INDUSTRIAL APPLICATIONS OF CARBON NANOTUBES

INDUSTRIAL APPLICATIONS OF CARBON NANOTUBES

Edited by

HUISHENG PENG
Fudan University, Shanghai, China

QINGWEN LI
Chinese Academy of Sciences, Suzhou, China

TAO CHEN
Tongji University, Shanghai, China

ELSEVIER

Amsterdam • Boston • Heidelberg • London • New York • Oxford
Paris • San Diego • San Francisco • Singapore • Sydney • Tokyo

Elsevier
Radarweg 29, PO Box 211, 1000 AE Amsterdam, Netherlands
The Boulevard, Langford Lane, Kidlington, Oxford OX5 1GB, United Kingdom
50 Hampshire Street, 5th Floor, Cambridge, MA 02139, United States

Library of Congress Cataloging-in-Publication Data
A catalog record for this book is available from the Library of Congress

British Library Cataloguing-in-Publication Data
A catalogue record for this book is available from the British Library

ISBN: 978-0-323-41481-4

Publisher: Matthew Deans
Acquisition Editor: Simon Holt
Editorial Project Manager: Sabrina Webber
Production Project Manager: Julie-Ann Stansfield
Designer: Greg Harris

Typeset by TNQ Books and Journals

CONTENTS

CONTRIBUTORS

Z.A. ALOthman
King Saud University, Riyadh, Saudi Arabia

J. An
Nanyang Technological University, Singapore, Singapore

M. Batmunkh
The Flinders University of South Australia, Bedford Park, SA, Australia; The University of Adelaide, Adelaide, SA, Australia

M.J. Biggs
The University of Adelaide, Adelaide, SA, Australia; Loughborough University, Loughborough, United Kingdom

X. Bo
Northeast Normal University, Changchun, China

P. Chen
Fudan University, Shanghai, China

T. Chen
Tongji University, Shanghai, China

L. Cheng
Northwestern Polytechnical University, Xi'an, China

S. Fang
Nanjing University of Aeronautics and Astronautics, Nanjing, China

T. Grace
The Flinders University of South Australia, Bedford Park, SA, Australia

L.-H. Guo
Chinese Academy of Science, Beijing, China

B. Hao
The Xinjiang Technical Institute of Physics and Chemistry, Chinese Academy of Sciences, Urumqi, China

S. He
Fudan University, Shanghai, China

J. Hou
Chinese Academy of Science, Beijing, China

G. Hu
Chinese Academy of Sciences, Urumqi, China; Umea University, Umea, Sweden

M.Q. Jian
Tsinghua University, Beijing, People's Republic of China

K. Jiang
Tsinghua University, Beijing, China

L.X. Kang
Chinese Academy of Sciences, Suzhou, China

D. Li
Chinese Academy of Sciences, Suzhou, China

N. Li
Tongji University, Shanghai, China

Q. Li
Chinese Academy of Sciences, Suzhou, China

T. Liu
Northeast Normal University, Changchun, China

T. Lv
Tongji University, Shanghai, China

P.C. Ma
The Xinjiang Technical Institute of Physics and Chemistry, Chinese Academy of Sciences, Urumqi, China

H. Peng
Fudan University, Shanghai, China

D. Qin
Chinese Academy of Sciences, Urumqi, China; University of Chinese Academy of Sciences, Beijing, People's Republic of China

H. Qiu
University of Shanghai for Science and Technology, Shanghai, China

J.G. Shapter
The Flinders University of South Australia, Bedford Park, SA, Australia

C. Shearer
The Flinders University of South Australia, Bedford Park, SA, Australia

H. Shen
Chinese Academy of Sciences, Urumqi, China; University of Chinese Academy of Sciences, Beijing, People's Republic of China

L. Shen
Nanjing University of Aeronautics and Astronautics, Nanjing, China

X. Sun
Fudan University, Shanghai, China

D. Tune
The Flinders University of South Australia, Bedford Park, SA, Australia; Karlsruhe Institute of Technology, Karlsruhe, Germany

T. Wagberg
Umea University, Umea, Sweden

B. Wan
Chinese Academy of Science, Beijing, China

F. Wang
Chinese Academy of Sciences, Urumqi, China

L. Wang
Chinese Academy of Sciences, Urumqi, China; University of Chinese Academy of Sciences, Beijing, People's Republic of China

X. Wang
Nanyang Technological University, Singapore, Singapore

Z. Wang
Nanyang Technological University, Singapore, Singapore

H. Wu
Northwestern Polytechnical University, Xi'an, China

K.L. Xia
Tsinghua University, Beijing, People's Republic of China

H.H. Xie
Tsinghua University, Beijing, People's Republic of China

Y. Xue
Wenzhou Medical University, Wenzhou, China

J. Yang
University of Shanghai for Science and Technology, Shanghai, China

Y. Yao
Tongji University, Shanghai, China

Z.Z. Yong
Chinese Academy of Sciences, Suzhou, China

L. Yu
The Flinders University of South Australia, Bedford Park, SA, Australia

W. Yuan
Northwestern Polytechnical University, Xi'an, China

Q. Yuan
Chinese Academy of Sciences, Urumqi, China

Z. Zhan
Chongqing Institute of Green and Intelligent Technology, Chinese Academy of Sciences, Chongqing, People's Republic of China

X. Zhang
Nanjing University of Aeronautics and Astronautics, Nanjing, China

X.H. Zhang
Chinese Academy of Sciences, Suzhou, China

Y. Zhang
Northwestern Polytechnical University, Xi'an, China

Y.Y. Zhang
Tsinghua University, Beijing, People's Republic of China

L. Zheng
Khalifa University of Science, Technology and Research, Abu Dhabi, United Arab Emirates

M. Zhou
Northeast Normal University, Changchun, China

PREFACE

Due to its unique structure and remarkable physical and chemical properties, carbon nanotubes (CNTs) have drawn great attention in both education and industry since they were discovered in 1991. As a typical one-dimensional carbon nanomaterial, CNTs represent one of the most attractive candidates for a variety of application fields, ranging from electronics, sensing, energy, and biomedical science to structure materials. However, for most proposed electronic applications, it remains a great challenge for CNTs to replace silicon because it is currently almost impossible to synthesize CNTs with uniform chiral structure and semiconducting or metallic property at a large scale, although a CNT computer with its central processor totally based on CNTs has been demonstrated. A promising way to promote its practical applications is to assemble them into macroscopic materials in formats of fibers, films, and arrays, based on which a broad spectrum of electronic, sensing, and energy devices and structure materials have been developed. Energy-storage devices and touchscreens based on CNT-conducting films are approaching commercialization, and CNT-reinforced structural materials have already been used for some high-end products, such as those in sports and automobiles. It can be expected that there will be more and more products based on CNTs in the near future.

This book aims to give an overview of the recent progress on CNTs in the areas of synthesis, property, and application, with a focus on the practical applications that are available or will appear soon. Chapter 1 summarizes the controllable growth of CNTs, and Chapter 2 then discusses the structure and property of CNTs. Based on the high mechanical property of CNTs, Chapter 3 demonstrates the use of CNTs as composite and structural materials. Chapter 4 reviews the CNT electrodes used for displays and touchscreens. Chapter 5 generally demonstrates their promise for sensing applications. Chapter 6 further highlights the response of CNTs to a variety of environmental stimulations such as light irradiation, temperature, and mechanical force by carefully assembling them, while Chapter 7 describes the use of CNTs as sensors in wearable devices. Carbon nanotubes have also been widely explored for energy harvesting and storage devices due to their large specific surface areas and high electrical conductivities. The use of CNTs as photoactive and/or electrodes materials in solar cells is presented in Chapter 8, and the advancements of CNTs as electrode materials for lithium-ion batteries and supercapacitors are carefully discussed in Chapters 9 and 10, respectively. Carbon nanotubes have also been extensively investigated for biomedical applications, as discussed in Chapter 11. Recently, the alignment of CNTs has been studied as a general and effective strategy to extend their excellent properties such as high electrical conductivity and high mechanical strength to a macroscopic scale. For example, aligned CNT fiber has been developed as electrodes for wearable electronics as discussed in Chapter 12, and aligned CNT film and array are explored for structural materials and electronic devices in Chapter 13. Of course, it is always relevant to study the safety of CNTs in practical applications, and Chapter 14 thus shows the main development of this issue. This book concludes with a brief review of the challenges and opportunities of CNTs in Chapter 15.

We sincerely hope that this book is beneficial for the scientists and engineers involved or/and interested in CNTs. It may also be used as a textbook for both undergraduate and graduate students in a variety of related fields including chemistry, physics, biology, engineering, and textiles. Lastly, we greatly appreciate the authors who contributed these wonderful chapters. It has been our great pleasure to work with them in the past year.

Huisheng Peng
Qingwen Li
Tao Chen

CHAPTER 1

Controllable Synthesis of Carbon Nanotubes

J. An[1], Z. Zhan[2], L. Zheng[3]
[1]Nanyang Technological University, Singapore, Singapore; [2]Chongqing Institute of Green and Intelligent Technology, Chinese Academy of Sciences, Chongqing, People's Republic of China; [3]Khalifa University of Science, Technology and Research, Abu Dhabi, United Arab Emirates

Contents

Industrial Applications of Carbon Nanotubes
ISBN 978-0-323-41481-4

1.1 INTRODUCTION

Carbon nanotubes (CNTs) have attracted intense attention since their discovery *(1)*. The past two decades have witnessed their spread in a range of applications including building blocks in modern electronics, field emission sources, photodetectors, nonvolatile memory, ultrasensitive chemical and biosensors, and transparent conductive membranes *(2–10)*.

Gaining controllability over CNT synthesis is critical in realizing CNT's potentials, because each particular application may require a specific type of CNT. For example, when developing nanoelectronics based on single-walled carbon nanotubes (SWNTs), the chirality and diameter essentially determine their electronic properties, which further direct their availability as either functional devices or interconnects in integrated circuits *(11–13)*. The orientation and alignment of the SWNTs are important for optimizing device performance by exploiting the anisotropic nature of SWNTs *(14,15)*. The morphology, in terms of length and density, should be easily tuned according to specific applications—thin-film-type transistors require dense and parallel SWNTs to sustain large currents *(2,16)*, while their ultralong length enables easy fabrication of numerous devices out of one individual SWNT and the construction of logic circuits at a single nanotube level *(17,18)*. Other applications may use various macroscopic CNT assemblies, such as 1D CNT fiber/bundle, 2D CNT film/sheet, 2D/3D CNT network, and 3D-aligned CNT arrays or foams *(19)*, to take advantage of CNT's high specific surface area, superior mechanical properties, and 1D electrical conductivity. The processability and performance of these assemblies in a particular application directly depend on the nature or morphology of raw materials that these assemblies are derived from.

Therefore, for the purpose of achieving high-performance materials/devices for industrial applications, no effort should be spared to produce CNTs with control over nanotube geometrical structure, chirality, orientation, alignment, and morphology. Considering that chemical vapor deposition (CVD) possesses advantages in terms of scalability and controllability, we focus our interest in this chapter on the controllable synthesis by the CVD method for two typical types of CNTs: individual SWNTs (horizontal aligned) and CNT arrays (vertically aligned).

1.2 CONTROLLABLE SYNTHESIS OF INDIVIDUAL SWNTS ON SUBSTRATES

Compared to bulk synthesis and postassembly of SWNTs, which involve cumbersome processes and thereby degrade the quality of nanotubes because of the inevitable introduction of defects and contamination *(20)*, in situ CVD methodology permits single-step preparation of horizontally aligned SWNTs on surfaces. High-quality SWNTs grown in such a manner have the qualities of narrow diameter distribution, enriched chirality, well-defined orientation, and tunable length and density *(21–24)*. Here we elaborate recent progress on surface synthesis of horizontally aligned SWNTs.

1.2.1 Diameter and Chirality Control

The main hurdle in the development of SWNT-based technologies is the lack of control over their structure, more precisely their diameter and chirality. Tremendous efforts have been made aimed at controlling the nanotube's structure directly during the growth process.

1.2.1.1 Diameter and Chirality Control Through Catalyst

Catalytic chemical vapor deposition is the prevailing route for controllable CNT synthesis. Various metals, such as Fe, Mo, Co, Ni, Pd, Cu, Au, Ag, Mn, Cr, Sn, and Al, and their combinations, have been employed as catalysts in the form of nanoparticles, nanoclusters, or single crystals *(25–30)*. According to the vapor–liquid–solid (VLS) (Fig. 1.1A–C) mechanism, the growth process includes the dissociation of gaseous hydrocarbon on the catalyst, the diffusion of adsorbed carbon atoms into the bulk of the nanoparticle, carbon precipitation, and the nucleation of cap structures of nanotubes *(24,31)*. Since the

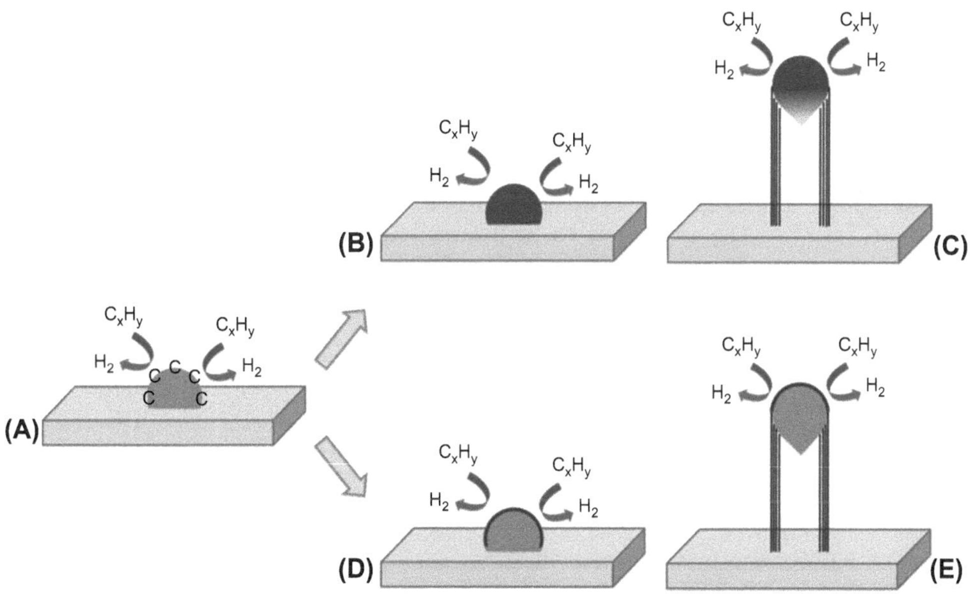

Figure 1.1 Schematic demonstration of growth dynamics of carbon nanotubes (CNTs). Vapor–liquid–solid (VLS) growth model (A–C) versus Vapor–solid–solid (VSS) growth model (A, D, and E). *(Adapted from Tessonnier, J.-P.; Su, D. S. Recent Progress on the Growth Mechanism of Carbon Nanotubes: A Review. ChemSusChem* ***2011****, 4 (7), 824–847 with permission.)*

caps that are initially formed around the catalyst define the diameter and the chirality of the growing nanotube, it is extremely effective to control the diameter and chirality by tuning the size, shape, and composition of the catalyst nanoparticles.

A considerable number of routes have been developed. Iron-group metals (Fe, Co, Ni), because of their strong activity in catalyzing the decomposition of carbon feedstock and sufficient carbon solubility, are the most widely investigated catalyst *(32)*. Monodispersed Fe–Mo nanoparticles with uniform size were prepared by thermal decomposition of metal carbonyl complexes. Long-chain amine and long-chain carboxylic acid were used as the capping agents to prevent particle aggregation. By changing the experimental conditions, the particle size could be tuned from 3 to 14 nm. However, only nanoparticles smaller than 8 nm succeeded in SWNT growth *(33)*. Similarly, Ni nanoparticles with controllable size and a narrow size distribution were obtained on a flat Si/SiO_2 surface from thermal decomposition of the nickel precursor. The monodispersed Ni nanoparticles resulted in the formation of CNTs with well-controllable diameters *(34)*.

Based on the catalyst size–CNT diameter relationship *(35,36)*, discrete nanoparticles with identical size are highly desired for producing SWNTs with narrow diameter distribution. By placing controllable numbers of Fe atoms into the cores of apoferritin, Li et al. reported the synthesis of discrete Fe_2O_3 nanoparticles with tunable sizes *(37)*. Isolated SWNTs were grown from the discrete catalyst nanoparticles. Transmission electron microscopy (TEM) characterization (Fig. 1.2A) confirmed that the SWNT diameters were determined by the particle sizes. Apoferritins were also adopted as a Co storage unit. Discrete Co nanoparticles were formed after calcination in air, which catalyzed the growth of suspended SWNTs with narrow diameter distribution *(38)*. In addition to the metal-contained protein, block copolymer micelles and dendrimers have also been used as the synthetic templates to derive discrete catalyst nanoparticles *(39–41)*, and metal-based inorganic molecular nanoclusters were proposed as an alternative catalyst precursor. The route for preparing uniform catalysts is facile, and the diameter distribution of the synthesized SWNTs fell in a narrow range *(42,43)*.

Conventional CVD usually produces a mixture of metallic (m-) and semiconducting (s-) SWNTs. Their coexistence has become a major obstacle to SWNT's application in high-performance nanoelectronics. Although postsynthesis sorting enables separation of the m- and s-SWNTs *(44–46)*, it cannot allow the selection of a specific chiral index in the as-grown aggregates *(21)*. Thus direct selective growth of SWNTs has been regarded as the best choice to solve this problem.

By tuning the shape of Fe nanoparticles, Harutyunyan et al. managed to increase the fraction of m-SWNTs to 91%. They found that the rearrangement of Fe nanoparticles could be dramatically influenced by the gas environment. Faceted particles with sharp corners, which were annealed in He/H_2O ambient, were considered to be responsible

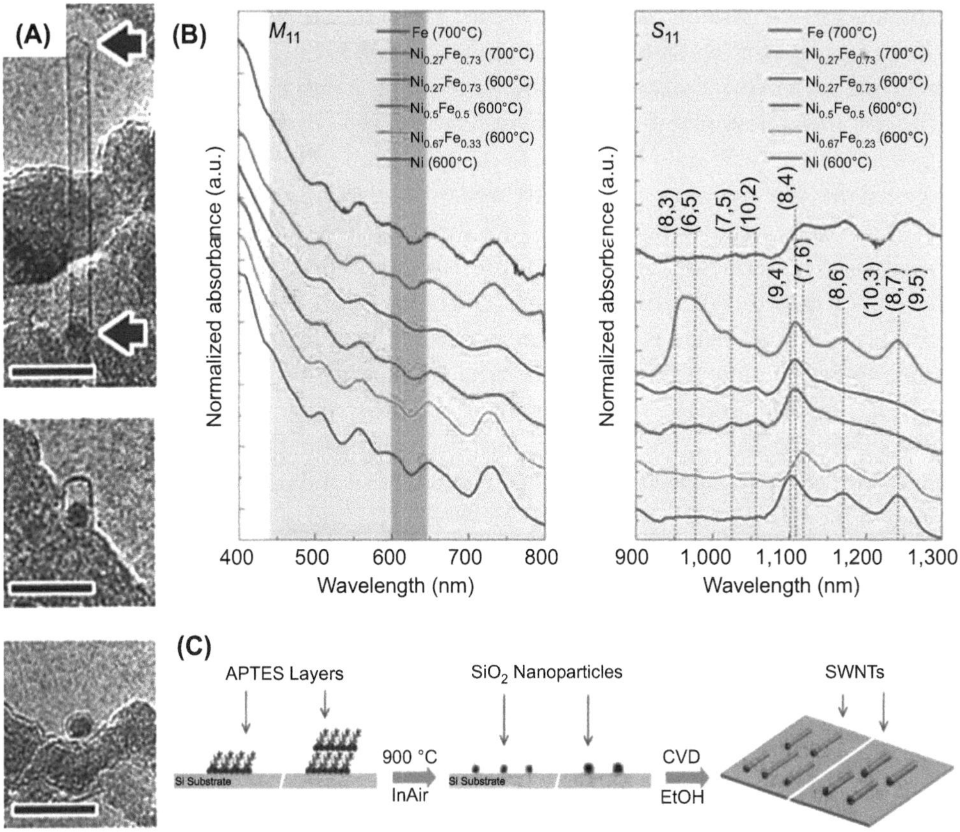

Figure 1.2 (A) Transmission electron microscopy (TEM) images showing the growth of single-walled carbon nanotubes (SWNTs). Scale bars: 10 nm. *(Adapted from Li, Y.; Kim, W.; Zhang, Y.; Rolandi, M.; Wang, D.; Dai, H. Growth of Single-walled Carbon Nanotubes from Discrete Catalytic Nanoparticles of Various Sizes. Journal of Physical Chemistry B* ***2001****, 105 (46), 11424–11431 with permission.)* (B) UV–Vis–NIR absorbance spectra of SWNTs grown with Ni_xFe_{1-x} nanoparticles. *(Adapted from Chiang, W.-H.; Mohan Sankaran, R. Linking Catalyst Composition to Chirality Distributions of as-Grown Single-walled Carbon Nanotubes by Tuning Ni_xFe_{1-x} Nanoparticles. Nature Materials* ***2009****, 8 (11), 882–886 with permission.)* (C) Schematic illustration of the preparation of SiO_2 nanoparticles and the diameter-controlled growth of SWNTs. *(Adapted from Chen, Y.; Zhang, J. Diameter Controlled Growth of Single-walled Carbon Nanotubes from SiO_2 Nanoparticles. Carbon* ***2011****, 49 (10), 3316–3324 with permission.)*

for the preferential growth *(47)*. Inert coinage metals, such as Cu, Ag, and Au, have served as the catalysts to produce SWNTs with narrow diameter and chirality distributions *(48,49)*. Based on systematic investigations, it has been found that the binding energy of hydrocarbons on the Au surface is much weaker than on the Fe surface, which inhibits the formation of large-diameter nanotubes. Within the limited diameter range, the number of stable caps is also restricted, leading to fewer chirality possibilities *(48)*.

Binary catalyst systems, such as Co–Mo, Fe–Ru, Co–Pt, Ni–Fe catalysts, have shown advantages in chiral-preferential growth of SWNTs *(50–56)*. In the bimetallic system, the second metal may impede the aggregation against high-temperature sintering of the first metal, resulting in small catalyst nanoparticles. Bachilo et al. reported the selective growth of (6,5) and (7,5) SWNTs using bimetallic Co–Mo catalysts *(50)*. It was suggested that Co catalysts were first stabilized by interacting with the Mo oxides. When the catalysts were exposed to CO, the Co^{2+} ions were reduced to small clusters, leading to small diameter SWNTs. Similarly, an Fe–Ru nanocluster was also reported to facilitate SWNT growth with narrow diameter and chirality distribution *(53)*. Small nanoparticles in the 1–2 nm range were formed because of the intimate alloying of Fe–Ru and the strong Fe–Ru interaction, and they were responsible for good chirality selectivity. He et al. prepared an MgO-supported Fe–Cu catalyst system to facilitate the growth of small diameter SWNTs with predominantly (6,5) nanotubes *(57)*. They found that the addition of Cu to Fe promoted the reducibility of Fe and inhibited their aggregation, leading to the enhancement in the yield of SWNTs. To bridge the gap between the catalyst composition and the chirality distribution of SWNTs, Chiang et al. developed a continuous gas-phase process to precisely tune the composition of Ni_xFe_{1-x} nanoparticles at constant size *(56)*. It has been shown that, with the increase of Fe content in the bimetallic nanocatalysts, the number of metallic nanotubes was decreased (Fig. 1.2B, left panel). The narrowest (n,m) distribution—predominantly (8,4) nanotubes with a small fraction of (7,5), (6,5), (7,6), and (8,3) nanotubes—was obtained by $Ni_{0.27}Fe_{0.73}$ nanoclusters (Fig. 1.2B, right panel). It was suggested that the perturbations to the crystal structure caused by compositional tuning affected the lattice match of the catalyst with certain chiralities and led to chirality selectivity.

Attention has recently been given to nonmetal catalysts, because of their advantage in diameter and chirality control *(58–62)*. The nonmetal catalysts, such as metal oxide and diamond, have high melting points so that their structure and morphology remain unchanged at high growth temperature. Such stable catalysts are superior templates for cap nucleation, thereby defining the diameter and chirality of the growing nanotube. Additionally, these solid nonmetal catalysts have little activity of carbon feed stock decomposition, resulting in the tendency of carbon atoms to diffuse along the particle surface (surface diffusion) instead of within the nanoparticle (bulk diffusion). Accordingly, a vapor–solid–solid (VSS, illustrated in Fig. 1.1A, D, and E) growth mechanism has been proposed for nonmetal catalysts *(31)*.

SiO_2 nanoparticles are the most widely investigated nonmetal catalysts. Huang et al. prepared SiO_2 nanoparticles by simply scratching the surface of quartz or Si/SiO_2 substrate. SWNTs grown from the SiO_2 nanoparticles had a narrow diameter distribution of 50% in the range of 1.0–1.2 nm *(61)*. Chen et al. obtained SiO_2 nanoparticles with size distributed from 1.22 to 1.98 nm by the thermal oxidation of 3-aminopropyltriethoxysilane. Ethanol-CVD was then carried out for the synthesis of

SWNTs, as depicted in Fig. 1.2C. The diameter distribution falls in the range of 0.9−1.82 nm, indicating a direct correlation to size distribution of the SiO_2 nanoparticles. In situ Raman spectroscopy and X-ray photoelectron spectroscopy revealed that no Si−C alloy was formed; the SiO_2 nanoparticles remained in a solid state and only played the role as nucleation centers, confirming the VSS growth mechanism *(58)*.

The latest improvement has been demonstrated by Yang et al. utilizing W-based bimetallic alloys as catalysts. These metal alloys possess a high melting point so that they maintain the crystalline structure with well-faceted shapes at the high growth temperature. A perfect geometrical match between (12,6) SWNTs and (0012) plane of W_6Co_7 nanoparticles has been achieved during the growth process, resulting in chirality selectivity toward (12,6) nanotubes higher than 92% *(30)*.

In addition to the catalysts preparation procedure, the method of loading catalysts onto the substrates is also essential. Several strategies have been developed to deposit monodispersed and uniform catalyst nanoparticles. A thin film of Fe catalyst precursor was sputtered on sapphire and then annealed in ultrahigh vacuum *(63)*. It has been found that thermal annealing reduces Fe nanoparticle size and narrows its size distribution. Under optimized conditions, SWNTs with diameters in the range of 1.3−1.4 nm were successfully produced. Kim et al. later simplified the procedure to atmospheric annealing for diameter-specified growth of SWNTs *(64)*. Barzegar et al. dip coated silicon wafer with a mixed solution of Fe and Co *(65)*. The size and distribution of the catalyst could be well tuned by changing the solution's concentration. By changing the ratio of Fe/Co, selective growth toward dominantly (70%) four types of nanotubes was achieved. Similarly, Ishgami et al. adopted a dip-coating process to pattern ethyl alcohol solution of CoMo mixture on sapphire *(66)*. Atomic force microscopy (AFM) characterization revealed that the mean diameter of catalyst nanoparticles was dependent on the crystal plane of the sapphire. Preferential growth of near-zigzag tubes was obtained on the A-plane while the R-plane showed dominant near-armchair tubes. The growth mechanism was proposed as "epitaxial nanotube growth" because of the matching of atomic arrangements of the sapphire surface and the specific cap structures.

1.2.1.2 Diameter and Chirality Control Through Growth Conditions

Growth conditions, such as temperature, carbon source, and gas composition and flow rate, notably influence the kinetics of SWNT growth and also cause structural changes in the metal nanoparticles. It has been observed that higher growth temperature resulted in larger SWNT diameter and wider diameter distribution *(67−71)*. Li et al. managed to tune the diameter of SWNTs from 0.6−0.8 nm to 1.8−2.0 nm by increasing the reaction temperature from 550°C to 950°C *(68)*. Dervishi et al. studied the thermal effect from an MgO-supported Fe−Co catalyst system. It was found that the diameter distribution was broadened when raising the growth temperature from 800°C to 1000°C *(71)*. These results are normally explained by particle coalescence or Ostwald ripening.

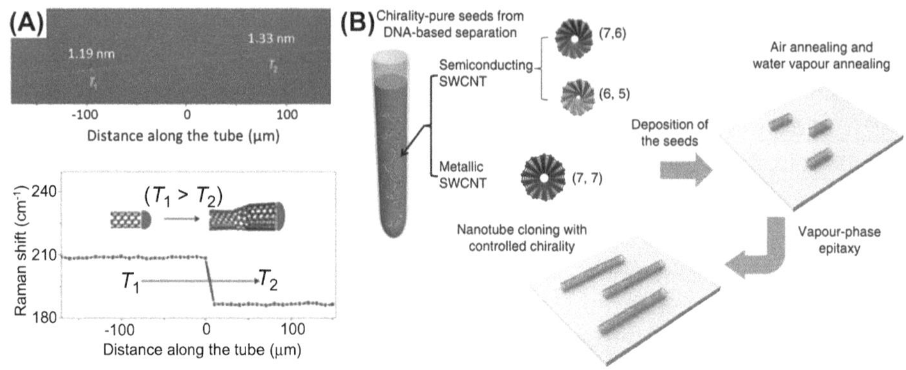

Figure 1.3 (A) Tuning single-walled carbon nanotube (SWNT) diameter by varying the temperature during growth. *(Adapted from Yao, Y.; Li, Q.; Zhang, J.; Liu, R.; Jiao, L.; Zhu, Y. T.; Liu, Z. Temperature-Mediated Growth of Single-walled Carbon-Nanotube Intramolecular Junctions. Nature Materials* ***2007****, 6 (4), 283–286 with permission.)* (B) Schematic illustration of SWNT cloning process. *(Adapted from Liu, J.; Wang, C.; Tu, X.; Liu, B.; Chen, L.; Zheng, M.; Zhou, C. Chirality-Controlled Synthesis of Single-wall Carbon Nanotubes Using Vapour-Phase Epitaxy. Nature Communications* ***2012****, 3, 1199 with permission.)*

However, the opposite effect of temperature is also true in the case of single catalyst nanoparticles. Yao et al. varied the growth temperature to tune the diameter and found that SWNTs grown from the same catalyst particle became thicker when decreasing the temperature, as demonstrated in Fig. 1.3A *(72)*. They later discovered that the diameter change was highly related to the initial nanotube diameter and the range of temperature-tuned diameter variation was limited *(73)*.

Feeding gas is another factor that affects diameter and chirality. It is believed that the diameter is tightly related to the carbon feeding rate because of the selective activation of catalyst nanoparticles. Lu et al. reported that small diameter SWNTs were preferentially obtained at a low carbon feeding rate *(74)*. In contrast to this, Saito et al. found that increasing the flow rate of the carbon source would lead to a smaller diameter *(75)*. The type of carbon source may lead to the change in chirality distribution. Wang et al. utilized four different carbon precursors, CO, C_2H_5OH, CH_3OH, and C_2H_2, to synthesize SWNTs on CoMo catalysts *(52)*. It was discovered that narrowly (n,m) distributed SWNTs could be obtained only under high-pressure CO or vacuumed C_2H_5OH and CH_3OH. The same group also reported that SWNTs with enriched three dominant chiralities could be obtained by changing the pressure of the carbon precursor *(76)*.

To enrich the percentage of s-SWNTs in production, a number of etchants have been employed for selective destruction of m-SWNTs. Ding et al. introduced methanol with ethanol to preferentially synthesize s-SWNTs (over 95%) *(77)*. The semiconducting/metallic ratio of the SWNTs grew with the increase of methanol concentration. It was proposed that the OH radical from methanol selectively etched m-SWNTs because of the smaller ionization potential for m-SWNTs. The interaction between lattice and nanotubes promoted the narrowing of diameter distribution and the enrichment of s-SWNTs. It has also been found that H_2 gas and water vapor have a similar effect on the selective growth of s-SWNTs *(48,78,79)*. More recently, Li et al. demonstrated that by controlling the diameter distribution and adjusting the concentration of water vapor, the selectivity of s-SWNTs could reach as high as 95% *(79)*. The group discovered that the etching of SWNTs was both electronic-type dependent and diameter dependent. Narrowing the diameter distribution of as-synthesized SWNTs could promote "selective etching," which increases the percentage of semiconducting nanotubes in production.

1.2.1.3 Diameter and Chirality Control Through Cap Engineering and SWNT Cloning

To specifically control the diameter and chirality of SWNTs, researchers epitaxially grew SWNTs out of cap structures or seeds with predefined diameter and chirality. Thermally opened C_{60} molecules were exploited as hemispherical caps *(80)*. Cap engineering was carried out by tuning the temperature of thermal oxidation, which determines the diameter distribution of the SWNTs. It was found that stronger oxidation treatment was favorable for the growth of thinner SWNTs, while weaker oxidation promoted the growth of SWNTs with larger diameters. However, such C_{60} cap engineering could only affect the diameter, thereby SWNT "cloning" was proposed to control the diameter and chirality simultaneously *(81–85)*. Generally, the "cloning" procedures include cutting long SWNTs into short segments, docking catalysts at the open ends, and using short "parent" segments as seeds to grow long SWNTs. Because of epitaxial growth, the duplicated SWNTs inherit the same diameter and chirality from the seeded SWNTs. Yao et al. freed the use of metal catalysts in the seeded growth by directly cloning from the parent segments via an open-ended growth mechanism *(84)*. The cloning yield was relatively low (9%), but could be greatly improved to 40% by using quartz substrates. Liu et al. later demonstrated SWNT cloning from chirality-pure nanotube seeds *(85)*. They initially obtained chirality-pure semiconducting (7,6), (6,5), and metallic (7,7) nanotube seeds using DNA-based chromatographic purification. The nanotube seeds were then deposited on substrate for subsequent catalyst-free, chirality-control cloning process, as illustrated in Fig. 1.3B. Cloned SWNTs obtained on ST-cut quartz substrates showed alignment along the crystal orientation, while random orientation was observed on Si/SiO_2 substrates. Raman characterization on both parent and duplicated SWNTs suggested that the diameter and chirality of the parent SWNTs were well preserved by the cloned nanotubes.

1.2.2 Orientation and Alignment Control

As an anisotropic material, SWNT possesses better properties in the tube axial direction as compared to the transverse direction. This necessitates the utilization of well-aligned SWNTs with controlled location and orientation in SWNT-based nanodevices, for the purpose of achieving high device performance.

1.2.2.1 Electric-Field Directed Orientation and Alignment Control

External forces, for instance the electric field, have been applied to align the SWNTs into ordered structures *(86–88)*. As early as 2001, Zhang et al. introduced electric field into the CVD process *(86)*. Poly-Si trenches were patterned on quartz to form capacitors in series; voltages were applied to establish electric filed across all of the trenches. As exhibited in the scanning electron microscope (SEM) image of Fig. 1.4A, suspended SWNTs were grown across the trenches with good alignment. The alignment was

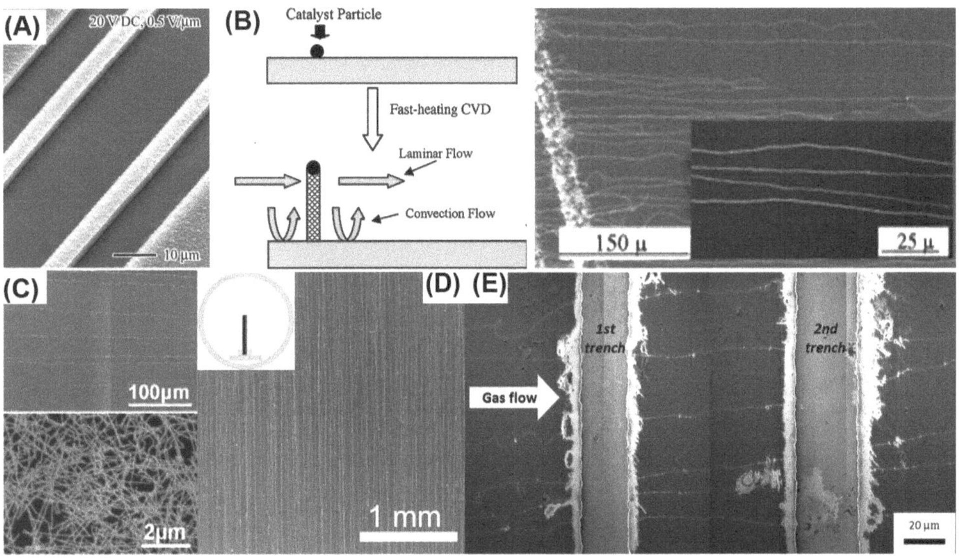

Figure 1.4 (A) Scanning electron microscope (SEM) image of suspended single-walled carbon nanotubes (SWNTs) aligned by electric field. *(Adapted from Zhang, Y.; Chang, A.; Cao, J.; Wang, Q.; Kim, W.; Li, Y.; Morris, N.; Yenilmez, E.; Kong, J.; Dai, H. Electric-Field-Directed Growth of Aligned Single-walled Carbon Nanotubes. Applied Physics Letters **2001**, 79 (19), 3155–3157 with permission.)* (B) Left panel: Schematics of the "kite mechanism." Right panel: SEM of aligned SWNTs grown by the "fast-heating" process. *(Adapted from Huang, S.; Woodson, M.; Smalley, R.; Liu, J. Growth Mechanism of Oriented Long Single Walled Carbon Nanotubes Using "Fast-Heating" Chemical Vapor Deposition Process. Nano Letters **2004**, 4 (6), 1025–1028 with permission.)* (C) Comparison of SWNTs catalyzed by Cu (upper panel) and Fe (lower panel) nanoparticles. *(Adapted from Zhou, W.; Han, Z.; Wang, J.; Zhang, Y.; Jin, Z.; Sun, X.; Zhang, Y.; Yan, C.; Li, Y. Copper Catalyzing Growth of Single-walled Carbon Nanotubes on Substrates. Nano Letters **2006**, 6 (12), 2987–2990 with permission.)* (D) SEM image of long and oriented SWNTs grown with ultralow feeding gas flow. *(Adapted from Jin, Z.; Chu, H.; Wang, J.; Hong, J.; Tan, W.; Li, Y. Ultralow Feeding Gas Flow Guiding Growth of Large-Scale Horizontally Aligned Single-walled Carbon Nanotube Arrays. Nano Letters **2007**, 7 (7), 2073–2079 with permission.)* (E) SEM showing improvement of alignment by the trench structure-induced "zipping effect." *(Adapted from An, J.; Zhan, Z.; Salila Vijayalal Mohan, H. K.; Sun, G.; Hansen, R. V.; Zheng, L. Trench Structure Assisted Alignment in Ultralong and Dense Carbon Nanotube Arrays. Journal of Materials Chemistry C **2015**, 3 (10), 2215–2222 with permission.)*

improved as the bias voltage increased. The field-alignment effect was attributed to the high polarization of SWNTs under electric field. The same group also managed to grow aligned SWNTs on Si/SiO_2 substrates with the assistance of electric fields established across Mo electrodes *(87)*. Through the calculations of the electric field distribution, the authors pointed out that the SWNTs were aligned up along electric field lines during their elongation. A similar study was reported by Joselevich et al. later on *(88)*. They demonstrated that the dipole moments induced by the electric field would lead to aligning torques, forcing the SWNTs to grow parallel to the electric field. The drawbacks of this method are very obvious—the process of patterning electrode pairs and the use of electric field in growth bring difficulties to the experiments.

1.2.2.2 Gas Flow-Directed Orientation and Alignment Control

It has been considered that the strong interaction between the catalyst and substrates, which is generally described as van der Waals (vdW) interaction, tends to hold the growing nanotubes in place and hinder their alignment. To overcome this interaction, a "fast-heating" technique was developed, where the catalyst-loaded substrate was rapidly transferred to the center zone of a preheated CVD furnace *(89,90)*. A convection flow was induced as a result of the different heating-up speeds of the solid substrate and its surrounding gas. The convection flow helped to lift up the catalyst and the growing nanotube tail. As depicted in Fig. 1.4B left panel, the active end of the nanotube, with the catalyst attached, floats above the substrate like a flying kite. During the growth process, the floating nanotube will be carried and aligned by the horizontal laminar gas flow. This "kite mechanism" was evidenced by successfully growing SWNTs over barriers on substrate and across trenches. Two-dimensional, well-defined, crossed-network SWNTs were also fabricated via a two-step growth process. Centimeter-long, well-aligned SWNTs could be grown by the "fast-heating" CVD, as shown in Fig. 1.4B right panel. Apparently, to successfully leverage the gas flow as the alignment force, the key is to avoid the vdW interaction and thus ensure that the growing SWNTs float up into the gas phase.

Many other approaches have also been tried to lift up the catalyst. Yu et al. elevated the catalyst using a metal under a layer to prevent the interface from the substrate *(91)*. By using methane as carbon source, they obtained 1.5 mm-long and straight SWNTs. Huang et al. found that intentionally raising the catalyst side of the substrate would benefit the alignment of long SWNTs *(92)*. Cu has been recognized as a superior catalyst for the high-efficiency growth of aligned SWNTs *(27)*, because it exhibits weaker interaction with the SiO_2 surface. Zhou et al. carried out SWNT growth using Cu catalyst and methane, ethanol, and 2-propanol as feeding gas on Si/SiO_2 substrates. They obtained centimeter-long, aligned SWNTs with a density of 2–3 SWNTs/10 μm. Under similar growth conditions, however, only a network of random SWNTs was produced from the Fe–Mo catalysts, as compared in Fig. 1.4C.

In early reported studies, high flow rate feeding gas was thought essential to guide the elongating of SWNTs *(90)*. However, the high flow rate normally induces turbulence to the growth environment, disturbs the stability of the laminar flow, and as a result degrades the alignment. Hong et al. designed a reactor with tube-in-tube configuration to stabilize the gas flow *(93)*. Laminar flow was created in the inner tube, confirmed by the small Reynolds number inside the growth system ($Re \approx 50$). By placing catalyst-loaded substrate in the hot zone in the middle of the inner quartz tube, extremely long CNTs with aligned geometries were grown. It has therefore been found that high gas flow rate is not necessary for aligning the growing SWNTs. Jin et al. developed a "normal heating" CVD system to prepare aligned SWNTs with an ultralow gas flow rate of 1.5 sccm *(94)*. They discovered that a buoyancy effect was induced by vertical gas density gradient, which would help to uplift the catalyst and the SWNTs. The buoyancy effect became dominant in the case of a small gas flow flux. It was experimentally confirmed by growing SWNTs on a vertically placed Si/SiO_2 substrate. As shown in Fig. 1.4D, straight and dense SWNTs can be obtained under a growth condition of 2 sccm CH_4 and 4 sccm H_2 at 970°C. Although lying on vertical-clamped substrate, the SWNTs were well aligned without obvious obliquities to the horizontal direction.

Problems still exist because of the crowding effect or the local turbulence of the gas flow *(95,96)*. An et al. described a trench structure-assisted approach to in situ improve the alignment while maintaining their length and density *(97)*. Trench structures were prefabricated on the Si/SiO_2 surface and the CVD process was carried out on the patterned substrates. Although the SWNTs initially appeared with wiggles and curves, as displayed in Fig. 1.4E, they were greatly straightened and aligned after growing past the first trench. Raman characterization suggested that tensile strain was induced to the nanotubes growing across trenches because of the "zipping effect" at the trench edge. The strain was responsible for stretching the slack nanotube into a taut and straight body. At the same time, the "apparent" elongation rate was slowed down in the presence of the trenches, which promotes the gas flow guiding of the floating nanotubes.

1.2.2.3 Surface-Directed Orientation and Alignment Control

Although the nanotube—substrate interaction may hinder the smooth growth of SWNTs, exceptional studies have demonstrated the controlled alignment by taking advantage of the vdW interaction by using substrates with miscut atomic steps or anisotropic lattice *(98—101)*. Ismach et al. reported that highly aligned SWNTs could be grown along 0.2 nm-high atomic steps of a miscut C-plane sapphire surface, as shown in Fig. 1.5A *(98)*. They suggested that the surface steps may template the formation of periodic SWNTs. In their subsequent work, the sapphire substrate was annealed at 1100°C in air for 5—10 h to force unstable atomic steps to bunch together into faceted nanosteps. After CVD growth, well-aligned SWNTs were separated over substrate *(102)*. The step edge-induced alignment was also found in other substrates.

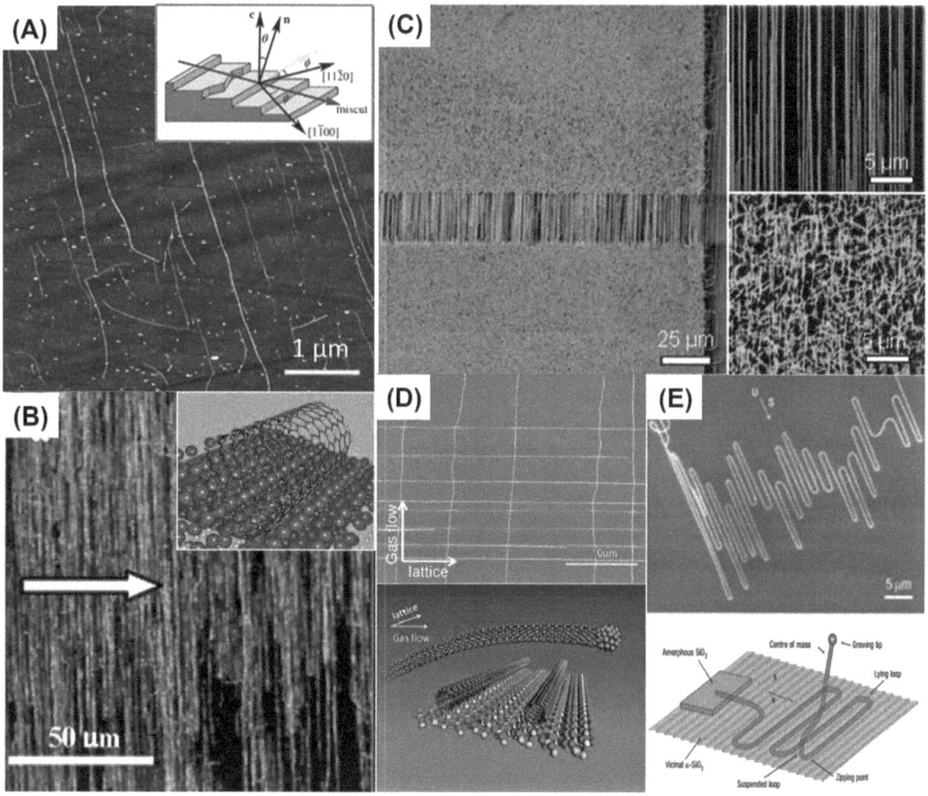

Figure 1.5 (A) Atomic force microscopy (AFM) image of single-walled carbon nanotubes (SWNTs) aligned along faceted atomic steps. Inset: Schematic representation of the atomic steps on C-plane sapphire. *(Adapted from Ismach, A.; Segev, L.; Wachtel, E.; Joselevich, E. Atomic-Step-Templated Formation of Single Wall Carbon Nanotube Patterns. Angewandte Chemie International Edition* ***2004****, 43 (45), 6140–6143 with permission.)* (B) Dense SWNTs aligned on A-plane sapphire. Inset: Atomic illustration of an SWNT. *(Adapted from Han, S.; Liu, X.; Zhou, C. Template-Free Directional Growth of Single-walled Carbon Nanotubes on a- and r-Plane Sapphire. Journal of the American Chemical Society* ***2005****, 127 (15), 5294–5295 with permission.)* (C) Scanning electron microscope (SEM) image of SWNTs grown on ST-cut quartz substrate. *(Adapted from Kocabas, C.; Shim, M.; Rogers, J. A. Spatially Selective Guided Growth of High-Coverage Arrays and Random Networks of Single-walled Carbon Nanotubes and Their Integration into Electronic Devices. Journal of the American Chemical Society* ***2006****, 128 (14), 4540–4541 with permission.)* (D) SEM image of SWNT crossbars (upper panel), and schematic illustration showing the growth of SWNTs (lower panel). *(Adapted from Zhang, B.; Hong, G.; Peng, B.; Zhang, J.; Choi, W.; Kim, J. M.; Choi, J.-Y.; Liu, Z. Grow Single-walled Carbon Nanotubes Cross-Bar in One Batch. Journal of Physical Chemistry C* ***2009****, 113 (14), 5341–5344 with permission.)* (E) SEM image of serpentine SWNTs (upper panel) and schematic illustration of the growth mechanism. *(Adapted from Geblinger, N.; Ismach, A.; Joselevich, E. Self-Organized Nanotube Serpentines. Nature Nanotechnology* ***2008****, 3 (4), 195–200 with permission.)*

For example, Kocabas et al. performed growth on miscut single-crystal quartz substrates *(103)*. Well-aligned, densely packed, horizontal SWNTs were produced over a large area. To organize SWNT structures on a silicon wafer oxidized with SiO_2 layer, Yoshihara et al. intentionally modified the surface with CF_4 plasma *(104)*. Radial steps were fabricated. The produced SWNTs resembled the radial steps in the alignment, forming flower-like morphology. Inside each "petal," the SWNTs were highly aligned.

Some specific lattice structures also serve as the driving force to control the alignment. Han et al. synthesized SWNTs on sapphire substrates with ferritin as the catalyst and a mixture of C_2H_4, CH_4, and H_2 as the feeding gas *(105)*. High-throughput growth of highly aligned SWNTs was obtained on the A-plane and R-plane (Fig. 1.5B); in contrast, the C-plane and M-plane sapphire produced predominantly random SWNTs. Control over orientation and alignment was attributed to the anisotropic interactions between the SWNTs and the topmost layer of atoms on the substrate. Almost at the same time, growth of highly aligned SWNTs on sapphire was reported in another study *(106)*. Because of the trigonal crystal structure of sapphire, SWNTs obtained on different planes exhibited varied degrees of alignment. AFM characterization revealed that the highest degree of alignment was obtained on the R-plane, while aligned SWNTs with highest density were formed on the A-plane. On the C-face, however, randomly oriented SWNTs were observed. It has been suggested the Al atoms on the R-face and A-face surfaces interact with the SWNTs and thus guide their alignment. Yu et al. investigated the mechanism specifically on the R-plane sapphire *(107)*. It was believed that Al atoms in the $[1\bar{1}0\bar{1}]$ direction on the oxygen-depleted sapphire were responsible for the alignment in the middle part of the wafers, while atomic steps or nanosteps played a superior role at the edges of the wafers.

On the surface of thermally annealed ST-cut quartz substrates, Kocabas et al. patterned ferritin catalyst as stripes at specific areas to grow SWNTs with controlled geometries *(108)*. As displayed in Fig. 1.5C, the process produced highly aligned SWNTs between the catalyst-patterned areas, while the catalyst regions were highly covered by random networks. Later on, the same group further optimized the growth conditions and produced SWNTs with levels of alignment >99.9% *(16)*. The average tube length and density were also improved in the study. Ding et al. patterned line-shaped copper catalysts on oxygen plasma-treated ST-cut quartz substrate *(100)*. With ethanol as the carbon source, they also managed to obtain high-density, uniform, perfectly aligned long SWNTs. The patterned catalysts efficiently prevented the aggregation of nanoparticles, thus promoting the elongation of straight SWNTs with high density. Xiao et al. systematically explored the alignment mechanism on single-crystal quartz substrates with different crystal directions *(14)*. Their studies revealed that the angle-dependent vdW interactions governed the alignment of tubes on quartz with X, Y, Z, and ST cuts.

1.2.2.4 Synergetic Effect of Different Directional Modes for Complex Architectures

By combining the previously mentioned techniques, various complex and coherent nanotube configurations can be constructed. For example, Ismach et al. assembled SWNTs in orthogonal configuration on sapphire substrate patterned with amorphous SiO_2 stripes *(102)*. On a miscut C-plane sapphire surface, the SWNTs were graphoepitaxially grown along the faceted nanosteps, while the electric field guided the SWNT growth in the perpendicular direction between the electrodes on SiO_2 stripes. Zhang et al. demonstrated a similar approach to fabricate SWNT crossbars on ST-cut quartz wafers *(109)*. By perpendicularly patterning the catalyst stripes, growth of SWNT crossbars was activated based on two growth modes simultaneously. As displayed in Fig. 1.5D, in one direction the SWNTs were guided by the gas flow, while the lattice of the quartz oriented the tubes in the second direction. Growth temperature and adhesion of catalyst nanoparticles on quartz were found to be critical in affecting the length and density. Under optimized conditions, the node density of the SWNT crossbars could reach up to $10^7/cm^2$. Another example is the organization of serpentine SWNTs on substrates. Geblinger et al. succeeded in forming such serpentine geometry with combinational driving forces of gas flow and step orientation *(110)*. As observed in the SEM image of Fig. 1.5E, the straight segments of the tubes followed existing steps of the vicinal α-SiO_2 $[1\bar{1}01]$ surface, while the serpentines propagated in the gas flow direction. A "falling spaghetti mechanism" has been proposed to depict the growth: the SWNT is first lifted up and grown following the gas flow-directed mode, and it falls down onto the substrate in an oscillatory fashion. Because of the competition between the flow-induced tension and tube—substrate adhesion, the nanotube is adsorbed on the surface in serpentine configuration (Fig. 1.5E lower panel). Yao et al. further verified the growth mechanism later, and they improved the growth conditions to controllably crinkle ultra-long SWNTs into serpentine geometries *(111)*.

1.2.3 Length and Density Control

Control over the morphology of SWNTs, in terms of length and density, represents a significant step toward the realization of practically applying SWNTs in nanoelectronics. The length and density of CVD-synthesized SWNTs are inherently related to the nucleation kinetics and growth mechanism.

1.2.3.1 Growth Kinetics and Mechanism for Horizontally Aligned SWNTs

Depending on the location of the nanoparticles on the growing nanotubes, the growth mechanism can generally be described as "tip growth" and "base growth." In the tip-growth mode, the catalyst always locates at the tip of a growing SWNT and is pushed forward by the newly grown segment. While in the case of the base-growth mode, the catalyst is stationed at its original position, and the SWNT grows up with the catalyst rooted on its base. The growth mode for horizontally aligned SWNTs mostly depends on the interaction between catalyst and the substrates. When the interaction is weak, the nanoparticles are easily lifted up and remain suspended in the gas stream together with the growing nanotubes, as shown in Fig. 1.6A upper panel. On the contrary, the catalyst tends to anchor on the substrate if the interaction is strong. As a result, the nanotube body crawls along the substrate, as shown in Fig. 1.6A lower panel.

The tip-growth mechanism has been taken as the preferred explanation for gas flow-directed SWNTs. The direct evidence was provided by Huang et al., in that the catalyst situates at the floating SWNT tip *(90)*. It has extensively been discovered that SWNTs succeed in crossing wide trenches and growing over obstacles, which indicates the floating of SWNTs in the gas flow during the growth *(90,94,96,97,112,113)*. Only the active end of the nanotube is floating above the substrate, while the portions grown earlier come in contact with the surface. Other studies have argued that the whole nanotube was floating in the gas flow, following either a tip- or base-growth mechanism, while the upstream end is pinned down on the surface *(15,114)*.

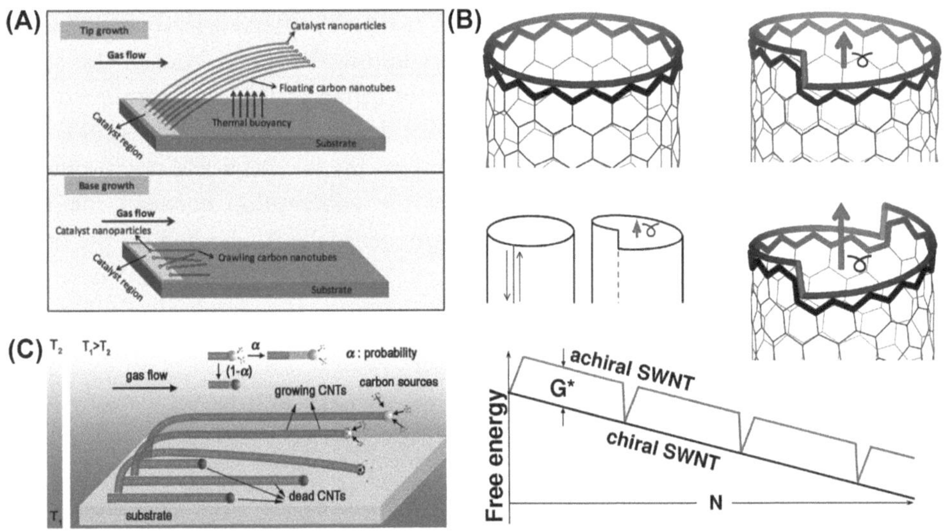

Figure 1.6 (A) "Tip-growth" and "base-growth" mechanisms for horizontally aligned single-walled carbon nanotubes (SWNTs). *(Adapted from Zhang, R.; Xie, H.; Zhang, Y.; Zhang, Q.; Jin, Y.; Li, P.; Qian, W.; Wei, F. The Reason for the Low Density of Horizontally Aligned Ultralong Carbon Nanotube Arrays. Carbon* ***2013****, 52, 232–238 with permission.)* (B) The "screw dislocation-like" (SDL) kinetic mechanism. *(Adapted from Ding, F.; Harutyunyan, A. R.; Yakobson, B. I. Dislocation Theory of Chirality-Controlled Nanotube Growth. Proceedings of the National Academy of Sciences of the United States of America* ***2009****, 106 (8), 2506–2509 with permission.)* (C) Illustration of carbon nanotube (CNT) growth obeying the Schulz–Flory distribution. *(Adapted from Zhang, R.; Zhang, Y.; Zhang, Q.; Xie, H.; Qian, W.; Wei, F. Growth of Half-Meter Long Carbon Nanotubes Based on Schulz–Flory Distribution. ACS Nano* ***2013****, 7 (7), 6156–6161 with permission.)*

Tip growth is also applicable to surface-directed SWNTs. Yu et al. proposed a "raised-head" growth model for SWNTs grown horizontally aligned on R-plane sapphire *(107)*. In this case, only the catalyst and the short nanotube segment close to it hovered above the substrate, the rest of the SWNT nestled tightly on the surface. Ding et al. obtained direct proof on the tip-growth mechanism for SWNTs grown on ST-cut quartz, based on the SEM and AFM observations of individual nanotube ends after growth *(100)*. They found catalyst at the terminated end of the SWNTs, and the nanoparticle had larger diameters than the nanotube. However, many reports have reviewed that the base-growth mechanism dominates over most of the SWNT growth. Ago et al. utilized an isotope-labeled growth method to visualize the growth process *(115)*. Through switching carbon feedstock from $^{13}CH_4$ to $^{12}CH_4$, a gradient distribution of carbon isotopes was induced along the tube axis. It was observed that ^{12}C was near the catalyst region, suggesting a base-growth mechanism. Feng et al. also reported the base-growth mechanism on a quartz surface by using nanotube barriers to terminate the growth *(101)*. On the one hand, the lattice-oriented SWNTs were terminated when hitting the nanobarriers, while on the other hand, AFM characterization showed that the catalyst could only be identified from the growth "tail" but not from the growth "head." These findings strongly implied the base-growth mechanism for the nanotubes oriented by quartz lattice.

Ding et al. recently contended a "screw dislocation-like" (SDL) mechanism *(116)*. As illustrated in Fig. 1.6B, a zigzag $(n,0)$ tube is viewed as a perfect crystal, while a chiral tube can be viewed as having a screw dislocation along the axis. The model predicted the overall dominance of nearly-armchair nanotubes in mass production. This kinetic mechanism has been remarkably testified by scores of experimental studies *(60,117–119)*. Marchand et al. directly observed the growth of individual SWNTs by field emission microscopy, and discovered that the nanotubes rotate axially during growth, thereby supporting the SDL kinetic mechanism *(117)*. Wang et al. uncovered the mechanism by single C atom addition and chirality change by C_2 dimer incorporation, based on density functional theory calculations *(60)*.

Various influential factors, including growth temperature, catalyst types and composition, carbon source, substrate class, etc., would determine the nature of nanotube growth and the resultant morphology of CNTs *(26,96,100,120,121)*. The formation of horizontally aligned CNTs obeys the SDL kinetic mechanism, according to which the carbon dimers repetitively integrate to the growing edge, forming a spiral linear carbon "polymer" with high molecular weight *(116,118)*. In analogy with the kinetics-controlled polymerization process, the growth of CNTs is comprised of three primary reactions: initiation by the active catalysts, chain propagation driven by axial screw dislocation, and termination by disturbance. Therefore it is rational to apply Schulz–Flory distribution to describe the relative ratios of CNTs of different lengths, based on catalyst activity/deactivation probability. Zhang et al. specifically interpreted the length distribution of gas flow-directed CNTs with the Schulz–Flory distribution *(122)*. As shown in Fig. 1.6C, each catalyst nanoparticle has the probability α to maintain active during nanotube growth, which is defined as "catalyst activity probability." According to the Schulz–Flory theory, the percentage (P_L) of CNTs with length L can be expressed as $P_L = \alpha^{(L-1)}(1-\alpha)$. Therefore all factors affecting CNT growth could be ascribed to the catalyst activity probability. With sufficient carbon feedstock, high catalyst activity probability is the requisite toward long CNTs with high density.

1.2.3.2 Length Control of SWNTs

In general, Si/SiO_2 substrates, because of their weak interaction with the catalyst nanoparticles and the nanotubes, are more appealing for the growth of SWNTs under gas flow-directed mode. However, even by following the tip-growth mechanism, the length of horizontally aligned SWNTs had been restricted within several millimeters for a long period of time, until the pioneer work reported by Zheng et al. in 2004, where they synthesized 4 cm-long SWNTs at a high growth rate of 11 μm/s (Fig. 1.7A) *(25)*. In their study, Fe nanoparticles derived from $FeCl_3$ ethanol solution were utilized as the catalysts, ethanol vapor was chosen as the source material, and the growth temperature was set as 900°C. It was suggested that the high catalytic activity of Fe together with the high reactivity and dissociation property of the ethanol molecules were responsible for the huge improvement in SWNT length. With similar growth conditions, the length of CNTs was further increased up to 10 cm, only limited by the size of silicon wafers or the length of the hot zone of the furnace *(93,123)*. Adding a trace amount of water in the gas feed was found to greatly facilitate the elongation of horizontally aligned SWNTs *(92,124)*. Water served as a mild oxidizer in the reaction to remove or inhibit the formation of amorphous carbon on the catalyst and thus to maintain the activity for a longer time. The average growth rate was tripled and the length of the CNTs was increased to 20 cm *(124)*.

The properties of catalysts are highly sensitive to growth temperature, thus it is highly desired to maintain a spatially homogeneous temperature during the growth process. Wang et al. used an 86 cm-long electrical thermal furnace, which gave a constant temperature zone of 25 cm in length *(125)*. A layer of superaligned CNT film was used to support the monodispersed Fe–Mo catalysts, which allowed for the "catalyst residual-free" production of SWNTs, facilitating subsequent fabrication of SWNT-based devices. Different feed gas mixtures were tried and ultralong SWNTs up to 18.5 cm in length were successfully obtained. Zhang et al. have developed a "furnace-moving" method where the furnace was moved at the same speed as the CNT growth speed (Fig. 1.6B). Half-meter-long CNTs were synthesized, and the defined "catalyst activity probability" was up to 0.995 *(122)*.

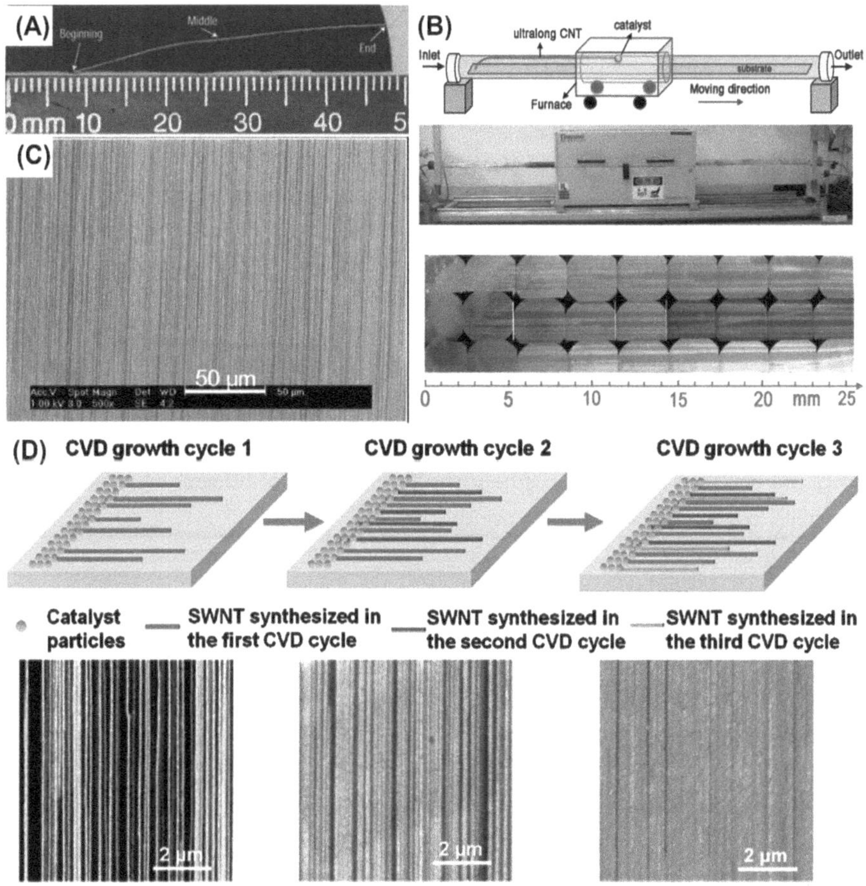

Figure 1.7 (A) Growth of 4 cm-long individual single-walled carbon nanotubes (SWNTs). *(Adapted from Zheng, L. X.; O'Connell, M. J.; Doorn, S. K.; Liao, X. Z.; Zhao, Y. H.; Akhadov, E. A.; Hoffbauer, M. A.; Roop, B. J.; Jia, Q. X.; Dye, R. C.; Peterson, D. E.; Huang, S. M.; Liu, J.; Zhu, Y. T. Ultralong Single-wall Carbon Nanotubes. Nature Materials* ***2004****, 3 (10), 673–676 with permission.)* (B) "Furnace-moving" method for producing half-meter-long carbon nanotubes (CNTs). *(Adapted from Zhang, R.; Zhang, Y.; Zhang, Q.; Xie, H.; Qian, W.; Wei, F. Growth of Half-Meter Long Carbon Nanotubes Based on Schulz–Flory Distribution. ACS Nano* ***2013****, 7 (7), 6156–6161 with permission.)* (C) Ultradense SWNTs grown on quartz substrate. *(Adapted from Ding, L.; Yuan, D.; Liu, J. Growth of High-Density Parallel Arrays of Long Single-walled Carbon Nanotubes on Quartz Substrates. Journal of the American Chemical Society* ***2008****, 130 (16), 5428–5429 with permission.)* (D) Growth of aligned SWNTs by three-cycle chemical vapor deposition (CVD) method. *(Adapted from Zhou, W.; Ding, L.; Yang, S.; Liu, J. Synthesis of High-Density, Large-Diameter, and Aligned Single-walled Carbon Nanotubes by Multiple-Cycle Growth Methods. ACS Nano* ***2011****, 5 (5), 3849–3857 with permission.)*

1.2.3.3 Density Control of SWNTs

The main obstacle for the bulk production of horizontally aligned ultralong SWNTs is their extremely low density, not to mention that the density drops rapidly along the tube axial direction *(94,125)*. As described by the Schulz–Flory distribution, the key for increasing CNT density is to raise the "catalyst activity probability." Several routes have been proposed. Liu et al. annealed the Fe catalyst nanoparticles in air at 750°C for 5 min *(96)*, and the thickness of Fe film for catalyzing the synthesis was optimized in the gas flow process. As a result, the density of the SWNTs reached >60 nanotubes/100 μm, although the length of the SWNTs was only a few millimeters. An et al. systematically investigated the effect of growth parameters on the activity and lifetime of catalysts *(121)*. It was suggested that smaller nanoparticles were more fluid than the larger ones and they were more active at higher temperature. Besides, a pretreatment procedure with a proper duration was required to obtain uniform nanoparticles and also to prevent the particles coarsening. Their results demonstrated that the length and density could be well controlled by growth conditions.

The high growth temperature provides sufficient energy for nucleation; however, not all of the nucleated CNTs could attain macroscopic length. Zhang et al. discovered that the reason was caused by the aggregation of catalyst nanoparticles at high temperature *(126)*. The merged nanoclusters produced multiwalled carbon nanotubes (MWNTs) with large diameters, which tend to follow the base-growth mechanism, and hence hinder the elongation of nanotubes. Therefore the efficient way to improve the density of ultralong SWNTs is to prevent catalyst agglomeration on substrates. Xie et al. respectively employed silica nanospheres and grapheme/graphite sheet for catalyst nanoparticle anchoring *(127,128)*. The coarse surface of silica nanospheres and the wrinkles on the grapheme/graphite sheet both functioned to suppress small catalyst nanoparticles agglomerating into large ones, thereby leading to production of dense SWNTs with macroscopic length.

Another major category of obtaining super-dense, superaligned SWNTs is to grow nanotubes on crystalline substrates. It was suggested that catalyst and reactive carbon species would preferentially accumulate on the dense atomic steps and edges and high atomic density of a specific crystal lattice. The enhanced reactivity and the strong interactions between the growing nanotubes and the substrates would lead to SWNTs with high density and large scale *(15)*. Groundbreaking experiments of growing SWNTs on sapphire or quartz substrates have been carried out; the average density of the produced SWNTs could reach as high as 10 SWNTs/μm *(2,16,103,129)*. Ding et al. improved the density on quartz substrate up to 50 SWNTs/μm by employing copper as catalyst, as shown in Fig. 1.7C *(100)*. To give the catalyst more opportunities to nucleate SWNTs, a multicycle process was adopted *(130,131)*. As shown in Fig. 1.7D, Zhou et al. repeated the CVD growth process three times to improve the density. SWNTs with densities of 20–40 SWNTs/μm were successfully obtained over a large area. Most recently, Hu et al.

developed a new type of Trojan catalyst system, where the catalysts were gradually released and thus the catalytic efficiency was increased. The record for the density of SWNTs is set at 130 SWNTs/μm *(132)*.

1.3 SYNTHESIS OF VERTICALLY ALIGNED CARBON NANOTUBE ARRAYS

In addition to nanoelectronics, there are many other applications *(133—140)*, including e-textile, artificial muscle, composites, field emission, etc., that prefer another type of CNT product: the vertically aligned carbon nanotube arrays (CNTAs). It is reported that 1D (bundle and fiber), 2D (film or sheet), and 3D (array and foam) assemblies could be directly derived from CNTAs.

1.3.1 Wall Number and Diameter Control

Many applications require distinctive properties that are determined by the number of walls of CNTs. Consequently, the development of wall number controllable growth of CNTAs will be of great significance in both fundamental study and innovative applications. Many strategies have been developed to grow CNTAs with defined wall number, by varying the catalyst size and composition.

1.3.1.1 Catalyst Size Optimization

It is very straightforward to control diameter and wall number of nanotubes in CNTAs by tuning the catalyst size. As shown in Fig. 1.8, the number of graphitic walls in a CNT is dependent on CNT diameter, which is dependent on catalyst thickness. The mean CNT diameter is proportional to the thickness of the Fe thin film, and thin CNTs have a strong tendency to be SWNTs while thick CNTs tend to be double-walled carbon nanotubes (DWNTs) or MWNTs. SWNTs occupied the majority of the nanotube population within a diameter range of 1.0—2.5 nm, and DWNTs account for the majority of CNTs with a diameter in the range of 3.0—4.5 nm, while MWNTs have a higher percentage for the diameter larger than 4.5 nm. Maximum DWNT selectivity was reached at an Fe thin film thickness of 1.69 nm. It is generally believed that during catalyst annealing, thin (thick) metal films break up into smaller (larger) catalytic nanoparticles, which, in turn, produce smaller (larger) tubes. Empirically, it is well established that a thin catalytic metal film (around 1 nm) is required to grow SWNTs *(141)*.

To achieve a better wall-selective growth of CNTAs, Chiang et al. *(142)* presented direct wall number control by engineering Fe catalysts with a low-deposition-rate sputter deposition, which allows for the control of Fe catalyst thickness with small variations *(142)*. AFM characterization on the annealed Fe catalysts confirmed that catalyst particles with high density were formed and the average particle size varied in proportion to the

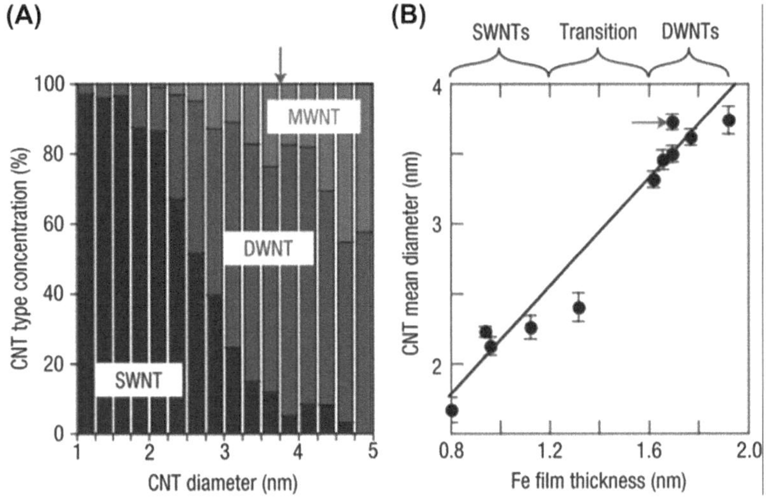

Figure 1.8 Trends in carbon nanotube (CNT) type and diameter. (A) Concentrations of single-walled carbon nanotube (SWNT), double-walled carbon nanotube (DWNT), and multiwalled carbon nanotube (MWNT) as a function of the CNT diameter; (B) CNT mean diameter as a function of Fe film thickness. *(Adapted from Yamada, T.; Namai, T.; Hata, K.; Futaba, D. N.; Mizuno, K.; Fan, J.; Yudasaka, M.; Yumura, M.; Iijima, S. Size-Selective Growth of Double-walled Carbon Nanotube Forests from Engineered Iron Catalysts. Nature Nanotechnology* ***2006****, 1 (2), 131–136 with permission.)*

initial Fe film thickness. TEM images revealed that Fe catalyst films with a mean thickness of 1.5 nm produce primarily SWNTs; as the catalyst film thickness increased to 2.0 nm, a larger portion of DWNTs were found with an outer and inner diameter of 5.2 and 4.9 nm, respectively; for catalyst film thickness of 3.0 nm, a higher fraction of CNTs with a typical wall number of three were observed. The same group also proposed a method to quantitatively characterize the thickness of Fe thin film and controlled the wall number of CNTs by electrical resistance measurement *(143)*. Combining electrical characterization and TEM analysis, they found that SWNT, DWNT, and triple-walled nanotube CNTAs were grown on Fe catalyst films with mean sheet resistances of 646.63, 75.40, and 27.84 MΩ/sq, respectively; and the average wall number and outer diameter of CNTs were found to be linearly proportional to the logarithm of the mean sheet resistances of Fe catalyst films. All these experimental findings demonstrate that precisely controlling the catalyst film thickness is a key factor for wall number-controlled growth.

For a given thickness of catalyst films, the particle size also changes according to the catalyst annealing and CNT growth processes. At elevated temperature, many processes happened simultaneously, such as Ostwald ripening *(17)* and catalyst aggregation *(144,145)*. Thurakitseree et al. *(146)* have systematically studied the influence of catalyst annealing procedures on the catalyst size and average diameter of SWNTs. Raman radial breathing mode spectra indicated a decrease in the average SWNT diameter for lower reduction temperatures, and the reason was believed to be the less agglomeration of catalyst particles. But when temperatures are lower than 500°C, SWNTs were randomly aligned. This demonstrated that such a reduction was too mild to reach the minimum areal density of active catalyst particles. The first optical band gap E_{11} of SWNTs synthesized from low-temperature-annealed catalysts showed a significant blue shift, indicating a small average SWNT diameter. In addition, the extended annealing time also caused the aggregation of catalyst particles and, in turn, increased average SWNT diameters.

Subsurface diffusion is another key factor affecting the catalyst size and thus wall number and diameter of CNTs. Fe could directly migrate into Si substrate or diffuse across the native oxide into Si to form silicide at temperatures higher than 450°C *(147)*. The aluminum oxide (AlO_x) layer was then widely used as the supporting layer to control the subsurface diffusion of the catalyst. The effect of AlO_x depends on its nanostructures: a porous supporting layer may not inhibit Fe diffusion away from the surface. Therefore at high temperature, there exist two competing processes: Ostwald ripening that increases particle size and subsurface diffusion that decreases particle size. The catalyst particle size and, in turn, the diameter of CNTs will be determined by comprehensive effects of these two processes, as schematically described in Fig. 1.9 *(144)*.

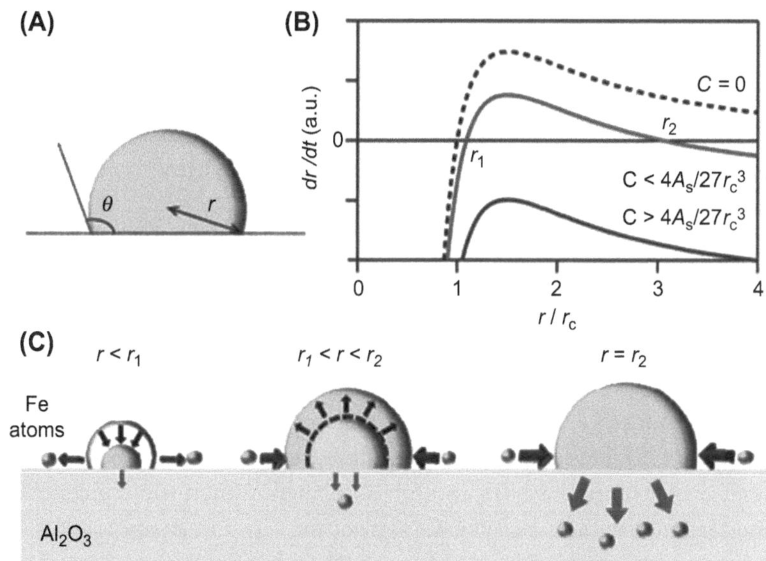

Figure 1.9 (A) The Fe nanoparticle modeled in Eq. (1.1). (B) dr/dt values versus r/r_c. (C) Schematic representation of the growth behavior of Fe nanoparticles. *(Adapted from Sakurai, S.; Nishino, H.; Futaba, D. N.; Yasuda, S.; Yamada, T.; Maigne, A.; Matsuo, Y.; Nakamura, E.; Yumura, M.; Hata, K. Role of Subsurface Diffusion and Ostwald Ripening in Catalyst Formation for Single-walled Carbon Nanotube Forest Growth. Journal of the American Chemical Society* ***2012****, 134 (4), 2148–2153 with permission.)*

Catalyst nanoparticle formation was modeled on the basis of the balance between these two opposing processes. The growth rate of a particle with radius r could be described as Eq. (1.1) *(144)*:

$$\frac{dr}{dt} = -\frac{A_s}{r^3}\left(1 - \frac{r}{r_c}\right) - C \tag{1.1}$$

where A_s is constant, r_c is critical nucleus size, and C is the subsurface diffusion rate of Fe in a catalyst particle.

It is important to note that nanoparticles cannot exist in the size range if dr/dt is negative. Any particle with a size larger than r_c can survive stably, because the growth rate is positive, as shown in Fig. 1.9B. Considering the subsurface diffusion effect, growth rate is positive only in a specific range (between r_1 and r_2) if $C < 4A_s/(27r_c^3)$, see middle solid line in Fig. 1.9B. This is fully consistent with the experimental observation. Because of Ostwald ripening, small particles with $r < r_1$ become smaller and eventually extinguish. On the other end, the particles stop growing at $r = r_2$ where effects of subsurface diffusion and Ostwald ripening are equalized (Fig. 1.9C). When subsurface diffusion is strong $C > 4A_s/(27r_c^3)$, the growth rate of catalyst particles is always negative, and thus nanoparticles of any diameter will disappear. By controlling the Ostwald ripening and subsurface diffusion, the catalyst particle size could be tuned. Sakurai et al. found that the SWNT arrays with similar average diameters (2.8–3.1 nm) could be grown from Fe/AlO_x catalysts regardless of the iron precursors used to prepare the catalyst. This observation highlighted the competing mechanisms of Ostwald ripening and subsurface diffusion of Fe catalyst *(144)*. A compact and dense supporting layer is able to effectively increase subsurface diffusion resistance, for example, oxygen plasma treatment on the AlO_x supporting layer could greatly densify the AlO_x supporting layer, and benefit the growth of CNTAs *(148)*.

1.3.1.2 Catalyst Composition Optimization

Other secondary species were introduced to form bimetal or even multimetal catalysts to inhibit/slow Ostwald ripening. Mo itself is inactive, but it is highly effective in controlling CNT diameters once being alloyed with commonly used catalyst metals (Fe, Co, and Ni) *(149–151)*. Xiang et al. reported wide-range diameter modulation using a bimetal Co–Mo catalyst *(145)*. By carefully adjusting the catalyst reduction conditions, CNTAs with a diameter of 1.6 nm SWNTs could be obtained. When the concentration of Co was decreased in the solution, a clear blue shift of the E_{11} and E_{22} peaks was seen in optical absorbance as displayed in Fig. 1.10A, suggesting that the average diameter decreased from 2 to 1.6 nm when the Co concentration was decreased by two orders in magnitude. Fig. 1.10B shows optical absorbance spectra of SWNTs grown using various Mo amounts while keeping the Co concentration constant. The E_{11} and E_{22} peaks of the SWNTs because of the increase of Mo concentration exhibit more significant shifts than in the

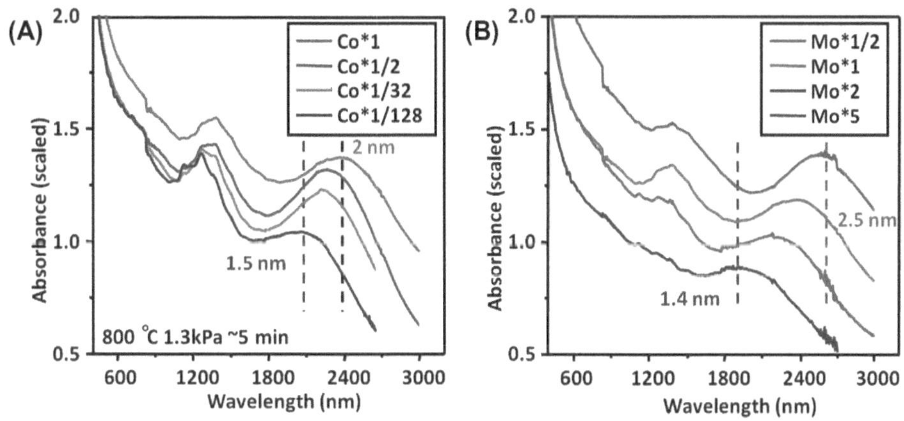

Figure 1.10 UV–Vis–NIR optical absorbance spectra of aligned single-walled carbon nanotubes (SWNTs) synthesized using different catalyst recipes. The concentration of the catalyst solution is normalized by a standard value of 0.01 wt% for each of the metal species in (A) cobalt acetate and (B) molybdenum acetate. *(Adapted from Xiang, R.; Einarsson, E.; Murakami, Y.; Shiomi, J.; Chiashi, S.; Tang, Z.; Maruyama, S. Diameter Modulation of Vertically Aligned Single-walled Carbon Nanotubes. ACS Nano* ***2012****, 6 (8), 7472–7479 with permission.)*

reduced Co case, indicating the average diameter decreases from 2.5 to 1.4 nm. This diameter decrease was obtained by increasing the Mo concentration by only 10 times, yet was much more significant than that obtained by reducing the Co concentration by more than 100 times. This agrees well with the hypothesis that Mo forms compounds (MoO_x or MoC_3) *(152)* that strongly interact with Co, thereby reducing the mobility of Co nanoparticles. Mo in a bimetal catalyst could also improve the diameter distribution of CNTs *(149)*. Sugime et al. found that with the nominal thickness ratio Co/Mo decreasing from 0.11/0.28, 0.22/0.14, 0.070/0.047, and 0.78/0, the diameter distribution of CNTs was transited from unimodal to bimodal shape, which suggests that Mo suppresses the surface diffusion of Co adatoms and the Ostwald ripening process, resulting in a Co catalyst assembly with a narrow size distribution *(149)*. Many other binary systems have also been developed for growth of CNTAs *(146,151,153)*. However, a thorough understanding of the detailed mechanism of the interactions between different components in these alloyed catalysts is still lacking.

1.3.2 Areal Density-Controlled Growth of CNTAs

CNTAs, if used as interconnect materials or thermal interfacial materials in electronics or used as field emission display, are required to have particular areal density. For example, to replace copper in interlayer vias, a target density of CNTAs of at least $2 \times 10^{13}/cm^2$ is needed *(154)*. Currently achieved maximum densities for CNTAs in vias do not exceed $10^{11}/cm^2$. The low areal density is one of the key reasons why CNT interconnects have not yet been implemented by electronics manufacturers.

1.3.2.1 Control the Catalyst Loading

Controlling the density of CNTAs could be realized by adjusting the density of catalyst nanoparticles. Huh et al. *(155)* have controllably grown CNTs using uniformly distributed Co nanoparticles, which were distributed on substrates by spin coating colloidal solutions of Co nanoparticles of different concentrations (1, 20, and 80 mM). They found that Co nanoparticles were well distributed on the substrates and their density was determined by concentration of nanoparticles in colloid. By pyrolysis of C_2H_2 gas at 850°C, CNTAs were obtained. In the case of high-density Co nanoparticles, the high-density CNTAs were synthesized with good alignment, while the low-density CNTAs were produced with poor alignment when the low-density Co nanoparticles were used. The alignment and density of CNTAs can be simply controlled by adjusting the Co concentration in colloidal solution. However, special care should be taken if the Co concentration is too high; Co nanoparticle coarsening will dominate during catalyst annealing and may decrease the density of CNTs.

Similarly, Bennett et al. *(156)* demonstrated a route to utilize iron oxide nanoclusters that utilize micelles formed by the amphiphilic block copolymer, polystyrene-*b*-poly(acrylic acid) to create arrays of uniform diameter iron oxide nanoclusters, with quantifiable areal densities that can be varied over more than an order of magnitude. This catalyst system enables the creation of nanocluster arrays with independent control of the nanocluster diameter and areal density. The density of catalyst nanoclusters could be well tuned by adding varied amounts of polystyrene into copolymer micelles. A 50 nm-thick micellar thin film with multilayers of iron oxide nanoclusters resulted in an iron oxide areal density of 1.5×10^{11} cm^{-2} (Fig. 1.11A). CNT growth using this catalyst (Fig. 1.11F) resulted in uniform, dense, and vertically aligned CNTAs. By decreasing the catalyst areal density to 6×10^{10} cm^{-2} (Fig. 1.11B), a uniform, dense, and vertical CNTA could also be obtained (Fig. 1.11G). However, an areal density of nanoclusters of 2.5×10^{10} cm^{-2} (Fig. 1.11C and H) resulted in an irregular oriented array, where the average thickness of the CNT thin film was significantly reduced. These results suggest that the minimum areal density of iron oxide nanoclusters required to grow uniform and thick CNTAs is in the range between 2.5×10^{10} cm^{-2} and 6.0×10^{10} cm^{-2}. When the areal density of nanoclusters was further decreased to 1×10^{10} cm^{-2} (Fig. 1.11D and I), the overall density of the CNTs decreased significantly. By combining the length of the CNTs with the diameter and the number of concentric walls, the mass of an individual CNT can be estimated. The ratio of the mass of the CNT film to the mass of an individual CNT gives the number of CNTs, which can be converted to the CNT areal density by dividing by the substrate area. The CNT areal density was estimated to be 5.2×10^{9} cm^{-2} (0.879 mg cm^{-2}), 2.2×10^{9} cm^{-2} (0.363 mg cm^{-2}) for catalyst clusters with a density of 1.5×10^{11} cm^{-2} and 6×10^{10} cm^{-2}, respectively. By comparing with the known iron oxide nanocluster density, the catalytic activity for the multilayer nanocluster catalyst was determined to be approximately 4%.

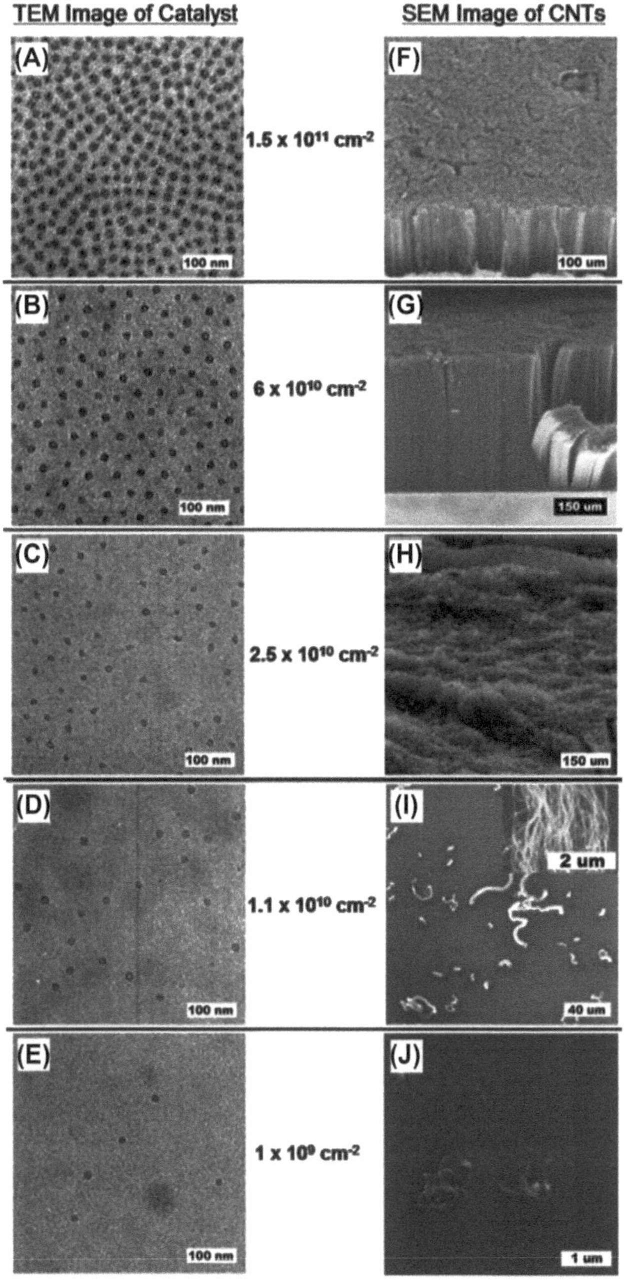

Figure 1.11 Effects of varying nanocluster areal density on the morphology of the carbon nanotube (CNT) films. The left column (A—E) shows transmission electron microscopy (TEM) images of the iron oxide nanocluster arrays with varying areal density prior to CNT growth. The right column (F—45 degree tilt, G—85 degree tilt, H—70 degree tilt, I—0 degree tilt, J—0 degree tilt) shows scanning electron microscope (SEM) images of CNTs. The figures correspond to the following iron oxide nanocluster densities: (A, F) 1.5×10^{11} cm^{-2}; (B, G) 6×10^{10} cm^{-2}; (C, H) 2.5×10^{10} cm^{-2}; (D, I) 1.1×10^{10} cm^{-2}. The inset in I shows a higher magnification image of one vertically grown CNT column; and (E, J) approximately 1×10^{9} cm^{-2}. *(Adapted from Bennett, R. D.; Hart, A. J.; Cohen, R. E. Controlling the Morphology of Carbon Nanotube Films by Varying the Areal Density of Catalyst Nanoclusters Using Block-Copolymer Micellar Thin Films. Advanced Materials* ***2006****, 18 (17), 2274 with permission.)*

1.3.2.2 Tune Catalyst Effectiveness

Comparing the density of catalyst particles and the density of CNTAs, one can find that only 1−10% of the catalysts produce CNTs. It is obvious that catalyst effectiveness is more critical in obtaining high density CNTAs. A complicated strategy was then introduced during catalyst treatment and CNT synthesis.

Nesssim et al. *(157)* have examined the effects of pretreatment on the subsequent growth of CNTAs. They found that controlling only the timing of the introduction of hydrogen led to tunability of the CNT diameter by a factor of 2, and a corresponding increase of the areal density by a factor of 10. Furthermore, introduction of H_2 at different times after ethylene was introduced, rather than following the standard practice of simultaneously introducing the reactant and reducing gases, led to significantly enhanced CNT growth rates. The time/temperature/flow sequence is illustrated in Fig. 1.12A. Delaying H_2 introduction leads to better-aligned and taller CNTAs. The average CNT diameter decreased from 14 nm (7 walls) when H_2 is introduced 5 min before carbon source (H_2 at −5) to only 7 nm (3 walls) when H_2 was introduced 10 min after carbon source (H_2 at +10). Concomitantly, the areal density increased by an order of magnitude from 4×10^9 cm^{-2} for H_2 at −5 to 5×10^{10} cm^{-2} for H_2 at +10 (as shown in the table in Fig. 1.12B). During catalyst pretreatment, the clusters will initially be small, but they start

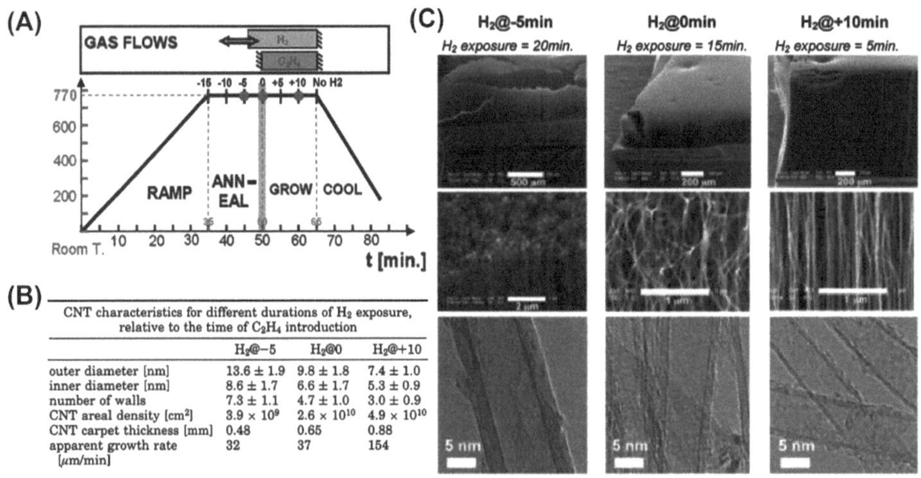

CNT characteristics for different durations of H_2 exposure, relative to the time of C_2H_4 introduction

	H_2@−5	H_2@0	H_2@+10
outer diameter [nm]	13.6 ± 1.9	9.8 ± 1.8	7.4 ± 1.0
inner diameter [nm]	8.6 ± 1.7	6.6 ± 1.7	5.3 ± 0.9
number of walls	7.3 ± 1.1	4.7 ± 1.0	3.0 ± 0.9
CNT areal density [cm^2]	3.9×10^9	2.6×10^{10}	4.9×10^{10}
CNT carpet thickness [mm]	0.48	0.65	0.88
apparent growth rate [μm/min]	32	37	154

Figure 1.12 (A) Thermal schedule used for chemical vapor deposition (CVD) growth. (B) Measurements of carbon nanotubes (CNTs) from the three processes indicated by the dots in (A). (C) Scanning electron microscope (SEM) and high-resolution transmission electron microscopy (HRTEM) images of CNT carpets after processing when H_2 is introduced 5 min before ethylene (H_2 at −5), at the same time as ethylene (H_2 at 0), and 10 min after ethylene (H_2 at +10). *(Adapted from Nessim, G. D.; Hart, A. J.; Kim, J. S.; Acquaviva, D.; Oh, J.; Morgan, C. D.; Seita, M.; Leib, J. S.; Thompson, C. V. Tuning of Vertically-Aligned Carbon Nanotube Diameter and Areal Density through Catalyst Pre-treatment. Nano Letters* ***2008****, 8 (11), 3587−3593 with permission.)*

to ripen driven by surface diffusion. In this process, small clusters tend to shrink and disappear with their mass being redistributed to larger clusters, resulting in an increase in the average cluster size as well as a decrease in cluster density. Therefore catalyst pretreatment could be an effective tool to tune the diameter and areal density of CNTAs.

1.3.3 Alignment Controlled Growth of CNTAs

In many applications, CNTs must be effectively assembled into macroscopic products, so that their excellent properties can be retained at a larger scale. CNTAs could be used to derive 1D CNT fibers, 2D CNT sheets/films, and 3D foams, and the processability or spinnability of converting a CNTA into a macroscopic assembly is found strongly dependent on the alignment of the CNTA. A good alignment in CNTAs is believed to be the prerequisite to full exploitation of CNT's potentials.

1.3.3.1 Crowding Effect

During CNT growth, if the interspacing between CNTs is decreased, there will exist less space for CNTs to become wavy, and, in turn, the alignment will be improved. Futaba et al. found that the alignment of wavy CNTs could be greatly improved after liquid densification *(139)*. Xu et al. demonstrated a general approach to control the alignment of few walled CNTAs from nearly random to ideally aligned by tuning the density of active catalysts. Experimentally, the catalyst density and the degree of alignment were inseparably related because of a crowding effect *(158)*. They also proposed a simple, first-order model based on the flexural modulus of the CNTs that could quantitatively describe the relationship between the degree of alignment (Herman's orientation factor, HOF) and CNT crowding effect (Fig. 1.13A and B). HOF could be derived from this angular deviation according to a classic mechanical model *(158)*:

$$\mathrm{HOF} \approx 1 - \mathrm{C}\left[\left(E^{-1/3}\right)\right] D^{2/3} \tag{1.2}$$

where C is a constant representing geometric factors, D is catalyst spacing, and E is the flexural modulus, representing the CNT stiffness.

The crowding effect describes how CNTs physically interact with their neighbors to determine the growth direction. Specifically, at low CNT density, the lateral confinement was weak and allowed the CNTs to experience minimal interaction with neighboring CNTs, and thus the CNTs grew in arbitrary directions away from the substrate, resulting in a nonaligned forest. In contrast, at high CNT density, the lateral confinement increased where the "crowding" of neighboring CNTs becomes prominent, and thus the CNTs could no longer grow in arbitrary directions but primarily in the vertical direction where the disturbance from neighboring CNTs was minimal, as shown in Fig. 1.13C.

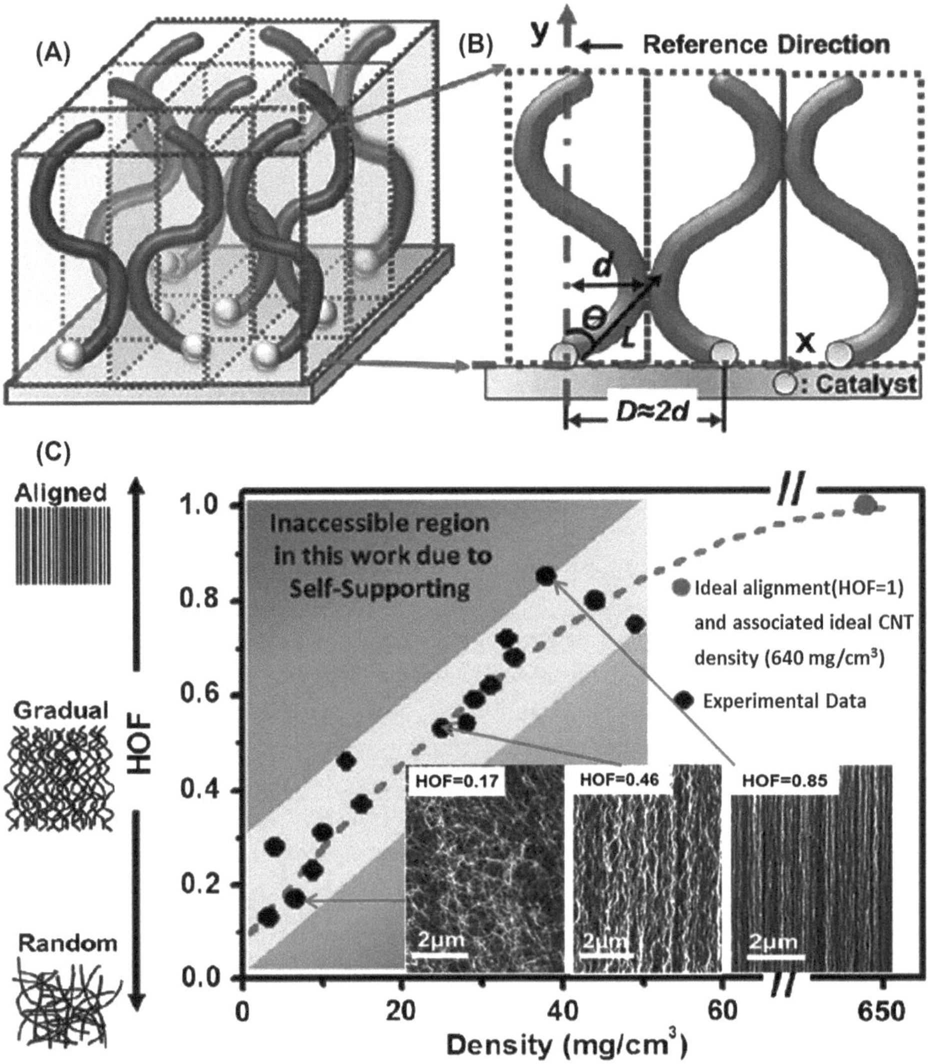

Figure 1.13 (A) Schematic of the assumption to treat the carbon nanotube (CNT) forest "many-body" problem to a more solvable "single-CNT" problem. (B) Graph of alignment as a function of catalyst spacing. (C) Graph of the alignment as a function of density; inset shows the morphology of CNT arrays with varied Herman's orientation factor (HOF). *(Adapted from Zhang, Y.; Zou, G.; Doorn, S. K.; Htoon, H.; Stan, L.; Hawley, M. E.; Sheehan, C. J.; Zhu, Y.; Jia, Q. Tailoring the Morphology of Carbon Nanotube Arrays: From Spinnable Forests to Undulating Foams. ACS Nano* ***2009****, 3 (8), 2157–2162 with permission.)*

First, this relationship indicates that CNT interspacing and HOF should fall within a limited window as experimentally observed (Fig. 1.13C). Second, the model presents a quantitative description of the crowding effect where the alignment (HOF) is predicted to increase with decreasing CNT spacing (D). The model could be expanded to predicate the degree of alignment of arbitrary CNTs provided that the flexural modulus could be estimated. With HOF increasing from 0 to 1, the CNT morphology will become nearly perfectly aligned from nearly random, as shown in the inset of Fig. 1.13C *(158)*.

The crowding effect exerted by the areal density provides a new strategy to tune the alignment of CNTs. The mechanism of crowding effect realized by catalyst pretreatment could be schematically described in Fig. 1.14. Generally, there are two main stages involved in the preparation of CNTs: the formation of catalyst particles and the nucleation/growth of CNTs from the catalyst. The H_2 in the forming gas first reduces the catalyst film and cracks the film into densely packed catalyst particles with narrow size

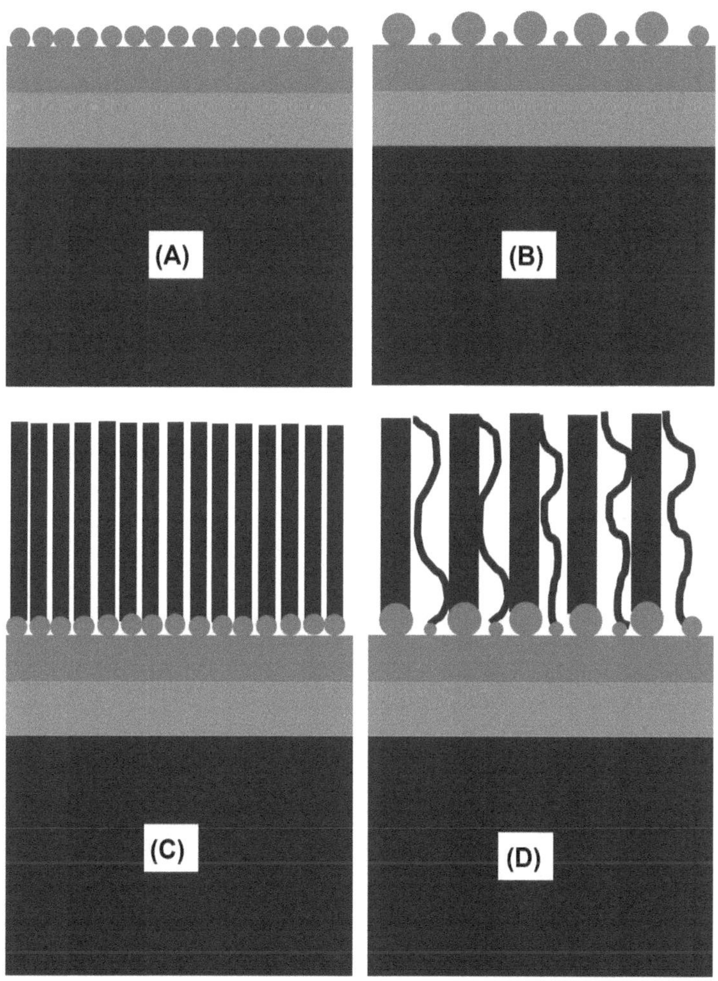

Figure 1.14 Schematic diagram of catalyst evolution and growth of carbon nanotubes (CNTs). (A) Reduction of catalyst film into fine particles. (B) Coarsening of catalyst particles. (C) Well-aligned CNTs grown on fine catalyst. (D) Nonuniform and wavy CNTs grown on coarse catalyst.

distribution, as shown in Fig. 1.14A. Meanwhile, the prolonged exposure to H_2 at CNT growth temperature will cause the ripening of catalyst particles, resulting in nonuniform particle size and an increase in particle spacing as displayed in Fig. 1.14B. The CNTAs grown from the densely packed catalyst particle assembly and scattered catalyst particle array are shown in Fig. 1.14C and D, respectively. These densely packed fine catalyst particles have the tendency to promote the growth of CNT arrays with high areal density and good alignment, because there is not enough space for CNTs to become wavy. In addition, the narrow size distribution of fine particles enables the CNTs to grow with the same speed, which also contributes to good alignment. While the catalyst particles that experienced a longer pretreating time become nonuniform and less dense, the small particles are more efficient in promoting CNT growth; and whereas the coarse particles will have lower catalytic activity and the CNTs will grow slower, the nonuniform growth will then contribute to the wavy and random CNTs, as shown in Fig. 1.14D.

1.3.3.2 Alignment From Uniform Growth

As illustrated in Fig. 1.14, good alignment could be obtained from dense and uniform CNT growth, in which each CNT tends to grow with the same speed, and the physical interaction between CNTs will not disturb the growth of CNTs in the vertical direction. Zhang et al. *(159)* have tuned the catalyst particle size distribution and studied its effect on CNT alignment: the shortest catalyst pretreatment time led to CNTAs with the best alignment, while prolonged pretreatment resulted in wavy CNTAs. The CNTAs grown by introducing C_2H_4 at the temperature ramping-up stage were straight and parallel; in contrast curly CNTAs were observed in samples grown by introducing carbon source after 10 min catalyst pretreatment. This trend became much more obvious with longer catalyst pretreatment time; most of the CNTAs grown from 60-min pretreated catalysts are curly and show random orientations.

Zheng et al. have observed that the CNTA morphologies could be easily tuned by introducing hydrogen and/or oxygen during CNTA growth *(138)*, and there exists a strong correlation between the array morphologies and the fiber (spun from the arrays) properties: well-aligned arrays yield high performance in CNT fibers, while wavy arrays give poor performance. Fig. 1.15A shows a CNTA grown by the oxygen-assisted-growth array with very wavy and entangled CNTs. In contrast, the arrays synthesized in the hydrogen-assisted-growth array show uniformly distributed and well-aligned CNTs (Fig. 1.15C). A small amount of oxygen is able to change the well-aligned morphology into the wavy one, even with hydrogen cointroduction, suggesting that hydrogen and oxygen play different roles. The typical morphology of the normally grown CNTAs (neither hydrogen nor oxygen introduced) is somewhere between these two morphologies, as shown in Fig. 1.15B. Fig. 1.15D–F shows SEM images for fibers spun from according arrays, indicating that the CNTs in fibers share a similar alignment with their original arrays.

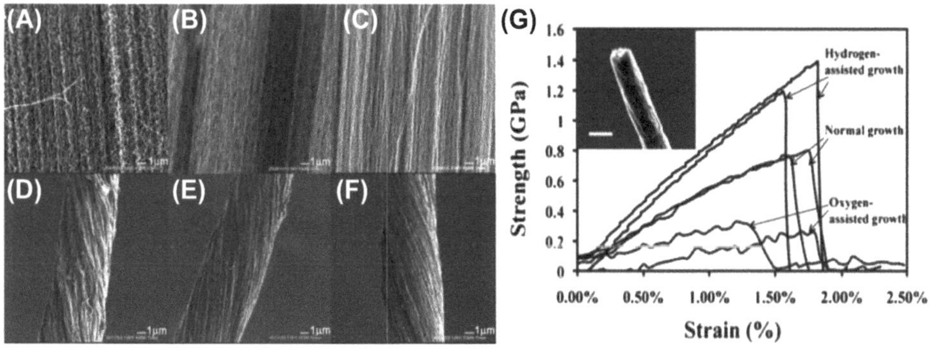

Figure 1.15 Typical scanning electron microscope (SEM) images of (A) the oxygen assistance-grown carbon nanotube arrays (CNTAs); (B) the normally grown CNTAs; (C) the hydrogen assistance-grown CNTAs; (D) carbon nanotube (CNT) fibers spun from oxygen assistance-grown CNTAs; (E) CNT fibers spun from normally grown CNTAs; (F) CNT fibers spun from hydrogen assistance-grown CNTAs; (G) stress–strain plots for CNT fibers spun from different arrays; the inset shows a fractured fiber (scale bar is 5 mm). *(Adapted from Zheng, L.; Sun, G.; Zhan, Z. Tuning Array Morphology for High-Strength Carbon-Nanotube Fibers. Small* ***2010****, 6 (1), 132–137 with permission.)*

In the presence of hydrogen, the growth of CNTAs and the formation of amorphous carbon from/on the catalyst particles will eventually reach a steady state, resulting in a uniform growth rate for all individual CNTs. However, in the presence of oxygen, the formation and the removal of amorphous carbon on/from the catalysts happen randomly, which results in nonuniform growth rates. The alignment in the arrays directly determines the mechanical performance of CNT fibers. The average tensile strengths are 0.29, 0.78, and 1.3 GPa for the fibers spun from the oxygen-assisted-growth, the normal growth, and the hydrogen-assisted-growth arrays, respectively. With a slight change in growth condition (from 150 ppm oxygen addition to 2% hydrogen addition), the strength of resultant fibers has been improved by 4.5 times, suggesting that the alignment of CNTs is critical to the mechanical performance of CNT fibers, as shown in Fig. 1.15G.

1.3.4 Length-Controlled Growth of CNTAs

It is found that the CNT fibers spun from long CNTAs show higher mechanical performance compared to those spun from short ones *(160)*, so the strategies to enhance CNT growth in the vertical direction have also been intensively studied. General ideas of growing ultralong CNTAs include the enhancement of growth rate and elongation of catalyst lifetime.

1.3.4.1 Grow Long CNTAs Through Enhancing the Growth Rate

The growth rate of CNTs is strongly dependent on the size of catalyst particles and growth temperature. Small catalyst particles tend to have high catalytic activity, while high temperature is prone to enhance CNT growth rate. However, it is challenging to keep small catalyst and high temperature growth at the same time. Thus decoupling the catalyst pretreatment (to get a small catalyst) and high temperature growth (to get a high growth rate) is an effective method to preserve both high catalytic activity of catalyst particles and also maximize the temperature-dependent growth rate. Zhan et al. have developed a two-step method that separates catalyst pretreatment from the subsequent growth stage *(161)*. They first optimized the catalyst pretreatment at 760°C, and then increase the temperature to 800°C for fast CNT growth. The CNTAs could grow to 1.5 mm high in 10 min.

Preheating the carbon source materials externally (at T_p) and then growing CNTs in a reactor at another temperature (T_s) is another way for decoupling the growth from catalyst treatment. Meshot et al. have found that the substrate temperature (T_s) primarily determined the CNT diameter, but independent (outside reactor) thermal treatment (T_p) of the C_2H_4/H_2 reactant mixture significantly changed the growth rate and CNTA height. This approach allowed preheating of the carbon source at a much higher temperature than growth temperature. By balancing the growth rate and catalyst lifetime, 3.5 mm-tall CNTAs were obtained by determining the substrate temperature at 825°C and precursor treatment temperature at 1020°C *(162)*.

Catalyst activity is also sensitive to carbon source concentration. Hasegawa et al. found that the partial pressure of carbon source was a critical factor controlling the activity of catalyst particles *(163)*. The growth rate increased linearly with the pressure of carbon source, although the growth lifetime decreased more rapidly. The key for millimeter-scale growth of CNTAs was to limit the carbon source below a certain partial pressure to retain the catalyst activity during the growth. By implementing this strategy, millimeter-tall SWNT CNTAs on Fe-supported AlO_x/Si were obtained.

The growth rate tends to decease with the increase of CNT density and height because of the slower mass delivery of carbon precursor across the CNTAs; and the diffusion issue becomes more serious on large substrates. Yasuda et al. proposed a top-flow growth process by introducing a gas shower system to improve the transport of feedstock (including carbon precursors, carrier gas, and week oxidizer) through the densely packed CNTAs to catalyst particles. This approach solves one limiting factor by changing the gas flow direction and directly projecting the reagents to the catalytic sites. They demonstrated synthesis of a 1 cm-tall forest with 1×1 cm size, and the approach was even employed to grow A4 size CNTAs to highlight the scalability *(164)*.

1.3.4.2 Elongate Catalyst Lifetime Using Supergrowth

During the growth process, the carbon coating on the catalyst would gradually degrade catalyst activity. Hata et al. found that some weak oxidizers could selectively remove amorphous carbon but would not damage the CNTs *(141)*. The catalysts could be effectively cleaned and reactivated by water vapor; and catalyst lifetime and activity were dramatically enhanced. The water-stimulated catalytic activity results in the growth of dense and vertically aligned single-walled CNTAs with 2.5 mm height in a 10-min growth time. After the pioneering work by Hata et al., the supergrowth that introduces a certain amount of weak oxidizer has been widely exploited *(138,139,142,165–167)*.

The role of weak oxidizer (water vapor, etc.) was also manifested in supergrowth of MWNTAs. Li et al. have demonstrated the growth of 4.7 mm-long MWNTs at 750°C in 2 h, and found that a certain amount of water vapor did not have a great effect on the CNT growth rate at the lower temperature (730–750°C). However, at a higher temperature of 780°C, it significantly enhanced the apparent growth rate. Without water vapor the growth rate increased with temperature in the range of 730–750°C, but increased slightly when the temperature increased to 780°C. In contrast, water vapor accelerated remarkably the apparent growth rate even at higher temperatures. Raman spectroscopy revealed that CNTAs grown with supergrowth had less amorphous carbon, indicating that water vapor promoted CNT growth by decreasing the formation of amorphous carbon and thus prolonging the catalyst lifetime *(136)*. Cho et al. found that though the catalyst lifetime was drastically prolonged by introducing a certain amount of water vapor, the growth rate decreased as the water concentration increased *(168)*, which was in contrast to the observation made by Li et al. *(136)*. The discrepancy in the understanding of the effect of water vapor on growth rate may be because of the different calculation methods used by these groups. Despite the debatable effect of water vapor on growth rate of CNTAs, the role in enhancing catalyst lifetime was once again confirmed. They obtained 21.7 mm-high CNTAs and a catalyst lifetime of 790 min *(168)*.

Though water vapor could significantly prolong catalyst lifetime, the concentration of water vapor should be precisely controlled. Zheng et al. have demonstrated the growth of 4.2 cm-long individual SWNTs by using ethanol as a carbon source, suggesting that ethanol and other oxygen containing carbon sources be able to maintain high activity and long lifetime of catalyst even at a temperature as high as 900°C *(25)*. Li et al. extended this strategy to the growth of MWNT arrays. By introducing a certain amount of ethanol into the feedstock and separating the growth zone from the carbon thermal annealing zone, up to 7 mm-long CNTAs were successively obtained within 45 min, with a maximal growth rate of ~280 μm/min, as shown in Fig. 1.16A *(169)*. It is found that with alcohol vapor, the growth of CNTAs is quick and stable. While CNTA height decreased dramatically to 600 μm and growth could only last for <5 min without ethanol introduction (Fig. 1.16B), the catalyst could only survive a very short time at a high pyrolysis temperature without ethanol addition. Considering the small amount of alcohol vapor, it is very unlikely that it will be used as a carbon source in comparison to its main function as a weak oxidizer removing the impure carbon.

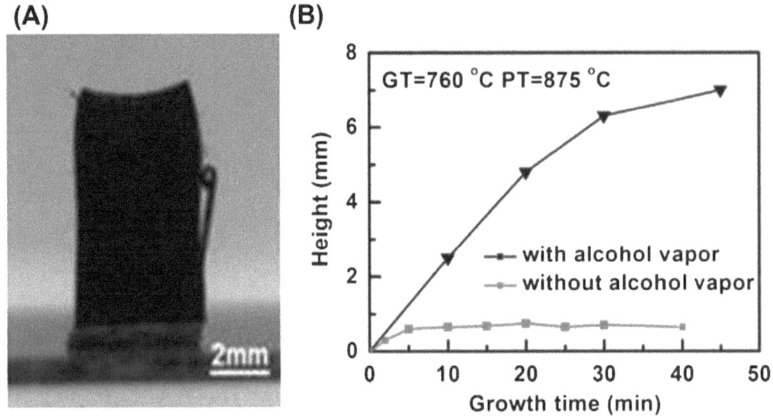

Figure 1.16 (A) Photograph images of a 7 mm-long carbon nanotube array (CNTA) grown by alcohol-assisted chemical vapor deposition (CVD) process within 45 min; (B) dependence of CNTA height on growth time. *(Adapted from Li, Y.; Xu, G.; Zhang, H.; Li, T.; Yao, Y.; Li, Q.; Dai, Z. Alcohol-Assisted Rapid Growth of Vertically Aligned Carbon Nanotube Arrays. Carbon **2015**, 91, 45–55 with permission.)*

REFERENCES

1. Baughman, R. H.; Zakhidov, A. A.; de Heer, W. A. Carbon Nanotubes-the Route Toward Applications. *Science* **2002,** *297* (5582), 787–792.
2. Kang, S. J.; Kocabas, C.; Ozel, T.; Shim, M.; Pimparkar, N.; Alam, M. A.; Rotkin, S. V.; Rogers, J. A. High-performance Electronics Using Dense, Perfectly Aligned Arrays of Single-walled Carbon Nanotubes. *Nature Nanotechnology* **2007,** *2* (4), 230–236.
3. Shulaker, M. M.; Hills, G.; Patil, N.; Wei, H.; Chen, H.-Y.; Wong, H. S. P.; Mitra, S. Carbon Nanotube Computer. *Nature* **2013,** *501* (7468), 526–530.
4. de Heer, W. A.; Châtelain, A.; Ugarte, D. A Carbon Nanotube Field-Emission Electron Source. *Science* **1995,** *270* (5239), 1179–1180.
5. He, X.; Fujimura, N.; Lloyd, J. M.; Erickson, K. J.; Talin, A. A.; Zhang, Q.; Gao, W.; Jiang, Q.; Kawano, Y.; Hauge, R. H.; Léonard, F.; Kono, J. Carbon Nanotube Terahertz Detector. *Nano Letters* **2014,** *14* (7), 3953–3958.
6. Zhan, Z.; Liu, C.; Zheng, L.; Sun, G.; Li, B.; Zhang, Q. Photoresponse of Multi-walled Carbon Nanotube-Copper Sulfide (MWNT-cus) Hybrid Nanostructures. *Physical Chemistry Chemical Physics* **2011,** *13* (45), 20471–20475.
7. Sun, G.; Liu, J.; Zheng, L.; Huang, W.; Zhang, H. Preparation of Weavable, All-carbon Fibers for Non-volatile Memory Devices. *Angewandte Chemie International Edition* **2013,** *52* (50), 13351–13355.
8. Salila Vijayalal Mohan, H. K.; An, J.; Zheng, L. Sequence-dependent Electrical Response of sDNA-decorated Carbon Nanotube, Field-effect Transistors to Dopamine. *Beilstein Journal of Nanotechnology* **2014,** *5,* 2113–2121.
9. Salila Vijayalal Mohan, H. K.; An, J.; Zhang, Y.; Wong, C. H.; Zheng, L. Effect of Channel Length on the Electrical Response of Carbon Nanotube Field-effect Transistors to Deoxyribonucleic Acid Hybridization. *Beilstein Journal of Nanotechnology* **2014,** *5,* 2081–2091.
10. Wu, Z.; Chen, Z.; Du, X.; Logan, J. M.; Sippel, J.; Nikolou, M.; Kamaras, K.; Reynolds, J. R.; Tanner, D. B.; Hebard, A. F.; Rinzler, A. G. Transparent, Conductive Carbon Nanotube Films. *Science* **2004,** *305* (5688), 1273–1276.
11. Tans, S. J.; Verschueren, A. R. M.; Dekker, C. Room-Temperature Transistor Based on a Single Carbon Nanotube. *Nature* **1998,** *393* (6680), 49–52.

12. Javey, A.; Guo, J.; Wang, Q.; Lundstrom, M.; Dai, H. Ballistic Carbon Nanotube Field-effect Transistors. *Nature* **2003,** *424* (6949), 654–657.
13. Li, J.; Ye, Q.; Cassell, A.; Ng, H. T.; Stevens, R.; Han, J.; Meyyappan, M. Bottom-up Approach for Carbon Nanotube Interconnects. *Applied Physics Letters* **2003,** *82* (15), 2491–2493.
14. Xiao, J.; Dunham, S.; Liu, P.; Zhang, Y.; Kocabas, C.; Moh, L.; Huang, Y.; Hwang, K.-C.; Lu, C.; Huang, W.; Rogers, J. A. Alignment Controlled Growth of Single-walled Carbon Nanotubes on Quartz Substrates. *Nano Letters* **2009,** *9* (12), 4311–4319.
15. Huang, L.; Jia, Z.; O'Brien, S. Orientated Assembly of Single-walled Carbon Nanotubes and Applications. *Journal of Materials Chemistry* **2007,** *17* (37), 3863–3874.
16. Kocabas, C.; Kang, S. J.; Ozel, T.; Shim, M.; Rogers, J. A. Improved Synthesis of Aligned Arrays of Single-walled Carbon Nanotubes and Their Implementation in Thin Film Type Transistors. *Journal of Physical Chemistry C* **2007,** *111* (48), 17879–17886.
17. Chen, Z.; Appenzeller, J.; Lin, Y.-M.; Sippel-Oakley, J.; Rinzler, A. G.; Tang, J.; Wind, S. J.; Solomon, P. M.; Avouris, P. An Integrated Logic Circuit Assembled on a Single Carbon Nanotube. *Science* **2006,** *311* (5768), 1735.
18. Wang, S.; Zhang, Z.; Ding, L.; Liang, X.; Shen, J.; Xu, H.; Chen, Q.; Cui, R.; Li, Y.; Peng, L.-M. A Doping-free Carbon Nanotube CMOS Inverter-based Bipolar Diode and Ambipolar Transistor. *Advanced Materials* **2008,** *20* (17), 3258–3262.
19. Zhang, Y.; Sun, G.; Zheng, L. Fabrication and Properties of Macroscopic Carbon Nanotube Assemblies Transforming from Aligned Nanotubes. *Journal of Nanomaterials* **2015,** 907954.
20. Ma, Y.; Wang, B.; Wu, Y.; Huang, Y.; Chen, Y. The Production of Horizontally Aligned Single-walled Carbon Nanotubes. *Carbon* **2011,** *49* (13), 4098–4110.
21. Zhang, Y.; Zheng, L. Towards Chirality-pure Carbon Nanotubes. *Nanoscale* **2010,** *2* (10), 1919–1929.
22. Liu, B.; Wang, C.; Liu, J.; Che, Y.; Zhou, C. Aligned Carbon Nanotubes: from Controlled Synthesis to Electronic Applications. *Nanoscale* **2013,** *5* (20), 9483–9502.
23. Ibrahim, I.; Bachmatiuk, A.; Warner, J. H.; Büchner, B.; Cuniberti, G.; Rümmeli, M. H. CVD-grown Horizontally Aligned Single-walled Carbon Nanotubes: Synthesis Routes and Growth Mechanisms. *Small* **2012,** *8* (13), 1973–1992.
24. Chen, Y.; Zhang, J. Chemical Vapor Deposition Growth of Single-walled Carbon Nanotubes with Controlled Structures for Nanodevice Applications. *Accounts of Chemical Research* **2014,** *47* (8), 2273–2281.
25. Zheng, L. X.; O'Connell, M. J.; Doorn, S. K.; Liao, X. Z.; Zhao, Y. H.; Akhadov, E. A.; Hoffbauer, M. A.; Roop, B. J.; Jia, Q. X.; Dye, R. C.; Peterson, D. E.; Huang, S. M.; Liu, J.; Zhu, Y. T. Ultralong Single-wall Carbon Nanotubes. *Nature Materials* **2004,** *3* (10), 673–676.
26. Zheng, L.; Liao, X.; Zhu, Y. T. Parametric Study of Carbon Nanotube Growth via Cobalt-catalyzed Ethanol Decomposition. *Materials Letters* **2006,** *60* (16), 1968–1972.
27. Zhou, W.; Han, Z.; Wang, J.; Zhang, Y.; Jin, Z.; Sun, X.; Zhang, Y.; Yan, C.; Li, Y. Copper Catalyzing Growth of Single-walled Carbon Nanotubes on Substrates. *Nano Letters* **2006,** *6* (12), 2987–2990.
28. Bhaviripudi, S.; Mile, E.; Steiner, S. A.; Zare, A. T.; Dresselhaus, M. S.; Belcher, A. M.; Kong, J. CVD Synthesis of Single-walled Carbon Nanotubes from Gold Nanoparticle Catalysts. *Journal of the American Chemical Society* **2007,** *129* (6), 1516–1517.
29. Yuan, D.; Ding, L.; Chu, H.; Feng, Y.; McNicholas, T. P.; Liu, J. Horizontally Aligned Single-walled Carbon Nanotube on Quartz from a Large Variety of Metal Catalysts. *Nano Letters* **2008,** *8* (8), 2576–2579.
30. Yang, F.; Wang, X.; Zhang, D.; Yang, J.; Luo, D.; Xu, Z.; Wei, J.; Wang, J.-Q.; Xu, Z.; Peng, F.; Li, X.; Li, R.; Li, Y.; Li, M.; Bai, X.; Ding, F.; Li, Y. Chirality-specific Growth of Single-walled Carbon Nanotubes on Solid Alloy Catalysts. *Nature* **2014,** *510* (7506), 522–524.
31. Tessonnier, J.-P.; Su, D. S. Recent Progress on the Growth Mechanism of Carbon Nanotubes: A Review. *ChemSusChem* **2011,** *4* (7), 824–847.
32. Li, Y.; Cui, R.; Ding, L.; Liu, Y.; Zhou, W.; Zhang, Y.; Jin, Z.; Peng, F.; Liu, J. How Catalysts Affect the Growth of Single-walled Carbon Nanotubes on Substrates. *Advanced Materials* **2010,** *22* (13), 1508–1515.

33. Li, Y.; Liu, J.; Wang, Y.; Wang, Z. L. Preparation of Monodispersed Fe–Mo Nanoparticles as the Catalyst for CVD Synthesis of Carbon Nanotubes. *Chemistry of Materials* **2001,** *13* (3), 1008–1014.
34. Geng, J.; Li, H.; Golovko, V. B.; Shephard, D. S.; Jefferson, D. A.; Johnson, B. F. G.; Hofmann, S.; Kleinsorge, B.; Robertson, J.; Ducati, C. Nickel Formate Route to the Growth of Carbon Nanotubes. *Journal of Physical Chemistry B* **2004,** *108* (48), 18446–18450.
35. Dai, H.; Rinzler, A. G.; Nikolaev, P.; Thess, A.; Colbert, D. T.; Smalley, R. E. Single-wall Nanotubes Produced by Metal-catalyzed Disproportionation of Carbon Monoxide. *Chemical Physics Letters* **1996,** *260* (3–4), 471–475.
36. Nasibulin, A. G.; Pikhitsa, P. V.; Jiang, H.; Kauppinen, E. I. Correlation between Catalyst Particle and Single-walled Carbon Nanotube Diameters. *Carbon* **2005,** *43* (11), 2251–2257.
37. Li, Y.; Kim, W.; Zhang, Y.; Rolandi, M.; Wang, D.; Dai, H. Growth of Single-walled Carbon Nanotubes from Discrete Catalytic Nanoparticles of Various Sizes. *Journal of Physical Chemistry B* **2001,** *105* (46), 11424–11431.
38. Jeong, G.-H.; Yamazaki, A.; Suzuki, S.; Yoshimura, H.; Kobayashi, Y.; Homma, Y. Cobalt-filled Apoferritin for Suspended Single-walled Carbon Nanotube Growth with Narrow Diameter Distribution. *Journal of the American Chemical Society* **2005,** *127* (23), 8238–8239.
39. Lu, J.; Yi, S. S.; Kopley, T.; Qian, C.; Liu, J.; Gulari, E. Fabrication of Ordered Catalytically Active Nanoparticles Derived from Block Copolymer Micelle Templates for Controllable Synthesis of Single-walled Carbon Nanotubes. *Journal of Physical Chemistry B* **2006,** *110* (13), 6655–6660.
40. Lu, J. Q.; Moll, N.; Fu, Q.; Liu, J. Iron Nanoparticles Derived from Iron-complexed Polymethylglutarimide to Produce High-quality Lithographically Defined Single-walled Carbon Nanotubes. *Chemistry of Materials* **2005,** *17* (9), 2237–2240.
41. Choi, H. C.; Kim, W.; Wang, D.; Dai, H. Delivery of Catalytic Metal Species onto Surfaces with Dendrimer Carriers for the Synthesis of Carbon Nanotubes with Narrow Diameter Distribution. *Journal of Physical Chemistry B* **2002,** *106* (48), 12361–12365.
42. An, L.; Owens, J. M.; McNeil, L. E.; Liu, J. Synthesis of Nearly Uniform Single-walled Carbon Nanotubes Using Identical Metal-containing Molecular Nanoclusters as Catalysts. *Journal of the American Chemical Society* **2002,** *124* (46), 13688–13689.
43. Fu, Q.; Huang, S.; Liu, J. Chemical Vapor Depositions of Single-walled Carbon Nanotubes Catalyzed by Uniform Fe_2O_3 Nanoclusters Synthesized Using Diblock Copolymer Micelles. *Journal of Physical Chemistry B* **2004,** *108* (20), 6124–6129.
44. Arnold, M. S.; Green, A. A.; Hulvat, J. F.; Stupp, S. I.; Hersam, M. C. Sorting Carbon Nanotubes by Electronic Structure Using Density Differentiation. *Nature Nanotechnology* **2006,** *1* (1), 60–65.
45. Liu, H.; Nishide, D.; Tanaka, T.; Kataura, H. Large-scale Single-chirality Separation of Single-wall Carbon Nanotubes by Simple Gel Chromatography. *Nature Communications* **2011,** *2,* 309.
46. Hong, G.; Zhou, M.; Zhang, R.; Hou, S.; Choi, W.; Woo, Y. S.; Choi, J.-Y.; Liu, Z.; Zhang, J. Separation of Metallic and Semiconducting Single-walled Carbon Nanotube Arrays by "Scotch Tape". *Angewandte Chemie International Edition* **2011,** *50* (30), 6819–6823.
47. Harutyunyan, A. R.; Chen, G.; Paronyan, T. M.; Pigos, E. M.; Kuznetsov, O. A.; Hewaparakrama, K.; Kim, S. M.; Zakharov, D.; Stach, E. A.; Sumanasekera, G. U. Preferential Growth of Single-walled Carbon Nanotubes with Metallic Conductivity. *Science* **2009,** *326* (5949), 116–120.
48. Ghorannevis, Z.; Kato, T.; Kaneko, T.; Hatakeyama, R. Narrow-Chirality Distributed Single-walled Carbon Nanotube Growth from Nonmagnetic Catalyst. *Journal of the American Chemical Society* **2010,** *132* (28), 9570–9572.
49. Yazyev, O. V.; Pasquarello, A. Effect of Metal Elements in Catalytic Growth of Carbon Nanotubes. *Physical Review Letters* **2008,** *100* (15), 156102.
50. Bachilo, S. M.; Balzano, L.; Herrera, J. E.; Pompeo, F.; Resasco, D. E.; Weisman, R. B. Narrow (n,m)-Distribution of Single-walled Carbon Nanotubes Grown Using a Solid Supported Catalyst. *Journal of the American Chemical Society* **2003,** *125* (37), 11186–11187.
51. Lolli, G.; Zhang, L.; Balzano, L.; Sakulchaicharoen, N.; Tan, Y.; Resasco, D. E. Tailoring (n,m) Structure of Single-walled Carbon Nanotubes by Modifying Reaction Conditions and the Nature of the Support of CoMo Catalysts. *Journal of Physical Chemistry B* **2006,** *110* (5), 2108–2115.

52. Wang, B.; Poa, C. H. P.; Wei, L.; Li, L.-J.; Yang, Y.; Chen, Y. (n,m) Selectivity of Single-walled Carbon Nanotubes by Different Carbon Precursors on Co–Mo Catalysts. *Journal of the American Chemical Society* **2007,** *129* (29), 9014–9019.
53. Li, X.; Tu, X.; Zaric, S.; Welsher, K.; Seo, W. S.; Zhao, W.; Dai, H. Selective Synthesis Combined with Chemical Separation of Single-walled Carbon Nanotubes for Chirality Selection. *Journal of the American Chemical Society* **2007,** *129* (51), 15770–15771.
54. Liu, B.; Ren, W.; Li, S.; Liu, C.; Cheng, H.-M. High Temperature Selective Growth of Single-walled Carbon Nanotubes with a Narrow Chirality Distribution from a CoPt Bimetallic Catalyst. *Chemical Communications* **2012,** *48* (18), 2409–2411.
55. Chiang, W.-H.; Sankaran, R. M. In-flight Dimensional Tuning of Metal Nanoparticles by Microplasma Synthesis for Selective Production of Diameter-Controlled Carbon Nanotubes. *Journal of Physical Chemistry C* **2008,** *112* (46), 17920–17925.
56. Chiang, W.-H.; Mohan Sankaran, R. Linking Catalyst Composition to Chirality Distributions of as-Grown Single-walled Carbon Nanotubes by Tuning Ni_xFe_{1-x} Nanoparticles. *Nature Materials* **2009,** *8* (11), 882–886.
57. He, M.; Chernov, A. I.; Fedotov, P. V.; Obraztsova, E. D.; Sainio, J.; Rikkinen, E.; Jiang, H.; Zhu, Z.; Tian, Y.; Kauppinen, E. I.; Niemelä, M.; Krause, A. O. I. Predominant (6,5) Single-walled Carbon Nanotube Growth on a Copper-Promoted Iron Catalyst. *Journal of the American Chemical Society* **2010,** *132* (40), 13994–13996.
58. Chen, Y.; Zhang, J. Diameter Controlled Growth of Single-walled Carbon Nanotubes from SiO_2 Nanoparticles. *Carbon* **2011,** *49* (10), 3316–3324.
59. Takagi, D.; Kobayashi, Y.; Homma, Y. Carbon Nanotube Growth from Diamond. *Journal of the American Chemical Society* **2009,** *131* (20), 6922–6923.
60. Wang, Q.; Ng, M.-F.; Yang, S.-W.; Yang, Y.; Chen, Y. The Mechanism of Single-walled Carbon Nanotube Growth and Chirality Selection Induced by Carbon Atom and Dimer Addition. *ACS Nano* **2010,** *4* (2), 939–946.
61. Huang, S.; Cai, Q.; Chen, J.; Qian, Y.; Zhang, L. Metal-Catalyst-Free Growth of Single-walled Carbon Nanotubes on Substrates. *Journal of the American Chemical Society* **2009,** *131* (6), 2094–2095.
62. Liu, B.; Ren, W.; Gao, L.; Li, S.; Pei, S.; Liu, C.; Jiang, C.; Cheng, H.-M. Metal-Catalyst-Free Growth of Single-walled Carbon Nanotubes. *Journal of the American Chemical Society* **2009,** *131* (6), 2082–2083.
63. Ago, H.; Ayagaki, T.; Ogawa, Y.; Tsuji, M. Ultrahigh-Vacuum-Assisted Control of Metal Nanoparticles for Horizontally Aligned Single-walled Carbon Nanotubes with Extraordinary Uniform Diameters. *Journal of Physical Chemistry C* **2011,** *115* (27), 13247–13253.
64. Jin-Ju, K.; Byeong-Joo, L.; Seung-Hwan, L.; Goo-Hwan, J. Size Engineering of Metal Nanoparticles to Diameter-Specified Growth of Single-walled Carbon Nanotubes with Horizontal Alignment on Quartz. *Nanotechnology* **2012,** *23* (10), 105607.
65. Barzegar, H. R.; Nitze, F.; Sharifi, T.; Ramstedt, M.; Tai, C. W.; Malolepszy, A.; Stobinski, L.; Wågberg, T. Simple Dip-Coating Process for the Synthesis of Small Diameter Single-walled Carbon Nanotubes—Effect of Catalyst Composition and Catalyst Particle Size on Chirality and Diameter. *Journal of Physical Chemistry C* **2012,** *116* (22), 12232–12239.
66. Ishigami, N.; Ago, H.; Imamoto, K.; Tsuji, M.; Iakoubovskii, K.; Minami, N. Crystal Plane Dependent Growth of Aligned Single-walled Carbon Nanotubes on Sapphire. *Journal of the American Chemical Society* **2008,** *130* (30), 9918–9924.
67. Wang, W. L.; Bai, X. D.; Xu, Z.; Liu, S.; Wang, E. G. Low Temperature Growth of Single-walled Carbon Nanotubes: Small Diameters with Narrow Distribution. *Chemical Physics Letters* **2006,** *419* (1–3), 81–85.
68. Li, N.; Wang, X.; Ren, F.; Haller, G. L.; Pfefferle, L. D. Diameter Tuning of Single-walled Carbon Nanotubes with Reaction Temperature Using a Co Monometallic Catalyst. *Journal of Physical Chemistry C* **2009,** *113* (23), 10070–10078.
69. Zoican Loebick, C.; Podila, R.; Reppert, J.; Chudow, J.; Ren, F.; Haller, G. L.; Rao, A. M.; Pfefferle, L. D. Selective Synthesis of Subnanometer Diameter Semiconducting Single-walled Carbon Nanotubes. *Journal of the American Chemical Society* **2010,** *132* (32), 11125–11131.

70. He, M.; Chernov, A.; Obraztsova, E.; Sainio, J.; Rikkinen, E.; Jiang, H.; Zhu, Z.; Kaskela, A.; Nasibulin, A.; Kauppinen, E.; Niemelä, M.; Krause, O. Low Temperature Growth of SWNTs on a Nickel Catalyst by Thermal Chemical Vapor Deposition. *Nano Research* **2011,** *4* (4), 334–342.
71. Dervishi, E.; Li, Z.; Watanabe, F.; Xu, Y.; Saini, V.; Biris, A. R.; Biris, A. S. Thermally Controlled Synthesis of Single-wall Carbon Nanotubes with Selective Diameters. *Journal of Materials Chemistry* **2009,** *19* (19), 3004–3012.
72. Yao, Y.; Li, Q.; Zhang, J.; Liu, R.; Jiao, L.; Zhu, Y. T.; Liu, Z. Temperature-Mediated Growth of Single-walled Carbon-Nanotube Intramolecular Junctions. *Nature Materials* **2007,** *6* (4), 283–286.
73. Yao, Y.; Dai, X.; Liu, R.; Zhang, J.; Liu, Z. Tuning the Diameter of Single-walled Carbon Nanotubes by Temperature-Mediated Chemical Vapor Deposition. *Journal of Physical Chemistry C* **2009,** *113* (30), 13051–13059.
74. Lu, C.; Liu, J. Controlling the Diameter of Carbon Nanotubes in Chemical Vapor Deposition Method by Carbon Feeding. *Journal of Physical Chemistry B* **2006,** *110* (41), 20254–20257.
75. Saito, T.; Ohshima, S.; Okazaki, T.; Ohmori, S.; Yumura, M.; Iijima, S. Selective Diameter Control of Single-walled Carbon Nanotubes in the Gas-Phase Synthesis. *Journal of Nanoscience and Nanotechnology* **2008,** *8* (11), 6153–6157.
76. Wang, B.; Wei, L.; Yao, L.; Li, L.-J.; Yang, Y.; Chen, Y. Pressure-Induced Single-walled Carbon Nanotube (n,m) Selectivity on Co–Mo Catalysts. *Journal of Physical Chemistry C* **2007,** *111* (40), 14612–14616.
77. Ding, L.; Tselev, A.; Wang, J.; Yuan, D.; Chu, H.; McNicholas, T. P.; Li, Y.; Liu, J. Selective Growth of Well-Aligned Semiconducting Single-walled Carbon Nanotubes. *Nano Letters* **2009,** *9* (2), 800–805.
78. Li, W.-S.; Hou, P.-X.; Liu, C.; Sun, D.-M.; Yuan, J.; Zhao, S.-Y.; Yin, L.-C.; Cong, H.; Cheng, H.-M. High-quality, Highly Concentrated Semiconducting Single-wall Carbon Nanotubes for Use in Field Effect Transistors and Biosensors. *ACS Nano* **2013,** 7 (8), 6831–6839.
79. Li, J.; Ke, C.-T.; Liu, K.; Li, P.; Liang, S.; Finkelstein, G.; Wang, F.; Liu, J. Importance of Diameter Control on Selective Synthesis of Semiconducting Single-walled Carbon Nanotubes. *ACS Nano* **2014,** *8* (8), 8564–8572.
80. Yu, X.; Zhang, J.; Choi, W.; Choi, J.-Y.; Kim, J. M.; Gan, L.; Liu, Z. Cap Formation Engineering: From Opened C_{60} to Single-walled Carbon Nanotubes. *Nano Letters* **2010,** *10* (9), 3343–3349.
81. Wang, Y.; Kim, M. J.; Shan, H.; Kittrell, C.; Fan, H.; Ericson, L. M.; Hwang, W.-F.; Arepalli, S.; Hauge, R. H.; Smalley, R. E. Continued Growth of Single-walled Carbon Nanotubes. *Nano Letters* **2005,** *5* (6), 997–1002.
82. Smalley, R. E.; Li, Y.; Moore, V. C.; Price, B. K.; Colorado, R.; Schmidt, H. K.; Hauge, R. H.; Barron, A. R.; Tour, J. M. Single Wall Carbon Nanotube Amplification: En Route to a Type-Specific Growth Mechanism. *Journal of the American Chemical Society* **2006,** *128* (49), 15824–15829.
83. Ren, Z. Nanotube Synthesis: Cloning Carbon. *Nature Nanotechnology* **2007,** *2* (1), 17–18.
84. Yao, Y.; Feng, C.; Zhang, J.; Liu, Z. "Cloning" of Single-walled Carbon Nanotubes via Open-End Growth Mechanism. *Nano Letters* **2009,** *9* (4), 1673–1677.
85. Liu, J.; Wang, C.; Tu, X.; Liu, B.; Chen, L.; Zheng, M.; Zhou, C. Chirality-Controlled Synthesis of Single-wall Carbon Nanotubes Using Vapour-Phase Epitaxy. *Nature Communications* **2012,** *3,* 1199.
86. Zhang, Y.; Chang, A.; Cao, J.; Wang, Q.; Kim, W.; Li, Y.; Morris, N.; Yenilmez, E.; Kong, J.; Dai, H. Electric-Field-Directed Growth of Aligned Single-walled Carbon Nanotubes. *Applied Physics Letters* **2001,** *79* (19), 3155–3157.
87. Ural, A.; Li, Y.; Dai, H. Electric-Field-Aligned Growth of Single-walled Carbon Nanotubes on Surfaces. *Applied Physics Letters* **2002,** *81* (18), 3464–3466.
88. Joselevich, E.; Lieber, C. M. Vectorial Growth of Metallic and Semiconducting Single-wall Carbon Nanotubes. *Nano Letters* **2002,** *2* (10), 1137–1141.
89. Huang, S.; Cai, X.; Liu, J. Growth of Millimeter-Long and Horizontally Aligned Single-walled Carbon Nanotubes on Flat Substrates. *Journal of the American Chemical Society* **2003,** *125* (19), 5636–5637.
90. Huang, S.; Woodson, M.; Smalley, R.; Liu, J. Growth Mechanism of Oriented Long Single Walled Carbon Nanotubes Using "Fast-Heating" Chemical Vapor Deposition Process. *Nano Letters* **2004,** *4* (6), 1025–1028.

91. Yu, Z.; Li, S.; Burke, P. J. Synthesis of Aligned Arrays of Millimeter Long, Straight Single-walled Carbon Nanotubes. *Chemistry of Materials* **2004,** *16* (18), 3414–3416.
92. Huang, L.; White, B.; Sfeir, M. Y.; Huang, M.; Huang, H. X.; Wind, S.; Hone, J.; O'Brien, S. Cobalt Ultrathin Film Catalyzed Ethanol Chemical Vapor Deposition of Single-walled Carbon Nanotubes. *Journal of Physical Chemistry B* **2006,** *110* (23), 11103–11109.
93. Hong, B. H.; Lee, J. Y.; Beetz, T.; Zhu, Y.; Kim, P.; Kim, K. S. Quasi-Continuous Growth of Ultralong Carbon Nanotube Arrays. *Journal of the American Chemical Society* **2005,** *127* (44), 15336–15337.
94. Jin, Z.; Chu, H.; Wang, J.; Hong, J.; Tan, W.; Li, Y. Ultralow Feeding Gas Flow Guiding Growth of Large-Scale Horizontally Aligned Single-walled Carbon Nanotube Arrays. *Nano Letters* **2007,** *7* (7), 2073–2079.
95. Murakami, Y.; Chiashi, S.; Miyauchi, Y.; Hu, M.; Ogura, M.; Okubo, T.; Maruyama, S. Growth of Vertically Aligned Single-walled Carbon Nanotube Films on Quartz Substrates and Their Optical Anisotropy. *Chemical Physics Letters* **2004,** *385* (3–4), 298–303.
96. Liu, H.; Takagi, D.; Chiashi, S.; Homma, Y. The Controlled Growth of Horizontally Aligned Single-walled Carbon Nanotube Arrays by a Gas Flow Process. *Nanotechnology* **2009,** *20* (34), 345604.
97. An, J.; Zhan, Z.; Salila Vijayalal Mohan, H. K.; Sun, G.; Hansen, R. V.; Zheng, L. Trench Structure Assisted Alignment in Ultralong and Dense Carbon Nanotube Arrays. *Journal of Materials Chemistry C* **2015,** *3* (10), 2215–2222.
98. Ismach, A.; Segev, L.; Wachtel, E.; Joselevich, E. Atomic-Step-Templated Formation of Single Wall Carbon Nanotube Patterns. *Angewandte Chemie International Edition* **2004,** *43* (45), 6140–6143.
99. Ismach, A.; Kantorovich, D.; Joselevich, E. Carbon Nanotube Graphoepitaxy: Highly Oriented Growth by Faceted Nanosteps. *Journal of the American Chemical Society* **2005,** *127* (33), 11554–11555.
100. Ding, L.; Yuan, D.; Liu, J. Growth of High-Density Parallel Arrays of Long Single-walled Carbon Nanotubes on Quartz Substrates. *Journal of the American Chemical Society* **2008,** *130* (16), 5428–5429.
101. Feng, C.; Yao, Y.; Zhang, J.; Liu, Z. Nanobarrier-Terminated Growth of Single-walled Carbon Nanotubes on Quartz Surfaces. *Nano Research* **2009,** *2* (10), 768–773.
102. Ismach, A.; Joselevich, E. Orthogonal Self-Assembly of Carbon Nanotube Crossbar Architectures by Simultaneous Graphoepitaxy and Field-Directed Growth. *Nano Letters* **2006,** *6* (8), 1706–1710.
103. Kocabas, C.; Hur, S.-H.; Gaur, A.; Meitl, M. A.; Shim, M.; Rogers, J. A. Guided Growth of Large-Scale, Horizontally Aligned Arrays of Single-walled Carbon Nanotubes and Their Use in Thin-Film Transistors. *Small* **2005,** *1* (11), 1110–1116.
104. Yoshihara, N.; Ago, H.; Imamoto, K.; Tsuji, M.; Ikuta, T.; Takahashi, K. Horizontally Aligned Growth of Single-walled Carbon Nanotubes on a Surface-Modified Silicon Wafer. *Journal of Physical Chemistry C* **2009,** *113* (19), 8030–8034.
105. Han, S.; Liu, X.; Zhou, C. Template-Free Directional Growth of Single-walled Carbon Nanotubes on a- and r-Plane Sapphire. *Journal of the American Chemical Society* **2005,** *127* (15), 5294–5295.
106. Ago, H.; Nakamura, K.; Ikeda, K.-i.; Uehara, N.; Ishigami, N.; Tsuji, M. Aligned Growth of Isolated Single-walled Carbon Nanotubes Programmed by Atomic Arrangement of Substrate Surface. *Chemical Physics Letters* **2005,** *408* (4–6), 433–438.
107. Yu, Q.; Qin, G.; Li, H.; Xia, Z.; Nian, Y.; Pei, S.-S. Mechanism of Horizontally Aligned Growth of Single-wall Carbon Nanotubes on R-Plane Sapphire. *Journal of Physical Chemistry B* **2006,** *110* (45), 22676–22680.
108. Kocabas, C.; Shim, M.; Rogers, J. A. Spatially Selective Guided Growth of High-Coverage Arrays and Random Networks of Single-walled Carbon Nanotubes and Their Integration into Electronic Devices. *Journal of the American Chemical Society* **2006,** *128* (14), 4540–4541.
109. Zhang, B.; Hong, G.; Peng, B.; Zhang, J.; Choi, W.; Kim, J. M.; Choi, J.-Y.; Liu, Z. Grow Single-walled Carbon Nanotubes Cross-Bar in One Batch. *Journal of Physical Chemistry C* **2009,** *113* (14), 5341–5344.
110. Geblinger, N.; Ismach, A.; Joselevich, E. Self-Organized Nanotube Serpentines. *Nature Nanotechnology* **2008,** *3* (4), 195–200.
111. Yao, Y.; Dai, X.; Feng, C.; Zhang, J.; Liang, X.; Ding, L.; Choi, W.; Choi, J.-Y.; Kim, J. M.; Liu, Z. Crinkling Ultralong Carbon Nanotubes into Serpentines by a Controlled Landing Process. *Advanced Materials* **2009,** *21* (41), 4158–4162.

112. Zhang, Y.; Zhang, J.; Son, H.; Kong, J.; Liu, Z. Substrate-Induced Raman Frequency Variation for Single-walled Carbon Nanotubes. *Journal of the American Chemical Society* **2005,** *127* (49), 17156–17157.
113. Zhang, Y.; Son, H.; Zhang, J.; Dresselhaus, M. S.; Kong, J.; Liu, Z. Raman Spectra Variation of Partially Suspended Individual Single-walled Carbon Nanotubes. *Journal of Physical Chemistry C* **2007,** *111* (5), 1983–1987.
114. Hofmann, M.; Nezich, D.; Reina, A.; Kong, J. In-Situ Sample Rotation as a Tool to Understand Chemical Vapor Deposition Growth of Long Aligned Carbon Nanotubes. *Nano Letters* **2008,** *8* (12), 4122–4127.
115. Ago, H.; Ishigami, N.; Yoshihara, N.; Imamoto, K.; Akita, S.; Ikeda, K.-i.; Tsuji, M.; Ikuta, T.; Takahashi, K. Visualization of Horizontally-aligned Single-walled Carbon Nanotube Growth with $^{13}C/^{12}C$ Isotopes. *Journal of Physical Chemistry C* **2008,** *112* (6), 1735–1738.
116. Ding, F.; Harutyunyan, A. R.; Yakobson, B. I. Dislocation Theory of Chirality-Controlled Nanotube Growth. *Proceedings of the National Academy of Sciences of the United States of America* **2009,** *106* (8), 2506–2509.
117. Marchand, M.; Journet, C.; Guillot, D.; Benoit, J.-M.; Yakobson, B. I.; Purcell, S. T. Growing a Carbon Nanotube Atom by Atom: "And Yet It Does Turn". *Nano Letters* **2009,** *9* (8), 2961–2966.
118. Morin, S. A.; Bierman, M. J.; Tong, J.; Jin, S. Mechanism and Kinetics of Spontaneous Nanotube Growth Driven by Screw Dislocations. *Science* **2010,** *328* (5977), 476–480.
119. Rao, R.; Liptak, D.; Cherukuri, T.; Yakobson, B. I.; Maruyama, B. In Situ Evidence for Chirality-Dependent Growth Rates of Individual Carbon Nanotubes. *Nature Materials* **2012,** *11* (3), 213–216.
120. Zheng, L.; Satishkumar, B. C.; Gao, P.; Zhang, Q. Kinetics Studies of Ultralong Single-walled Carbon Nanotubes. *Journal of Physical Chemistry C* **2009,** *113* (25), 10896–10900.
121. An, J.; Zhan, Z.; Hari Krishna, S. V.; Zheng, L. Growth Condition Mediated Catalyst Effects on the Density and Length of Horizontally Aligned Single-walled Carbon Nanotube Arrays. *Chemical Engineering Journal* **2014,** *237,* 16–22.
122. Zhang, R.; Zhang, Y.; Zhang, Q.; Xie, H.; Qian, W.; Wei, F. Growth of Half-Meter Long Carbon Nanotubes Based on Schulz–Flory Distribution. *ACS Nano* **2013,** *7* (7), 6156–6161.
123. Wen, Q.; Qian, W.; Nie, J.; Cao, A.; Ning, G.; Wang, Y.; Hu, L.; Zhang, Q.; Huang, J.; Wei, F. 100 mm Long, Semiconducting Triple-walled Carbon Nanotubes. *Advanced Materials* **2010,** *22* (16), 1867–1871.
124. Wen, Q.; Zhang, R.; Qian, W.; Wang, Y.; Tan, P.; Nie, J.; Wei, F. Growing 20 cm Long DWNTs/TWNTs at a Rapid Growth Rate of 80–90 μm/s. *Chemistry of Materials* **2010,** *22* (4), 1294–1296.
125. Wang, X.; Li, Q.; Xie, J.; Jin, Z.; Wang, J.; Li, Y.; Jiang, K.; Fan, S. Fabrication of Ultralong and Electrically Uniform Single-walled Carbon Nanotubes on Clean Substrates. *Nano Letters* **2009,** *9* (9), 3137–3141.
126. Zhang, R.; Xie, H.; Zhang, Y.; Zhang, Q.; Jin, Y.; Li, P.; Qian, W.; Wei, F. The Reason for the Low Density of Horizontally Aligned Ultralong Carbon Nanotube Arrays. *Carbon* **2013,** *52,* 232–238.
127. Xie, H.; Zhang, R.; Zhang, Y.; Li, P.; Jin, Y.; Wei, F. Growth of High-Density Parallel Arrays of Ultralong Carbon Nanotubes with Catalysts Pinned by Silica Nanospheres. *Carbon* **2013,** *52,* 535–540.
128. Xie, H.; Zhang, R.; Zhang, Y.; Zhang, W.; Jian, M.; Wang, C.; Wang, Q.; Wei, F. Graphene/Graphite Sheet Assisted Growth of High-Areal-Density Horizontally Aligned Carbon Nanotubes. *Chemical Communications* **2014,** *50* (76), 11158–11161.
129. Ago, H.; Nakamura, Y.; Ogawa, Y.; Tsuji, M. Combinatorial Catalyst Approach for High-Density Growth of Horizontally Aligned Single-walled Carbon Nanotubes on Sapphire. *Carbon* **2011,** *49* (1), 176–186.
130. Hong, S. W.; Banks, T.; Rogers, J. A. Improved Density in Aligned Arrays of Single-walled Carbon Nanotubes by Sequential Chemical Vapor Deposition on Quartz. *Advanced Materials* **2010,** *22* (16), 1826–1830.
131. Zhou, W.; Ding, L.; Yang, S.; Liu, J. Synthesis of High-Density, Large-Diameter, and Aligned Single-walled Carbon Nanotubes by Multiple-Cycle Growth Methods. *ACS Nano* **2011,** *5* (5), 3849–3857.
132. Hu, Y.; Kang, L.; Zhao, Q.; Zhong, H.; Zhang, S.; Yang, L.; Wang, Z.; Lin, J.; Li, Q.; Zhang, Z.; Peng, L.; Liu, Z.; Zhang, J. Growth of High-Density Horizontally Aligned SWNT Arrays Using Trojan Catalysts. *Nature Communications* **2015,** *6.*

133. Cao, A. Y.; Dickrell, P. L.; Sawyer, W. G.; Ghasemi-Nejhad, M. N.; Ajayan, P. M. Super-Compressible Foamlike Carbon Nanotube Films. *Science* **2005,** *310* (5752), 1307–1310.
134. Ci, L.; Manikoth, S. M.; Li, X.; Vajtai, R.; Ajayan, P. M. Ultrathick Freestanding Aligned Carbon Nanotube Films. *Advanced Materials* **2007,** *19* (20), 3300.
135. Jiang, K. L.; Li, Q. Q.; Fan, S. S. Nanotechnology: Spinning Continuous Carbon Nanotube Yarns – Carbon Nanotubes Weave Their Way into a Range of Imaginative Macroscopic Applications. *Nature* **2002,** *419* (6909), 801.
136. Li, Q. W.; Zhang, X. F.; DePaula, R. F.; Zheng, L. X.; Zhao, Y. H.; Stan, L.; Holesinger, T. G.; Arendt, P. N.; Peterson, D. E.; Zhu, Y. T. Sustained Growth of Ultralong Carbon Nanotube Arrays for Fiber Spinning. *Advanced Materials* **2006,** *18* (23), 3160.
137. Zhang, X.; Li, Q.; Holesinger, T. G.; Arendt, P. N.; Huang, J.; Kirven, P. D.; Clapp, T. G.; DePaula, R. F.; Liao, X.; Zhao, Y.; Zheng, L.; Peterson, D. E.; Zhu, Y. Ultrastrong, Stiff, and Lightweight Carbon-Nanotube Fibers. *Advanced Materials* **2007,** *19* (23), 4198.
138. Zheng, L.; Sun, G.; Zhan, Z. Tuning Array Morphology for High-Strength Carbon-Nanotube Fibers. *Small* **2010,** *6* (1), 132–137.
139. Futaba, D. N.; Hata, K.; Yamada, T.; Hiraoka, T.; Hayamizu, Y.; Kakudate, Y.; Tanaike, O.; Hatori, H.; Yumura, M.; Iijima, S. Shape-Engineerable and Highly Densely Packed Single-walled Carbon Nanotubes and Their Application as Super-Capacitor Electrodes. *Nature Materials* **2006,** *5* (12), 987–994.
140. Hinds, B. J.; Chopra, N.; Rantell, T.; Andrews, R.; Gavalas, V.; Bachas, L. G. Aligned Multiwalled Carbon Nanotube Membranes. *Science* **2004,** *303* (5654), 62–65.
141. Hata, K.; Futaba, D. N.; Mizuno, K.; Namai, T.; Yumura, M.; Iijima, S. Water-Assisted Highly Efficient Synthesis of Impurity-Free Single-Waited Carbon Nanotubes. *Science* **2004,** *306* (5700), 1362–1364.
142. Chiang, W.-H.; Futaba, D. N.; Yumura, M.; Hata, K. Direct Wall Number Control of Carbon Nanotube Forests from Engineered Iron Catalysts. *Journal of Nanoscience and Nanotechnology* **2013,** *13* (4), 2745–2751.
143. Chiang, W.-H.; Futaba, D. N.; Yumura, M.; Hata, K. Growth Control of Single-walled, Double-walled, and Triple-walled Carbon Nanotube Forests by a Priori Electrical Resistance Measurement of Catalyst Films. *Carbon* **2011,** *49* (13), 4368–4375.
144. Sakurai, S.; Nishino, H.; Futaba, D. N.; Yasuda, S.; Yamada, T.; Maigne, A.; Matsuo, Y.; Nakamura, E.; Yumura, M.; Hata, K. Role of Subsurface Diffusion and Ostwald Ripening in Catalyst Formation for Single-walled Carbon Nanotube Forest Growth. *Journal of the American Chemical Society* **2012,** *134* (4), 2148–2153.
145. Xiang, R.; Einarsson, E.; Murakami, Y.; Shiomi, J.; Chiashi, S.; Tang, Z.; Maruyama, S. Diameter Modulation of Vertically Aligned Single-walled Carbon Nanotubes. *ACS Nano* **2012,** *6* (8), 7472–7479.
146. Thurakitseree, T.; Einarsson, E.; Xiang, R.; Zhao, P.; Aikawa, S.; Chiashi, S.; Shiomi, J.; Maruyama, S. Diameter Controlled Chemical Vapor Deposition Synthesis of Single-walled Carbon Nanotubes. *Journal of Nanoscience and Nanotechnology* **2012,** *12* (1), 370–376.
147. Chemelli, C.; D'Angelo, D.; Girardi, G.; Pizzini, S. Effects of the Presence of Native Silicon Oxide at the Fe/Si Interface on the Formation of Silicides Studied by Auger Spectroscopy. *Applied Surface Science* **1993,** *68* (2), 173–177.
148. Zhong, G.; Warner, J. H.; Fouquet, M.; Robertson, A. W.; Chen, B.; Robertson, J. Growth of Ultrahigh Density Single-walled Carbon Nanotube Forests by Improved Catalyst Design. *ACS Nano* **2012,** *6* (4), 2893–2903.
149. Sugime, H.; Noda, S.; Maruyama, S.; Yamaguchi, Y. Multiple "Optimum" Conditions for Co–Mo Catalyzed Growth of Vertically Aligned Single-walled Carbon Nanotube Forests. *Carbon* **2009,** *47* (1), 234–241.
150. Yang, H. S.; Zhang, L.; Dong, X. H.; Zhu, W. M.; Zhu, J.; Nelson, B. J.; Zhang, X. B. Precise Control of the Number of Walls Formed during Carbon Nanotube Growth Using Chemical Vapor Deposition. *Nanotechnology* **2012,** *23* (6).

151. Zhou, L. P.; Ohta, K.; Kuroda, K.; Lei, N.; Matsuishi, K.; Gao, L. Z.; Matsumoto, T.; Nakamura, J. Catalytic Functions of Mo/Ni/MgO in the Synthesis of Thin Carbon Nanotubes. *Journal of Physical Chemistry B* **2005,** *109* (10), 4439–4447.
152. Balbuena, P. B.; Zhao, J.; Huang, S. P.; Wang, Y. X.; Sakulchaicharoen, N.; Resasco, D. E. Role of the Catalyst in the Growth of Single-wall Carbon Nanotubes. *Journal of Nanoscience and Nanotechnology* **2006,** *6* (5), 1247–1258.
153. Cheng, H.-C.; Lin, K.-C.; Tai, H.-C.; Juan, C.-P.; Lai, R.-L.; Liu, Y.-S.; Chen, H.-W.; Syu, Y.-Y. Growth and Field Emission Characteristics of Carbon Nanotubes Using Co/Cr/Al Multilayer Catalyst. *Japanese Journal of Applied Physics Part 1-Regular Papers Brief Communications & Review Papers* **2007,** *46* (7A), 4359–4363.
154. Robertson, J.; Zhong, G.; Hofmann, S.; Bayer, B. C.; Esconjauregui, C. S.; Telg, H.; Thomsen, C. Use of Carbon Nanotubes for VLSI Interconnects. *Diamond and Related Materials* **2009,** *18* (5–8), 957–962.
155. Huh, Y.; Green, M. L. H.; Kim, Y. H.; Lee, J. Y.; Lee, C. J. Control of Carbon Nanotube Growth Using Cobalt Nanoparticles as Catalyst. *Applied Surface Science* **2005,** *249* (1–4), 145–150.
156. Bennett, R. D.; Hart, A. J.; Cohen, R. E. Controlling the Morphology of Carbon Nanotube Films by Varying the Areal Density of Catalyst Nanoclusters Using Block-Copolymer Micellar Thin Films. *Advanced Materials* **2006,** *18* (17), 2274.
157. Nessim, G. D.; Hart, A. J.; Kim, J. S.; Acquaviva, D.; Oh, J.; Morgan, C. D.; Seita, M.; Leib, J. S.; Thompson, C. V. Tuning of Vertically-Aligned Carbon Nanotube Diameter and Areal Density through Catalyst Pre-treatment. *Nano Letters* **2008,** *8* (11), 3587–3593.
158. Xu, M.; Futaba, D. N.; Yumura, M.; Hata, K. Alignment Control of Carbon Nanotube Forest from Random to Nearly Perfectly Aligned by Utilizing the Crowding Effect. *ACS Nano* **2012,** *6* (7), 5837–5844.
159. Zhang, Y.; Zou, G.; Doorn, S. K.; Htoon, H.; Stan, L.; Hawley, M. E.; Sheehan, C. J.; Zhu, Y.; Jia, Q. Tailoring the Morphology of Carbon Nanotube Arrays: From Spinnable Forests to Undulating Foams. *ACS Nano* **2009,** *3* (8), 2157–2162.
160. Zhang, X.; Li, Q.; Tu, Y.; Li, Y.; Coulter, J. Y.; Zheng, L.; Zhao, Y.; Jia, Q.; Peterson, D. E.; Zhu, Y. Strong Carbon-Nanotube Fibers Spun from Long Carbon-Nanotube Arrays. *Small* **2007,** *3* (2), 244–248.
161. Zhan, Z.-Y.; Zhang, Y.-N.; Sun, G.-Z.; Zheng, L.-X.; Liao, K. The Effects of Catalyst Treatment on Fast Growth of Millimeter-Long Multi-walled Carbon Nanotube Arrays. *Applied Surface Science* **2011,** *257* (17), 7704–7708.
162. Meshot, E. R.; Plata, D. L.; Tawfick, S.; Zhang, Y.; Verploegen, E. A.; Hart, A. J. Engineering Vertically Aligned Carbon Nanotube Growth by Decoupled Thermal Treatment of Precursor and Catalyst. *ACS Nano* **2009,** *3* (9), 2477–2486.
163. Hasegawa, K.; Noda, S. Millimeter-Tall Single-walled Carbon Nanotubes Rapidly Grown with and without Water. *ACS Nano* **2011,** *5* (2), 975–984.
164. Yasuda, S.; Futaba, D. N.; Yamada, T.; Satou, J.; Shibuya, A.; Takai, H.; Arakawa, K.; Yumura, M.; Hata, K. Improved and Large Area Single-walled Carbon Nanotube Forest Growth by Controlling the Gas Flow Direction. *ACS Nano* **2009,** *3* (12), 4164–4170.
165. Amama, P. B.; Pint, C. L.; Mirri, F.; Pasquali, M.; Hauge, R. H.; Maruyama, B. Catalyst-Support Interactions and Their Influence in Water-Assisted Carbon Nanotube Carpet Growth. *Carbon* **2012,** *50* (7), 2396–2406.
166. Hasegawa, K.; Noda, S. Moderating Carbon Supply and Suppressing Ostwald Ripening of Catalyst Particles to Produce 4.5-mm-Tall Single-walled Carbon Nanotube Forests. *Carbon* **2011,** *49* (13), 4497–4504.
167. Zhang, G. Y.; Mann, D.; Zhang, L.; Javey, A.; Li, Y. M.; Yenilmez, E.; Wang, Q.; McVittie, J. P.; Nishi, Y.; Gibbons, J.; Dai, H. J. Ultra-High-Yield Growth of Vertical Single-walled Carbon Nanotubes: Hidden Roles of Hydrogen and Oxygen. *Proceedings of the National Academy of Sciences of the United States of America* **2005,** *102* (45), 16141–16145.

168. Cho, W.; Schulz, M.; Shanot, V. Growth and Characterization of Vertically Aligned Centimeter Long CNT Arrays. *Carbon* **2014,** *72,* 264–273.
169. Li, Y.; Xu, G.; Zhang, H.; Li, T.; Yao, Y.; Li, Q.; Dai, Z. Alcohol-Assisted Rapid Growth of Vertically Aligned Carbon Nanotube Arrays. *Carbon* **2015,** *91,* 45–55.
170. Yamada, T.; Namai, T.; Hata, K.; Futaba, D. N.; Mizuno, K.; Fan, J.; Yudasaka, M.; Yumura, M.; Iijima, S. Size-Selective Growth of Double-walled Carbon Nanotube Forests from Engineered Iron Catalysts. *Nature Nanotechnology* **2006,** *1* (2), 131–136.

CHAPTER 2

Structure and Properties of Carbon Nanotubes

H. Qiu, J. Yang
University of Shanghai for Science and Technology, Shanghai, China

Contents

2.1 INTRODUCTION

Carbon nanotubes (CNTs) are a kind of tubular structure typically having nanometer-scaled diameter and micrometer-scaled length. Because of their tiny size, this fascinating new class of materials was not noticed until 1991, when Iijima observed the raw soot produced in the arc-discharge synthesis of fullerenes by transmission electronic microscope *(1)*. Ever since then, the unique structure of nanotubes has been attracting the attention of the nanoscience communities. It was anticipated that this nanostructure would have interesting and exceptional properties, and thus find many but irreplaceable applications in nanotechnology. Shortly after the discovery of multiwalled CNTs (MWNTs) in 1991, single-walled carbon nanotubes (SWNTs) as small as 1 nm were catalytically synthesized using arc-discharge methods by Bethune et al. *(2)* and Iijima and Ichihashi *(3)* independently in 1993. Different from MWNTs, SWNTs generally exist in the form of

Industrial Applications of Carbon Nanotubes
ISBN 978-0-323-41481-4

crystalline ropes containing tens to hundreds of tubes of similar diameter and chirality, which are closely packed into a triangular lattice.

It is well known that CNTs are intrinsically composed of pure carbon atoms that arrange and interact with each other by the strong sp^2 carbon—carbon chemical bonds and form the unique geometric structure of a carbon network; this gives CNTs fascinating and attractive properties, such as electronic, mechanical, and thermal properties. These properties have convinced scientists that CNTs and related nanostructures may play crucial but irreplaceable roles in many different kinds of applications by either replacing the traditional materials or exploring novel functions.

In this chapter we mainly focus on the introduction of the geometric structure of CNTs and present an overview of the knowledge on typical aspects of basic properties of CNTs. These properties are not only strongly dependent on the structure of nanotubes, but also are all interrelated and influenced. Typically, the strong chemical bonding in the carbon network endows CNTs with strong mechanical modulus, high thermal transport, as well as remarkable electrical properties.

In this chapter, Section 2.2 is devoted to the introduction of the geometric and basic structure of nanotubes on which their consequent properties essentially depend. In Section 2.3, the current state of knowledge of electronic behaviors of CNTs and the correlation to the structures are presented. Section 2.4, as one of most important parts of this chapter, specifies the mechanical properties of nanotubes. We first discuss how the mechanical properties of a nanoscaled substance can be characterized from a theoretical perspective. Second, developments achieved both in theoretical and experimental studies are extensively reviewed. Subsequently, thermal transport behaviors of CNTs and their affecting factors are reviewed in Section 2.5. Finally, the chapter concludes with a brief summary and gives an outlook on future developments in the field.

2.2 GEOMETRIC STRUCTURE AND SYMMETRY OF SINGLE-WALLED CARBON NANOTUBES

An SWNT can be viewed as a conformal mapping of the two-dimensional hexagonal lattice of a single sheet of graphene onto the surface of a cylinder, forming a seamless tubular structure. As pointed out, the proper boundary conditions around the cylinder can only be satisfied if one of the Bravais lattice vectors of the graphene sheet maps to a circumference around the cylinder *(4)*. Therefore each lattice vector of the two-dimensional lattice defines a different way of rolling up the graphene sheet into a tube. Since the properties of nanotubes intimately depend on their diameter and chirality, for a better understanding, a vector, $c_h = na_1 + ma_2$, is theoretically introduced to determine the geometric structure and chirality of a nanotube, where a_1 and a_2 represent the graphene lattice vectors and n and m are integers, respectively (Fig. 2.1).

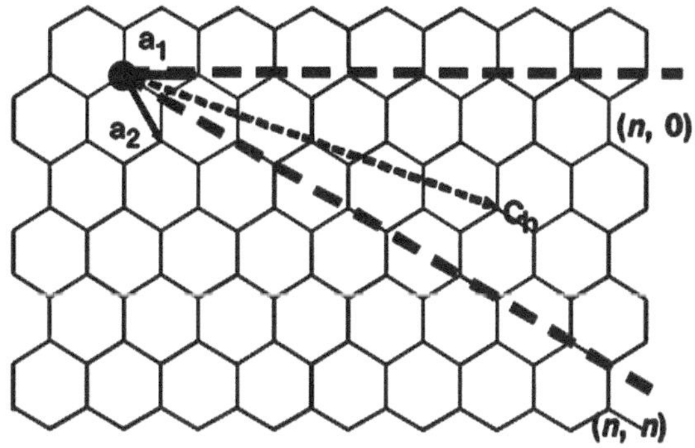

Figure 2.1 Schematic of a two-dimensional graphene sheet illustrating lattice vectors a_1 and a_2, and the roll-up vector $c_h = na_1 + ma_2$. The vectors representing cases of (n, 0) zigzag and (n, n) armchair tubes are indicated with *dashed lines. (Adapted with permission from Odom, T. W.; Huang, J. L.; Kim, P.; Lieber, C. M. Atomic Structure and Electronic Properties of Single-walled Carbon Nanotubes. Nature* ***1998****, 391, 62–64. Copyright 1998, Nature Publishing Group.)*

An SWNT formed with a vector c_h has an axial direction perpendicular to the lattice vector. The construction of SWNTs includes the following types according to the values of (n, m) indices in the vector *(5)*. Nanotubes constructed by rolling up a graphene sheet along a direction equivalent to lattice indices of $(n, 0)$, or (n, m) where $n = m$, are termed zigzag SWNTs or armchair SWNTs, respectively. While constructions mapped in directions equivalent to transition indices from $(n, 0)$ to (n, m) by rotating 30 degrees will be defined as chiral (n, m) SWNTs, where $n \neq m \neq 0$. The schematic illustrations of these types of SWNTs are shown in Fig. 2.2. Typical examples of three types of nanotubes, SWNTs of (5, 5), (9, 0), and (8, 2), have been schematically drawn to illustrate the structures of armchair, zigzag, and chiral nanotubes, respectively. According to the number of graphene sheets, CNTs can be classified into the following types. CNTs, consisting of only one graphene sheet, are termed SWNTs, whereas those comprising more than one graphene sheet are defined as MWNTs. By convention, MWNTs containing just two graphene sheets are named double-walled CNTs with an abbreviation of DWNTs.

2.3 ELECTRONIC PROPERTIES OF SINGLE-WALLED CARBON NANOTUBES

The nanometer scale and the unique geometric structure of CNTs make their electronic properties highly unusual. The small difference in the structure will sensitively affect the electronic properties of nanotubes. In this section, the main electronic properties, as well as the correlation with the geometric structure of CNTs, are reviewed.

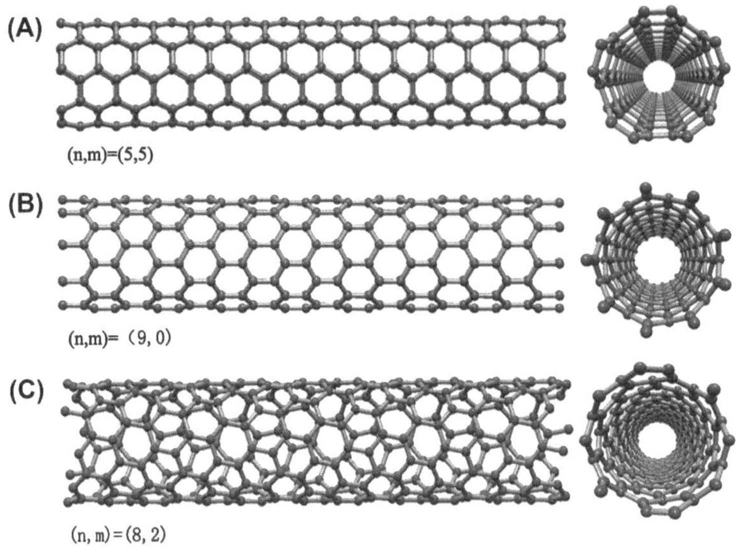

Figure 2.2 Schematic illustration of carbon nanotubes with different chirality: (A) armchair (5,5) single-walled carbon nanotube (SWNT), (B) zigzag (9,0) SWNT, and (C) chiral (8,2) SWNT.

2.3.1 Electronic Structure of Nanotubes

Theoretical calculations predicted that the properties, particularly the electronic behaviors of SWNTs, are strongly correlated to their geometric structure *(7—11)*. Though graphene behaves as a zero-energy band gap semiconductor, CNTs formed by rolling up of graphene can be either metals or semiconductors with tiny energy gaps induced by energy band folding, depending very sensitively on their diameter and chirality, that is, the (n, m) indices of vector c_h.

The dependence of the electronic properties of CNTs on their structure can be physically understood in a band-folding picture *(4)*. Graphene has a unique band structure in which states cross the Fermi level at only two inequivalent points in κ-space [see the figure in Ref. *(4)*]. The electronic structure of an isolated graphene sheet is depicted near the Fermi energy by an occupied π-band and an empty π^*-band both of which have linear dispersion and meet at the K-point in the Brillouin zone. The Fermi surface of an idea graphene sheet consists of the six corner K-points. However, when rolling up a graphene sheet and forming tubes, only a set of κ-states of the planar graphene sheet is allowed, because of the periodic boundary conditions imposed in the circumferential direction. The values of the allowed κ-vectors of tubes are dependent on the diameter and chirality of tubes. The allowed κ-vectors of (5, 5), (7, 1), and (8, 0) tubes indicated by solid lines [see the figure in Ref. *(4)*] are illustrated as the metallic and semiconducting samples, respectively. Once the allowed κ contains the K-point as shown by (5, 5), (7, 1) tubes, the system behaves as a metal with an almost nonzero density of states at the

Fermi level. While the K-point is not contained in the allowed κ-states, taking the (8, 0) tube as an example, the system exhibits a behavior of a semiconductor with different energy band gaps. It is worth noting that for both metallic and semiconducting tubes, the allowed states near the Fermi level are all becoming closest to the K-point, which determines the main electronic and transport properties of tubes.

Extensive calculation has been carried out to deduce a general principle for evaluating the metallicity of a given tube *(6–11)*. Depending on their diameter and chirality of the arrangement of graphitic rings in the walls, CNTs are predicted to be metallic or semiconducting. The general principles deduced by calculation are as follows. For zigzag SWNTs with lattice vectors of (n, 0), when n is completely divided by 3, zigzag (n, 0) SWNTs will behave like a metal, and otherwise a semiconductor. When the vector c_h rotates 30 degrees relative to (n, 0), where n equates to m, armchair (n, n) SWNTs will then form. All armchair tubes are expected to be approximately metallic. As c_h rotates away from (n, 0) to (n, n), chiral SWNTs, transition forms from zigzag to armchair tubes, with various n and m indices are possible. Similarly, as zigzag tubes, chiral tubes have two different behaviors of either metallic or semiconducting depending on the relative values of n and m. When the value of $(2n + m)/3$ is an integer, the tubes behave like a metal, otherwise they are semiconductors. Besides the dependence of chirality on electronic properties of SWNTs, diameter also affects the electronic behaviors of nanotubes. Typically, the energy of the electronic band gap of semiconducting tubes is inversely proportional to diameters. In general, as the tube radius, R, increases, the band gaps of semiconducting tubes will decrease with dependence of $1/R$ and $1/R$ *(2)*, respectively, because of the tube curvature effects *(4)*.

2.3.2 Metallicity of Single-Walled Carbon Nanotubes

Experimental measurements of the electronic properties of CNTs have been extensively studied over the past years and provided solid evidence of promising electronic behaviors, and is in good accordance with the theoretical predictions *(12–16)*. The correlation of the atomic structure of SWNTs and their corresponding electronic properties was first established by virtue of scanning tunneling microscopy (STM) *(6,17)*. In Thess's work *(18)*, the electronic properties of a given metallic or semiconducting tube were simultaneously measured when they were resolved by STM. An SWNT produced by laser vaporization method was identified by STM, showing the expected honeycomb lattice with a C–C spacing of 0.14 ± 0.02 nm. The chiral angle is determined by identifying the zigzag tube axis direction relative to the sample tine axis. It is clearly shown that the tube is chiral with an axis orientated at an angle of -8.0 ± 0.5 degrees relative to that for a zigzag tube. Based on this angle and the measured diameter of 1.0 ± 0.05 nm, the (n, m) indices of the tube can be assigned either (11, 2) or (12, 2) for which the angle/diameter are $-8.2/0.95$ nm and $-7.6/1.03$ nm, respectively.

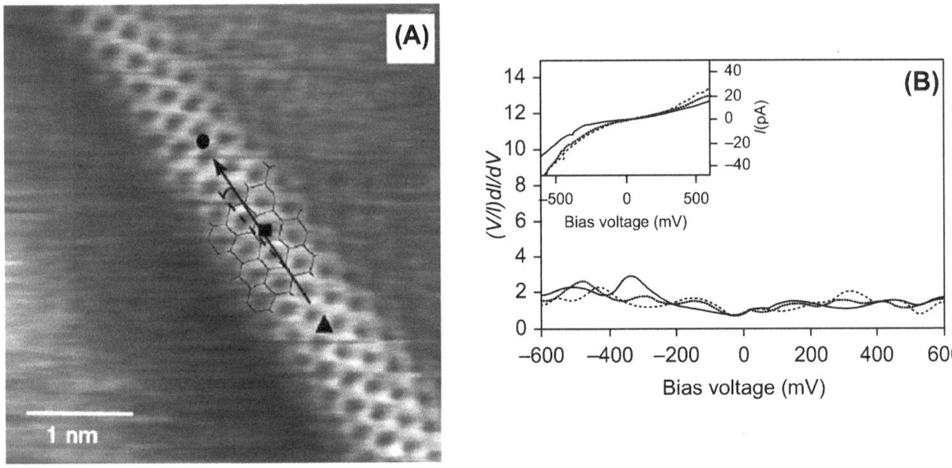

Figure 2.3 Identifying atomic structure of a single-walled carbon nanotube (SWNT) by scanning tunneling microscopy (STM). (A) the STM image was recorded in the constant-current mode with bias voltages of 50 and 150 mV, and tunneling current of 150 pA. The tube axes in the image are indicated with a *solid, black arrow*, and the zigzag direction is highlighted by *dashed lines*. A portion of a two-dimensional graphene layer is overlaid to highlight the atomic structure. (B) shows the calculated normalized conductance, (*V*/*I*)d*I*/d*V*, and measured *I*–*V* behavior. *(Adapted with permission from Odom, T. W.; Huang, J. L.; Kim, P.; Lieber, C. M. Atomic Structure and Electronic Properties of Single-walled Carbon Nanotubes. Nature* ***1998****, 391, 62–64. Copyright 1998, Nature Publishing Group.)*

According to the principles for evaluating the metallicity of a tube, (11, 2) tube is expected to be metallic, whereas (12, 2) is expected to be semiconducting. The tight panel of Fig. 2.3 shows consequent *I*–*V* behaviors of the resolved SWNT shown in the left panel of this figure. The gradual increase in current in the *I*–*V* data demonstrates that the tube is metallic. The metallic behavior measured suggests that the indices for the tube in the figure is (11, 2) tube rather than (12, 2). Also in this work, a number of semiconducting SWNTs were studied in the same way. The results accord well with the predictions.

2.3.3 Summary

Besides the fundamental electronic properties, including the intrinsically metallic or semiconducting behaviors, because of the fascinating one-dimensional structures, SWNTs possess exceptional transport properties by which many potential nanodevices may be fabricated *(19)*. Most of these envisaged electric or electronic applications of nanotubes, such as high-performance transistors, logic integrated circuits, and other various nanodevices, are related in one way or another to the nature of semiconductor and transport properties of SWNTs. More details about these properties and applications based on transport properties of nanotubes can be found in Chapter 3 of this book contributed by Prof. Yutaka Ohno, as well as in the literature *(4,10)*.

2.4 MECHANICAL PROPERTIES OF CARBON NANOTUBES

It is widely accepted that the sp^2 C—C covalent bond in the graphitic hexagonal ring is probably the strongest chemical bond among the known systems in nature. Therefore CNTs completely comprised of the pure carbon atoms by rolling up of graphene sheets have naturally been expected to possess exceptional mechanical properties, and many nanotube applications are strongly desired to be implemented in wide areas.

2.4.1 Derivative Definition of Young's Modulus of Nanotubes

The mechanical properties of a material are defined as its response when subjected to an external stress that may change its shape or volume. From the perspective of classical theory, a series of traditionally defined parameters, including Young's modulus, elastic constants, and Poisson ratio, are employed to characterize a material *(20)*. The Young's modulus of a macroscopic material, Y, along a given direction is defined as:

$$Y = \frac{1}{V_{eq}}\left(\frac{\partial^2 E}{\partial \epsilon^2}\right)_{\epsilon=0} \tag{2.1}$$

where E is the total energy of the system, ϵ is the strain, and V_{eq} is the equilibrium volume of the system. Since the system energy is an extensive property, and it relates to the dimensions of the material, the parameter V_{eq} is introduced in this formula so as to make Y size independent. However, because of the unusual features of nanotubes, the definition of V_{eq} becomes problematic. Though the length of SWNTs can be macroscopic, up to a few millimeters, the thickness of the wall is only one atom thick, of which the value is quite uncertain and difficult to be determined. Since there is no unambiguous way to define V_{eq}, a more generally useful definition would be:

$$\tilde{Y} = \frac{1}{S_{eq}}\left(\frac{\partial^2 E}{\partial \epsilon^2}\right)_{\epsilon=0} \tag{2.2}$$

where S_{eq} is the surface area of the nanotube at zero strain, which can be determined unambiguously *(21)*. The advantage of this definition lies in that it can be directly used without considering the actual geometry of the system. A frequent convention can be realized by replacing V_{eq} in Eq. (2.1) with $S_{eq}h$, where h has the value of 0.34 nm, that is, the interlayer spacing in graphite.

It worth noting that to avoid any confusion when predicting the modulus of nanotubes, it is strongly recommended to use the latter definition.

2.4.2 Mechanical Properties in Elastic Deformation Regime

Numerous theoretical and experimental studies have been extensively carried out to determine the mechanical properties of CNTs ever since the discovery of CNTs *(22)*.

However, there were some differences in these achieved results because of differences in the simulation and experiment methods employed. Especially for experimental measurements of mechanical properties of CNTs, the tiny size of CNTs makes the positioning of nanotubes on an appropriate testing platform extremely difficult, as it does for the operating and measuring processes. Scientists face great challenges when applying a desired loading and detecting slight deformations at nanometer scales. The slight difference in the experimental values measured so far can be attributed to a series of side effects induced by the nanoscale dimension of CNTs, the inappropriate testing techniques for nanoscaled matters, inadequacy in the preparation techniques of testing specimens, and lack of better control over the alignment and distribution of nanotubes. Despite all this, tremendous improvements have been achieved in theoretical prediction and experimental measurement for mechanical properties of both MWNTs and SWNTs.

The first experimental measurement of the Young's modulus of MWNTs was reported by Treacy et al. *(23)* in 1996. In the study, one of the ends of the tube was fixed, while the other one stood in free space. It was then possible by transmission electron microscopy (TEM) to monitor the amplitude of the thermally induced oscillations of the free-standing tip as a function of temperature. Supposing that the tube behaves as a hollow cylindrical structure with a given wall thickness, the authors achieved a Young's modulus of 0.4—4.15 TPa for 11 different nanotubes, with an average value of 1.8 TPa. Although the scatter of the data is relatively large, reflecting the difficulties in operation and measurement process, the study provided the first solid evidence that nanotubes indeed possessed exceptional mechanical properties. A subsequent study by the same experimental approach on a larger sample of SWNTs (27 in total) was carried out by Krishnan et al. *(24)*. A Young's modulus with an average value of 1.25 TPa was obtained, much closer to the value of the C_{11} elastic modulus of the graphite basal plane. Almost at the same time, Chopra and Zettl *(25)* obtained the value of 1.22 TPa of the Young's modulus of boron nitride (BN) nanotubes. As the first prediction of BN modulus, the result indicates that the Young's modulus of BN nanotubes is as high as CNTs.

Shortly after the report of Treacy and coworkers, a different experimental procedure to probe the Young's modulus of nanotubes was employed by Wong et al. *(26)*. Different to the previously reported approach, the authors employed the tip of an atomic force microscope (AFM) to laterally bend the cantilevered nanotubes, and simultaneously record the force exerted by the bent nanotube on the AFM tip as a function of the nanotube deflection. In this way, a Young's modulus with an average value of 1.28 ± 0.5 TPa was extracted. With a deeper understanding of AFM tip-assisted measurement of mechanical properties, Falvo et al. *(27)* succeeded in measuring the extreme flexibility and toughness of nanotubes in their experiments by laterally bending nanotubes on a substrate using an AFM tip until they twisted over themselves, but no failure of the nanotube structure was observed, identifying the exceptional capacity for reversible deformation of nanotubes. Salvetat et al. *(28,29)* also probed the mechanical properties of nanotubes using an AFM tip. In their work, bundles of SWNTs *(28)* or individual MWNTs *(29)* deposited and suspended across the holes on a filtration membrane with a regular pattern were probed by an AFM tip. When a load was applied by AFM tip on the suspended length of the tube, the restoring force was measured as a function of the deflection. Assuming that the tubes were pinned to the substrate by attractive dispersion force, these nanotubes could not slide in the surface of the membrane when the load was applied. It was reported that the MWNTs as synthesized by the arc-discharge method have an average value of Young's modulus of 0.87 TPa, and that nanotubes catalytically grown by the chemical vapor deposition (CVD) method have an average value of Y as low as 27 GPa. The values for MWNTs are comparable with or slight smaller than the previously obtained Young's modulus, and might be caused by the concentration of structural defects in the nanotubes used in the experiments. The much smaller value for catalytically grown SWNTs compared to those produced by laser ablation or arc-discharge methods could only be explained by the much higher extent of structural defects of the nanotubes, clearly showing how much the presence of structural defects affected the mechanical behaviors of the nanotubes.

Poncharal et al. *(30)* measured what they call an effective bending modulus for a series of MWNTs of different diameters by a TEM-assisted technique. While bending, stretching in the outer shell occurs more seriously than in the inner one. Especially in the case of measured nanotubes having a diameter smaller than 12 nm, the effective bending modulus corresponds to the standard Young's modulus and is found to have a value of c. 1 TPa, in agreement with previously obtained Young's modulus results. While for MWNTs of larger diameters, it was reported in this study that the effective bending modulus reduced quickly to a value of c. 100 GPa. High-resolution TEM revealed that large-diameter nanotubes were easily prone to rippled or buckling distortion when bending, as might be attributed to the onset of a strain—relaxation mechanism observed previously, which are more variable for wide nanotubes than for narrow ones *(31—33)*. The experiments by Poncharal et al. indicate that MWNTs can be easily bent laterally than stretched axially, and moreover they can reversibly withstand large lateral distortions.

In subsequent reports, Yu et al. *(34,35)* measured the direct axial tensile strain for both MWNTs *(34)* and SWNTs *(35)* by employing an experimental set-up assembled from two opposing AFM tips and using these to stretch the nanotube beyond the elastic limit. The experiments reached a tensile strength of between 11 and 63 GPa for samples of 19 MWNTs, corresponding the calculated Young's modulus between 0.27 and 0.95 TPa. In the same manner, the tensile strength for 15 SWNT ropes was measured to be between 13 and 52 GPa, with an average value of 30 GPa. The corresponding Young's modulus for individual SWNTs was estimated to be between 0.32 and 1.47 TPa, with an average value of 1.002 TPa. The slight differences reflected in the various samples of both the studied MWNT and SWNT bundles can be attributed to the presence of defects and thus their influence on the mechanical properties.

Along with the aforementioned experimental studies on the mechanical properties of CNTs, simulation and computation studies from many research groups were performed, aimed at guiding and complementing the experimental activities. Theoretical calculations on models of SWNTs using empirical potentials *(36)*, semiempirical tight binding *(37,38)*, and ab initio methods *(39,40)* were performed on estimating the strain energy of nanotubes. The strain energy achieved by these calculations, defined as the energy difference between a given nanotube even with a very small radius and an infinite flat graphene sheet, is in good agreement with what has been predicted by classical theory *(41)*. Considering the Young's modulus, the calculation reveals that the stiffness of a tube with limit radius corresponds directly to the C_{11} elastic constant of graphite calculated by the same model. The empirical potential of Brenner *(42)* provided a value of C_{11} that agreed well with the accepted experimental value of 1.06 TPa *(43,44)*. The other result of this study shows that the predicted modulus should decrease with radius of a narrow tube. This can be explained by understanding the stability of sp^2 C—C bonds. For a much narrower tube, the increased curvature strain would induce the weakening of C—C bonds, thus affecting the mechanical properties.

Results reported by Lu et al. *(45)* were believed to be the first detailed study of the mechanical properties of CNTs. They employed an empirical force field to calculate mechanical properties, including Young's modulus, Poisson ratio (defined as minus the radial strain divided by the axial strain), and some elastic constants of MWNTs and SWNTs. It was found that the Young's modulus of isolated SWNTs has a value of c. 1.0 TPa, which was insensitive to the diameter and chirality of the tube. The result somewhat contradicted with the early report by Robertson et al. whose finding revealed that the Young's modulus of CNTs was affected by diameter and chirality of the tubes. The contradiction between the results may be induced by the simplicity of the model employed by Lu. In Lu's study, harmonic springs were used to represent C—C bonds, and pairwise Lennard—Jones-type potentials were used to model the interlayer interaction in MWNTs. It was also found for the bundles of SWNTs that the Young's modulus decreases with the increase of diameters. However, this does not indicate the weakening of the C—C bonds in the bundles but a fact of geometry effects that the unit cell volume grows more rapidly than the number of carbon atoms. The Young's modulus in Lu's work was 1.11 TPa for MWNTs and 0.97 TPa for isolated SWNTs. Subsequent work by Hernández et al. *(37,38)* compared the Young's modulus and Poisson ratio for SWNTs comprising C, BN, BC2N, C3N4, and BC3. Results revealed that C nanotubes with a diameter of 2 nm have a Young's modulus of 1.26 TPa, while BN nanotubes have a Young's modulus of 0.8 TPa, confirming that C nanotubes have the highest Young's modulus among those studied, although this value was insensitive to the tube diameter except for narrower C nanotubes, which have much higher curvature, resulting in weakening of C—C bonds and consequently reduction in Young's modulus. Both Lu and Hernández et al. concluded the accordant results that Young's modulus is insensitive to the diameter and chirality of nanotubes except for much narrower ones and provides almost the same values of Young's modulus when the thickness of carbon atoms is defined in the same convention. More importantly, these theoretical achievements match well with the experimental findings by Wong et al. *(26)* and Krishnan et al. *(24)* for either MWNTs or SWNTs.

2.4.3 Mechanical Properties in Plastic Deformation Regime

What we discussed earlier concerned the measurements of Young's modulus of CNTs within the axial, linear, elastic regime. However, it is of great significance to estimate and characterize the mechanical properties beyond the elastic regime, namely, when plastic deformation occurs, particularly for practical applications, since in most cases it is necessary for us to know how much stress is needed to break a tube, or how much strain it will sustain before complete failure. To estimate the strength of nanotubes by simulation is a big challenge, because of the wide varying scales involved in fracture, both in time and length domains. This inherent limitation in computation makes simulation impossible to span the macroscopic timescale relevant to fracture experiments. It is therefore understandable that so far no substantial progress has been achieved on theoretically predicting mechanical properties of plastically deformed CNTs. Even so, a theoretical simulation by Samsonidze et al. *(46)* was conducted by studying the energies of the different possible transition states in the formation of the 5/7/7/5 Stone—Wales defect. By using this transition state theory, they determined the yield strain of nanotubes as a function of the chirality of the nanotube and expected failure time. The results showed that chiral tubes have much lower yield strain than zigzag or armchair tubes, and the applied strain until failure was predicted to be c. 17%, in good accordance with experimental results. Subsequently, a brittle yield to tension via a bond-breaking mechanism was suggested by Dumitrică et al. to be more relevant at room temperature, and thus may serve as the acting yield mechanism instructing the tensile-test experiments.

The systematic and solid experiments intended to measure the breaking stress and strain of MWNTs were conducted by Yu et al. *(34)*. In the experimental set-up, two opposing AFM tips were employed to exert tensile load on the nanotubes. The two ends were bound to the tips with carbonaceous solid film produced by the electron beam of a scanning electron microscope (SEM). Once the two ends of the nanotube had been positioned by the AFM tips as shown in the figure, tensile loading was exerted on the tube by moving apart the tips until the nanotube was fractured and failed. Yu et al.

succeeded in carrying out load experiments of 19 MWNTs, and successfully monitored the failure of the used tubes. With the gradual pulling apart of the tips, the attached nanotubes having 6.9 μm become stretched and lengthened, and after complete breaking the total length of the two fractured tubes reached 12.5 μm (6.6 plus 5.9) [for details, please refer to series images in the figure in Ref. *(34)*]. It was shown in other series images that MWNTs having a section length of 11 μm became S-shaped just after fracture, and the length reached 23 μm, indicating the partial pullout of the two nanotube fragments after breaking. This may be because the outer layers of MWNTs directly attached to AFM tips were of tensile strain and fractured, while the inner layers were seemingly unaffected, behaving like a sword being pulled from its sheath once the outer tube failed. This study achieved a Young's modulus of the outer tube between 270 and 950 GPa, a breaking strain between 3% and 12%, and a strength from 11 to 63 GPa. The wide distribution in these data reflects, to some extent, the difference in contents of structural defects in outer carbon shells of MWNTs. Yu and his coworkers also investigated the related mechanical properties of the bundles of SWNTs by similar methods *(35)*. Fig. 2.4 shows the SEM images demonstrating how SWNT bundle tensile-loading experiments were conducted. A tensile load between an SWNT bundle attached to an AFM tip and an SWNT "paper" sample was gradually applied until it broke. It was found in their work that the tensile load was essentially transmitted to the outermost nanotubes comprising the bundle, so that the inner tubes remained intact. Assuming that the SWNT bundle was composed by hexagonally close-packed SWNTs, the measured values were as follows: an average value of 30 GPa for breaking strength, and an average value of 3.1% for breaking strain.

Not only MWNTs but also bundles of SWNTs have been demonstrated by Yu et al. to be brittle, especially under ambient conditions. What about the tensile strength and strain of CNTs at elevated temperatures? And how are they affected by the temperature?

Simulations by Yakobson et al. *(47)* have been conducted to predict the strength of nanotubes under tension. It was revealed that when nanotube was stretched and almost fractured, chains of atoms formed between the two fragments to link them together. Nardelli et al. *(48)* further conformed the formation of a dislocation with a 5/7/7/5 Stone–Wales pair in their work by means of molecular dynamics simulations. At sufficient strain, the energetic barrier to the formation of a Stone–Wales defect is reduced, therefore this defect forms more easily. The presence of this defect definitely facilitates the further relaxation of the tube, which can proceed in the form of either brittle fracture or superplastic tensile deformation, depending on the strain rate, the temperature, and mainly the structure of the nanotubes. The results of these simulations conclude that under conditions of low temperature and high tensile strain rate, nanotubes become brittle and easily break, while at high temperature and low applied tensile strain rate, nanotubes narrower than 1.1 nm are ductile and experience a plastic deformation, although the stress–strain behaviors of those with larger diameters largely depend on their structure.

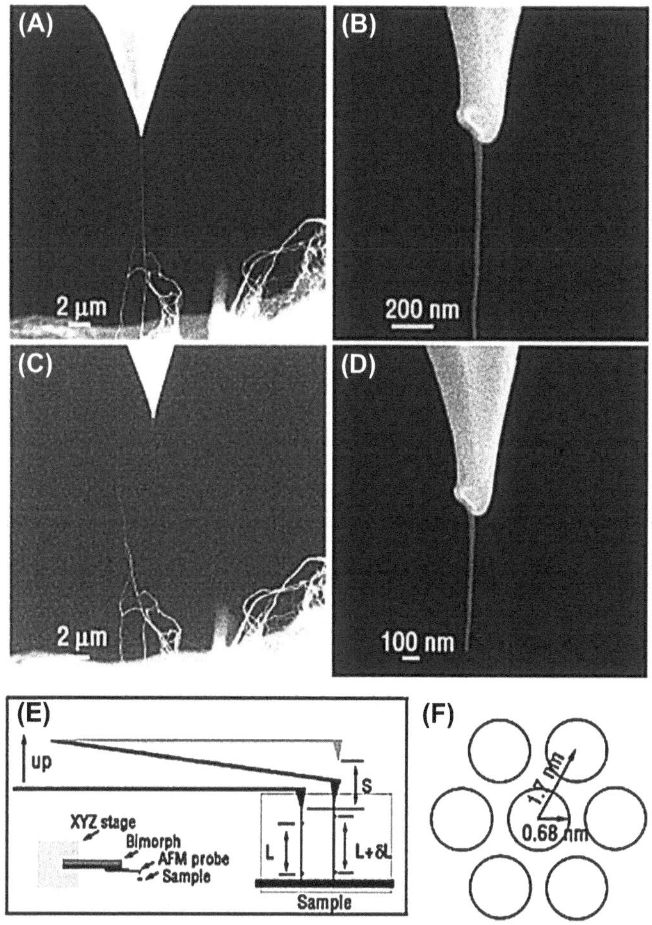

Figure 2.4 Scanning electron microscope (SEM) images showing a single-walled carbon nanotube (SWNT) rope tensile loading experiment, before and after the SWNT rope was broken. (A) A tensile loaded SWNT rope between an atomic force microscope (AFM) tip and an SWNT "paper" sample. (B) Close-up view showing the attachment (carbonaceous deposit) of the end of the SWNT rope to the AFM tip. (C) The same SWNT rope after being loaded to the point where it broke. The image shows that one rope fragment was about 1 mm from the attachment region on the AFM tip. (D) Another close-up view of the attachment area after the rope was broken, showing the deposit was still robust. (E) A schematic showing an overview of the tensile-loading experiment. In the schematic, the *gray cantilever* indicates where the cantilever would be if no rope were attached on the AFM tip. (F) The assumed close-packed SWNT rope with hexagonal cross-section used in the paper for the purpose of calculating the cross-sectional area from the diameter measured with SEM. *(Adapted with permission from Yu, M.-F.; Files, B. S.; Arepalli, S.; Ruoff, R. S. Tensile Loading of Ropes of Single-wall Carbon Nanotubes and Their Mechanical Properties. Physical Review Letters **2000**, 84, 5552–5555. Copyright 2000, American Physical Society.)*

Dumitrică and coworkers *(49)* have developed a strength theory of nanotubes combining both the brittle and plastic yield mechanisms, demonstrating the dependence of these different mechanisms on the symmetric structure of nanotubes, imposed strain, and temperature. These simulations convince us that nanotubes presumably exhibit quite different behavior at elevated temperatures or low strain rates and are helpful to understand the experimental results. To demonstrate these issues, series experiments have been reported. In Huang's work *(50)*, an SWNT was stretched at temperatures as high as 2000°C, and exhibited a superplastic deformation behavior, that is, sustaining strains of up to 280% of the original length and with a consequent reduction of diameter from 12 to 0.8 nm. Furthermore, DWNTs were also reported to have similar mechanical behavior as MWNTs *(51)*. The superelastic deformation behavior exhibited by both MWNTs and DWNTs at elevated temperatures was well explained by considering the formation of dislocation kinks at these conditions, as predicted by Yakobson et al. *(47)* and Nardelli et al. *(48)*.

2.4.4 Factors Affecting Mechanical Properties of Carbon Nanotubes

Though a large number of experimental approaches and various simulation methods have been employed by researchers to measure and predict the mechanical properties of both MWNTs and SWNTs, as discussed earlier, the resulting values were quite inconsistent among those predicted and measured. The possible reasons taken into account to explain these disparities may be as follows. First, the experimental set-up or measuring techniques employed in each investigation were more or less diverse, to a great extent resulting in the slight difference in values of certain mechanical properties. Similarly, the simulation or calculation predictions for mechanical values also varied from each other, because of the employment of various theoretical *(39,40)*, empirical *(36)*, or semiempirical *(37,38)* simulation approaches each of which involves many variables or parameters. So it is not for us to understand the discrepancies of mechanical properties caused by using different measuring techniques or theoretical methods. However, besides the methods employed on mechanical properties, more factors or conditions from CNTs, including the number of defects, types of defects, arrangement of SWNTs in a bundle, and number of walls, have significant influence on the mechanical properties. Therefore the prediction or measurement of mechanical properties often varies with some uncertainties, especially because of the unavoidable defects induced during the production, postpurification, and functionalization processes *(35,46,48,52–54)*.

Both theoretically and experimentally, many types of defects are possible in the CNT structure, such as the Stone–Wales defects *(55)*, vacancies, pentagons, heptagons, and lattice-trapped states. The presence of each type of defect has more or less influence on the mechanical properties of CNTs. Among these defects, the effects of the Stone–Wales defect on the properties of CNTs have been extensively studied by many researchers *(46,56–58)*. The basic structure of the Stone–Wales defect contains two pentagons and two heptagons, which can be formed by appropriate rotation of C–C bonds in hexagonal rings of graphitic sheets. Taking most of the research into account, it was widely believed that the presence of the Stone–Wales defect accounted for the reduction of failure strength and ductility of CNTs. However, the chemical reactivities were enhanced for defective zigzag SWNTs compared to a planer graphene *(59)*. The effects of vacancy defect on mechanical properties of CNTs were also investigated by many researchers. Haskins et al. studied the role of defects on the elastic moduli and failure behavior of CNTs by using tight-binding molecular dynamics simulation *(60)*. In Pozrikidis' report, it was shown that inclined, axial, and circumferential defect orientations have a strong influence on the mechanical response of zigzag and armchair SWNTs *(57)*. Tunvir et al. reported in their work the effects of two neighboring defects on the mechanical properties of CNTs in detail *(58)*. They compared the tensile behavior of SWNTs having both vacancy and Stone–Wales defects positioned next to each other. The influence of the spatial arrangement of defects on the mechanical properties of CNTs was simulated by using classical molecular dynamics at the atomic scale. Two neighboring vacancy defects reduced the failure strength by as much as 46% and the failure strain by as much as 80% when compared to those of pristine SWNTs, while two neighboring Stone–Wales defects reduced them by as much as 34% and 70%, respectively. SWNTs having two defects in the loading (axial) direction showed higher failure strength than SWNTs with defects perpendicular to the loading direction. For both types of defect, the closer the defects, the weaker the SWNTs. The effects of wall numbers of MWNTs on their buckling behavior and mechanical properties were examined by Zhang et al. via molecular dynamics simulations *(61)*. Similar results to previous reports concluded that the buckling strain and Young's modulus of MWNTs increase as the number of tubes is increased while keeping the outermost tube diameter constant, whereas Poisson's ratio was observed to decrease. In the research of Huang and Cao, the effects of structure, temperature, and strain rate on the mechanical properties of a group of polygonal carbon nanotubes (P-CNTs) were investigated by classical molecular dynamics simulations *(62)*. The results indicate that the Young's modulus of P-CNTs is lower than those of circumcircle carbon nanotubes (C-CNTs). With an increase in the number of sides to the polygons, the Young's modulus increases and is much closer to that of C-CNTs. The effects of temperature and strain rate on the mechanical properties of P-CNTs show that the higher temperature and slower strain rate result in a lower critical strain and weaker tensile strength, and is in good agreement with related reports.

Various simulation investigations have been conducted so far to clarify the effects of different defects on the mechanical properties of both SWNTs and MWNTs, which are believed to be helpful for predicting the values of mechanical properties of certain CNTs and instructing experimental measurements under appropriate conditions.

2.4.5 Summary of Mechanical Properties

The remarkable mechanical properties of CNTs have thus far been evidenced by tremendous theoretical simulation and experimental measurement, convincing us that they may find important applications as a reinforcement component in high-performance composites or other related application fields. The most recent measurements of mechanical properties of CNTs have always been involved in evaluating the performance of CNT-reinforced composites *(63,64)*. It is revealed that not only the inherent Young's modulus and tensile strength, but also the quality and purity that insensitively depend on the type and number of defects determine the performance of CNT-based applications. Typically, when MWNTs were used as reinforcement phase to enhance the tensile strength of a specimen, the performance of the specimen will be largely dependent on the nature of the "sword out of a sheath" mechanism, that is, the susceptibility of the inner tubes to be pulled out of the outer tubes upon tensile stresses. For this reason, both theoretical and experimental studies imposed on CNTs will better direct the use of CNTs, broaden the application fields, and value the effect of CNTs on improving the mechanical performance of samples used.

2.5 THERMAL PROPERTIES OF CARBON NANOTUBES

Except for their outstanding electronic and mechanical properties, CNTs possess excellent thermal behaviors, especially for their thermal transport properties, because of the strong carbon−carbon chemical bonding. In the early 2000s, it had been theoretically predicted that CNTs had an unusual high value of thermal conductivity, approximately 6600 W/m K for an isolated (10, 10) nanotube at room temperature, much higher than values of a hypothetical isolated graphene monolayer or diamond that has the highest 3D thermal conductivity *(65)*. The unusual thermal conductivity of CNTs renders them promising applications in areas involving heat dissipation, such as heat management and dissipation in highly integrated circuits based on CNTs *(66,67)*. In recent years thermal properties of CNTs have therefore attracted much attention from the scientific communities working on both theoretical and experimental studies of CNTs. However, the thermal properties have not been as extensively investigated as the electronic and mechanical properties, because of the technical difficulties in manipulating and measuring nanotubes at a single-nanotube level. Generally, the basic thermal properties of CNTs, including specific heat, thermal expansion, and thermal conductivity, result from their relationship to a 2D graphitic sheet and their 1D tubular structure with a nanoscaled size. Therefore the specific heat, as well as thermal expansion, should be similar to those of 2D graphene particularly at normal temperatures, about which knowledge can be found in the literature *(4)*. In this subsection, we present a comprehensive review of recent advances in thermal transport behaviors of CNTs and the influencing factors on thermal transport.

2.5.1 Thermal Transport Behaviors of Carbon Nanotubes

In view of the geometrical structure of CNTs, rolling up of 2D graphene layers composed of carbon atomic hexagons forms 1D nanotubes, thus CNTs should have a similar mechanism of thermal transport to graphite, namely, ballistic phonons primarily contribute to the thermal conductivity. However, CNTs have more apparent long-range crystallinity and much longer phonon mean free path (MFP) compared to graphite, which results in the extraordinary longitudinal thermal transport behavior of CNTs than those of the in-plane structural graphite *(33)*.

The phonon MFP of SWNTs is predicted to be of the order of micrometers at low temperatures *(68,69)*. And the typical lengths of nanotubes used in devices are normally shorter than the value of phonon MFP. Therefore the thermal transport of SWNTs can be thought of as quasiballistic phonon conductive behavior at low temperatures. Since the low-frequency photons serve as the dominant carriers of thermal energy during thermal transport, the thermal conductivity of SWNTs is sensitive to the phonon density of states, namely, both the sub- and acoustic bands, velocity of phonons, temperature, and value of MFP *(4)*. According to experiments performed on crystalline ropes of SWNTs by Hone *(70)*, the measured thermal conductance $\kappa_{ph}(T)$ varies linearly with temperature T at low temperatures and goes to zero at 0K, implying the quantum transport in a 1D system. Subsequently, measurements of thermal conductivity of SWNT bundles *(71)* and individual MWNTs *(72)* were reported by connecting both ends of the SWNT bundle or MWNTs with the two heating pads/thermal contacts, respectively. The thermal conduction measurements of the SWNT bundle samples with diameters of 10–100 nm were performed between 15K and 330K and with a peak at 310K. Based on the measurements, the thermal conductance exhibits a $T^{1.6}$ dependence, clearly lower than that for graphite exhibiting a $T^{2.3}$ dependence, but still greater than the linear T dependence expected for a 1D system and reported for "mat" SWNT samples *(73)*. The coupling to band phonons and tube–tube scattering may account for the discrepancies of a linear T dependence. The maximum of thermal conductivity observed at 310K suggests that phonon–phonon scattering becomes important near room temperature for SWNT bundle samples, but this effect is much weaker than in graphite, possibly reflecting the much smaller number of available scattering states for a 1D system. Thermal conductivity measurements for an MWNT of 14 nm in diameter exhibit a maximum value of 3000 W/m K, as shown in Fig. 2.5, indicating a strong scattering of phonon–phonon at high temperatures *(72)*. The function form of $\kappa(T)$ is 2.0 for the low T, while around 2.5 at larger T, corresponding to a 2D system and a quasi-2D system when compared to a $T^{2.3}$ dependence for graphite. The differences in magnitudes of $\kappa(T)$ for either SWNT bundles or an MWNT or "mat" CNTs samples reflect the influence of sample structure on phonon quantization.

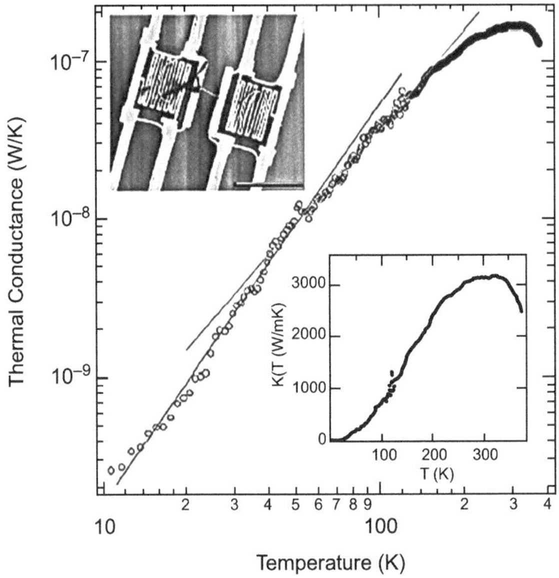

Figure 2.5 Measured thermal conductance of an individual multiwalled carbon nanotube (MWNT) of diameter 14 nm. The *solid lines* represent linear fits of the data in a logarithmic scale at different temperature ranges. The slopes of the line fits are 2.50 and 2.01, respectively (lower inset). Thermal conductivity of the MWNT (upper inset). Scanning electron microscope (SEM) image of the suspended islands with the individual MWNT. *(Adapted with permission from Kim, P.; Shi, L.; Majumdar, A.; McEuen, P. L. Thermal Transport Measurements of Individual Multiwalled Nanotubes. Physical Review Letters* ***2001****, 87, 215502. Copyright 2001, American Physical Society.)*

2.5.2 Influencing Factors on Thermal Conductance

It can be seen in the aforementioned studies of thermal transport behaviors of both SWNT bundles and an MWNT that with increasing temperature the thermal conductance deviates upward from the *T*-linear dependence, indicating that the optical phonons begin to contribute to the heat transport. Typically, for a defect-free (8, 8) SWNT, at lower *T*, where the optical photons are frozen out, the thermal conductance exhibits a *T*-linear behavior with a gradient of $4 \times (9.465 \times 10^{-13})$ W/K where the factor 4 represents the number of acoustic branches of phonons *(74)*. However, with increasing temperatures, phonons make a gradually increasing contribution to thermal conductance. The most recent study on temperature-dependent thermal properties of SWNT thin films shows that the thermal conductance reaches a saturation plateau when temperatures increase to 410K *(75)*. Funding may contribute to a better understanding of the thermal behaviors of CNT thin films.

The length dependence of thermal conductivity of SWNTs was extensively investigated, both theoretically and experimentally. In principle, the phonon-derived thermal conductivity behaves linearly in relation to the tube length *L* because the phonon

MFP is bound to tube length L. The length dependence of thermal conductivity of a (5, 5) SWNT with finite length at 300K was studied in Maruyama's work by the molecular dynamics simulation method based on two different thermostats *(76)*. The results show that the room temperature thermal conductivity of SWNT exhibits linear length dependence when tube lengths are below around 20 nm, but a power law length dependence when tube lengths are longer than 100 nm. However, it is believed that the exponent α is not always the same but variable with regard to the tube length L. Mingo et al. reported the first calculations of finite length carbon nanotube thermal conductivity that extend from the ballistic to the diffusive regime *(77)*. According to theoretical analyses based on the Boltzmann—Peierls phonon transport equation, thermal conductivity saturates to a finite value in the thermodynamic limit $L \to \infty$ if 3-phonon scattering processes are considered to second or higher order. Namely, the exponent α in this case becomes zero in the thermodynamic limit $L \to \infty$. Though the dependence of thermal conduc tivity on tube length was extensively studied *(76—78)*, there is no widely accepted theory so far to demonstrate the correlation of exponent α to the length of SWNTs. But it is believed that the exponent α is closely related to the phonon-scattering mechanism, which is not well understood yet and requires further study.

Another factor affecting the thermal behaviors of CNTs is defects contained in the chemical structure of tubes. Although it is theoretically suggested that the room temperature thermal conductivity of CNTs could be higher than that of graphene and diamond *(65)*, the real thermal conductivity is greatly diminished because of the presence of various defects, including vacancy defects, Stone—Wales defects, or isotope impurities, induced during the processes of synthesis and posttreatment. Many simulation studies show that the thermal conductivity of CNTs is sensitively decreased because of defects concentration *(22,79—81)*. This can be explained by considering that the structural defects locally disrupt the carbon hexagonal network of CNTs and thus dramatically lower the ability of phonons to transport heat. Even though heat treatment techniques have been developed to "heal" the chemical structure by rebonding the dangling bonds in the vicinity of defect location and have recovered to some extent the thermal properties of CNTs, high-quality CNTs are desirable to retain the intrinsic extraordinary thermal conductance.

2.5.3 Summary

Fundamental knowledge about thermal transport has been introduced by reviewing the research in the past a few years. The thermal transport behavior of CNTs is based on the phonon conduction mechanism. With temperatures increasing from relatively low to around room temperature, phonon conduction varies correspondingly from ballistic/quasiballistic to diffusive behavior. In addition, factors, such as temperature, tube length, and defects, affect the thermal transport behavior of CNTs. The thermal transport dependence of these factors is briefly discussed in this subsection.

2.6 SUMMARY AND OUTLOOK

In this chapter, we summarized the important achievements regarding the structure and intriguing properties of CNTs. Because of their unique structure and special properties, CNTs have attracted much scientific interest in the field of nanomaterials, and undoubtedly will become the most eye-catching new material. Based on the fascinating electronic properties of SWNTs, nanoelectronics possessing distinguishing functions and performance will be fabricated and bring about revolutionary effects on the miniaturization of devices and integrated circuits used today. The remarkable mechanical properties of CNTs offer a huge potential for nanoreinforcing composites, is believed to be the most easily realized application of CNTs among all properties, and may serve us in the very near future. The outstanding thermal properties of CNTs give them great potential for applications involving heat-removal nanodevices. Although we have carried out extensive studies on CNTs for the past two decades and indeed have learnt more about nanotubes, we are not exactly convinced whether there are any other undiscovered properties of CNTs, or how much they can change our way of life.

Furthermore, we must realize that what we have achieved in CNTs up to now, both theoretically and experimentally, is far from what we expected or anticipated. On the one hand, characterizations of the properties of CNTs still intimately rely on the development of modern instrument technology. On the other hand, to exploit the usefulness and application field of nanotubes strongly depends on the ability of nanoengineering and manipulating CNTs by researchers at the nanoscale level, and the depth of understanding novel theories suitable for microscopic substances. Fortunately, materials scientists and engineers are currently working on CNTs with unprecedented passion. Therefore everyone has reason to believe that it is just a matter of time before the promise of CNTs is fully realized.

REFERENCES

1. Ijima, S. Helical Microtubules of Graphitic Carbon. *Nature* **1991,** *354,* 56–58.
2. Bethune, D. S.; Kiang, C. H.; de Vries, M. S.; Gorman, C.; Savoy, R.; Vazquez, J.; Beyers, R. Cobalt-Catalysed Growth of Carbon Nanotubes with Single Atomic Layer Walls. *Nature* **1993,** *363,* 605–607.
3. Iijima, S.; Ichihashi, T. Single Shell Carbon Nanotubes of 1-nm Diameter. *Nature* **1993,** *363,* 603–605.
4. Dresselhaus, M. S.; Dresselhaus, G.; Charlier, J. C.; Hernández, E. Electronic, Thermal and Mechanical Properties of Carbon Nanotubes. *Philosophical Transactions of the Royal Society of London A* **2004,** *362,* 2065–2098.
5. Rafii-Tabar, H. Computational Modelling of Thermo-mechanical and Transport Properties of Carbon Nanotubes. *Physics Reports* **2004,** *390,* 235–452.
6. Odom, T. W.; Huang, J. L.; Kim, P.; Lieber, C. M. Atomic Structure and Electronic Properties of Single-walled Carbon Nanotubes. *Nature* **1998,** *391,* 62–64.
7. Hamada, N.; Sawada, S.; Oshiyama, A. New One-Dimensional Conductors: Graphitic Microtubules. *Physical Review Letters* **1992,** *68,* 1579–1581.

8. Saito, R.; Fujita, M.; Dresselhaus, G.; Dresselhaus, M. S. Electronic Structure of Chiral Graphene Tubules. *Applied Physics Letters* **1992,** *60,* 2204–2206.
9. Mintmire, J. W.; Dunlap, B. I.; White, C. T. Are Fullerene Tubules Metallic? *Physical Review Letters* **1992,** *68,* 631–634.
10. Mintmire, J. W.; White, C. T. Electronic and Structural Properties of Carbon Nanotubes. *Carbon* **1995,** *33* (7), 893–902.
11. Saito, R.; Fujita, M.; Dresselhaus, G.; Dresselhaus, M. S. Electronic Structure of Graphene Tubules Based on C60. *Physical Review B* **1992,** *46,* 1804–1811.
12. Zhou, W.; Bai, X.; Wang, E.; Xie, S. Synthesis, Structure, and Properties of Single walled Carbon Nanotubes. *Advanced Materials* **2009,** *21,* 4565–4583.
13. Zhang, Z.; Lieber, C. M. Nanotube Structure and Electronic Properties Probed by STM. *Applied Physics Letters* **1993,** *62,* 2972–2974.
14. Olk, C. H.; Heremans, J. P. Scanning Tunneling Spectroscopy of Carbon Nanotubes. *Journal of Materials Research* **1994,** *9,* 259–262.
15. Ge, M.; Sattler, K. Vapor-Condensation Generation and STM Analysis of Fullerene Tubes. *Science* **1993,** *260,* 515–518.
16. Carroll, D. L.; Redlich, P.; Ajiayan, P. M. Electronic Structure and Localized States at Carbon Nanotube Tips. *Physical Review Letters* **1997,** *78,* 2811–2814.
17. Ge, M.; Sattler, K. STM of Single-Shell Nanotubes of Carbon. *Applied Physics Letters* **1994,** *65,* 2284–2286.
18. Thess, A.; Lee, R.; Nikoraev, P.; Dai, H. J.; Petit, P.; Robert, J.; Xu, C.; Lee, Y. X.; Kim, S. G.; Rinzler, A. G.; et al. Crystalline Ropes of Metallic Carbon Nanotubes. *Science* **1996,** *273,* 483–487.
19. Biercuk, M. J.; Ilani, S.; Marcus, C. M.; McEuen, P. L. Electrical Transport in Single-wall Carbon Nanotubes. In *Carbon Nanotubes;* Jorio, A., Dresselhaus, G., Dresselhaus, M. S., Eds.; *Topics in Applied Physics,* Vol. 111; Springer-Verlag: Berlin Heidelberg, 2008; pp 455–493.
20. Landau, L. D.; Lifshitz, E. M. *Theory of Elasticity;* Pergamon: Oxford, 1986.
21. Yakobson, B. I.; Avouris, P. Mechanical Properties of Carbon Nanotubes. In *Carbon Nanotubes: Synthesis, Structure, Properties and Applications;* Dresselhaus, M. S., Dresselhaus, G., Avouris, P., Eds.; *Topics in Applied Physics,* Vol. 80; Springer, 2001; pp 287–327.
22. Yamamoto, T.; Watanabe, K.; Hernandez, E. R. Mechanical Properties, Thermal Stability and Heat Transport in Carbon Nanotubes. In *Carbon Nanotubes;* Jorio, A., Dresselhaus, G., Dresselhaus, M. S., Eds.; *Topics in Applied Physics,* Vol. 111; Springer-Verlag: Berlin Heidelberg, 2008; pp 165–194.
23. Treacy, M. M.; Ebbesen, T. W.; Gibson, J. M. Exceptionally High Young's Modulus Observed for Individual Carbon Nanotubes. *Nature (London)* **1996,** *381,* 678–680.
24. Krishnan, A.; Dujardin, E.; Ebbesen, T. W.; Yianilos, P. N.; Treacy, M. M. J. Young's Modulus of Single-walled Nanotubes. *Physical Review B* **1998,** *58,* 14013.
25. Chopra, N. G.; Zettl, A. Measurement of the Elastic Modulus of a Multi-wall Boron Nitride Nanotube. *Solid State Communications* **1998,** *105,* 297–300.
26. Wong, E. W.; Sheehan, P. E.; Lieber, C. M. Nanobeam Mechanics: Elasticity, Strength and Toughness of Nanorods and Nanotubes. *Science* **1997,** *277,* 1971–1975.
27. Falvo, M. R.; Clary, G. J.; Taylor, R. M.; Chi, V.; Brooks, F. P.; Washburn, S.; Superfine, R. Bending and Buckling of Carbon Nanotubes Under Large Strain. *Nature (London)* **1997,** *389,* 582–584.
28. Salvetat, J. P.; Briggs, G. A. D.; Bonard, J. M.; Bacsa, R. R.; Kulik, A. J.; Stoeckli, T.; Burnham, N. A.; Forró, L. Elastic and Shear Moduli of Singlewalled Carbon Nanotube Ropes. *Physical Review Letters* **1999,** *82,* 944–947.
29. Salvetat, J. P.; Kulik, A. J.; Bonard, J. M.; Briggs, G. A. D.; Stoeckli, T.; Méténier, K.; Bonnamy, S.; Béguin, F.; Burnham, N. A.; Forró, L. Elastic Modulus of Ordered and Disordered Multiwalled Carbon Nanotubes. *Advanced Materials* **1999,** *11,* 161–165.
30. Poncharal, P.; Wang, Z. L.; Ugarte, D.; de Heer, W. A. Electrostatic Deflections and Electromechanical Resonances of Carbon Nanotubes. *Science* **1999,** *283,* 1513–1516.
31. Iijima, S.; Brabec, C.; Maiti, A.; Bernholc, J. Structural Flexibility of Carbon Nanotubes. *Journal of Chemical Physics* **1996,** *104,* 2089–2092.

32. Lourie, O.; Cox, D. M.; Wagner, H. D. Buckling and Collapse of Embedded Carbon Nanotubes. *Physical Review Letters* **1998,** *81,* 1638–1641.
33. Ruoff, R. S.; Lorents, D. C. Mechanical and Thermal Properties of Carbon Nanotubes. *Carbon* **1995,** *33,* 925–930.
34. Yu, M.-F.; Lourie, O.; Dyer, M. J.; Moloni, K.; Kelly, T. F.; Ruoff, R. S. Strength and Breaking Mechanism of Multiwalled Carbon Nanotubes Under Tensile Load. *Science* **2000,** *287,* 637–640.
35. Yu, M.-F.; Files, B. S.; Arepalli, S.; Ruoff, R. S. Tensile Loading of Ropes of Single-wall Carbon Nanotubes and Their Mechanical Properties. *Physical Review Letters* **2000,** *84,* 5552–5555.
36. Robertson, D. H.; Brenner, D. W.; Mintmire, J. W. Energetics of Nanoscale Graphitic Tubules. *Physical Review B* **1992,** *45,* 12592–12595.
37. Hernández, E.; Goze, C.; Bernier, P.; Rubio, A. Elastic Properties of C and $B_xC_yN_z$ Composite Nanotubes. *Physical Review Letters* **1998,** *80,* 4502–4505.
38. Hernández, E.; Goze, C.; Bernier, P.; Rubio, A. Elastic Properties of Single-wall Nanotubes. *Applied Physics A* **1999,** *68,* 287–292.
39. Miyamoto, Y.; Rubio, A.; Louie, S. G.; Cohen, M. L. Chiral Tubules of Hexagonal BC_2N. *Physical Review B* **1994,** *50,* 4976–4979.
40. Miyamoto, Y.; Rubio, A.; Cohen, M. L.; Louie, S. G. Electronic Properties of Tubule Forms of Hexagonal BC_3. *Physical Review B* **1994,** *50,* 18360–18366.
41. Tibbets, G. G. Why Are Carbon Filaments Tubular. *Journal of Crystal Growth* **1983,** *66,* 632–638.
42. Brenner, D. W. Empirical Potential for Hydrocarbons for Use in Simulating the Chemical Vapor Deposition of Diamond Films. *Physical Review B* **1990,** *42,* 9458–9471.
43. Kelly, B. T. *Physics of Graphite;* Applied Science Publishers: London and New Jersey, 1981; p 477.
44. Dresselhaus, M. S.; Dresselhaus, G.; Sugihara, K.; Spain, I. L.; Goldberg, H. A. *Graphite Fibers and Filaments;* Springer: Berlin, Heidelberg, 1988; pp 12–34.
45. Lu, J. P. Elastic Properties of Nanotubes and Nanoropes. *Physical Review Letters* **1997,** *79,* 1297–1300.
46. Samsonidze, G. G.; Samsonidze, G. G.; Yakobson, B. I. Kinetic Theory of Symmetry Dependent Strength in Carbon Nanotubes. *Physical Review Letters* **2002,** *88,* 065501.
47. Yakobson, B. I. *In Proceedings of the Recent Advances in the Chemistry and Physics of Fullerenes and Related Materials,* **97**; Electrochemical Society: Pennington, 1997; p 549.
48. Nardelli, M.; Yakobson, B. I.; Bernholc, J. Brittle and Ductile Behavior in Carbon Nanotubes. *Physical Review Letters* **1998,** *81,* 4656–4659.
49. Dumitrică, T.; Hua, M.; Yakobson, B. I. Symmetry, Time and Temperature Dependent Strength of Carbon Nanotubes. *Proceedings of the National Academy of Sciences of the United States of America* **2006,** *103,* 6105–6109.
50. Huang, J. Y.; Chen, S.; Wang, Z. Q.; Kempa, K.; Wang, Y. M.; Jo, S. H.; Chen, G.; Dresselhaus, M. S.; Ren, Z. F. Superplastic Carbon Nanotubes-Conditions Have Been Discovered that Allow Extensive Deformation of Rigid Single-walled Nanotubes. *Nature (London)* **2006,** *439,* 281.
51. Huang, J. Y.; Chen, S.; Ren, Z. F.; Wang, Z. Q.; Wang, D. Z.; Vaziri, M.; Suo, Z.; Chen, G.; Dresselhaus, M. S. Kink Formation and Motion in Carbon Nanotubes at High Temperatures. *Physical Review Letters* **2006,** *97,* 075501.
52. Batra, R. C.; Sears, A. Uniform Radial Expansion/Contraction of Carbon Nanotubes and Their Transverse Elastic Moduli. *Modelling and Simulation in Material Science and Engineering* **2007,** *15* (8), 835–844.
53. Chou, T. W.; Gao, L.; Thostenson, E. T.; Zhang, Z.; Byun, J. H. An Assessment of the Science and Technology of Carbon Nanotube-Based Fibers and Composites. *Composites Science and Technology* **2010,** *70,* 1–19.
54. Talukdar, K.; Mitra, A. K. Molecular Dynamics Simulation Study on the Mechanical Properties and Fracture Behavior of Single-wall Carbon Nanotubes. In *Carbon Nanotubes – Synthesis, Characterization, Applications;* Yellampalli, S., Ed.; InTech: Rijeka, Croatia, 2011; pp 305–338.
55. Stone, A. J.; Wales, D. J. Theoretical Studies of Icosahedral C_{60} and Some Related Species. *Chemical Physics Letters* **1986,** *128,* 501–503.
56. Belytschko, T.; Xiao, S. P.; Schatz, G. C.; Ruoff, R. S. Atomistic Simulations of Nanotube Fracture. *Physical Review B* **2002,** *65,* 235430–235438.
57. Pozrikidis, C. Effect of the Stone-Wales Defect on the Structure and Mechanical Properties of Single-wall Carbon Nanotubes in Axial Stretch and Twist. *Archive of Applied Mechanics* **2009,** *79,* 113–123.

58. Tunvir, K.; Kim, A.; Nahm, S. H. The Effect of Two Neighboring Defects on the Mechanical Properties of Carbon Nanotubes. *Nanotechnology* **2008,** *19,* 065703.
59. Picozzi, S.; Santucci, S.; Lozzi, L.; Valentini, L.; Delley, D. Ozone Adsorption on Carbon Nanotubes: the Role of Stone-Wales Defects. *Journal of Chemical Physics* **2004,** *120,* 7147–7152.
60. Haskins, R. W.; Maier, R. S.; Ebeling, R. M.; Marsh, C. P.; Majure, D. L.; Bednar, A. J.; et al. Tight-Binding Molecular Dynamics Study of the Role of Defects on Carbon Nanotube Moduli and Failure. *Journal of Chemical Physics* **2007,** *127,* 074708.
61. Zhang, Y. Y.; Wang, C. M.; Tan, V. B. C. Examining the Effects of Wall Numbers on Buckling Behavior and Mechanical Properties of Multiwalled Carbon Nanotubes via Molecular Dynamics Simulations. *Journal of Applied Physics* **2008,** *103,* 053505.
62. Huang, L.; Cao, D. P. Mechanical Properties of Polygonal Carbon Nanotubes. *Nanoscale* **2012,** *4,* 5420–5424.
63. Moghadam, A. D.; Omrani, E.; Menezes, P. L.; Rohatgi, P. K. Mechanical and Tribological Properties of Self-Lubricating Metal Matrix Nanocomposites Reinforced by Carbon Nanotubes (CNTs) and Graphene-A Review. *Composites Part B* **2015,** *77,* 402–420.
64. Shokrieh, M. M.; Rafiee, R. A Review of the Mechanical Properties of Isolated Carbon Nanotubes and Carbon Nanotube Composites. *Mechanics of Composite Materials* **2010,** *46,* 1–18.
65. Berber, S.; Kwon, Y. K.; Tománek, D. Unusually High Thermal Conductivity of Carbon Nanotubes. *Physical Review Letters* **2000,** *84,* 4613–4616.
66. Fasano, M.; Bigdeli, M. B.; Sereshk, M. R. V.; Chiavazzo, E.; Asinari, P. Thermal Transmittance of Carbon Nanotube Networks: Guidelines for Novel Thermal Storage Systems and Polymeric Material of Thermal Interest. *American Journal of Neuroradiology* **2014,** *41* (3), 1028–1036.
67. Balandin, A. A. Thermal Properties of Graphene and Nanostructured Carbon Materials. *Nature Materials* **2011,** *10* (8), 569–581.
68. Mingo, N.; Broido, D. A. Carbon Nanotube Ballistic Thermal Conductance and Its Limits. *Physical Review Letters* **2005,** *95,* 096105.
69. Yu, C.; Shi, L.; Yao, Z.; Li, D.; Majumdar, A. Thermal Conductance and Thermopower of an Individual Single-wall Carbon Nanotube. *Nano Letters* **2005,** *5,* 1842–1846.
70. Hone, J.; Whitney, M.; Piskoti, C.; Zettl, A. Thermal Conductivity of Single-walled Carbon Nanotubes. *Physical Review B* **1999,** *59,* R2514.
71. Li, D. *Thermal Transport in Individual Nanowires and Nanotubes* (Ph.D. thesis); Department of Mechanical Engineering, University of California: Berkeley, CA, USA, 2002.
72. Kim, P.; Shi, L.; Majumdar, A.; McEuen, P. L. Thermal Transport Measurements of Individual Multi-walled Nanotubes. *Physical Review Letters* **2001,** *87,* 215502.
73. Hone, J. Phonons and Thermal Properties of Carbon Nanotubes. In *Carbon Nanotubes: Synthesis, Structure, Properties and Applications;* Dresselhaus, M. S., Dresselhaus, G., Avouris, P., Eds.; *Topics in Applied Physics*, Vol. 80; Springer-Verlag: Berlin Heidelberg, 2001; pp 273–286.
74. Yamamoto, T.; Watanabe, K. Nonequilibrium Green's Function Approach to Photon Transport in Defective Carbon Nanotubes. *Physical Review Letters* **2006,** *96,* 255503.
75. Duzynska, A.; Taube, A.; Korona, K. P.; Judek, J.; Zdrojek, M. Temperature-Dependent Thermal Properties of Single-walled Carbon Nanotube Thin Films. *Applied Physics Letters* **2015,** *106,* 183108.
76. Maruyama, S. A Molecular Dynamics Simulation of Heat Conduction in Finite Length SWNTs. *Physica B Condensed Matter* **2002,** *323* (1–4), 193–195.
77. Mingo, N.; Broido, D. A. Length Dependence of Carbon Nanotube Thermal Conductivity and the "Problem of Long Waves". *Nano Letters* **2005,** *5,* 1221–1225.
78. Wang, J.; Wang, J. S. Carbon Nanotube Thermal Transport: Ballistic to Diffuse. *Applied Physics Letters* **2006,** *88,* 111909.
79. Maruyama, S.; Igarashi, Y.; Taniguchi, Y.; Shiomi, J. Anisotropic Heat Transfer of Single-walled Carbon Nanotubes. *Journal of the Thermal Science and Technology* **2006,** *1,* 138–148.
80. Chen, H.; Mcgaughey, A. J. H. Thermal Conductivity of Carbon Nanotubes with Defects. *Micro/Nano Scale Phenomena & Thermal Properties* **2011,** 7 (2), 138–148. T30075–T30075-4.
81. Peng, J.; Feng, Y. H.; Li, W.; Zhang, X. X. Thermal Conductivity of Carbon Nanotubes with Stone-Wales Defects. *Journal of Heat Transfer* **2010,** *134* (9), 507–511.

CHAPTER 3

Carbon Nanotubes for Defect Monitoring in Fiber-Reinforced Polymer Composites

B. Hao, P.C. Ma
The Xinjiang Technical Institute of Physics and Chemistry, Chinese Academy of Sciences, Urumqi, China

Contents

3.1 INTRODUCTION

Fiber-reinforced polymers (FRPs), composite materials consisting of a polymer matrix and continuous fibers, have become one of the most important materials in the field of lightweight constructions because of their high specific stiffness and strength as well as their outstanding fatigue performance *(1,2)*. In FRPs, long fibers ensure longitudinal stiffness and strength, while the polymer matrix is responsible for the fracture toughness, delamination strength, and out-of-plane strength and stiffness of the materials. Lamination and sandwich are the common build-ups for FRPs, such specific discrete structures make maintenance and damage analysis of these materials a challenging task. Because these composites are sensitive to intrinsic damages, such as delamination, matrix cracking, fiber debonding, and breakage, a reliable method for the structural health monitoring (SHM) of FRPs is critical to optimize the maintenance schedules, reduce downtime, and mitigate the risk of implementation *(2)*. Furthermore, an in situ monitoring of damage development in FRP composites could be a useful tool to evaluate their reliability and lifetime.

Industrial Applications of Carbon Nanotubes
ISBN 978-0-323-41481-4

Various methods have been adopted to investigate the defects in FRPs. Among them, scanning acoustic microscopy is a highly developed one. The principle of this method is that the velocity and attenuation of an ultrasonic pulse passing through composites provide information about structural defects *(2)*. This technique suffers from the incapability of illustrating the initiation and propagation of cracks in composites under mechanical load. By embedding conventional health-monitoring sensors, such as optical fibers *(3)* and shape memory alloys *(4)*, it is possible to monitor the structural defects in composites. However, there are large stress perturbations caused by either weak interfacial adhesion with the polymer matrix or the large diameters of fibers.

The emergence of nanotechnology as a major field of research, especially based on carbon nanotubes (CNTs), has impacted almost every scientific discipline. This is because of the unique structure and excellent properties of CNTs, such as tube-like morphology with aspect ratio higher than 500, extremely high modulus and strength (Young's modulus 1.0 TPa, strength >10 GPa), remarkable thermal conductivity (3500 W/m K) and electrical conductivity (10^6 S/cm), and so on *(5)*. Therefore CNTs have been widely studied and employed in various polymer matrices to develop polymer nanocomposites with multifunctional properties. The use of CNTs as a sensory material has emerged as one of the most promising fields for practical applications, and the employment of CNTs for defect monitoring in FRPs is a relatively new research area. Indeed, this concept was proposed by Fiedler and coworkers in 2004 whereby introducing CNTs into the polymer matrix, the structural damage in FRPs can be monitored by measuring the resistance change of materials *(6)*. Moreover, because CNTs can improve the mechanical, thermal, and electrical properties of the matrix with an extremely low content (generally below 1 wt%) *(7)*, the utilization of CNTs in FRPs brings additional benefit for the composites. Given these points, application of CNTs in health sensing of composites has been extensively studied and significant progress has been made so far. It is noted that the bulk of previous and ongoing sensor development studies is mainly based on CNT/polymer nanocomposites, where the structural deformation is measured by monitoring the deformation of the matrix. Coupling of electrical properties and mechanical deformation in CNTs within a matrix material produces composites with sensory capabilities. Optical methods can also be utilized to localize the strain or stress distribution in composites because certain bands of CNTs can be shifted under mechanical force, which are related to the applied load transferred to CNTs, and this process can be characterized using Raman spectrum.

This chapter reviews the state-of-the-art understanding of CNTs for defect monitoring in FRPs, with two particular topics: (1) the mythologies and principles of CNTs for SHM of FRPs, and (2) the effects of CNT distribution and dispersion on the reliability and sensitivity for analyzing the structural defects in FRPs. In view of the progress made so far, some suggestions paving the way for the practical application of CNTs for defect monitoring in FRPs are presented.

3.2 UTILIZATION OF CNTS FOR MONITORING STRUCTURAL DEFECTS IN FRPS

Based on the components of FRPs, efforts aiming at monitoring structural defects using CNTs can be made on modifying the properties of either polymer matrix or fiber *(8)*. Specifically, CNTs can be incorporated into the materials for FRPs via the following two ways. (1) By adding CNTs into polymer resins through proper dispersion and processing, and the materials are employed as a matrix for FRPs (Route A in Fig. 3.1). This strategy can be further extended depending on the physical states of CNTs, such as direct dispersion of a large number individual CNTs into the matrix, incorporation of a polymer matrix into CNT foam/aerogel, and embedding of CNT film/paper into the matrix. (2) Growth or deposition of CNTs into fiber/yarn or on the fiber surface (Route B in Fig. 3.1), the large surface area and excellent electrical properties of nanomaterials make it possible not only to improve the interfacial interactions between the fibers and the matrix, but also to change the insulating inorganic fiber [such as glass fiber (GF), silicon carbide fiber] to the conductive ones, thus offering an alternative method for defect monitoring in FRPs.

3.2.1 Dispersing CNTs into Polymer Matrix

Dispersing individual CNTs into the polymer matrix is a general way to prepare CNT/polymer nanocomposites with enhanced electrical properties. One of the common features among the sensory applications of such nanocomposites is that the mechanical load applied to the material will result in changes in the electrical properties of nanocomposites, which can be quantified by measuring electrical resistance or impedance. In fact, this idea is a continuation of employing an electrical technique as a nondestructive way to monitor the damage in carbon fiber reinforced polymers under static and dynamic conditions. The principle of this method is that carbon fibers are conductive and Raman sensitive; the fracture of fibers results in the change in the electrical/Raman response of composites.

CNT dispersed in the matrix should form a stable conductive network to utilize electrical properties of nanocomposites to evaluate the health states of FRP structure. Good dispersion and uniform distribution of CNTs are the prerequisites to achieve this goal. The formation of a conductive network is based on the percolation theory, which says that when the content of fillers reaches a critical value (percolation threshold), the conductivity of nanocomposites jumps by several orders of magnitude and the material becomes conductive *(11)*. The excellent electrical properties and high aspect ratio of CNTs lead to a very low percolation threshold (generally <0.5 wt%) in nanocomposites. For example, Thostenson and coworkers *(12)* prepared CNT/epoxy nanocomposites using a high shear stress field. This technique maintained the tube-like structure of CNTs and resulted in a percolation threshold below 0.1 wt% (Fig. 3.2A). The tensile test

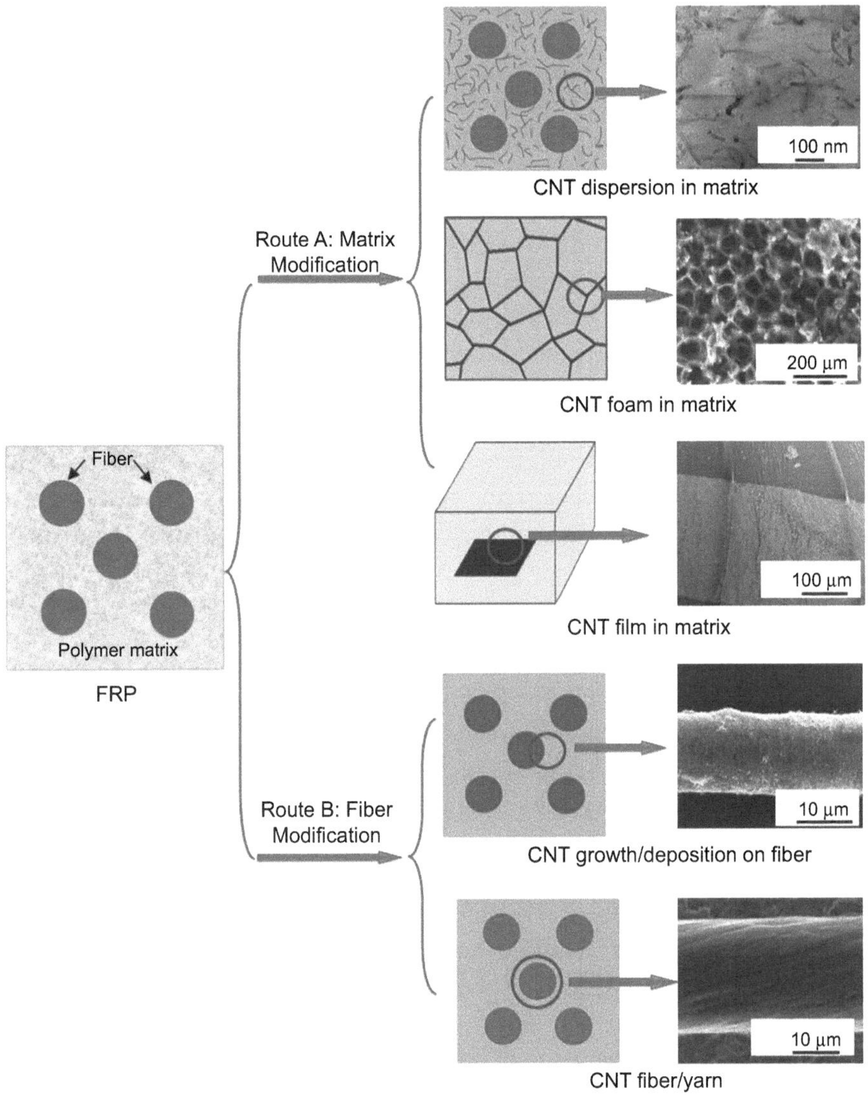

Figure 3.1 Schematic showing the strategies of using CNTs for monitoring structural defects in fiber-reinforced polymers (FRPs). *(Adapted with permission from Ma, P.-C.; Zhang, Y. Perspectives of Carbon Nanotubes/Polymer Nanocomposites for Wind Blade Materials. Renewable & Sustainable Energy Reviews* ***2014****, 30, 651–660. Copyright (2014) Elsevier, from Rein, M. D.; Breuer, O.; Wagner, H. D. Sensors and Sensitivity: Carbon Nanotube Buckypaper Films as Strain Sensing Devices. Composites Science and Technology* ***2011****, 71, 373–381. Copyright (2011) Elsevier, from Liu, K.; Sun, Y.; Lin, X.; Zhou, R.; Wang, J.; Fan, S.; et al. Scratch-Resistant, Highly Conductive, and High-Strength Carbon Nanotube-Based Composite Yarns. ACS Nano* ***2010****, 4, 5827–5834. Copyright (2010) American Chemical Society.)*

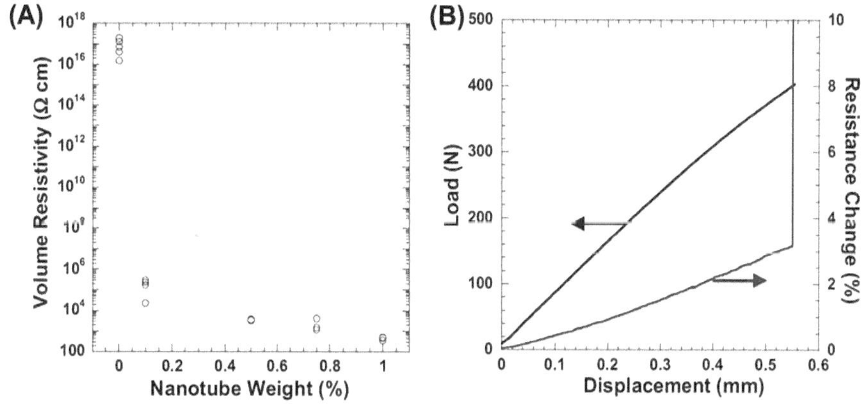

Figure 3.2 (A) Electrical percolation behavior in carbon nanotube (CNT)/epoxy nanocomposites. (B) Resistance change with deformation for a 0.5 wt% CNT/epoxy nanocomposite loaded in tension. *(Adapted with permission from Thostenson, E. T.; Chou, T.-W. Carbon Nanotube Networks: Sensing of Distributed Strain and Damage for Life Prediction and Self Healing. Advanced Materials* ***2006****, 18, 2837–2841. Copyright (2006) John Wiley and Sons.)*

performed on the sample showed a linear curve when plotting the resistance change of nanocomposites as a function of mechanical deformation until the fracture of material where the resistance turned into infinity simultaneously (Fig. 3.2B). From these results, a relationship between the stress and electrical properties of nanocomposites can be established, which provides the fundamental issues for the health evaluation of FRPs using CNT/polymer nanocomposites.

As confirmed by numerous reports, CNT/polymer nanocomposites showed a typical piezoresistive effect under mechanical load, that is, the electrical resistance of samples increases with an increasing strain. The sensitivity to strain can be quantitatively defined by a gauge factor, K, according to Eq. (3.1) *(13)*:

$$K = \frac{\Delta R}{R_0} \Big/ \varepsilon = \frac{R_\varepsilon - R_0}{R_0} \Big/ \varepsilon \tag{3.1}$$

where ε is the applied strain, R_0 is the resistance of nanocomposites before deformation, R_ε is the resistance of nanocomposites at the strain of ε, and ΔR is the resistance change of nanocomposites.

The conductive networks in nanocomposites are made of numerous individual CNTs (good dispersion) or agglomerates (poor dispersion) contacting with each other. Thus the resistance of networks, R_0, is governed not only by the intrinsic resistance of CNTs but also by the intertube resistance. The latter originated from the tunneling effect of non-contacted CNTs separated by small gaps and the contact resistance between two connected CNTs. Since the deformation of nanocomposites generally did not change the structure of CNTs, the variation on the intrinsic resistance of CNTs can be negligible. In other words, the resistance change of nanocomposites under deformation, ΔR, could be ascribed to the variation on the tunneling and contact resistance of CNTs in the matrix. It is known that the contact distance and the one arising from the tunneling effect are in subnanoscale; very small deformations to the nanocomposites will affect these parameters significantly, thus leading to a remarkable change to the resistance of bulky material.

Based on this analysis, Chou et al. *(14)* developed a way for the real-time in situ sensing of damage evolution in FRPs using CNT networks, which were fabricated by dispersing CNTs (0.5 wt%) into epoxy and then incorporated into the nanotube/epoxy with GFs (Fig. 3.3A). The sample in the damage accumulation test was subjected to the cyclic loadings with increasing strain peak. The results showed that during the first and second cycle, the resistance change of nanocomposites was in a linear way with increasing strain, and the resistance could return to the original value after the unloading (Fig. 3.3B), which was attributed to the low level of strain below the threshold where transverse

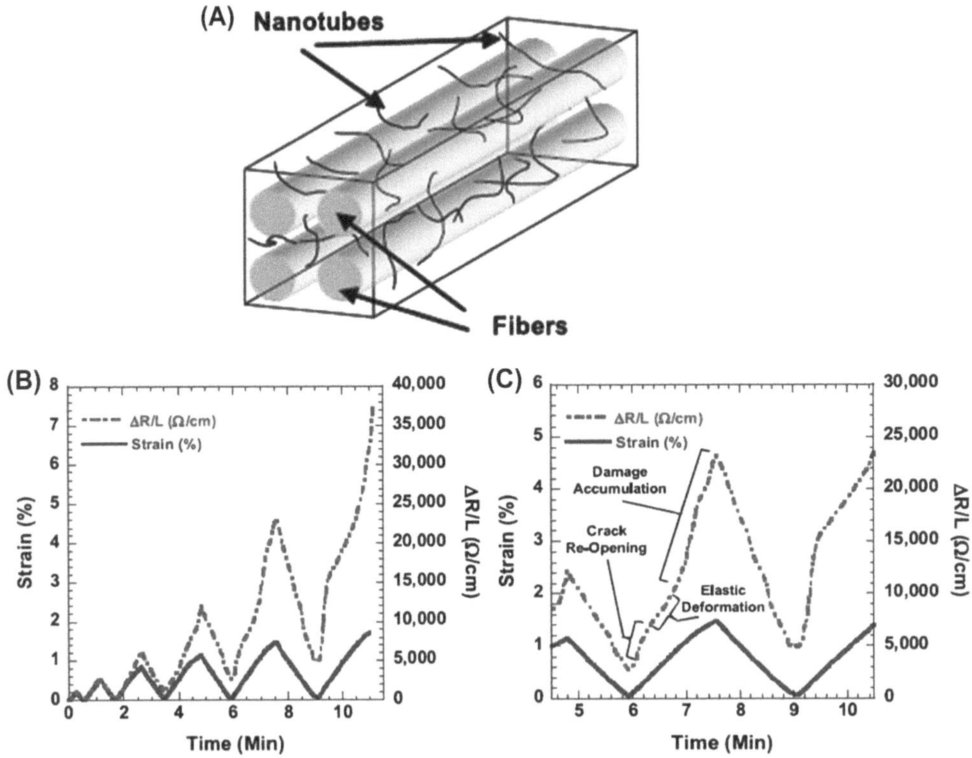

Figure 3.3 Dispersed carbon nanotubes (CNTs) for monitoring structural defects in FRPs. (A) A schematic showing the penetration of CNTs in fiber-reinforced polymers (FRPs). (B) Transient resistance and strain responses of FRPs under cyclic loadings. (C) Expanded view of the fifth cycle showing the reopening and formation of cracks in FRP structure. *(Adapted with permission from Thostenson, E. T.; Chou, T. W. Real-Time in Situ Sensing of Damage Evolution in Advanced Fiber Composites Using Carbon Nanotube Networks. Nanotechnology* ***2008****, 19, 215713. Copyright (2006) IOP Publishing.)*

cracking happened in the nanocomposites. In the third cycle, the resistance of nanocomposites went up significantly, and permanent change on resistance was observed after unloading, which was attributed to the formation of a permanent microcrack or deformation in the material. The resistance curve in the following cycles displayed more pronounced deviation from the strain response with increasing residual resistance change.

More information on damage evolution of FRPs under mechanical loading can be obtained by studying the response of material in the fifth cycle under an expanded view, as shown in Fig. 3.3C. At the beginning of the cycle, a large initial slope on the resistance curve was observed, which was caused by the reopening of previously formed cracks and elastic deformation of CNT networks. The lower slope of the curve was related to the elastic deformation without further accumulation. Then a larger slope in the resistance curve occurred with a larger strain, denoting the emergence of new damage in the material. The crack features during unloading also exhibited abrupt changes in slope, suggesting the closing of generated cracks in FRPs.

While the resistance of material can function in an indirect way to illustrate the damage accumulation and evolution in FRPs, the curve on resistance change of material under mechanical load is not always linear, possibly because of the complexity of the issues arising from crack initiation, reopening, and closing. In addition, factors such as noise signal, generation of voids in FRPs, and isotropy of material should be considered when interpreting the obtained results. Yu and coworkers *(15)* studied the feasibility of CNT/polyester nanocomposites for detecting matrix failure and analyzing damage behavior of three-dimensional (3D) braided FRPs. The results showed that the resistance change of material consisted of five stages in the whole strain range with obvious nonlinear behavior (Fig. 3.4A). In stage 1, the resistance displayed almost no change. This was followed by a significant increase in resistance in stages 2−4. The different slopes of the curve at these stages implied the various damaging mechanisms involved, and at a critical strain after stage 4, the resistance change decreased (stage 5). By combining direct experiment characterization using X-ray computed tomography (CT), the authors concluded that CNTs dispersed in the polymer matrix enable microcracks to be sensed and the damage modes of composites can be analyzed. This conclusion was supported by the CT images of composites in the thickness direction (Fig. 3.4B). Before deformation, no obvious damage was checked. In stage 2, transverse cracks and the initiation of microdelamination were observed. More transverse cracks were noticed in stage 3 and the microdelamination became obvious in terms of size and number, as shown by an elliptical void area between the yarns appearing along their boundary. Although the empty region caused by microcracks and microdelamination was reduced, the number of microcracks (crack density) seemed to increase in stage 4. It is interesting to note that these microcracks and microdelamination were less obvious in stage 5, implying closing of the microcracks. This observation can explain the reduced electrical resistance of the sample.

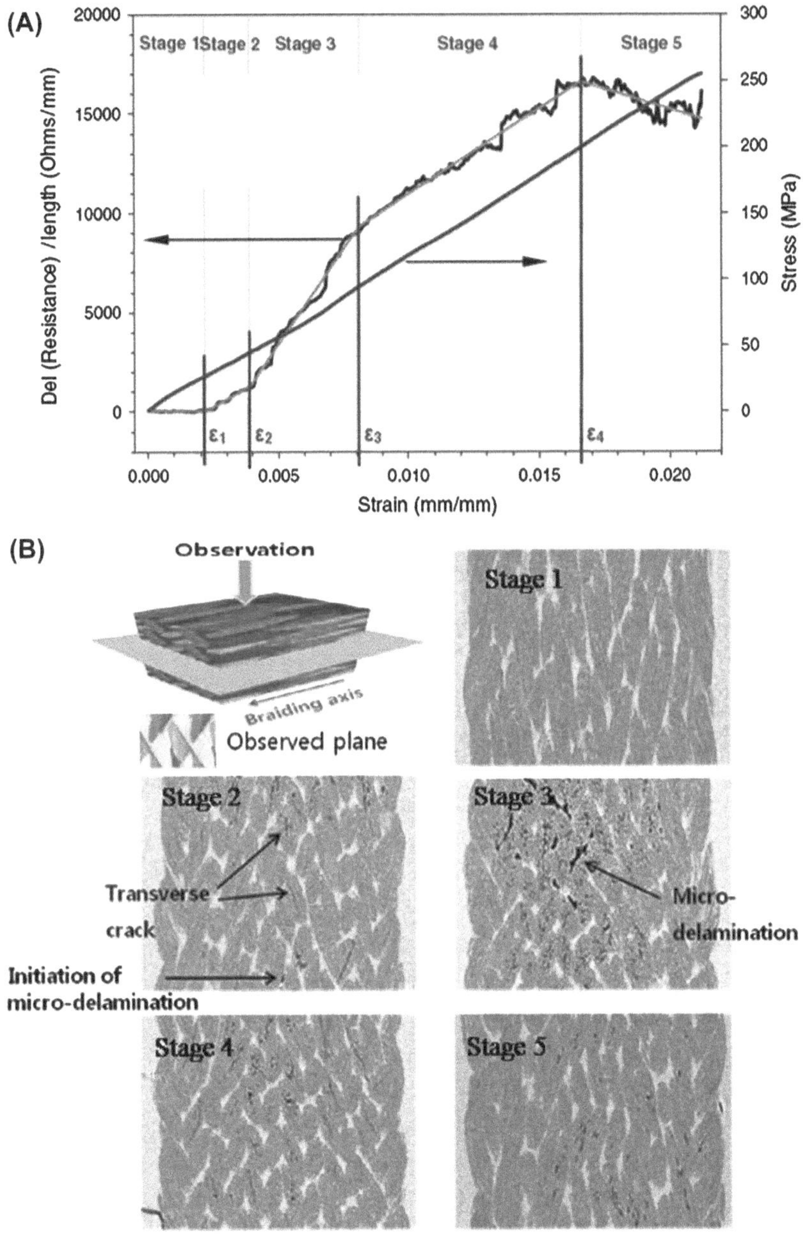

Figure 3.4 (A) Piezoresistive behavior of 3-D braided fiber-reinforced polymers (FRPs) during tensile testing (where ΔR_0 and L represent the resistance change and length of samples). (B) Computed tomography (CT) images of 3-D composites in different stages. *(Adapted with permission from Kim, K. J.; Yu, W.-R.; Lee, J. S.; Gao, L.; Thostenson, E. T.; Chou, T.-W.; et al. Damage Characterization of 3D Braided Composites Using Carbon Nanotube-Based in Situ Sensing. Composites Part A: Applied Science and Manufacturing **2010**, 41, 1531–1537. Copyright (2010) Elsevier.)*

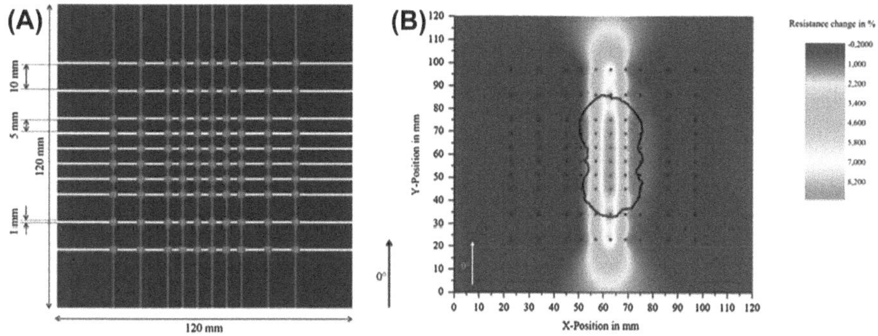

Figure 3.5 (A) Schematic illustrations showing the prototype sensor made from CNT/epoxy nanocomposites in fiber-reinforced polymers (FRPs), and (B) comparison between impact damage map of a glass fiber-reinforced plastic (GFRP) specimen by resistance and delamination contour (the *black curve*) derived from an ultrasonic C-scan. *(Adapted with permission from Viets, C.; Kaysser, S.; Schulte, K. Damage Mapping of GFRP via Electrical Resistance Measurements Using Nanocomposite Epoxy Matrix Systems. Composites Part B: Engineering* ***2014****, 65, 80–88. Copyright (2014) Elsevier.)*

The majority of nanocomposite sensors are based on the analysis of the resistance change of FRP structures under mechanical strain or time. In contrast, Viets et al. *(16)* reported a sensor containing CNTs that provided the resistance signal versus space by mapping the resistance change of composites to achieve spatial distribution of damage sites. The prototype was fabricated by incorporating CNTs/epoxy nanocomposites with GF fabrics and introducing point-electrodes (red circle dots, the gray and white lines were silver lines) in an array on the bottom and top surface in the thickness direction of FRPs, as shown in Fig. 3.5A. The sample was impacted by using a drop weight tower with an impact energy of 7.65 J (impactor diameter 20 mm). The results showed that the maximum resistance increased by 9.2% in the central part of the sample because of the damage to the conductive network, and there was less resistance change around the red area (Fig. 3.5B). The black line was the outer contour of the delamination obtained using ultrasonic C-scans, a complementary tool to evaluate the damaged parts in FRPs. It was clear that the most damaged region from mapping resistance change was roughly included with that from ultrasonic C-scans, suggesting the reliability of this methodology for the health monitoring of FRPs. The unique advantage of this method is that it offers information on the location and distribution of damage in FRPs, which cannot be achieved by measuring the electrical responses of FRPs under mechanical loading. It is expected that a higher resolution on the damage in FRPs can be realized by placing more electrodes in the FRPs.

3.2.2 Foam-Like CNTs in Polymer Matrix

CNT/polymer nanocomposites hold the promise of delivering exceptional mechanical properties and multifunctional characteristics. The potential of employing CNTs as reinforcements has, however, been severely limited because of the difficulties associated with dispersion of entangled CNTs during processing. To ensure effective reinforcements for nanocomposites, proper dispersion of CNTs in the polymer matrix has to be guaranteed.

The foam-like structure of CNTs provides an alternative to address this issue. Such a structure can be sponge, aerogel, xerogel, and so on. The continuous integrity, structural stability, and high porosity of material make it an ideal filler for polymer nanocomposites by infiltrating different polymers into the foam structure, ultimately eliminating the problems associated with dispersion and reagglomeration of CNTs in the polymer matrix.

Various methods have been developed to fabricate foam-like CNTs, including dip coating *(17)*, dispersion-aerogel process *(18)*, the sol–gel method *(19)*, and chemical vapor deposition (CVD) *(20)*. Among them, the CVD method has unique advantages like easy control, high efficiency, and the ability to realize high quantity production of CNT foam at an industrial scale. The CNT foam prepared from CVD could be CNT array *(21,22)* in which CNTs are aligned vertically on a substrate such as silicon wafer, or in a random orientation by optimizing experimental conditions *(23)*. The random approach is more suitable for sensory applications because of the easier fabrication procedure, isotropic properties, and the possibility of being processed into a large sample size. Generally, foam-like CNTs can be achieved directly on metal templates like nickel or copper foam *(20)*.

By utilizing the high porosity of CNT foam, CNT/polymer nanocomposites can be prepared using matrix impregnation/infiltration. In this process, there is no need to consider the dispersion of CNTs, because CNT foam exhibited high thermal and structural stability. While the distribution of CNTs in a matrix can be a CNT-rich region and a CNT-poor one, this will not be a problem as the conductive networks were formed before the incorporation of polymer matrix, and such networks can remain largely after the preparation of nanocomposites. For example, Ma and coworkers *(23)* used commercial melamine sponge as a template to prepare CNT foam by CVD. CNTs were fully covered on the skeleton of carbonized polymer substrate with a high porosity (Fig. 3.6A). During the process, polymer can be partly carbonized, forming a substrate for the growth of CNTs with the assistance of metal catalysts, whereas the majority of polymer template can be decomposed at a high temperature of 800°C. The prepared foam displayed a high electrical conductivity of $\sim 10^{-2}$ S/cm, and this can remain after incorporating polydimethylsiloxane (PDMS). The piezoresistivity of CNT/PDMS nanocomposites was studied, and the results showed that at the beginning of applied strain (S1 in Fig. 3.6B), the resistance displayed little change, suggesting there was not much damage or deformation in the conductive networks. In the second stage (S2 in Fig. 3.6B), the resistance–strain curve exhibited a higher slope with a gauge factor of ~ 6, denoting more pronounced deformation in the conductive networks, and such

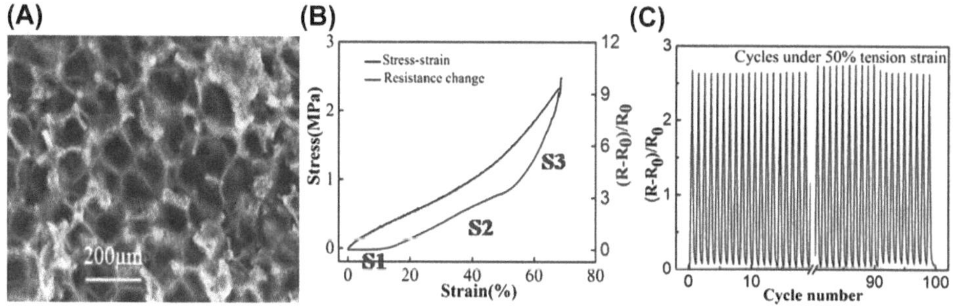

Figure 3.6 Foam-like carbon nanotubes (CNTs)/polymer nanocomposites for sensory application. (A) Morphology of CNT foam grown on carbonized polymer template. (B) Mechanical and electrical responses of CNT foam/ polydimethylsiloxane (PDMS) nanocomposites under tension. (C) Cyclic performance of nanocomposite sensor with a maximum strain of 50%. *(Adapted from Hao, B.; Cui, L. F.; Ma, P. C. Largely Stretchable and Compressible Strain Sensor Based on Carbon Nanotubes with 3-D Networks, 2015 Under review.)*

behavior can survive at a maximum strain of 50%. The following stage (S3 in Fig. 3.6B) was related to a drastic resistance change as a function of strain, and the obtained gauge factor was 14, suggesting the accumulation of damaged networks in nanocomposites. The sensor based on CNT/PDMS nanocomposites showed excellent durability, as confirmed by the marginal variations on the resistance change after 100 cycles (Fig. 3.6C). The 3-D CNT foam in hierarchical structure provides another new way for the fabrication of CNT-based nanocomposite sensors.

3.2.3 CNT Film/Paper in Polymer Matrix

Two-dimensional (2-D) assembly of CNTs, like CNT paper or film, has also been studied in recent years for strain sensing and health monitoring in composites. In a typical experiment for the preparation of CNT film, CNT dispersions were first deposited on various substrates by using techniques such as spray coating, spinning coating, dip coating, electrophoretic deposition (EPD), vacuum filtration, rod coating, drop casting, inkjet printing, and so on *(24)*. This was followed by a solvent evaporation process and free-standing CNT films can be obtained directly by peeling off from the substrates.

There are several ways to incorporate CNT film into polymers for healthy monitoring, and direct embedding is one of the most convenient ways for this application. For example, Wagner et al. *(9)* prepared CNT paper with a thickness around 70 μm and embedded it into epoxy as a sensor to measure strains in polymers with different elastic properties. The results showed that the CNT paper was penetrated by epoxy matrix (Fig. 3.7A) because of its loose structure, thus leading to the formation of CNT film/ polymer nanocomposites. The electrical properties of nanocomposites were studied under mechanical strain, and the results showed that the resistance of the sample increased linearly with increasing strain, which was caused by the enlarging gaps among CNTs by

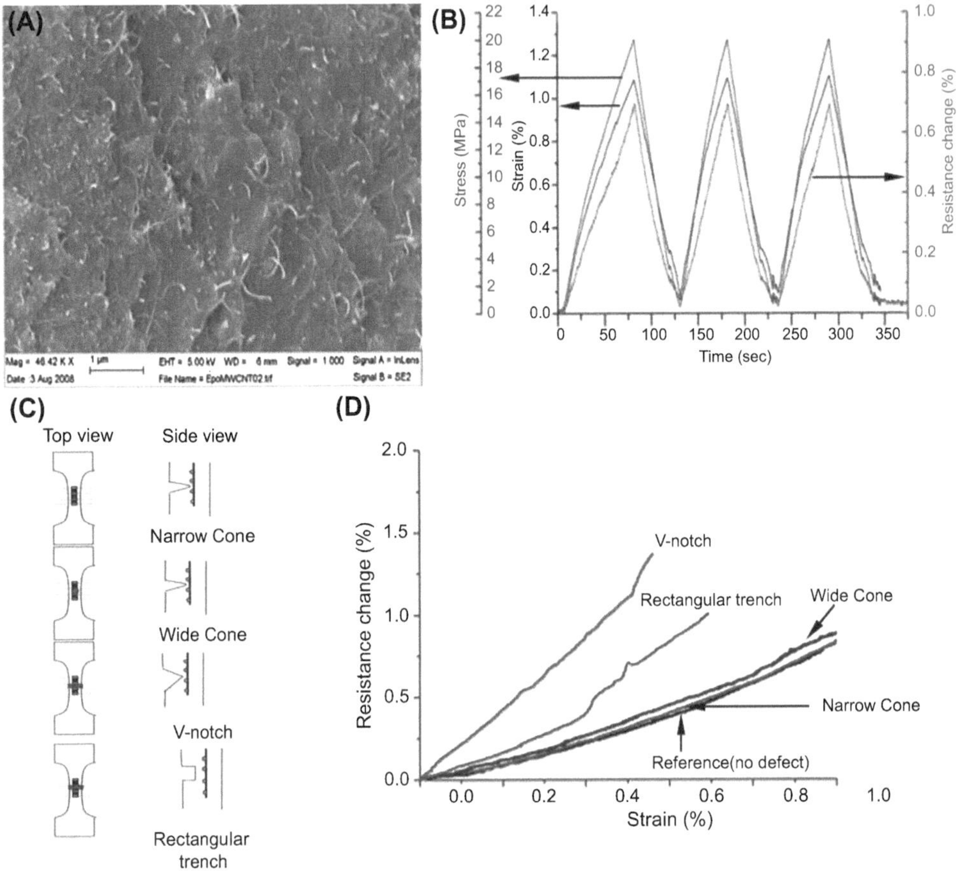

Figure 3.7 Carbon nanotube (CNT) paper embedded in polymer matrix for sensory application. (A) Impregnation of polymer matrix into the loose networks of CNT paper. (B) Strain, stress, and resistance change as a function of time of CNT paper/epoxy nanocomposites under cyclic tests. (C) Top view and side view of the sensors embedded into epoxy to monitor different local defects in composites. (D) Resistance response of samples with different defects under mechanical loading. *(Adapted with permission from Rein, M. D.; Breuer, O.; Wagner, H. D. Sensors and Sensitivity: Carbon Nanotube Bucky-paper Films as Strain Sensing Devices. Composites Science and Technology* ***2011****, 71, 373–381. Copyright (2011) Elsevier.)*

the deformation (Fig. 3.7B). When the strain was large enough, residual resistance was seen, denoting permanent damage to the CNT conductive networks.

To assess the sensitivity of CNT paper to the geometry of defects in composites, the authors introduced four types of designed local defects (narrow cone, wide cone, V-notch, and rectangular trench) (Fig. 3.7C) in the prepared samples, and found that CNT paper was capable of reflecting these local defects with different sensitivities. The responses of the sensor in the presence of a V-notch and a trench differ significantly from the sensor response when no defect is present, whereas the effect of the presence of elliptical defects seems negligible (Fig. 3.7D). The cause of such behavior may be related to the nature of defects in composites.

CNT film can also be attached to the surface of a testing object for sensory applications. In most cases, CNT film was bonded onto the outside of materials and utilized as a strain gauge. The main mechanisms involve simultaneous deformation of CNT film with substrate or the stress transfer from the structural materials to the CNT film, which changes the electrical resistance of CNT film. This kind of CNT film was mostly employed by combining with polymer to form nanocomposites in practical applications to enhance the reliability of the sensor. Mainly, such nanocomposites can be classified into two groups based on the rigidity of matrix material: (1) a rigid sensor with a thermoplastic or thermosetting matrix, such as polyolefin and epoxy, and (2) a flexible sensor with an elastomer matrix, such as PDMS.

Srivastava et al. *(25)* reported a strain sensor made from a nanocomposite film consisting of polystyrene (PS) and 6 wt% of CNTs. The sensor was attached to a brass specimen with the assistance of epoxy to measure the microstrain applied on the metal sample. Fig. 3.8 shows the different modes of strain applied on brass, and the electrical responses of nanocomposites tracked closely the change of strain of metal substrate without time delay. An interesting observation was that for three loading conditions, that is, ramp,

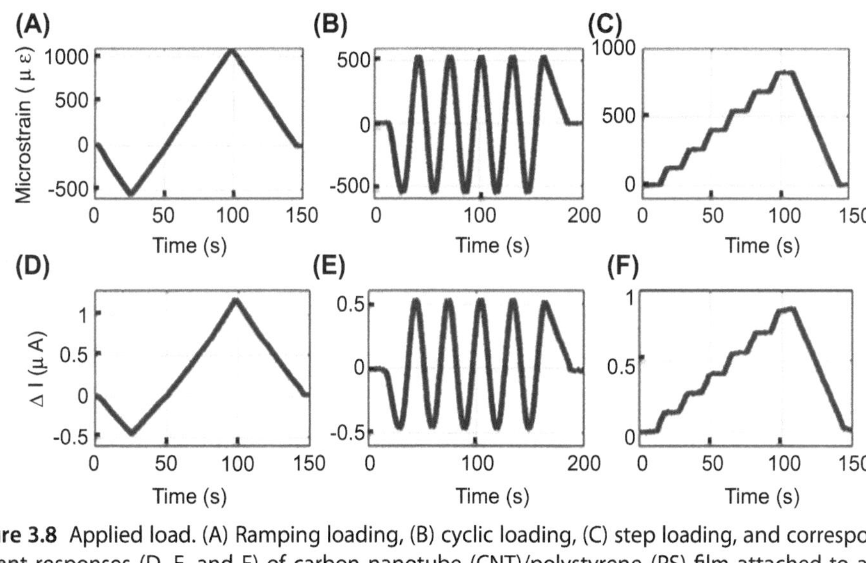

Figure 3.8 Applied load. (A) Ramping loading, (B) cyclic loading, (C) step loading, and corresponding current responses (D, E, and F) of carbon nanotube (CNT)/polystyrene (PS) film attached to a brass specimen (ΔI in D, E, and F represents the current change of sample). *(Adapted with permission from Srivastava, R. K.; Vemuru, V. S. M.; Zeng, Y.; Vajtai, R.; Nagarajaiah, S.; Ajayan, P. M.; et al. The Strain Sensing and Thermal–Mechanical Behavior of Flexible Multi-Walled Carbon Nanotube/Polystyrene Composite Films. Carbon* ***2011****, 49, 3928–3936. Copyright (2011) Elsevier.)*

cyclic, and stepped tensile loadings, the electrical responses of the sensor were noise free, instant, and stable, and could return to its original condition when the brass specimen was unloaded. In addition, the current change measured in the sensor was linear, suggesting the negligible strain variations caused by the elastic deformation of CNT networks. While this sensor was incapable of evaluating defects in composites, it offers a way to mapping the strain states of structural materials. Additionally, it should be emphasized here that when the strain applied on the brass was large, the mode and efficiency of load transfer to the strain sensor should be considered seriously.

Flexible strain sensors using 2-D CNT film also displayed remarkable performance, especially in some novel applications. Bao et al. *(26)* fabricated conductive, transparent, and stretchable nanotube film by spray coating directly onto an ozone-treated PDMS substrate. The results showed that the film exhibited a memory effect on strain history under cyclic operations, that is, the film can be "programmed" by the first cycle of strain and release, to be reversibly stretchable within the range defined by the first strain. Hata and coworkers *(27)* reported a wearable and stretchable sensory device based on aligned CNT films on PDMS substrate. The sample was capable of measuring strain by up to 280%. The mechanism showed that once stretched, the nanotube film would fracture into gaps and islands, and bundles bridging the gaps

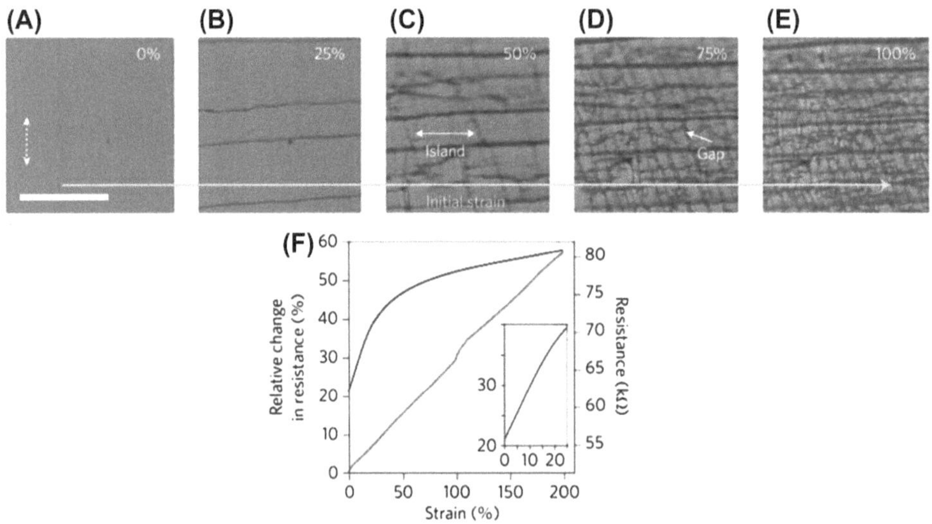

Figure 3.9 Carbon nanotube (CNT) film deposited onto polydimethylsiloxane (PDMS) for sensory application. (A–E) Images showing the progressive agglomeration of CNT bundles under the different strain states from 0% to 100%, scale bar = 100 mm. (F) The resistance change of sample in loading (*red line* (gray in print versions)) and unloading (*blue line* (dark gray in print versions)) conditions. *(Adapted with permission from Yamada, T.; Hayamizu, Y.; Yamamoto, Y.; Yomogida, Y.; Izadi-Najafabadi, A.; Futaba, D. N.; et al. A Stretchable Carbon Nanotube Strain Sensor for Human-Motion Detection. Nature Nanotechnology* ***2011****, 6, 296–301. Copyright (2011) Nature Publishing Group.)*

(Fig. 3.9A–E), leading to a large strain sustained by the film. This mechanism allows the films to act as strain sensors capable of measuring strains with high durability, fast response, and low creep (Fig. 3.9F). The CNT sensors assembled on stockings, bandages, and gloves can detect different types of human motion, including movement, typing, breathing, and speech.

While flexible CNT film exhibited piezoresistivity with a very high extension ratio (most of them can be higher than 50%), excellent transparency, and stretchability, few reports were found in the literature that described their practical application for sensing the strain state in structural materials. This was possibly because of the small deformation of FRPs before catastrophic failure, as well as the dimension of the sensor, which was much smaller than the bulky FRPs. Anyway, the methodology and results obtained so far present us with new concepts for designing novel materials for health monitoring in composites.

3.2.4 CNT Coating on Fiber Surface

The bulk of previous and ongoing studies on the sensory application of CNTs is mainly based on the modification or reinforcement of polymers by CNTs, which can be used as a matrix for FRPs, and the structural deformation can be measured by monitoring the deformation of the matrix. Coupling the electrical properties and mechanical deformation of material makes composites with a sensory capability. A major drawback based on this approach is that one cannot predict the damage in the reinforcements, which actually sustain the majority of the load applied to the structure *(28)*.

As discussed before, optical fibers, shape memory alloys, or polymers have been employed to study the defects and failure behavior of FRPs. However, these conventional materials cause large stress perturbations in the FRP structure because of their large diameters and weak interfacial adhesion between the fibers and the polymer matrix. As an alternative, CNT as a fiber coating/sizing has been developed in recent years, and the CNT-coated fibers exhibited comparable mechanical properties to the structural reinforcements (fibers without coating/sizing), making them an integral part of composite structures without any negative effects on their performance. Such implementation offers a new route for the advanced warning of defects in composites by integrating the interphase modification concept into the FRPs.

Introduction of CNTs onto a fiber surface can be realized by spray coating, fiber sizing, dip coating, EPD, and direct growth. Among these methods, dip coating is easiest and most commonly used method. During the process, the fiber was vertically dipped into a CNT solution for a certain time and then taken out slowly from the solution; the CNT can then be deposited onto the fiber surface because of electrostatic interactions. By optimizing experimental conditions, CNTs or CNT/polymer nanocomposites can be coated homogeneously onto GFs by using this method *(28)*.

EPD is based on the process in which charged particles (CNTs could be charged with the aid of surfactants) can move and deposit on electrodes under electric fields, and this method can be used to coat CNTs as well as graphene on GFs *(29)*. Fig. 3.10 shows that the EPD process produced a fairly uniform coating layer on a fiber surface. Graphene exhibits a full coverage to the GF (Fig. 3.10A), and CNTs form networks along the fiber direction (Fig. 3.10C). These morphological differences originated from the structures of coatings: graphene has a plate-like structure and can be easily assembled layer by layer on the GF surface, whereas CNTs are one-dimensional (1-D) materials with high aspect ratio, and the connectivity of individual CNTs leads to continuous networks on the fiber surface. A close examination of the cross-section of coated fibers shows that the thicknesses of graphene and CNT coating are 250 and 300 nm, respectively (Fig. 3.10B and D).

Single-coated fibers embedded in polymer matrix as the sensor have been studied by some groups. The results presented by Zhang et al. *(30)* showed that the resistance change

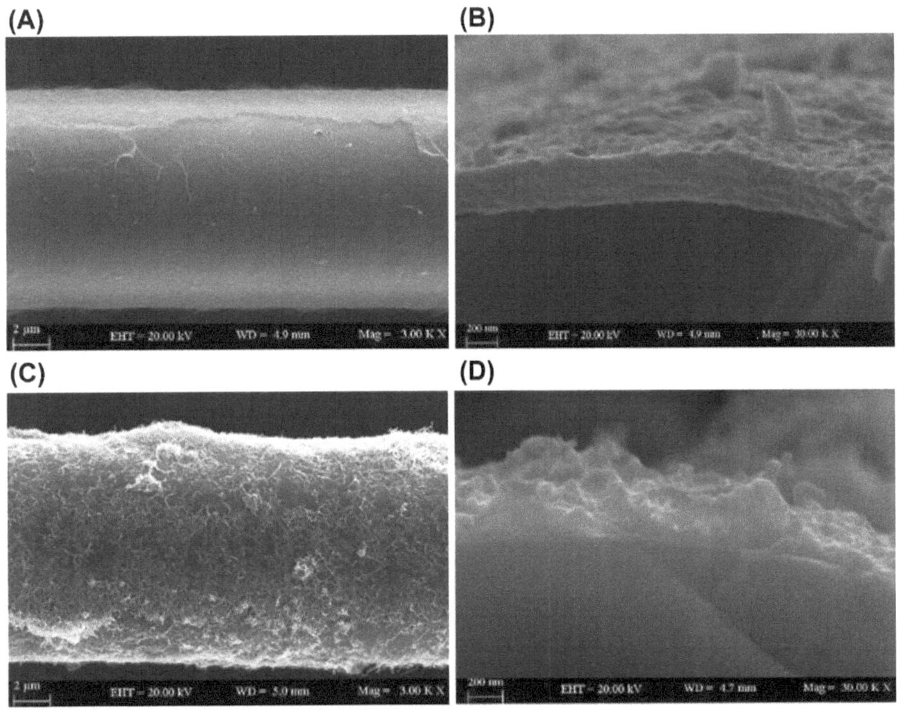

Figure 3.10 Scanning electron microscope (SEM) images showing the in-plane and cross-section morphologies of glass fibers (GFs) with different coatings produced by the electrophoretic deposition (EDP) method. (A and B) Graphene coating. (C and D) Carbon nanotube (CNT) coating. *(Adapted from Hao, B.; Ma, Q.; Yang, S. D.; Mäder, E.; Ma, P. C. Comparative study on monitoring structural damage in fiber-reinforced polymers using glass fibers with carbon nanotubes and graphene coating. Composites Science and Technology 2016, 129, 38–45. Copyright (2016) Elsevier.)*

of corresponding FRPs exhibited three stages (Fig. 3.11A): (1) a linear stage where the electrical resistance of FRPs increased with increasing strain, which was attributed to the elastic deformation of the interphase formed between the polymer matrix and the fiber; (2) a nonlinear stage where the resistance varied exponentially with increased strain; this process is a reflection of the structural defects in FRPs, which were caused by the increasing intertube distance, loss of junction points among CNTs in networks, as well as stress concentration before fiber breakage; (3) abrupt changes on resistance of FRPs, suggesting the conductive network failed completely and the resistance jumped to "infinity" suddenly before the complete failure of the composites.

It is noted that the abrupt change behavior is closely related to the interphase structure. Ma and coworkers *(29)* compared the piezoresistivity of FRPs containing fibers with CNT and graphene coating, and found that for the case of CNT-coated GFs

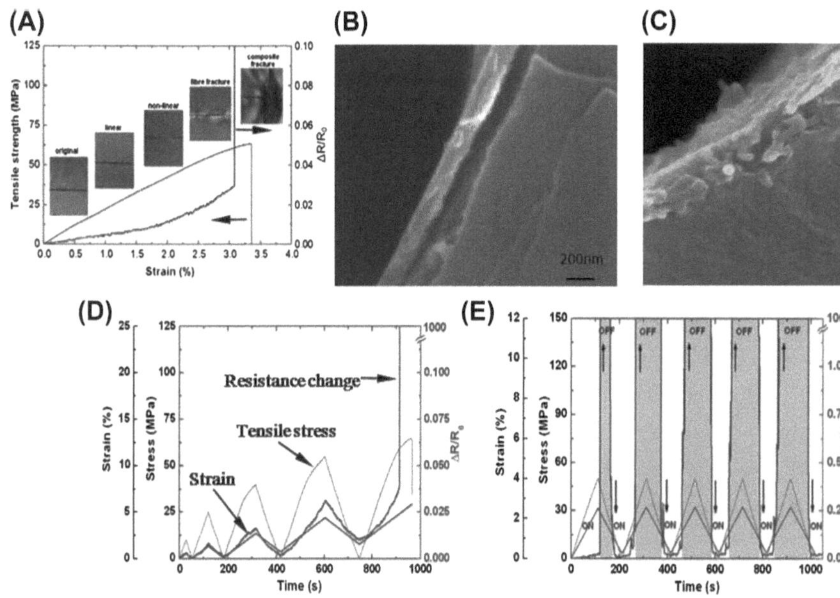

Figure 3.11 The performance of single carbon nanotube (CNT)-coated fiber embedded in epoxy as a strain sensor. (A) Simultaneous change of electrical resistance and stress as a function of strain. (B and C) Fractured morphology of fiber-reinforced polymers (FRPs) with fibers with graphene and CNT coating. (D) Variation of electrical resistance with progressive increase of peak values of tensile strain/stress in a single CNT-coated fiber model composite. (E) Electromechanical switch acts during the cyclic tensile of CNT-coated fiber model composites. *(Adapted with permission from Hao, B.; Ma, Q.; Yang, S. D.; Mäder, E.; Ma, P. C. Comparative study on monitoring structural damage in fiber-reinforced polymers using glass fibers with carbon nanotubes and graphene coating. Composites Science and Technology 2016, 129, 38–45. Copyright (2016) Elsevier; from Zhang, J.; Zhuang, R.; Liu, J.; Mäder, E.; Heinrich, G.; Gao, S. Functional Interphases with Multi-Walled Carbon Nanotubes in Glass Fibre/Epoxy Composites. Carbon* ***2010****, 48, 2273–2281. Copyright (2010) Elsevier, from Zhang, J.; Liu, J.; Zhuang, R.; Mäder, E.; Heinrich, G.; Gao, S. Single MWNT-Glass Fiber as Strain Sensor and Switch. Advanced Materials* ***2011****, 23, 3392–3397. Copyright (2011) John Wiley and Sons.)*

embedded into epoxy, the conductive networks originated from the nanocomposites composed of epoxy and CNTs (Fig. 3.11C), whereas for graphene coating, the conductivity of FRPs could be sustained until the failure of the whole composite. The conductive networks of graphene coating were in the form of a compacted layer (Fig. 3.11B), thus preventing the penetration of matrix onto the fiber surface, which could deform easily by sliding with each other, which is beneficial to sustaining a large strain.

The introduction of CNTs onto fibers made it possible to analyze the damage accumulation of FRPs under cyclic loadings, thus providing a way to assess the life of composites in service conditions. The results presented in Fig. 3.11D by Gao and coworkers *(31)* showed that when the strain peak applied to FRPs was below 1.5% (in the first and second cycle), the resistance change of composites was linear and could return to the original value after unloading. When the strain exceeded a critical value (in the third cycle), the composite displayed residual resistance after unloading, suggesting the existence of permanent structural deformation or damage of CNT networks. With a further increase of strain peak, the residual resistance became more pronounced until the failure of conductive CNT networks in FRPs (in the fourth and fifth cycle), confirming the damage evolution by mechanical strain. Considering CNT network failure before the breakage of the matrix, the authors studied the feature of the local microcrack of a CNT network, and found that when the gap of the microcrack exceeded a critical value of 2.7 μm during the loading process, the material turned into a nonconductivity or "OFF" state. When the size of the gap was below this critical value, the composite turned into a conductivity or "ON" state; such a change in behavior between the ON and OFF states gave the material an interesting "switch effect" (Fig. 3.11E).

It should be noted that the resistance change of fibers originated from the structure and chemical states of the coating layer rather than the fiber. Ma and coworkers reported the preparation of GFs with CNTs and graphene coating by using the EPD process. The introduction of carbon-based coatings can modify the electrical properties of individual fibers significantly, and the thermal treatment after EPD played an important role in controlling the electrical resistance of coated GFs (Table 3.1). For example, the resistance of CNT-GF decreased by more than 60% after thermal treatment, making the GF more

Table 3.1 Electrical resistivity of glass fibers coated by carbon-based nanoparticles

Sample name	Resistance (length of 4 cm, MΩ)
CNT-GF before thermal treatment	25.4 + 8.0(18.1 ~ 42.4)
CNT-GF	7.4 + 2.3(4.9 ~ 13.5)
GO-GF	Insulating (>2 GΩ)
Graphene-GF	7.9 + 4.1 (4.2 ~ 16.5)

Data in brackets: minimum and maximum resistance of sample.
CNT, carbon nanotube; *GF*, glass fiber; *GO*, graphene oxide.
Adapted from Hao, B.; Ma, Q.; Yang, S. D.; Mäder, E.; Ma, P. C. Comparative study on monitoring structural damage in fiber-reinforced polymers using glass fibers with carbon nanotubes and graphene coating. Composites Science and Technology 2016, 129, 38–45. Copyright (2016) Elsevier.

conductive. This ameliorating effect became more pronounced for the sample with graphene coating. The resistance of graphene oxide-GF exceeded the measurement range of a digital multimeter (>2 GΩ). In sharp contrast, the electrical resistance of graphene-GF was comparable with that of a CNT coated one after thermal treatment (7.4 MΩ), suggesting the removal of heterogeneous atoms in the graphene. Anyway, the enhanced electrical property of coated fibers after thermal treatment was caused by the removal of defects and oxygen-containing functional groups, as well as the recovery of aromatic structures in CNTs or graphene in the coating layer, which decreased the contact resistance among the individual nanofillers on the fiber surface.

Fig. 3.12 shows the effect of temperature on the resistance change (R/R_0) of samples (R_0 is the resistance at 228K). It was obvious that both fibers showed a typically negative temperature coefficient behavior, that is, the electrical resistance of the sample decreases with increasing temperature. This behavior suggested that raising the temperature made more charge carriers of higher energy level in the sample cross the barriers between the localized states in the material and contact between adjacent particles, leading to a lower resistivity for the sample. Additionally, graphene-GF exhibited a higher sensitivity to the temperature than CNT-GF, possibly because of the higher overlapped area in the graphene layer than the point-to-point contact among CNTs, which facilitates the transport of electrons in the coating layer. Considering the sensitivity and size of the sample (diameter ~12 μm), the fibers with CNTs or graphene coating can be used for temperature sensing at the microscale.

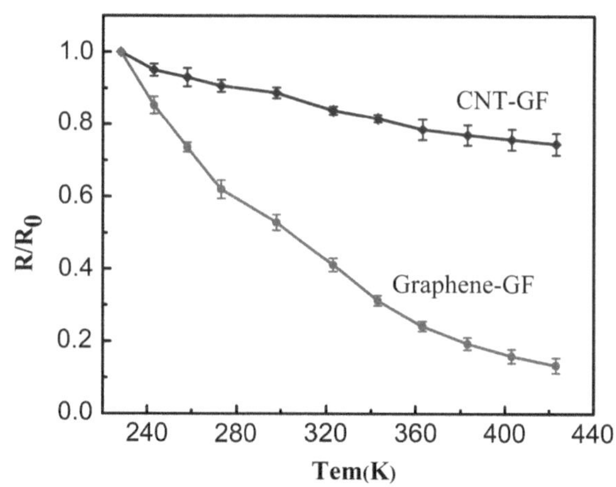

Figure 3.12 Temperature-dependent electrical resistance of glass fibers (GFs) with different coatings, suggesting the fibers can be used for temperature sensing at the microscale. *(Adapted from Hao, B.; Ma, Q.; Yang, S. D.; Mäder, E.; Ma, P. C. Comparative study on monitoring structural damage in fiber-reinforced polymers using glass fibers with carbon nanotubes and graphene coating. Composites Science and Technology 2016, 129, 38–45. Copyright (2016) Elsevier.)*

Besides the coating method for the modification of fibers, direct growth of CNT on a fiber surface using the CVD method has also been developed in recent years. Different to the coating methods that can be accomplished at room temperature and in liquid states, CVD requires relatively harsh conditions for the growth and deposition of CNTs on fiber. In a typical experiment *(32,33)*, a metal catalyst, such as Fe, Ni nanoparticles, was deposited onto the fiber substrate. This was followed by the introduction of hydrocarbon gases into a CVD chamber with a temperature of more than 600°C. Under such circumstances, the hydrocarbon gases will be decomposed, offering a carbon source for the growth of CNTs with assistance from the metal catalyst and inert atmosphere. The fibers with CNTs can be used directly for defect monitoring in structural materials. In most cases, the prepared fibers were in fuzzy states, as shown in Fig. 3.13A. Boehle et al. *(34)* reported that CNT-covered fiber sensors exhibited similar sensitivity to conventional strain gauges, having a gauge factor of 1.6–2.3, and were more easily integrated into FRP structures as the sensor itself was a composite. The fuzzy fiber strain gauges can be used to sense strain within composite structures and can be readily integrated into the structural laminate to provide sensing over large sections and in previously inaccessible locations. Measurement of the sensor response in longitudinal, transverse, and off-axis orientations (Fig. 3.13B) showed that the sensor does indeed respond to the

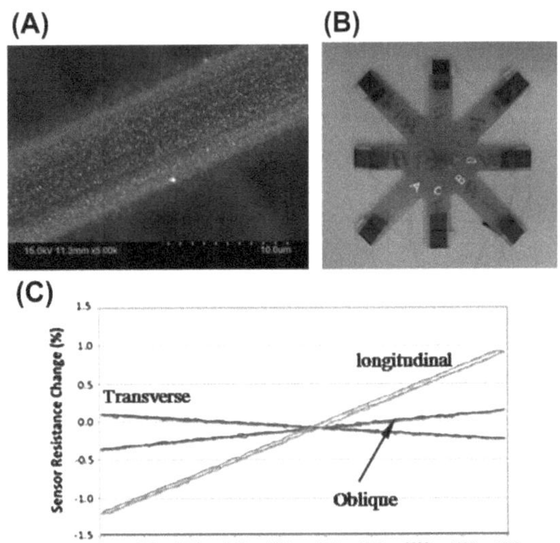

Figure 3.13 Carbon nanotube (CNT) grown on glass fibers (GFs) for sensory application in fiber-reinforced polymers (FRPs). (A) Morphology of fiber with CNT growth. (B) Specimen embedded with CNT-grafted fibers in different directions. (C) Typical off-axis sensor response under tension along the longitudinal direction. *(Adapted with permission from Sebastian, J.; Schehl, N.; Bouchard, M.; Boehle, M.; Li, L.; Lagounov, A.; et al. Health Monitoring of Structural Composites with Embedded Carbon Nanotube Coated Glass Fiber Sensors. Carbon* ***2014****, 66, 191–200. Copyright (2014) Elsevier.)*

mechanical stimuli, and the one in the longitude direction showed the highest sensitivity (Fig. 3.13C). This phenomenon may be caused by the compressive stress causing a combined effect wherein the resistance of the sensing element decreases because of Poisson effects while the resistance simultaneously increases because of stress on the conductive CNT network. By better understanding these effects, it may be possible to tailor the fabrication of the sensor so that its performance is predictable under both tensile and compressive stresses.

It should be noted that the number of reports on grafting CNTs on fibers for SHM was much less than that found by coating CNTs to achieve the same goal. A possible reason was that few fibers can maintain their mechanical properties during the high temperature of the CVD process. Direct growth of CNTs onto carbon fiber during fiber carbonization and graphitization may be more attractive because the CNTs on the surface of carbon fibers will significantly enhance the interfacial interactions between the fiber and the polymer matrix, with the promise of improving the sensitivity of fibers for monitoring structural defects in corresponding FRPs.

3.2.5 Utilization of CNT Fiber/Yarn

CNT fiber, a type of microscale fiber/yarn consisting of 1-D assembly CNTs (Fig. 3.14A), has attracted considerable interest by researchers since its first production by Vigolo et al. *(35)*. The individual CNTs in the fiber come together thanks to van der Waals interactions and entanglements of long and continuous CNTs caused by twisting during preparation. CNT fibers can be prepared by spinning continuously from CNT solutions, aerogels, and arrays *(10,35,36)*. It should be noted that the CNTs fibers produced using different methods have numerous voids and porosity in their

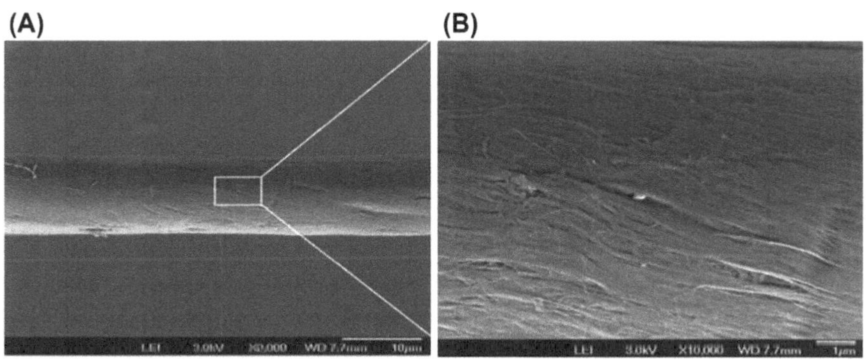

Figure 3.14 Scanning electron microscope (SEM) images showing the morphology of carbon nanotube (CNT) fibers under different magnifications. *(Adapted with permission from Zu, M.; Li, Q.; Wang, G.; Byun, J.-H.; Chou, T.-W. Carbon Nanotube Fiber Based Stretchable Conductor. Advanced Functional Materials* ***2013****, 23, 789–793. Copyright (2013) John Wiley and Sons.)*

microstructures (Fig. 3.14B), which renders a lower strength (0.1−0.2 GPa) and modulus (10−250 GPa) compared to individual CNTs *(37)*.

Similar to that found in CNT/polymer nanocomposites, CNT fibers also presented electromechanical behavior, for example, gauge factors of 3.0−4.4 were obtained in a tensile test of a single fiber prepared by an aerogel-spinning process *(39)*. Different to the trend on the resistance change as a function of strain in CNT/polymer nanocomposites, the variation on electrical properties of CNT fibers consists of two parts. One is because of the deformation of individual CNTs, and the other is because of the competitiveness between the improvement on the radial contact and that from axial contact caused by intersliding among CNTs *(37)*. Given this point, utilization of CNT fibers for monitoring the structural defects in composites by averaging the electrical resistance change is feasible. Yuan et al. *(40)* studied CNT fiber as a strain sensor in epoxy and achieved a gauge factor of 0.38 in the longitudinal direction over the whole strain range, while the transverse gauge factor located from 0.02 to 0.04. The yarn sensors embedded in the polymer matrix can be used to monitor the initiation and propagation of cracks in composites in real time, as well as to work as a general strain gauge if top mounted on the structural components.

Utilization of CNT fiber in FRP for SHM was also reported by Abot and coworkers *(41)*. The researchers designed a sensor made from CNT threads, and glass/epoxy laminated samples with a precrack between the central plies were instrumented with sensor threads. Fig. 3.15A shows a schematic of the sample. Three sensor threads were placed between the central plies perpendicular to the crack propagation direction. Another sensor thread was placed between the central plies, in front of the initial crack tip and

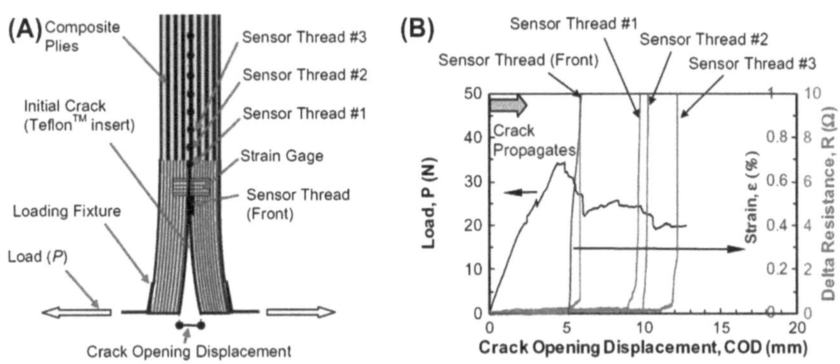

Figure 3.15 Monitoring evolution of mode I delamination in laminated fiber-reinforced polymers (FRPs) using carbon nanotube (CNT) yarn. (A) Side view schematic of samples for mode I delamination test. (B) Load and resistance behavior versus crack opening displacement during the test. *(Adapted with permission from Abot, J. L.; Song, Y.; Vatsavaya, M. S.; Medikonda, S.; Kier, Z.; Jayasinghe, C.; et al. Delamination Detection with Carbon Nanotube Thread in Self-Sensing Composite Materials. Composites Science and Technology* ***2010****, 70, 1113−1119. Copyright (2010) Elsevier.)*

parallel to the crack propagation direction. A strain gauge was placed near the front of the crack tip to detect any transverse strain in the laminate prior to and during the crack propagation. The sample was then tested in a double cantilever beam configuration. The results showed that once the crack reaches the sensor thread, the resistance increases significantly, and a slight increase in the resistance was observed in all of the sensor threads before the crack reaches the sensor, thus providing a warning of a crack approaching a specific location (Fig. 3.15B). The sensor threads in such a configuration can capture the cracks that specifically reach their location, warn about a nearby crack, and can be upgraded with continuous sensor threads to capture any cracks that may develop between the plies in the laminated FRPs.

While CNT fiber exhibited outstanding performance for defect monitoring in FRPs, few reports are available in the literature so far, possibly because of the difficulties associated with the preparation of the fiber sample. In addition, the structural instability of CNT yarn, that is, the radial and axial contraction of fiber caused by intersliding among CNTs under mechanical loading, makes analysis of the obtained results a challenging topic. Definitely much effort should be made in this field for further research.

3.2.6 Utilization of CNT for SHM Using Optical Methods

The methodologies described earlier for damage sensing and monitoring in FRPs are based on the electrical resistance of samples under mechanical deformation, which causes the change in the conductive networks in the composites. Other than depending on the electrical properties of composites, an optical method can also be employed for damage analysis in FRPs, and Raman spectrum is one of the most commonly used techniques.

Raman spectroscopy is an inelastic photon scattering spectrum, which reflects unique information related to the vibration of materials *(42)*. For CNTs, various nanotubes possess their own featured Raman peaks (Fig. 3.16A), that is, the D band ("D" for disorder, located at $\sim 1350\ cm^{-1}$), G band ("G" for graphite, located at $\sim 1600\ cm^{-1}$), and G′ band (the first overtone of the D band, located at $\sim 2700\ cm^{-1}$), which can be used to characterize their intrinsic properties, such as diameter, defects in the structure, degree of functionalization, and so on. More importantly, the positions of these bands were sensitive to the chemical states of C atoms, and displayed peak shifts when the length of the C–C bond was changed under the deformation. For example, Young et al. *(43)* studied the position of the G′ band in glass fiber-reinforced plastics (GFRPs) where GFs were coated with CNT/silane nanocomposites; the results showed that the G′ peak shifted to a lower wavenumber and the position change versus fiber strain was linear (Fig. 3.16B). Similarly, this technique was adopted to investigate the strain or stress subjected to CNTs. In other words, Raman spectrum is a powerful tool to evaluate the interactions between the nanotubes and polymer matrix, and can be further extended to monitor defects in composites.

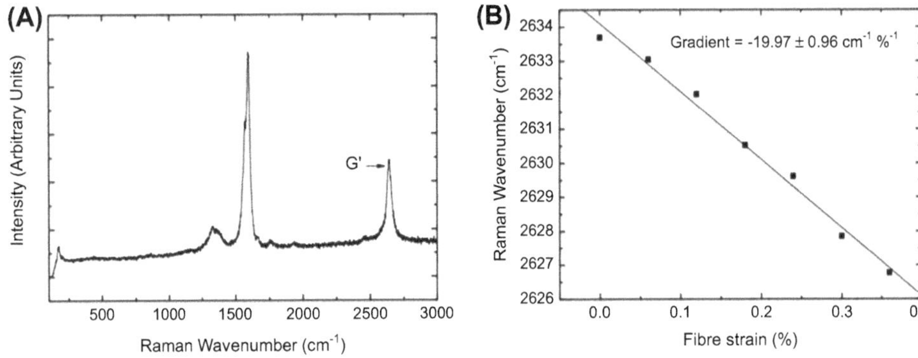

Figure 3.16 (A) Raman spectrum of glass fiber (GF) sized by carbon nanotubes (CNTs). (B) G′ band shift of sample under strain. *(Adapted with permission from Sureeyatanapas, P.; Hejda, M.; Eichhorn, S. J.; Young, R. J. Comparing Single-Walled Carbon Nanotubes and Samarium Oxide as Strain Sensors for Model Glass-Fibre/Epoxy Composites. Composites Science and Technology* ***2010****, 70, 88–93. Copyright (2010) Elsevier.)*

By using Raman spectrum, Liu and Ma *(28)* studied the strain and stress distribution along the fiber by embedding GF coated with CNT/epoxy nanocomposites into a polymer matrix. Fig. 3.17A shows that the G′ band of CNTs shifted downward when applying strain to the GF, and reduction in wavenumber was linear to the strain with a slope of 15.2 and 16.6 cm^{-1}/strain% for the samples with 0.12 and 0.50 wt% CNTs, respectively. Based on the relationship between the Raman G′ band shift and fiber strain, the strain on the fiber could be deduced in FRPs. Raman spectroscopy and electrical response of GFs under mechanical load were coupled by the authors for in situ sensing of deformation in composites. The results showed that the fibers with nanocomposite coating exhibited efficient stress transfer across the fiber/matrix interface. A relationship between the fiber stress and the change in electrical resistance against the fiber strain either in air or embedded in composites was established (Fig. 3.17B). In addition, the authors studied the birefringence patterns of FRP containing a single fiber with CNT coating under the tension state. Four different stages are identified, as shown in Fig. 3.17C. Stage A represented the linear relationship between the change in resistance and fiber strain. Stage B corresponded to the period when the resistance changed abruptly because of the disruption in the CNT networks on the fiber surface. This observation was accompanied by a weak birefringence pattern around the stretched coating layer arising from the partially unstable CNT networks. Clear birefringence patterns were observed in stage C, where the fiber was fractured, resulting in a surge in electrical resistance to infinity at a net fiber strain of about 0.9%. The image at stage D was obtained after matrix fracture. It is worth noting that the fiber was broken before matrix fracture, suggesting that the fiber with a nanocomposite coating can be employed as a sensor to offer an early warning of fracture in FRP structures.

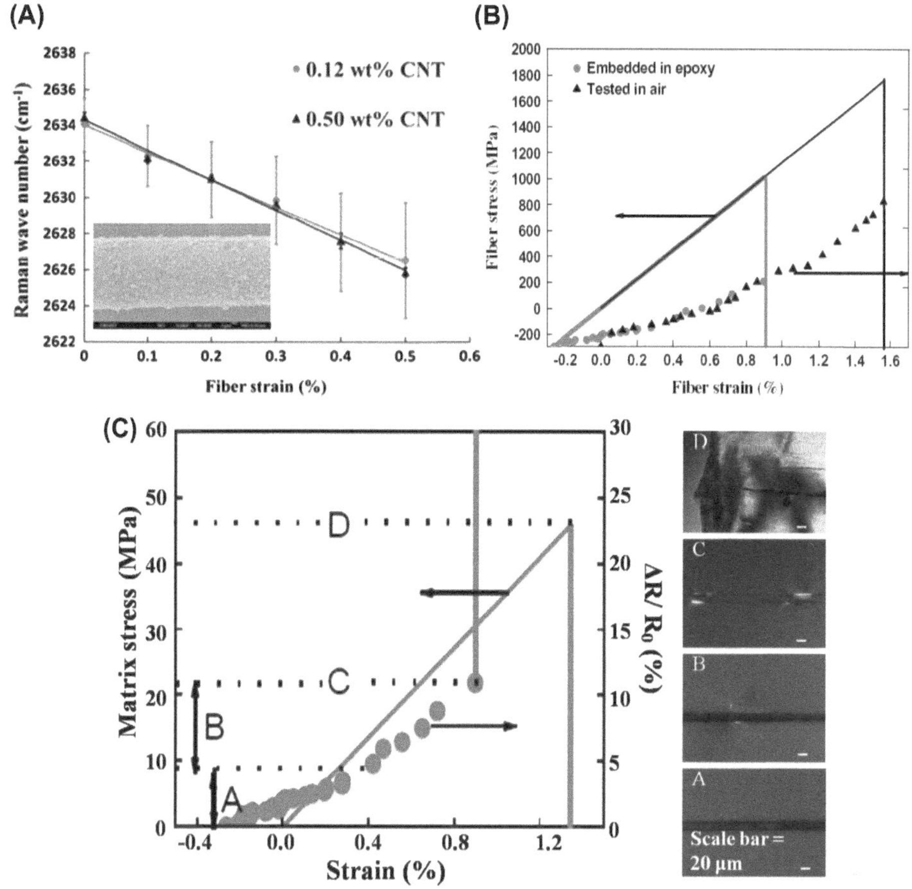

Figure 3.17 Application of Raman spectrum for monitoring structural defects in fiber-reinforced polymers (FRPs). (A) G′ band shift of fiber coated by carbon nanotube (CNT)/epoxy nanocomposites under strain. (B) Changes in fiber stress and electrical resistance as a function of fiber tensile strain for fibers with nanocomposite coating. (C) Birefringence patterns of single glass fiber (GF) coated with CNT/epoxy in fiber-reinforced polymer (FRP) under different loading conditions and corresponding electrical resistance change. *(Adapted with permission from Ma, P.-C.; Zhang, Y. Perspectives of Carbon Nanotubes/Polymer Nanocomposites for Wind Blade Materials. Renewable & Sustainable Energy Reviews* ***2014****, 30, 651–660. Copyright (2014) Elsevier, from Liu, L.; Ma, P. C.; Xu, M.; Khan, S. U.; Kim, J. K. Strain-Sensitive Raman Spectroscopy and Electrical Resistance of Carbon Nanotube-Coated Glass Fibre Sensors. Composites Science and Technology* ***2012****, 72, 1548–1555. Copyright (2012) Elsevier.)*

It should be noted that when applying Raman spectroscopy to study nanocomposites or FRPs containing CNTs, the laser size on the specimen surface should be optimized, and the laser beam should be polarized using a polarizer along the direction of mechanical load to maximize the signals from CNTs parallel to this direction *(44)*. In addition, laser power and scanning time should be controlled for each Raman spectrum collection to minimize stress relaxation of composites and the heating effect of the laser. The Raman spectra should be collected from exactly the same spot of the specimen at different strains to detect the in situ deformation of CNTs in the nanocomposites, making analysis a challenging task to achieve reliable results. Anyway, mapping of Raman data offers a way to display the strain and stress distribution in nanocomposites or FRPs at the micro- or nanoscale.

3.3 SUMMARY AND CONCLUSIONS

With unique structural and transport properties, CNTs have attracted much interest as reinforcements for polymer nanocomposites, and the developed materials hold promise for delivering exceptional mechanical and multifunctional properties for various applications. The use of CNTs and CNT/polymer nanocomposites for health monitoring in FRPs is a very new research area. The principle behind such application is that the electrical properties of materials containing CNTs will be changed under mechanical loading. Coupling of electrical properties and mechanical deformation of material with CNTs produces a composite with sensory capabilities.

Based on the constituents of FRPs, CNTs can be incorporated into materials for FRPs via the following two ways: (1) by adding CNTs with forms of dispersed states, foam, aerogel, paper and film, into the polymer resins through proper dispersion and processing, and the materials can be employed as a matrix for FRPs, and (2) by growth or deposition of CNTs into fiber/yarn or on the fiber surface; the large surface area and excellent electrical properties of nanomaterials give the fiber a sensory capability for defect monitoring in FRPs.

While significant progress has been made in developing CNTs and corresponding nanocomposites for defect monitoring in FRPs, there are still many issues to be addressed to optimize the structural, morphological, and applicable properties of such materials. For example, the structural defects in FRPs under deformation can be monitored using electrical or Raman techniques; however, a major drawback based on this approach is that one cannot predict the damage in the reinforcements that sustain the majority of the load applied to the structure. Additionally, a process on how to prepare CNT-decorated microfibers using a facile method, especially for bulk-scale production, is highly desirable.

Besides CNTs, graphene has been utilized as a novel functional coating for fibers in recent years. However, few reports were found in the literature on controlling the structure and morphology of coating layers, and less effort has been made to compare the role of CNT and graphene coatings on the properties of fibers and structural damage process of corresponding FRPs under mechanical loading. It is known that CNTs have a tube-like morphology, whereas graphene exhibits a plate-like structure, such a significant difference in dimension may result in the different functions of coatings in governing the electrical behavior of FRPs. Therefore comparative studies to address these issues are highly desirable.

ACKNOWLEDGMENTS

This project was supported by the National Natural Science Foundation of China (Project No. 51203180, 11472294), the 1000-Talent Program (Recruitment Program of Global Expert, In Chinese: Qian-Ren-Ji-Hua), as well as the Funds for Distinguished Young Scientists of Xinjiang Uyghur Autonomous Region (Project No. 2013711005).

REFERENCES

1. Kim, J. K.; Mai, Y. W. *Engineered Interfaces in Fiber Reinforced Composites;* Elsevier: Oxford, 1998.
2. *Engineering Composite Materials,* 2nd ed.; Maney Publishing: London, 1999.
3. Yeh, T.-S.; Chu, C.-S.; Lo, Y.-L. Highly Sensitive Optical Fiber Oxygen Sensor Using Pt(II) Complex Embedded in Sol–Gel Matrices. *Sensors Actuators B-Chemical* **2006,** *119,* 701–707.
4. Nagai, H.; Oishi, R. Shape Memory Alloys as Strain Sensors in Composites. *Smart Materials & Structures* **2006,** *15,* 493–498.
5. Ma, P. C.; Kim, J. K. *Carbon Nanotubes for Polymer Reinforcement;* Taylor & Francis Group: Boca Raton, 2011.
6. Fiedler, B.; Gojny, F. H.; Wichmann, M. H. G.; Bauhofer, W.; Schulte, K. Can Carbon Nanotubes Be Used to Sense Damage in Composites? *Annales De Chimie-Science Des Materiaux* **2004,** *29,* 81–94.
7. Ma, P. C.; Siddiqui, N. A.; Marom, G.; Kim, J. K. Dispersion and Functionalization of Carbon Nanotubes for Polymer-Based Nanocomposites: A Review. *Composites Part A: Applied Science and Manufacturing* **2010,** *41,* 1345–1367.
8. Ma, P.-C.; Zhang, Y. Perspectives of Carbon Nanotubes/Polymer Nanocomposites for Wind Blade Materials. *Renewable & Sustainable Energy Reviews* **2014,** *30,* 651–660.
9. Rein, M. D.; Breuer, O.; Wagner, H. D. Sensors and Sensitivity: Carbon Nanotube Buckypaper Films as Strain Sensing Devices. *Composites Science and Technology* **2011,** *71,* 373–381.
10. Liu, K.; Sun, Y.; Lin, X.; Zhou, R.; Wang, J.; Fan, S.; et al. Scratch-Resistant, Highly Conductive, and High-Strength Carbon Nanotube-Based Composite Yarns. *ACS Nano* **2010,** *4,* 5827–5834.
11. Tjong, S. C. Polymer Composites with Graphene Nanofillers: Electrical Properties and Applications. *Journal of Nanoscience and Nanotechnology* **2014,** *14,* 1154–1168.
12. Thostenson, E. T.; Chou, T.-W. Carbon Nanotube Networks: Sensing of Distributed Strain and Damage for Life Prediction and Self Healing. *Advanced Materials* **2006,** *18,* 2837–2841.
13. Zhao, J.; He, C.; Yang, R.; Shi, Z.; Cheng, M.; Yang, W.; et al. Ultra-Sensitive Strain Sensors Based on Piezoresistive Nanographene Films. *Applied Physics Letters* **2012,** *101,* 063112.
14. Thostenson, E. T.; Chou, T. W. Real-Time in Situ Sensing of Damage Evolution in Advanced Fiber Composites Using Carbon Nanotube Networks. *Nanotechnology* **2008,** *19,* 215713.
15. Kim, K. J.; Yu, W.-R.; Lee, J. S.; Gao, L.; Thostenson, E. T.; Chou, T.-W.; et al. Damage Characterization of 3D Braided Composites Using Carbon Nanotube-Based in Situ Sensing. *Composites Part A: Applied Science and Manufacturing* **2010,** *41,* 1531–1537.
16. Viets, C.; Kaysser, S.; Schulte, K. Damage Mapping of GFRP via Electrical Resistance Measurements Using Nanocomposite Epoxy Matrix Systems. *Composites Part B: Engineering* **2014,** *65,* 80–88.
17. Chen, W.; Rakhi, R.; Hu, L.; Xie, X.; Cui, Y.; Alshareef, H. High-Performance Nanostructured Supercapacitors on a Sponge. *Nano Letters* **2011,** *11,* 5165–5172.
18. Zou, J.; Liu, J.; Karakoti, A. S.; Kumar, A.; Joung, D.; Li, Q.; et al. Ultralight Multiwalled Carbon Nanotube Aerogel. *ACS Nano* **2010,** *4,* 7293–7302.
19. Worsley, M. A.; Satcher, J. H., Jr.; Baumann, T. F. Synthesis and Characterization of Monolithic Carbon Aerogel Nanocomposites Containing Double-Walled Carbon Nanotubes. *Langmuir* **2008,** *24,* 9763–9766.

20. Fang, Y.; Jiang, F.; Liu, H.; Wu, X.; Lu, Y. Free-Standing Ni-Microfiber-Supported Carbon Nanotube Aerogel Hybrid Electrodes in 3D for High-Performance Supercapacitors. *RSC Advances* **2012,** *2,* 6562–6569.
21. Eom, K.; Nam, K.; Jung, H.; Kim, P.; Strano, M. S.; Han, J.-H.; et al. Controllable Viscoelastic Behavior of Vertically Aligned Carbon Nanotube Arrays. *Carbon* **2013,** *65,* 305–314.
22. Yaglioglu, O.; Cao, A.; Hart, A. J.; Martens, R.; Slocum, A. H. Wide Range Control of Microstructure and Mechanical Properties of Carbon Nanotube Forests: A Comparison between Fixed and Floating Catalyst CVD Techniques. *Advanced Functional Materials* **2012,** *22,* 5028–5037.
23. Hao, B.; Cui, L. F.; Ma, P. C. *Largely Stretchable and Compressible Strain Sensor Based on Carbon Nanotubes with 3-D Networks,* 2015. Under review.
24. Kanoun, O.; Mueller, C.; Benchirouf, A.; Sanli, A.; Trong Nghia, D.; Al-Hamry, A.; et al. Flexible Carbon Nanotube Films for High Performance Strain Sensors. *Sensors* **2014,** *14,* 10042–10071.
25. Srivastava, R. K.; Vemuru, V. S. M.; Zeng, Y.; Vajtai, R.; Nagarajaiah, S.; Ajayan, P. M.; et al. The Strain Sensing and Thermal–Mechanical Behavior of Flexible Multi-Walled Carbon Nanotube/Polystyrene Composite Films. *Carbon* **2011,** *49,* 3928–3936.
26. Lipomi, D. J.; Vosgueritchian, M.; Tee, B. C. K.; Hellstrom, S. L.; Lee, J. A.; Fox, C. H.; et al. Skin-Like Pressure and Strain Sensors Based on Transparent Elastic Films of Carbon Nanotubes. *Nature Nanotechnology* **2011,** *6,* 788–792.
27. Yamada, T.; Hayamizu, Y.; Yamamoto, Y.; Yomogida, Y.; Izadi-Najafabadi, A.; Futaba, D. N.; et al. A Stretchable Carbon Nanotube Strain Sensor for Human-Motion Detection. *Nature Nanotechnology* **2011,** *6,* 296–301.
28. Liu, L.; Ma, P. C.; Xu, M.; Khan, S. U.; Kim, J. K. Strain-Sensitive Raman Spectroscopy and Electrical Resistance of Carbon Nanotube-Coated Glass Fibre Sensors. *Composites Science and Technology* **2012,** *72,* 1548–1555.
29. Hao, B.; Ma, Q.; Yang, S. D.; Mäder, E.; Ma, P. C. Comparative study on monitoring structural damage in fiber-reinforced polymers using glass fibers with carbon nanotubes and graphene coating. *Composites Science and Technology* **2016,** *129,* 38–45.
30. Zhang, J.; Zhuang, R.; Liu, J.; Mäder, E.; Heinrich, G.; Gao, S. Functional Interphases with Multi-Walled Carbon Nanotubes in Glass Fibre/Epoxy Composites. *Carbon* **2010,** *48,* 2273–2281.
31. Zhang, J.; Liu, J.; Zhuang, R.; Mäder, E.; Heinrich, G.; Gao, S. Single MWNT-Glass Fiber as Strain Sensor and Switch. *Advanced Materials* **2011,** *23,* 3392–3397.
32. Rahmanian, S.; Suraya, A. R.; Zahari, R.; Zainudin, E. S. Synthesis of Vertically Aligned Carbon Nanotubes on Carbon Fiber. *Applied Surface Science* **2013,** *271,* 424–428.
33. Garcia, E. J.; Wardle, B. L.; John Hart, A.; Yamamoto, N. Fabrication and Multifunctional Properties of a Hybrid Laminate with Aligned Carbon Nanotubes Grown in Situ. *Composites Science and Technology* **2008,** *68,* 2034–2041.
34. Sebastian, J.; Schehl, N.; Bouchard, M.; Boehle, M.; Li, L.; Lagounov, A.; et al. Health Monitoring of Structural Composites with Embedded Carbon Nanotube Coated Glass Fiber Sensors. *Carbon* **2014,** *66,* 191–200.
35. Vigolo, B.; Pénicaud, A.; Coulon, C.; Sauder, C.; Pailler, R.; Journet, C.; et al. Macroscopic Fibers and Ribbons of Oriented Carbon Nanotubes. *Science* **2000,** *290,* 1331–1334.
36. Zhong, X. H.; Li, Y. L.; Liu, Y. K.; Qiao, X. H.; Feng, Y.; Liang, J.; et al. Continuous Multilayered Carbon Nanotube Yarns. *Advanced Materials* **2010,** *22,* 692–696.
37. Wu, A. S.; Chou, T.-W. Carbon Nanotube Fibers for Advanced Composites. *Materials Today* **2012,** *15,* 302–310.
38. Zu, M.; Li, Q.; Wang, G.; Byun, J.-H.; Chou, T.-W. Carbon Nanotube Fiber Based Stretchable Conductor. *Advanced Functional Materials* **2013,** *23,* 789–793.
39. Wu, A. S.; Nie, X.; Hudspeth, M. C.; Chen, W. W.; Chou, T.-W.; Lashmore, D. S.; et al. Strain Rate-Dependent Tensile Properties and Dynamic Electromechanical Response of Carbon Nanotube Fibers. *Carbon* **2012,** *50,* 3876–3881.
40. Zhao, H.; Zhang, Y.; Bradford, P. D.; Zhou, Q.; Jia, Q.; Yuan, F.-G.; et al. Carbon Nanotube Yarn Strain Sensors. *Nanotechnology* **2010,** *21,* 305502.

41. Abot, J. L.; Song, Y.; Vatsavaya, M. S.; Medikonda, S.; Kier, Z.; Jayasinghe, C.; et al. Delamination Detection with Carbon Nanotube Thread in Self-Sensing Composite Materials. *Composites Science and Technology* **2010,** *70,* 1113−1119.
42. Dresselhaus, M. S.; Dresselhaus, G.; Saito, R.; Jorio, A. Raman Spectroscopy of Carbon Nanotubes. *Physics Reports-Review Section of Physics Letters* **2005,** *409,* 47−99.
43. Sureeyatanapas, P.; Hejda, M.; Eichhorn, S. J.; Young, R. J. Comparing Single-Walled Carbon Nanotubes and Samarium Oxide as Strain Sensors for Model Glass-Fibre/Epoxy Composites. *Composites Science and Technology* **2010,** *70,* 88−93.
44. Ma, P. C.; Zheng, Q. B.; Mäder, E.; Kim, J. K. Behavior of Load Transfer in Functionalized Carbon Nanotube/Epoxy Nanocomposites. *Polymer* **2012,** *53,* 6081−6088.

CHAPTER 4

Carbon Nanotubes for Displaying

K. Jiang
Tsinghua University, Beijing, China

Contents

4.1 BRIEF INTRODUCTION ABOUT DISPLAY TECHNOLOGIES

Vision may be our most important sense, and about 60–80% of the information we receive is acquired by visual sense *(1,2)*. With the development of information technology (IT), more and more information exchange between human beings and machines will depend on display devices. Display technology is an important part of modern IT. The global sale value of display technology has been estimated to exceed $150 billion *(3)*. Display technology influences people's lives deeply at the same time. Every cell phone or smart terminal is assembled with a display panel. Besides the dynamic image display, the control part of an instrument is usually assembled with high fidelity display panels. The display devices have also been further incorporated with a control function in the form of a touch panel (TP). In the foreseeable future, displays will still be the most important information acquisition device for human beings.

According to the characteristics of display devices, display devices can be categorized into different types. Based on the light-emitting manner, display devices can be categorized into light-emitting and light-manipulating types. Based on the geometric shape of the display devices, they can be categorized into flat-panel display and nonflat-panel display, or flexible display and rigid display. According to the display content, the display devices can be categorized into image display and character display, or dynamics display and static display. There are also projection display, stereo display, helmet display, etc. depending on the different display manners.

As a kind of technology bridge between the outer world and the human self, display is a combined technology that involves psychology, ergonomics, optesthesia engineering, optics and photonics, material science, electronics, data processing, etc. Display is

Industrial Applications of Carbon Nanotubes
ISBN 978-0-323-41481-4

intensely related with manufacturing technology. The development of display technology will also intensely depend on these different technologies.

Two kinds of prevalent display device, liquid crystal display (LCD) and organic light-emitting diode (OLED) display, are the main products of the display market. At the same time, there are also vacuum fluorescent display, field emission display (FED), electric light-emitting display, projection display, etc. Traditional cathode ray tube (CRT) display is still necessary in some special applications. However, it is difficult to say which kind of technology is the best. The development history of LCD can vividly prove this. LCD was first suggested in the 1960s in the RCA lab *(4,5)*. Although LCD holds almost 90% of the market *(3)* nowadays, during the development of LCD, it was criticized for having a narrow view angle, slow response, low brightness, and other problems. With the development of the IT industry, LCD has met the requirements of people and overcome its own problems step by step. The development and trends of the next generation of display technologies may also follow this mode. To find applications of the newly emerging materials and technology in the display market, research should meet the people's requirements and follow the trends of technology development.

Carbon nanotubes (CNTs) have found applications in different kinds of displays. CNTs can be used as the field emitter in FEDs, the polarizer in LCDs, and the transparent electrode and thin film transistor (TFT) in OLEDs. CNTs have also been used in some new types of displays, such as incandescence displays, thermochromic displays, and flexible fiber displays. This chapter will give a brief review of the research of CNT in the different kinds of display devices.

4.2 CNT IN FED

The traditional CRT display has occupied our daily life for more than 70 years. The CRT display has the advantages of the best color reproduction, fast response, and wide view angle. However, the traditional CRT is heavy and thick, and cannot follow the requirement of modern IT development. It is hoped that a thin and light display with CRT performance can be fabricated. FED is one kind of promising technology.

Field emission is a quantum mechanics phenomenon, which was discovered in 1929 by Fowler and Nordheim *(6)*. It can be simply described by the schematic illustration shown in Fig. 4.1.

Because of the applied intense electric field on the emitter surface, the barrier becomes a triangle shape and tunneling probability increases. The emission current can be described by Eq. (4.1) *(7)*:

$$j = \frac{1.54 \times 10^{-6} F^2}{\varphi t^2 \left(3.79 \times 10^{-4} \frac{F^{\frac{1}{2}}}{\varphi}\right)} \exp\left(- 6.83 \times 10^{7} \frac{\varphi^{\frac{3}{2}}}{F} v\left(3.79 \times 10^{-4} \frac{F^{\frac{1}{2}}}{\varphi}\right)\right) \tag{4.1}$$

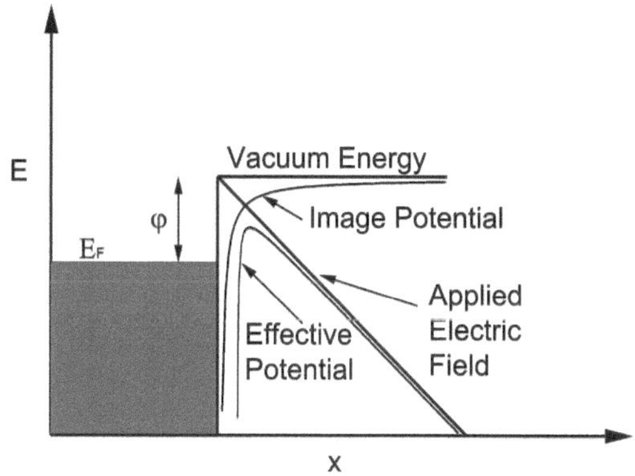

Figure 4.1 Theory model of field emission.

Here, j is the emission current density, F is the local electric field intensity, φ is the work function, t and v are the Nordheim functions *(7–9)*, and F is the electric field above the emitter surface. For an emitter with a given geometry, the surface electric field is correlated with the macroscopic electric field E with $F = \beta E$. β is the electric field enhancement factor.

In the early period of FED development, Spindt-type emitters (silicon or molybdenum tips) were used as field emitters *(10–12)*. FEDs from 6″ to 15″ diagonal have been fabricated successfully *(13–15)*. Fabrication of the Spindt emitter is complicated and device stability is not satisfactory. The commercialized Spindt FED is still not available.

From the Fowler–Nordheim formula, we can see that the field enhancement factor will influence emission greatly. In the 1990s, with intense studies, CNTs became potential field emitters because of their large field enhancement and excellent stability *(16–19)*. An FED with a CNT field emitter has attracted intense attention from both academia and industry *(20,21)*. The application of a CNT emitter was first realized in an element tube *(22,23)*. A field emission lamp has also been fabricated successfully *(24,25)*.

To fabricate an FED, mass production of a uniform CNT emitter over a large area is a basic technical requirement. In situ chemical vapor deposition (CVD) and postformation methods such as screen printing or solution deposition are the main methods of fabricating CNT field emitters *(17,18,26)*. Except for thermal CVD, plasma enhanced CVD has also been used to synthesize CNT emitters *(27–29)*. Although the morphology of a CVD CNT emitter can be well controlled, the postformation method is more popular because of its simplicity, low temperature process, etc. Screen printing can fabricate the emitter over a large area, but the CNT emitter fabricated by screen printing usually needs posttreatment to acquire excellent emission performance *(20)*.

The CNT FED has adopted a similar structure to the Spindt-type emitter, as shown in Fig. 4.2A and B *(20,30)*. The structure is the so-called vertical structure. The CNT emitter locates at the bottom, the gate electrode in the middle, and the anode electrode at the top. Emission is controlled by the gate electrode. The anode electrode is applied

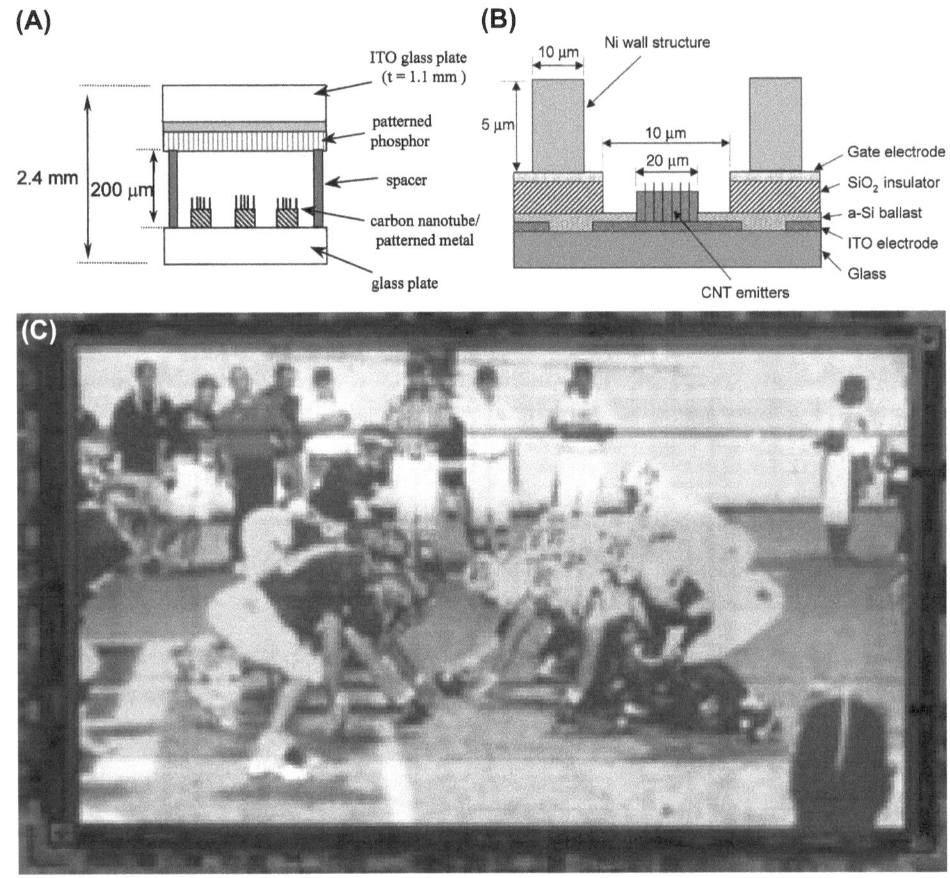

Figure 4.2 Carbon nanotube (CNT) field emission display (FED). (A) Schematic structure of a diode-type structure. *(Reproduced with permission from Choi, W. B.; Chung, D. S.; Kang, J. H.; Kim, H. Y.; Jin, Y. W.; Han, I. T.; Lee, Y. H.; Jung, J. E.; Lee, N. S.; Park, G. S.; Kim, J. M. Fully Sealed, High-Brightness Carbon-Nanotube Field-Emission Display. Applied Physics Letters* ***1999,*** *75, 3129–3131, Copyright 1999, AIP Publishing LLC.)* (B) Schematic structure of a triode-type structure (anode not shown). *(Reproduced with permission from Jung, J. E.; Jin, Y. W.; Choi, J. H.; Park, Y. J.; Ko, T. Y.; Chung, D. S.; Kim, J. W.; Jang, J. E.; Cha, S. N.; Yi, W. K.; Cho, S. H.; Yoon, M. J.; Lee, C. G.; You, J. H.; Lee, N. S.; Yoo, J. B.; Kim, J. M. Fabrication of Triode-Type Field Emission Displays with High-Density Carbon-Nanotube Emitter Arrays. Physica B: Condensed Matter* ***2002,*** *323, 71–77, Copyright 2002, Elsevier.)* It seems that the size of the CNT emitter should be 10 μm, and the distance between the two SiO_2 insulators should be 20 μm. (C) Image of a 38-inch lateral-gate FED manufactured by Samsung (2003). *(Reproduced with permission from Kim, Y. C.; Han, I. T.; Kim, J. M. Screen-Printed Carbon Nanotube Field Emitters for Display Applications. In Carbon Nanotube and Related Field Emitters: Fundamentals and Applications; Yahachi, S., Ed.; Wiley-VCH Verlag GmbH & Co. KGaA: Weinheim,* ***2010****; pp. 287–309, Copyright 2010, John Wiley and Sons.)*

with a high voltage to accelerate the electrons to illuminate the phosphor. Fig. 4.2C shows an image of a dynamically displayed structure using a triode CNT FED *(31)*.

Because the fabrication of an FED requires the field emitter and structure to be mass produced uniformly, the superaligned CNTs (SACNTs) are also very suitable candidate materials because the CNTs are parallel to each other and are easily mass produced *(32,33)*. The SACNT can be drawn out from the as-grown array in the form of a film, as shown in Fig. 4.3A, and the film can be further shrunk or spun into a yarn.

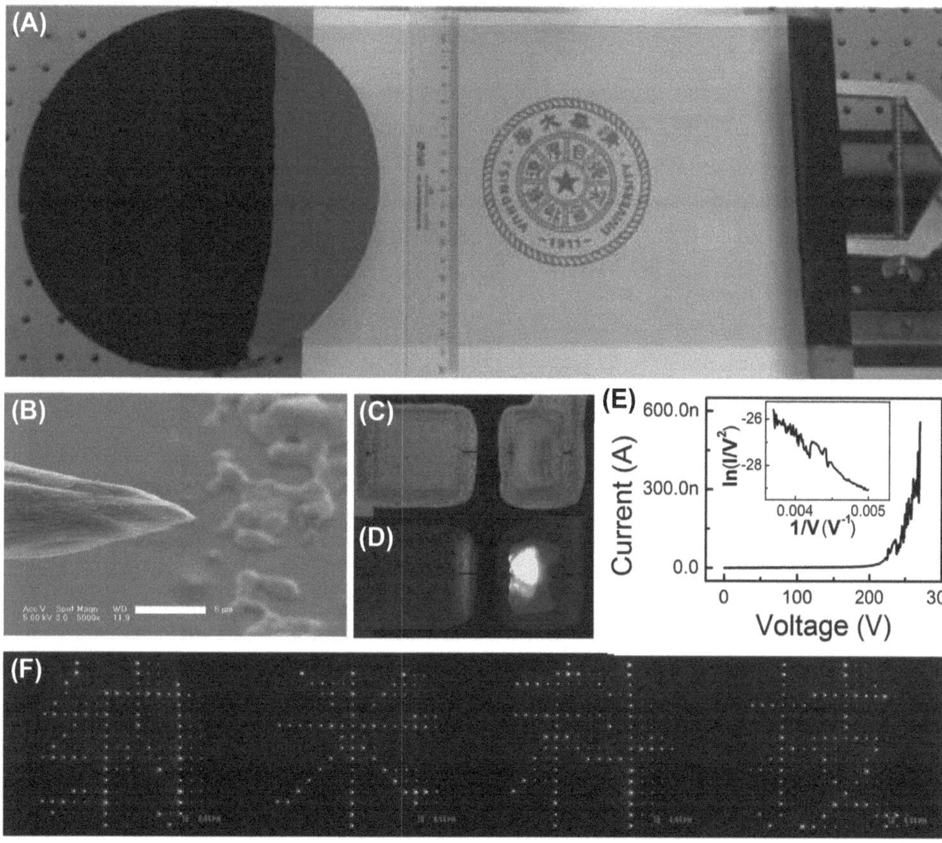

Figure 4.3 Superaligned CNTs (SACNTs) and a planar-type field emission display prototype. (A) SACNT film drawn out from a carbon nanotube (CNT) array. *(Reproduced with permission from Jiang, K. L.; Wang, J. P.; Li, Q. Q.; Liu, L. A.; Liu, C. H.; Fan, S. S. Superaligned Carbon Nanotube Arrays, Films, and Yarns: A Road to Applications. Advanced Materials* ***2011,*** *23, 1154–1161, Copyright 2011, John Wiley and Sons.)* (B) The CNT yarn emitter. (C) The planar structure and (D) the image during emission. (E) The emission current–voltage curves of a pixel. (F) The dynamic Chinese characters with the prototype. *((B–F) Reproduced with permission from Liu, P.; Wei, Y.; Liu, K.; Liu, L.; Jiang, K. L.; Fan, S. S. New-Type Planar Field Emission Display with Superaligned Carbon Nanotube Yarn Emitter. Nano Letters,* ***2012,*** *12, 2391–2396, Copyright 2012, American Chemical Society.)*

Experiments have shown that the CNT yarn emitters are excellent and suitable for FED applications *(34)*. The CNT yarn emitter can be fabricated by both mechanical cutting and electrical breakdown in vacuum *(34–36)*. A very large field enhancement factor has been observed in a plate–hole structure, which is estimated at about 400,000 and attributed to a two-level enhancement model. A CNT yarn field emission element tube has also been demonstrated *(34)*. With the CNT yarn linear cathode, a SACNT emitter lamp has also been fabricated. Micro-SACNT yarn tips can be made with the shrunk CNT film *(35)*. Double-walled CNT (DWCNT) and single-walled CNT (SWCNT) yarn tip emitters can be made with the correspondent SWCNT or DWCNT bundle with a joule heating breakdown method *(36)*.

Since SACNTs are easy to form a film or yarn shape, it is especially suitable to fabricate a planar-type emission structure. The planar-type field emission structure in the early 1990s adopted the silicon emitter fabricated by microfabrication technologies *(37)*. To acquire a large field enhancement factor, the silicon emitter is suspended above the substrate and is kept as thin as possible. A planar-type field emission element prototype of the SACNT has been demonstrated in *(38)*. The emitter is fabricated by joule heating breakdown in a vacuum and assembled with a phosphor-coated metal rod as the anode. A mass production method to fabricate the CNT emitter has also been proposed *(39)*. The two kinds of macroscopic structure, the yarn and the film, can both be used as the emitter after laser cutting. With the screen printing electrode structure, a planar-type CNT FED prototype has been fabricated. The emitter geometry can be optimized by control over laser-cutting parameters, as shown in Fig. 4.3B. The FED prototype has also been vacuum sealed in a glass unit with a low-melting point glass frit. Getters are used to maintain the vacuum. Fig. 4.3C–E shows the emission characteristics of an individual CNT yarn emitter. Driven with a specially designed circuit, the dynamic display of the Chinese characters has been demonstrated, as shown in Fig. 4.3F.

Brightness can reach about 30 cd/m^2 at 0.5 μA emission current in a diode structure with a low-voltage ZnO phosphor. High-voltage phosphor can improve the lumen efficiency of the devices. Furthermore, the CNT field emitter can be laser treated again, even after vacuum sealing. It is convenient to improve the emission uniformity by post-treatment after device fabrication. The mass production of SACNT has been a great convenience for the fabrication of CNT FED. It is also possible to fabricate a triode-type FED structure with the planar-type emitter structure.

Another application of the CNT field emission structure in display is the backlight unit (BLU) in the LCD *(40–42)*. The BLU is the main power-dissipating part in the LCD and has obvious influence on display uniformity. A traditional cold cathode fluorescence lamp (CCFL) has now been replaced by the light-emitting diode (LED) to realize the dynamical control of the backlight part to lower energy consumption and improve contrast. Since a CNT emitter has been used to fabricate a lamp-like field emission illumination unit with a brightness of 30,000 cd/m^2 at 26.4 lm/W *(43)* and

Figure 4.4 Display appearance of a thin film transistor-liquid crystal display (TFT-LCD) screen using a carbon nanotube (CNT) field emission backlight unit (BLU), compared to that of a cold cathode fluorescence lamp (CCFL) and light-emitting diode (LED) BLU. (A) In indoor lighting conditions and (B) in outdoor sunlight conditions. *(Reproduced with permission from Zhang, Y.; Deng, S. Z.; Xu, N. S.; Chen, J. Fully Sealed Carbon Nanotube Flat-Panel Light Source and Its Application as Thin Film Transistor-Liquid-Crystal Display Backlight. Journal of Vacuum Science & Technology B* ***2008,*** *26, 1033–1037, Copyright 2008, AIP Publishing LLC.)*

10,000 cd/m^2 at 40 lm/W *(24)* in the different reports, respectively, a CNT field emission BLU can also be used in an LCD. Compared with the FED, the BLU is much simpler. A diode structure can meet the requirement. Another advantage of CNT BLU is that the FED backlight is a plane light source itself and can realize the patterned light emission with a triode structure. The light guide plate used in both of the CCFL and LED BLU can be removed. The addressable triode CNT field emission BLU can realize much more accurate control than the LED BLU. The performance of the LCD with CNT field emission BLU has also been improved *(40,41)* as a result, as shown in Fig. 4.4.

However, both for the illumination and BLU applications, a problem is that the lumen efficiency of the present phosphor is low compared with the phosphor in the CCFL and LED. If there were a breakthrough to improve phosphor efficiency, it would be possible to realize the mass applications of CNT field emission lighting devices.

4.3 CNT IN LCD

LCD is the mainstream product in the present display market. The basic principle of LCD is that liquid crystal (LC) molecules work as a light valve to control the light. The LC

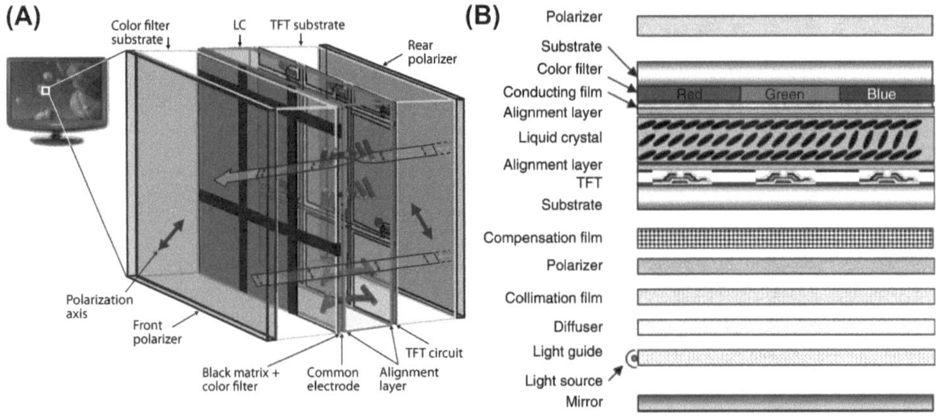

Figure 4.5 Schematic structure of a liquid crystal display (LCD). (A) Schematic diagram showing the construction of a twisted nematic-mode thin film transistor (TFT) LCD panel. The twisted liquid crystal (LC) structure rotates the polarization of incident light (upper pixel) without an applied electric field, aligning the polarization with the front polarizer (bright state). When an electric field is applied (lower pixel), the LC layer becomes homeotropically aligned and the polarization of light is no longer aligned with the front polarizer (dark state). *(Reproduced with permission from Kim, K. H.; Song, J. K. Technical Evolution of Liquid Crystal Displays. NPG Asia Materials* ***2009****, 1, 29–36, Copyright 2009, Nature Publishing Group.)* (B) Schematic diagram of the layer structure of an active matrix (AM) LCD. *(Reproduced with permission from Yang, D. K., Wu, S. T., Eds. Fundamentals of Liquid Crystal Devices; John Wiley & Sons, Ltd: Chichester,* ***2015,*** *Copyright 2015, John Wiley and Sons.)*

molecule is usually a rod-like (calamitic) or disk-like (discotic) shape and can rotate under an electric field *(44)*. A typical LCD structure is shown in Fig. 4.5 *(44,45)*. It can be seen that the whole structure is composed of an LC molecule cell, transparent electrode, backlight module, TFT, color filter, polarizer, connection wire, etc. Correspondingly, we can find the applications of CNTs in the different parts of the LCD.

The CNT itself can be viewed as a large rod-like molecule and shows an LC-like behavior *(46–50)*. LC molecules have been mixed or integrated with CNTs, and it is found that the LC can align CNTs *(51–53)*, or CNTs can also be used to align LC molecules *(54)*. The LC with CNTs also shows a better performance over the original ones *(55)*.

The CNT can emit polarized light because of its one-dimensional electron movement *(32,56–58)*. The polarity can be as large as 0.33, 0.72, and 0.75 for a multiwalled CNT (MWCNT) bundle, film, and individual CNT according to the respective reports. Fig. 4.6A and B show the polarized light emission of CNTs. The incandescent brightness of CNT has reached about 6000 cd/m^2 *(59)*, and it is possible that CNTs can be used as a polarized BLU. At the same time, the CNT is also an excellent polarized light absorber *(32,58)*. Fig. 4.6C and D show the experimental results of a CNT polarizer. It is easy to fabricate a CNT polarizer with the SACNT film.

Two-thirds of CNTs are semiconducting and one-third are metallic, according to their chiral indices. A field effect transistor (FET) made with semiconducting CNTs

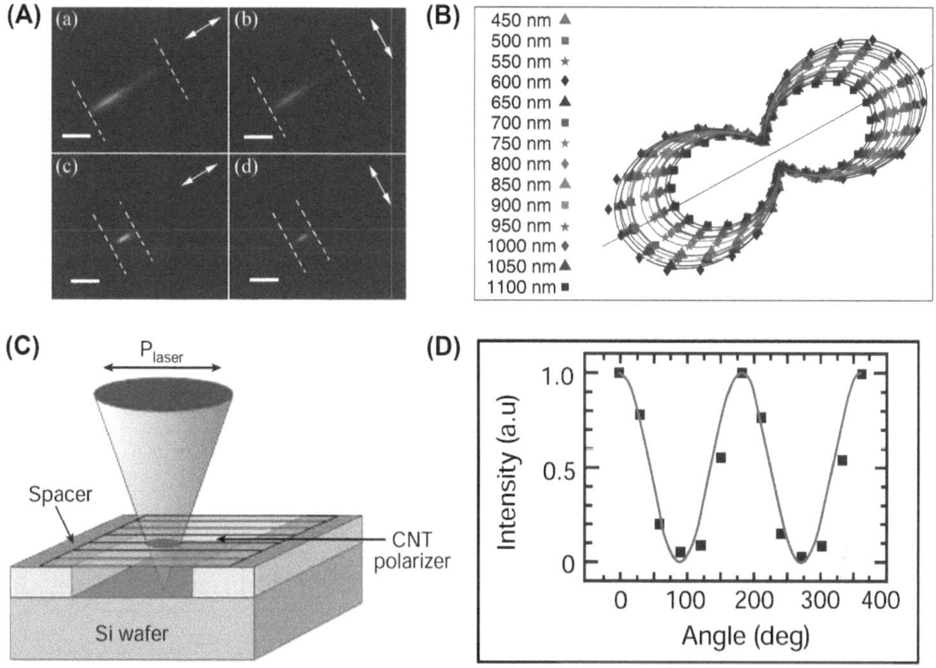

Figure 4.6 Carbon nanotubes (CNTs) as the incandescence polarized light emitter and as the polarizer. (A) The result for the multiwalled carbon nanotube (MWCNT) bundle. *(Reproduced with permission from Li, P.; Jiang, K. L.; Liu, M.; Li, Q. Q.; Fan, S. S.; Sun, J. L. Polarized Incandescent Light Emission from Carbon Nanotubes. Applied Physics Letters* ***2003,*** *82, 1763–1765, Copyright 2003, AIP Publishing LLC.)* (B) Individual MWCNT. *(Reproduced with permission from Singer, S. B.; Mecklenburg, M.; White, E. R.; Regan, B. C. Polarized Light Emission from Individual Incandescent Carbon Nanotubes. Physical Review B* ***2011,*** *83, 233404, Copyright 2011, American Physical Society.)* (C) and (D) A CNT polarizer. Polarized laser light (P_{laser}) is focused on the surface of a silicon wafer and reflected back, passing the CNT polarizer twice before being collected by a charge-coupled device detector. Inset, normalized intensity (*squares*, ratio of the current I to I_{max}; *a.u.*, arbitrary units) plotted against the angle between the polarization direction of the laser light and the CNT polarizer, compared with values calculated from Malus's law ($I = I_0 \cos^2\theta$; red line (gray in print versions)). *(Reproduced with permission from Jiang, K. L.; Li, Q. Q.; Fan, S. S. Nanotechnology: Spinning Continuous Carbon Nanotube Yarns – Carbon Nanotubes Weave Their Way into a Range of Imaginative Macroscopic Applications. Nature* ***2002,*** *419, 801, Copyright 2002, Nature Publishing Group.)*

has been studied for a long time. The on/off ratio of the present α-Si TFTs over a large area in an LCD is 10^6 with mobility ~1 $cm^2/(V \cdot s)$ *(44,60,61)*. With the development of the CNT FET, FETs made with SWCNTs are possible for use in LCDs *(62–64)*. At the same time, the CNT TFT is also flexible. A flexible and stretchable display can be made based on the CNT TFT.

The mechanism of the LCD is the LC molecular rotation under the applied electric field. After electrode formation on a glass substrate, a conventional process is to coat a polyimide (PI) film onto the glass. Then the PI film is rubbed under a given angle. After such treatment, the LC molecule in the LC cell can be well aligned and used to modulate the light by the applied electric field. This is called the prealignment of LC molecules. Different mechanisms have been adopted to explain the prealignment phenomena. It has been found that the grooves are one of the reasons for the alignment effect. However, the rubbing process has disadvantages such as the introduction of broken debris, surface electrostatic charge, and damage to the film. Because CNTs have the linear molecule shape, it is found that the SACNT substrate can also be used to align the LC molecule *(65)*.

Fig. 4.7 shows the result of a CNT film as the alignment layer of an LCD. The CNT film is placed on the back plate, and a 50 nm-thick SiO_2 layer is coated as the insulating layer. The front plate is an indium–tin oxide (ITO) glass. The LC molecule is capsuled between the two plates. Spacer beads of 5 μm diameter are used. After applying a voltage of 15 V, the light can be switched off in the prototype device. CNT films can also be used as a heater. The device has been demonstrated to be able to work under a temperature lower than −50°C. The SWCNT network embedded in a block copolymer matrix after rubbing is also found to be able to improve the switching speed of LC *(66)*. The speed is

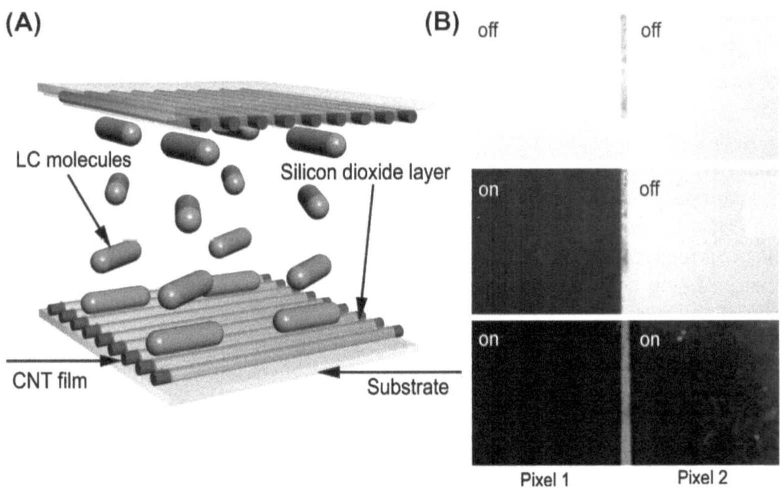

Figure 4.7 Carbon nanotube (CNT) as the alignment layer of a liquid crystal display (LCD). (A) A schematic illustration of the aligning of liquid crystal (LC) molecules by a CNT film. The upper lower panels have the same structure and were placed perpendicular to each other. A layer of silicon dioxide was grown by plasma-enhanced chemical vapor deposition (CVD) to improve the stability of the CNT film. (B) Optical image of a two-pixel LCD cell taken behind the crossed polarizer. Upper: "Off" state; middle: the pixel on the left is at the "on" state; lower: "On" state. The working voltage of the "on" state is 15 V "On" means the voltage between the two aligning layers is applied. *(Reproduced with permission from Fu, W. Q.; Liu, L.; Jiang, K. L.; Li, Q. Q.; Fan, S. S. Super-Aligned Carbon Nanotube Films as Aligning Layers and Transparent Electrodes for Liquid Crystal Displays. Carbon **2010,** 48, 1876–1879, Copyright 2010, Elsevier.)*

about four times faster than that on a commercial PI layer. Furthermore, the LCD cells containing SWCNT nanocomposite alignment layers exhibited low power operation at an effective switching voltage amplitude of approximately 1.3 V without capacitance hysteresis.

Another necessary part in an LCD is the transparent electrode. ITO film is now widely used. It is a mixture of In_2O_3 and SnO_2 with a ratio of 9:1. However, indium is scarce and expensive, and the formation of ITO film needs a thin film process in a sputter or evaporation system. The control of uniform sputtering of ITO over a large area is difficult *(67)*. Because of the ultrasmall diameter of CNT and porous property of the CNT film, it is also a good candidate as a transparent electrode. The square resistance of CNT film is about tens to hundreds of Ω/sq at an optical transparency of 50–90% *(67,68)*. There are different methods that can be used to improve transparency and conductivity. For example, after surface coating of metal particles, the square resistance can be greatly reduced, while the transparency of CNT film is still kept above 50% *(69)*. The advantages of the CNT transparent electrode are the low price for a rich resource of carbon and the simple fabrication technology. However, it still needs to meet the industrial requirement of 100 Ω/sq and 90% transparency using a cheap and reliable method *(70,71)*.

Many LCDs are incorporated with a TP, especially in a cell phone or pad smart terminal. According to the different working principles, the TP can be divided into capacitance, resistance, surface acoustic wave, infrared light, etc. The resistance and capacitance panel are the two most important technologies. The principle of the TP is that the detector addresses the touched x and y position based on different principles in different devices. Because a TP is always incorporated with an LCD, the electrode must be transparent, and the electrode should also be isotropic to realize detection along the x and y directions.

Because CNTs can meet these two requirements, CNT film is a very suitable candidate for TP *(69)*. Fig. 4.8 shows a TP made with SACNT film. Mass production of CNT TP has been realized and the product has been used in many different cell phone and touchpad products. The CNT TP shows a similar performance to ITO ones but with much better flexibility. At the same time, the fabrication of CNT TP does not need the thin film process such as sputtering; the lithography process is also not required to form an isotropic conductance. As a result, the cost can be greatly lowered. The original size of CNT film can be as wide as 12 inches. It can also realize a larger size by merging small pieces according to requirements. Another advantage of CNT TP is its flexibility, which can be used in the future flexible TP in the flexible display as OLED.

4.4 CNT IN OLED

OLED was first proposed by Dr. Tang *(72)*. Because of its unique properties, such as active light emitting, ultrasmall thickness, nonlinear $I-V$ characteristics, fast response,

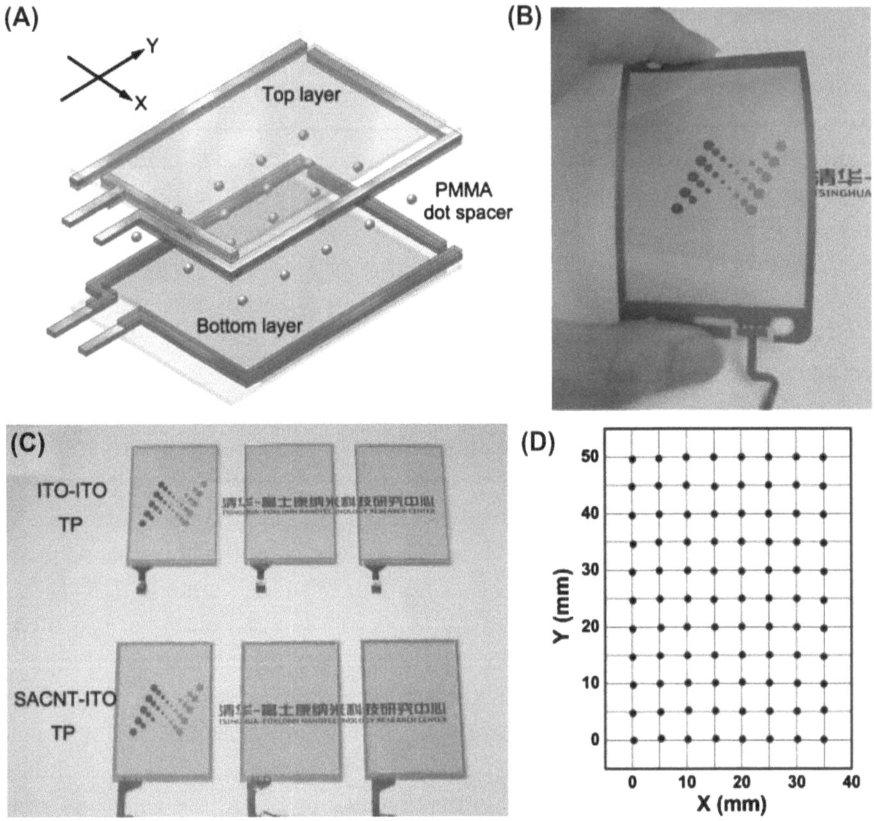

Figure 4.8 Carbon nanotube touch panel (CNT TP). (A) Illustration of the structure of an assembled resistive four-line TP. Two groups of parallel linear electrodes [silver color (gray in print versions)] are coated on the bottom layer and the top layer [yellow color (light gray in print versions)], perpendicular to the *x* and *y* directions, respectively. These two conducting layers are separated by poly(-methyl methacrylate) dot spacers (40 mm in diameter). If the conducting layer is superaligned CNT/poly(ethyleneterephthalate) (SACNT/PET) film, the *x* or *y* direction also indicates the drawing direction of the SACNT film on the bottom or top layer, respectively. (B) Transparent and flexible SACNT–SACNT TP. (C) Comparison of indium–tin oxide–indium–tin oxide (ITO–ITO) TP and SACNT–ITO TP. (D) Linearity test of an SACNT–SACNT TP. The intersections of the *lines* indicate the practical contact points, and the black circles indicate the detected points. *(Reproduced with permission from Feng, C.; Liu, K.; Wu, J. S.; Liu, L.; Cheng, J. S.; Zhang, Y. Y.; Sun, Y. H.; Li, Q. Q.; Fan, S. S.; Jiang, K. L. Flexible, Stretchable, Transparent Conducting Films Made from Superaligned Carbon Nanotubes. Advanced Functional Materials* ***2010,*** *20, 885–891, Copyright 2010, John Wiley and Sons.)*

low working voltage, low-temperature fabrication, and potential application in flexible displays, OLED is believed to be the next generation display, and can now be found in cell phones and TVs. OLED has a multilayered structure. The principle of OLED is the recombination of electron and hole.

A key part of OLED structure is the transparent electrode. It currently still uses the ITO electrode. However, except for similar cost and fabrication problems with LCD, it has also been found that ITO can migrate to the active layer in OLED and deteriorate the device *(73—79)*. The microcavity effect of high reflecting electrode will also limit the emission spectral and angle range *(80)*. OLED is believed to be the best possible candidate for flexible display. ITO is easily broken or cracked according to experimental trials *(81)*. Therefore many endeavors have been made to search for other transparent electrode materials *(82—84)*. As we have described in the previous section, CNT can be used as a flexible transparent electrode. Besides this, the work function of CNT is especially suitable for use as the hole injection layer *(58,85,86)*. It has been pointed out that the ultralarge aspect ratio will also benefit the performance of OLED *(85)*. There have been many attempts to utilize CNTs as the transparent electrode of OLED.

A solution based SWCNT of 500 Ω/sq and 90% transparency has also been used in a top emission OLED structure as the anode on poly(3,4-ethylenedioxythiophene): poly(4-styrenesulfonate) (PEDOT:PSS). With the active layer of poly(vinylcarbazole), 2-(4-biphenylyl)-5-(4-*tert*-butylphenyl)-1,3,4-oxadiazole, *tris*(2-phenylpyridinato) iridium, and *N,N′*-diphenyl-*N,N′*-*bis*(3-methylphenyl)-1,1′-biphenyl-4,4′-diamine, luminance can reach 3588 cd/m^2 at an efficiency of 1.24 cd/A *(87)*. The device with a solution-processed SWCNT electrode of 60 Ω/sq and 44% transparency as the bottom electrode with copper phthalocyanine, *N,N′*-*bis*-(1-naphthyl)-*N,N′*-diphenyl-1,1-biphenyl-4,4′-diamine, *tris*(8-hydroxyquinoline) aluminum (Alq3) LiF/Al shows a luminance of 2800 cd/m^2 at 1.4 cd/A *(88)*. The SWCNT electrode formed by a dry transfer technique with as-grown SWCNT aerogel can reach 110 Ω/sq and 90% transparency after HNO_3 doping, and the OLED of blue-emitting polyfluorene copolymer poly[(9,9-dioctylfluorenyl-2,7-diyl)-*co*-(4,4-(*N*-(4-*sec*-butylphenyl)) diphenylamine)] and poly [9,9-dioctylfluorene] with the PEDOT:PSS-coated SWCNT anode shows a luminance of 2000 cd/m^2 *(89)*. The SWCNT anode with a structure of SWCNT/PEDOT-PSS: MeOH shows 3500 cd/m^2 and a current efficiency of 1.6 cd/A *(90)*. After the HNO_3 modification and polymer encapsulation, the device of the CNT anode shows a maximum luminescence of 9000 cd/m^2, which is close to ITO-based OLED device performance, and a high efficiency of 10 cd/A, similar to the ITO-based OLED device *(91)*. The results are shown in Fig. 4.9A and B. The study of the SWCNT poly(*m*-phenylene vinylene-*co*-2,5-dioctoxy-*p*-phenylene) (PmPV) composite electrode indicated that the SWCNT in the PmPV matrix acts as a hole-blocking material *(92)*. The SWCNT/ PEDOT:PSS polymeric nanocomposite anode has shown a lower turn-on voltage and higher luminous intensity than those fabricated on ITO/poly(ethyleneterephthalate) (PET) anodes in a flexible OLED *(81)*. ITO modification of SWCNT can also improve the performance of flexible OLED to a maximum light output of 8900 cd/m^2 with a current efficiency of 4.5 cd/A *(93)*. The device with SWCNT as both anode and cathode fabricated by the roll lamination method shows a low turn-on voltage of 3.8 V, high

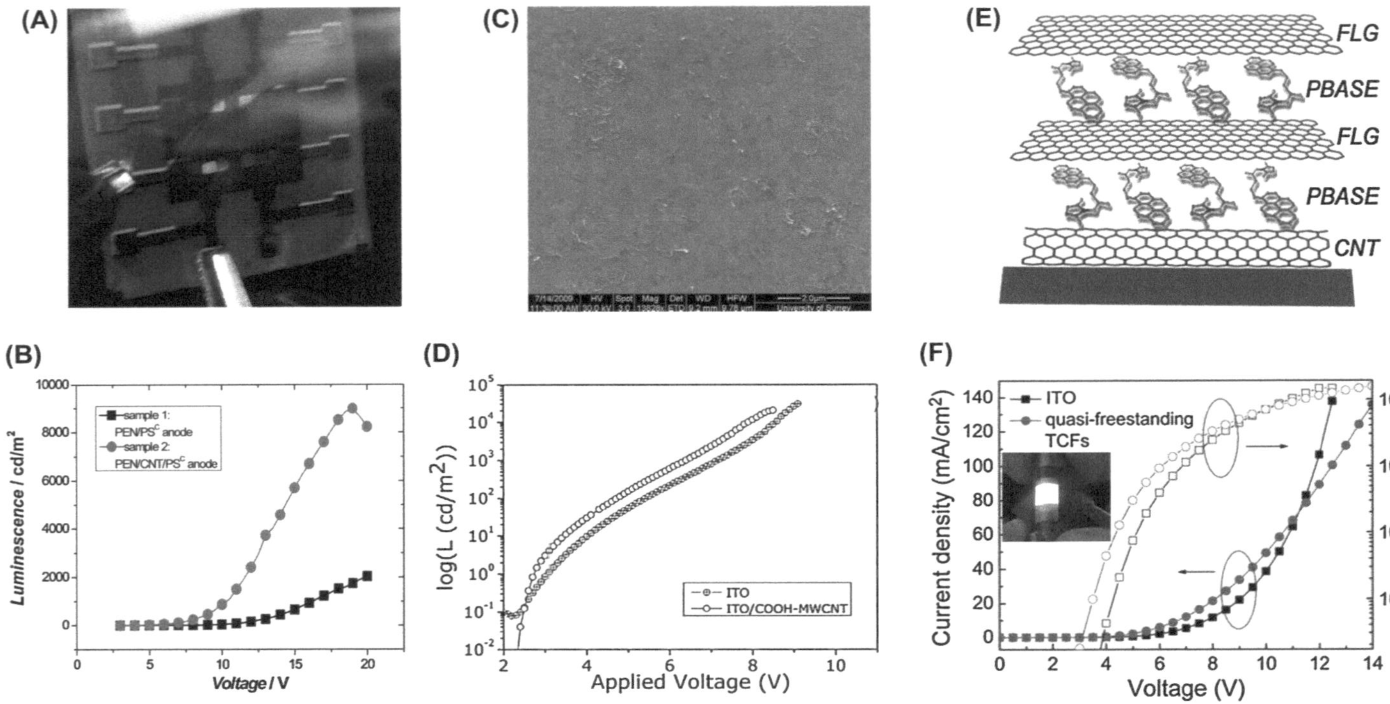

Figure 4.9 (A) and (B) Photograph of working organic light-emitting diode (OLED) device with surface-modified carbon nanotube (CNT) anodes and the luminescence versus voltage of OLEDs built on poly(3,4-ethylenedioxythiophene):poly(4-styrenesulfonate) (PEN/PEDOT: PSS) composite (sample 1) and versus polyethylene naphthalate (PEN)/CNT/PEDOT:PSS composite layer (sample 2). *((A and B) reproduced with permission from Ou, E. C. W.; Hu, L. B.; Raymond, G. C. R.; Soo, O. K.; Pan, J. S.; Zheng, Z.; Park, Y.; Hecht, D.; Irvin, G.; Drzaic, P.; et al. Surface-Modified Nanotube Anodes for High Performance Organic Light-Emitting Diode. ACS Nano* ***2009,*** *3, 2258–2264, Copyright 2009, American Chemical Society.)* (C) Scanning electron microscope (SEM) images of a functionalized multiwalled carbon nanotube (MWCNT) spin cast at 1000 rpm for 1 min from aqueous solution on indium–tin oxide (ITO)-coated glass substrates. (D) Luminance versus voltage characteristics of reference ITO and ITO/COOH-MWCNT electrode OLED. *((C and D) reproduced with permission from Inigo, A. R.; Underwood, J. M.; Silva, S. R. P. Carbon Nanotube Modified Electrodes for Enhanced Brightness in Organic Light Emitting Devices. Carbon* ***2011,*** *49, 4211–4217, Copyright 2011, Elsevier.)* (E) Full carbon transparent single-walled carbon nanotube (SWCNT)/1-pyrenebutyric acid N-hydroxysuccinimide ester (PBASE)/few layer graphene (FLG)/PBASE/FLG/PBASE electrodes are assembled in a layer-by-layer fashion. (F) Luminance–voltage–current density characteristics of devices using ITO and full carbon transparent electrode, respectively; inset is the flexibility test of devices using full carbon transparent electrode. *((E and F) reproduced with permission from Liu, Y. P.; Jung, E.; Wang, Y.; Zheng, Y.; Park, E. J.; Cho, S. M.; Loh, K. P. "Quasi-Freestanding" Graphene-on-Single Walled Carbon Nanotube Electrode for Applications in Organic Light-Emitting Diode. Small* ***2014,*** *10, 944–949, Copyright 2013, John Wiley and Sons.)*

brightness of 1400 cd/m^2 at 10 V, and maximum efficiency of 2.2 cd/A at 480 cd/m^2 *(94)*. By using PEDOT passivation for better surface smoothness and using $SOCl_2$ doping for lower sheet resistance, the SWCNT film can show 160 Ω/sq at 87% transparency and has been used successfully to fabricate OLEDs with high stabilities and long lifetimes *(95)*. SWCNT with poly[2-methoxy-5-(2-ethylhexyloxy)-1,4-phenylenevinylene] composite as the anode can significantly enhance the current density, but the device performance will deteriorate because of the discrepancy of two kinds of carrier density *(96)*. SWCNTs are mainly believed to trap the holes in the hole conducting layer by the comparison of electroluminescence, photoluminescence, and $I-V$ curves *(97)*.

It was discovered that the relative efficiencies of using MWCNT PmPV composite as an electron transport layer have been improved compared to the poly(2,5-dimethoxy-1,4-phenylene-vinylene-2-methoxy-5(2′-ethylhexyloxy)-1,4-phenylene-vinylene) single layer devices *(98)*. The OLED with the MWCNT PmPV composite electrode shows better stability than that of polymer only *(99)*. The microcavity effect can be avoided with MWCNT as the top electrode in the structure *(80)*. MWCNT has successfully formed a stable ink with PEDOT:PSS to be used as the charge transport layer in OLED *(100)*. Nanocomposite layers consisting of MWCNT-doped poly(3-hexylthiophene) (P3HT) were employed as a hole-injection layer (HIL) in OLEDs with the structure of glass/Al-doped zinc oxide/MWCNTs-doped P3HT/*N,N′*-*bis*(3-methylphenyl)-*N,N′*-*bis*(phenyl) benzidine/Alq3/LiF/Al. The obtained current efficiency of OLED is as high as 10 cd/A, which is about 30% higher than that of a conventional device without an MWCNT-doped HIL *(101)*. The nanocomposite of acid-treated MWCNT and PEDOT:PSS can show an improved luminance and lower turn-on voltage as the HIL over the pristine PEDOT:PSS electrode *(102)*. MWCNT film as the HIL after PEDOT:PSS planarization can show a brightness of 4500 cd/m^2 and a current efficiency near 2.5 cd/A *(85)*. Acid-oxidized MWCNT as a hole-injection buffer layer provides higher brightness in excess of 20,000 cd/m^2 (Fig. 4.9C and D) and the reason is attributed to enhanced hole injection *(103)*.

Besides the CNT and CNT polymer composite electrode, there are also other kinds of materials being integrated with CNT and used as the electrode. The gold grid transparent conducting electrodes with regioregular poly(3-hexylthiophene) (πP3HT)-wrapped SWCNT OLEDs have a luminance of 8260 cd/m^2 at 9 V *(78)*. A nanocomposite electrode comprising SWCNT and silver nanowires on polymer substrate can show maximum current efficiency of 118 cd/A at 10,000 cd/m^2 with a calculated external quantum efficiency of 38.9%. Nanoparticles of barium strontium titanate are dispersed in the substrate to enhance light extraction efficiency *(79)*. ITO-modified SWCNT fabricated on flexible PET substrates shows a maximum light output of 8900 cd/m^2 with a current efficiency of 4.5 cd/A *(93)*. It is found that an SWCNT and graphene electrode exhibits 88% power efficiency of OLEDs fabricated on an ITO anode with a low turn-on voltage of 3.1 V, high luminance up to 29,490 cd/m^2, and current efficiency of 14.7 cd/A (Fig. 4.9E and F) *(104)*.

It has been demonstrated that CNTs can be used as an electron injection part in an OLED device with a cavity field emission structure. The CNT paste cathode is vacuum encapsulated with an OLED emissive layer as the anode. After an Al film is coated, the light emission uniformity and lifetime of the device are both improved *(105)*.

Similar to the LCD, the active matrix (AM) OLED also needs TFT to control each individual pixel. The CNTs can be used to fabricate the TFT of OLED, and it is possible to use this in a flexible display panel. Because CNTs are naturally two-thirds semiconducting and one-third metallic, two kinds of strategy have been adopted in the fabrication of CNT FETs. The first uses the separated semiconducting CNTs. Its advantage is that the devices are more uniform, while the problem of separation of CNT is still a big challenge. The second strategy uses the mixed CNTs by controlling the percolation. Its advantage is that the fabrication is very simple, while the problem of control over percolation of CNTs is difficult. Both kinds of CNT TFT devices have been used in OLED.

Zhou *(106)* has demonstrated that TFT with separated 95% semiconducting SWCNT shows small sheet resistance (25 kΩ/sq), high current density (10 μA/μm), and superior mobility [52 cm^2/(V·s)]. On/off ratios $>10^4$ are achieved with channel length $L > 20$ μm. The TFT can be used to drive OLED at a modulation of output light intensity if it exceeds 10^4. In 2011, with the separated 98% semiconducting SWCNT, TFT made with the solution deposition technique *(107)* had been used in a 20 × 25 pixel AM-OLED display, as shown in Fig. 4.10A; the individual device showed a mobility of 31.65 cm^2/(V·s), an on/off ratio $>10^4$, and current driving capability of ~0.8 μA/μm. With the random grown SWCNT network, a transparent TFT with SWCNT has been demonstrated with an on/off ratio of 10^3, transparency of 80%, and mobility of 16–22 cm^2/(V·s). This method is also ideal for large-scale fabrication and the TFT is used in a prototype of OLED *(108)*. With a TFT made with a CVD SWCNT network, a dynamic AM-OLED display with 6 × 6 pixels has also been successfully demonstrated in *(64)*, as shown in Fig. 4.10B.

One of the biggest advantages of OLED is that it can be used as a flexible display. With the CNT TFT, a flexible OLED display has been fabricated. With 99% semiconducting SWCNT, TFT has been fabricated on PI substrate. After OLED integration and a pressure-sensitive layer lamination, a pressure-sensitive OLED display has been fabricated. The drive current can be adjusted by the stretchability of CNT TFT and electrode. The intensity of emitted light quantifies the magnitude of the applied pressure. The skin-like device can be used to indicate the pressure of the surface, as shown in Fig. 4.10C and D *(109)*. SWCNTs have also been used as a stretchable electrode with a conductivity of more than 100 S/cm; the OLED display shows about 50% stretchability *(110)*.

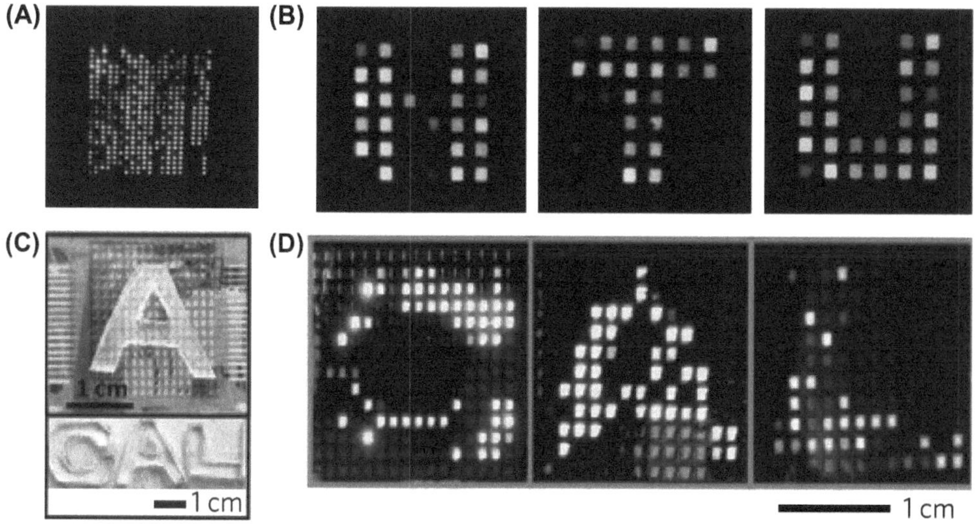

Figure 4.10 (A) Photograph of a 20 × 25 pixel active matrix organic light-emitting diode (AM-OLED)-made single-walled carbon nanotube-thin film transistor (SWCNT-TFT) with solution deposition technique, $V_{DATA} = 5$ V, $V_{SCAN} = 5$ V, and $V_{DD} = 8$ V. *(Reproduced with permission from Zhang, J. L.; Fu, Y.; Wang, C.; Chen, P. C.; Liu, Z. W.; Wei, W.; Wu, C.; Thompson, M. E.; Zhou, C. W. Separated Carbon Nanotube Macroelectronics for Active Matrix Organic Light-Emitting Diode Displays. Nano Letters* ***2011****, 11, 4852–4858, Copyright 2011, American Chemical Society.)* (B) Display demonstration of the 6 × 6 AM-OLED pixels driven by 72 SWCNT-TFTs, $V_{SCAN} = 10$ V, $V_{DATA} = 10$ V, and $V_{DD} = 5$ V, the letters "N," "T," and "U" are displayed sequentially on this OLED display. *(Reproduced with permission from Zou, J. P.; Zhang, K.; Li, J. Q.; Zhao, Y. B.; Wang, Y. L.; Pillai, S. K. R.; Demir, H. V.; Sun, X. W.; Chan-Park, M. B.; Zhang, Q. Carbon Nanotube Driver Circuit for 6 × 6 Organic Light Emitting Diode Display. Scientific Reports-UK* ***2015****, 5, 11755.)* (C) and (D) The application of a carbon nanotube (CNT) TFT-OLED used as an e-skin. (C) The polydimethylsiloxane (PDMS) slabs with C, A, and L shapes are prepared and used to apply pressure onto the sensor array. (D) Green (lightest gray in print versions), blue (light gray in print versions) and red (gray in print versions) color interactive e-skins are used to spatially map and display the pressure applied with C- (left), A- (center), and L- (right) shaped PDMS slabs, respectively. *(Reproduced with permission from Wang, C.; Hwang, D.; Yu, Z. B.; Takei, K.; Park, J.; Chen, T.; Ma, B. W.; Javey, A. User-Interactive Electronic Skin for Instantaneous Pressure Visualization. Nature Materials* ***2013****, 12, 899–904, Copyright 2013, Nature Publishing Group.)*

Different from the TFT in the LCD, the drive circuit of OLED is the current-drive type. CNTs have been used to enable dinaphtho-[2,3-*b*:2′,3′-*f*]thieno[3,2-*b*]-thiophene and to fabricate vertical FET. Current modulation is realized by the gate field modulation of the Schottky barrier between the nanotubes and organic channel material. This structure realized sufficient on-currents to drive OLED pixels at low operating voltages without high-resolution patterning. This CNT vertical-FET has been used to drive an RGB OLED pixel in a prototype device *(111,112)*.

4.5 CNT IN OTHER KINDS OF DISPLAY

CNT has also been used in other kinds of new display devices. As is well known, CNT is very thin with a diameter of about several nanometers and with a hollow core. The intrinsic heat capacity of CNT is very small. The heat capacity of SACNT film is also ultrasmall as a result. Based on this property, thermal acoustic loudspeakers and earphones have been fabricated *(113,114)*. The thermal response of CNT film has been studied in a vacuum chamber with a current heating method *(59)*. It has been found that the CNT film can be heated to incandescence in less than 1 ms and cooled down to room temperature with almost the same speed. Fig. 4.11 shows the thermal response of CNT films.

The temperature changes are recorded with a photodiode by monitoring the incandescence signal. Simulation of the heating process shows good accordance with the experiment. Because the CNT film is heated to incandescence in vacuum, thermal radiation is the major thermal dissipation term and the convective term can be omitted. It is found that the temperature can respond to a periodic heating pulse up to 10 kHz. An

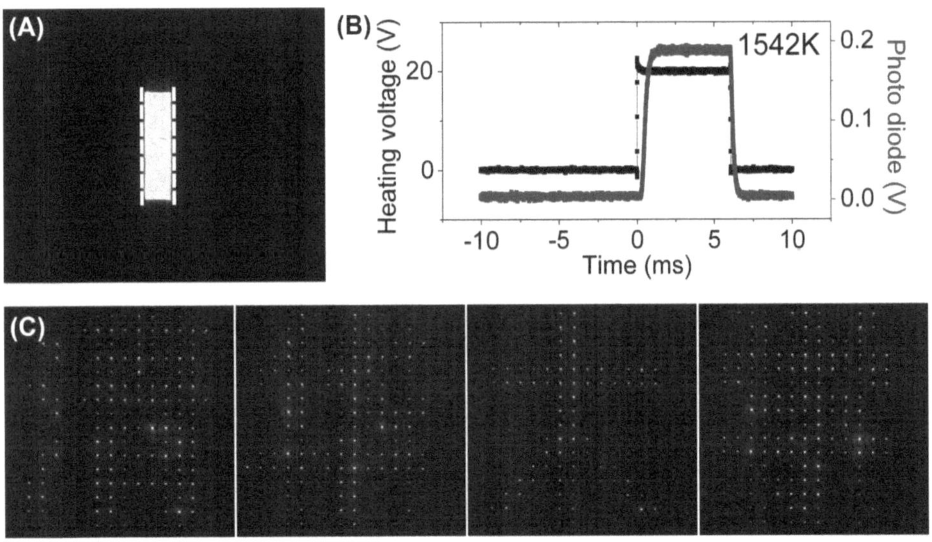

Figure 4.11 (A) Incandescence image of a carbon nanotube (CNT) film 2.5 mm long and 8 mm wide; dashed lines are the position of the nickel rod electrodes. (B) Incandescence response of the CNT film to heating pulse. (C) Image of displayed Chinese characters, which are the Chinese characters for Tsinghua University. *(Reproduced with permission from Liu, P.; Liu, L.; Wei, Y.; Liu, K.; Chen, Z.; Jiang, K. L.; Li, Q. Q.; Fan, S. S. Fast High-Temperature Response of Carbon Nanotube Film and Its Application as an Incandescent Display. Advanced Materials* ***2009,*** *21, 3563–3566, Copyright 2009, John Wiley and Sons.)*

incandescence display, which was first reported in Ref. *(115)*, can be fabricated with the CNT film. The electrode structure is fabricated with screen printing technologies. SACNT film is placed on the electrode structure and cut to the desired size with an infrared laser. Then the device is vacuum enveloped. The 16 × 16 pixel device can display Chinese characters dynamically. The lifetime of the device is estimated to be about 1000 h.

Successively, an incandescence display of CNT thin yarn has also been fabricated *(116)*. It is hoped that the thin yarn devices will have better stability. Except for the pure CNT film, the thermal response of CNT film loaded with graphite powder has also been studied *(117)*. The thermal response—temperature relation is similar to that of pure CNT film.

The thermal response of CNT film in atmosphere has also been studied. The CNT film is heated to about only 100°C to avoid oxidation damage. Different to a vacuum, convection is the major heat dissipation term in the atmosphere. The temperature is monitored with an infrared thermometer or infrared imager. The results show that the thermal response can be as fast as about <0.1 s. The thermal response with a stable thermal load, for example, graphite powder, has also been studied. The increased specific heat per unit area will slow down the thermal response. Although the speed is low for dynamical display of a movie, it still satisfies some applications such as electronic paper. As an example, thermochromic pigments have been loaded on the CNT film using a spraying method. The thermochromic pigments are capsulated spheres of a mixture of different components. With increasing temperature, the mixture experiences phase change and chemical reaction, and color transition happens as a result. Compared with the graphite powder load, the thermochromic pigment will experience phase change and a latent heat process. The thermal response characterization clearly shows this. It is found that the latent heat will slow down the "on and off" delay. Even so, the on and off delay with phase change load can reach less than 0.2 s at an appropriate load quantity. A thermochromic display has been fabricated on a glass substrate. To sustain the free-standing state of CNT film, the substrate is dug to form a concave pit array. Electrodes are also screen printed. The thermochromic display prototype can be used to display dynamic Chinese characters, as shown in Fig. 4.12A and B. The all-on power consumption of the 16 × 16 array is about 6 W. The brightness contrast of black thermochromic pigment at on and off state is about 6.

The CNT film can be used to fabricate an addressable microheater array *(118)*. Because the CNT film is flexible at the same time, it is possible to fabricate a flexible and bendable microheater array. Both PET and PI substrates work well. The electrodes are also formed by screen printing after adopting low-temperature silver paste. CNT film is adhered to the polymer substrate. The microheater array can be heated separately with a driving circuit, as shown in Fig. 4.12C. A flexible thermochromic display has also been demonstrated with the microheater array on PET (Fig. 4.12D). The response is slower

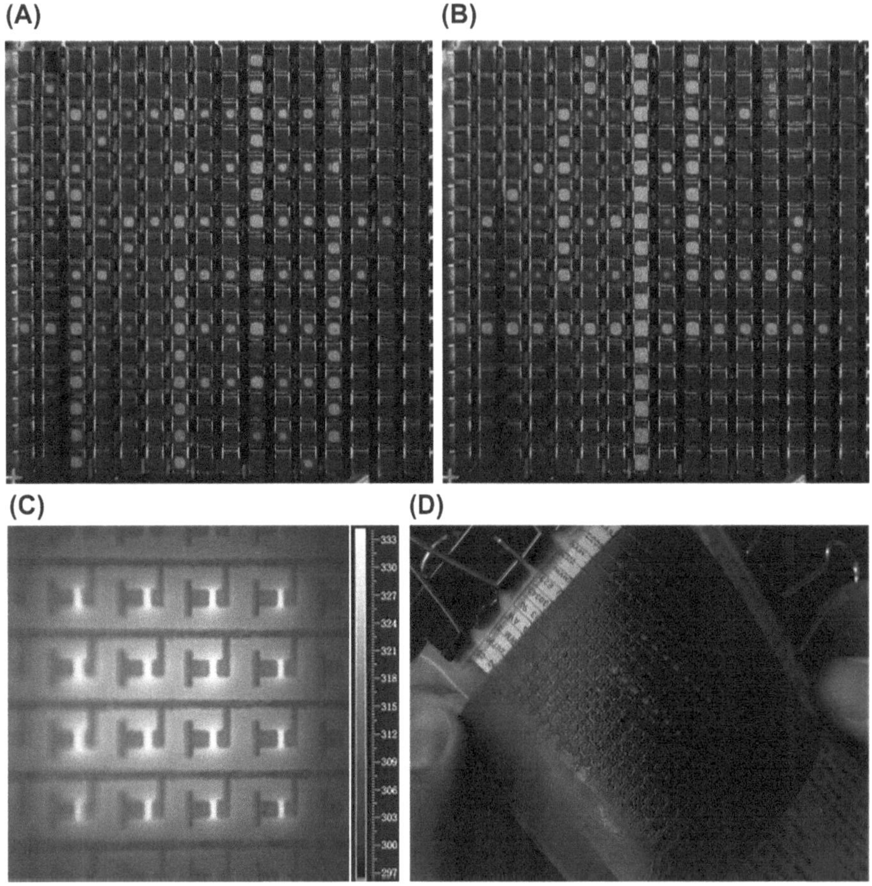

Figure 4.12 Thermochromic display of carbon nanotubes (CNTs). (A) and (B) Dynamic Chinese characters displayed by the suspended 16 × 16 pixel CNT film thermochromic display. *((A and B) reproduced with permission from Liu, P.; Zhou, D. L.; Wei, Y.; Jiang, K. L.; Wang, J. P.; Zhang, L. N.; Li, Q. Q.; Fan, S. S. Load Characteristics of a Suspended Carbon Nanotube Film Heater and the Fabrication of a Fast-Response Thermochromic Display Prototype. ACS Nano* ***2015,*** *9, 3753–3759, Copyright 2015, American Chemical Society.)* (C) The addressable microheater array and (D) the flexible thermochromic display. *(Reproduced with permission from Liu, P.; Liu, L. A.; Jiang, K. L.; Fan, S. S. Carbon-Nanotube-Film Microheater on a Polyethylene Terephthalate Substrate and Its Application in Thermochromic Displays. Small* ***2011,*** *7, 732–736, Copyright 2011, John Wiley and Sons.)*

than that at the free-standing state. The on and off delays are 2 and 5 s, respectively. The brightness contrast induced by thermochromic pigment color transition is also about 6.

With the development of wearable electronics, it is hoped that the display function can be incorporated simultaneously. It has been found that the CNT fibers can be integrated with functional polymers. Incorporation with functional polymer materials such as polydiacetylene, the CNT fiber can rapidly (about 1 s) and reversibly respond to electrical current, with the resulting color change being readily observable with the naked eye *(119)*. With CNT as anode, a weavable fiber-shaped polymer light-emitting electrochemical cell (PLEC) has also been demonstrated. The fiber-shaped PLEC is fabricated using all-solution-based processes that can be scaled up for practical applications. The design has a coaxial structure comprising a modified metal wire cathode and a conducting aligned carbon nanotube sheet anode, with an electroluminescent polymer layer sandwiched between them *(120)*. These experiments show that it is possible to fabricate new types of flexible display devices based on CNTs.

4.6 SUMMARY AND OUTLOOK

Information display involves a multidisciplinary technology. Resolution at tens of micrometers is enough to the vision feeling of human beings. The contemporary IC technique is capable of fabricating devices at such size. However, many other factors need to be considered than just the resolution for the fabrication of real display devices. Nanomaterials and nanotechnology can possibly find chances from the viewpoint of bottom-up methodology. CNT is one of the nanomaterials that has been most frequently studied and has realized mass production. It may be a suitable candidate to improve the performance of display devices or bring a revolution to display technologies. Of course, it will need a new design and structure. Display is a technology that relates with human subjective feeling, convenience, simplicity, low cost and low power, high performance and high robustness, which will be the trends of technology development. Application research into nanomaterials is just following these requirements. It is optimistic that the CNTs will find more real applications in the future display technologies.

REFERENCES

1. Myers, R. L., Ed. *Display Interfaces;* John Wiley & Sons, Ltd.: Chichester, 2002.
2. Ying, G., Tu, Y., Wan, B., Eds. *Handbook of Application Technologies in Flat Panel Display;* Publishing House of Electronics Industry: Beijing, 2007.
3. Hu, C., Gao, H., Eds. *Yearbook of the Flat Panel Display Development in China;* Publishing House of Electronics Industry: Beijing, 2012.
4. Heilmeie, G.; Zanoni, L. A. Guest-Host Interactions in Nematic Liquid Crystals. A New Electro-Optic Effect. *Applied Physics Letters* **1968,** *13,* 91–92.
5. Heilmeie, G.; Zoanoni, L. A.; Barton, L. A. Dynamic Scattering in Nematic Liquid Crystals. *Applied Physics Letters* **1968,** *13,* 46–47.
6. Fowler, R. H.; Nordheim, L. Electron Emission in Intense Electric Fields. *Proceedings of the Royal Society of London, Series A* **1928,** *119,* 173–181.
7. Flugge, S., Ed. Handbuch der Physik. Field Emission, Eds. R.H. Good and E.W. Muller, Springer-Verlag: Berlin, 1956; p. 187.
8. Nordhlim, L. W. The Effect of the Image Force on the Emission and Reflexion of Electrons by Metals. *Proceedings of the Royal Society of London, Series A* **1928,** *121,* 626–639.

9. Burgess, R. E.; Kroemer, H.; Houston, J. M. Corrected Values of Fowler-Nordheim Field Emission Functions V(Y) and S(Y). *Physical Review* **1953,** *90,* 515.
10. Spindt, C. A. A Thin-Film Field-Emission Cathode. *Journal of Applied Physics* **1968,** *39,* 3504–3505.
11. Thomas, R. N.; Wickstro, R.; Schroder, D. K.; Nathanso, H. Fabrication and Some Applications of Large-Area Silicon Field-Emission Arrays. *Solid State Electron* **1974,** *17,* 155–157.
12. Zhu, W., Ed. *Vacuum Microelectronics;* John Wiley & Sons, Inc.: New York, 2001.
13. Vaudaine, P.; Meyer, R. 'Microtips' Fluorescent Display. In *International Electron Devices Meeting 1991. Technical Digest (Cat. No.91CH3075-9);* 1991; pp 197–200.
14. Talin, A. A.; Dean, K. A.; Jaskie, J. E. Field Emission Displays: A Critical Review. *Solid State Electron* **2001,** *45,* 963–976.
15. Meyer, R. 6″ Diagonal Microtips Fluorescent Display for TV Applications. In Eurodisplay, 1990.
16. de Jonge, N.; Lamy, Y.; Schoots, K.; Oosterkamp, T. H. High Brightness Electron Beam from a Multi-Walled Carbon Nanotube. *Nature* **2002,** *420,* 393–395.
17. Fan, S. S.; Chapline, M. G.; Franklin, N. R.; Tombler, T. W.; Cassell, A. M.; Dai, H. J. Self-Oriented Regular Arrays of Carbon Nanotubes and Their Field Emission Properties. *Science* **1999,** *283,* 512–514.
18. Deheer, W. A.; Chatelain, A.; Ugarte, D. A Carbon Nanotube Field-Emission Electron Source. *Science* **1995,** *270,* 1179–1180.
19. Rinzler, A. G.; Hafner, J. H.; Nikolaev, P.; Lou, L.; Kim, S. G.; Tomanek, D.; Nordlander, P.; Colbert, D. T.; Smalley, R. E. Unraveling Nanotubes – Field-Emission from an Atomic Wire. *Science* **1995,** *269,* 1550–1553.
20. Choi, W. B.; Chung, D. S.; Kang, J. H.; Kim, H. Y.; Jin, Y. W.; Han, I. T.; Lee, Y. H.; Jung, J. E.; Lee, N. S.; Park, G. S.; et al. Fully Sealed, High-Brightness Carbon-Nanotube Field-Emission Display. *Applied Physics Letters* **1999,** *75,* 3129–3131.
21. Wang, Q. H.; Setlur, A. A.; Lauerhaas, J. M.; Dai, J. Y.; Seelig, E. W.; Chang, R. P. H. A Nanotube-Based Field-Emission Flat Panel Display. *Applied Physics Letters* **1998,** *72,* 2912–2913.
22. Saito, Y.; Hamaguchi, K.; Uemura, S.; Uchida, K.; Tasaka, Y.; Ikazaki, F.; Yumura, M.; Kasuya, A.; Nishina, Y. Field Emission from Multi-Walled Carbon Nanotubes and Its Application to Electron Tubes. *Applied Physics A* **1998,** *67,* 95–100.
23. Saito, Y.; Uemura, S.; Hamaguchi, K. Cathode Ray Tube Lighting Elements with Carbon Nanotube Field Emitters. *Japanese Journal of Applied Physics Part 2-Letters* **1998,** *37,* L346–L348.
24. Croci, M.; Arfaoui, I.; Stockli, T.; Chatelain, A.; Bonard, J. M. A Fully Sealed Luminescent Tube Based on Carbon Nanotube Field Emission. *Microelectronics Journal* **2004,** *35,* 329–336.
25. Wei, Y.; Xiao, L.; Zhu, F.; Liu, L.; Tang, J.; Liu, P.; Fan, S. S. Cold Linear Cathodes with Carbon Nanotube Emitters and Their Application in Luminescent Tubes. *Nanotechnology* **2007,** 18.
26. Deheer, W. A.; Bacsa, W. S.; Chatelain, A.; Gerfin, T.; Humphreybaker, R.; Forro, L.; Ugarte, D. Aligned Carbon Nanotube Films – Production and Optical and Electronic-Properties. *Science* **1995,** *268,* 845–847.
27. Milne, W. I.; Teo, K. B. K.; Amaratunga, G. A. J.; Legagneux, P.; Gangloff, L.; Schnell, J. P.; Semet, V.; Binh, V. T.; Groening, O. Carbon Nanotubes as Field Emission Sources. *Journal of Materials Chemistry* **2004,** *14,* 933–943.
28. Teo, K. B. K.; Chhowalla, M.; Amaratunga, G. A. J.; Milne, W. I.; Hasko, D. G.; Pirio, G.; Legagneux, P.; Wyczisk, F.; Pribat, D. Uniform Patterned Growth of Carbon Nanotubes without Surface Carbon. *Applied Physics Letters* **2001,** *79,* 1534–1536.
29. Ren, Z. F.; Huang, Z. P.; Xu, J. W.; Wang, J. H.; Bush, P.; Siegal, M. P.; Provencio, P. N. Synthesis of Large Arrays of Well-Aligned Carbon Nanotubes on Glass. *Science* **1998,** *282,* 1105–1107.
30. Jung, J. E.; Jin, Y. W.; Choi, J. H.; Park, Y. J.; Ko, T. Y.; Chung, D. S.; Kim, J. W.; Jang, J. E.; Cha, S. N.; Yi, W. K.; et al. Fabrication of Triode-Type Field Emission Displays with High-Density Carbon-Nanotube Emitter Arrays. *Physica B: Condensed Matter* **2002,** *323,* 71–77.
31. Kim, Y. C.; Han, I. T.; Kim, J. M. Screen-Printed Carbon Nanotube Field Emitters for Display Applications. In *Carbon Nanotube and Related Field Emitters: Fundamentals and Applications;* Yahachi, S., Ed.; Wiley-VCH Verlag GmbH & Co. KGaA: Weinheim, 2010; pp 287–309.
32. Jiang, K. L.; Li, Q. Q.; Fan, S. S. Nanotechnology: Spinning Continuous Carbon Nanotube Yarns – Carbon Nanotubes Weave Their Way into a Range of Imaginative Macroscopic Applications. *Nature* **2002,** *419,* 801.

33. Jiang, K. L.; Wang, J. P.; Li, Q. Q.; Liu, L. A.; Liu, C. H.; Fan, S. S. Superaligned Carbon Nanotube Arrays, Films, and Yarns: A Road to Applications. *Advanced Materials* **2011,** *23,* 1154−1161.
34. Wei, Y.; Weng, D.; Yang, Y. C.; Zhang, X. B.; Jiang, K. L.; Liu, L.; Fan, S. S. Efficient Fabrication of Field Electron Emitters from the Multiwalled Carbon Nanotube Yarns. *Applied Physics Letters* **2006,** *89,* 063101.
35. Wei, Y.; Liu, L.; Liu, P.; Xiao, L.; Jiang, K. L.; Fan, S. S. Scaled Fabrication of Single-Nanotube-Tipped Ends from Carbon Nanotube Micro-Yarns and Their Field Emission Applications. *Nanotechnology* **2008,** *19,* 475707.
36. Wei, Y.; Liu, P.; Jiang, K. L.; Liu, L.; Fan, S. S. Breaking Single-Walled Carbon Nanotube Bundles by Joule Heating. *Applied Physics Letters* **2008,** *93,* 023118.
37. Busta, H. H. Vacuum Microelectronics-1992. *Journal of Micromechanics and Microengineering* **1992,** *2,* 43−74.
38. Wei, Y.; Jiang, K. L.; Liu, L.; Chen, Z.; Fan, S. S. Vacuum-Breakdown-Induced Needle-Shaped Ends of Multiwalled Carbon Nanotube Yarns and Their Field Emission Applications. *Nano Letters* **2007,** *7,* 3792−3797.
39. Liu, P.; Wei, Y.; Liu, K.; Liu, L.; Jiang, K. L.; Fan, S. S. New-Type Planar Field Emission Display with Superaligned Carbon Nanotube Yarn Emitter. *Nano Letters* **2012,** *12,* 2391−2396.
40. Choi, Y. C.; Lee, J. W.; Lee, S. K.; Kang, M. S.; Lee, C. S.; Jung, K. W.; Lim, J. H.; Moon, J. W.; Hwang, M. I.; Kim, I. H.; et al. The High Contrast Ratio and Fast Response Time of a Liquid Crystal Display Lit by a Carbon Nanotube Field Emission Backlight Unit. *Nanotechnology* **2008,** *19,* 235306.
41. Zhang, Y.; Deng, S. Z.; Xu, N. S.; Chen, J. Fully Sealed Carbon Nanotube Flat-Panel Light Source and Its Application as Thin Film Transistor-Liquid-Crystal Display Backlight. *Journal of Vacuum Science & Technology B* **2008,** *26,* 1033−1037.
42. Kobayashi, S., Mikoshiba, S., Lim, S., Eds. *LCD Backlights;* John Wiley & Sons Ltd: Chichester, 2009.
43. Fu, W.; Liu, P.; Tang, J.; Liu, L.; Fan, S. Spherical Field Emission Cathode Based on Carbon Nanotube Paste and Its Application in Luminescent Bulbs. *Journal of Vacuum Science & Technology B* **2008,** *26,* 1404−1406.
44. Yang, D. K., Wu, S. T., Eds. *Fundamentals of Liquid Crystal Devices;* John Wiley & Sons, Ltd.: Chichester, 2015.
45. Kim, K. H.; Song, J. K. Technical Evolution of Liquid Crystal Displays. *NPG Asia Materials* **2009,** *1,* 29−36.
46. Song, W. H.; Windle, A. H. Isotropic-Nematic Phase Transition of Dispersions of Multiwall Carbon Nanotubes. *Macromolecules* **2005,** *38,* 6181−6188.
47. Song, W. H.; Kinloch, I. A.; Windle, A. H. Nematic Liquid Crystallinity of Multiwall Carbon Nanotubes. *Science* **2003,** *302,* 1363.
48. Badaire, S.; Zakri, C.; Maugey, M.; Derre, A.; Barisci, J. N.; Wallace, G.; Poulin, P. Liquid Crystals of DNA-Stabilized Carbon Nanotubes. *Advanced Materials* **2005,** *17,* 1673−1676.
49. Zhang, S. J.; Kumar, S. Carbon Nanotubes as Liquid Crystals. *Small* **2008,** *4,* 1270−1283.
50. Moulton, S. E.; Maugey, M.; Poulin, P.; Wallace, G. G. Liquid Crystal Behavior of Single-Walled Carbon Nanotubes Dispersed in Biological Hyaluronic Acid Solutions. *Journal of the American Chemical Society* **2007,** *129,* 9452−9457.
51. Dierking, I.; Scalia, G.; Morales, P.; LeClere, D. Aligning and Reorienting Carbon Nanotubes with Nematic Liquid Crystals. *Advanced Materials* **2004,** *16,* 865−869.
52. Lagerwall, J.; Scalia, G.; Haluska, M.; Dettlaff-Weglikowska, U.; Roth, S.; Giesselmann, F. Nanotube Alignment Using Lyotropic Liquid Crystals. *Advanced Materials* **2007,** *19,* 359−364.
53. Lynch, M. D.; Patrick, D. L. Organizing Carbon Nanotubes with Liquid Crystals. *Nano Letters* **2002,** *2,* 1197−1201.
54. Russell, J. M.; Oh, S. J.; LaRue, I.; Zhou, O.; Samulski, E. T. Alignment of Nematic Liquid Crystals Using Carbon Nanotube Films. *Thin Solid Films* **2006,** *509,* 53−57.
55. Chen, H. Y.; Lee, W.; Clark, N. A. Faster Electro-Optical Response Characteristics of a Carbon-Nanotube-Nematic Suspension. *Applied Physics Letters* **2007,** *90,* 033510.
56. Singer, S. B.; Mecklenburg, M.; White, E. R.; Regan, B. C. Polarized Light Emission from Individual Incandescent Carbon Nanotubes. *Physical Review B* **2011,** *83,* 233404.

57. Li, P.; Jiang, K. L.; Liu, M.; Li, Q. Q.; Fan, S. S.; Sun, J. L. Polarized Incandescent Light Emission from Carbon Nanotubes. *Applied Physics Letters* **2003,** *82,* 1763–1765.
58. Zhang, M.; Fang, S. L.; Zakhidov, A. A.; Lee, S. B.; Aliev, A. E.; Williams, C. D.; Atkinson, K. R.; Baughman, R. H. Strong, Transparent, Multifunctional, Carbon Nanotube Sheets. *Science* **2005,** *309,* 1215–1219.
59. Liu, P.; Liu, L.; Wei, Y.; Liu, K.; Chen, Z.; Jiang, K. L.; Li, Q. Q.; Fan, S. S. Fast High-Temperature Response of Carbon Nanotube Film and Its Application as an Incandescent Display. *Advanced Materials* **2009,** *21,* 3563–3566.
60. Yamamoto, Y. Technological Innovation of Thin-Film Transistors: Technology Development, History, and Future. *Japanese Journal of Applied Physics* **2012,** *51,* 060001.
61. Kagan, C. R., Andry, P., Eds. *Thin-Film Transistors;* Marcel Dekker: New York, 2003.
62. Schindler, A.; Brill, J.; Fruehauf, N.; Novak, J. P.; Yaniv, Z. Solution-Deposited Carbon Nanotube Layers for Flexible Display Applications. *Physica E* **2007,** *37,* 119–123.
63. Hu, L. B.; Hecht, D. S.; Gruner, G. Carbon Nanotube Thin Films: Fabrication, Properties, and Applications. *Chemical Reviews* **2010,** *110,* 5790–5844.
64. Zou, J. P.; Zhang, K.; Li, J. Q.; Zhao, Y. B.; Wang, Y. L.; Pillai, S. K. R.; Demir, H. V.; Sun, X. W.; Chan-Park, M. B.; Zhang, Q. Carbon Nanotube Driver Circuit for 6 × 6 Organic Light Emitting Diode Display. *Scientific Reports-UK* **2015,** *5,* 11755.
65. Fu, W. Q.; Liu, L.; Jiang, K. L.; Li, Q. Q.; Fan, S. S. Super-Aligned Carbon Nanotube Films as Aligning Layers and Transparent Electrodes for Liquid Crystal Displays. *Carbon* **2010,** *48,* 1876–1879.
66. Lee, W. K.; Choi, Y. S.; Kang, Y. G.; Sung, J.; Seo, D. S.; Park, C. Super-Fast Switching of Twisted Nematic Liquid Crystals on 2D Single Wall Carbon Nanotube Networks. *Advanced Functional Materials* **2011,** *21,* 3843–3850.
67. Hecht, D. S.; Hu, L. B.; Irvin, G. Emerging Transparent Electrodes Based on Thin Films of Carbon Nanotubes, Graphene, and Metallic Nanostructures. *Advanced Materials* **2011,** *23,* 1482–1513.
68. Hecht, D. S.; Heintz, A. M.; Lee, R.; Hu, L. B.; Moore, B.; Cucksey, C.; Risser, S. High Conductivity Transparent Carbon Nanotube Films Deposited from Superacid. *Nanotechnology* **2011,** *22,* 075201.
69. Feng, C.; Liu, K.; Wu, J. S.; Liu, L.; Cheng, J. S.; Zhang, Y. Y.; Sun, Y. H.; Li, Q. Q.; Fan, S. S.; Jiang, K. L. Flexible, Stretchable, Transparent Conducting Films Made from Superaligned Carbon Nanotubes. *Advanced Functional Materials* **2010,** *20,* 885–891.
70. De Volder, M. F. L.; Tawfick, S. H.; Baughman, R. H.; Hart, A. J. Carbon Nanotubes: Present and Future Commercial Applications. *Science* **2013,** *339,* 535–539.
71. De, S.; Coleman, J. N. The Effects of Percolation in Nanostructured Transparent Conductors. *MRS Bulletin* **2011,** *36,* 774–781.
72. Tang, C. W.; Vanslyke, S. A. Organic Electroluminescent Diodes. *Applied Physics Letters* **1987,** *51,* 913–915.
73. Lee, S. T.; Gao, Z. Q.; Hung, L. S. Metal Diffusion from Electrodes in Organic Light-Emitting Diodes. *Applied Physics Letters* **1999,** *75,* 1404–1406.
74. Sheats, J. R.; Roitman, D. B. Failure Modes in Polymer-Based Light-Emitting Diodes. *Synthetic Metals* **1998,** *95,* 79–85.
75. Lo, M. F.; Ng, T. W.; Lai, S. L.; Fung, M. K.; Lee, S. T.; Lee, C. S. Stability Enhancement in Organic Photovoltaic Device by Using Polymerized Fluorocarbon Anode Buffer Layer. *Applied Physics Letters* **2011,** *99,* 033302.
76. Lo, M. F.; Ng, T. W.; Mo, H. W.; Lee, C. S. Direct Threat of a UV-Ozone Treated Indium-Tin-Oxide Substrate to the Stabilities of Common Organic Semiconductors. *Advanced Functional Materials* **2013,** *23,* 1718–1723.
77. Aziz, H.; Popovic, Z. D. Degradation Phenomena in Small-Molecule Organic Light-Emitting Devices. *Chemistry of Materials* **2004,** *16,* 4522–4532.
78. Sam, F. L. M.; Dabera, G. D. M. R.; Lai, K. T.; Mills, C. A.; Rozanski, L. J.; Silva, S. R. P. Hybrid Metal Grid-Polymer-Carbon Nanotube Electrodes for High Luminance Organic Light Emitting Diodes. *Nanotechnology* **2014,** *25,* 345202.

79. Li, L.; Liang, J. J.; Chou, S. Y.; Zhu, X. D.; Niu, X. F.; Yu, Z. B.; Pei, Q. B. A Solution Processed Flexible Nanocomposite Electrode with Efficient Light Extraction for Organic Light Emitting Diodes. *Scientific Reports-UK* **2014,** *4,* 4307.
80. Freitag, P.; Zakhidov, A. A.; Luessem, B.; Zakhidov, A. A.; Leo, K. Lambertian White Top-Emitting Organic Light Emitting Device with Carbon Nanotube Cathode. *Journal of Applied Physics* **2012,** *112,* 114505.
81. Wang, G. F.; Tao, X. M.; Wang, R. X. Flexible Organic Light-Emitting Diodes with a Polymeric Nanocomposite Anode. *Nanotechnology* **2008,** *19,* 145201.
82. Gaynor, W.; Hofmann, S.; Christoforo, M. G.; Sachse, C.; Mehra, S.; Salleo, A.; McGehee, M. D.; Gather, M. C.; Lussem, B.; Muller-Meskamp, L.; et al. Color in the Corners: ITO-Free White OLEDs with Angular Color Stability. *Advanced Materials* **2013,** *25,* 4006–4013.
83. Wu, J. B.; Agrawal, M.; Becerril, H. A.; Bao, Z. N.; Liu, Z. F.; Chen, Y. S.; Peumans, P. Organic Light-Emitting Diodes on Solution-Processed Graphene Transparent Electrodes. *ACS Nano* **2010,** *4,* 43–48.
84. Gruner, G. Carbon Nanotube Films for Transparent and Plastic Electronics. *Journal of Materials Chemistry* **2006,** *16,* 3533–3539.
85. Williams, C. D.; Robles, R. O.; Zhang, M.; Li, S.; Baughman, R. H.; Zakhidov, A. A. Multiwalled Carbon Nanotube Sheets as Transparent Electrodes in High Brightness Organic Light-Emitting Diodes. *Applied Physics Letters* **2008,** *93,* 183506.
86. Hu, L. B.; Li, J. F.; Liu, J.; Gruner, G.; Marks, T. Flexible Organic Light-Emitting Diodes with Transparent Carbon Nanotube Electrodes: Problems and Solutions. *Nanotechnology* **2010,** *21.*
87. Chien, Y. M.; Lefevre, F.; Shih, I.; Izquierdo, R. A Solution Processed Top Emission OLED with Transparent Carbon Nanotube Electrodes. *Nanotechnology* **2010,** *21,* 134020.
88. Aguirre, C. M.; Auvray, S.; Pigeon, S.; Izquierdo, R.; Desjardins, P.; Martel, R. Carbon Nanotube Sheets as Electrodes in Organic Light-Emitting Diodes. *Applied Physics Letters* **2006,** *88,* 183104.
89. Kaskela, A.; Nasibulin, A. G.; Timmermans, M. Y.; Aitchison, B.; Papadimitratos, A.; Tian, Y.; Zhu, Z.; Jiang, H.; Brown, D. P.; Zakhidov, A.; et al. Aerosol-Synthesized SWCNT Networks with Tunable Conductivity and Transparency by a Dry Transfer Technique. *Nano Letters* **2010,** *10,* 4349–4355.
90. Li, J.; Hu, L.; Wang, L.; Zhou, Y.; Gruner, G.; Marks, T. J. Organic Light-Emitting Diodes Having Carbon Nanotube Anodes. *Nano Letters* **2006,** *6,* 2472–2477.
91. Ou, E. C. W.; Hu, L. B.; Raymond, G. C. R.; Soo, O. K.; Pan, J. S.; Zheng, Z.; Park, Y.; Hecht, D.; Irvin, G.; Drzaic, P.; et al. Surface-Modified Nanotube Anodes for High Performance Organic Light-Emitting Diode. *ACS Nano* **2009,** *3,* 2258–2264.
92. Woo, H. S.; Czerw, R.; Webster, S.; Carroll, D. L.; Ballato, J.; Strevens, A. E.; O'Brien, D.; Blau, W. J. Hole Blocking in Carbon Nanotube-Polymer Composite Organic Light-Emitting Diodes Based on Poly(m-Phenylene Vinylene-co-2,5-Dioctoxy-p-Phenylene Vinylene). *Applied Physics Letters* **2000,** *77,* 1393–1395.
93. Li, J. F.; Hu, L. B.; Liu, J.; Wang, L.; Marks, T. J.; Gruner, G. Indium Tin Oxide Modified Transparent Nanotube Thin Films as Effective Anodes for Flexible Organic Light-Emitting Diodes. *Applied Physics Letters* **2008,** *93,* 083306.
94. Yu, Z. B.; Hu, L. B.; Liu, Z. T.; Sun, M. L.; Wang, M. L.; Gruner, G.; Pei, Q. B. Fully Bendable Polymer Light Emitting Devices with Carbon Nanotubes as Cathode and Anode. *Applied Physics Letters* **2009,** *95,* 203304.
95. Zhang, D. H.; Ryu, K.; Liu, X. L.; Polikarpov, E.; Ly, J.; Tompson, M. E.; Zhou, C. W. Transparent, Conductive, and Flexible Carbon Nanotube Films and Their Application in Organic Light-Emitting Diodes. *Nano Letters* **2006,** *6,* 1880–1886.
96. Ha, Y. G.; You, E. A.; Kim, B. J.; Choi, J. H. Fabrication and Characterization of OLEDs Using MEH-ppv and SWCNT Nanocomposites. *Synthetic Metals* **2005,** *153,* 205–208.
97. Woo, H. S.; Czerw, R.; Webster, S.; Carroll, D. L.; Park, J. W.; Lee, J. H. Organic Light Emitting Diodes Fabricated with Single Wall Carbon Nanotubes Dispersed in a Hole Conducting Buffer: the Role of Carbon Nanotubes in a Hole Conducting Polymer. *Synthetic Metals* **2001,** *116,* 369–372.

98. Fournet, P.; Coleman, J. N.; Lahr, B.; Drury, A.; Blau, W. J.; O'Brien, D. F.; Horhold, H. H. Enhanced Brightness in Organic Light-Emitting Diodes Using a Carbon Nanotube Composite as an Electron-Transport Layer. *Journal of Applied Physics* **2001,** *90,* 969–975.
99. Curran, S. A.; Ajayan, P. M.; Blau, W. J.; Carroll, D. L.; Coleman, J. N.; Dalton, A. B.; Davey, A. P.; Drury, A.; McCarthy, B.; Maier, S.; et al. A Composite from Poly(m-phenylenevinylene-co-2,5-dioctoxy-p-phenylenevinylene) and Carbon Nanotubes: A Novel Material for Molecular Optoelectronics. *Advanced Materials* **1998,** *10,* 1091–1093.
100. Mills, C. A.; Sam, F. L. M.; Alshammari, A. S.; Rozanski, L. J.; Emerson, N. G.; Silva, S. R. P. Storage Lifetime of Polymer-Carbon Nanotube Inks for Use as Charge Transport Layers in Organic Light Emitting Diodes. *Journal of Display Technology* **2014,** *10,* 125–131.
101. Chun, J. Y.; Han, J. W.; Kim, T. W.; Seo, D. S. Enhancement of Organic Light-Emitting Diodes Efficiency Using Carbon Nanotube Doped Hole-Injection Layer on the Al-Doped ZnO Anode. *ECS Solid State Letters* **2012,** *1,* R13–R15.
102. Wang, G. F.; Tao, X. M.; Wang, R. X. Fabrication and Characterization of OLEDs Using PEDOT: PSS and MWCNT Nanocomposites. *Composites Science and Technology* **2008,** *68,* 2837–2841.
103. Inigo, A. R.; Underwood, J. M.; Silva, S. R. P. Carbon Nanotube Modified Electrodes for Enhanced Brightness in Organic Light Emitting Devices. *Carbon* **2011,** *49,* 4211–4217.
104. Liu, Y. P.; Jung, E.; Wang, Y.; Zheng, Y.; Park, E. J.; Cho, S. M.; Loh, K. P. "Quasi-Freestanding" Graphene-on-Single Walled Carbon Nanotube Electrode for Applications in Organic Light-Emitting Diode. *Small* **2014,** *10,* 944–949.
105. Li, C. S.; Su, S. H.; Chi, H. Y.; Li, Z. B.; Yokoyama, M. Enhancing the Lifetime of Field-Emission Organic Light-Emitting Diodes Using a Cavity Structure. *Thin Solid Films* **2009,** *517,* 5330–5332.
106. Wang, C.; Zhang, J. L.; Ryu, K. M.; Badmaev, A.; De Arco, L. G.; Zhou, C. W. Wafer-Scale Fabrication of Separated Carbon Nanotube Thin-Film Transistors for Display Applications. *Nano Letters* **2009,** *9,* 4285–4291.
107. Zhang, J. L.; Fu, Y.; Wang, C.; Chen, P. C.; Liu, Z. W.; Wei, W.; Wu, C.; Thompson, M. E.; Zhou, C. W. Separated Carbon Nanotube Macroelectronics for Active Matrix Organic Light-Emitting Diode Displays. *Nano Letters* **2011,** *11,* 4852–4858.
108. Kim, S.; Kim, S.; Park, J.; Ju, S.; Mohammadi, S. Fully Transparent Pixel Circuits Driven by Random Network Carbon Nanotube Transistor Circuitry. *ACS Nano* **2010,** *4,* 2994–2998.
109. Wang, C.; Hwang, D.; Yu, Z. B.; Takei, K.; Park, J.; Chen, T.; Ma, B. W.; Javey, A. User-Interactive Electronic Skin for Instantaneous Pressure Visualization. *Nature Materials* **2013,** *12,* 899–904.
110. Sekitani, T.; Nakajima, H.; Maeda, H.; Fukushima, T.; Aida, T.; Hata, K.; Someya, T. Stretchable Active-Matrix Organic Light-Emitting Diode Display Using Printable Elastic Conductors. *Nature Materials* **2009,** *8,* 494–499.
111. McCarthy, M. A.; Liu, B.; Rinzler, A. G. High Current, Low Voltage Carbon Nanotube Enabled Vertical Organic Field Effect Transistors. *Nano Letters* **2010,** *10,* 3467–3472.
112. McCarthy, M. A.; Liu, B.; Donoghue, E. P.; Kravchenko, I.; Kim, D. Y.; So, F.; Rinzler, A. G. Low-Voltage, Low-Power, Organic Light-Emitting Transistors for Active Matrix Displays. *Science* **2011,** *332,* 570–573.
113. Xiao, L.; Chen, Z.; Feng, C.; Liu, L.; Bai, Z. Q.; Wang, Y.; Qian, L.; Zhang, Y. Y.; Li, Q. Q.; Jiang, K. L.; et al. Flexible, Stretchable, Transparent Carbon Nanotube Thin Film Loudspeakers. *Nano Letters* **2008,** *8,* 4539–4545.
114. Wei, Y.; Lin, X. Y.; Jiang, K. L.; Liu, P.; Li, Q. Q.; Fan, S. S. Thermoacoustic Chips with Carbon Nanotube Thin Yarn Arrays. *Nano Letters* **2013,** *13,* 4795–4801.
115. Hochberg, F.; Seitz, H. K.; Brown, A. V. A Thin-Film Integrated Incandescent Display. *IEEE Transactions on Electron Devices* **1973,** *20,* 1002–1005.
116. Wei, Y.; Liu, P.; Jiang, K. L.; Fan, S. S. A Display Module Implemented by the Fast High-Temperature Response of Carbon Nanotube Thin Yarns. *Nano Letters* **2012,** *12,* 2548–2553.
117. Liu, P.; Zhou, D. L.; Wei, Y.; Jiang, K. L.; Wang, J. P.; Zhang, L. N.; Li, Q. Q.; Fan, S. S. Load Characteristics of a Suspended Carbon Nanotube Film Heater and the Fabrication of a Fast-Response Thermochromic Display Prototype. *ACS Nano* **2015,** *9,* 3753–3759.

118. Liu, P.; Liu, L. A.; Jiang, K. L.; Fan, S. S. Carbon-Nanotube-Film Microheater on a Polyethylene Terephthalate Substrate and Its Application in Thermochromic Displays. *Small* **2011,** *7,* 732–736.
119. Peng, H. S.; Sun, X. M.; Cai, F. J.; Chen, X. L.; Zhu, Y. C.; Liao, G. P.; Chen, D. Y.; Li, Q. W.; Lu, Y. F.; Zhu, Y. T.; et al. Electrochromatic Carbon Nanotube/Polydiacetylene Nanocomposite Fibres. *Nature Nanotechnology* **2009,** *4,* 738–741.
120. Zhang, Z. T.; Guo, K. P.; Li, Y. M.; Li, X. Y.; Guan, G. Z.; Li, H. P.; Luo, Y. F.; Zhao, F. Y.; Zhang, Q.; Wei, B.; et al. A Colour-Tunable, Weavable Fibre-Shaped Polymer Light-Emitting Electrochemical Cell. *Nature Photonics* **2015,** *9,* 233–238.

CHAPTER 5

Carbon Nanotubes for Sensing Applications

M. Zhou[1], Z. Wang[2], X. Wang[2]
[1]Northeast Normal University, Changchun, China; [2]Nanyang Technological University, Singapore, Singapore

Contents

5.1 INTRODUCTION

Carbon nanotubes (CNTs) have attracted lots of attention since their discovery by Iijima *(1−3)*. Because of their interesting and unique properties, more than 135,000 scientific reports have been published on the topic of CNTs from 1991 to 2015. Among them, over 12,000 papers are specifically focused on the topic of CNT-based sensors *(4−12)*.

CNTs exhibit excellent mechanical properties because of their σ-bonds and tubular structure. Analytical calculation indicated that the Young's modulus of single-walled carbon nanotubes (SWCNTs) is at least 1 TPa *(14)*, which has been confirmed by experiments *(15)*. CNTs have also shown high tensile strength of up to 100 GPa without any form of plastic deformation, brittle-like behavior, or rupture, which is caused by the extra energy absorption required for the hollow structures of CNTs *(7)*. Because of the high electrochemically accessible surface area of porous nanotube arrays, combined with their high electronic conductivity and useful mechanical properties, CNTs are recognized as a kind of attractive electrode material for device construction that explored electrochemical double-layer charge injection *(4)*. In addition, the easy protein immobilization on CNTs with the retention of protein activity nominates CNTs for electrochemical sensing and biosensing applications *(4)*. The relationship between the electronic property and the structure of the CNTs implies that mechanical deformation can alter the band structure

Industrial Applications of Carbon Nanotubes
ISBN 978-0-323-41481-4

of CNTs *(13,16)*. Such an effect was introduced into the CNT field for strain sensor construction *(16)*.

The objective of this chapter is to highlight recent developments and advances in the field of sensors that have been developed by using CNTs as the active and key sensing element. We will discuss and focus on different types of CNT-based sensors including biosensors, gas sensors, strain and pressure sensors, flow sensors, and mass and pH sensors. The outlook and recommendations of future directions of CNT-based sensors will be proposed at the end.

5.2 BIOSENSORS

A biological sensor (i.e., biosensor) is an analytical device used for the detection of an analyte, which combines a biological component with a physicochemical detector *(17,18)*. The concept of the biosensor was first proposed in 1962 by Clark and Lyons from the Children's Hospital of Cincinnati for glucose level measurement *(19)*. This glucose biosensor was composed of an oxygen electrode, an inner oxygen semipermeable membrane, a thin layer of glucose oxidase (GOx), and an outer dialysis membrane. The enzyme of GOx was immobilized at an electrochemical detector to form an enzyme electrode. A decrease in the measured oxygen concentration was proportional to the glucose concentration. The biosensors differ from the classical chemical sensors in the following two ways: (1) the sensing element in biosensors consists of biological materials such as proteins, oligo- or polynucleotides, microorganisms, or even whole biological tissues, and (2) the biosensors applied to monitor the biological processes or the recognition of biomolecules *(20)*. Usually, two types of biosensors are applied for real applications. One is the in vitro biosensor, in which the solution with samples (such as blood serum, urine, and milk products) is dropped onto the biosensor and the output signals give the analyte-related information. The other type is called the in vivo biosensor, which mainly aims to in situ measure the rate of uptake or efflux of relevant species or to estimate the spatial distribution of the concentration of analyte in living organisms *(20)*. CNTs are a kind of promising material for biosensor construction because of several intriguing properties *(4)*. For example, the CNTs exhibit excellent solubility and biocompatibility, which make them attractive for the fabrication of a biological material—CNT composite with high biological activities *(20)*. In addition, the fast electron-transfer kinetics for a wide range of electroactive species is observed on the CNT surface *(20)*. Furthermore, CNTs can be easily functionalized, suggesting that it could be relatively easy to attach almost any desired chemical species to the CNT surface *(21)*. Therefore sp^2 hybridization, small size, and outstanding electronic properties of CNTs coupled with specific recognition properties of immobilized biosystems make CNTs extremely attractive for a wide range of biosensor construction *(4,22)*. The integration of CNTs with biomolecules could result in hybrid systems. In hybrid systems, the CNTs are used as nanoscale electrode elements *(23)*, electronic elements, and platforms upon which biomolecules can be attached. At present, different biomolecules or biomolecule-related molecules, including glucose *(23–39)*, phenolic compounds *(40–46)*, organophosphorus compounds *(47–51)*, cholesterol *(52–55)*, proteins *(21,56–58)*, NADH *(32,59–61)*, nitric oxide *(62–67)*, and DNA *(23,68–76)*, have been detected by employing CNTs. Here, CNT-based glucose and DNA sensors will be introduced in detail as two examples of CNT-based biosensors.

It is well known that biosensors for glucose play a leading role in the monitoring of blood glucose levels, which is important for the diagnosis and management of diabetes mellitus *(24,26–29,32,39)*. The dramatic decrease in the overpotential of hydrogen peroxidase as well as the direct electron transfer of GOx observed at CNT-based electrodes pave the way for the construction of glucose biosensors. Wang's group found that the CNTs/Nafion/GOx-coated electrode was useful to detect hydrogen peroxide *(38)*. Accordingly, the resulting CNTs/Nafion/GOx biosensor offered a highly selective low-potential (−0.05 V vs. Ag/AgCl) biosensing of glucose, which greatly benefits from the coupling of the efficient electrocatalytic action of CNTs toward hydrogen peroxide. In addition to the random distribution of CNTs, the Ni-coordinated, vertically aligned CNTs on a flexible graphite substrate were used as a biosensor to detect glucose *(39)*. The electrode exhibited enhanced electrocatalytic activity toward the oxidation of glucose in alkaline solution. The resulting biosensor exhibited a sensitivity of 950.6 $\mu A/mM\ cm^2$ to glucose measurement with a linear range of 0.05–1.0 mM and a detection limit of 30 μM. This type of biosensor will offer exciting opportunities for wearable, real-time, and/or on-site applications in fields ranging from clinical analysis to industry.

It is known that the redox center of GOx is electrically insulated by a protein shell. Because of this glycoprotein shell, the enzyme cannot be oxidized or reduced at an electrode at any potential. Therefore the direct electron transfer between CNTs and GOx is vital for the construction of amperometric glucose biosensors. Such electrical communication with GOx would obviate the need for a cosubstrate and allow efficient transduction of the biorecognition event *(4)*. Willner's group demonstrated the aligned reconstitution of a redox flavoenzyme (GOx) on the edge of CNTs that are linked to an electrode surface *(25)*. The SWCNTs acts as a nanoconnector that electrically contacts the active site of the enzyme and the electrode. The electrons are transported along distances greater than 150 nm and the rate of electron transport is controlled by the length of the SWCNTs. It has been estimated that the interfacial electron-transfer rate constants would be 83, 42, 19, and 12 s^{-1} for assemblies that include standing SWCNTs of 25, 50, 100, and 150 nm average length, respectively. Such enzyme reconstitution on the end of CNTs represents an extremely efficient approach for "plugging" an electrode into GOx *(4)*.

Besides the glucose biosensor, the CNT-based DNA biosensor is another important application. The DNA biosensor is based on the nucleic acid recognition processes *(72)*. It is crucial to develop effective DNA biosensing systems for the diagnosis and treatment of genetic diseases, pathogen identification, drug discovery, identification in forensic cases, or antibioterrorism efforts. Currently, it is rapidly being developed toward the goal of a rapid, simple, and inexpensive test of genetic and infectious diseases. The DNA biosensor can greatly benefit from the use of CNTs. For example, surface-confined multiwalled carbon nanotubes (MWCNTs) facilitate the adsorptive accumulation of the guanine nucleon bases and accordingly greatly enhance their oxidation signal *(77)*. The advantages of the CNT/glassy carbon (GC) electrode were illustrated from its comparison to the common unmodified GC, carbon paste, and graphite pencil electrodes. The dramatic amplification of the guanine signal was combined with a label-free electrical detection of DNA hybridization. Subsequently, several methods have been adopted to capture the DNA probes onto the transducer surface such as chemical functionalization either by covalent binding *(23,78)* or by noncovalent binding *(79)* (physical adsorption, electrostatic attraction, $\pi-\pi$ stacking interactions). Baker et al. *(78)* developed a multistep route to the formation of covalently linked adducts of SWCNTs and DNA. The resulting DNA-SWCNT adducts hybridize selectively with complementary sequence, with only minimal interaction with noncomplementary sequences, which strongly leads to the successful synthesis of covalently linked DNA-SWCNT adducts. In another work, a nanoelectrode array based on vertically aligned MWCNTs embedded in SiO_2 has been successfully fabricated *(23)* (Fig. 5.1). These aligned MWCNTs provide a new bottom-up scheme for fabricating reliable nanoelectrode arrays. Such electrode platforms are then combined with $Ru(bpy)_3^{2+}$-mediated guanine oxidation. Such MWCNTs were applied in the label-free detection of DNA polymerase chain reaction amplification. In another work, by combining the $Ru(bpy)_3^{2+}$-mediated guanine oxidation and the aligned MWCNTs (Fig. 5.2), the detection limit of the aligned MWCNT-based DNA biosensor is lower than a few attomoles of oligonucleotide targets.

Unfortunately, the covalent functionalization of CNTs may disrupt the unique one-dimensional electronic structure and desired optical properties of CNTs *(80)*. Therefore it would be efficient to both preserve CNTs' electronic structure (as no bonds are broken) and provide sites for selective binding by modifying the CNTs with electroactive species by the noncovalent method, although it is difficult to control *(80)*. For example, Barone et al. reported the synthesis and successful testing of solution-phase, near-infrared sensors, with β-D-glucose sensing as a model system. Here, SWCNTs have been used to modulate their emission in response to the adsorption of specific biomolecules *(80)*. Therefore new types of noncovalent functionalization using electron-withdrawing molecules can provide sites for transferring electrons in and out of the nanotube.

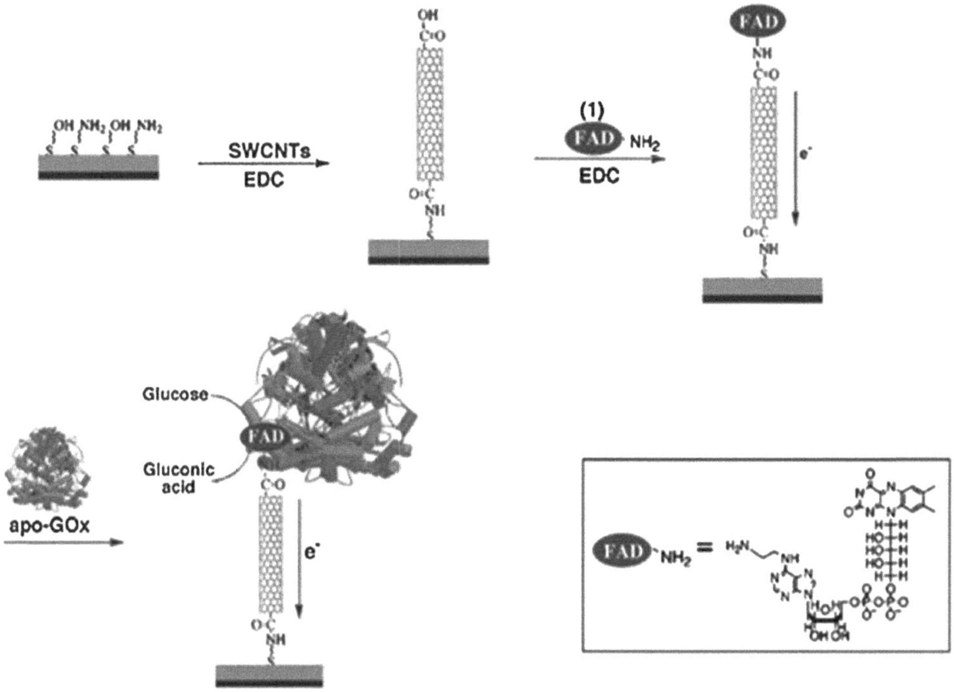

Figure 5.1 Assembly of the single-walled carbon nanotube (SWCNT) electrically contacted glucose oxidase electrode. *EDC*, N-(3-Dimethylaminopropyl)-N′-ethylcarbodiimide; *FAD*, flavin adenine dinucleotide. *(Adapted with permission from Patolsky, F.; Weizmann, Y.; Willner, I. Long-Range Electrical Contacting of Redox Enzymes by SWCNT Connectors. Angewandte Chemie International Edition* ***2004****, 43, 2113–2117. Copyright 2004 Wiley.)*

5.3 GAS SENSORS

The development of gas sensors is critical to environmental monitoring, control of chemical processes, space missions, agricultural, medical applications, etc. *(81)*. It has been reported that the adsorption of gaseous molecules either donates or withdraws electrons from CNTs, resulting in the change of CNTs' electrical properties. Especially, CNTs change their electrical resistance drastically when they are exposed to alkalis, halogens, or other gases like O_2, NO_2, or NH_3 at room temperature *(73,82–91)*, which raises the hopes for developing gas sensors with excellent performance. Subsequently, efforts have been made to exploit the development of new gas sensors based on CNTs *(81,89,91)*.

In 2000, Dai's group developed an interesting chemical sensor based on individual SWCNTs. They found that the electrical resistance of a semiconducting SWCNT dramatically increased or decreased upon exposure to gaseous molecules such as NO_2 or NH_3 *(81)*. After the SWCNT sample was exposed to NH_3, the $I-V_g$ curve exhibits

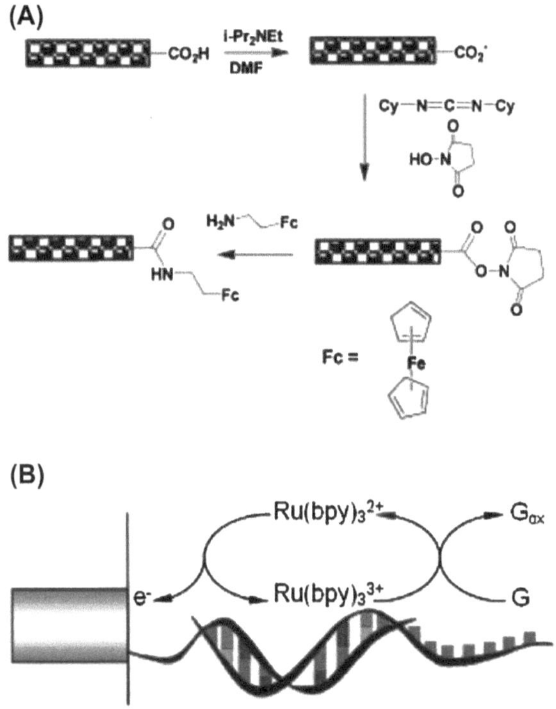

Figure 5.2 (A) The functionalization process of the amine-terminated ferrocene derivative to carbon nanotubes (CNTs) ends by carbodiimide chemistry and (B) the schematic mechanism of $Ru(bpy)_3^{2+}$-mediated guanine oxidation. *DMF*, N,N-dimethylformamide. *(Adapted with permission from Li, J.; Ng, H. T.; Cassell, A.; Fan, W.; Chen, H.; Ye, Q.; Koehne, J.; Han, J.; Meyyappan, M. Carbon Nanotube Nanoelectrode Array for Ultrasensitive DNA Detection. Nano Letters **2003**, 3, 597–602. Copyright 2003 American Society.)*

a shift of −4 V (Fig. 5.3, left curve). In contrast, the $I-V_g$ curve was shifted by +4 V after NO_2 exposure (Fig. 5.3, right curve). Exposure to NH_3 effectively shifts the valence band of the nanotube away from the Fermi level, which results in hole depletion and reduced conductance. However, for the NO_2 case, exposure of the initially depleted sample to NO_2 resulted in the nanotube Fermi level shifting closer to the valence band, which caused enriched hole carriers in the nanotube and enhanced sample conductance. These results indicate that the Fermi level of SWCNTs can be shifted by the molecular gating effects and thus its resistance can be modulated by orders of magnitude. The response time of the SWCNT sensor to 200 ppm NO_2 was a few seconds, and the sensitivity (defined as the ratio between the device resistance after gas exposure and the one before gas exposure) was ~100−1000. Sensor reversibility can be achieved by slow recovery (~12 h) at room temperature or by relatively fast recovery (~1 h) at high temperature. The response time to ~1% NH_3 was a few minutes with sensitivity between 10 and 100.

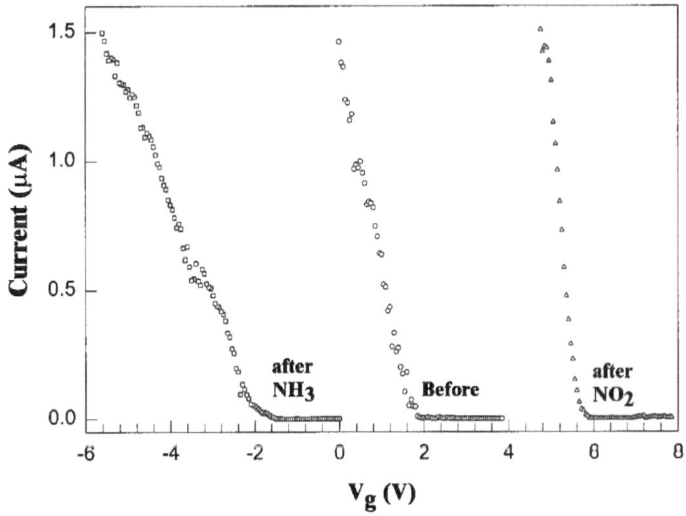

Figure 5.3 Chemical gating effects to the semiconducting single-walled carbon nanotubes (SWCNTs). Current versus gate voltage curves before NO_2 (*circles*), after NO_2 (*triangles*), and after NH_3 (*squares*) exposures. The measurements with NH_3 and NO_2 were carried out successively after sample recovery. *(Adapted with permission from Kong, J.; Franklin, N. R.; Zhou, C.; Chapline, M. G.; Peng, S.; Cho, K.; Dai, H. Nanotube Molecular Wires as Chemical Sensors. Science* ***2000****, 287, 622–625. Copyright 2000 Science.)*

The effects of NO_2 and NH_3 exposure have been explained. For NO_2, it was explained by the removal of electronic density from the SWCNTs at a rate of ~0.1 electrons per adsorbed NO_2 molecule. However, the device response to NH_3 is still unclear. The authors suggested that NH_3 may have indirectly influenced the SWCNTs electronic structure through the interaction with hydroxyl groups on the SiO_2 substrate, which leads to partial neutralization of negative charges.

This amazing report has attracted a lot of interesting as soon as it was published. Then, a floodgate of interest in the development of CNT-based gas sensors was generated by this initial report of CNT gas sensing *(33,82–91)*. The mechanism for the interaction between NO_2 (or NH_3) with CNTs has been conducted by theoretical investigations *(92–99)*. However, the theoretical and experimental reports are not consistent. Despite the controversy over physisorption versus chemisorption and charge transfer versus Schottky barrier modification, great strides have been made in the sensitivity and selectivity of CNT-based sensors for these species. Several groups focused on improving the response of sensors composed of CNT networks without any chemical functionalization *(100–103)*. In 2003, Li et al. reported the detection of NO_2 on a bare SWCNT electrode *(103)*. In this platform, SWCNTs form a network or mesh on interdigitated electrodes because of a solution casting process, providing a large enough density of nanotubes on the sensor surface *(103)*. The interdigitated electrode configuration enables effective electric contact between SWCNTs and the electrodes over large areas while providing good accessibility for gas/vapor adsorption to all SWCNTs including semiconducting tubes. The sensor responses are linear for concentrations of sub ppm to hundreds of ppm with a detection limit of 44 ppb for NO_2 or 262 ppb for nitrotoluene. The extended detection capability from gas to organic vapors is attributed to the direct charge transfer on individual semiconducting SWCNT conductivity with additional electron-hopping effects on intertube conductivity through physically adsorbed molecules on SWCNTs.

Besides the bare CNTs, organic polymers *(104–107)*, catalytic metal *(91,108–111)*, and metal-oxide nanoparticles *(112–116)* have also been used to functionalize CNTs and applied in gas sensors. These materials serve as sensing elements, exhibiting a broad range of electronic, chemical, physical, and biological recognition properties that are often highly sensitive to changes in their chemical environments. However, CNTs are the support that physically responds to stimuli from the molecular world around these functionalizations *(117)*. For example, theoretical simulations *(108,118,119)* and experimental investigations *(91,109–111)* both indicate that metallic clusters strongly interact with CNTs, and such interaction produces significant charge reorganization. Therefore the CNT–metal cluster assemblies can be used to detect the molecules that have a weak or repulsive interaction with CNTs. Moreover, the clusters can be mechanically and chemically robust and stable. Therefore these metal–CNT sensors will be able to operate at higher temperatures or under other harsher environments *(109)*. Since different metallic particle-functionalized CNTs make the composites sensitive to several gases at room temperature, multiple sensor elements can be prepared by the decoration of CNTs with different metallic particles. Star et al. reported the decoration of SWCNTs with 18 metal nanoparticles by chemical vapor deposition *(109)*. The resulting composite-based sensor can detect H_2, CH_4, CO, or H_2S gases. The chip based on metal–SWCNTs demonstrates comparative selectivity to traditional metal evaporation techniques but exhibits a much smaller spatial footprint. The employment of metal/metal oxide nanoparticles as gas recognition layers makes the CNT sensor arrays robust and suitable to real-life applications. Applications for these small-size and low-power nanoelectronic sensor arrays are in detection and identification of toxic/explosive gases for personal safety and air pollution monitoring.

5.4 STRAIN AND PRESSURE SENSORS

Experimental investigations indicate that the mechanical deformation can change the electrical conductance of metallic and semiconducting CNTs *(120–124)*, which provides the foundation for the application of CNTs as high-sensitivity electromechanical pressure sensors. In 2004, Sazonva et al. designed a tunable CNT electromechanical oscillator *(125)*. In this report, the lateral vibration frequency can be adjusted by introducing electrical force, which causes extra tension strain inside the CNTs. In addition, Dereli et al. found that the vibration frequency of CNTs in the radial direction is sensitive to the strain rate, whereas vibration frequency decreases with increasing strain rate *(126)*. Bao et al. reported that the transparent, conducting, spray-deposited films of SWCNTs can be rendered stretchable by applying strain along each axis, and then releasing this strain *(16)*. These CNT films were produced by spray coating (nanotube length = 2–3 μm) directly onto a substrate of poly(dimethylsiloxane) (PDMS, activated by exposure to ultraviolet/O_3) from a solution in *N*-methylpyrrolidone. Fig. 5.4A presents the evolution of the change in resistance ($\Delta R/R_0$) as a function of strain for seven stages of applied strain and relaxation. Fig. 5.4B shows four cycles of strain from 0% to 50%, in which the resistance increases reversibly by $\leq$10% because the film was previously strained to a maximum of 50%. Measurement of resistance (R) over the course of 12,500 cycles of

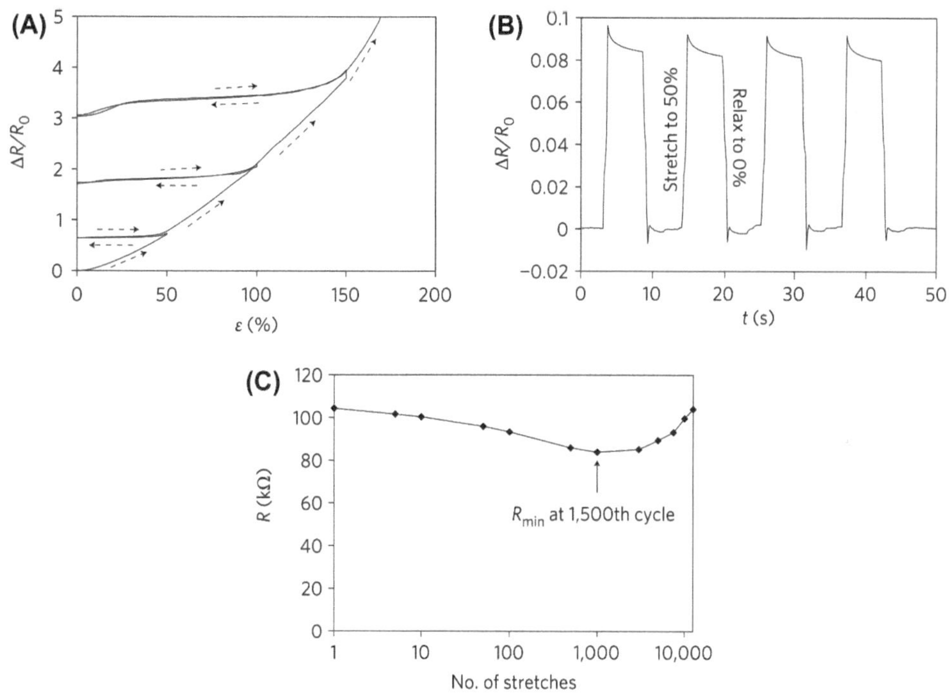

Figure 5.4 Effect of applied strain on films of spray-coated carbon nanotubes (CNTs) on poly(dimethylsiloxane) (PDMS) substrates. (A) Change in resistance $\Delta R/R_0$ versus strain ε for a nanotube film on a PDMS substrate. When the film is strained (*arrow*, bottom left), $\Delta R/R_0$ increases, and remains constant as the strain is released. When the strain is increased again, $\Delta R/R_0$ remains constant, and then increases when ε exceeds the value at which the strain was released before. This sequence is repeated up to $\Delta R/R_0 \approx 5$ and $\varepsilon \approx 150\%$. (B) $\Delta R/R_0$ versus time in response to four cycles of stretching from 0% to 50%. (C) Resistance versus number of stretches (on a log scale) over 12,500 cycles of stretching to 25%. *(Adapted with permission from Lipomi, D. J.; Vosgueritchian, M.; Tee, B. C.-K.; Hellstrom, S. L.; Lee, J. A.; Fox, C. H.; Bao, Z. Skin-Like Pressure and Strain Sensors Based on Transparent Elastic Films of Carbon Nanotubes. Nature Nanotechnology* ***2011****, 6, 788–792. Copyright 2011 Nature Publishing Group.)*

stretching produced the plot shown in Fig. 5.4C. Over the course of the experiment, the resistance decreased by 22% after the 1500th cycle, and then increased linearly. It was observed that the same minimum in resistance at ~1000 cycles of stretching in three similar experiments. This can be attributed to a period in which the nanotube bundles adopted their optimum morphology. Subsequent cycles of stretching possibly decreased the number of conductive junctions between bundles. Based on the experimental results, the nanotube films were used as electrodes in arrays of transparent, stretchable capacitors, which behave as pressure and strain sensors *(127–129)*.

Besides, Cao's group used conventional twist-spinning process to get a highly twisted yarn-derived double-helix structure *(127)*. Pulling the yarn ends completely resolved the entanglement in an elastic and reversible way, yielding large tensile strains with linear change in electrical resistance *(128)*, indicating that these entanglements can serve as elastic strain sensors over a wide strain range. Moreover, Zhou's group fabricated an active fiber-based strain sensor by coiling a fiber-based generator around a stretchable silicone fiber *(129)*. This strain sensor showed excellent sensitivity and stability. Up to 25% strain could be detected and it could be demonstrated to detect finger motion states.

5.5 FLOW SENSORS

Flow sensors are devices used for measuring the flow rate or quantity of a moving liquid or gas *(22)*. New materials have been introduced into flow sensors to improve their performance. Among them, CNTs were introduced into flow sensors by Ghosh et al. in *(130)*. It was found that current/voltage in a bundle of SWCNTs will be generated along the direction of the flow when SWCNTs were used in the flow sensor. The authors suggested that the dominant mechanism responsible for this highly nonlinear response involves a direct forcing of the free charge carriers in the nanotubes by the fluctuating coulombic field of the liquid flowing past the nanotubes. However, pristine SWCNTs contain carbonaceous and metallic impurities that influence their electrochemical performance. Later, Escarpa et al. used the high near-infrared purity index SWCNTs to design microfluidic chips *(131)*. In this work, a set of 0.1 mg/mL SWCNT dispersions with different degrees of purity and different dispersants was carefully characterized by near-infrared spectroscopy (Fig. 5.5). These dispersions were utilized to modify microfluidic chip electrodes for electrochemical measurement of dopamine and catechol. In comparison with a non-SWCNT-based electrode, the sample with the highest purity exhibited the best analytical performance in terms of improved sensitivity, very good signal-to-noise ratio, high resistance to fouling, and enhanced resolution. Such work is the first demonstration of the sensitivity–purity relationship in SWCNT microfluidic chips.

Flow sensors based on SWCNT thin films for liquids were also reported *(132)*. They proved that the flow-induced current on the surface of CNT thin films depended closely on the angular velocity, concentration, properties, and temperature of the liquids.

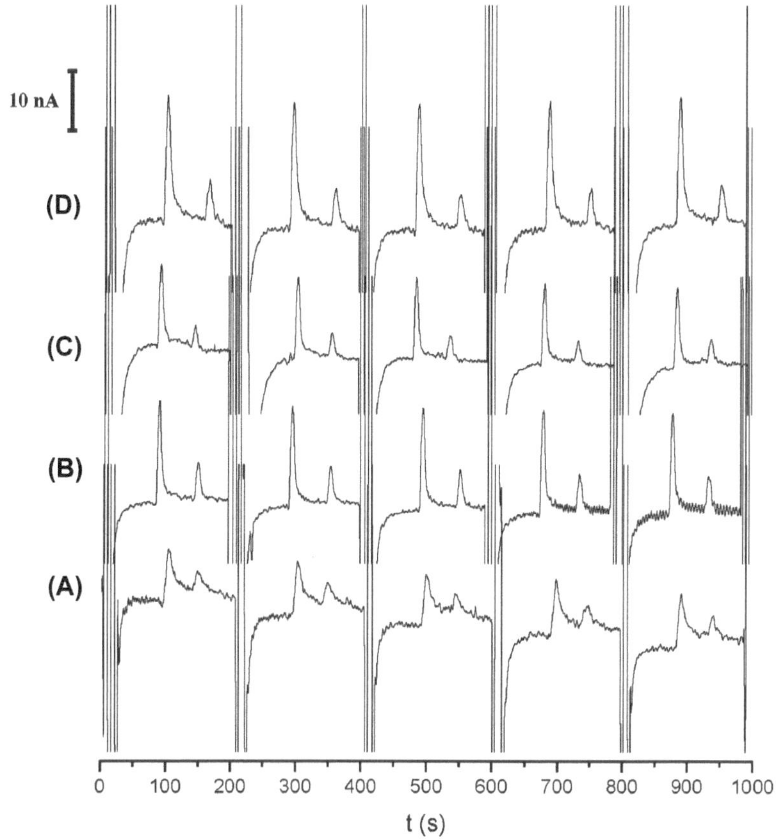

Figure 5.5 Microchip electropherograms corresponding to a mixture of dopamine and catechol (100 mM each) with electrode materials studied. (A) Carbon screen-printed electrodes, (B) single-walled carbon nanotubes (SWCNTs) as grown, (C) centrifuged SWCNTs, and (D) oxidized and centrifuged SWCNTs. Conditions: MES buffer 25 mM pH = 6.5, separation voltage 1.5 kV, injection voltage 1.5 kV for 5 s, detection voltage +0.7 V. *(Adapted with permission from Vilela, D.; Anson-Casaos, A.; Martinez, M. T.; Gonzalez, M. C.; Escarpa, A. High NIR-Purity Index Single-Walled Carbon Nanotubes for Electrochemical Sensing in Microfluidic Chips. Lab on a Chip* ***2012****, 12, 2006–2014. Copyright 2012 Royal Society of Chemistry.)*

Unlike normal flow sensors, a novel SWCNT press-transfer electrode was fabricated for microfluidic sensing *(133)*. In this work, the SWCNTs are press-transferred on poly(methyl methacrylate) substrate and it was easily coupled to microfluidic chips to act as the exclusive transducer in electrochemical sensing, which resulted in an electrode with high stability. The obtained sensor showed good detection results, excellent repeatability, interelectrode reproducibility, and an extreme resistance to fouling. Importantly, this type of sensor can be afforded in any lab since it does not require clean-room facilities. Similarly, a novel microfluidic pH-sensing chip was developed based on pH-sensitive SWCNTs *(134)*. In this study, the SWCNT thin film acted as both an electrode and a pH-sensitive membrane. The potentiometric pH response was observed by electronic structure changes in these semiconducting SWCNTs in response to the pH level. In a microfluidic chip consisting of a SWCNT pH-sensing working electrode and an Ag/AgCl reference electrode, the calibration plot exhibited promising pH-sensing performance with an ideal Nernstian response of 59.71 mV/pH between pH 3 and 11 (standard deviation of the sensitivity is 1.5 mV/pH, $R^2 = 0.985$). Moreover, the SWCNTs electrode in the microfluidic device showed no significant variation at any pH value in the range of the flow rate between 0.1 and 15 mL/min. The selectivity coefficients of the SWCNT electrode revealed good selectivity against common interfering ions. In addition, MWCNTs have also been used as microflow sensors on polymer substrates and polymer-based devices *(135)*, which might be an alternative sensor for silicon-based flow sensors with the requirements of biocompatibility and low-cost applications.

In addition, Dai's group found that flow-induced voltage can be significantly enhanced by aligning the nanotubes along the flow direction, increasing the flow velocity and/or the ionic strength of the flowing liquid *(136)*. Seung et al. investigated the flow-induced voltage generation of SWCNTs, comparing metallic and semiconducting types, flow velocity, and different ionic concentration solutions *(137)*. They found that the voltage generated for semiconducting nanotubes was three times greater than that for metallic nanotubes and that both types of SWCNTs showed an unexpected reversal in signal sign, likely caused by the switching of the major carrier between holes and electrons. These generated voltages increased proportionally for both types of SWCNTs as functions of flow velocity and ionic concentration.

5.6 MASS SENSORS

Micromechanical resonators such as microcantilevers are based on the detection principle that the resonant frequency of the cantilever depends on the inverse square root of the cantilever mass *(22)*. Therefore the operation of a nanomechanical mass sensor relies on monitoring how the resonance frequency of a nanomechanical resonator changes when additional mass is adsorbed onto its surface. The small size and extraordinary mechanical properties of CNTs make them promising candidates for replacing cantilever structures in a mass sensor *(22)*.

The idea of using individual CNTs as high-sensitivity nanobalances was first proposed by Poncharal et al. *(138)*. In this work, they successfully measured masses in the pictogram to femtogram mass range. Marc Bockrath's group realized atomic scale mass sensing using doubly clamped suspended CNT nanomechanical resonators *(139)*. They used the detection of shifts in the resonance frequency of the nanotubes to sense and determine the inertial mass of atoms as well as the mass of the nanotube. Bachtold et al. developed a

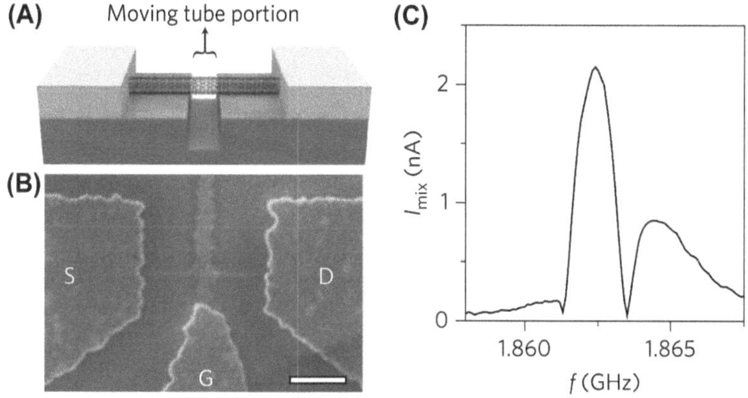

Figure 5.6 Device and characterization. Schematic (A) and scanning electron microscope image (B) of the device (diameter of nanotube, $d = 1.7$ nm; length of suspended nanotube section, $L_{tube} \approx 150$ nm, measured with an atomic-force microscopy). Scale bar, 300 nm. *S*, source; *D*, drain; *G*, gate. (C) Mechanical resonance obtained by measuring the mixing current I_{mix} as a function of driving frequency *f* using the frequency modulation (FM) mixing technique at 6K (amplitude of applied FM voltage, 4 mV; integration time of the lock-in amplifier used to measure I_{mix}, 30 ms). The separation in frequency between two data points is 100 kHz. The curve has not been smoothed. The resonance is asymmetric (the lobe on the right side of the central peak is larger than the lobe on the left side); this asymmetry is attributed to the Duffing force, because it is less pronounced at lower driving forces. Hysteresis jumps are not observed even at higher driving forces. This resonance probably corresponds to the fundamental eigenmode, as it is the resonance with the lowest frequency and the highest I_{mix} (we detected a second resonance at 3.4 GHz). *(Adapted with permission from Chaste, J.; Eichler, A.; Moser, J.; Ceballos, G.; Rurali, R.; Bachtold, A. A Nanomechanical Mass Sensor with Yoctogram Resolution. Nature Nanotechnology* ***2012****, 7, 301–304. Copyright 2012 Nature Publishing Group.)*

shrinking mechanical resonator to submicrometer dimensions (~100 nm), which tremendously improved capabilities in sensing applications *(140)*. They used a 1 nm diameter CNT as a mechanical resonator for mass sensing. The mass responsivity was measured to be 11 Hz/yg and mass resolution is 25 zg at room temperature ($1\ yg = 10^{-24}$ g and $1\ zg = 10^{-21}$ g). Later, they improved the mass sensing system and achieved a resolution of 1.7 yg *(141)* (Fig. 5.6). They combined all the ingredients known to favor the detection of small masses, such as the use of a short nanotube, low-noise motion detection, and ultrahigh vacuum. The obtained resonator was a CNT with a length of ~150 nm that vibrated at a frequency of almost 2 GHz. This unprecedented level of sensitivity realizes the adsorption detection events of naphthalene molecules and the measurement of the binding energy of a xenon atom on the nanotube surface. In addition, Zhu et al. proposed a scheme of an optical sensor to weigh the masses of nanoparticles via a doubly clamped suspended CNT resonator *(142)*. By measuring the resonance frequency shift of the nanotube in the probe absorption spectrum, the masses of external particles landing onto the surface of the nanotube can be easily determined.

5.7 pH Sensors

pH sensing is important in the fields of real-time biomedical, clinical, in situ environmental monitoring and industrial process control *(134)*. Solid-state pH electrodes have attracted much attention because of their advantages over conventional glass pH electrodes including robustness, cost-effective fabrication, flexibility, no need for maintenance, and compatibility with microfluidic systems known as lab-on-chip and micrototal analytical systems *(134)*. Normally, solid-state pH electrodes employ metal/metal oxides or polymer membranes as sensing layers *(143)*. However, the brittleness of metal oxide and the complicated process restrict the application of metal oxide-based pH electrodes. Moreover, the conventional glass membrane or ion-selective field-effect transistor pH sensing technologies both require a reference electrode, which may suffer from leakage of electrolytes and potential contamination *(144)*. The low-cost, rugged properties of CNT-based sensors and the excellent electronic properties of SWCNTs offer great potential for the development of stable, miniaturized, and

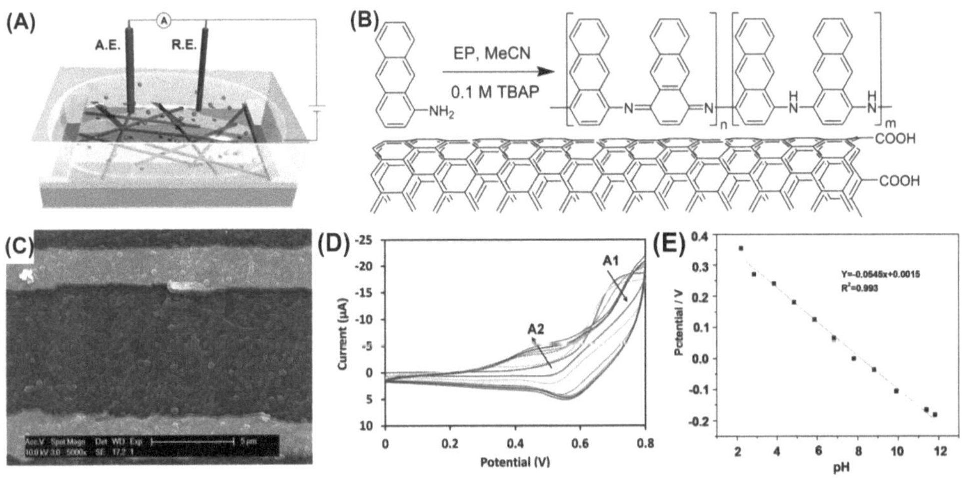

Figure 5.7 Poly(1-aminoanthracene) (PAA)/oxidized single-walled carbon nanotube (ox-SWCNT) device fabrication and characterization. (A) Generalized schematic of electropolymerization (EP) of PAA onto ox-SWCNTs using the nanotube network as the working electrode with an Ag/AgCl quasireference electrode (R.E.) and a Pt wire auxiliary electrode (A.E.) (B) Synthesis of PAA from 1-aminoanthracene (AA) by electropolymerization on ox-SWCNTs. *MeCN*, acetonitrile; *TBAP*, tetrabutylammonium phosphate. (C) Scanning electron microscopy image of the ox-SWCNT/PAA network in between two finger electrodes. (D) Cyclic voltammetry curves of EP of PAA onto ox-SWCNTs. As the number of cycles increases, from 1 [*red* (dark gray in print versions)] to 30 cycles [*purple* (gray in print versions)], the oxidation peak for the AA monomeric unit (A1) disappears and the peak for PAA (A2) appears. (E) Calibration curve derived from potentiometry measurement of ox-SWCNTs/PAA exposed to various pH solutions (vs Ag/AgCl electrode). *(Adapted with permission from Gou, P.; Kraut, N. D.; Feigel, I. M.; Bai, H.; Morgan, G. J.; Chen, Y.; Tang, Y.; Bocan, K.; Stachel, J.; Berger, L.; Mickle, M.; Sejdic, E.; Star, A. Carbon Nanotube Chemiresistor for Wireless PH Sensing. Scientific Reports **2014**, 4, 4468. Copyright 2014 Nature Publishing Group.)*

implantable pH sensors *(144)*. Several reports were published on CNT-based pH sensing *(145–147)*; however, the selectivity between hydronium ions (H_3O^+) and metal cations is still not satisfied. In addition, pristine SWNTs are only sensitive to high concentrations of metal ions (>1 mM), and strong base (pH > 12) or acid (pH < 2). To achieve the required specificity to H_3O^+, a new system incorporating SWCNTs modified with polymers or proteins was developed to create a cost-effective, easy-to-use pH sensing platform *(134,143,144)*. In the typical work reported by Gou et al., oxidized SWCNTs were functionalized with the conductive polymer poly(1-aminoanthracene) (PAA) to build a solid-state sensor (Fig. 5.7) *(144)*. This device had a Nernstian response over a wide pH range (2–12) and retained sensitivity over 120 days. The sensor was also attached to a passively powered radiofrequency identification tag that transmits pH data through simulated skin. As a chemiresistor, such a device was able to achieve the sensitive pH response without the need of a reference electrode. This is very important to relieve the size limitation of the device. Therefore a smaller device can be made, and it could be easily implanted at the location of interest inside the body or become incorporated with other implanted devices such as prosthetics. Thus such a battery-less, reference electrode-free, and wirelessly transmitting sensor platform shows potential for biomedical applications as an implantable sensor, adjacent to surgical implants detecting for infection.

5.8 SUMMARY AND OUTLOOK

CNTs present significant opportunities to construct a wide range of sensors because of their distinguished properties such as good electronic properties, high surface-to-volume ratios, and excellent mechanical properties. The CNT-based sensors benefit from the use of CNT support platforms because of the enhanced performance for the detection of the product or the target. However, the commercial applications of CNTs for sensing are limited. For example, the efficient growth of macroscopic-length carbon nanotubes, controlled growth of nanotubes on desired substances, durability of nanotube-based sensors and actuators, effective dispersion of nanotubes in required solution or matrix and the control of their alignment, and the scale-up in fabrication of sensors and actuators based upon CNTs. Furthermore, for the development of the CNT-based sensor, a fundamental knowledge of their structure/property relationship is also necessary. In a macroscopic material system the characteristic length scales span several orders of magnitude and highlight the need to develop both experimental and analytical techniques to bridge these scales toward the optimization of these systems. In addition, the slow recovery of sensors remains a major drawback for CNT-based chemical sensors. In the future, the synthesis of CNTs should be developed to simplify the preparation process and achieve high quality so that it will be cost effective. At the same time, it is important to apply the pristine CNTs to sensors since the pristine CNTs can maintain their properties well. In addition, detailed studies on the toxicological effects of CNTs are required for their application in living cells. Last but not least, studies on the mechanism of the sensors using CNTs are also important, which will pave the way to the design and construction of new sensors with distinguished properties.

ACKNOWLEDGMENT

M.Z. is grateful to the support from Northeast Normal University startup funds, Jilin Provincial Department of Education, and the Recruitment Program of Global Youth Experts. M. Zhou, Z. Wang, and X. Wang are the corresponding authors.

REFERENCES

1. Iijima, S. Helical Microtubules of Graphitic Carbon. *Nature* **1991,** *354,* 56–58.
2. Zhu, Z. G.; Garcia-Gancedo, L.; Flewitt, A. J.; Xie, H.; Moussy, F.; Milne, W. I. A Critical Review of Glucose Biosensors Based on Carbon Nanomaterials: Carbon Nanotubes and Graphene. *Sensors* **2012,** *12,* 5996–6022.
3. Baughman, R. H.; Zakhidov, A. A.; de Heer, W. A. Carbon Nanotubes: The Route Toward Applications. *Science* **2002,** *297,* 787–792.
4. Wang, J. Carbon-Nanotube Based Electrochemical Biosensors: A Review. *Electroanalysis* **2005,** *17,* 7–14.
5. Kauffman, D. R.; Star, A. Carbon Nanotube Gas and Vapor Sensors. *Angewandte Chemie International Edition* **2008,** *47,* 6550–6570.
6. Jacobs, C. B.; Peairs, M. J.; Venton, B. J. Review: Carbon Nanotube Based Electrochemical Sensors for Biomolecules. *Analytica Chimica Acta* **2010,** *662,* 105–127.
7. Obitayo, W.; Liu, T. A Review: Carbon Nanotube-Based Piezoresistive Strain Sensors. *Journal of Sensors* **2012,** *2012,* 1–15.
8. Bondavalli, P. Carbon Nanotubes Based Transistors Composed of Single-Walled Carbon Nanotubes Mats as Gas Sensors: A Review. *Comptes Rendus Physique* **2010,** *11,* 389–396.
9. Goldoni, A.; Petaccia, L.; Lizzit, S.; Larciprete, R. Sensing Gases with Carbon Nanotubes: A Review of the Actual Situation. *Analytica Chimica Acta* **2008,** *626,* 119–129.
10. Bondavallia, P.; Legagneuxa, P.; Pribatb, D. Carbon Nanotubes Based Transistors as Gas Sensors: State of the Art and Critical Review. *Sensors and Actuators B: Chemical* **2009,** *140,* 304–318.
11. Mahar, B.; Laslau, C.; Yip, R.; Sun, Y. Development of Carbon Nanotube-Based Sensors-A Review. *IEEE Sensors Journal* **2007,** *7,* 266–284.
12. Agui, L.; Yanez-Sedeno, P.; Pingarron, J. Role of Carbon Nanotubes in Electroanalytical Chemistry-A Review. *Analytica Chimica Acta* **2008,** *622,* 11–47.
13. Balasubramanian, K.; Burghard, M. Chemically Functionalized Carbon Nanotubes. *Small* **2005,** *1,* 180–192.
14. Yakobson, B. I.; Brabec, C. J.; Bernholc, J. Nanomechanics of Carbon Tubes: Instabilities beyond Linear Response. *Physical Review Letters* **1996,** *76,* 2511–2514.
15. Wong, E. W.; Sheehan, P. E.; Lieber, C. M. Nanobeam Mechanics: Elasticity, Strength, and Toughness of Nanorods and Nanotubes. *Science* **1997,** *277,* 1971–1975.
16. Lipomi, D. J.; Vosgueritchian, M.; Tee, B. C.-K.; Hellstrom, S. L.; Lee, J. A.; Fox, C. H.; Bao, Z. Skin-Like Pressure and Strain Sensors Based on Transparent Elastic Films of Carbon Nanotubes. *Nature Nanotechnology* **2011,** *6,* 788–792.
17. Anthony, T.; George, W.; Isao, K. *Biosensors: Fundamentals and Applications;* Oxford University Press: Oxford, UK, ISBN 0198547242; 1987.
18. Bănică, F.-G. *Chemical Sensors and Biosensors: Fundamentals and Applications;* John Wiley & Sons: Chichester, UK, ISBN 9781118354230; 2012; p 576.
19. Clark, L. C. J.; Lyons, C. Electrode Systems for Continuous Monitoring in Cardiovascular Surgery. *Annals of the New York Academy of Sciences* **1962,** *102,* 29–45.

20. Balasubramanian, K.; Burghard, M. Biosensors Based on Carbon Nanotubes. *Analytical and Bioanalytical Chemistry* **2006,** *385,* 452–468.
21. Ali, M. A.; Solanki, P. R.; Srivastava, S.; Singh, S.; Agrawal, V. V.; John, R.; Malhotra, B. D. Protein Functionalized Carbon Nanotubes-Based Smart Lab-on-a-Chip. *ACS Applied Materials & Interfaces* **2015,** *7,* 5837–5846.
22. Varghese, S. H.; Nair, R.; Nair, B. G.; Hanajiri, T.; Maekawa, T.; Yoshida, Y.; Kumar, D. S. Sensors Based on Carbon Nanotubes and Their Applications: A Review. *Current Nanoscience* **2010,** *6,* 331–346.
23. Li, J.; Ng, H. T.; Cassell, A.; Fan, W.; Chen, H.; Ye, Q.; Koehne, J.; Han, J.; Meyyappan, M. Carbon Nanotube Nanoelectrode Array for Ultrasensitive DNA Detection. *Nano Letters* **2003,** *3,* 597–602.
24. Jia, J.; Guan, W.; Sim, M.; Li, Y.; Li, H. Carbon Nanotubes Based Glucose Needle-Type Biosensor. *Sensors* **2008,** *8,* 1712–1718.
25. Patolsky, F.; Weizmann, Y.; Willner, I. Long-Range Electrical Contacting of Redox Enzymes by SWCNT Connectors. *Angewandte Chemie International Edition* **2004,** *43,* 2113–2117.
26. Wang, S. G.; Zhang, Q.; Wang, R.; Yoon, S. F. A Novel Multi-Walled Carbon Nanotube-Based Biosensor for Glucose Detection. *Biochemical and Biophysical Research Communications* **2003,** *311,* 572–576.
27. Jeykumari, D. R. S.; Narayanan, S. S. A Novel Nanobiocomposite Based Glucose Biosensor Using Neutral Red Functionalized Carbon Nanotubes. *Biosensors and Bioelectronics* **2008,** *23,* 1404–1411.
28. Zhao, H.; Ju, H. Multilayer Membranes for Glucose Biosensing via Layer-by-Layer Assembly of Multiwall Carbon Nanotubes and Glucose Oxidase. *Analytical Biochemistry* **2006,** *350,* 138–144.
29. Wang, J.; Musameh, M. Enzyme-Dispersed Carbon-Nanotube Electrodes: A Needle Microsensor for Monitoring Glucose. *Analyst* **2003,** *128,* 1382–1385.
30. Wang, L.; Yuan, Z. B. Direct Electrochemistry of Glucose Oxidase at a Gold Electrode Modified with Single-Wall Carbon Nanotubes. *Sensors* **2003,** *3,* 544–554.
31. Liu, J.; Chou, A.; Rahmat, W.; Paddon-Row, M. N.; Gooding, J. J. Achieving Direct Electrical Connection to Glucose Oxidase Using Aligned Single Walled Carbon Nanotube Arrays. *Electroanalysis* **2005,** *17,* 38–45.
32. Wang, J.; Musameh, M. Carbon Nanotube/Teflon Composite Electrochemical Sensors and Biosensors. *Analytical Chemistry* **2003,** *75,* 2075–2079.
33. Tsai, Y.-C.; Li, S.-C.; Chen, J.-M. Cast Thin Film Biosensor Design Based on a Nafion Backbone, a Multiwalled Carbon Nanotube Conduit, and a Glucose Oxidase Function. *Langmuir* **2005,** *21,* 3653–3658.
34. Dai, Y.-Q.; Shiu, K.-K. Glucose Biosensor Based on Multi-Walled Carbon Nanotube Modified Glassy Carbon Electrode. *Electroanalysis* **2004,** *16,* 1697–1703.
35. Cella, L. N.; Chen, W.; Myung, N. V.; Mulchandani, A. Single-Walled Carbon Nanotube-Based Chemiresistive Affinity Biosensors for Small Molecules: Ultrasensitive Glucose Detection. *Journal of the American Chemical Society* **2010,** *132,* 5024–5026.
36. Claussen, J. C.; Franklin, A. D.; Haque, A.; Porterfield, M.; Fisher, T. S. Electrochemical Biosensor of Nanocube-Augmented Carbon Nanotube Networks. *ACS Nano* **2009,** *3,* 37–44.
37. Korani, A.; Salimi, A.; Hadadzadeh, H. Nickel-Phendione Complex Covalently Attached onto Carbon Nanotube/Cross Linked Glucose Dehydrogenase as Bioanode for Glucose/Oxygen Compartment-Less Biofuel Cell. *Journal of Power Sources* **2015,** *282,* 586–595.
38. Wang, J.; Musameh, M.; Lin, Y. Solubilization of Carbon Nanotubes by Nafion toward the Preparation of Amperometric Biosensors. *Journal of the American Chemical Society* **2003,** *125,* 2408–2409.
39. Kim, W.-S.; Lee, G.-J.; Ryu, J.-H.; Park, K.; Park, H.-K. A Flexible, Nonenzymatic Glucose Biosensor Based on Ni-Coordinated, Vertically Aligned Carbon Nanotube Arrays. *RSC Advances* **2014,** *4,* 48310–48316.
40. Wang, J.; Deo, R. P.; Musameh, M. Stable and Sensitive Electrochemical Detection of Phenolic Compounds at Carbon Nanotube Modified Glassy Carbon Electrodes. *Electroanalysis* **2003,** *15,* 1830–1834.
41. Zhao, Q.; Guan, L.; Gu, Z.; Zhuang, Q. Determination of Phenolic Compounds Based on the Tyrosinase-Single Walled Carbon Nanotubes Sensor. *Electroanalysis* **2005,** *17,* 85–88.

42. Rivas, G. A.; Rubianes, M. D. Enzymatic Biosensors Based on Carbon Nanotubes Paste Electrodes. *Electroanalysis* **2005,** *17,* 73–78.
43. Xu, Z.; Chen, X.; Qu, X.; Dong, S. Electrocatalytic Oxidation of Catechol at Multi-Walled Carbon Nanotubes Modified Electrode. *Electroanalysis* **2004,** *16,* 684–687.
44. Qi, H.; Zhang, C. Simultaneous Determination of Hydroquinone and Catechol at a Glassy Carbon Electrode Modified with Multiwall Carbon Nanotubes. *Electroanalysis* **2005,** *17,* 832–838.
45. Liu, N.; Cai, X.; Zhang, Q.; Lei, Y.; Chan-Park, M. B. Real-Time Nitrophenol Detection Using Single-Walled Carbon Nanotube Based Devices. *Electroanalysis* **2008,** *20,* 558–562.
46. Cao, W.; Hu, S. S.; Ye, L. H.; Cao, J.; Xu, J. J.; Pang, X. Q. Trace-Chitosan-Wrapped Multi-Walled Carbon Nanotubes as a New Sorbent in Dispersive Micro Solid-Phase Extraction to Determine Phenolic Compounds. *Journal of Chromatography A* **2015,** *1390,* 13–21.
47. Joshi, K. A.; Tang, J.; Haddon, R.; Wang, J.; Chen, W.; Mulchandani, A. A. Disposable Biosensor for Organophosphorus Nerve Agents Based on Carbon Nanotubes Modified Thick Film Strip Electrode. *Electroanalysis* **2005,** *17,* 54–58.
48. Liu, G.; Lin, Y. Biosensor Based on Self-Assembling Acetylcholinesterase on Carbon Nanotubes for Flow Injection/Amperometric Detection of Organophosphate Pesticides and Nerve Agents. *Analytical Chemistry* **2006,** *78,* 835–843.
49. Kandimalla, V. B.; Ju, H. Binding of Acetylcholinesterase to Multiwall Carbon Nanotube-Cross-Linked Chitosan Composite for Flow-Injection Amperometric Detection of Organophosphorous Insecticide. *Chemistry – A European Journal* **2006,** 1074–1080.
50. Deo, R. P.; Wang, J.; Block, I.; Mulchandani, A.; Joshi, K. A.; Lin, Y. Determination of Organophosphate Pesticides at a Carbon Nanotube/Organophosphorus Hydrolase Electrochemical Biosensor. *Analytica Chimica Acta* **2005,** *530,* 185–189.
51. Pereira, N. D.; de Oliveira, F. M.; Pereira, N. R.; Verly, R. M.; Souto, D. E. P.; Kubota, L. T.; Tanaka, A. A.; Damos, F. S.; Luz, R. D. S. Ultrasensitive Biosensor for Detection of Organophosphorus Pesticides Based on a Macrocycle Complex/Carbon Nanotubes Composite and 1-Methyl-3-octylimidazolium Tetrafluoroborate as Binder Compound. *Analytical Sciences* **2015,** *31,* 29–35.
52. Tan, X.; Li, M.; Cai, P.; Lou, L.; Zhou, X. An Amperometric Cholesterol Biosensor Based on Multiwalled Carbon Nanotubes and Organically Modified Sol-Gel/Chitosan Hybrid Composite Film. *Analytical Biochemistry* **2005,** *337,* 111–120.
53. Roy, S.; Vedala, H.; Choi, W. Vertically Aligned Carbon Nanotube Probes for Monitoring Blood Cholesterol. *Nanotechnology* **2006,** *17,* S14–S18.
54. Shi, Q.; Peng, T.; Zhu, Y.; Catherine, F. Y. An Electrochemical Biosensor with Cholesterol Oxidase/Sol-Gel Film on a Nanoplatinum/Carbon Nanotube Electrode. *Electroanalysis* **2005,** *17,* 857–861.
55. Poulsen, S. S.; Saber, A. T.; Mortensen, A.; Szarek, J.; Wu, D. M.; Williams, A.; Andersen, O.; Jacobsen, N. R.; Yauk, C. L.; Wallin, H. Changes in Cholesterol Homeostasis and Acute Phase Response Link Pulmonary Exposure to Multi-Walled Carbon Nanotubes to Risk of Cardiovascular Disease. *Toxicology and Applied Pharmacology* **2015,** *283,* 210–222.
56. Gooding, J. J.; Wibowo, R.; Liu, J.; Yang, W.; Losic, D.; Orbons, S.; Mearns, F. J.; Shapter, J. G.; Hibbert, D. B. Protein Electrochemistry Using Aligned Carbon Nanotube Arrays. *Journal of the American Chemical Society* **2003,** *125,* 9006–9007.
57. Kam, N. W. S.; Dai, H. Carbon Nanotubes as Intracellular Protein Transporters Generality and Biological Functionality. *Journal of the American Chemical Society* **2005,** *127,* 6021–6026.
58. Chen, R. J.; Zhang, Y.; Wang, D.; Dai, H. Noncovalent Sidewall Functionalization of Single-Walled Carbon Nanotubes for Protein Immobilization. *Journal of the American Chemical Society* **2001,** *123,* 3838–3839.
59. Devadoss, A.; Lee, J. W.; Terashima, C.; Fujishima, A.; Kim, Y. P.; Kang, J. K.; Paik, U. Synergistic Oxidation of NADH on Bimetallic CoPt Nanoparticles Decorated Carbon Nitride Nanotubes. *Sensors and Actuators B: Chemical* **2015,** *208,* 204–211.
60. Yuan, J. H.; Chen, J. R.; Wu, X. H.; Fang, K. M.; Niu, L. A NADH Biosensor Based on Diphenylalanine Peptide/Carbon Nanotube Nanocomposite. *Journal of Electroanalytical Chemistry* **2011,** *656,* 120–124.
61. Musameh, M.; Wang, J.; Merkoci, A.; Lin, Y. Low-Potential Stable NADH Detection at Carbon-Nanotube-Modified Glassy Carbon Electrodes. *Electrochemistry Communications* **2002,** *4,* 743–746.

62. Fu, D.; Xu, Y.; Li, L.-J.; Chen, Y. Electrical Detection of Nitric Oxide Using Single Walled Carbon Nanotube Network Devices. *Carbon* **2007,** *45,* 1911–1920.
63. Zhao, G. C.; Wu, F. X.; Wei, X. W. Catalytic Activity of Multi Walled Carbon Nanotubes for the Oxidation of Nitric Oxide. *Chemistry Letters* **2002,** *31,* 520–521.
64. Wang, Y.; Li, Q.; Hu, S. A Multiwall Carbon Nanotubes Film-Modified Carbon Fiber Ultramicroelectrode for the Determination of Nitric Oxide Radical in Liver Mitochondria. *Bioelectrochemistry* **2005,** *65,* 135–142.
65. Zhang, L.; Fang, Z.; Zhao, G. C.; Wei, X. W. Electrodeposited Platinum Nanoparticles on the Multi-Walled Carbon Nanotubes and its Electrocatalytic for Nitric Oxide. *International Journal of Electrochemical Science* **2008,** *3,* 746–754.
66. Kuzmych, O.; Allen, B. L.; Star, A. Carbon Nanotube Sensors for Exhaled Breath Components. *Nanotechnology* **2007,** *18,* 375502 (7 pp).
67. Wu, F. H.; Zhao, G. C.; Wei, X. W. Electrocatalytic Oxidation of Nitric Oxide at Multi-Walled Carbon Nanotubes Modified Electrode. *Electrochemistry Communications* **2005,** *4,* 690–694.
68. Baek, Y.-K.; Yoo, S. M.; Kim, J.-H.; Jung, D.-H.; Choi, Y.-K.; Kim, Y. S.; Lee, S. Y.; Jung, H.-T. The Effect of Network Density on the DNA-Sensing Performance of Single-Walled Carbon Nanotubes. *Journal of Physical Chemistry C* **2009,** *113,* 21566–21571.
69. Tang, X.; Bansaruntip, S.; Nakayama, N.; Yenilmez, E.; Chang, Y.-L.; Wang, Q. Carbon Nanotube DNA Sensor and Sensing Mechanism. *Nano Letters* **2006,** *6,* 1632–1636.
70. Münzer, A. M.; Morgan, G. J.; Michael, Z. P.; Zhao, Y.; Melzer, K.; Scarpa, G.; Star, A. Sensing Reversible Protein-Ligand Interactions with Single-Walled Carbon Nanotube Field-Effect Transistors. *Journal of Physical Chemistry C* **2014,** *118,* 17193–17199.
71. Jung, D.-H.; Kim, B. H.; Ko, Y. K.; Jung, M. S.; Jung, S.; Lee, S. Y.; Jung, H.-T. Covalent Attachment and Hybridization of DNA Oligonucleotides on Patterned Single-Walled Carbon Nanotube Films. *Langmuir* **2004,** *20,* 8886–8891.
72. Nie, H.; Yang, Z.; Huang, S.; Wu, Z.; Wang, H.; Yu, R.; Jiang, J. DNA-Wrapped Carbon Nanotubes as Sensitive Electrochemical Labels in Controlled-Assembly-Mediated Signal Transduction for the Detection of Sequence-Specific DNA. *Small* **2012,** *8,* 1407–1414.
73. Singh, R.; Pantarotto, D.; McCarthy, D.; Chaloin, O.; Hoebeke, J.; Partidos, C. D.; Briand, J.-P.; Prato, M.; Bianco, A.; Kostarelos, K. Binding and Condensation of Plasmid DNA onto Functionalized Carbon Nanotubes: Toward the Construction of Nanotube-Based Gene Delivery Vectors. *Journal of the American Chemical Society* **2005,** *127,* 4388–4396.
74. Roy, S.; Vedala, H.; Roy, A. D.; Kim, D-h.; Doud, M.; Mathee, K.; Shin, H-k.; Shimamoto, N.; Prasad, V.; Choi, W. Direct Electrical Measurements on Single-Molecule Genomic DNA Using Single-Walled Carbon Nanotubes. *Nano Letters* **2008,** *8,* 26–30.
75. Huang, Y.; Zhao, S.; Liu, Y.-M.; Chen, J.; Chen, Z.-F.; Shia, M.; Liang, H. An Amplified Single-Walled Carbon Nanotube-Mediated Chemiluminescence Turn-on Sensing Platform for Ultrasensitive DNA Detection. *Chemical Communications* **2012,** *48,* 9400–9402.
76. Li, J.; Lee, E. C. Carbon Nanotube/Polymer Composite Electrodes for Flexible, Attachable Electrochemical DNA Sensors. *Biosensors and Bioelectronics* **2015,** *71,* 414–419.
77. Wang, J.; Kawde, A.-N.; Musameh, M. Carbon-Nanotube-Modified Glassy Carbon Electrodes for Amplified Label-Free Electrochemical Detection of DNA Hybridization. *Analyst* **2003,** *128,* 912–916.
78. Baker, S. E.; Cai, W.; Lasseter, T. L.; Weidkamp, K. P.; Hamers, R. J. Covalently Bonded Adducts of Deoxyribonucleic Acid (DNA) Oligonucleotides with Wingle-Wall Carbon Nanotubes: Synthesis and Hybridization. *Nano Letters* **2002,** *2,* 1413–1417.
79. Chen, Y.; Liu, H.; Ye, T.; Kim, J.; Mao, C. DNA-Directed Assembly of Single-Wall Carbon Nanotubes. *Journal of the American Chemical Society* **2007,** *129,* 8696–8697.
80. Barone, P. W.; Balk, S.; Heller, D. A.; Strano, M. S. Near-Infrared Optical Sensors Based on Single-Walled Carbon Nanotubes. *Nature Materials* **2005,** *4,* 86–92.
81. Kong, J.; Franklin, N. R.; Zhou, C.; Chapline, M. G.; Peng, S.; Cho, K.; Dai, H. Nanotube Molecular Wires as Chemical Sensors. *Science* **2000,** *287,* 622–625.
82. Cantalini, C.; Valentini, L.; Armentano, I.; Kenny, J. M.; Lozzi, L.; Santucci, S. Carbon Nanotubes as New Materials for Gas Sensing Applications. *Journal of the European Ceramic Society* **2004,** *24,* 1405–1408.

83. Modi, A.; Koratkar, N.; Lass, E.; Wei, B.; Ajayan, P. M. Miniaturized Gas Ionization Sensors Using Carbon Nanotubes. *Nature* **2003,** *424,* 171–174.
84. Ganzhorn, M.; Vijayaraghavan, A.; Dehm, S.; Hennrich, F.; Green, A. A.; Fichtner, M.; Voigt, A.; Rapp, M.; von Lohneysen, H.; Hersam, M. C.; et al. Hydrogen Sensing with Diameter- and Chirality-Sorted Carbon Nanotubes. *ACS Nano* **2011,** *5,* 1670–1676.
85. Iakoubovskii, K.; Minami, N.; Karthigeyan, A. Optical Gas Sensing with Dip-Coated Carbon Nanotubes through the Modulation of Photoluminescence and Optical Absorption. *Journal of Materials Chemistry* **2012,** *22,* 4716–4719.
86. Chen, G.; Paronyan, T. M.; Pigos, E. M.; Harutyunyan, A. R. Enhanced Gas Sensing in Pristine Carbon Nanotubes under Continuous Ultraviolet Light Illumination. *Scientific Reports* **2012,** *2,* 343.
87. Ventura, D. N.; Li, S.; Baker, C. A.; Breshike, C. J.; Spann, A. L.; Strouse, G. F.; Kroto, H. W.; Acquah, S. F. A. A Flexible Cross-Linked Multi-Walled Carbon Nanotube Paper for Sensing Hydrogen. *Carbon* **2012,** *50,* 2666–2674.
88. Boyd, A.; Dube, I.; Fedorov, G.; Paranjape, M.; Barbara, P. Gas sensing Mechanism of Carbon Nanotubes: From Single Tubes to High-Density Networks. *Carbon* **2014,** *69,* 417–423.
89. Du, N.; Zhang, H.; Yu, J.; Wu, P.; Zhai, C.; Xu, Y.; Wang, J.; Yang, D. General Layer-by-Layer Approach to Composite Nanotubes and Their Enhanced Lithium-Storage and Gas-Sensing Properties. *Chemistry of Materials* **2009,** *21,* 5264–5271.
90. Lee, C. Y.; Baik, S.; Zhang, J.; Masel, R. I.; Strano, M. S. Charge Transfer from Metallic Single-Walled Carbon Nanotube Sensor Arrays. *Journal of Physical Chemistry B* **2006,** *110,* 11055–11061.
91. Kong, J.; Chapline, M. G.; Dai, H. Functionalized Carbon Nanotubes for Molecular Hydrogen Sensors. *Advanced Materials* **2001,** *13,* 1384–1386.
92. Chang, H.; Lee, J. D.; Lee, S. M.; Lee, Y. H. Adsorption of NH_3 and NO_2 Molecules on Carbon Nanotubes. *Applied Physics Letters* **2001,** *79,* 3863–3865.
93. Zhao, J.; Buldum, A.; Han, J.; Lu, J. P. Gas Molecule Adsorption in Carbon Nanotubes and Nanotube Bundles. *Nanotechnology* **2002,** *13,* 195–200.
94. Shi, W.; Johnson, J. K. Gas Adsorption on Heterogeneous Single-Walled Carbon Nanotube Bundles. *Physical Review Letters* **2003,** *91,* 015504.
95. Andzelm, J.; Govind, N.; Maiti, A. Nanotube-Based Gas Sensors-Role of Structural Defects. *Chemical Physics Letters* **2006,** *421,* 58–62.
96. Seo, K.; Park, K. A.; Kim, C.; Han, S.; Kim, B.; Lee, Y. H. Chirality- and Diameter-Dependent Reactivity of NO_2 on Carbon Nanotube Walls. *Journal of the American Chemical Society* **2005,** *127,* 15724–15729.
97. Yim, W. L.; Gong, X. G.; Liu, Z. F. Chemisorption of NO_2 on Carbon Nanotubes. *Journal of Physical Chemistry B* **2003,** *107,* 9363–9369.
98. Zhang, Y.; Suc, C.; Liu, Z.; Li, J. Carbon Nanotubes Functionalized by NO_2: Coexistence of Charge Transfer and Radical Transfer. *Journal of Physical Chemistry B* **2006,** *110,* 22462–22470.
99. Ricca, A.; Baushlicher, C. W. The Adsorption of NO_2 on (9,0) and (10,0) Carbon Nanotubes. *Chemical Physics Letters* **2006,** *323,* 511–518.
100. Quang, N. H.; Trinh, M. V.; Lee, B. H.; Huh, J. S. Effect of NH_3 Gas on the Electrical Properties of Single-Walled Carbon Nanotube Bundles. *Sensors and Actuators B* **2006,** *113,* 341–346.
101. Valentini, L.; Armentano, I.; Kenny, J. M.; Cantalini, C.; Lozzi, L.; Santucci, S. Sensors for Sub-ppm NO_2 Gas Detection Based on Carbon Nanotube Thin Films. *Applied Physics Letters* **2003,** *82,* 961–963.
102. Penza, M.; Cassano, G.; Rossi, R.; Rizzo, A.; Signore, M. A.; Alvisi, M.; Lisi, L.; Serra, E.; Giorgi, R. Effect of Growth Catalysts on Gas Sensitivity in Carbon Nanotube Film Based Chemiresistive Sensors. *Applied Physics Letters* **2007,** *90,* 103101.
103. Li, J.; Lu, Y.; Ye, Q.; Cinke, M.; Han, J.; Meyyappan, M. Carbon Nanotube Sensors for Gas and Organic Vapor Detection. *Nano Letters* **2003,** *3,* 929–933.
104. Qi, P.; Vermesh, O.; Grecu, M.; Javey, A.; Wang, Q.; Dai, H. Toward Large Arrays of Multiplex Functionalized Carbon Nanotube Sensors for Highly Sensitive and Selective Molecular Detection. *Nano Letters* **2003,** *3,* 347–351.

105. Zhang, T.; Nix, M. B.; Yoo, B.-Y.; Deshusses, M. A.; Myung, N. V. Electrochemically Functionalized Single-Walled Carbon Nanotube Gas Sensor. *Electroanalysis* **2006,** *18,* 1153–1158.
106. Gao, M.; Huang, S.; Dai, L.; Wallace, G.; Gao, R.; Wang, Z. Aligned Coaxial Nanowires of Carbon Nanotubes Sheathed with Conducting Polymers. *Angewandte Chemie International Edition* **2000,** *39,* 3664–3667.
107. Hea, L.; Jiaa, Y.; Meng, F.; Li, M.; Liu, J. Gas Sensors for Ammonia Detection Based on Polyaniline-Coated Multi-Wall Carbon Nanotubes. *Materials Science and Engineering: B* **2009,** *163,* 76–81.
108. Yeung, C. S.; Liu, L. V.; Wang, Y. A. Adsorption of Small Gas Molecules onto Pt-Doped Single-Walled Carbon Nanotubes. *Journal of Physical Chemistry C* **2008,** *112,* 7401–7411.
109. Star, A.; Joshi, V.; Skarupo, S.; Thomas, D.; Gabriel, J.-C. P. Gas Sensor Array Based on Metal-Decorated Carbon Nanotubes. *Journal of Physical Chemistry B* **2006,** *110,* 21014–21020.
110. Kumar, M. K.; Ramaprabhu, S. Nanostructured Pt Functionalized Multiwalled Carbon Nanotube Based Hydrogen Sensor. *Journal of Physical Chemistry B* **2006,** *110,* 11291–11298.
111. Mubeen, S.; Zhang, T.; Yoo, B.; Deshusses, M. A.; Myung, N. V. Palladium Nanoparticles Decorated Single-Walled Carbon Nanotube Hydrogen Sensor. *Journal of Physical Chemistry C* **2007,** *111,* 6321–6327.
112. Liu, J.; Guo, Z.; Meng, F.; Jia, Y.; Liu, J. A Novel Antimony-Carbon Nanotube-Tin Oxide Thin Film: Carbon Nanotubes as Growth Guider and Energy Buffer. Application for Indoor Air Pollutants Gas Sensor. *Journal of Physical Chemistry C* **2008,** *112,* 6119–6125.
113. Ionescu, R.; Espinosa, E. H.; Leghrib, R.; Felten, A.; Pireaux, J. J.; Erni, R.; Van Tendeloo, G.; Bittencourt, C.; Cañellas, N.; Llobet, E. Novel Hybrid Materials for Gas Sensing Applications Made of Metal-Decorated MWCNTs Dispersed on Nano-Particle Metal Oxides. *Sensors and Actuators B* **2008,** *131,* 174–182.
114. Du, N.; Zhang, H.; Maa, X.; Yang, D. Homogeneous Coating of Au and SnO_2 Nanocrystals on Carbon Nanotubes via Layer-by-Layer Assembly: A New Ternary Hybrid for a Room-Temperature CO Gas Sensor. *Chemical Communications* **2008,** *46,* 6182–6184.
115. Willinger, M.-G.; Neri, G.; Bonavita, A.; Micali, G.; Rauwel, E.; Herntricha, T.; Pinna, N. The Controlled Deposition of Metal Oxides onto Carbon Nanotubes by Atomic Layer Deposition: Examples and a Case Study on the Application of V_2O_4 Coated Nanotubes in Gas Sensing. *Physical Chemistry Chemical Physics* **2009,** *11,* 3615–3622.
116. Xu, Y.; Zhou, X.; Sorensen, O. T. Oxygen Sensors Based on Semiconducting Metal Oxides: An Overview. *Sensors and Actuators B* **2000,** *65,* 2–4.
117. Goldoni, A.; Petaccia, L.; Lizzit, S.; Larciprete, R. Sensing Gases with Carbon Nanotubes: A Review of the Actual Situation. *Journal of Physics: Condensed Matter* **2010,** *22,* 013001.
118. Wang, J.; Liu, G.; Jan, M. R. Ultrasensitive Electrical Biosensing of Proteins and DNA: Carbon-Nanotube Derived Amplification of the Recognition and Transduction Events. *Journal of the American Chemical Society* **2004,** *126,* 3010–3011.
119. He, P.; Dai, L. Aligned Carbon Nanotube-DNA Electrochemical Sensors. *Chemical Communications* **2004,** 348–349.
120. Cao, J.; Wang, Q.; Dai, H. Electromechanical Properties of Metallic, Quasimetallic, and Semiconducting Carbon Nanotubes Under Stretching. *Physical Review Letters* **2003,** *90,* 157601.
121. Grujicic, M.; Gao, G.; Roy, W. N. A Computational Analysis of the Carbon-Nanotube-Based Resonant-Circuit Sensors. *Applied Surface Science* **2004,** *229,* 316–323.
122. Zhao, Q.; Wood, J. R.; Wagner, H. D. Stress Fields Around Defects and Fibers in a Polymer Using Carbon Nanotubes as Sensors. *Applied Physics Letters* **2001,** *78,* 1748–1750.
123. Zhao, Q.; Frogley, M. D.; Wagner, H. D. Direction-Sensitive Strain-Mapping with Carbon Nanotube Sensors. *Composites Science and Technology* **2002,** *62,* 147–150.
124. Tombler, T. W.; Zhou, C. W.; Alexseyev, L.; Kong, J.; Dai, H. J.; Lei, L.; Jayanthi, C. S.; Tang, M. J.; Wu, S. Y. Reversible Electromechanical Characteristics of Carbon Nanotubes under Local-Probe Manipulation. *Nature* **2000,** *405,* 769–772.
125. Sazonova, V.; Yaish, Y.; Ustunel, H.; Roundy, D.; Arias, T. A.; McEuen, P. L. A Tunable Carbon Nanotube Electromechanical Oscillator. *Nature* **2004,** *431,* 284–287.

126. Dereli, G.; Ozdogan, C. Structural Stability and Energetics of Single-Walled Carbon Nanotubes under Uniaxial Strain. *Physical Review B* **2003,** *67,* 035416.
127. Shang, Y.; Li, Y.; He, X.; Du, S.; Zhang, L.; Shi, E.; Wu, S.; Li, Z.; Li, P.; Wei, J.; et al. Highly Twisted Double-Helix Carbon Nanotube Yarns. *ACS Nano* **2013,** *7,* 1446–1453.
128. Li, Y.; Shang, Y.; He, X.; Peng, Q.; Du, S.; Shi, E.; Wu, S.; Li, Z.; Li, P.; Cao, A. Overtwisted, Resolvable Carbon Nanotube Yarn Entanglement as Strain Sensors and Rotational Actuators. *ACS Nano* **2013,** *7,* 8128–8135.
129. Zhong, J.; Zhong, Q.; Hu, Q.; Wu, N.; Li, W.; Wang, B.; Hu, B.; Zhou, J. Stretchable Self-Powered Fiber-Based Strain Sensor. *Advanced Functional Materials* **2015,** *25,* 1798–1803.
130. Ghosh, S.; Sood, A. K.; Kumar, N. Carbon Nanotube Flow Sensors. *Science* **2003,** *299,* 1042–1044.
131. Vilela, D.; Anson-Casaos, A.; Martinez, M. T.; Gonzalez, M. C.; Escarpa, A. High NIR-Purity Index Single-Walled Carbon Nanotubes for Electrochemical Sensing in Microfluidic Chips. *Lab on a Chip* **2012,** *12,* 2006–2014.
132. Cao, C. L.; Fang, L.; Liao, K. J.; Wei, F. Y.; Li, L. A Flow Sensor for Liquids Based on Single-Walled Carbon Nanotube Thin Films. *Diffusion and Defect Data Solid State Phenomena Part B* **2007,** *121–123,* 75–79.
133. Vilela, D.; Garoz, J.; Colina, A.; Gonzalez, M. C.; Escarpa, A. Carbon Nanotubes Press-Transferred on PMMA Substrates as Exclusive Transducers for Electrochemical Microfluidic Sensing. *Analytical Chemistry* **2012,** *84,* 10838–10844.
134. Cheng, A. L.; Han, K. N.; Pham, X.-H.; Seong, G. H. A Single-Walled Carbon Nanotube Thin Film-Based PH-Sensing Microfluidic Chip. *Analyst* **2014,** *139,* 2011–2015.
135. Fung, C. K. M.; Sin, M. L. Y.; Lei, T. K. F.; Chow, W. W. Y.; Lai, K. W. C.; Li, W. J. Flow Rate Measurement Inside Polymer Microfluidic Systems Using Carbon Nanotube Sensors. In *Proceedings of the IEEE International Conference on Sensors (IEEE-Sensors), Irvine, CA;* October 31–November 03, 2005; pp 541–544.
136. Liu, J. W.; Dai, L.; Baur, J. W. Multiwalled Carbon Nanotubes for Flow-Induced Voltage Generation. *Journal of Applied Physics* **2007,** *101,* 064312.
137. Lee, S. H.; Kim, D.; Kim, S.; Han, C.-S. Flow-Induced Voltage Generation in High-Purity Metallic and Semiconducting Carbon Nanotubes. *Applied Physics Letters* **2011,** *99,* 104103.
138. Poncharal, P.; Wang, Z. L.; Ugarte, D.; de Heer, W. A. Electrostatic Deflections and Electromechanical Resonances of Carbon Nanotubes. *Science* **1999,** *283,* 1513–1516.
139. Chiu, H.-Y.; Hung, P.; Postma, H. W.; Bockrath, M. Atomic-Scale Mass Sensing Using Carbon Nanotube Resonators. *Nano Letters* **2008,** *8,* 4342–4346.
140. Lassagne, B.; Garcia-Sanchez, D.; Aguasca, A.; Bachtold, A. Ultrasensitive Mass Sensing with a Nanotube Electromechanical Resonator. *Nano Letters* **2008,** *8,* 3735–3738.
141. Chaste, J.; Eichler, A.; Moser, J.; Ceballos, G.; Rurali, R.; Bachtold, A. A Nanomechanical Mass Sensor with Yoctogram Resolution. *Nature Nanotechnology* **2012,** *7,* 301–304.
142. Li, J.-J.; Jiang, C.; Chen, B.; Zhu, K.-D. Optical Mass Sensing with a Carbon Nanotube Resonator. *Journal of the Optical Society of America B* **2012,** *29,* 965–969.
143. Tsai, W.-L.; Huang, B.-T.; Wang, K.-Y.; Huang, Y.-C.; Yang, P.-Y.; Cheng, H.-C. Functionalized Carbon Nanotube Thin Films as the PH Sensing Membranes of Extended-Gate Field-Effect Transistors on the Flexible Substrates. *IEEE Transactions on Nanotechnology* **2014,** *13,* 760–766.
144. Gou, P.; Kraut, N. D.; Feigel, I. M.; Bai, H.; Morgan, G. J.; Chen, Y.; Tang, Y.; Bocan, K.; Stachel, J.; Berger, L.; et al. Carbon Nanotube Chemiresistor for Wireless PH Sensing. *Scientific Reports* **2014,** *4,* 4468.
145. Li, P.; Martin, C.; Yeung, K. K.; Xue, W. Dielectrophoresis Aligned Single-Walled Carbon Nanotubes as PH Sensors. *Biosensors* **2011,** *1,* 23–35.
146. Muenzer, A. M.; Melzer, K.; Heimgreiter, M.; Scarpa, G. Random CNT Network and Regioregular Poly(3-hexylthiophen) FETs for PH Sensing Applications: A Comparison. *Biochimica et Biophysica Acta (BBA) – General Subjects* **2013,** *1830,* 4353–4358.
147. Takeda, S.; Nakanura, M.; Ishii, A.; Subagyo, A.; Hosoi, H.; Sueoka, K.; Mukasa, K. A PH Sensor Based on Electric Properties of Nanotubes on a Glass Substrate. *Nanoscale Research Letters* **2007,** *2,* 207–212.

CHAPTER 6

Stimuli-Responsive Materials From Carbon Nanotubes

S. He, P. Chen, X. Sun, H. Peng
Fudan University, Shanghai, China

Contents

6.1 INTRODUCTION

The discovery of carbon nanotubes (CNTs) by Iijima in 1991 *(1)* has attracted great research interest for their light weight, high strength, and high electrical and thermal conductivity *(2)*. In 1999, Kim et al. first used nanometer-diameter CNTs to create robust nanotweezers *(3)*. In the same year, Baughman et al. reported that the single-walled CNT (SWNT)-based actuator could exhibit bending actuations in an electrolyte solution *(4)*. Then the actuators from CNTs attracted increasing attention for their potential applications in intelligent robots, sensors, biomedical devices, and so on *(5−7)*. Apart from the change of shape, the CNT-based materials can also change their colors electrochemically *(8)*. The chromic devices from CNTs are promising in various applications including smart windows, optical shutters, antiglare mirrors, sunglasses, and displays. In addition, because of the excellent electrical property, great efforts have been made to incorporate CNTs into other stimuli-responsive materials to enhance their performances, such as response speed and magnitude as well as endow the devices

Industrial Applications of Carbon Nanotubes
ISBN 978-0-323-41481-4

with new sensitivities. In this chapter, we will introduce the basic mechanisms and performances of these responsive materials with actuating and chromic properties based on CNTs.

6.2 ACTUATORS FROM CARBON NANOTUBES

Actuators from CNTs, including pure CNTs and their composites, can be mainly classified into three categories according to the different external stimuli: electricity, solvent, and light. Because of the nanoscaled size, unique electrical and mechanical properties, and high chemical and thermal stability of CNTs, these actuators show superior actuation performances to the conventional actuators including shape memory alloys, piezoelectric ceramics, and polymers.

6.2.1 Electrically Driven Actuators

Among the actuators, the electromechanical actuator is mostly explored because of the easy controllability of electricity and promising application in various areas such as intelligent robots. CNTs could serve as good candidates for electricity-based actuators because of their inherent advantages of light weight and high electrical conductivity compared with the other electroactive materials including shape memory alloys, piezoelectric ceramics, and polymers *(9–11)*. Four types of CNT-based actuators in response to electricity are widely explored and described here: electrostatic, electrochemical, electrothermal, and electromagnetic actuators.

6.2.1.1 Electrostatically Driven Actuator

In 1999, Kim et al. first used CNTs to fabricate robust nanotweezers on the basis of the electrostatic mechanism *(3)*. The nanotweezer was made by attaching two CNTs with a length of ~4 μm on two sides of a tapered glass micropipette. Both sides of the micropipettes were deposited with metal electrodes and then connected to the external circuit. The resulting nanotweezers bent closer to each other upon application of a voltage. In 2001, Akita et al. reported a nanotweezer that was composed of two CNTs with a length of 1–5 μm and operated by atomic force microscopy *(12)*. The nanotweezer could repeatedly grab, lift, and release submicron-sized objects upon the use and removal of a voltage. The driving force of the nanotweezer was derived from the electrostatic force *(12–14)*. CNTs could serve as an ideal building block for a nanoelectromechanical system. By fixing the CNTs at both ends to suspend over counter electrodes, CNTs could be used for a force measurement system *(15)*. Besides, it could be used for electromechanical data storage, where CNTs serve as both molecular device elements and molecular wires for reading and writing information *(16–17)*.

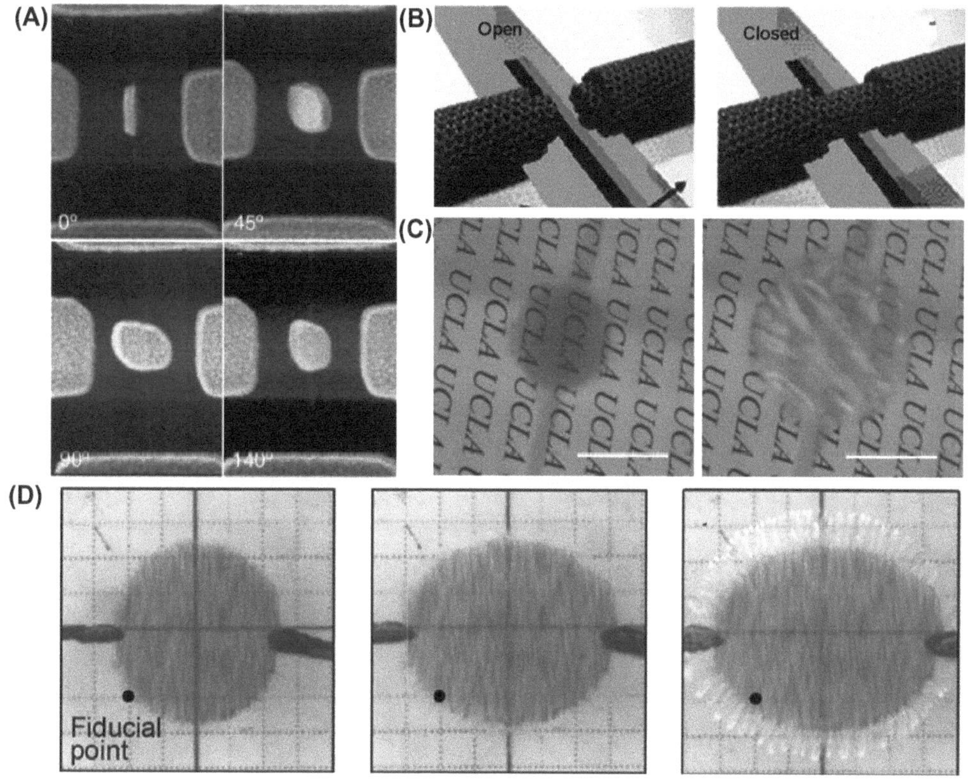

Figure 6.1 ***Electrostatically driven actuators based on carbon nanotubes (CNTs).*** (A) Scanning electron microscope (SEM) images of the CNT actuator rotor plate at increasing angles under 5 V. Scale bars, 300 nm *(Reproduced with permission from Fennimore, A.; Yuzvinsky, T.; Han, W.-Q.; Fuhrer, M.; Cumings, J.; Zettl, A. Rotational Actuators Based on Carbon Nanotubes.* Nature ***2003,*** *424, 408–410, Copyright 2003 Macmillan Publishers Ltd.)* (B) Schematic illustration of a CNT linear bearing nanoswitch producing a conducting state at a voltage above 4.5 V *(Reproduced with permission from Deshpande, V.; Chiu, H.-Y.; Postma, H. C.; Miko, C.; Forro, L.; Bockrath, M. Carbon Nanotube Linear Bearing Nanoswitches.* Nano Letters ***2006,*** *6, 1092–1095, Copyright 2006 American Chemical Society.)* (C) Expansion of the VHB 4910 film with single-walled CNT (SWNT) electrodes at 5 kV *(Reproduced with permission from Yuan, W.; Hu, L. B.; Yu, Z. B.; Lam, T. L.; Biggs, J.; Ha, S. M.; Xi, D. J.; Chen, B.; Senesky, M. K.; Gruner, G.; Pei, Q. B. Fault-Tolerant Dielectric Elastomer Actuators Using Single-Walled Carbon Nanotube Electrodes.* Advanced Materials ***2008,*** *20, 621–625, Copyright 2008 Wiley-VCH.)* (D) Actuation of VHB film with aligned CNT sheet as electrodes at increasing electrical field from 0, 80, and 100 V/μm *(Reproduced with permission from Cakmak, E.; Fang, X. M.; Yildiz, O.; Bradford, P. D.; Ghosh, T. K. Carbon Nanotube Sheet Electrodes for Anisotropic Actuation of Dielectric Elastomers.* Carbon ***2015,*** *89, 113–120, Copyright 2015 Elsevier Ltd.)*

After incorporating into a rotatable metal plate, the CNT could serve as a fully synthetic nanoscale electromechanical actuator (Fig. 6.1A) *(5,18)*. A metal rotor was suspended on a CNT, and an electrode was located around the metal rotor. The voltages applied to the electrode and metal rotor could induce a rotation of the CNT, and the rotation output could be tuned by varying the electric field. The devices from an SWNT showed superior actuation performance. The SWNT could turn back to the initial position upon switching off the voltage even though the pendulum was turned by 180 degrees

Other successes in electrostatic actuation devices are nanoelectromechanical switches or servo motors for electromechanical electronic circuits *(19–21)*. This is because CNTs exhibit low friction-bearing capabilities. For instance, to prepare the switch, a CNT was suspended between two electrical contacts and a gap was generated after application of a high voltage to produce an insulating state (i.e., OFF). When a bias voltage was used between the ruptured ends, the switch was actuated by electrostatic forces and underwent linear bearing motion that telescoped the inner shell of the CNT. As a result, the gap was bridged and a conducting state was produced (i.e., ON) (Fig. 6.1B) *(19)*. Though prototype devices were built using the current-driving method, the capability to modify the structure of CNTs in parallel is still a big challenge for eventual large-scale fabrication and commercialization. To solve this problem, a novel biasing approach was explored to ensure high yield and low cycle time. This method was also used to fabricate a single CNT with a sophisticated shell architecture and high-density bearing *(22)*.

CNTs have also been explored to be promising electrodes for other electrostatically driven actuators *(23–25)*. This is because CNTs are capable of forming electrodes with fault-tolerant behavior *(26)*. This kind of actuator is typically composed of a dielectric elastomer film sandwiched between two electrodes. Upon application of a voltage, the bending motion is produced because of the coulombic force on the elastomer film, resulting from the rearrangement of electric charges on the surfaces of the dielectric layer *(27)*. A wide variety of polymers including acrylic, silicones, polyurethanes, and fluoroelastomers are mainly investigated as dielectric layers *(28–30)*. For instance, the actuator prepared from 3M VHB 4910 acrylic film (dielectric layer) and SWNT electrodes could generate a large area strain up to 200% under an electric field of 160 kV/mm (Fig. 6.1C) *(23)*. The actuation performance of the elastomer with SWNT electrodes was even better than that with silver nanowires *(31)*.

Aligned CNT sheet was further used as electrodes for enhancing the actuation performances (Fig. 6.1D) *(32)*. The resulting actuator showed a strain of 40% under a relatively low electric field of 100 kV/mm. Recently, a fiber-shaped electrostatic actuator was developed by twisting two CNT sheets wrapped in rubber fibers. The resulting actuator could exhibit rotation under an electric field. The torsional stroke of the fiber-shaped actuator was 45 degrees/cm with a rotary speed of 3000 rpm under an electric field of 12 MV/m.

6.2.1.2 Electrochemically Driven Actuator

Although electrostatically driven actuators exhibit large actuation strain combined with a rapid responsiveness, they still suffer from the disadvantage of high driving electric field at hundreds of kilovolts per millimeter, which hinders their practical applications *(27)*. To this end, electrochemically driven actuators have attracted increasing attention because of the low operation voltage *(33)*.

The electrochemical actuator is usually composed of a semipermeable ion-exchange polymer membrane sandwiched between two electrodes. The bending actuation is generally produced through the volume changes of electrodes caused by ion migration under the applied electrical field *(34)*. The electrode plays an important role in actuation performance. However, the metal electrodes suffer from many disadvantages including heavy weight, inflexibility, and electrochemical instability. The intercalation and deintercalation of electrolyte ions into the metal electrodes during the actuation process induced partially irreversible phase changes in metal *(34)*. However, CNTs could serve as a promising candidate to endow electrochemical actuators with stable actuation performances. Therefore a lot of effort has been made to develop an actuator from CNTs with two general and effective strategies, such as incorporating CNTs into the electrode layer and directly using CNTs as the electrode.

The SWNT sheets combined with electrolyte were used as electrodes to fabricate the electrochemical actuator *(4)*. The SWNT sheets, laminated together with double-sided Scotch tape, showed a bending actuation in NaCl aqueous solution. A stress of 0.75 MPa was generated when a 0.1-Hz square voltage between −0.5 and 1.5 V was used. The actuating stress was higher than that of natural muscle. The actuation mechanism was quantum chemical-based expansion because of electrochemical double-layer charging, without requiring ion intercalation. However, the high cost and dramatically decreased properties after chemical modification of SWNTs limited their practical applications. A free-standing CNT mat was also used to fabricate electrochemical actuators *(4,35–37)*. The actuator exhibited actuation strains of 0.06–0.2%.

In recent years, as an alternative to disordered CNTs, the highly aligned CNT sheet has been directly used as electrodes *(38)*. The strain level (>8% strain under 4 V) and actuation speed (>10% strain/s) were greatly improved by using vertically aligned CNT/Nafion composite electrodes (Fig. 6.2A). The enhanced performance was attributed to the following three reasons: creating continuous pathways through aligned channels among aligned CNTs for rapid ion migration, minimizing electrical resistance caused by the formation of continuous CNTs, and anisotropically tailoring elastic modulus to enhance actuation strain.

In addition, organic electrolytes and ionic liquids have been mainly used in this kind of actuator *(37,39)*. However, these actuators cannot be operated in air and their performances are also unstable. To this end, all-solid-state actuators based on solid electrolytes were developed to allow rapid and long-life operation in air with high performances

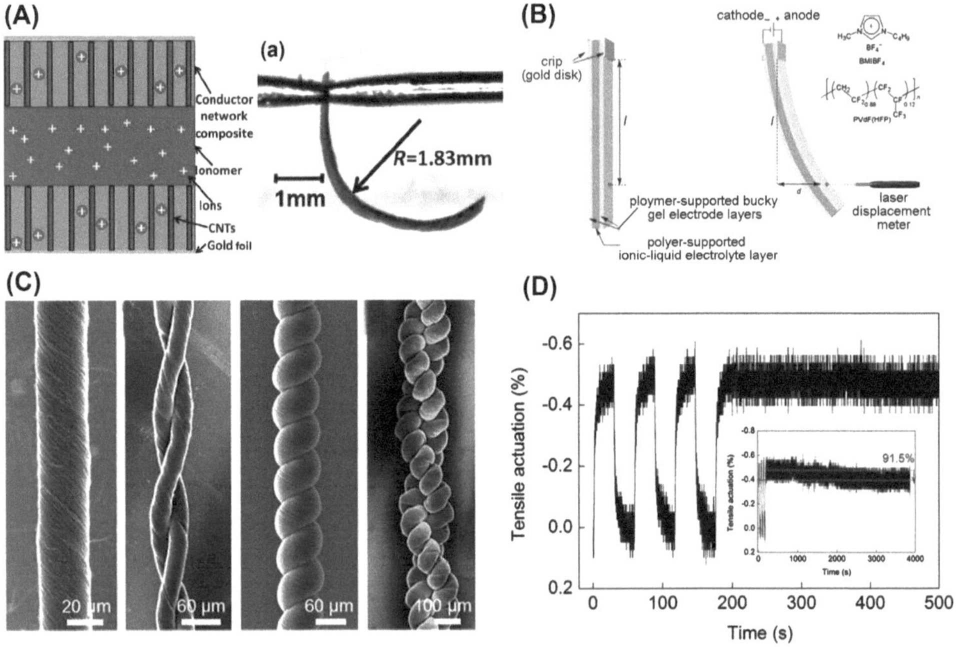

Figure 6.2 ***Electrochemically driven actuators based on carbon nanotube (CNT) materials.*** (A) Schematic illustration of an aligned CNT electrode-based actuator and bending actuation of the actuator at (a) under 4 V *(Reproduced with permission from Liu, S.; Liu, Y.; Cebeci, H.; de Villoria, R. G.; Lin, J. H.; Wardle, B. L.; Zhang, Q. High Electromechanical Response of Ionic Polymer Actuators With Controlled-morphology Aligned Carbon Nanotube/Nafion Nanocomposite Electrodes.* Advanced Functional Materials ***2010,*** *20, 3266–3271, Copyright 2010 Wiley-VCH.)* (B) Bending actuation of the actuator with the electrode layer containing single-walled CNT (SWNT) at an applied voltage of 3.5 V *(Reproduced with permission from Fukushima, T.; Asaka, K.; Kosaka, A.; Aida, T. Fully Plastic Actuator through Layer-by-Layer Casting With Ionic-Liquid-Based Bucky Gel.* Angewandte Chemie International Edition ***2005,*** *44, 2410–2413, Copyright 2005 Wiley-VCH.)* (C) The structures of an all-solid-state twisted CNT fiber actuator. (D) Actuation at (C) with the use of a square-wave voltage (1 V for the magnitude) and constant voltage of 1 V (*inserted graph*) *(Reproduced with permission from Lee, J. A.; Kim, Y. T.; Spinks, G. M.; Suh, D.; Lepró, X.; Lima, M. D.; Baughman, R. H.; Kim, S. J. All-Solid-State Carbon Nanotube Torsional and Tensile Artificial Muscles.* Nano Letters ***2014,*** *14, 2664–2669. Copyright 2014 American Chemical Society.)*

(40,41). The all-solid-state actuator using CNTs was composed of a polymer-supported internal ionic electrolyte layer sandwiched between SWNTs *(40)*. It was found that SWNTs created a physical gel (called bucky-gel) when grounded with imidazolium-based ionic liquids *(42)*. This phenomenon may be derived from the specific interaction between the imidazolium ion component and the π-electronic CNT surface *(40,43)*. The actuator with the electrode layer containing 13 wt% SWNT generated a strain of 0.9% and stress of 0.1 MPa at an applied voltage of ±3.5 V (0.01 Hz) (Fig. 6.2B) *(40)*. The performance of the actuator could be further improved by increasing the SWNT content in the electrode layer. The actuator with 22 wt% SWNT performed the maximal

strain and stress of 4.7 MPa and 1.9% at an operating voltage of ±3.5 V, respectively *(44)*. To exclude the underlying polymer, which would greatly lower the conductivities and capacitances of the electrode layers in the dry actuators, the SWNT electrode-based actuator without polymer was then designed to show a high strain speed of 2.3%/s and high stress rate of 3.26 MPa/s at a voltage of 2.5 V *(45)*. Later, Chen et al. prepared an SWNT/chitosan composite film that could exhibit high electromechanical actuation performance at a low voltage of 5 V *(46)*. This actuator was realized through the use of free-standing SWNT film instead of dispersed CNTs as additives in the electrode. In addition, the actuator exhibited superfast response (19 ms) and an incredibly large stress generation rate (1080 MPa/s). The better actuation performance can be further achieved by using functionalized CNT electrode *(47)*. Because the functioned CNTs had much higher specific surface area and pore volume, and better wettability than the nonactivated CNT electrode, a corresponding increase of the real area of the built-up double layer was achieved. Another effective way to increase the actuation performance is by introducing other capacitive materials, for example, RuO_2 and MnO_2, into the CNT electrode *(48,49)*.

The aforementioned electrochemical actuators are typically in a planer structure so that they mainly generate the bending actuation under the applied voltage. CNT fibers twisted from CNT sheets could also be used to fabricate the fiber-shaped electrochemical actuators with multiple actuations. Mirfakhrai et al. investigated the performance of an electrochemical actuator made of CNT fibers *(39)*. The actuator produced a contraction strain up to 0.5% at an applied voltage of 2.5 V. The electrochemical actuator prepared from CNT fiber exhibited complex and tunable actuations. An electrochemically driven CNT fiber in liquid electrolyte could simultaneously generate a contraction of ~1% and reversible rotation with a speed of 590 rpm *(50)*. The simultaneous occurrence of lengthwise contraction and torsional rotation was explained by the increase in fiber volume induced by an electrochemical double-layer charge injection. The disadvantage of this fiber actuator is exploiting the liquid electrolyte, which was much heavier than the fiber and caused low volumetric and gravimetric performances for the actuating system. Therefore the solid-state CNT fiber actuator without the use of three-electrode eletromechanical setup, liquid electrolyte, and package was developed later (Fig. 6.2C) *(51)*. For instance, the actuator was fabricated by twisting the solid electrolyte-infiltrated CNT fibers together. The performance was nearly unchanged through the introduction of the solid electrolyte. A torsional stroke of 53 degrees/mm and a tensile contraction of 0.52% were achieved by applying a voltage of 5 V (Fig. 6.2D).

6.2.1.3 Electrothermally Driven Actuator

The electrothermally driven actuators based on CNTs have also been widely explored. The CNT sheet was demonstrated as an actuator with large stroke as a result of electrothermal heating *(6)*. The CNT aerogel sheet drawn from CNT forests was electrically conductive, and a rapid electrothermal heating could be generated by the CNT sheet when a current was passed through. Because of the gas-like density and large Poisson's ratio, the CNT sheet generated a lengthwise contraction and widthwise elongation. The generated elongation and elongation rate reached 220% and (3.7×10^4)% per second with increasing operating temperatures from 80 to 1900 K at a voltage of 5 kV, respectively.

CNTs can also function as effective fillers for electrothermal polymers to overcome the high applied voltage constraint of the electrothermal polymers besides enhancing the actuation performances. The electrothermal CNT/polymer composite actuators can be divided into two types. One is a uniform film that can expand or contract with a change under heating and cooling induced by electricity. The actuation is attributed to the thermal expansion and contraction of the composite film. The other is a double-layer film consisting of electrically conductive and polymer layers. The two layers have different coefficients of thermal expansion, and asymmetrical volume expansions in the two layers were generated to cause a bending motion.

The uniform electrothermal CNT/polymer composite actuator was introduced first. For instance, the actuator based on the CNT/polyimide composite with 0.5 vol% SWNT exhibited a reversible strain of 4% under an electric field of 10 V/mm (Fig. 6.3A) *(52)*. However, it needs a relatively high voltage. To this end, more CNTs (5 wt%) were added into the silicone elastomer, and the resulting composite actuator could be operated under a low voltage of 1.5 V/mm *(53)*. Meanwhile, it could generate a maximal strain of 4.4% along with a maximal stress of 0.08 MPa. The biocompatible chitosan could also be used to composite with CNTs to fabricate actuators *(54,55)*. The CNT/chitosan composite actuators showed tunable vibrational motions, including frequency and waveform, which were controlled by the applied wave voltage (<0.9 V/mm) *(55)*.

Among the electrothermally driven actuator with a double-layer structure, the CNT/polydimethylsiloxane (PDMS) bimorph composite actuator was mostly reported (Fig. 6.3B) *(7,56)*. The aligned CNT sheet was embedded into one side of polymer matrix, which could reduce the coefficients of thermal expansion of the polymer by two orders of magnitude (from 3.1×10^{-4} down to 6×10^{-6}/K). The CNT/PDMS composite actuator generated a large bending actuation at a low voltage (<0.7 V/mm) because of the significant thermal expansion difference between the CNT-rich and PDMS layers (Fig. 6.3C). A large-area and highly aligned CNT paper was further used to fabricate the double-layer composite actuator *(56)*. By cutting the composite film in different directions relative to the aligned direction of CNTs, the composite could achieve sophisticated and tunable deformations at a low voltage, accompanied by a large bending magnitude (Fig. 6.3D).

Recently, the electrothermally driven actuators based on CNT fibers have attracted increasing attention for their unique contractive and rotary actuations. Because of the

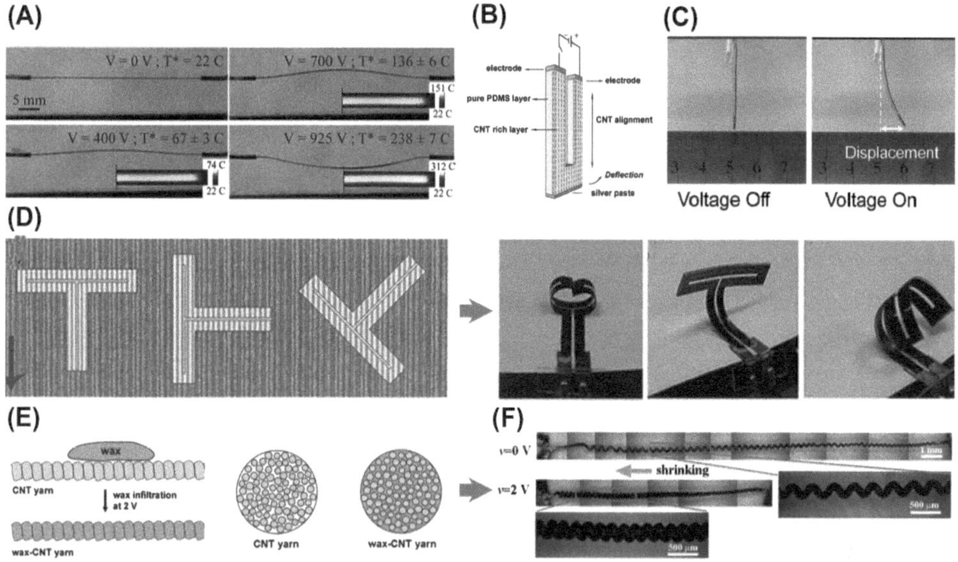

Figure 6.3 ***Electrothermally driven actuators based on carbon nanotubes (CNTs).*** (A) Deformation of a CNT/polyimide film with variation of the applied voltage *(Reproduced with permission from Sellinger, A. T.; Wang, D. H.; Tan, L. S.; Vaia, R. A. Electrothermal Polymer Nanocomposite Actuators.* Advanced Materials ***2010,*** *22, 3430–3435, Copyright 2010 Wiley-VCH.)* (B) Schematic illustration to the structure of the CNT/polydimethylsiloxane (PDMS) composite actuator. (C) The actuation of (B) without and with the application of 40 V *(Reproduced with permission from Chen, L.; Liu, C.; Liu, K.; Meng, C.; Hu, C.; Wang, J.; Fan, S. High-Performance, Low-Voltage, and Easy-Operable Bending Actuator Based on Aligned Carbon Nanotube/Polymer Composites.* ACS Nano ***2011,*** *5, 1588–1593, Copyright 2011 American Chemical Society.)* (D) Preparation and actuation of double-layer CNT/PDMS film actuators with the CNTs aligned vertically, horizontally, and diagonally *(Reproduced with permission from Li, Q.; Liu, C.; Lin, Y.-H.; Liu, L.; Jiang, K.; Fan, S. Large-Strain, Multiform Movements From Designable Electrothermal Actuators Based on Large Highly-Anisotropic Carbon Nanotube Sheets.* ACS Nano ***2015,*** *9, 409–418, Copyright 2015 American Chemical Society.)* (E) Schematic illustration to the preparation of CNT/wax composite CNT fiber. (F) Photographs of the fiber actuator showing the shrinking actuation at a voltage of 2 V *(Reproduced with permission from Shang, Y.; He, X.; Wang, C.; Zhu, L.; Peng, Q.; Shi, E.; Wu, S.; Yang, Y.; Xu, W.; Wang, R. Large-Deformation, Multifunctional Artificial Muscles Based on Single-Walled Carbon Nanotube Yarns.* Advanced Engineering Materials ***2015,*** *17, 14–20, Copyright 2015 Wiley-VCH.)*

massive pores among the aligned CNT fiber, it is beneficial to be further incorporated with guest materials with a high coefficient of thermal expansion to prepare electrothermally driven actuators with complex actuation. The expansion of the guest material increases the distance among the neighboring CNTs for actuations. Among these guest materials, paraffin wax is one of the most used candidates because of its large volume change, tenability of transition temperatures, and wettability to CNTs *(57)*. The wax was introduced into CNT fiber by a hot wire method *(57)*. Upon the use of a pulse current (20 Hz and 15 V/cm), the CNT/wax composite fiber generated a contraction of 10% under a tensile strength of 5.5 MPa. Besides, the two-end-tethered, half-wax-infiltrated, and multi-ply twisted fiber could accelerate a 16.5-times-heavier paddle to a rotation speed of 11,500 rotations per minute. SWNT fibers have also been used as actuators and artificial muscles. The SWNT/wax composite coiled fiber produced a large contraction of 90%, and thus could exhibit a variety of promising functionalities such as dragging an object, self-twisting, and tightening an object (Fig. 6.3E and F) *(58)*. However, the wax-infiltrated, electrothermally powered actuations were torsionally underdamped, so it experienced unwanted dynamic oscillations, which needed complicated control of the torsional and tensile actuation stroke when a rapid response was required.

Learning from the strategy of spiders to eliminate uncontrolled spinning at the end of dragline silk, a triblock copolymer (polystyrene-poly(ethylene—butylene)-polystyrene triblock copolymer, SEBS) was introduced into the paraffin to develop a hybrid CNT fiber without the oscillation *(59)*. The CNT composite fiber with the weight ratio between wax and SEBS of 8/2 generated a 9800 rpm rotation without obvious oscillation when excited by a 0.5-Hz square-wave voltage. As expected, the actuation could be also achieved by a heating source such as radiation from an incandescent lamp. For instance, a reversible torsional actuation (12.6 degrees/mm) could be produced by a half-wax-infiltrated CNT fiber after being exposed to an incandescent lamp *(57)*.

6.2.1.4 Electromagnetically Driven Actuator

Although the aforementioned actuators from CNTs have some advantages, there are many challenges for each of them. For instance, high voltages are required for the electrostatic actuators, and slow responsiveness and short cyclic life are still the main problems for electrothermal and electrochemical actuators. The macroscopic CNT fiber has been found to be an excellent actuator based on an electromagnetic mechanism *(60)*. The actuators are composed of pure CNTs without other guest materials such as electrolytes or other electroactive polymers. The actuators exhibited a high actuation output with a low applied voltage, rapid responsiveness, and long cyclic life.

The macroscopic CNT fibers by spinning from the CNT arrays had a helical structure and the fibers could simultaneously contract and rotate upon application of a small current of several microamperes (Fig. 6.4A) *(60)*. This generated an actuation strain of ~2% and rotation of 360 degrees. A mechanism based on Ampere's law among helically aligned CNTs was proposed to explain the phenomenon. In detail, a CNT fiber could be considered to be many parallel disks along the fiber, and the electric current along one disk could exert attractive and torsional electromagnetic forces simultaneously on the neighboring disk. The collected electromagnetic forces from all parallel disks made the whole CNT fiber contract and rotate (Fig. 6.4B). Particularly, the CNT fiber actuator based on the novel electromagnetic mechanism could be operated in various media including air, water, and organic solvents *(60)*. Besides, the additional electrothermal effect generated in the CNT fiber could allow for an efficient curing and reordering of thermosetting and thermoplastic polymers infiltrated within the fibers *(61)*.

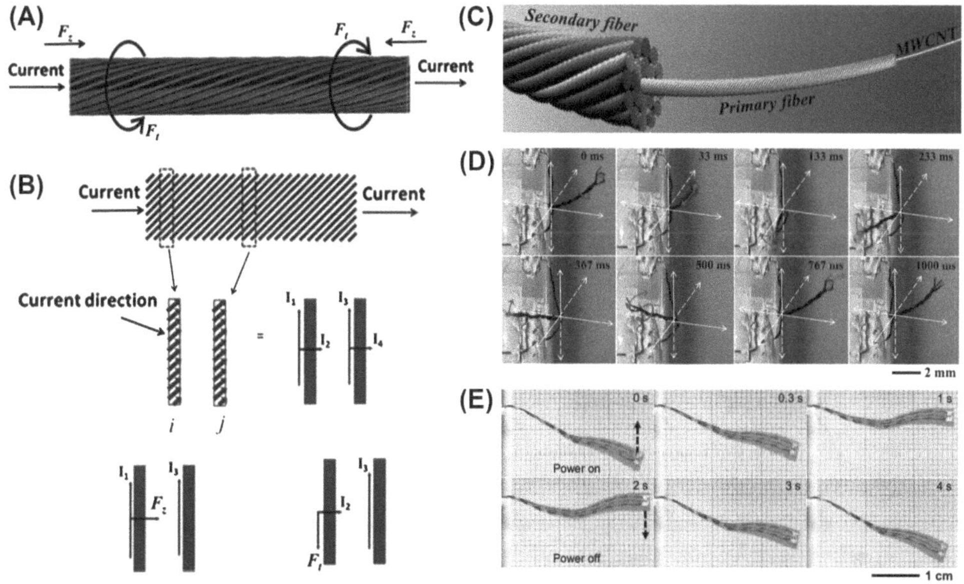

Figure 6.4 ***Electromagnetically driven actuations based on carbon nanotube (CNT) fibers.*** (A and B) Schematic illustration of the actuation and mechanism of an electromagnetically driven CNT fiber *(Reproduced with permission from Guo, W.; Liu, C.; Zhao, F.; Sun, X.; Yang, Z.; Chen, T.; Chen, X.; Qiu, L.; Hu, X.; Peng, H. A Novel Electromechanical Actuation Mechanism of a Carbon Nanotube Fiber.* Advanced Materials ***2012,*** *24, 5379–5384, Copyright 2012 Wiley-VCH.)* (C) Schematic illustration to the helical assembly of CNT fiber and (D) the waggling motion of an artificial tail from CNT fibers *(Reproduced with permission from Chen, P.; Xu, Y.; He, S.; Sun, X.; Guo, W.; Zhang, Z.; Qiu, L.; Li, J.; Chen, D.; Peng, H. Biologically Inspired, Sophisticated Motions from Helically Assembled, Conducting Fibers.* Advanced Materials ***2014,*** *27, 1042–1047. Copyright 2014 Wiley-VCH.)* (E) Photographs of the bending motions generated by a Kapton ribbon woven from the fiber upon application of a pulse current *(Reproduced with permission from Chen, P.; He, S.; Xu, Y.; Sun, X.; Peng, H. Electromechanical Actuator Ribbons Driven by Electrically Conducting Spring-Like Fibers.* Advanced Materials ***2015,*** *27, 4982–4988, Copyright 2015 Wiley-VCH.) MWCNT,* Multiwalled carbon nanotube.

The hierarchically helical CNT fibers were further assembled by twisting multi-ply primary fibers for more powerful electromechanical actuations (Fig. 6.4C) *(62)*. Compared with the single CNT fiber, the resulting secondary fiber exhibited improved actuation performances. For instance, the contractive stress of the secondary fibers prepared by 50 primary fibers was over nine times that of the single primary fiber. The actuating fiber could realize some sophisticated and programmable biomimetic motions such as the flapping motion of an artificial wing and the wagging motion of an artificial tail (Fig. 6.4D). The secondary CNT fiber could also be overtwisted into a spring-like configuration to enhance the contractive strain and rotation output. Upon application of a voltage, the spring-like fiber actuator could generate an actuation strain of 20% and rotary speed of 978 rpm, respectively. These electromagnetic actuations generated by the CNT fiber were rapid and reversible. They could be performed at low voltage (<10 V/cm) and be repeated for thousands of cycles without fatigue because of the excellent electrical and mechanical properties. In addition, a single-layer fabric ribbon with tunable electromechanical actuation could be obtained by integrating spring-like CNT fiber into a polymer film through a sewing method (Fig. 6.4E) *(63)*. The ribbon actuator could be further made into a crawler and helically three-dimensional structures through a heat-setting method. The resulting ribbon actuators could generate programmable electromechanical actuations such as bending, rotation, contraction, and elongation with high reversibility and stability.

6.2.2 Actuators Driven by Solvent

Polymer is the main investigated material to develop a solvent-induced actuator *(64–67)*. For conducting polymers, the incorporation of aligned CNTs can induce the orientation of polymers with conjugated backbones to achieve controllable actuation. Poly[1-phenyl-2-(*p*-trimethylsilyl)phenylacetylene] (PTP), for example, can perform a remarkable actuation because of the changed interchain distance upon absorption and evaporation of organic solvent (*68,69*). By spin coating PTP solution onto aligned CNT sheets paved on a PDMS substrate, the CNT/PTP composite film was prepared *(70)*. Consequently, the conjugated polymer chains could be oriented by the aligned CNTs, so the composite films exhibited anisotropic bending and unbending deformations with controlled direction along the perpendicular direction relative to the CNT length. The film could roll into a tube as it was exposed to organic solvent (Fig. 6.5A). A solvent-induced fiber actuator has been also developed by introducing silicone rubber in coiled CNT fiber *(71)*. The actuation of the composite fiber was driven by the swelling of rubber upon exposure to solvent. The actuator could generate a contractive strain of 50%, accompanied by a high-energy density of 1.2 kJ/kg. In addition, the conversion efficiency of chemical energy to mechanical energy could reach up to 4.3% (Fig. 6.5B).

These solvent-induced CNT actuators are based on the swelling of the guest polymers. However, they suffer from certain disadvantages such as low stress, slow responsiveness, and poor actuation stability, which limit their practical applications *(70)*. As solvent can be infiltrated into the void of the CNT fiber, the fiber with a helical structure could generate torsional actuation in acetonitrile and hexane. The actuation was shown as a function of immersion length in these solvents *(57)*. A general strategy has been developed to prepare a solvent-driving fiber actuator from the pure CNTs with high actuating performances *(72)*. The actuation performances in response to solvent and vapor could be greatly enhanced through a hierarchical and helical assembly of aligned CNTs. The hierarchically arranged helical fiber with massive nano- and micron-sized gaps was prepared by overtwisted multi-ply primary CNT fibers, and it exhibited a rapid response of ~8 MPa/s, a large actuation stroke of ~10%, and a rapid rotary speed of 6361 rpm

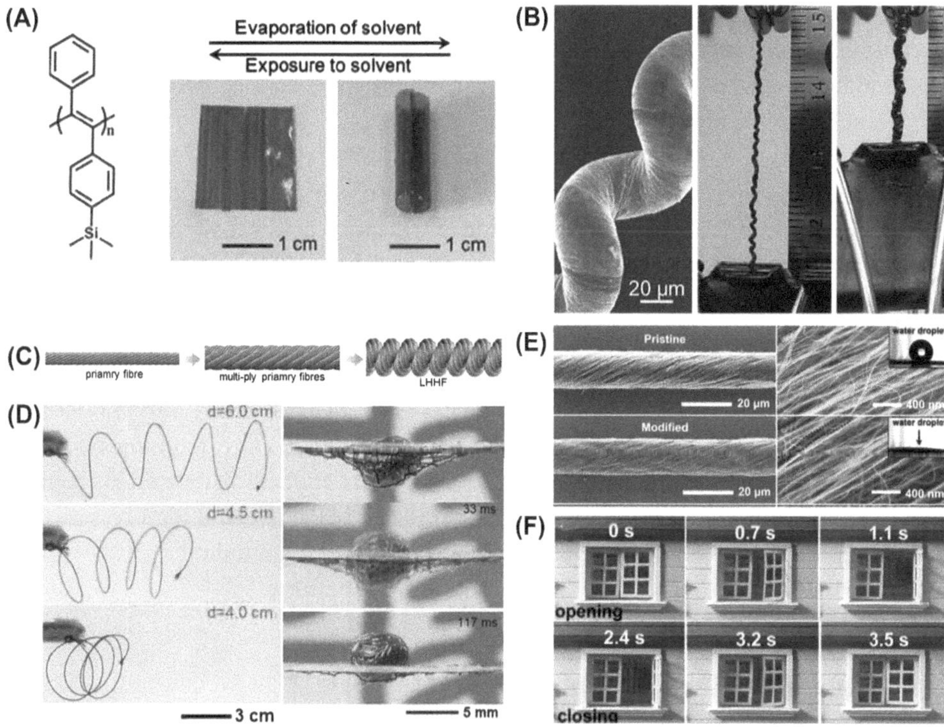

Figure 6.5 ***Actuators driven by solvents.*** (A) Molecular structure of poly[1-phenyl-2-(p-trimethylsilyl) phenylacetylene] (PTP) and photographs of a carbon nanotube (CNT)/PTP composite film being rolled and recovered upon evaporation and absorption of ethanol *(Reproduced with permission from Lu, X.; Zhang, Z.; Li, H.; Sun, X.; Peng, H. Conjugated Polymer Composite Artificial Muscle With Solvent-Induced Anisotropic Mechanical Actuation.* Journal of Materials Chemistry A ***2014,*** *2, 17272–17280, Copyright 2014 Royal Society of Chemistry.)* (B) Scanning electron microscope (SEM) image of a thin coiled CNT/silicone hybrid yarn and photographs of the fiber showing 50% contraction when exposed to hexane under a 2-MPa tensile stress *(Reproduced with permission from Lima, M. D.; Hussain, M. W.; Spinks, G. M.; Naficy, S.; Hagenasr, D.; Bykova, J. S.; Tolly, D.; Baughman, R. H. Efficient, Absorption-powered Artificial Muscles Based on Carbon Nanotube Hybrid Yarns.* Small ***2015,*** *11, 3113–3118, Copyright 2015 Wiley-VCH.)* (C) Schematic illustration of the creation of a hierarchically arranged helical fiber. It was prepared from multi-ply primary fibers through a twisting process, LHHF in the figure refers to the left-handed hierarchically arranged helical fiber. (D) Contractive deformations of smart springs and woven fabric prepared from the hierarchically arranged helical fiber in response to vapor and solution of ethanol *(Reproduced with permission from Chen, P.; Xu, Y.; He, S.; Sun, X.; Pan, S.; Deng, J.; Chen, D.; Peng, H. Hierarchically Arranged Helical Fibre Actuators Driven by Solvents and Vapours.* Nature Nanotechnology ***2015,*** *10, 1077–1083, Copyright 2015 Macmillan Publishers Ltd.)* (E) SEM images of a pristine primary CNT fiber and the CNT fiber after plasma modification at low and high magnification. (F) Schematic illustration of the smart window in response to water. *(Reproduced with permission from He, S.; Chen, P.; Qiu, L.; Wang, B.; Sun, X.; Xu, Y.; Peng, H. A Mechanically Actuating Carbon-Nanotube Fiber in Response to Water and Moisture.* Angewandte Chemie International Edition ***2015,*** *54, 14880–14884, Copyright 2015 Wiley-VCH.)*

(Fig. 6.5C). In addition, the compactly arranged coil formed inside contributed to the reversible and stable rotation when one end was not stabilized. These fibers were promising in many applications including energy-harvesting generators, deformable sensing springs, and smart textiles (Fig. 6.5D).

However, because of the hydrophobic nature of CNTs, the CNT fiber could mechanically respond to the organic solvents, which limited the practical applications. Water is a significant stimulus to trigger movements and transports in living systems to ensure life sustainability in nature. By transferring the surface of the CNT fiber to hydrophilic through oxygen plasma treatment, the modified fiber showed a rapid and powerful actuation in response to water (Fig. 6.5E) *(73)*. The oxygen plasma treatment was a readily, mild, and easily scaled-up method, which could maintain the helical alignment of CNTs well. The hydrophilic CNT fibers were further twisted into a multi-ply larger fiber with hierarchically helical channels to mimic the nature structure. The resulting fiber performed larger actuation output than that of the fiber only containing nanoscale channels. The fiber exhibited a high responding stress rate of 22 MPa/s with a high reversibility, which was promising for various applications in our daily life including a smart window and louver that responded to water and moisture (Fig. 6.5F).

6.2.3 Actuators Driven by Light

Because light can be controlled remotely, instantly, and precisely, a light-driving actuator can realize remote and wireless control in many fields, such as an artificial muscle, a photomobile soft actuator, and in microphotomechanical systems.

CNTs can absorb light over a wide range of wavelengths and convert the light to local heat *(74)*. Therefore many photoactuators have been developed based on CNTs and thermally driven materials. A number of polymers and ceramics have been used to fabricate the photoactuators by the incorporation of CNTs, such as PDMS *(75–78)*, polycarbonate (PC) *(79)*, liquid crystal (LC) polymers *(80–85)*, ethylene vinyl acetate *(86)*, and vanadium dioxide bilayers *(87)*.

A CNT/PDMS composite photoactuator was fabricated by dispersing a small content of CNTs (1 wt%) in a PDMS matrix *(75)*. The composite film showed compression and expansion in response to the infrared (IR) light irradiation. The direction of deformation was controlled by the external uniaxial prestrain applied to the composite, which was induced by the alignment of CNTs. The composite expanded under IR irradiation with small prestrains in the range of 2–8%. When the prestrain was higher than 10%, the composite contracted in response to IR radiation. This phenomenon was applicable for both MWNTs and SWNTs *(76)*. To improve the dispersion of CNTs in the polymer matrix, a polystyrene chain was grafted to the CNT surface and then dispersed in the linear triblock copolymer styrene-*b*-isoprene-*b*-styrene matrix to obtain the composite photoactuator *(88)*.

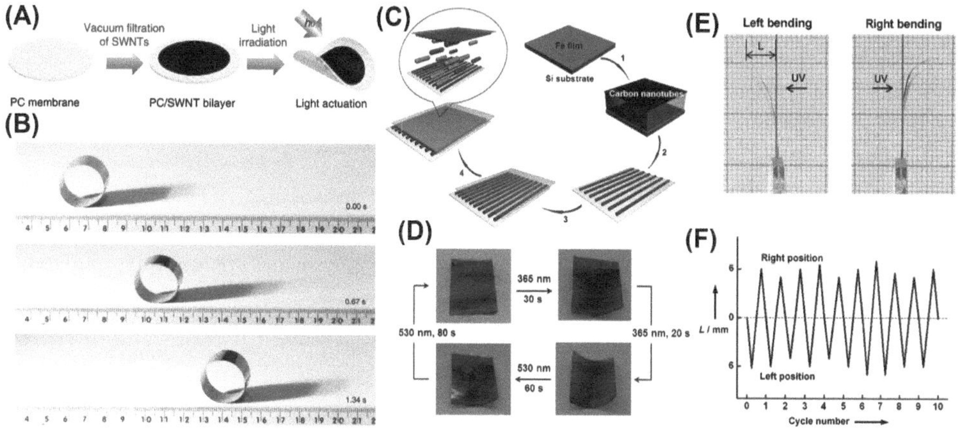

Figure 6.6 ***Light-driving actuations of carbon nanotube (CNT)/polymer composite materials.*** (A) Schematic illustration to the single-walled CNT/polycarbonate (SWNT/PC) composite film with a bilayer structure. (B) Photographs of a moving SWNT/PC motor driven by a halogen lamp *(Reproduced with permission from Zhang, X.; Yu, Z.; Wang, C.; Zarrouk, D.; Seo, J.-W. T.; Cheng, J. C.; Buchan, A. D.; Takei, K.; Zhao, Y.; Ager, J. W. Photoactuators and Motors Based on Carbon Nanotubes With Selective Chirality Distributions.* Nature Communications ***2014,*** *5, 3983, Copyright 2014 Macmillan Publishers Ltd.)* (C) Schematic illustration of the preparation of a CNT/liquid crystal (LC) polymer composite film. (D) Photographs of the CNT/LC polymer composite film under irradiation by UV and visible light with a bending behavior toward the light source *(Reproduced with permission from Wang, W.; Sun, X.; Wu, W.; Peng, H.; Yu, Y. Photoinduced Deformation of Crosslinked Liquid-Crystalline Polymer Film Oriented by a Highly Aligned Carbon Nanotube Sheet.* Angewandte Chemie International Edition ***2012,*** *124, 4722–4725, Copyright 2012 Wiley-VCH.)* (E) Photographs of the CNT/LC polymer composite under UV light with a bending behavior away from the light source. (F) Bent distances during reversible deformation at (E) *(Reproduced with permission from Sun, X.; Wang, W.; Qiu, L.; Guo, W.; Yu, Y.; Peng, H. Unusual Reversible Photomechanical Actuation in Polymer/Nanotube Composites.* Angewandte Chemie International Edition ***2012,*** *51, 8520–8524, Copyright 2012 Wiley-VCH.)*

Besides, a simple and interesting bilayer CNT/PC composite photoactuator was also fabricated by vacuum filtration of an SWNT solution on a PC membrane (Fig. 6.6A) *(79)*. The device performed a large deflection of ~100 degrees, fast response of 0.5 s, and a controllable wavelength response. The SWNTs converted the absorbed photon energy to heat, then the heat was transported to the PC layer. As the PC layer had a much larger coefficient of thermal expansion (~65 ppm per K) than that of SWNTs (<4 ppm per K), the film bent toward the SWNT side upon light irradiation. The CNT/PC bilayers could be used to make a rapid forward-moving motor, which was driven by light, as shown in Fig. 6.6B.

A CNT/LC polymer composite was prepared via mixing CNTs into the polymer matrix *(85)*. With a small concentration of CNTs (0.01%), the thermal property did not change but the LC polymer was sensitive to infrared-visible light, so the actuation stress (7.8 kPa) was greatly increased compared with a pure LC polymer (0.5 kPa). However, the stress was slightly decreased with a further increase of the CNT concentration to 1%. A cantilever made of CNT/LC composite performed dynamic bending upon exposure to light *(83)*. The movement was produced by inhomogeneous strain from exponential heat generation across the width of the cantilever. The bulk photothermal actuator with three dimensions could also be achieved by the combination of SWNTs with LC polymer, and could contract to 25% of the initial height under IR irradiation *(81)*. The contents of CNTs were very small (only 0.02 wt%) in the LC polymer because of the aggregation of CNTs in the matrix at higher concentrations. The use of poly(*p*-phenylene ethynylene) to functionalize SWNT favored a homogeneous dispersion of CNTs in the LC polymer matrix *(80)*. The resulting composite film achieved a large and reversible IR-induced strain of ~30% when the concentration of SWNT was 0.1−0.2 wt%.

In addition to the photothermal effect, LC polymers can also change their shapes based on the transition between nematic and isotropic phases derived from the isomerization of photoactive components, such as azobenezene. Therefore the LC polymer may reversibly change their bulk dimensions in response to light because of the *trans*- to *cis*-photoisomerization of azobenzene *(89)*. In this case, the aligned direction of LC mesogens dominates the deformation behavior of the LC films *(90)*. Aligned CNTs with parallel nanogrooves were then used to achieve the orientation of the LC polymer and a photodeformable CNT/LC polymer nanocomposite film was successfully prepared *(84)*. The composite film was prepared by injecting molten precursors into the LC cell that was made of two CNT sheet-covered glass slides, followed by photopolymerization (Fig. 6.6C). The aligned nanostructure of the CNT sheet could effectively orient the LC mesogens along the length of the CNTs instead of using the other aligning layer. The CNT/LC polymer composite film performed bending and unbending by alternating the irradiation with ultraviolet (UV) and visible light (Fig. 6.6D). The orientation of mesogens could be further tuned by changing the fabrication process. The mesogens preferred to orient perpendicularly to the aligned CNTs when the precursor was introduced to the aligned CNTs by the solution process *(91)*. The resulting composite film showed an unusual deformation by using the UV light alone. Specifically, it bent against the incident direction of UV light and was reversed by the opposite irradiation also by UV light (Fig. 6.6E and F). The mechanism is still unclear, and more effort is required to understand it. Moreover, the addition of the aligned CNTs greatly increased the mechanical strength and stability and also offered an electrically conductive property to the photodeformable film. Therefore the composite could be used as a remote switch for electrical circuits.

6.3 CHROMIC MATERIALS FROM CNTs

Another kind of responsive material from CNTs is chromic material. Chromic materials are mainly driven by electricity and can be classified into three kinds according to the chromatic mechanism: electrochromatic, electrothermal chromatic, and electrochemical chromatic.

6.3.1 Electrochromatic Material

Polydiacetylene (PDA), which can change color from blue to red under various external stimuli, has attracted increasing interest to fabricate a variety of chemically and physically based sensors *(92)*. The color change is derived from the conformation change of PDA induced by the stimuli. However, the stimulus for the chromatism of PDA is limited to temperature, pH, chemical, and mechanical stress. Although the electricity can be easily controlled and has been widely used at our life, it is difficult to achieve the color change of PDA by the electricity as PDA displays low electrical conductivity. In addition, most chromatic responses are irreversible, limiting their applications in colorimetric sensing. Much effort has been made to improve reversibility, sensitivity, and increase the scope of stimuli *(93—94)*. To develop the current-induced chromatic behavior in PDA, CNT fiber was introduced to PDA to realize the desired electrochromism.

The CNT/PDA composite fibers were prepared by directly coating diacetylenic precursors onto the fiber, followed by topochemical polymerization under UV light (Fig. 6.7A) *(95)*. The fiber rapidly changed color from blue to red upon passing a current, and reversed to blue after removal of the current. The minimum current to achieve chromatic transition was 10 mA for a fiber with a diameter of 11 μm. The colorimetric reversibility of CNT/PDA fibers could be controlled by varying the value of the current, that is, the color transition was reversible for tens of cycles when current was lower than 30 mA, but once the current was higher, the transition was irreversible. This mechanism was proposed on the basis of interactions between CNTs and PDA and the high conductivity with three-dimensional hopping conduction of the composite fiber. That is to say, electrons possibly hop from one CNT to another inside the fiber, resulting in electric fields among neighboring CNTs when passed with a current. These electric fields could induce the polarization of COOH groups at side chains and conjugated PDA backbones in neighboring CNTs *(96)*. Polarization could then induce the color change of the fiber, similar to the other stimuli. The composite fiber was very promising for applications in many fields such as sensors, actuators, and other electronic devices.

6.3.2 Electrothermal Chromatic Material

Because of the high conductivity and joule-heating effect, CNTs can be used as nanoheaters to induce a color change of thermochromic materials. Among them, PTP has been reported to show a reversible change in fluorescent intensity under the stimulus of heat, which can be used for fluorescent image patterning *(97)*. It was produced by an occurrence of a slight molecular perturbation and exciton deconfinement in the

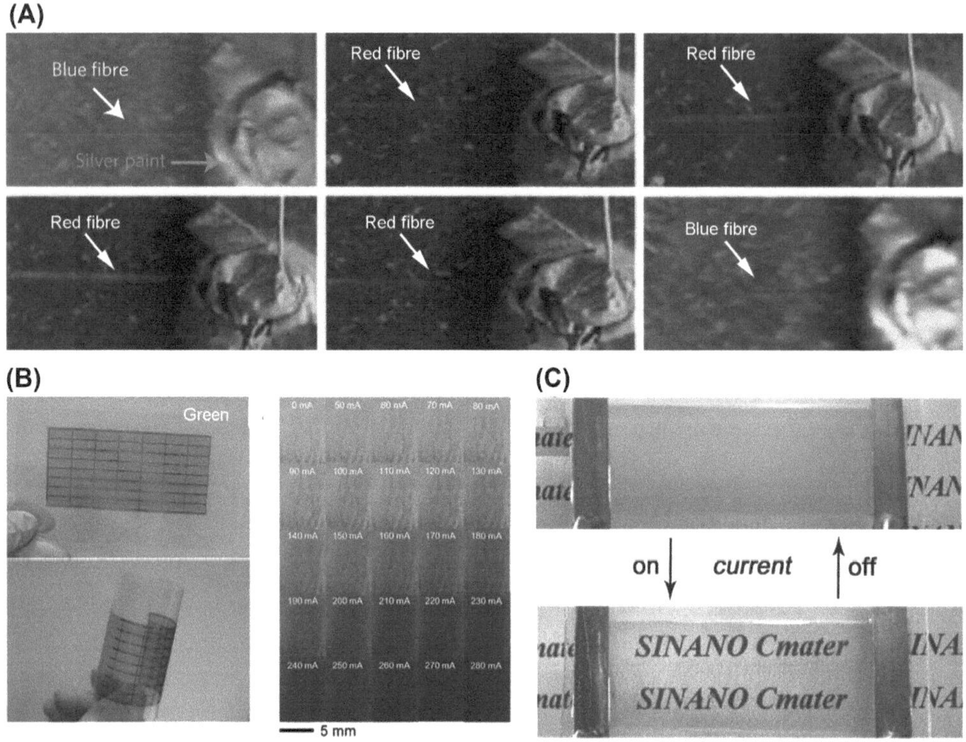

Figure 6.7 ***Electrochromatic carbon nanotube (CNT)/polymer composite materials.*** (A) Photographs of the electrochromatic CNT/polydiacetylene (PDA) composite fiber with reversible color change between blue and red *(Reproduced with permission from Peng, H.; Sun, X.; Cai, F.; Chen, X.; Zhu, Y.; Liao, G.; Chen, D.; Li, Q.; Lu, Y.; Zhu, Y. Electrochromatic Carbon Nanotube/Polydiacetylene Nanocomposite Fibres.* Nature Nanotechnology ***2009,*** *4, 738–741, Copyright 2009 Macmillan Publishers Ltd.)* (B) Photographs of the flexible electrochromatic CNT/poly[1-phenyl-2-(p-trimethylsilyl)phenylacetylene] (PTP) composite film with color change from yellowish green to dark green with increasing current from 0 to 280 mA *(Reproduced with permission from Sun, X.; Zhang, Z.; Lu, X.; Guan, G.; Li, H.; Peng, H. Electric Current Test Paper Based on Conjugated Polymers and Aligned Carbon Nanotubes.* Angewandte Chemie International Edition ***2013,*** *125, 7930–7934, Copyright 2013 Wiley-VCH.)* (C) Photographs of CNT/polyurethane (PU) composite film under opaque and transparent states with an electric current on and off, respectively *(Reproduced with permission from Meng, F.; Zhang, X.; Xu, G.; Yong, Z.; Chen, H.; Chen, M.; Li, Q.; Zhu, Y. Carbon Nanotube Composite Films With Switchable Transparency.* ACS Applied Materials & Interfaces ***2011,*** *3, 658–661, Copyright 2011 American Chemical Society.)*

main chain at high temperature *(98)*. Moreover, PTP exhibited a high thermal stability and its decomposition temperature was above 400°C. Because the conjugated PTP backbone was surrounded by a rigid "jacket" formed through the intra- and interchain molecular interactions of benzene and silicon moieties in the side chains, so the polymer main chain was protected from the thermal attack *(99)*. Thus PTP represents a promising candidate to fabricate electrothermal chromic devices.

Electrothermal chromic CNT/PTP composite film was prepared by casting a PTP solution onto the aligned CNT sheet, followed by evaporation of the solvent at room temperature *(100)*. The composite film changed color from yellowish green to dark green with the applied current increasing from 0 to 280 mA. The response was very fast (~1 s). The color transition was reversible when the current was below 130 mA, and could be repeated for over 1000 cycles. Therefore the electrochromatic films were promising for practical sensing applications as current test papers, similar to the pH test paper (Fig. 6.7B). What is more, the CNT/PTP composite material could be made into a fiber format, which is available in integrated or portable electronic devices. The composite fiber also showed a rapid and reversible fluorescent quenching when currents were applied but with much lower values, that is, the color was changed from yellowish green to dark green with the current increasing from 0 to 32 mA. The highly aligned CNTs may contribute to the conversion from electrical energy to thermal energy.

Polyurethane (PU) is another useful polymer for the thermal chromatic composite, showing a switchable transparency in response to temperature. It is produced by a thermal effect of the soft segment, which undergoes a crystal melting. The soft segment crystallized in randomly distributed large-size spherulites before melting, so PU is opaque because of the scattering of crystal grains and boundaries. After the temperature reaches the crystal melting temperature, the soft segments melt into an amorphous structure and refract, reflect, and transmit light, so it becomes transparent.

The electrochromic CNT/PU composite film was prepared by sandwiching an aligned CNT sheet between two PU films via a simple solution process *(101)*. Because of the joule heating of CNTs, the composite film became transparent when subjected to a voltage, and it reversed to opaque after the removal of the voltage (Fig. 6.7C). In addition, the spherulites were changed to smaller sizes with improved alignment compared with the case without the addition of CNTs, resulting in a decrease of the melting temperature from 52 to 47°C. Therefore the composite film exhibited a more rapid response when applied to a voltage than that of the pure PU under heating, and it also showed a higher thermal stability.

6.3.3 Electrochemical Chromatic Material

Electrochemical chromatic materials are the most studied chromatic materials. Since CNTs show excellent electrical and electrochemical properties, they can be directly used to fabricate electrochromic devices. Moreover, CNTs can be introduced to the electrochemical chromatic devices as electrodes or directly combined with other electrochemical chromatic materials to further improve their performances.

The CNTs can intrinsically show blue-green, magenta, and yellow colors by using metallic SWNTs with different average diameters of about 1.4, 1.0, and 0.84 nm, respectively. The electrochemical metallic SWNT films can change their colors under the three-electrode system *(8)*. The pure SWNTs played the role of both a working electrode and an electrochromic component with a Pt wire as the counter, an Ag/Ag^+ as reference electrode, and an ionic liquid [*N*,*N*,*N*-trimethyl-*N*-propylammonium *bis*(trifluoromethanesulfonyl)imide] as electrolyte solution. The blue-green-, magenta-, and yellow-colored metallic SWNT films changed to yellow, yellow-orange, and light yellow, respectively, when they were applied with a voltage of −3.0 V (Fig. 6.8A–C). The color

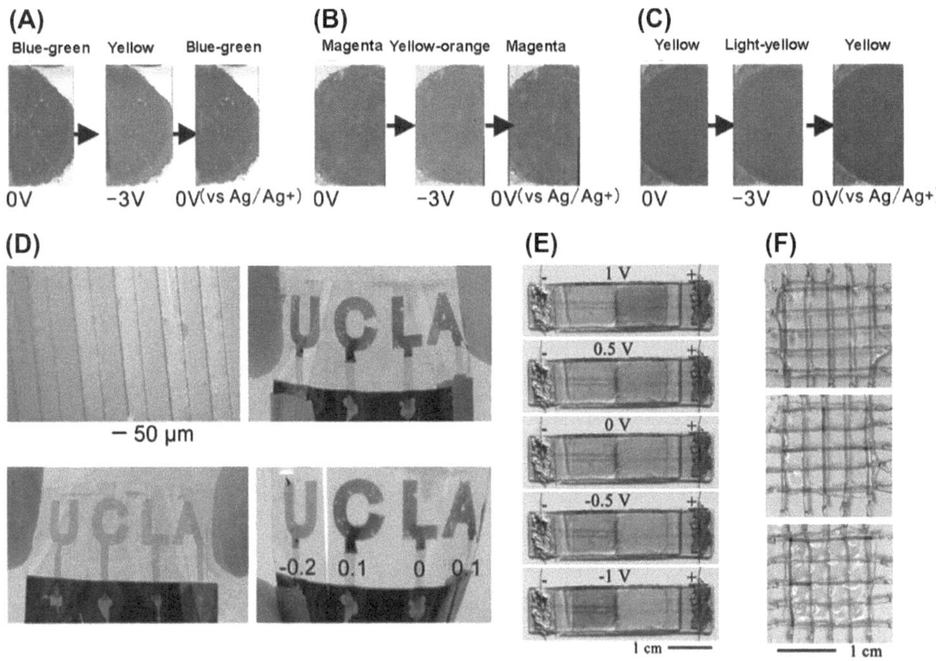

Figure 6.8 ***Electrochemical chromatic carbon nanotube (CNT) and CNT/polymer composite materials.*** (A–C) Photographs of color change generated by metallic single-walled CNT (SWNT) films with diameters of 1.4, 1.0, and 0.84 nm, respectively, when the voltages were changed from 0 to −3 V and then shifted to 0 V *(Reproduced with permission from Yanagi, K.; Moriya, R.; Yomogida, Y.; Takenobu, T.; Naitoh, Y.; Ishida, T.; Kataura, H.; Matsuda, K.; Maniwa, Y. Electrochromic Carbon Electrodes: Controllable Visible Color Changes in Metallic Single-Wall Carbon Nanotubes.* Advanced Materials ***2011,*** *23, 2811–2814, Copyright 2011 Wiley-VCH.)* (D) Photographs of color change generated by a patterned SWNT/poly(aniline) (PANI) composite on poly(ethylene glycol terephthalate) under different voltages *(Reproduced with permission from Hu, L.; Gruner, G.; Li, D.; Kaner, R.B.; Cech, J. Patternable Transparent Carbon Nanotube Films for Electrochromic Devices.* Journal of Applied Physics ***2007,*** *101, 016102, Copyright 2007 American Institute of Physics.)* (E) Photographs of color change generated by an electrochromic supercapacitor during the charge–discharge process *(Reproduced with permission from Chen, X.; Lin, H.; Chen, P.; Guan, G.; Deng, J.; Peng, H. Smart, Stretchable Supercapacitors.* Advanced Materials ***2014,*** *26, 4444–4449, Copyright 2014 Wiley-VCH.)* (F) Photographs of electrochromic fiber-shaped supercapacitors that were woven to display the signs of "+" and "F" *(Reproduced with permission from Chen, X.; Lin, H.; Deng, J.; Zhang, Y.; Sun, X.; Chen, P.; Fang, X.; Zhang, Z.; Guan, G.; Peng, H. Electrochromic Fiber-Shaped Supercapacitors.* Advanced Materials ***2014,*** *26, 8126–8132, Copyright 2014 Wiley-VCH.)*

changes were reversible and repeatable, and the response was also fast (about 1 s). The color change was produced from the injection of electrons and holes in the ionic liquid molecules on the surface of the SWNTs.

Apart from CNTs, conjugated polymers are the most investigated electrochromic materials including the derivatives of poly(thiophene), poly(pyrrole), and poly(aniline) (PANI). They share the advantages of easy processability, high optical contrasts, reversible color change, and rapid response. The color change of the conjugated polymer is derived from the change in the π-electronic character of the conjugated polymer, that is, the insertion and extraction of ions through the polymer during electrochemical oxidation and reduction. For instance, the original color of the PANI is green; it turns dark blue at the oxidized state and bright yellow at the reduced state *(102)*.

The CNT film with high electrical conductivity can be used as transparent and flexible electrodes in the electrochromic devices with the electrochromic polymer as the active material. A typical electrochromic device can be regarded as an electrochemical cell, which is composed of two electrodes with electrochromic materials separated by an ionic conducting layer. The colors are changed on the basis of the electrochemical reactions of the electrochromic materials. A flexible electrochemical chromic device was fabricated by paving the SWNT film on polyethylene glycol terephthalate as the electrode and PANI as the electrochromic material *(103)*. The electrochromic device displayed yellow, yellow-green, and green colors under different applied voltages (−0.2, 0.1, and 0 V), and the color stayed unchanged for ~20 s after removing the SWNT film from the device (Fig. 6.8D).

By introducing CNTs into the electrochemical chromic polymer, the mechanical strength, electrical conductivity, and electroactivity can be enhanced. Meanwhile, the incorporation of CNTs does not alter the electrochemical reaction of polymer (*104,105*). For instance, the CNTs can be incorporated into PANI and act as dopants to interact with the N−H groups and quinoid units, which increase the degree of electron delocalization, leading to a higher electrical conductivity (*106,107*). Hence CNTs were introduced to many polymers, such as poly(methylpyrrole) *(108)*, PANI (*109,110*), poly(3,4-ethylenedioxythiophene) (PEDOT) *(111)*, and polyschiff base *(112)*, to enhance the electrochromic properties.

Because of the easy preparation and high stability, PANI has been widely used as an anodically coloring electrochromic polymer and is usually used as an electrochromic electrode. For instance, the electrochromic device was fabricated by using the modified CNT/PANI and CNT/PEDOT thin films as anodically and cathodically coloring electrodes based on a liquid electrolyte, respectively *(84)*. The coloration efficiencies of the CNT/PANI and CNT/PEDOT films were 302 cm^2/C at 700 nm and 562 cm^2/C at 630 nm, respectively. The electrochromic devices shared a transmittance change of 35%. However, a much higher coloration efficiency of 836 cm^2/C at 650 nm had been achieved with the CNT, compared with 285 cm^2/C at 570 nm without the CNT. An all-solid-state electrochromic device was also developed by employing CNT/PANI as the anodically coloring electrode and poly(butyl viologen) as the cathodically coloring electrode *(113)*. It displayed a coloration efficiency of 151.3 cm^2/C at 550 nm and the transmittance could be changed reversibly from 54% (bleached) to 10% (darkened) with an applied voltage of −2.0 and 0.75 V, respectively.

However, the CNTs tend to agglomerate because of their van der Waals interactions, so they are poorly dispersed in most solvents and polymer matrices *(114)*. To increase the dispersion, surfactants are often introduced or the CNTs are further modified with functional groups *(115)*. However, it is difficult to remove the surfactant, and the excessive chemical functionalization decreases the property of CNTs. A mild modification of CNTs alleviates the problem, and the introduced functional groups may serve as dopants during the electropolymerization of conjugated polymers in some cases. For instance, the SWNTs are oxidized with concentrated nitric acid and sulfuric acid to yield carboxylic acid-functionalized SWNTs. PANI was then grafted onto SWNTs in the presence of poly(styrene sulfonate) (PSS) in an aqueous medium *(110)*. The resulting hybrid material exhibited a good dispersion of the SWNT in PANI/PSS through covalent bonds. With the addition of SWNTs, the charge-transfer resistance of the material was largely reduced, which led to the enhanced electrochromic contrast from 0.34 for PANI to 0.47 for PANI/SWNT composite with 0.8 wt% SWNT and higher coloration efficiency from 105 to 168 cm^2/C at 630 nm. Functionalized CNTs can also be introduced into PEDOT by electrochemical polymerization of the 3,4-ethylenedioxythiophene monomer in the presence of CNTs as cathode for the electrochromic device *(111)*. The CNT/PEDOT film displayed a higher coloration efficiency of 367 cm^2/C at 550 nm, compared with 251 cm^2/C for the PEDOT film. The high electrical conductivity and doping effect of the acid-modified CNTs led to an improvement in the electrochemical activity and coloration efficiency of PEDOT.

By combination of the electrochromatic PANI into an energy device such as a supercapacitor, a new family of smart energy storage devices can indicate working states such as the change of voltage by displaying different colors. The electrochromic supercapacitor was fabricated by electrodepositing PANI onto CNT sheets as two electrodes with a poly(vinyl alcohol)/H_3PO_4 gel electrolyte between the two electrodes *(116)*. The electrochromic supercapacitor could be stretched by 100% for 200 cycles or bent for 1000 cycles. During the charge and discharge processes, the two electrodes changed their colors. They switched to blue upon charging to 1 V, indicating the end of the charge process, returned to green after discharging to 0.5 V, and then switched to bright yellow after discharging to −0.5 V and −1 V (Fig. 6.8E). Moreover, it can be fabricated into a one-dimensional fiber shape *(117)*. These smart supercapacitor fibers could be further woven into a fabric to display various signs (Fig. 6.8F).

6.4 SUMMARY AND OUTLOOK

CNTs have been widely developed for responsive materials with high actuating properties and chromatic performances. As a broad spectrum of actuating and chromatic materials or devices with a variety of mechanisms and structures, they may generate high actuation stress with short response times and exhibit many colors with high coloration efficiency. However, macroscopic CNT materials show comparably lower electrical conductivity and mechanical strength in comparison to individual CNTs, and the intrinsic hydrophobic property of the CNTs also hinders their applications. There is a lot of room for improvement of the performance of the CNT-based responsive materials. Another challenge lies in the development of the commercial products. Currently, it remains difficult to produce these CNT-based sensing materials at an applicant scale. New preparation methods that are compatible to industrial production should be developed with low cost and high efficiency. Furthermore, increasing interest should be also stimulated to synthesize new materials and design new structures for enhanced sensing properties.

REFERENCES

1. Iijima, S. Helical Microtubules of Graphitic Carbon. *Nature* **1991,** *354,* 56–58.
2. De Volder, M. F. L.; Tawfick, S. H.; Baughman, R. H.; Hart, A. J. Carbon Nanotubes: Present and Future Commercial Applications. *Science* **2013,** *339,* 535–539.
3. Kim, P.; Lieber, C. M. Nanotube Nanotweezers. *Science* **1999,** *286,* 2148–2150.
4. Baughman, R. H.; Cui, C.; Zakhidov, A. A.; Iqbal, Z.; Barisci, J. N.; Spinks, G. M.; Wallace, G. G.; Mazzoldi, A.; De Rossi, D.; Rinzler, A. G. Carbon Nanotube Actuators. *Science* **1999,** *284,* 1340–1344.
5. Fennimore, A.; Yuzvinsky, T.; Han, W.-Q.; Fuhrer, M.; Cumings, J.; Zettl, A. Rotational Actuators Based on Carbon Nanotubes. *Nature* **2003,** *424,* 408–410.
6. Aliev, A. E.; Oh, J.; Kozlov, M. E.; Kuznetsov, A. A.; Fang, S.; Fonseca, A. F.; Ovalle, R.; Lima, M. D.; Haque, M. H.; Gartstein, Y. N. Giant-Stroke, Superelastic Carbon Nanotube Aerogel Muscles. *Science* **2009,** *323,* 1575–1578.
7. Chen, L.; Liu, C.; Liu, K.; Meng, C.; Hu, C.; Wang, J.; Fan, S. High-Performance, Low-Voltage, and Easy-Operable Bending Actuator Based on Aligned Carbon Nanotube/Polymer Composites. *ACS Nano* **2011,** *5,* 1588–1593.
8. Yanagi, K.; Moriya, R.; Yomogida, Y.; Takenobu, T.; Naitoh, Y.; Ishida, T.; Kataura, H.; Matsuda, K.; Maniwa, Y. Electrochromic Carbon Electrodes: Controllable Visible Color Changes in Metallic Single-Wall Carbon Nanotubes. *Advanced Materials* **2011,** *23,* 2811–2814.
9. Rodel, J.; Jo, W.; Seifert, K. T. P.; Anton, E. M.; Granzow, T.; Damjanovic, D. Perspective on the Development of Lead-Free Piezoceramics. *Journal of the American Ceramic Society* **2009,** *92,* 1153–1177.
10. Carpi, F.; Kornbluh, R.; Sommer-Larsen, P.; Alici, G. Electroactive Polymer Actuators as Artificial Muscles: Are They Ready for Bioinspired Applications? *Bioinspiration & Biomimetics* **2011,** *6,* 045006.
11. Abel, J.; Luntz, J.; Brei, D. Hierarchical Architecture of Active Knits. *Smart Materials and Structures* **2013,** *22,* 125001.
12. Akita, S.; Nakayama, Y.; Mizooka, S.; Takano, Y.; Okawa, T.; Miyatake, Y.; Yamanaka, S.; Tsuji, M.; Nosaka, T. Nanotweezers Consisting of Carbon Nanotubes Operating in an Atomic Force Microscope. *Applied Physics Letters* **2001,** *79,* 1691–1693.
13. Mølhave, K.; Hansen, T. M.; Madsen, D. N.; Bøggild, P. Towards Pick-and-Place Assembly of Nanostructures. *Journal of Nanoscience and Nanotechnology* **2004,** *4,* 279–282.

14. Lee, J.; Kim, S. Manufacture of a Nanotweezer Using a Length Controlled CNT Arm. *Sensors and Actuators A: Physical* **2005,** *120,* 193–198.
15. Sazonova, V.; Yaish, Y.; Üstünel, H.; Roundy, D.; Arias, T. A.; McEuen, P. L. A Tunable Carbon Nanotube Electromechanical Oscillator. *Nature* **2004,** *431,* 284–287.
16. Rueckes, T.; Kim, K.; Joselevich, E.; Tseng, G. Y.; Cheung, C.-L.; Lieber, C. M. Carbon Nanotube-Based Nonvolatile Random Access Memory for Molecular Computing. *Science* **2000,** *289,* 94–97.
17. Ke, C.; Espinosa, H. D. Feedback Controlled Nanocantilever Device. *Applied Physics Letters* **2004,** *85,* 681–683.
18. Meyer, J. C.; Paillet, M.; Roth, S. Single-Molecule Torsional Pendulum. *Science* **2005,** *309,* 1539–1541.
19. Deshpande, V.; Chiu, H.-Y.; Postma, H. C.; Miko, C.; Forro, L.; Bockrath, M. Carbon Nanotube Linear Bearing Nanoswitches. *Nano Letters* **2006,** *6,* 1092–1095.
20. Cumings, J.; Zettl, A. Low-Friction Nanoscale Linear Bearing Realized From Multiwall Carbon Nanotubes. *Science* **2000,** *289,* 602–604.
21. Dong, L.; Nelson, B. J.; Fukuda, T.; Arai, F. Towards Nanotube Linear Servomotors. *IEEE Transactions on Automation Science and Engineering* **2006,** *3,* 228–235.
22. Subramanian, A.; Dong, L.; Tharian, J.; Sennhauser, U.; Nelson, B. Batch Fabrication of Carbon Nanotube Bearings. *Nanotechnology* **2007,** *18,* 075703.
23. Yuan, W.; Hu, L. B.; Yu, Z. B.; Lam, T. L.; Biggs, J.; Ha, S. M.; Xi, D. J.; Chen, B.; Senesky, M. K.; Gruner, G.; et al. Fault-Tolerant Dielectric Elastomer Actuators Using Single-Walled Carbon Nanotube Electrodes. *Advanced Materials* **2008,** *20,* 621–625.
24. Yuan, W.; Hu, L. B.; Ha, S.; Lam, T. L.; Gruner, G.; Pei, Q. B. Self-Clearable Carbon Nanotube Electrodes for Improved Performance of Dielectric Elastomer Actuators. *Electroactive Polymer Actuators and Devices (Eapad)* **2008,** *2008,* 6927–69270P.
25. Hu, L. B.; Yuan, W.; Brochu, P.; Gruner, G.; Pei, Q. B. Highly Stretchable, Conductive, and Transparent Nanotube Thin Films. *Applied Physics Letters* **2009,** *94,* 161108.
26. Yu, Y. T.; Yuan, W. Z.; Xu, J. H. Design of a New MEMS-Based Tunable Grating With Electrostatic Comb Drive. *Design, Manufacturing, and Testing of Micro- and Nano-Optical Devices and Systems* **2007,** *6724,* 67241G.
27. Bar-Cohen, Y. Electroactive Polymer (EAP) as Artificial Muscles: Progress and Challenges. *SPIE Society of Photo-optical Instrumentation Engineers, Bellingham, WA* **2004,** *5385,* 10–16.
28. Pelrine, R. E.; Kornbluh, R. D.; Joseph, J. P. Electrostriction of Polymer Dielectrics With Compliant Electrodes as a Means of Actuation. *Sensors and Actuators A: Physical* **1998,** *64,* 77–85.
29. Krakovsky, I.; Romijn, T.; de Boer, A. P. A Few Remarks on the Electrostriction of Elastomers. *Journal of Applied Physics* **1999,** *85,* 628–629.
30. Pelrine, R.; Kornbluh, R.; Kofod, G. High-Strain Actuator Materials Based on Dielectric Elastomers. *Advanced Materials* **2000,** *12,* 1223–1225.
31. Shian, S.; Diebold, R. M.; McNamara, A.; Clarke, D. R. Highly Compliant Transparent Electrodes. *Applied Physics Letters* **2012,** *101,* 061101.
32. Cakmak, E.; Fang, X. M.; Yildiz, O.; Bradford, P. D.; Ghosh, T. K. Carbon Nanotube Sheet Electrodes for Anisotropic Actuation of Dielectric Elastomers. *Carbon* **2015,** *89,* 113–120.
33. Tiwari, R.; Garcia, E. The State of Understanding of Ionic Polymer Metal Composite Architecture: A Review. *Smart Materials and Structures* **2011,** *20,* 083001.
34. Kong, L. R.; Chen, W. Carbon Nanotube and Graphene-Based Bioinspired Electrochemical Actuators. *Advanced Materials* **2014,** *26,* 1025–1043.
35. Hughes, M.; Spinks, G. M. Multiwalled Carbon Nanotube Actuators. *Advanced Materials* **2005,** *17,* 443–446.
36. Fraysse, J.; Minett, A.; Jaschinski, O.; Duesberg, G.; Roth, S. Carbon Nanotubes Acting Like Actuators. *Carbon* **2002,** *40,* 1735–1739.
37. Spinks, G. M.; Wallace, G. G.; Fifield, L. S.; Dalton, L. R.; Mazzoldi, A.; De Rossi, D.; Khayrullin, I. I.; Baughman, R. H. Pneumatic Carbon Nanotube Actuators. *Advanced Materials* **2002,** *14,* 1728–1732.
38. Liu, S.; Liu, Y.; Cebeci, H.; de Villoria, R. G.; Lin, J. H.; Wardle, B. L.; Zhang, Q. High Electromechanical Response of Ionic Polymer Actuators With Controlled-morphology Aligned Carbon Nanotube/Nafion Nanocomposite Electrodes. *Advanced Functional Materials* **2010,** *20,* 3266–3271.

39. Mirfakhrai, T.; Oh, J.; Kozlov, M.; Fok, E. C. W.; Zhang, M.; Fang, S.; Baughman, R. H.; Madden, J. D. Electrochemical Actuation of Carbon Nanotube Yarns. *Smart Materials and Structures* **2007,** *16,* S243.
40. Fukushima, T.; Asaka, K.; Kosaka, A.; Aida, T. Fully Plastic Actuator through Layer-by-Layer Casting With Ionic-Liquid-Based Bucky Gel. *Angewandte Chemie International Edition* **2005,** *44,* 2410–2413.
41. Biso, M.; Ricci, D. Multi-Walled Carbon Nanotubes Plastic Actuator. *Phsica Satus Solidi* **2009,** *246,* 2820–2823.
42. Fukushima, T.; Kosaka, A.; Ishimura, Y.; Yamamoto, T.; Takigawa, T.; Ishii, N.; Aida, T. Molecular Ordering of Organic Molten Salts Triggered by Single-Walled Carbon Nanotubes. *Science* **2003,** *300,* 2072–2074.
43. Fukushima, T.; Aida, T. Ionic Liquids for Soft Functional Materials With Carbon Nanotubes. *Chemistry-A European Journal* **2007,** *13,* 5048–5058.
44. Mukai, K.; Asaka, K.; Kiyohara, K.; Sugino, T.; Takeuchi, I.; Fukushima, T.; Aida, T. High Performance Fully Plastic Actuator Based on Ionic-Liquid-Based Bucky Gel. *Electrochimica Acta* **2008,** *53,* 5555–5562.
45. Mukai, K.; Asaka, K.; Sugino, T.; Kiyohara, K.; Takeuchi, I.; Terasawa, N.; Futaba, D. N.; Hata, K.; Fukushima, T.; Aida, T. Highly Conductive Sheets From Millimeter-Long Single-Walled Carbon Nanotubes and Ionic Liquids: Application to Fast-Moving, Low-Voltage Electromechanical Actuators Operable in Air. *Advanced Materials* **2009,** *21,* 1582–1585.
46. Li, J.; Ma, W.; Song, L.; Niu, Z.; Cai, L.; Zeng, Q.; Zhang, X.; Dong, H.; Zhao, D.; Zhou, W. Superfast-Response and Ultrahigh-Power-Density Electromechanical Actuators Based on Hierarchal Carbon Nanotube Electrodes and Chitosan. *Nano Letters* **2011,** *11,* 4636–4641.
47. Terasawa, N.; Ono, N.; Mukai, K.; Koga, T.; Higashi, N.; Asaka, K. A Multi-Walled Carbon Nanotube/Polymer Actuator that Surpasses the Performance of a Single-Walled Carbon Nanotube/Polymer Actuator. *Carbon* **2012,** *50,* 311–320.
48. Terasawa, N.; Mukai, K.; Yamato, K.; Asaka, K. Superior Performance of Non-activated Multi-Walled Carbon Nanotube Polymer Actuator Containing Ruthenium Oxide Over a Single-Walled Carbon Nanotube. *Carbon* **2012,** *50,* 1888–1896.
49. Terasawa, N.; Mukai, K.; Yamato, K.; Asaka, K. Superior Performance of Manganese Oxide/Multi-Walled Carbon Nanotubes Polymer Actuator Over Ruthenium Oxide/Multi-Walled Carbon Nanotubes and Single-Walled Carbon Nanotubes. *Sensors and Actuators B: Chemical* **2012,** *171,* 595–601.
50. Foroughi, J.; Spinks, G. M.; Wallace, G. G.; Oh, J.; Kozlov, M. E.; Fang, S.; Mirfakhrai, T.; Madden, J. D.; Shin, M. K.; Kim, S. J. Torsional Carbon Nanotube Artificial Muscles. *Science* **2011,** *334,* 494–497.
51. Lee, J. A.; Kim, Y. T.; Spinks, G. M.; Suh, D.; Lepró, X.; Lima, M. D.; Baughman, R. H.; Kim, S. J. All-Solid-State Carbon Nanotube Torsional and Tensile Artificial Muscles. *Nano Letters* **2014,** *14,* 2664–2669.
52. Sellinger, A. T.; Wang, D. H.; Tan, L. S.; Vaia, R. A. Electrothermal Polymer Nanocomposite Actuators. *Advanced Materials* **2010,** *22,* 3430–3435.
53. Chen, L.-Z.; Liu, C.; Hu, C.; Fan, S. Electrothermal Actuation Based on Carbon Nanotube Network in Silicone Elastomer. *Applied Physics Letters* **2008,** *92,* 263104.
54. Lu, L.; Chen, W. Biocompatible Composite Actuator: A Supramolecular Structure Consisting of the Biopolymer Chitosan, Carbon Nanotubes, and an Ionic Liquid. *Advanced Materials* **2010,** *22,* 3745–3748.
55. Hu, Y.; Chen, W.; Lu, L.; Liu, J.; Chang, C. Electromechanical Actuation With Controllable Motion Based on a Single-Walled Carbon Nanotube and Natural Biopolymer Composite. *ACS Nano* **2010,** *4,* 3498–3502.
56. Li, Q.; Liu, C.; Lin, Y.-H.; Liu, L.; Jiang, K.; Fan, S. Large-Strain, Multiform Movements From Designable Electrothermal Actuators Based on Large Highly-Anisotropic Carbon Nanotube Sheets. *ACS Nano* **2015,** *9,* 409–418.
57. Lima, M. D.; Li, N.; De Andrade, M. J.; Fang, S.; Oh, J.; Spinks, G. M.; Kozlov, M. E.; Haines, C. S.; Suh, D.; Foroughi, J. Electrically, Chemically, and Photonically Powered Torsional and Tensile Actuation of Hybrid Carbon Nanotube Yarn Muscles. *Science* **2012,** *338,* 928–932.

58. Shang, Y.; He, X.; Wang, C.; Zhu, L.; Peng, Q.; Shi, E.; Wu, S.; Yang, Y.; Xu, W.; Wang, R. Large-Deformation, Multifunctional Artificial Muscles Based on Single-Walled Carbon Nanotube Yarns. *Advanced Engineering Materials* **2015,** *17,* 14−20.
59. Chun, K.-Y.; Kim, S. H.; Shin, M. K.; Kwon, C. H.; Park, J.; Kim, Y. T.; Spinks, G. M.; Lima, M. D.; Haines, C. S.; Baughman, R. H. Hybrid Carbon Nanotube Yarn Artificial Muscle Inspired by Spider Dragline Silk. *Nature Communications* **2014,** *5,* 3322.
60. Guo, W.; Liu, C.; Zhao, F.; Sun, X.; Yang, Z.; Chen, T.; Chen, X.; Qiu, L.; Hu, X.; Peng, H. A Novel Electromechanical Actuation Mechanism of a Carbon Nanotube Fiber. *Advanced Materials* **2012,** *24,* 5379−5384.
61. Meng, F.; Zhang, X.; Li, R.; Zhao, J.; Xuan, X.; Wang, X.; Zou, J.; Li, Q. Electro-Induced Mechanical and Thermal Responses of Carbon Nanotube Fibers. *Advanced Materials* **2014,** *26,* 2480−2485.
62. Chen, P.; Xu, Y.; He, S.; Sun, X.; Guo, W.; Zhang, Z.; Qiu, L.; Li, J.; Chen, D.; Peng, H. Biologically Inspired, Sophisticated Motions from Helically Assembled, Conducting Fibers. *Advanced Materials* **2014,** *27,* 1042−1047.
63. Chen, P.; He, S.; Xu, Y.; Sun, X.; Peng, H. Electromechanical Actuator Ribbons Driven by Electrically Conducting Spring-Like Fibers. *Advanced Materials* **2015,** *27,* 4982−4988.
64. Sidorenko, A.; Krupenkin, T.; Taylor, A.; Fratzl, P.; Aizenberg, J. Reversible Switching of Hydrogel-Actuated Nanostructures Into Complex Micropatterns. *Science* **2007,** *315,* 487−490.
65. Ma, M. M.; Guo, L.; Anderson, D. G.; Langer, R. Bio-inspired Polymer Composite Actuator and Generator Driven by Water Gradients. *Science* **2013,** *339,* 186−189.
66. de Haan, L. T.; Verjans, J. M. N.; Broer, D. J.; Bastiaansen, C. W. M.; Schenning, A. P. H. J. Humidity-Responsive Liquid Crystalline Polymer Actuators With an Asymmetry in the Molecular Trigger That Bend, Fold, and Curl. *Journal of the American Chemical Society* **2014,** *136,* 10585−10588.
67. Zhao, Q.; Dunlop, J. W. C.; Qiu, X. L.; Huang, F. H.; Zhang, Z. B.; Heyda, J.; Dzubiella, J.; Antonietti, M.; Yuan, J. Y. An Instant Multi-responsive Porous Polymer Actuator Driven by Solvent Molecule Sorption. *Nature Communications* **2014,** *5,* 3322.
68. Kwak, G.; Lee, W.-E.; Jeong, H.; Sakaguchi, T.; Fujiki, M. Swelling-Induced Emission Enhancement in Substituted Acetylene Polymer Film With Large Fractional Free Volume: Fluorescence Response to Organic Solvent Stimuli. *Macromolecules* **2008,** *42,* 20−24.
69. Lee, W. E.; Jin, Y. J.; Park, L. S.; Kwak, G. Fluorescent Actuator Based on Microporous Conjugated Polymer With Intramolecular Stack Structure. *Advanced Materials* **2012,** *24,* 5604−5609.
70. Lu, X.; Zhang, Z.; Li, H.; Sun, X.; Peng, H. Conjugated Polymer Composite Artificial Muscle With Solvent-Induced Anisotropic Mechanical Actuation. *Journal of Materials Chemistry A* **2014,** *2,* 17272 17280.
71. Lima, M. D.; Hussain, M. W.; Spinks, G. M.; Naficy, S.; Hagenasr, D.; Bykova, J. S.; Tolly, D.; Baughman, R. H. Efficient, Absorption-powered Artificial Muscles Based on Carbon Nanotube Hybrid Yarns. *Small* **2015,** *11,* 3113−3118.
72. Chen, P.; Xu, Y.; He, S.; Sun, X.; Pan, S.; Deng, J.; Chen, D.; Peng, H. Hierarchically Arranged Helical Fibre Actuators Driven by Solvents and Vapours. *Nature Nanotechnology* **2015,** *10,* 1077−1083.
73. He, S.; Chen, P.; Qiu, L.; Wang, B.; Sun, X.; Xu, Y.; Peng, H. A Mechanically Actuating Carbon-Nanotube Fiber in Response to Water and Moisture. *Angewandte Chemie International Edition* **2015,** *54,* 14880−14884.
74. Mizuno, K.; Ishii, J.; Kishida, H.; Hayamizu, Y.; Yasuda, S.; Futaba, D. N.; Yumura, M.; Hata, K. A Black Body Absorber from Vertically Aligned Single-Walled Carbon Nanotubes. *Proceedings of the National Academy of Sciences* **2009,** *106,* 6044−6047.
75. Ahir, S. V.; Terentjev, E. M. Photomechanical Actuation in Polymer-Nanotube Composites. *Nature Material* **2005,** *4,* 491−495.
76. Lu, S.; Panchapakesan, B. Photomechanical Responses of Carbon Nanotube/Polymer Actuators. *Nanotechnology* **2007,** *18,* 305502.
77. Ahir, S. T.; Terentjev, E. M. Fast Relaxation of Carbon Nanotubes in Polymer Composite Actuators. *Physical Review Letters* **2006,** *96,* 133902.
78. Ahir, S. V.; Squires, A. M.; Tajbakhsh, A. R.; Terentjev, E. M. Infrared Actuation in Aligned Polymer-Nanotube Composites. *Physical Review B* **2006,** *73,* 085420.

79. Zhang, X.; Yu, Z.; Wang, C.; Zarrouk, D.; Seo, J.-W. T.; Cheng, J. C.; Buchan, A. D.; Takei, K.; Zhao, Y.; Ager, J. W. Photoactuators and Motors Based on Carbon Nanotubes With Selective Chirality Distributions. *Nature Communications* **2014,** *5,* 3983.
80. Yang, L.; Setyowati, K.; Li, A.; Gong, S.; Chen, J. Reversible Infrared Actuation of Carbon Nanotube-Liquid Crystalline Elastomer Nanocomposites. *Advanced Materials* **2008,** *20,* 2271–2275.
81. Chen, Z. G.; Wang, J. G.; Wang, Y.; Wang, G. Q.; Li, S.; Cheng, G. X. Novel High-Power Subterahertz-Range Radial Surface Wave Oscillator. *Physics of Plasmas* **2015,** *22,* 063114.
82. Li, M.; Lv, S.; Zhou, J. Photo-Thermo-Mechanically Actuated Bending and Snapping Kinetics of Liquid Crystal Elastomer Cantilever. *Smart Materials and Structures* **2014,** *23,* 125012.
83. Torras, N.; Zinoviev, K. E.; Marshall, J. E.; Terentjev, E. M.; Esteve, J. Bending Kinetics of a Photo-Actuating Nematic Elastomer Cantilever. *Applied Physics Letters* **2011,** *99,* 254102.
84. Wang, W.; Sun, X.; Wu, W.; Peng, H.; Yu, Y. Photoinduced Deformation of Crosslinked Liquid-Crystalline Polymer Film Oriented by a Highly Aligned Carbon Nanotube Sheet. *Angewandte Chemie International Edition* **2012,** *124,* 4722–4725.
85. Marshall, J. E.; Ji, Y.; Torras, N.; Zinoviev, K.; Terentjev, E. M. Carbon-Nanotube Sensitized Nematic Elastomer Composites for IR-Visible Photo-Actuation. *Soft Matter* **2012,** *8,* 1570–1574.
86. Czaniková, K.; Torras, N.; Esteve, J.; Krupa, I.; Kasák, P.; Pavlova, E.; Racko, D.; Chodák, I.; Omastová, M. Nanocomposite Photoactuators Based on an Ethylene Vinyl Acetate Copolymer Filled With Carbon Nanotubes. *Sensors and Actuators B: Chemical* **2013,** *186,* 701–710.
87. Wang, T.; Torres, D.; Fernandez, F. E.; Green, A. J.; Wang, C.; Sepúlveda, N. Increasing Efficiency, Speed, and Responsivity of Vanadium Dioxide Based Photothermally Driven Actuators Using Single-Wall Carbon Nanotube Thin-Films. *ACS Nano* **2015,** *9,* 4371–4378.
88. Ilčíková, M.; Mrlík, M.; Sedláček, T.; Chorvát, D.; Krupa, I.; Šlouf, M.; Koynov, K.; Mosnáček, J. Viscoelastic and Photo-Actuation Studies of Composites Based on Polystyrene-Grafted Carbon Nanotubes and Styrene-b-Isoprene-b-Styrene Block Copolymer. *Polymer* **2014,** *55,* 211–218.
89. Li, M. H.; Keller, P.; Li, B.; Wang, X.; Brunet, M. Light-Driven SIDE-on Nematic Elastomer Actuators. *Advanced Materials* **2003,** *15,* 569–572.
90. Kondo, M.; Yu, Y.; Ikeda, T. How Does the Initial Alignment of Mesogens Affect the Photoinduced Bending Behavior of Liquid-Crystalline Elastomers? *Angewandte Chemie International Edition* **2006,** *118,* 1406–1410.
91. Sun, X.; Wang, W.; Qiu, L.; Guo, W.; Yu, Y.; Peng, H. Unusual Reversible Photomechanical Actuation in Polymer/Nanotube Composites. *Angewandte Chemie International Edition* **2012,** *51,* 8520–8524.
92. Lu, Y.; Yang, Y.; Sellinger, A.; Lu, M.; Huang, J.; Fan, H.; Haddad, R.; Lopez, G.; Burns, A. R.; Sasaki, D. Y. Self-Assembly of Mesoscopically Ordered Chromatic Polydiacetylene/Silica Nanocomposites. *Nature* **2001,** *410,* 913–917.
93. Peng, H.; Tang, J.; Yang, L.; Pang, J.; Ashbaugh, H. S.; Brinker, C. J.; Yang, Z.; Lu, Y. Responsive Periodic Mesoporous Polydiacetylene/Silica Nanocomposites. *Journal of the American Chemical Society* **2006,** *128,* 5304–5305.
94. Peng, H.; Tang, J.; Pang, J.; Chen, D.; Yang, L.; Ashbaugh, H. S.; Brinker, C. J.; Yang, Z.; Lu, Y. Polydiacetylene/Silica Nanocomposites With Tunable Mesostructure and Thermochromatism From Diacetylenic Assembling Molecules. *Journal of the American Chemical Society* **2005,** *127,* 12782–12783.
95. Peng, H.; Sun, X.; Cai, F.; Chen, X.; Zhu, Y.; Liao, G.; Chen, D.; Li, Q.; Lu, Y.; Zhu, Y. Electrochromatic Carbon Nanotube/Polydiacetylene Nanocomposite Fibres. *Nature Nanotechnology* **2009,** *4,* 738–741.
96. Cheng, Q.; Stevens, R. C. Charge-Induced Chromatic Transition of Amino Acid-Derivatized Polydiacetylene Liposomes. *Langmuir* **1998,** *14,* 1974–1976.
97. Kwak, G.; Fukao, S.; Fujiki, M.; Sakaguchi, T.; Masuda, T. Nanoporous, Honeycomb-Structured Network Fibers Spun From Semiflexible, Ultrahigh Molecular Weight, Disubstituted Aromatic Polyacetylenes: Superhierarchical Structure and Unique Optical Anisotropy. *Chemistry of Materials* **2006,** *18,* 5537–5542.

98. Kwak, G.; Fukao, S.; Fujiki, M.; Sakaguchi, T.; Masuda, T. Temperature-Dependent, Static, and Dynamic Fluorescence Properties of Disubstituted Acetylene Polymer Films. *Chemistry of Materials* **2006,** *18,* 2081–2085.
99. Tang, B. Z.; Kong, X.; Wan, X.; Peng, H.; Lam, W. Y.; Feng, X.-D.; Kwok, H. S. Liquid Crystalline Polyacetylenes: Synthesis and Properties of Poly {n-[((4′-cyano-4-biphenylyl) Oxy) Carbonyl]-1-alkynes}. *Macromolecules* **1998,** *31,* 2419–2432.
100. Sun, X.; Zhang, Z.; Lu, X.; Guan, G.; Li, H.; Peng, H. Electric Current Test Paper Based on Conjugated Polymers and Aligned Carbon Nanotubes. *Angewandte Chemie International Edition* **2013,** *125,* 7930–7934.
101. Meng, F.; Zhang, X.; Xu, G.; Yong, Z.; Chen, H.; Chen, M.; Li, Q.; Zhu, Y. Carbon Nanotube Composite Films With Switchable Transparency. *ACS Applied Materials & Interfaces* **2011,** *3,* 658–661.
102. Argun, A. A.; Aubert, P. H.; Thompson, B. C.; Schwendeman, I.; Gaupp, C. L.; Hwang, J.; Pinto, N. J.; Tanner, D. B.; MacDiarmid, A. G.; Reynolds, J. R. Multicolored Electrochromism Polymers: Structures and Devices. *Chemistry of Materials* **2004,** *16,* 4401–4412.
103. Hu, L.; Gruner, G.; Li, D.; Kaner, R. B.; Cech, J. Patternable Transparent Carbon Nanotube Films for Electrochromic Devices. *Journal of Applied Physics* **2007,** *101,* 016102.
104. Pradhan, B.; Batabyal, S. K.; Pal, A. J. Functionalized Carbon Nanotubes in Donor/Acceptor-Type Photovoltaic Devices. *Applied Physics Letters* **2006,** *88,* 093106.
105. Huang, J.-E.; Li, X.-H.; Xu, J.-C.; Li, H.-L. Well-Dispersed Single-Walled Carbon Nanotube/Polyaniline Composite Films. *Carbon* **2003,** *41,* 2731–2736.
106. Zengin, H.; Zhou, W.; Jin, J.; Czerw, R.; Smith, D. W., Jr.; Echegoyen, L.; Carroll, D. L.; Foulger, S. H.; Ballato, J. Carbon Nanotube Doped Polyaniline. *Advanced Materials* **2002,** *14,* 1480–1483.
107. Matsunaga, D.; Tamaki, T.; Akiyama, H.; Ichimura, K. Photofabrication of Micro-patterned Polarizing Elements for Stereoscopic Displays. *Advanced Materials* **2002,** *12,* 1477–1480.
108. Ahmad, S.; Singh, S. Electrochromic Device Based on Carbon Nanotubes Functionalized Poly (Methyl Pyrrole) Synthesized in Hydrophobic Ionic Liquid Medium. *Electrochemistry Communications* **2008,** *10,* 895–898.
109. Small, W. R.; Masdarolomoor, F.; Wallace, G. G.; Marc in het Panhuis. Inkjet Deposition and Characterization of Transparent Conducting Electroactive Polyaniline Composite Films With a High Carbon Nanotube Loading Fraction. *Journal of Materials Chemistry* **2007,** *17,* 4359.
110. Xiong, S.; Wei, J.; Jia, P.; Yang, L.; Ma, J.; Lu, X. Water-Processable Polyaniline With Covalently Bonded Single-Walled Carbon Nanotubes: Enhanced Electrochromic Properties and Impedance Analysis. *ACS Applied Materials & Interfaces* **2011,** *3,* 782–788.
111. Bhandari, S.; Deepa, M.; Srivastava, A. K.; Lal, C.; Kant, R. Poly(3,4-ethylenedioxythiophene) (PEDOT)-Coated MWCNTs Tethered to Conducting Substrates: Facile Electrochemistry and Enhanced Coloring Efficiency. *Macromolecular Rapid Communications* **2008,** *29,* 1959–1964.
112. Ma, L.; Cai, J.; Zhao, P.; Niu, H.; Wang, C.; Bai, X.; Wang, W. In Situ Preparation of Composite From Conjugated Polyschiff Bases and Multiwalled Carbon Nanotube: Synthesis, Electrochromic, Acidochromic Properties. *Materials Chemistry and Physics* **2012,** *133,* 333–339.
113. Chang, T.-H.; Hu, C.-W.; Vittal, R.; Ho, K.-C. Incorporation of Plastic Crystal and Transparent UV-Cured Polymeric Electrolyte in a Complementary Electrochromic Device. *Solar Energy Materials and Solar Cells* **2014,** *126,* 213–218.
114. Xie, X.; Mai, Y.; Zhou, X. Dispersion and Alignment of Carbon Nanotubes in Polymer Matrix: A Review. *Materials Science and Engineering: R: Reports* **2005,** *49,* 89–112.
115. Islam, M. F.; Rojas, E.; Bergey, D. M.; Johnson, A. T.; Yodh, A. G. High Weight Fraction Surfactant Solubilization of Single-Wall Carbon Nanotubes in Water. *Nano Letter* **2003,** *3,* 269–273.
116. Chen, X.; Lin, H.; Chen, P.; Guan, G.; Deng, J.; Peng, H. Smart, Stretchable Supercapacitors. *Advanced Materials* **2014,** *26,* 4444–4449.
117. Chen, X.; Lin, H.; Deng, J.; Zhang, Y.; Sun, X.; Chen, P.; Fang, X.; Zhang, Z.; Guan, G.; Peng, H. Electrochromic Fiber-Shaped Supercapacitors. *Advanced Materials* **2014,** *26,* 8126–8132.

CHAPTER 7

Wearable Carbon Nanotube Devices for Sensing

H. Shen[1,4,a], T. Liu[2,a], D. Qin[1,4], X. Bo[2], L. Wang[1,4], F. Wang[1], Q. Yuan[1], T. Wagberg[3], G. Hu[1,3], M. Zhou[2]
[1]Chinese Academy of Sciences, Urumqi, China; [2]Northeast Normal University, Changchun, China; [3]Umea University, Umea, Sweden; [4]University of Chinese Academy of Sciences, Beijing, People's Republic of China

Contents

7.1 INTRODUCTION

Wearable technology refers to mobile electronic devices that are comfortably worn on the user's body or attached to their clothes. As written by Jayson Derrick at Bezinga.com, "wearable devices will far surpass market expectations, and become the fastest ramping consumer technology device to date, in our view" *(1)*.

Looking back in history, it is not surprising to notice that innovation in this area is closely coupled with advancements in electronics *(2)*. Fig. 7.1 illustrates the evolution of sensing technologies from bulky mechanical equipment to electronic devices (bench-top, portable devices, wearable and skin print devices). The advance of material science and its assembly techniques has evolved from macroscale to microscale water buckets and vacuum tubes, discrete transistors, integrated circuits to carbon nanotube (CNT)/graphene/organic electronics. There is a clear trend that the devices are becoming smaller, lighter, and less obtrusive and thus more comfortable for our daily lives *(2)*.

Among all the units of the devices, the key mission of original data collection is allotted to the sensors *(3)*. Sensor systems may support home-based or mobile assessment of a person's state and the data generated may be used to identify detrimental constellations or emergency situations *(4)*. Recent research and development on wearable sensors

[a] These authors contributed equally to this work.

Industrial Applications of Carbon Nanotubes
ISBN 978-0-323-41481-4

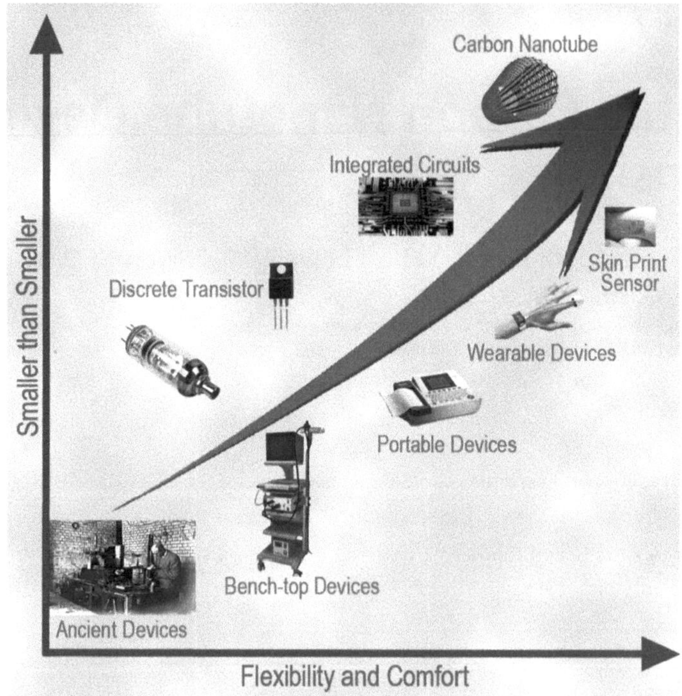

Figure 7.1 The evolution of electronic technology.

has demonstrated great potential in a wide range of applications, such as displays, robotics, in vitro diagnostics, advanced therapies, and energy harvesting *(5)*. In particular, acquisition of body-related information by wearable sensors indoors or outdoors is considered a cornerstone in health informatics. Besides capturing the information about the surroundings for health care and risk avoidance, certain weak human motions, including hand clenching, expression change, phonation, breath, blink, and pulse, can also be tested by smart wearable systems *(6–9)*.

An ideal wearable sensor would be energy harvesting, adaptable to follow human skin deformation, ultralight, transparent, and have relatively good sensitivity, high self-healing capabilities, superior physical robustness, switchable surfaces, and so on. It should thus neither constrain/affect users nor require maintenance *(8,10)*. As wireless technologies and ubiquitous and pervasive computing continue developing, so will wireless network sensors and intelligent wearable devices *(11)*. However, the door to futuristic devices is currently locked, and the key to open it lies within materials science, as Robert Byrne and Dermot Diamond report *(12)*. Because of the distinctive features of high sensitivity, superelasticity, and reversible extensibility, CNT-based materials have been not only demonstrated as prototypes reported in papers but also widely used in practical applications of wearable sensing and personal protection *(13)*. Their structural characteristics, which include high surface areas and unique electrical properties, make them the most suitable material for smart sensor applications *(14)*. The electronic properties of the CNT are determined by its structure, and any kind of mechanical deformation and surface functionalization is capable of changing the conductance of the CNT *(15)*.

In this chapter, we will discuss recent advances in the design and application of wearable devices based on CNT materials for sensing, especially in the fields of medical equipment, motion detection, and environmental sensors.

7.2 WEARABLE CNT SENSORS FOR HEALTH CARE

The future paths of population aging resulting from specific combinations of declining fertility and increasing life expectancies is a universal issue *(16)*. As Fig. 7.2A shows, it is predictable that the proportion of the global population 60+ years of age will increase nearly 11% every fifth decade from 2000 to 2100 *(17)*. The speed of aging over the coming years will reach its highest rate before 2035 (Fig. 7.2B), and before 2050 one-third of the population will be 60+ years of age in most world regions (Fig. 7.2C) *(17)*. The elderly suffers more health risks as a result of certain aspects of immune response decline with age *(18)*. Thus a large and diverse health care workforce that can effectively and efficiently diagnose and treat patients with complex medical conditions will be required to respond to the disease burden associated with a growing elderly population, as shown in Fig. 7.2D–F *(19)*. It is no surprise that the aging of the population really drives the demand for health care, as concluded by Uwe E. Reinhardt *(20)*. On the other hand, health data from 30 countries of the Organization for Economic Cooperation and Development shows that health care expenditures as a proportion of gross domestic product are at an all-time high *(21)*. In the traditional health care system, the patients, especially the elderly, are commonly restricted to length of hospital stay and thereby may suffer from loneliness or depression and even death, which imposes high costs and reduces the quality of life. In addition, those suffering from one or more functional impairments often prefer to stay in their familiar home environment rather than in an institutional care facility *(4)*. Smart technology and manufacture, telecare, telehealth, and telemedicine are new models of care already in use, bringing solutions to health care issues *(11)*. A small, flexible, wearable sensor system could support home-based or mobile assessment of a person's state of health and the data generated may be used to identify detrimental constellations or emergency situations. Thus timely intervention and prevention measures may be taken to avoid further deterioration. The wearable sensors enable medical professionals to monitor and survey an individual's state of health remotely and thereby allow them to work more efficiently so that the patient can enjoy an independent higher quality life (Fig. 7.2G) *(22)*.

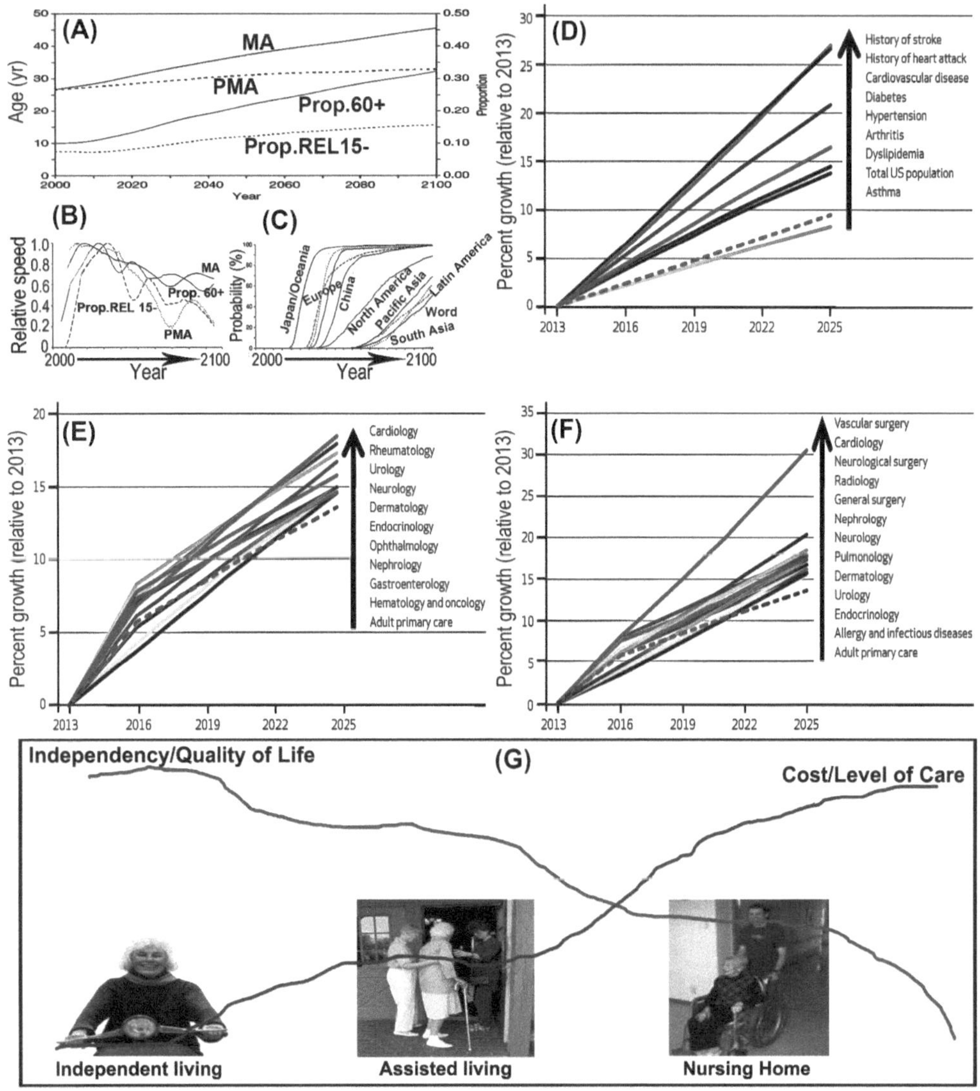

Figure 7.2 (A) Projected changes in the level of aging for the world population over the course of the century for four indicators of aging as defined in the text. (B) The changing speed of increase in selected indicators of aging. (C) Cumulative probabilities of reaching a proportion of 60+ of one-third or more of the world and selected world regions by calendar year *(17)*. (D) Projected growth from 2013 to 2025 in a population with chronic conditions, selected medical specialties (E), and demand for full-time equivalent physicians in selected specialties (F) *(19)*. (G) Supporting independent care and controlling the quality of life by wearable sensors *(23)*. *(Parts A, B, and C adapted with permission from Lutz, W.; Sanderson, W.; Scherbov, S. The Coming Acceleration of Global Population Ageing. Nature **2008**, 451 (7179), 716–719, Copyright 2008 Nature Publishing Group. Parts D and E adapted with permission from Dall, T. M.; Gallo, P. D.; Chakrabarti, R.; West, T.; Semilla, A. P.; Storm, M. V. An Aging Population and Growing Disease Burden Will Require a Large and Specialized Health Care Workforce by 2025. Health Affairs **2013**, 32 (11), 2013–2020, Copyright 2013 Project Hope. Part G adapted with permission from Marschollek, M.; Tian, F.; Koch, S.; Horsch, A.; Demiris, G. Smart Homes and Wearable Sensors in Healthcare–Infrastructures and Applications, Copyright 2011 European Federation for Medical Informatics.)*

Along with the cognition and exploitation of CNTs during the last 25 years *(24,25)*, various wearable CNT sensors have been invented and applied in our daily lives to trace physiological parameters in health informatics, such as heart rate *(26)*, core/surface body temperature *(27)*, blood pressure *(28)*, glucose concentration *(29,30)*, electroencephalogram *(13)*, and so forth.

As for homothermal animals, body temperature is a principal physiological parameter suggesting that the real-time monitoring of body temperature is one of the crucial ways to prevent illness (e.g., heat stroke, congestive heart failure, and many types of illnesses/diseases) *(31)*. Based on the linear relationship between temperature and electrical resistance variation of a composite ink of CNTs and conductive poly(3,4-ethylenedioxythiophene) and poly(styrenesulfonate) (PEDOT:PSS, 1.3 wt% in water), a printable resistive temperature sensor with a sensitivity of ~0.61%/°C between ~21 and ~50°C was reported by the Kuniharu Takei group *(32,33)*, as shown in Fig. 7.3A. Its performance is slightly better than that of the noble metal Pt-based temperature sensor *(34)*. The high-temperature sensitivity is most likely attributed to electron hopping at the interface between CNTs and PEDOT:PSS rather than the temperature coefficient of resistance for each material *(32,35)*. The mechanical flexibility of CNTs makes them suitable for flexible and wearable device applications, and it has been reported that a CNT sensor can be bent up to a 3.6-mm curvature radius without a resistance change, albeit with a little lower sensitivity and resolution (≈0.1°C) *(36)*. Thus it is clear that such sensors can be applicable to daily health care, and that the appeal of higher sensitivity of sensors promotes technical development. A sensor made of a highly percolative CNT/polymer network achieves a resistance sensitivity to temperature of 0.004/°C in the range 20–40°C, explained by the excellent dispersability of CNTs in poly(vinylbenzyl chloride) laminated with trimethylamine, which strongly increases the conductivity of the materials *(27)*. The use of CNTs improves the functionality of polymer composites by enhancing their strength and thermal or electrical conductivity *(37)*. Currently, there are still a number of limiting factors for the practical applications of wearable temperature sensors, such as their low stability during human motion with a retained sensitivity and resolution for a wide temperature range.

Fatigue, sleepiness, disturbed sleep, and insomnia are exacerbated among the elderly and these kinds of ailments are increasingly important factors affecting health and safety in modern society *(42–44)*. Hence it would be interesting to develop technologies for the routine ambulatory monitoring of related physiological indicators.

Electrophysiology, the measurement of electrical activity of biological origin, is a key technique for monitoring physiological parameters in several applications *(13,45)*. However, the specific procedures (such as skin preparation and the use of electrolytic gel) required by the high-quality low-amplitude measurements (such as electroencephalogram) would lead to longer preparation (up to several minutes per electrode) and longer stabilization time (diffusion of the electrolytic gel into the skin), which make it

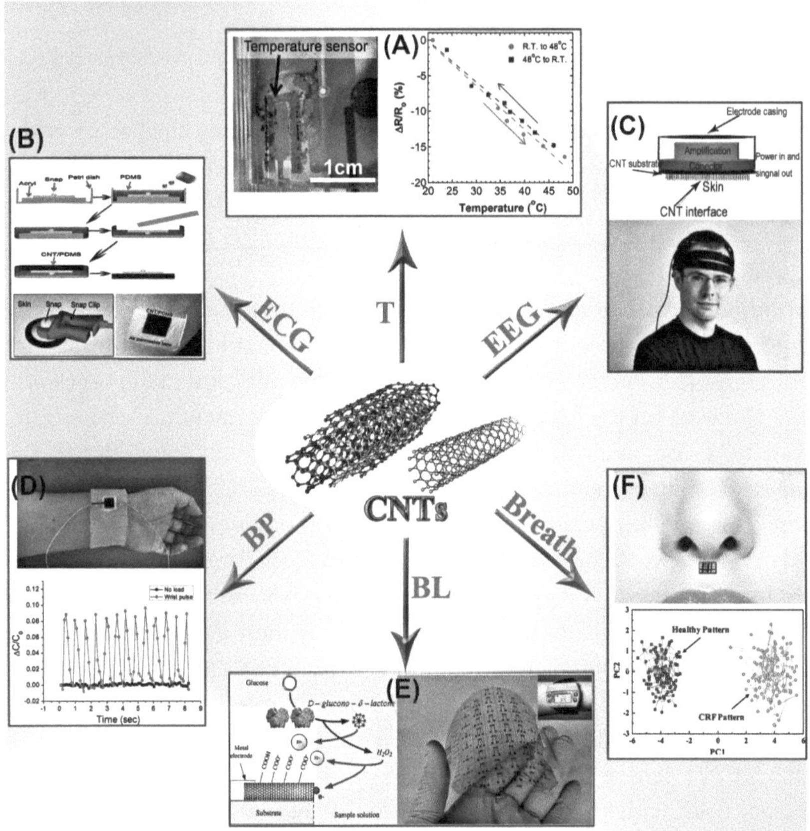

Figure 7.3 Stretchable wearable carbon nanotube (CNT) sensors. (A) Picture of a CNT-poly(3,4-ethylenedioxythiophene):poly(styrenesulfonate) (PEDOT:PSS) temperature sensor (T) on a Kapton substrate and relative resistance change range from 21°C to 50°C *(33)*. (B) Schematic diagram showing the fabrication process of CNT/polydimethylsiloxane composite electrocardiogram (ECG) electrodes connected to the snap clip and skin test for 7 days *(38)*. (C) Electrode design for an electroencephalogram (EEG) *(13,39)*. (D) Wrist bandage-type pulse sensor for real-time blood pulse (BP) monitoring *(28)*. (E) Glucose sensing mechanism in CNT ion-sensitive field-effect transistor sensors *(40)*. *BL*, blood components. (F) Detection of nonpolar molecules by means of carrier scattering in random networks of CNTs for lung cancer diagnosis via breath sample *(41)*. *CRF*, chronic renal failure. *(Part A adapted with permission from Honda, W.; Harada, S.; Arie, T.; Akita, S.; Takei, K. Wearable, Human-Interactive, Health-Monitoring, Wireless Devices Fabricated by Macroscale Printing Techniques. Advanced Functional Materials* ***2014****, 24 (22), 3299–3304, Copyright 2014 John Wiley and Sons. Part B adapted with permission from Ha-Chul, J.; Jin-Hee, M.; Dong-Hyun, B.; Jae-Hee, L.; Yoon-Young, C.; Joung-Sook, H.; et al. CNT/PDMS Composite Flexible Dry Electrodes for Long-Term ECG Monitoring. IEEE Transactions on Biomedical Engineering* ***2012****, 59 (5), 1472–1479, Copyright 2012 IEEE. Part C adapted with permission from Ruffini, G.; Dunne, S.; Fuentemilla, L.; Grau, C.; Farres, E.; Marco-Pallarés, J.; et al. First Human Trials of a Dry Electrophysiology Sensor Using a Carbon Nanotube Array Interface. Sensors and Actuators A: Physical* ***2008****, 144 (2), 275–279 and Casson, A.; Yates, D.; Smith, S.; Duncan, J.; Rodriguez-Villegas, E. Wearable Electroencephalography. IEEE Engineering in Medicine and Biology Magazine* ***2010****, 29 (3), 44–56, Copyright 2011 Elsevier and 2010 IEEE. Part D adapted with permission from Kwon, D.; Lee, T.-I.; Kim, M.; Kim, S.; Kim, T.-S.; Park, I., Eds. Porous Dielectric Elastomer Based Ultra-Sensitive Capacitive Pressure Sensor and Its Application to Wearable Sensing Device. 2015 Transducers-2015 18th International Conference on Solid-State Sensors, Actuators and Microsystems (TRANSDUCERS); IEEE, 2015, Copyright 2015 IEEE. Part E adapted with permission from Lee, D.; Cui, T. Low-Cost, Transparent, and Flexible Single-Walled Carbon Nanotube Nanocomposite Based Ion-Sensitive Field-Effect Transistors for PH/Glucose Sensing. Biosensors and Bioelectronics* ***2010****, 25 (10), 2259–2264, Copyright 2011 Elsevier. Part F adapted with permission from Peng, G.; Tisch, U.; Haick, H. Detection of Nonpolar Molecules by Means of Carrier Scattering in Random Networks of Carbon Nanotubes: Toward Diagnosis of Diseases via Breath Samples. Nano Letters* ***2009****, 9 (4), 1362–1368, Copyright 2009 American Chemical Society.)*

inconvenient to apply the sensors for measurements outside the laboratory or clinic *(13)*. Thanks to thriving materials science, such a dilemma was elegantly solved by the demonstration of the first human trials of a dry electrophysiology sensor using a CNT array interface in 2008 *(13)*. As depicted in Fig. 7.3B, a CNT was dispersed in polydimethylsiloxane (PDMS) via a two-step milling procedure to fabricate a flexible CNT/PDMS composite-based dry electrode for long-term wearable electrocardiogram (ECG) monitoring *(38)*. The sensor is further shown to be robust to environmental influences such as motion and sweat. It should be noted that ECG signal quality depends on the composition of the CNT/PDMS hybrid (less than 1.5 wt% CNT is desired *(46)*) and the size of the electrode. In bio- and skin-compatibility tests, the epidermal electrodes, which are directly printed onto the skin, did not generate any noticeable side effects (such as itching or irritation) *(47)*. Even after 1 week of continuous wearing, the signal quality did not show an obvious decrease over time, in contrast to conventional Ag/AgCl electrodes. Another example of a wearable monitoring device is the low-power wireless ECG necklace for reliable and on-the-move cardiac activity monitoring, which provides ease of use and comfort and allows flexibility in lead positioning and system functionality *(48)*.

Although the capacitive electrodes have been widely used for ECG, there is still a great challenge to use them for electroencephalogram (EEG) recording. The amplitude of EEG (10−100 μV) is much lower than that of ECG (0.05−3 mV), so such a low EEG signal can be easily affected by noise *(49)*. Giulio's group mounted CNT arrays on state-of-the-art commercially active electrodes and then connected them to commercial off-the-shelf research electrophysiology recording equipment (Fig. 7.3C) *(13)*. Compared to the existing "wet" sensors based on electrolytic gel *(50)*, such a nanostructure design provides a robust, stable, low-impedance, and low-noise electrical interface while barely penetrating the outer skin layer—the stratum corneum—resulting in a comfortable and pain-free interface, which is the result of the rational design based on the excellent mechanical (high modulus of elasticity, tensile strength *(51)*) and electrical properties *(52)* of CNTs. Also the elasticity and flexibility of the substrates within the dry EEG sensors allow them to adapt to irregular skin surfaces to maintain the low sensor-skin impedance *(53)*. Signal attenuation, phase distortion, and reduction of common mode rejection ratio in the hairy site are also challenges of EEG monitoring *(54)*. Based on a CNT/PDMS composite, a canal-type ear electrode for EEG recording has been developed with additional functions, triggering sound stimulation-like earphones and recording EEG simultaneously for auditory brain/computer interface *(55)*. Undoubtedly, these electrodes have a prosperous future.

The issue of age-related diseases is expected to rise in the coming decades because of aging societies worldwide *(56)*. Most age-related diseases, such as metabolic syndrome, usually exacerbate gradually for many years and there is an urgent need for strategies to prevent an emerging global epidemic *(57)*. The most effective method to solve the problem is a change of lifestyle and diet *(58)* in combination with the possibility of monitoring and controlling the characteristic index (e.g., blood pressure, glucose, cholesterol) at a healthy level in real time by wearable sensors *(4,33)*.

Another interesting area where CNT-based materials have found some applications in wearable sensors is CNT-based flexible pressure sensors, which exhibit the unique advantages of flexibility and low cost. The development of such a field is enhanced by introducing such a concept into inartificial intelligence systems and wearable health care devices *(59)*. A conductive, transparent, stretchable nanotube film was produced by spray coating CNTs directly onto a substrate of PDMS from a CNT solution in *N*-methylpyrrolidone; this was first demonstrated as a skin-like pressure sensor by Bao et al. *(60)*. The spring-like structures in the nanotubes are able to accommodate strains of up to 150%, even in the stretched state, and its conductivities can be as high as 2200 S/cm. It turned out that the capacitance of this sensor was in proportion to pressure from about 50 kPa to 1 MPa with high reversibility because of the programmable elastic networks. Compared to one another reported "skin-like" device *(61)*, though the sensor is less sensitive, the appealing features of transparency and stretchability are still attractive. In parallel, uniform semiconducting CNT networks are fabricated on polyimide substrates with a interesting honeycomb mesh structure and the networks are implemented as a highly robust conformal and sensitive active-matrix backplane for pressure mapping *(62)*. Inspired by the idea that employs CNT-elastomer nanocomposite thin film as the conducting electrode layer, Kwon et al. from the Korea Advanced Institute of Science and Technology developed a flexible capacitive-type pressure sensor with ultrahigh sensitivity and stability based on porous elastomeric dielectric *(28)*. Because of the presence of abundant micropores within the dielectric layer, their capacitive-type pressure sensor became highly deformable, even at very small applied pressure loads. Such pressure sensors have high performance with a ultrahigh sensitivity of 0.601/kPa at a low-pressure regime (<5 kPa) and wide working range of 0−130 kPa. Interestingly, the real-life test demonstrates that this sensor can be simply applied to wearable pressure-sensing devices as a form of bandage-type wrist pulse monitoring device (Fig. 7.3D) *(28)*.

Diabetes is a disease where the pancreas does not produce enough insulin or the cells of the body do not respond properly to the insulin produced, which can result in many complications *(63)*. The blood glucose monitor is a crucial reality in managing diabetes. Current glucose-level measurements usually require blood via the finger-pricking method and commercially available equipment to analyze the glucose level *(64−66)*. The GlucoWatch automatic glucose biographer (Cygnus Inc., Redwood City, California) provides a noninvasive means to measure glucose levels for as long as 12 h after one single blood glucose calibration *(67)*. The noninvasive method utilized by the GlucoWatch extracts glucose through the skin using an applied potential (a process known as iontophoresis) and measures the extracted sample using an electrochemical-enzymatic sensor. But iontophoresis usually causes transient mild skin irritation and the damage degree relies on the frequency with which the device is used *(68)*. Yoon et al.

fabricated a microneedle-based three-electrode integrated electrochemical sensor via the electrodeposition of platinum nanoparticles on CNT forest grown on the silicon microneedle array, and in vitro characterized this sensor for glucose-sensing applications. The high sensitivity ($17.73 \pm 3\ \mu A/mM\ cm^2$) of such a sensor could be attributed to the high surface area of the sensor increased by the CNT forest and dispersion of platinum nanoparticles *(29)*. However, even a minimally invasive technique involves pricking the fingers several times per day, which certainly impedes the quality of life of a person living with diabetes. Considering the invasiveness to acquire blood and the risk of infection, the birth of a noninvasive flexible wearable device can be considered as a milestone. Based on the principle of ion-sensitive field-effect transistors *(69)*, a transparent and flexible glucose sensor was reported using CNT and glucose oxidase layers (Fig. 7.3E). A detection limit can be achieved as low as 2 mM. It is anticipated that in the near future glucose can be monitored in real time by authentic noninvasive wearable devices and both patients with diabetes and health care professionals can use related information to make better decisions on all aspects of diabetes management *(67)*.

On the basis of the research on the molecular recognition of the CNT corona phase and the model of Sorensen *(70)*, researchers from Massachusetts Institute of Technology fabricated an interesting sensor that is able to measure the protein level in blood samples or the fibrinogen level changes in body tissues *(71)*.

Besides body liquids, breath is also a powerful indicator that provides useful information on body health *(72,73)*. For example, it is established that the exhaled nitric oxide (NO) level is markedly elevated from the breath of patients with asthma. Based on such a principle, Kuzmych et al. successfully constructed a CNT field-effect transistor for NO gas detection by utilizing the CNT conductivity change caused by the interaction between the gas and the chemically functionalized CNT *(74)*, which provides a potential method to enable self-diagnostics and home care for asthma sufferers. In another case, it was reported that the concentration of volatile organic compounds in the breath from patients with lung cancer or kidney diseases will be drastically different from those in the breath from individuals with no lung cancer or kidney diseases *(75)*. With this in mind, a conductive sensor array based on CNTs that can identify the change of volatile organic compounds was proposed *(41)*, with a lower cost than the normally used sensor that depends on gold nanoparticles *(76)* (Fig. 7.3F). However, it is still a challenge to miniaturize the sensor into a tip band on the region of the nasolabial angle or in the wall of the nostril together with wireless technology, so that the patient's condition can be detected in real time to prevent an emergency.

In addition, other body odors obtained from different regions and forms such as urine odor or armpit odor can also be probed for personal health status. Lately, a prototype of a wearable e-nose for monitoring human body odor was fabricated on the basis of a polymer/CNT-COOH nanocomposite gas sensor array *(77)*. The sensors were used to discriminate the human body odor from two people by testing various volatile compounds such as ammonium hydroxide, ethanol, pyridine, triethylamine, methanol, and acetone, which are commonly found in the waste released from the human body and are usually individually different (Fig. 7.4).

Health care is one of the most crucial factors benefiting human beings with improved quality of life. The application of wearable CNT sensors for health care would enable medical practitioners to perform an early diagnosis of potential disease issues, which in turn would lead to a better chance of full recovery and lower hospitalization expenditure, and enable people to live independently in the home environment for longer *(78)*.

7.3 WEARABLE CNT SENSORS FOR MOTION DETECTION

Human motion is a highly complex concept, which depends on and is, in turn, influenced by many factors, including physiological, anatomical, psychological, environmental, and social effects *(9,79)*. Monitoring human motion with wearable technology has been considered an effective approach to evaluate the human state; this technology is both accurate and able to do fast analyses *(23)*. Among the various types of techniques to detect and analyze periodic motion, the strain gauge is one of the most important and potential smart sensors *(80)*, which liberates the tracking motion from the space limitation (under the vision camera lens) of photographic technology *(81)*. Unique tenacity *(51)*,

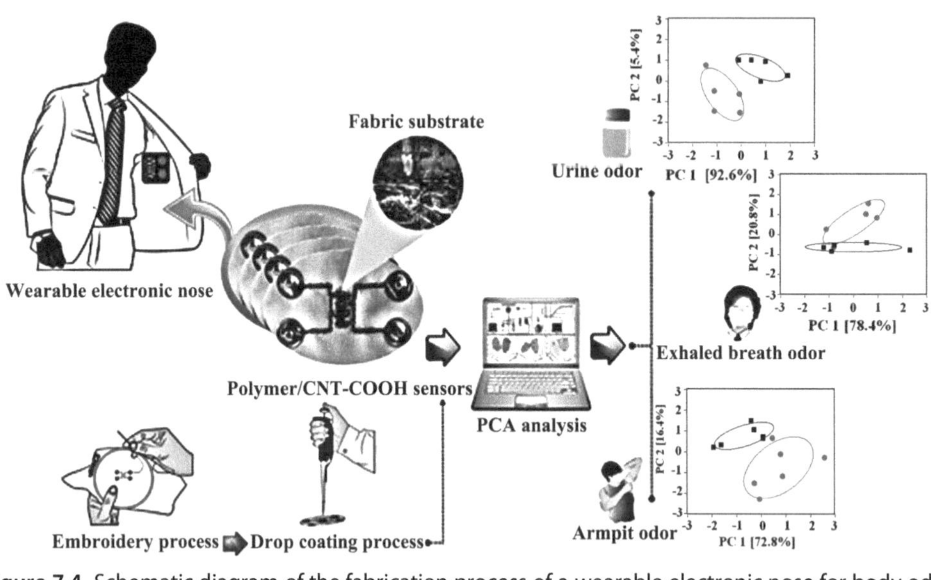

Figure 7.4 Schematic diagram of the fabrication process of a wearable electronic nose for body odor measurement *(77)*. PCA, principal component analysis. *(Adapted with permission from Seesaard, T.; Lorwongtragool, P.; Kerdcharoen, T. Development of Fabric-Based Chemical Gas Sensors for Use as Wearable Electronic Noses. Sensors* ***2015****, 15 (1), 1885–1902, Copyright 2015 Molecular Diversity Preservation International.)*

conductivity *(52)*, and resistance *(82)* make CNT an appropriate material for the construction of recoverable and stable strain sensors.

A class of wearable and stretchable devices based on aligned CNT thin films for a strain gauge was reported by Yamada and coworkers, Fig. 7.5A *(83)*. In the process of stretching, the alternate gaps and islands throughout the CNT film created by the irreversible fracture can lead to the monotonic increase of film resistance (Fig. 7.5B), with strain up to 280% before the PDMS substrate ruptured (50 times more than conventional

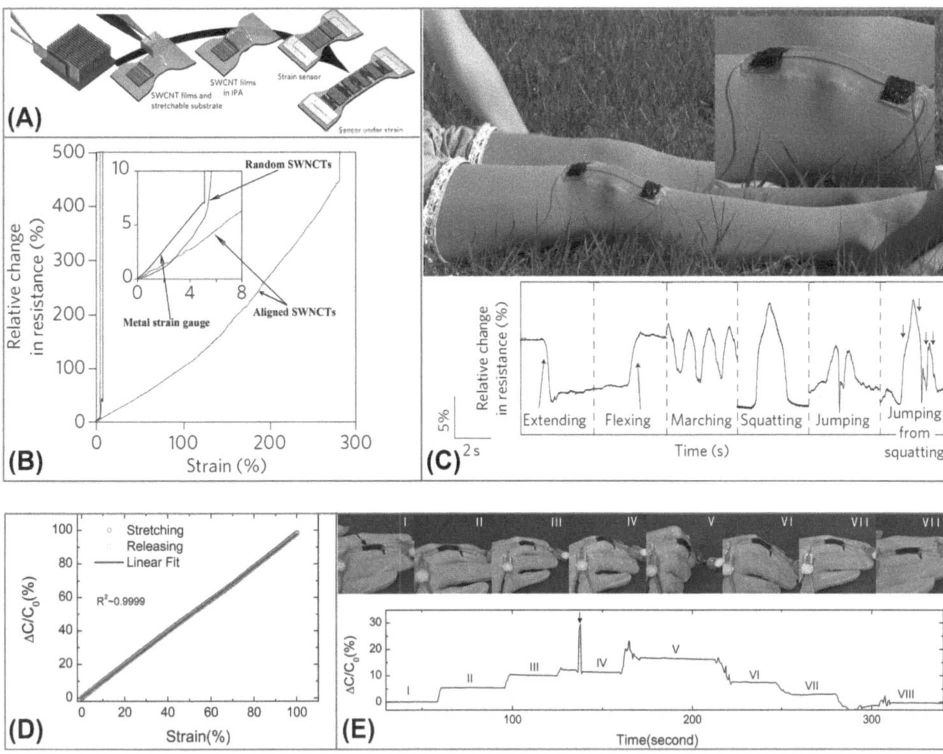

Figure 7.5 Demonstrations of using strain gauges to detect human motion. (A) Key steps in fabricating the single-walled carbon nanotube (SWCNT) strain sensor. (B) Relative change in resistance versus strain for the strain sensor [aligned carbon nanotube (CNT), *red* (gray in print versions)], randomly oriented CNT (*blue* (dark gray in print versions)), and conventional metal thin film (black). (C) A strain sensor fixed to a stocking and relative changes in resistance versus time for knee motion *(83)*. (D) Capacitive response of a strain gauge made from CNT/polydimethylsiloxane (PDMS) during both loading (*red circles* (gray in print versions)) and unloading (*green squares* (light gray in print versions)) of a strain of 100%, as well as the linear fit (*blue line* (dark gray in print versions)). (E) A prototypical data glove. Upper: Still pictures while the finger was gradually folded (I–V) and then unfolded (VI–VIII). Lower: Corresponding capacitive responses. *Arrow*: An accidental disrupt of the copper wire *(84)*. *(Parts A, B, and C adapted with permission from Yamada, T.; Hayamizu, Y.; Yamamoto, Y.; Yomogida, Y.; Izadi-Najafabadi, A.; Futaba, D. N.; et al. A Stretchable Carbon Nanotube Strain Sensor for Human-Motion Detection. Nature Nanotechnology* ***2011****, 6 (5), 296–301, Copyright 2011 Nature Publishing Group. Parts D and E adapted with permission from Cai, L.; Song, L.; Luan, P.; Zhang, Q.; Zhang, N.; Gao, Q.; et al. Super-Stretchable, Transparent Carbon Nanotube-Based Capacitive Strain Sensors for Human Motion Detection. Scientific Reports* ***2013****, 3, Copyright 2013 Nature Publishing Group.)*

metal strain gauges). Such extreme stretchability suggests the possibility of the application of a strain gauge for the detection of large-scale human motion. Eight films were conveniently connected together to assemble a large CNT strain sensor with extended sensing area, then assembled on a commercial stocking over a knee joint (Fig. 7.5C). Thus various human motions related to the extension and flexion of the knee, including bending, marching, squatting, and jumping, and combinations of these, can be easily detected and discriminated. Also the sensor on a glove was able to measure different motions of each finger or it can be made into adhesive bandages to detect fluctuation of the laryngeal prominence and chest. Similarly, a parallel plate-like capacitor was prepared by two layers of CNT films acting as electrodes laid on two sides of PDMS *(84)*. It was found that the change in capacitance of this capacitive sensor is linear to the stretch length (Fig. 7.5D). By integrating this strain sensor onto a rubber glove, a prototypical data glove was fabricated to detect the bending movements of the fingers, and such stretchable strain gauges can stably and reliably distinguish every single slight bending with ultrafast response at the subsecond scale (Fig. 7.5E). What is more, Hu et al. demonstrated that CNT-based films can accommodate strains of 700% with the use of specialized substrates *(85)*, which greatly expand the application of strain sensors.

Until now, a number of emerging materials including organic polymers and small molecules, inorganic nanowires, metallic nanowires, and graphene are in direct competition with CNTs in the field of wearable sensors for motion detection *(8,80,86–89)*. However, the unique advantages (such as high intrinsic conductivity, solution processability, flexibility, and the potential for production at low cost) of CNTs still make them one of the ideal candidates for various flexible strain sensors in the near future.

7.4 WEARABLE CNT SENSORS FOR ENVIRONMENT PROBER

As we discussed earlier, wearable CNT sensors have shown excellent and attractive performance in capturing some important physiological parameters of the human body. In addition, another intriguing application is to monitor the state or the change of the surrounding environments to prevent dangerous or toxic situations. Wearable CNT sensors can be used in various fields, such as detecting drug odors for police forces, discovering bombs used for terrorist activities, servicing special gases used in manufacturing, and even monitoring methanol or methane levels in rooms *(90,91)*.

Based on the band theory, a target gas that interacts with the surface of a film would result in a change of the charge carrier concentration for the material. Such a change may serve to alter the conductivity of the material *(92)*. By combining microlithography with lift-off techniques, CNT-based flexible sensors (Fig. 7.6A) were fabricated by assembling uniform CNT films on poly(dimethyldiallylammonium chloride) modified polyimide substrates and exhibited a high magnitude of response, reproducibility, linearity (Fig. 7.6B), selectivity (Fig. 7.6C), fast response, high stability, and very low detection

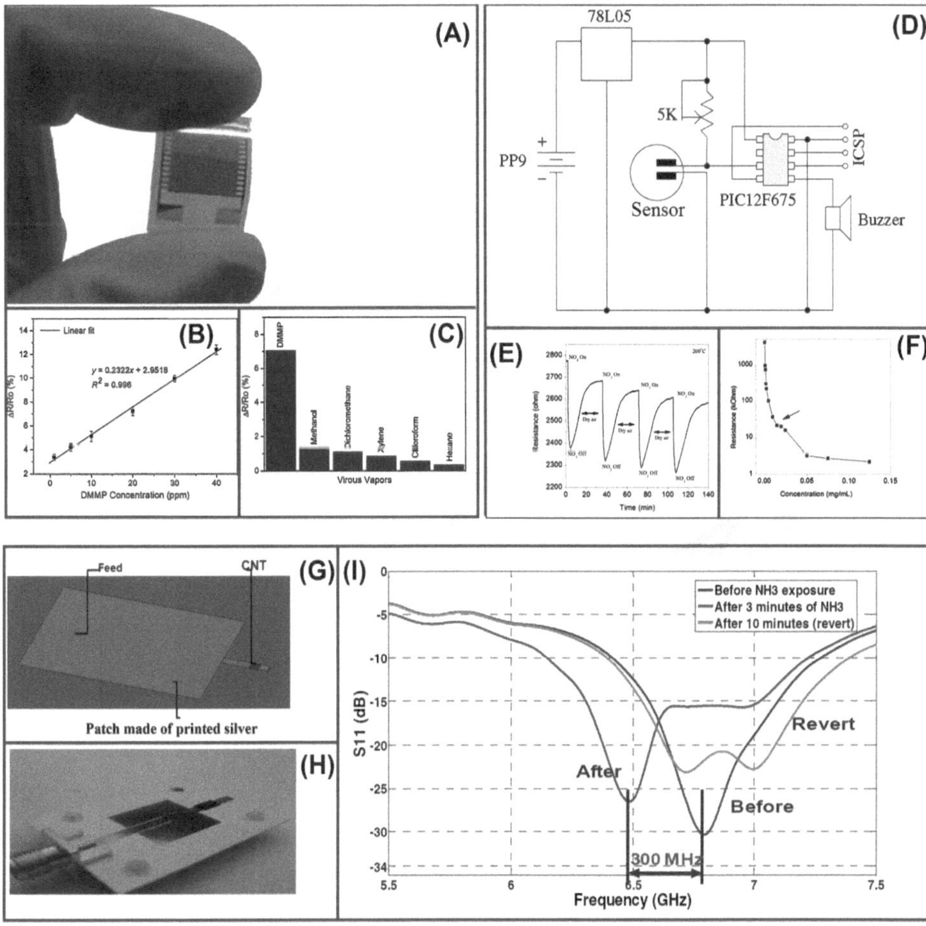

Figure 7.6 (A) The optical micrograph of the flexible assembled single-walled carbon nanotube (SWCNT) sensor on a polyimide (PI) substrate. (B) Calibration plot of the resistance responses of the flexible assembled SWCNT sensors versus the concentration of dimethyl methylphosphonate (DMMP) vapors. (C) Response of the flexible assembled SWCNT sensors to DMMP vapor and interferences diluted to 1% of saturated vapor concentrations *(93)*. (D) PIC12F675-based wearable sensor circuit. (E) Device performances when exposed to NO_2 at 200°C *(94)*. (F) Electrical resistance of different concentration of carbon nanotube (CNT) used to form networks films. The *arrow* indicates the approximate position of the conversion from 2D to 3D conduction *(95)*. (G) Sketch of the patch antenna with CNT loading. (H) Fabricated inset patch antenna with loaded polyaminobenzene sulfonic acid (PABS)-SWCNT. (I) Measured return loss with ammonia gas showing a resonance frequency shift of 300 MHz when exposed to 25 ppm ammonia *(96)*. *(Parts A, B, and C adapted with permission from Wang, Y.; Yang, Z.; Hou, Z.; Xu, D.; Wei, L.; Kong, E. S.-W.; et al. Flexible Gas Sensors with Assembled Carbon Nanotube Thin Films for DMMP Vapor Detection. Sensors and Actuators B: Chemical* ***2010****, 150 (2), 708–714, Copyright 2010 Elsevier. Parts D and E adapted with permission from Ramli, M. M.; Mat Isa, S. S.; Hasan, A. F.; Adnan, J., Eds. Wearable Gas Sensor Utilising Dye-Functionalized Single-Walled Carbon Nanotubes. 2014 2nd International Conference on Electronic Design (ICED); IEEE, 2014, Copyright 2014 IEEE. Part F adapted with permission from Ramli, M. M.; Isa, S. S. M.; Henley, S. J. Fabrication of Multi-Walled Carbon Nanotubes Hydrogen Sensor on Plastic. RSM2013 Proc., 2013; pp. 316–319, Copyright 2013 IEEE. Parts G, H, and I adapted with permission from Lee, H.; Naishadham, K.; Tentzeris, M.; Shaker, G., Eds. A Novel Highly-Sensitive Antenna-Based "Smart Skin" Gas Sensor Utilizing Carbon Nanotubes and Inkjet Printing. 2011 IEEE International Symposium on Antennas and Propagation (APSURSI); IEEE, 2011, Copyright 2011 IEEE.)*

limit to nerve agent stimulant dimethyl methylphosphonate (DMMP) vapors at room temperature. These make CNT-based flexible sensors very promising for portable and wearable on-site detection *(93)*. Fig. 7.6D illustrates a wearable sensor built on a model circuit of a PIC12F675 microcontroller. Such a sensor was constructed by depositing functionalized CNTs on plastic substrates via a vacuum filtration method, which allows reasonably good nanoscale control over the film thickness *(94)*. For possible real applications, a sensor based on dye-functionalized CNTs could be applied to monitor NO_2 (Fig. 7.6E) *(94)* and a Pt-modified CNT was used to detect H_2 (Fig. 7.6F) *(95)* for a corresponding field, respectively.

In addition to resistive-type sensors, CNTs were also utilized as the nanomaterial for fabricating antenna-based sensors *(96,97)*. Its chemical reactivity with ammonia or other analytes of interest to which the CNT is functionalized changes their effective permittivity, thereby shifting the resonant frequency. As depicted in Fig. 7.6F and G, the antenna and the sensor locations were separated to minimize the loss resulting from lower conductivity by coating a small portion of the antenna stub with a mesh of randomly oriented CNT. By doing so, the entire radiator is made of silver, and only a very small portion of the sensor consists of CNT. Thus the efficiency of the antenna remains high, and the frequency shift of the return loss is maximized by the change in impedance of the CNT load upon interaction with ammonia. As a result of being exposed to low levels of ammonia, a resonance frequency shift of 300 MHz for a patch antenna centered at ~6 GHz was observed (Fig. 7.6I), which is the highest frequency shift reported in the literature to the authors' knowledge *(96)*.

One of the most important problems in gas sensors is their selectivity, that is, the capability of providing different responses while exposed to different gaseous species *(98)*. Efforts have been focused on searching for new sensing materials to improve the selectivity of gas sensors. However, it is still a challenge to implement a smart gas sensor microsystem *(99)* for wearable devices as ultralight, switchable surfaces for energy harvesting.

7.5 CHALLENGE AND PERSPECTIVES

CNT exhibits its unique physical and chemical properties (such as high aspect ratio, ultralight weight, high mechanical strength, high electrical conductivity, and high thermal conductivity), which make CNTs an attractive nanomaterial with the potential for diverse wearable sensor fabrications. Nonetheless, several critical issues still need to be addressed before the wide applications of wearable CNT sensors can be realized. One of these issues is the biocompatibility of CNT. There have been plenty of papers reporting on the adverse effects of the CNT *(100–103)*. Smart et al. reviewed the nanotoxicology for some emerging carbon nanomaterials and discussed the toxicity of CNTs on lung, skin irritation, and especially the cytotoxicity of CNTs *(104)*. Table 7.1 summarizes several cytotoxicity test results carried out by different groups. Although it was never reported that the

Table 7.1 Cytotoxicity test for carbon nanotubes (CNTs)

Type	Affected area
CNT and graphene layers	Induce PC12 cell death *(110)*
Macrophages exposed to CNT	Necrosis, degeneration, and signs of apoptotic cell death *(111)*
CNT, graphene oxide, and nanodiamond	Dose- and time-dependent cytotoxicity to HeLa cells *(112)*
CNT	Cell of cultured human dermal fibroblast (HDF) death *(113)*
Purified CNT	Induces massive loss of cell viability through DNA damage and programmed cell death of HDF *(114)*
CNT	Elicits DNA damage and inflammation *(115)*
CNT	The cell of BEAS-2B (a human bronchial epithelial cell line) produced proinflammatory cytokines *(116)*
Metallic CNT	Loss of *Escherichia coli* viability was observed to increase with an increasing fraction of metallic single-walled carbon nanotubes *(117)*

CNT/PDMS film-based epidermal electrode, which was directly printed onto the skin, did not generate any noticeable side effects *(47)*, the majority of the test results still indicated CNT toxicity. The second challenge is how to create scalable and cost-effective means to deposit CNTs onto flexible substrates *(105)*. The processing conditions of chemical vapor deposition-grown CNTs are relatively laborious and difficult to implement to large-area fabrication, although the CNTs prepared by such a method generally yield better electrical performance. The third concern is the unsatisfied stability of the wearable sensors under ambient working conditions, which are more variable and complex than those in the test lab. A final consideration is system efficiency, durability, and unobtrusiveness, which are essential for the widespread application of devices and technologies. Wearable systems and devices must have efficient software and hardware and be unobtrusive *(11)*. For example, an ideal wearable CNT sensor system should continuously measure and wirelessly report all health-related information once it is deployed at one point on the human body.

Until now, there has been great progress in the field of nanomaterials given their great potential applications in daily life. At the fundamental level, insight into the property of CNT-based materials and their deformation mechanisms has already improved our understanding of CNT mechanical behavior *(83,106)*. On the practical side, there is a prospect of using wearable CNT sensors for many applications including health information monitoring, motion detection, and energy harvesting *(105,107)*. Meanwhile, the robust market demand of wearable technology is a stimulation for future advances of wearable CNT sensor systems *(108,109)*. With the development of such an exciting field, there will be lots of interesting science and technology experimentation to learn more. We hope that the coming groundbreaking advancements will bring forward conceptually different scope to this active area.

ACKNOWLEDGMENTS

G.H. acknowledges support from 1000-Talent Program (Recruitment Program of Global Expert, In Chinese: Qian-Ren-Ji-Hua) and the Special Talent Foundation of Director of Xinjiang Technical Institute of Physics & Chemistry, Chinese Academy of Sciences and National Natural Science Foundation of China (21505154). M.Z. is grateful to the support from Northeast Normal University startup funds, Jilin Provincial Department of Education, and the Recruitment Program of Global Youth Experts. G. Hu and M. Zhou are the corresponding authors.

REFERENCES

1. Jayson Derrick, B.S. *Morgan Stanley: Wearable Devices A Potential $1.6 Trillion Business.* http://wwwwearabledevicescom/2014/11/20/morgan-stanley-wearable-devices.
2. Ya-Li, Z.; Xiao-Rong, D.; Poon, C. C. Y.; Lo, B. P. L.; Heye, Z.; Xiao-Lin, Z.; et al. Unobtrusive Sensing and Wearable Devices for Health Informatics. *IEEE Transactions on Biomedical Engineering* **2014,** *61* (5), 1538−1554.
3. Wu, Z.; Xu, C.; Chen, H.; Wu, Y.; Yu, H.; Ye, Y.; et al. Mesoporous MgO Nanosheets: 1,6-hexanediamin-assisted Synthesis and Their Applications on Electrochemical Detection of Toxic Metal Ions. *Journal of Physics and Chemistry of Solids* **2013,** *74* (7), 1032−1038.
4. Marschollek, M.; Gietzelt, M.; Schulze, M.; Kohlmann, M.; Song, B.; Wolf, K.-H. Wearable Sensors in Healthcare and Sensor-Enhanced Health Information Systems: All Our Tomorrows? *Healthcare Informatics Research* **2012,** *18* (2), 97.
5. Pang, C.; Lee, C.; Suh, K. Y. Recent Advances in Flexible Sensors for Wearable and Implantable Devices. *Journal of Applied Polymer Science* **2013,** *130* (3), 1429−1441.
6. Aminian, K.; Najafi, B. Capturing Human Motion Using Body-Fixed Sensors: Outdoor Measurement and Clinical Applications. *Computer Animation and Virtual Worlds* **2004,** *15* (2), 79−94.
7. Wang, L.; Hu, W.; Tan, T. Recent Developments in Human Motion Analysis. *Pattern Recognition* **2003,** *36* (3), 585−601.
8. Wang, Y.; Wang, L.; Yang, T.; Li, X.; Zang, X.; Zhu, M.; et al. Wearable and Highly Sensitive Graphene Strain Sensors for Human Motion Monitoring. *Advanced Functional Materials* **2014,** *24* (29), 4666−4670.
9. Shany, T.; Redmond, S. J.; Narayanan, M. R.; Lovell, N. H. Sensors-Based Wearable Systems for Monitoring of Human Movement and Falls. *Sensors Journal, IEEE* **2012,** *12* (3), 658−670.
10. Pantelopoulos, A.; Bourbakis, N. G. A Survey on Wearable Sensor-Based Systems for Health Monitoring and Prognosis. *IEEE Transactions on Systems, Man, and Cybernetics, Part C: Applications and Reviews* **2010,** *40* (1), 1−12.
11. Chan, M.; Estève, D.; Fourniols, J.-Y.; Escriba, C.; Campo, E. Smart Wearable Systems: Current Status and Future Challenges. *Artificial Intelligence in Medicine* **2012,** *56* (3), 137−156.
12. Byrne, R.; Diamond, D. Chemo/Bio-Sensor Networks. *Nature Materials* **2006,** *5* (6), 421−424.
13. Ruffini, G.; Dunne, S.; Fuentemilla, L.; Grau, C.; Farres, E.; Marco-Pallarés, J.; et al. First Human Trials of a Dry Electrophysiology Sensor Using a Carbon Nanotube Array Interface. *Sensors and Actuators A: Physical* **2008,** *144* (2), 275−279.
14. Ponmozhi, J.; Frias, C.; Marques, T.; Frazão, O. Smart Sensors/Actuators for Biomedical Applications: Review. *Measurement* **2012,** *45* (7), 1675−1688.
15. Abraham, J. K., Aryasomayajula, L., Whitchurch, A., Varadan, V. K., Eds. *Carbon Nanotube Strain Sensors for Wearable Patient Monitoring Applications. The 15th International Symposium on: Smart Structures and Materials & Nondestructive Evaluation and Health Monitoring;* International Society for Optics and Photonics, 2008.
16. Anderson, G. F.; Hussey, P. S. Population Aging: A Comparison Among Industrialized Countries. *Health Affairs* **2000,** *19* (3), 191−203.
17. Lutz, W.; Sanderson, W.; Scherbov, S. The Coming Acceleration of Global Population Ageing. *Nature* **2008,** *451* (7179), 716−719.

18. Kiecolt-Glaser, J. K.; Glaser, R. Stress and Immunity: Age Enhances the Risks. *Current Directions in Psychological Science* **2001,** *10* (1), 18–21.
19. Dall, T. M.; Gallo, P. D.; Chakrabarti, R.; West, T.; Semilla, A. P.; Storm, M. V. An Aging Population and Growing Disease Burden Will Require a Large and Specialized Health Care Workforce by 2025. *Health Affairs* **2013,** *32* (11), 2013–2020.
20. Reinhardt, U. E. Does the Aging of the Population Really Drive the Demand for Health Care? *Health Affairs* **2003,** *22* (6), 27–39.
21. Huber, M.; Orosz, E. Health Expenditure Trends in OECD Countries, 1990–2001. *Health Care Financing Review* **2002,** *25* (1), 1–22.
22. Park, S.; Jayaraman, S. Enhancing the Quality of Life through Wearable Technology. *Engineering in Medicine and Biology Magazine, IEEE* **2003,** *22* (3), 41–48.
23. Marschollek, M.; Tian, F.; Koch, S.; Horsch, A.; Demiris, G. *Smart Homes and Wearable Sensors in Healthcare—Infrastructures and Applications*, 2011.
24. Hu, L.; Hecht, D. S.; Gruner, G. Carbon Nanotube Thin Films: Fabrication, Properties, and Applications. *Chemical Reviews* **2010,** *110* (10), 5790–5844.
25. Iijima, S. Helical Microtubules of Graphitic Carbon. *Nature* **1991,** *354* (6348), 56–58.
26. Gao, L.; Dong, D.; He, J.; Qiao, K.; Cao, F.; Li, M.; et al. Wearable and Sensitive Heart-Rate Detectors Based on PbS Quantum Dot and Multiwalled Carbon Nanotube Blend Film. *Applied Physics Letters* **2014,** *105* (15), 153702.
27. Giuliani, A.; Placidi, M.; Di Francesco, F.; Pucci, A. A New Polystyrene-Based Ionomer/MWCNT Nanocomposite for Wearable Skin Temperature Sensors. *Reactive and Functional Polymers* **2014,** *76,* 57–62.
28. Kwon, D., Lee, T.-I., Kim, M., Kim, S., Kim, T.-S., Park, I., Eds. *Porous Dielectric Elastomer Based Ultra-Sensitive Capacitive Pressure Sensor and Its Application to Wearable Sensing Device. 2015 Transducers-2015 18th International Conference on Solid-State Sensors, Actuators and Microsystems (TRANSDUCERS);* 2015; pp 299–302.
29. Yoon, Y.; Lee, G.; Yoo, K.; Lee, J.-B. Fabrication of a Microneedle/CNT Hierarchical Micro/Nano Surface Electrochemical Sensor and Its In-Vitro Glucose Sensing Characterization. *Sensors* **2013,** *13* (12), 16672–16681.
30. Lee, J.-Y.; Park, E.-J.; Lee, C.-J.; Kim, S.-W.; Pak, J. J.; Min, N. K. Flexible Electrochemical Biosensors Based on O_2 Plasma Functionalized MWCNT. *Thin Solid Films* **2009,** *517* (14), 3883–3887.
31. Takei, K.; Honda, W.; Harada, S.; Arie, T.; Akita, S. Toward Flexible and Wearable Human-Interactive Health-Monitoring Devices. *Advanced Healthcare Materials* **2015,** *4* (4), 487–500.
32. Harada, S.; Honda, W.; Arie, T.; Akita, S.; Takei, K. Fully Printed, Highly Sensitive Multifunctional Artificial Electronic Whisker Arrays Integrated with Strain and Temperature Sensors. *ACS Nano* **2014,** *8* (4), 3921–3927.
33. Honda, W.; Harada, S.; Arie, T.; Akita, S.; Takei, K. Wearable, Human-Interactive, Health-Monitoring, Wireless Devices Fabricated by Macroscale Printing Techniques. *Advanced Functional Materials* **2014,** *24* (22), 3299–3304.
34. Kim, D.-H.; Lu, N.; Ghaffari, R.; Kim, Y.-S.; Lee, S. P.; Xu, L.; et al. Materials for Multifunctional Balloon Catheters with Capabilities in Cardiac Electrophysiological Mapping and Ablation Therapy. *Nature Materials* **2011,** *10* (4), 316–323.
35. Yu, D.; Wang, C.; Wehrenberg, B. L.; Guyot-Sionnest, P. Variable Range Hopping Conduction in Semiconductor Nanocrystal Solids. *Physical Review Letters* **2004,** *92* (21), 216802.
36. Deng, Z.-S.; Liu, J. Mathematical Modeling of Temperature Mapping over Skin Surface and Its Implementation in Thermal Disease Diagnostics. *Computers in Biology and Medicine* **2004,** *34* (6), 495–521.
37. Sibinski, M.; Jakubowska, M.; Sloma, M. Flexible Temperature Sensors on Fibers. *Sensors* **2010,** *10* (9), 7934–7946.
38. Ha-Chul, J.; Jin-Hee, M.; Dong-Hyun, B.; Jae-Hee, L.; Yoon-Young, C.; Joung-Sook, H.; et al. CNT/PDMS Composite Flexible Dry Electrodes for Long-Term ECG Monitoring. *IEEE Transactions on Biomedical Engineering* **2012,** *59* (5), 1472–1479.

39. Casson, A.; Yates, D.; Smith, S.; Duncan, J.; Rodriguez-Villegas, E. Wearable Electroencephalography. *IEEE Engineering in Medicine and Biology Magazine* **2010,** *29* (3), 44–56.
40. Lee, D.; Cui, T. Low-Cost, Transparent, and Flexible Single-Walled Carbon Nanotube Nanocomposite Based Ion-Sensitive Field-Effect Transistors for PH/Glucose Sensing. *Biosensors and Bioelectronics* **2010,** *25* (10), 2259–2264.
41. Peng, G.; Tisch, U.; Haick, H. Detection of Nonpolar Molecules by Means of Carrier Scattering in Random Networks of Carbon Nanotubes: Toward Diagnosis of Diseases via Breath Samples. *Nano Letters* **2009,** *9* (4), 1362–1368.
42. Buysse, D. J. Insomnia, Depression and Aging. Assessing Sleep and Mood Interactions in Older Adults. *Geriatrics* **2004,** *59* (2), 47–51. quiz 2.
43. Ancoli-Israel, S. Sleep and Its Disorders in Aging Populations. *Sleep Medicine* **2009,** *10,* S7–S11.
44. Altevogt, B. M.; Colten, H. R. *Sleep Disorders and Sleep Deprivation: An Unmet Public Health Problem;* National Academies Press, 2006.
45. Huigen, E.; Peper, A.; Grimbergen, C. Investigation into the Origin of the Noise of Surface Electrodes. *Medical and Biological Engineering and Computing* **2002,** *40* (3), 332–338.
46. Lee, J. H.; Nam, Y. W.; Jung, H.-C.; Baek, D.-H.; Lee, S.-H.; Hong, J. S. Shear Induced CNT/PDMS Conducting Thin Film for Electrode Cardiogram (ECG) Electrode. *BioChip Journal* **2012,** *6* (1), 91–98.
47. Yeo, W. H.; Kim, Y. S.; Lee, J.; Ameen, A.; Shi, L.; Li, M.; et al. Multifunctional Epidermal Electronics Printed Directly onto the Skin. *Advanced Materials* **2013,** *25* (20), 2773–2778.
48. Penders, J., van de Molengraft, J., Altini, M., Yazicioglu, F., Van Hoof, C., Eds. *A Low-Power Wireless ECG Necklace for Reliable Cardiac Activity Monitoring On-the-Move. Proceedings of the International Conference of the IEEE Engineering in Medicine and Biology Society;* 2011.
49. Lee, S. M.; Kim, J. H.; Byeon, H. J.; Choi, Y. Y.; Park, K. S.; Lee, S.-H. A Capacitive, Biocompatible and Adhesive Electrode for Long-Term and Cap-Free Monitoring of EEG Signals. *Journal of Neural Engineering* **2013,** *10* (3), 036006.
50. Metting van Rijn, A. C.; Peper, A.; Grimbergen, C. High-Quality Recording of Bioelectric Events. Part 2. Low-Noise, Low-Power Multichannel Amplifier Design. *Medical & Biological Engineering & Computing* **1991,** *29* (4), 433–440.
51. Wong, E. W.; Sheehan, P. E.; Lieber, C. M. Nanobeam Mechanics: Elasticity, Strength, and Toughness of Nanorods and Nanotubes. *Science* **1997,** *277* (5334), 1971–1975.
52. Ebbesen, T.; Lezec, H.; Hiura, H.; Bennett, J.; Ghaemi, H.; Thio, T. Electrical Conductivity of Individual Carbon Nanotubes. *Nature* **1996,** *382.*
53. Liao, L.-D.; Chen, C.-Y.; Wang, I. J.; Chen, S.-F.; Li, S.-Y.; Chen, B.-W.; et al. Gaming Control Using a Wearable and Wireless EEG-Based Brain-Computer Interface Device with Novel Dry Foam-Based Sensors. *Journal of Neuroengineering and Rehabilitation* **2012,** *9* (1), 5.
54. Huang, Y.-J.; Wu, C.-Y.; Wong, A. M.-K.; Lin, B.-S. Novel Active Comb-Shaped Dry Electrode for EEG Measurement in Hairy Site. *IEEE Transactions on Biomedical Engineering* **2015,** *62* (1), 256–263.
55. Lee, J. H.; Lee, S. M.; Byeon, H. J.; Hong, J. S.; Park, K. S.; Lee, S.-H. CNT/PDMS-Based Canal-Typed Ear Electrodes for Inconspicuous EEG Recording. *Journal of Neural Engineering* **2014,** *11* (4), 046014.
56. Hartholt, K. A.; van Beeck, E. F.; Polinder, S.; van der Velde, N.; van Lieshout, E. M. M.; Panneman, M. J. M.; et al. Societal Consequences of Falls in the Older Population: Injuries, Healthcare Costs, and Long-Term Reduced Quality of Life. *The Journal of Trauma: Injury, Infection, and Critical Care* **2011,** *71* (3), 748–753.
57. Eckel, R. H.; Grundy, S. M.; Zimmet, P. Z. The Metabolic Syndrome. *The Lancet* **2005,** *365* (9468), 1415–1428.
58. Feldeisen, S. E.; Tucker, K. L. Nutritional Strategies in the Prevention and Treatment of Metabolic Syndrome. *Applied Physiology, Nutrition, and Metabolism* **2007,** *32* (1), 46–60.
59. Zang, Y.; Zhang, F.; Di, C.-a.; Zhu, D. Advances of Flexible Pressure Sensors toward Artificial Intelligence and Health Care Applications. *Materials Horizons* **2015,** *2* (2), 140–156.

60. Lipomi, D. J.; Vosgueritchian, M.; Tee, B. C.; Hellstrom, S. L.; Lee, J. A.; Fox, C. H.; et al. Skin-Like Pressure and Strain Sensors Based on Transparent Elastic Films of Carbon Nanotubes. *Nature Nanotechnology* **2011,** *6* (12), 788–792.
61. Mannsfeld, S. C.; Tee, B. C.; Stoltenberg, R. M.; Chen, C. V. H.; Barman, S.; Muir, B. V.; et al. Highly Sensitive Flexible Pressure Sensors with Microstructured Rubber Dielectric Layers. *Nature Materials* **2010,** *9* (10), 859–864.
62. Takahashi, T.; Takei, K.; Gillies, A. G.; Fearing, R. S.; Javey, A. Carbon Nanotube Active-Matrix Backplanes for Conformal Electronics and Sensors. *Nano Letters* **2011,** *11* (12), 5408–5413.
63. World Health Organisation. *Definition and Diagnosis of Diabetes Mellitus and Intermediate Hyperglycaemia,* 2006. http://www.who.int/%20diabetes/publications/diagnosis_diabetes2006/en/.
64. Schlagheck, J.; Steinberg, C.; Weishaupt, O.; Shaffer, G. H. *Glucose Meter,* 2003. Google Patents.
65. Clark, D.; Montgomery, P.; Brugler, J. S.; Parsay, S. *Blood Glucose Meter,* 1997. Google Patents.
66. Savage, D. L.; Montgomery, P.; Chapman, F. M. *Blood Glucose Meter,* 1996. Google Patents.
67. Tamada, J. A.; Garg, S.; Jovanovic, L.; Pitzer, K. R.; Fermi, S.; Potts, R. O.; et al. Noninvasive Glucose Monitoring: Comprehensive Clinical Results. *JAMA* **1999,** *282* (19), 1839–1844.
68. Van der Geest, R.; Elshove, D.; Danhof, M.; Lavrijsen, A.; Boddé, H. Non-invasive Assessment of Skin Barrier Integrity and Skin Irritation Following Iontophoretic Current Application in Humans. *Journal of Controlled Release* **1996,** *41* (3), 205–213.
69. Guo, J.; Koswatta, S. O.; Neophytou, N.; Lundstrom, M. Carbon Nanotube Field-Effect Transistors. *International Journal of High Speed Electronics and Systems* **2006,** *16* (04), 897–912.
70. Sorensen, J. T. *A Physiologic Model of Glucose Metabolism in Man and Its Use to Design and Assess Improved Insulin Therapies for Diabetes;* Massachusetts Institute of Technology, 1985.
71. Bisker, G.; Iverson, N. M.; Ahn, J.; Strano, M. S. A Pharmacokinetic Model of a Tissue Implantable Insulin Sensor. *Advanced Healthcare Materials* **2015,** *4* (1), 87–97.
72. Hunter, G. W.; Dweik, R. A. Applied Breath Analysis: an Overview of the Challenges and Opportunities in Developing and Testing Sensor Technology for Human Health Monitoring in Aerospace and Clinical Applications. *Journal of Breath Research* **2008,** *2* (3), 037020.
73. Kim, K.-H.; Jahan, S. A.; Kabir, E. A Review of Breath Analysis for Diagnosis of Human Health. *Trends in Analytical Chemistry* **2012,** *33,* 1–8.
74. Kuzmych, O.; Allen, B. L.; Star, A. Carbon Nanotube Sensors for Exhaled Breath Components. *Nanotechnology* **2007,** *18* (37), 375502.
75. Phillips, M.; Gleeson, K.; Hughes, J. M. B.; Greenberg, J.; Cataneo, R. N.; Baker, L.; et al. Volatile Organic Compounds in Breath as Markers of Lung Cancer: A Cross-Sectional Study. *The Lancet* **1999,** *353* (9168), 1930–1933.
76. Peng, G.; Tisch, U.; Adams, O.; Hakim, M.; Shehada, N.; Broza, Y. Y.; et al. Diagnosing Lung Cancer in Exhaled Breath Using Gold Nanoparticles. *Nature Nanotechnology* **2009,** *4* (10), 669–673.
77. Seesaard, T.; Lorwongtragool, P.; Kerdcharoen, T. Development of Fabric-Based Chemical Gas Sensors for Use as Wearable Electronic Noses. *Sensors* **2015,** *15* (1), 1885–1902.
78. Lymberis, A. Advanced Wearable Sensors and Systems Enabling Personal Applications. *Wearable and Autonomous Biomedical Devices and Systems for Smart Environment: Springer* **2010,** 237–257.
79. Godfrey, A.; Conway, R.; Meagher, D.; ÓLaighin, G. Direct Measurement of Human Movement by Accelerometry. *Medical Engineering & Physics* **2008,** *30* (10), 1364–1386.
80. Pang, C.; Lee, G.-Y.; Kim, T-i; Kim, S. M.; Kim, H. N.; Ahn, S.-H.; et al. A Flexible and Highly Sensitive Strain-Gauge Sensor Using Reversible Interlocking of Nanofibres. *Nature Materials* **2012,** *11* (9), 795–801.
81. Cutler, R.; Davis, L. S. Robust Real-Time Periodic Motion Detection, Analysis, and Applications. *IEEE Transactions on Pattern Analysis and Machine Intelligence* **2000,** *22* (8), 781–796.
82. Frank, S.; Poncharal, P.; Wang, Z.; de Heer, W. A. Carbon Nanotube Quantum Resistors. *Science* **1998,** *280* (5370), 1744–1746.
83. Yamada, T.; Hayamizu, Y.; Yamamoto, Y.; Yomogida, Y.; Izadi-Najafabadi, A.; Futaba, D. N.; et al. A Stretchable Carbon Nanotube Strain Sensor for Human-Motion Detection. *Nature Nanotechnology* **2011,** *6* (5), 296–301.

84. Cai, L.; Song, L.; Luan, P.; Zhang, Q.; Zhang, N.; Gao, Q.; et al. Super-Stretchable, Transparent Carbon Nanotube-Based Capacitive Strain Sensors for Human Motion Detection. *Scientific Reports* **2013,** *3*.
85. Hu, L.; Yuan, W.; Brochu, P.; Gruner, G.; Pei, Q. Highly Stretchable, Conductive, and Transparent Nanotube Thin Films. *Applied Physics Letters* **2009,** *94* (16), 161108.
86. Zhou, J.; Gu, Y.; Fei, P.; Mai, W.; Gao, Y.; Yang, R.; et al. Flexible Piezotronic Strain Sensor. *Nano Letters* **2008,** *8* (9), 3035–3040.
87. Xu, F.; Zhu, Y. Highly Conductive and Stretchable Silver Nanowire Conductors. *Advanced Materials* **2012,** *24* (37), 5117–5122.
88. Li, X.; Zhang, R.; Yu, W.; Wang, K.; Wei, J.; Wu, D.; et al. Stretchable and Highly Sensitive Graphene-on-Polymer Strain Sensors. *Scientific Reports* **2012,** *2*.
89. Park, J. J.; Hyun, W. J.; Mun, S. C.; Park, Y. T.; Park, O. O. Highly Stretchable and Wearable Graphene Strain Sensors with Controllable Sensitivity for Human Motion Monitoring. *ACS Applied Materials & Interfaces* **2015,** 7 (11), 6317–6324.
90. Röck, F.; Barsan, N.; Weimar, U. Electronic Nose: Current Status and Future Trends. *Chemical Reviews* **2008,** *108* (2), 705–725.
91. Gardner, J. W.; Bartlett, P. N. A Brief History of Electronic Noses. *Sensors and Actuators B: Chemical* **1994,** *18* (1), 210–211.
92. Fine, G. F.; Cavanagh, L. M.; Afonja, A.; Binions, R. Metal Oxide Semi-conductor Gas Sensors in Environmental Monitoring. *Sensors* **2010,** *10* (6), 5469–5502.
93. Wang, Y.; Yang, Z.; Hou, Z.; Xu, D.; Wei, L.; Kong, E. S.-W.; et al. Flexible Gas Sensors with Assembled Carbon Nanotube Thin Films for DMMP Vapor Detection. *Sensors and Actuators B: Chemical* **2010,** *150* (2), 708–714.
94. Ramli, M. M., Mat Isa, S. S., Hasan, A. F., Adnan, J., Eds. *Wearable Gas Sensor Utilising Dye-Functionalized Single-Walled Carbon Nanotubes. 2014 2nd International Conference on Electronic Design (ICED);* 2014; pp 301–304.
95. Ramli, M. M.; Isa, S. S. M.; Henley, S. J. *Fabrication of Multi-Walled Carbon Nanotubes Hydrogen Sensor on Plastic. IEEE Regional Symposium on Micro and Nanoelectronics*, 2013, pp 316–319.
96. Lee, H., Naishadham, K., Tentzeris, M., Shaker, G., Eds. *A Novel Highly-Sensitive Antenna-Based "Smart Skin" Gas Sensor Utilizing Carbon Nanotubes and Inkjet Printing. 2011 IEEE International Symposium on Antennas and Propagation. IEEE Antennas and Propagation Society International Symposium;* 2011; pp 1593–1596.
97. Lee, H.; Shaker, G.; Naishadham, K.; Song, X.; McKinley, M.; Wagner, B.; et al. Carbon-Nanotube Loaded Antenna-Based Ammonia Gas Sensor. *Transactions on Microwave Theory and Techniques, IEEE* **2011,** *59* (10), 2665–2673.
98. Sayago, I.; Terrado, E.; Aleixandre, M.; Horrillo, M.; Fernández, M.; Lozano, J.; et al. Novel Selective Sensors Based on Carbon Nanotube Films for Hydrogen Detection. *Sensors and Actuators B: Chemical* **2007,** *122* (1), 75–80.
99. Hagleitner, C.; Hierlemann, A.; Lange, D.; Kummer, A.; Kerness, N.; Brand, O.; et al. Smart Single-Chip Gas Sensor Microsystem. *Nature* **2001,** *414* (6861), 293–296.
100. Zhao, Y.; Xing, G.; Chai, Z. Nanotoxicology: Are Carbon Nanotubes Safe? *Nature Nanotechnology* **2008,** *3* (4), 191–192.
101. Hussain, S. M.; Braydich-Stolle, L. K.; Schrand, A. M.; Murdock, R. C.; Yu, K. O.; Mattie, D. M.; et al. Toxicity Evaluation for Safe Use of Nanomaterials: Recent Achievements and Technical Challenges. *Advanced Materials* **2009,** *21* (16), 1549–1559.
102. Firme, C. P.; Bandaru, P. R. Toxicity Issues in the Application of Carbon Nanotubes to Biological Systems. *Nanomedicine: Nanotechnology, Biology and Medicine* **2010,** *6* (2), 245–256.
103. Liu, Y.; Zhao, Y.; Sun, B.; Chen, C. Understanding the Toxicity of Carbon Nanotubes. *Accounts of Chemical Research* **2012,** *46* (3), 702–713.
104. Smart, S. K.; Cassady, A. I.; Lu, G. Q.; Martin, D. J. The Biocompatibility of Carbon Nanotubes. *Carbon* **2006,** *44* (6), 1034–1047.
105. Park, S.; Vosguerichian, M.; Bao, Z. A Review of Fabrication and Applications of Carbon Nanotube Film-Based Flexible Electronics. *Nanoscale* **2013,** *5* (5), 1727–1752.

106. Lu, N.; Yang, S. Mechanics for Stretchable Sensors. *Current Opinion in Solid State and Materials Science* **2015,** *19* (3), 149–159.
107. Xie, K.; Wei, B. Materials and Structures for Stretchable Energy Storage and Conversion Devices. *Advanced Materials* **2014,** *26* (22), 3592–3617.
108. Harrop, D. P.; Hayward, M. J.; Das, R.; Holland, G. *Wearable Technology 2015–2025: Technologies, Markets, Forecasts;* IDTechEx Ltd, 2015.
109. Ballve, M. Wearable Gadgets Are Still Not Getting the Attention They Deserve—here's Why They Will Create a Massive New Market. *Business Insider* **2013,** *1.*
110. Zhang, Y.; Ali, S. F.; Dervishi, E.; Xu, Y.; Li, Z.; Casciano, D.; et al. Cytotoxicity Effects of Graphene and Single-Wall Carbon Nanotubes in Neural Phaeochromocytoma-Derived PC12 Cells. *ACS Nano* **2010,** *4* (6), 3181–3186.
111. Tymecki, L.; Zwierkowska, E.; Koncki, R. Screen-Printed Reference Electrodes for Potentiometric Measurements. *Analytica Chimica Acta* **2004,** *526* (1), 3–11.
112. Zhang, X.; Hu, W.; Li, J.; Tao, L.; Wei, Y. A Comparative Study of Cellular Uptake and Cytotoxicity of Multi-Walled Carbon Nanotubes, Graphene Oxide, and Nanodiamond. *Toxicology Research* **2012,** *1* (1), 62–68.
113. Sayes, C. M.; Liang, F.; Hudson, J. L.; Mendez, J.; Guo, W.; Beach, J. M.; et al. Functionalization Density Dependence of Single-Walled Carbon Nanotubes Cytotoxicity in Vitro. *Toxicology Letters* **2006,** *161* (2), 135–142.
114. Patlolla, A.; Knighten, B.; Tchounwou, P. Multi-Walled Carbon Nanotubes Induce Cytotoxicity, Genotoxicity and Apoptosis in Normal Human Dermal Fibroblast Cells. *Ethnicity & Disease* **2010,** *20* (1 Suppl. 1), S1.
115. Yamashita, K.; Yoshioka, Y.; Higashisaka, K.; Morishita, Y.; Yoshida, T.; Fujimura, M.; et al. Carbon Nanotubes Elicit DNA Damage and Inflammatory Response Relative to Their Size and Shape. *Inflammation* **2010,** *33* (4), 276–280.
116. Hirano, S.; Fujitani, Y.; Furuyama, A.; Kanno, S. Uptake and Cytotoxic Effects of Multi-Walled Carbon Nanotubes in Human Bronchial Epithelial Cells. *Toxicology and Applied Pharmacology* **2010,** *249* (1), 8–15.
117. Vecitis, C. D.; Zodrow, K. R.; Kang, S.; Elimelech, M. Electronic-Structure-Dependent Bacterial Cytotoxicity of Single-Walled Carbon Nanotubes. *ACS Nano* **2010,** *4* (9), 5471–5479.

CHAPTER 8

Use of Carbon Nanotubes in Third-Generation Solar Cells

T. Grace[1], C. Shearer[1], D. Tune[1,2], L. Yu[1], M. Batmunkh[1,3], M.J. Biggs[3,4], Z.A. ALOthman[5], J.G. Shapter[1]

[1]The Flinders University of South Australia, Bedford Park, SA, Australia; [2]Karlsruhe Institute of Technology, Karlsruhe, Germany; [3]The University of Adelaide, Adelaide, SA, Australia; [4]Loughborough University, Loughborough, United Kingdom; [5]King Saud University, Riyadh, Saudi Arabia

Contents

8.1 INTRODUCTION

Over recent years, the world's energy consumption has dramatically increased because of the rapidly growing global population, improving standards of living, and development of modern technologies. The US Department of Energy predicts that the world's energy demands will double by 2050 and triple by 2100 *(1)*. The current world energy production of ~16.4 TW is principally met by burning fossil fuels *(2)*. Damage to the environment by the fuel-burning process has become a serious problem. This has the potential

Industrial Applications of Carbon Nanotubes
ISBN 978-0-323-41481-4

for disastrous consequences, and solutions need to be pursued with urgency. Considerable developments have been made in this respect, including energy technologies based on wind power *(3)*, biofuels *(4)*, solar cells *(5)*, and fuel cells *(6)*.

Only solar energy is present across the entire world, thus harnessing it is the only really worldwide solution to the global energy challenge. One option is to use this energy in solar thermal applications that use large mirrors to concentrate the solar energy onto a material such as molten salts that are capable of storing significant energy. While relatively simple, these solar thermal installations are hindered by the need to convert that thermal energy stored in the molten salts into electricity using a steam turbine and the requirements of a steady water supply.

Another solar energy option is photovoltaic (PV) cells, which convert sunlight directly into electricity. PV cells are generally classified into three main generations. In brief: (1) the first-generation solar cells based on crystalline (poly + single) silicon, which make up ~90% of the commercial production at present, are estimated to deliver power with approximately 20% efficiency; but they suffer from high manufacturing, installation, and material costs, (2) the second-generation cells (referred as a thin film technology) are more cost-effective than the traditional PV cells, but their lower performance is the main concern and still the subject of considerable research *(7)*, and (3) the third-generation PV cells, which include, for example, organic solar cells and dye-sensitized solar cells (DSSCs), are designed to further lower the costs achieved for second-generation cells (US$0.8−1.0/W at present) to at least US$0.5/W, potentially to US$0.2/W, by maintaining the economic and environmental aspects while increasing the performance *(8)*. This drive to lower energy costs has led to a keen interest in finding solutions to improve efficiencies in all solar cell architectures. Additionally, the latest generation of cells offers the potential to be very cheap, flexible, semitransparent, and still productive in low light conditions.

Despite the first deliberate fabrication of a PV device by Fritts in 1883 *(9)*, it was not until 1954, when Chapin and coworkers at Bell Labs developed a 6% efficient silicon *p−n* junction solar cell *(10)*, that the PV industry was born. Initially, PVs were only utilized in specialist applications where conventional power sources were unavailable or undesirable such as satellites and high-end electronic devices. In 1985, with the development of the first ~20% efficient solar cells such as by Green and coworkers *(11)*, the field of PVs established itself as a serious energy resource capable of supplying domestic demand. These devices were so-called first-generation solar cells. The high conversion efficiency (up to ~25% in the lab, up to ~22% for commercial modules), stability of high-purity silicon, excellent charge transport properties, and the mature processing technologies have led to silicon dominating the PV market *(12−14)*. However, the manufacturing process of high-efficiency silicon solar cells suffers from low throughput and thus these solar cells are costly, preventing silicon PV from contributing significantly as a source in the world's energy.

To solve some of these issues, thin film solar cells, second-generation solar cells, such as CdTe, $CuIn_xGa_{1-x}S(Se)_2$ (CIGS), Cu_2ZnSnS_4 (CZTS), and thin film silicon have become the subject of intense research *(13–15)*. The heavy metal Cd is toxic and the metal In is expensive, thus limiting the development of CdTe and CIGS-based solar cells, respectively *(16,17)*. Low-cost materials such as CZTS were recently introduced but is still underdeveloped *(18)*. For thin film silicon solar cell technology, the low efficiency and instability from the Staebler–Wronski effect may restrict its long-term usage in solar cells *(14)*. The tradeoff between the cost and the performance of these solar cells is still a great barrier to wide-scale commercial application. Therefore it is essential to search for alternative materials to be applied in various components of the many types of solar cells now commercially available or under development in research labs around the world. Carbon nanotubes (CNTs) are an attractive option because of their exceptional conductivity in a lightweight form, which can be incorporated into many materials at low loadings and still significantly change the material properties to improve the required performance.

This chapter will thus focus on third-generation solar cells and the application of CNTs to improve their performance. These cells have the potential to be cheap, efficient, flexible, and deployable in virtually any location or situation. In particular, DSSCs and the related perovskite cells, organophotovoltaic (OPV) solar cells and CNT/Si or nanotube/silicon heterojunction (NSH) solar cells will be discussed.

8.2 CARBON NANOTUBES

Ever since Iijima's seminal report, CNTs have attracted considerable attention from the scientific as well as industrial communities because of their remarkable physical and optoelectronic properties *(19,20)*. Much of this effort has been in the development of solar cells because of the CNT's unique structure, excellent conductivity, good transparency, high catalytic activity, low cost, and abundance. There are many very good reviews of CNTs in the literature and hence a summary of their properties, production, etc. will not be provided in this chapter *(21–23)*.

8.3 TRANSPARENT CONDUCTING ELECTRODES

Most solar cells, and many other electronic devices, require an electrode that is both transparent and highly conductive as an integral part of the structure. Because of the excellent conductivity and high degree of transparency of bulk CNT materials, they have been employed in a variety of approaches to build these electrodes and this work is summarized in the following section before specific solar cell architectures are discussed. Transparent conducting electrodes (TCEs) are used in both the photoelectrode and counter electrode of DSSCs, as electrodes in OPVs, and as an integral part of the CNT/Si junction in NSH solar cells.

8.3.1 ITO and FTO

TCEs are a necessary component in a wide range of modern optoelectronic devices such as organic light-emitting diodes, e-paper, touch screens, liquid-crystal displays, and thin film PV cells, all of which are growing rapidly in popularity *(24–26)*. Currently, the two most common materials used to meet the transparency and conducting requirements of the electrodes are indium tin oxide (ITO) (preferred) and fluorine tin oxide (FTO) (less effective but cheaper). An ideal TCE should possess a very low sheet resistance (R_s) with high transmittance and be composed of cheap and abundant materials. ITO and FTO are the most widely used TCEs because of their high conductivity and optical transparency but there are a number of issues with these materials including cost and a lack of flexibility. These shortcomings of the current TCEs have inspired researchers to seek other easily processable and sustainable electrode materials that could exhibit comparable sheet resistance at high transmittance while still being readily available, cheap, highly flexible, and lightweight. In the past few years, a number of alternative materials, including conductive polymers *(27)*, metal grids *(28)*, metallic nanowires *(29)*, graphene *(30)*, and CNTs *(31)*, have been developed to replace ITO and FTO. For a broad perspective of the field, there are several comprehensive reviews of TCEs available *(32–36)*.

Among the aforementioned alternative materials, CNTs are considered a promising replacement for conventional TCOs because they exhibit properties such as low resistivity, a high specular transmittance in a very broad spectral range from UV to mid-infrared, superior flexibility, and the electrodes can be made using simple fabrication techniques *(37–45)*. CNTs also form sheets, often called bucky paper, with a mesh-like three-dimensional topology with morphology similar to a bird's nest, which allows charge collection from a large surface area, not just a planar interface as from usual TCO electrodes. The high thermal conductivities that provide heat dissipation combined with stability during exposure to light make CNTs an ideal electrode material *(46)*. CNTs have a remarkably high level of electronic mobility and work function similar to that of the TCOs, which makes them an ideal candidate for use in the hole-extraction layer (or hole-transporting layer) of PV cells. Hence CNTs are generally, but certainly not exclusively, used for hole transport *(47,48)*.

8.3.2 CNTs for TCEs

Typically, the preparation methods of transparent conductive CNT films can be separated into dry processes and wet processes depending on the sample condition *(49)*. In the dry processing method, CNT films can be prepared by the following two routes: (1) depositing a CNT aerosol made in a chemical vapor deposition (CVD) reactor to form a thin film and transferring onto the substrate using the dry-transfer technique *(50–53)* and (2) drawing from a superaligned CNT forest *(54–56)*. Because of the excellent performance of these TCFs, transparent conductive CNT electrodes produced by the dry-processing methods have been used in a variety of applications. Though the dry processed CNT films possess good conductivity, they have some disadvantages in film production such as complicated experimental control, requirement of high temperature, limited substrates, and high cost.

Preparing transparent conductive CNT films using the wet processing method is of great interest because the solution processable techniques are simple, they are compatible with low-temperature processes, and are applicable for various substrates as well as being cost-effective. Since the preparation method of CNT-based TCFs using filtration-transfer and dip-coating techniques was first reported by Wu et al. *(31)* and Saran et al. *(57)* in 2004, several leading research groups and companies have put considerable effort into this area. Many solution processing methods such as vacuum filtration followed by transfer printing *(31,44,58)*, dip coating *(57,59)*, spin coating *(40,41)*, spray coating *(58,41,42,60,61)*, electrophoretic deposition (EPD) *(62,63)*, printing (rod coating, inkjet printing, and slot coating) *(38,64,65)*, and recently brush painting *(66)* have been developed to fabricate CNT-based TCFs. Of course all of these methods have advantages and disadvantages. Comprehensive reviews of the solution processing techniques for the fabrication of CNT-TCFs can be found elsewhere *(34,35,67,68)*.

8.4 DYE-SENSITIZED SOLAR CELLS

DSSCs have gained much attention not only because of the simple fabrication method, low manufacturing cost, and eco-friendliness, but also because of their relatively high efficiency (13% *(69)*–14% *(70)* for DSSCs) compared to OPVs (~12% *(71)*), although we should point out that these efficiencies are continuously increasing. Since the breakthrough work on DSSCs published in the early 1990s *(72)*, there has been a great deal of research to maximize the power conversion efficiency (PCE) of DSSCs but progress beyond 12% has been very slow *(73,74)*. Several companies (namely, Solaronix from Switzerland, Dyesol from Australia, and Dyenamo from Sweden) have been established to contribute to DSSC commercialization. Moreover, in 2009, G24-Power (the world leader in the design and manufacture of DSSCs) introduced the world's first commercial production of DSSCs using a roll-to-roll manufacturing process *(75)*.

Fig. 8.1 shows a typical DSSC geometry and some of the components where CNTs have been used. CNTs have been used in each component of the DSSC design. They have been used as the conducting films in either the counter electrode or the photoanode. CNTs have also been used to make stand-alone photoactive elements as well as in hybrids to make better conducting photoanodes. Indeed, the best efficiency to date (>10%) of carbon-based DSSCs (among sole materials or nonhybrids) was obtained by

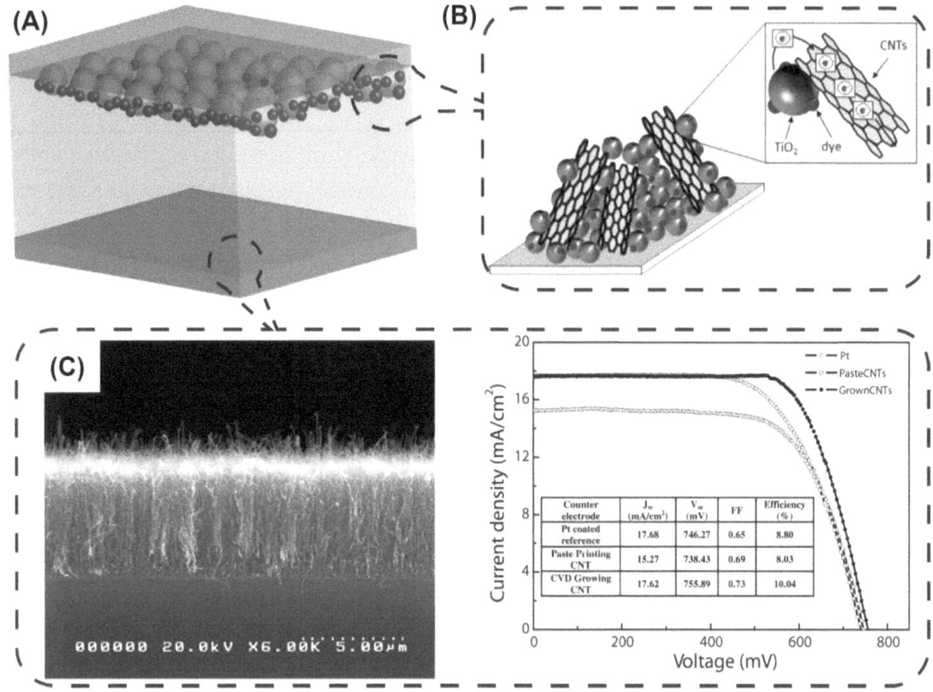

Figure 8.1 (A) Schematic of a typical dye-sensitized solar cell (DSSC) structure. (B) Schematic illustrating the introduction of carbon nanotubes (CNTs) into the photoanode. (C) Scanning electron microscope (SEM) image and *J–V* curve of a vertically grown CNT film-based DSSC. *(Reproduced with permission from Nam, J. G.; Park, Y. J.; Kim, B. S.; Lee, J. S. Enhancement of the Efficiency of Dye-Sensitized Solar Cell by Utilizing Carbon Nanotube Counter Electrode. Scripta Materialia* ***2010,*** *62 (3), 148–150. © Elsevier.)*

a vertically aligned CNT (VACNT) counter electrode-based cell (see Fig. 8.1C) *(76)*. On the other hand, among carbonaceous materials incorporating TiO_2 photoelectrode-based cells, DSSCs fabricated with a CNT/TiO_2 electrode are the champion cell.

8.4.1 CNTs-TCFs for DSSCs

Even though considerable advancements have been made in CNTs-TCFs, reports on the application of these films as the conducting electrodes in DSSCs are rare. There are very few reports available on the application of CNTs as the conducting component of the counter electrode of DSSCs *(77–80)*. Aitola et al. *(77)* synthesized single-walled carbon nanotubes (SWCNTs) using an aerosol CVD method and deposited a film on a polyethylene terephthalate (PET) plastic substrate. The resulting CNT film with an R_s of 60 Ω/sq was used as a counter electrode (TCO free and Pt free) in a DSSC. However, this DSSC fabricated with a CNT counter electrode had energy conversion efficiency at a light intensity of 100 mW/cm^2 of only 0.74%, which was far less than that (4.4%)

obtained with the reference cell. Tantang et al. *(80)* were able to improve the performance of Pt-free and TCO-free counter electrode-based DSSCs by using nitrogen-doped (N-doped) CNTs. Doped CNT films were prepared from solution processable N-doped CNT samples using an air-spray technique. The N-doping on CNTs forms one free electron, which can easily be supplied to the reduction of electrolyte and thus improve the performance of DSSCs. After N doping of the CNTs, the efficiency of DSSC was improved to 1.53%, which was better than that of undoped CNT counter electrodes reported by Aitola et al. *(77)* but still ~2.5-fold lower than the reference Pt-based device. Furthermore, Liu et al. *(78)* used CNTs as an alternative to TCO material in the Pt-based counter electrode of DSSCs. The dispersant [poly(oxyethylene)-backboned polyimide] was vital for preparing a good dispersion of the debundled CNTs and platinum salts in liquid. Highly dispersed Pt/CNT hybrids were drop coated on the substrates to fabricate the FTO-free counter electrode. The efficiency of the DSSC fabricated with the Pt/CNT counter electrode was as high as 6.96% at a standard light intensity. This performance was very close to that (7.02%) of the cell fabricated with a conventional Pt/FTO counter electrode and is the highest efficiency yet reported for TCO-free counter electrode-based DSSCs.

To our knowledge, only three papers have been published using CNTs as transparent and conductive component of the substrate for DSSC photoelectrodes *(81–83)*. Among them, the authors of two papers *(81,82)* have used ZnO nanoparticles, while the other one *(83)* used TiO_2 nanoparticles as a semiconducting layer. Wei et al. *(82)* were first to report a TCO-free photoelectrode in DSSCs. They fabricated a flexible DSSC composed of CNTs/PET as the TCE, ZnO nanoparticles as dye adsorbing semiconductor, black dye, iodide electrolyte, and Au as catalyst. Because of the suitable energy levels of the materials used in this study, a high PCE was expected. The CNTs-TCF fabricated in this study showed an R_s of 250 Ω/sq at 65% transparency and its cell was able to deliver a short circuit current (J_{sc}) of only 2.23 mA/cm^2 and open circuit voltage (V_{oc}) of only 0.23 V. This poor performance was caused by the catalytic properties of the CNTs for the redox reaction, which limited the continuous electron transfer. Du et al. *(81)* made further improvement to this DSSC and obtained a PCE of 2.5% by applying electrodeposited CNT film with a relatively high R_s (470 Ω/sq) at high transparency. On the other hand, Kyaw et al. *(83)* were able to reduce the charge recombination of CNTs with electrolyte by using a TiO_x layer. In their study, the DSSC was fabricated with a CNT film–TiO_2 semiconducting layer–TiO_x blocking layer-based photoelectrode and delivered an efficiency of 1.8%.

8.4.2 Semiconducting Layers

8.4.2.1 Nanostructured TiO_2 Materials

The heart of the DSSC is the semiconducting electron transport layer that collects and transfers the photogenerated electrons from the dye molecules and diffuses to the

TCE, which is connected to the external circuit. A mesoporous TiO_2 structure made with 20 nm diameter nanoparticles is used as the conventional semiconductor in the photoelectrode of DSSCs *(72)*. Many alternative oxide semiconductors such as ZnO *(84,85)*, SnO_2 *(86,87)*, Nb_2O_3 *(88)*, $SrTiO_3$ *(89)*, and $BaSnO_3$ *(90)* have been studied widely as the electron transport layer in DSSCs because of their suitable energy levels and high conductivity. A number of reviews of these materials for DSSCs have been published *(91–94)*. However, none of these materials have outperformed the conventional TiO_2 nanostructures because of their low surface area-to-volume ratio compared to the TiO_2 nanoparticles *(94–96)*. Indeed, TiO_2 nanoparticles still remain the material of choice for the DSSC photoelectrode. Unfortunately, nanostructured semiconducting materials used in the photoelectrodes have three main disadvantages: (1) significant recombination of the injected electrons with both the dye and oxidized species of the electrolyte, (2) slow electron transport caused by large number of grain boundaries, and (3) ineffective light harvesting. To solve these drawbacks, several strategies have been developed and they are briefly discussed in one of our previous reviews *(97)*.

8.4.2.2 Semiconducting Layers with CNTs

The use of CNTs in the semiconducting layer of DSSCs was first reported by Jang et al. *(98)* who incorporated acid-functionalized SWCNTs in nanocrystalline TiO_2 film. These authors observed a good enhancement (25%) in the J_{sc} and a slight improvement in the V_{oc} after the incorporation of SWCNTs in the unmodified (TiO_2-only-based) film. This improvement was believed to be because of the fact that CNTs significantly reduce the resistance of the films and facilitate the electron transport within the cell. Therefore during the electron transport through TiO_2 film, the photogenerated electrons can directly be transferred to the TCO without having to cross numerous grain boundaries. Moreover, the authors stated that the light scattering ability of the TiO_2 film was enhanced by adding SWCNTs and thereby improved the cell performance. However, the DSSCs fabricated by these authors showed J_{sc} and V_{oc} of only ~2.3 mA/cm^2 and ~0.44 V, respectively. Although the performance of DSSCs fabricated in this study was very low, this work has inspired many studies in this area *(99–102)*.

By incorporating SWCNTs, the amount of adsorbed dye molecules into the film fabricated in Jang et al. *(98)* increased by up to only 2%, which is considered to be negligible (no effect) for cell performance. Several authors have seen a similar result and concluded that no significant changes in the dye loading of the films can be observed by adding CNT structures (Table 8.1) *(101,103–121)*. These authors argued that the performance improvement achieved by adding CNTs does not result from an increased amount of adsorbed dye molecules and the improved performance must be caused by the improved electron transport and enhanced light scattering efficiency. On the other hand, Bakhshayesh et al. *(105)* reported that the amount of adsorbed dye onto the film

Table 8.1 Characteristics of DSSCs with a CNT/TiO_2 photoelectrode compared to TiO_2-only layer

Semiconducting layer	Cell area (cm^2)	Type of dye	Dye adsorption	Remark	CNT content (wt%)	PCE (%)	References
CNT/TiO_2	0.16	N719	Unchanged	With surfactant	—	5.3	*(104)*
TiO_2						3.9	
s-CNT/TiO_2	0.25	N719	Unchanged	With SDS surfactant	—	7.17	*(103)*
TiO_2						6.02	
CNT/TiO_2	0.25	N719	Unchanged	Acid functionalized	0.1	6.77	*(106)*
TiO_2						4.9	
CNT/TiO_2	0.25	N719	Decreased	Nonfunctionalized	0.025	7.38	*(105)*
TiO_2						6.53	
CNT/TiO_2	0.25	N719	Increased	O_2 plasma or acid treated	—	6.34	*(107)*
TiO_2						3.63	
CNT/TiO_2	—	N719	Increased	Acid functionalized	—	3.89	*(108)*
TiO_2						3.02	
CNT/TiO_2	—	N719	Increased	Acid functionalized	—	4.35	*(110)*
TiO_2						3.1	
CNT/TiO_2	0.36	N3	—	Functionalized	0.1	4.97	*(101)*
TiO_2	0.33					3.32	
CNT/TiO_2	0.25	N3	—	Acid functionalized	0.1	5.02	*(112)*
TiO_2						4.15	
CNT/TiO_2	0.25	N719	—	Acid functionalized	0.3	4.62	*(114)*
TiO_2						2.87	
CNT/TiO_2	0.36	N719	—	Non-functionalized	0.025	10.29	*(113)*

Continued

Table 8.1 Characteristics of DSSCs with a CNT/TiO_2 photoelectrode compared to TiO_2-only layer—cont'd

Semiconducting layer	Cell area (cm^2)	Type of dye	Dye adsorption	Remark	CNT content (wt%)	PCE (%)	References
TiO_2						6.31	
CNT/TiO_2	0.16	N719	—	Acid functionalized	0.2	4.52	*(120)*
TiO_2						3.78	
CNT/TiO_2	0.25	N3	—	Functionalized	0.3	9.05	*(115)*
TiO_2						5.86	
CNT/TiO_2	0.25	N719	—	Nonfunctionalized	0.03	4.1	*(118)*
TiO_2						3.7	
CNT/TiO_2	0.25	N719	—	Nonfunctionalized	0.01	9.0	*(119)*
TiO_2						7.0	
CNT/TiO_2	0.5	N719	—	Acid functionalized	0.172	5.15	*(116)*
TiO_2						4.34	
s-CNT/TiO_2	0.16	N719	—	Virus templated	0.1	10.6	*(117)*
TiO_2						8.3	
s-CNT/TiO_2	0.25	N719	—	DNA assembled	0.15	5.32	*(121)*
TiO_2						4.37	

In the dye adsorption column, "not changed," "increased," and ' decreased" are based on the dye loading of CNT/TiO_2 film compared to the TiO_2-only layer of each study. In the PCE (%) column, two different efficiencies are listed (the upper one is the PCE of CNTs/TiO_2, while the bottom is for the benchmark cells).
CNT, Carbon nanotube; *DSSCs*, dye-sensitized solar cells; *PCE*, power conversion efficiency; *s-CNT*, semiconducting carbon nanotube.

decreased when CNTs were added, whereas some others *(110, 107, 108)* claimed that the dye adsorption of the nanocrystalline TiO_2 film improved with the presence of CNT material. For instance, Zhang et al. *(107)* compared the dye loading of three different photoelectrode films (namely, TiO_2-only, acid-functionalized CNTs, and O_2 plasma-treated CNTs). The result revealed that both acid-treated and O_2 plasma-treated CNT-incorporated TiO_2 films adsorbed a higher amount of dye molecules than the pure TiO_2-based films. Because of the improved dye loading, these authors achieved a DSSC efficiency improvement (compared to the efficiency of TiO_2-only-based cell) of 28% and 75% by acid-treated CNTs/TiO_2 and O_2 plasma-treated CNT/TiO_2 films, respectively. They suggested the improvement in the dye adsorption of films is caused by the fact that the high surface area-to-volume ratio of CNTs provides more active sites, which enable adsorption of more dye. If this is the reason for greater dye loading, the adsorption of dye molecules should be improved in all studies that used CNT materials. Based on the available literature, we summarize that the adsorption mechanism of dye into CNT-incorporated TiO_2 films cannot be explained by the high surface area-to-volume ratio of CNTs. After careful examination of Table 8.1, it could be hypothesized that the dye adsorption of the films can be improved after incorporating only functionalized or chemically treated CNTs in TiO_2 films because the functional groups generated on CNTs would be capable of increasing adsorption of dye molecules through hydrogen bonding to the CNT functional groups *(109)*. In contrast, when nonfunctionalized CNTs were added into the TiO_2 films, the dye loading is reduced as compared to the TiO_2-only film. This is associated with the absence of the CNT active sites for dye adsorption. In addition to this hypothesis, another possibility is that dye molecules can be adsorbed on CNTs by strong $\pi-\pi$ interaction *(111)*. If our explanation is correct, highly functionalized CNTs may result in higher amount of dye adsorption. However, this may be offset by the fact that, in general, functionalization is well known to lower conductivity.

Although the incorporation of CNTs significantly increases the electron transport within the film, too high a concentration of CNTs decreases the performance of DSSCs. Therefore it is necessary to find an optimum content of CNTs for high-performance cells. Lee et al. *(112)* incorporated various CNT concentrations (0–0.5 wt%) in nanocrystalline TiO_2 films and compared the cell performance. The result showed that the highest PV efficiency (5.02%) of DSSCs was achieved by using a 0.1 wt% CNT/TiO_2 photoelectrode-based cell. Furthermore, Yen et al. *(114)* also compared the influence of different concentrations of CNTs on the TiO_2 film-based DSSCs. In their work, the best DSSC with an energy conversion efficiency of 4.62% was obtained by incorporating 0.3 wt% CNTs in TiO_2 film. This optimized efficiency was significantly higher than that (2.87%) recorded for a typical cell. The optimized concentration of CNTs in several other studies was also within this range (0.1–0.3 wt%) *(101, 115, 120)*. Very high energy conversion efficiency (10.29%) was recorded for the CNT/TiO_2

photoelectrode-based DSSC fabricated by Sawatsuk et al. *(113)*, while their benchmark cell with TiO_2-only photoelectrode achieved a 6.31% efficiency. In this CNT/TiO_2 composite film, the concentration of CNTs was optimized at 0.025 wt%. Furthermore, several authors have made a similar observation on this optimized value of CNT content and suggested that a concentration range between 0.01 and 0.03 wt% CNTs in TiO_2 films is suitable to achieve high-performance DSSCs *(105,118,119)*. It is surprising that this optimized CNT concentration is an order of magnitude smaller than that found in some other work (0.1–0.3 wt%) *(101,115,120)*. After a careful examination of the experimental method associated with each work (Table 8.1), we once again attribute this difference in the optimized CNT content to the functionalization of the CNTs. It seems that the optimal concentration of the functionalized CNTs ranges from 0.1 to 0.3 wt%, while that of the nonfunctionalized CNTs is between 0.01 and 0.03 wt%. Indeed, the authors of all these studies have seen dramatic decreases in the photocurrent and efficiency of DSSCs when the concentration of CNTs exceeds the optimized value. In these studies, this phenomenon was believed to be caused by the light-harvesting competition that affects the light absorption of the dye sensitizers. Chen et al. *(116)* recently argued that the explanation, attributed to the competition between light harvesting with dye, is an incomplete elucidation and is unsatisfactory. In their work, several DSSCs were fabricated with functionalized CNT/TiO_2 photoelectrodes with varying CNT concentrations (0–0.35 wt%) and have been studied to explain the role of CNTs on the PV characteristics. As a result, the DSSC assembled with 0.172 wt% CNT/TiO_2 photoelectrode exhibited the highest PV efficiency of 5.15%, which was 18.6% (relative) higher than the efficiency obtained by the benchmark cell. However, this high efficiency (5.15%) declined when the concentration of CNTs was further increased. These authors attributed their findings to (1) the effect of Schottky barrier height on the electron transport through the CNT/TiO_2 network and (2) electron leakage via the naked CNTs. Though CNTs can facilitate the electron transport, the performance of DSSCs fabricated with CNTs will decrease when too high a concentration is used because the catalytic activity of CNTs causes significant charge recombination at the interface of the CNT-containing photoelectrode and liquid electrolyte. It is well known that carbonaceous materials including CNTs possess excellent catalytic activity for the reduction of the electrolyte *(122)*. However, the application of CNTs in the photoelectrode brings undesirable loss of photogenerated electrons by reducing I_3^- to I^- because the CNTs in the composite cannot be fully covered by the TiO_2 particles.

To our knowledge, Dang et al. *(117)* were the first to report using s-SWCNTs in TiO_2 films and studied the PV characteristics of their DSSC devices. In their work, s-SWCNT/TiO_2 nanocrystal core–shell nanocomposites were synthesized with the support of a genetically engineered virus (called M13) as a template. By using this superior photoelectrode, these authors achieved an energy conversion efficiency of 10.6%. This result is the highest efficiency reported to date of carbonaceous material-incorporated

TiO_2 film-based DSSCs. Moreover, they also explored the effect of metallic SWCNT (m-SWCNT)-added TiO_2 films on DSSC performance. Dang et al. *(117)* concluded that the s-SWCNTs can significantly improve the power conversion efficiency of DSSCs, whereas the m-SWCNTs decrease the performance of typical TiO_2 photoelectrode-based cells. Furthermore, at least two other studies have used s-SWCNTs in TiO_2 photoelectrodes of DSSCs and made a similar observation *(103,121)*. The reason for this interesting phenomenon was explained by these authors as follows: s−SWCNTs increased the electron diffusion length, which leads to improved electron collection and resulted in an improved cell performance. The s-SWCNTs have a certain band gap energy that can be adjusted by several parameters including a diameter of nanotubes. Therefore, s-SWCNTs collect the electrons from the conduction band of TiO_2 and transfer them quickly to the external circuit without significant recombination or back transfer. Contrarily, m-SWCNTs have no band gap. So, though the m-SWCNTs can transport the electrons more quickly because of better conductivity than s-SWCNTs, the charge transport is disrupted with an accelerated recombination of electrons to the dye or increased back reaction to the electrolyte. In addition, we assume that in s-SWCNT/TiO_2 photoelectrodes, the photogenerated electrons would not be injected to only the TiO_2 conduction band, they could also be injected to the conduction band of s-SWCNTs from the photoexcited dyes at the same time because the functionalized s-SWCNTs themselves may have adsorbed some dye molecules. This may mean that electron collection from the dye molecules to the semiconductors in such s-SWCNT/TiO_2 composite electrodes could be very good and thereby results in high photocurrent density.

8.4.3 Catalyst Layers

8.4.3.1 Platinum and Other Catalysts

In DSSCs, the counter electrode serves an essential role in collecting electrons from the external circuit and reducing I_3^- to I^- for the regeneration of dye sensitizer. For the optimized DSSC, the counter electrode material should possess the highest possible electrical conductivity and excellent electrocatalytic activity for the reduction of the redox electrolyte. Other environmental and economic considerations include material abundance and stability, ease of preparation, and chemical nontoxicity. In a typical cell, Pt deposited on FTO is used as the counter electrode because of its simple preparation and high electrochemical activity for the reduction of triiodide *(123,124)*. In the past few years, a wide range of Pt architectures has been developed and used to improve the cell performance *(125−129)*. However, Pt is a very rare, expensive material. In addition, the slow dissolution of the Pt-deposited film in the corrosive redox electrolyte deteriorates cell stability *(130)*. Therefore substitution of Pt with other cost-effective materials that are readily available for use and can exhibit comparable or higher catalytic property for the reduction of electrolyte in DSSCs is required to lower the production cost of devices. Since the first fabrication of a Pt-free counter electrode for DSSCs reported by Grätzel's group in 1996 *(131)*, various materials have been employed as alternatives to Pt counter electrodes in DSSCs but this chapter will just consider CNTs.

8.4.3.2 CNTs

CNTs are very promising candidates for the counter electrode of DSSCs. It should be noted that just a brief overview of CNTs as a catalyst layer for DSSCs is provided here, as there are a number of excellent reviews available on this topic *(32,121,124,132–137)*.

The first successful application of CNTs as a counter electrode material of DSSCs was reported in 2003 by Suzuki et al. *(138)* who deposited SWCNTs on a Teflon membrane filter substrate for DSSC fabrication. Despite the fact that no optimization was done on the cell fabricated in this study, SWCNT counter electrode-based DSSCs exhibited a very promising result (PCE of 4.5%), which was even comparable to the Pt sputtered cell (PCE of 5.4%). This early work proved that CNT materials naturally possess excellent electrocatalytic activity and high conductivity and have the potential to perform a catalyzing role in DSSCs. Furthermore, Lee and coworkers were able to significantly improve this result to obtain a PCE of over 7% by using multiwalled carbon nanotubes (MWCNTs)/FTO *(139,140)*. The improved performance of DSSCs in their studies was mainly ascribed to the low charge transfer resistance (R_{ct}) of MWCNT films, which was comparable to that of Pt electrodes. In particular, these authors prepared a bamboo-like-structured MWCNT film on an FTO substrate for use in the counter electrode of DSSCs *(139)*. Defect-rich edge planes of CNTs can facilitate the electron-transfer kinetics and enhance the catalytic rate at the counter electrode–electrolyte interface. Raman spectroscopy (the intensity ratio of the D band to the G band was 0.9825) in this study confirmed that in this bamboo-like CNT structure the CNTs possess plenty of defects. As for the result, the R_{ct} of the defect-rich MWCNT-based counter electrodes was more than two times lower than that of Pt-based film. Even though the MWCNT film showed a lower R_{ct} than the Pt-based electrode, the Pt counter electrode-based DSSC (7.83%) exhibited a slightly higher PV efficiency than that (7.67%) of the DSSC fabricated with bamboo-like-structured CNT-based electrode. This was because of the following reasons: (1) the quasitransparent Pt film slightly reflects the unabsorbed portion of light incident toward the photoelectrode and (2) smaller diffusion impedance of redox species. Nevertheless, the energy conversion efficiency achieved by this MWCNT counter electrode-based DSSC was comparable to the benchmark DSSC.

Based on this study, it is clear that the defect density of CNTs plays a critical role in DSSCs when they are used in the counter electrode. For instance, it is well established in the case of graphene oxide that highly functionalized graphene structures possess excellent catalytic properties for I_3^- reduction *(141–143)*. So, it is not surprising that acid treatment on nanotubes would also be a reasonable strategy to improve the catalytic properties of CNT films because carbonaceous materials such as CNTs, graphite, and graphene share the same chemical structure *(144)*. However, high levels of functional groups on the CNTs significantly decrease the electrical conductivity of materials. Since the pioneering work of the Lee group *(139,140)*, many studies have attempted to improve the efficiency of CNT counter electrode-based DSSCs and the results are tabulated in Table 8.2.

Highly purified CNTs have been well aligned on an FTO substrate using a catalytic CVD method to make the counter electrode of I^-/I_3^- redox couple-based DSSCs *(76)*. This well-aligned CNTs counter electrode-based DSSC exhibited an energy conversion efficiency of 10.04%, which was considerably higher than those of the cells fabricated with screen printed CNTs (8.03%) and benchmark Pt counter electrodes (8.80%). These authors hypothesized that this high performance [especially the high fill factor (FF)] of DSSCs with a direct-grown CNT counter electrode could be caused by the faster redox reaction on the cathode. Therefore they suggested that the thermal CVD process, used to prepare the well-aligned CNT electrodes, also provides a significant improvement in the electron conduction path between the electrolyte and the FTO glass. DSSCs assembled with the VACNT counter electrodes show very comparable and/or even higher PCE than the expensive Pt electrode-based cells. However, reports on the exploration of the VACNT counter electrodes for DSSCs are still sparse. Therefore it is reasonable to expect further improvement in the DSSC performance by developing these vertically grown CNT structures.

Other work has used horizontally aligned films generally based on CNT yarns or fibers. Such alignment enhances conductivity and hence counter electrodes based on these free-standing aligned CNT structures have been used to produce DSSCs with high efficiencies *(155–162)*. The twisting of CNTs into fibers or yarns was pioneered over 10 years ago *(163)*. This approach has also been applied in DSSCs to produce fiber and wire-shaped DSSCs *(164–171)*; this has meant that the traditional planar structure of the cells can be expanded into new geometries. These cells have also received considerable attention because of their flexible and lightweight format, which is required in modern electronic devices. CNT-based materials have been proven to be promising candidates for this class of DSSCs because of their highly aligned structure *(172,173)*.

8.4.4 CNTs in Perovskite Solar Cells

Perovskite solar cells are a novel type of device that was first fabricated in 2009 *(174)* and several structures have been reported *(175)*. Excitons are created after light absorption in the perovskite material, which is then separated into holes and electrons to be collected at the metal electrode and the FTO-coated glass electrode, respectively. There are two possible ways to separate the excitons, either by thermal energy in the perovskite material or at the interface between perovskite material and TiO_2 or hole transport material. TCFs

Table 8.2 PV characteristics of DSSCs With CNT film-based counter electrodes and comparisons to Pt film-based cells

Counter electrode	J_{sc} (mA/cm^2)	V_{oc} (V)	FF	PCE (%)	Note and R_{ct} of CNTs	Year	References
SWCNTs	—	—	—	4.5	Teflon membrane filtered (FTO free)	2003	*(138)*
Pt				5.4			
MWCNTs	15.64	0.78	0.62	7.59	Spray coated, $R_{ct} = 2.34\ \Omega/cm^2$	2008	*(140)*
Pt	—	—	—	—			
MWCNTs	16.20	0.74	0.64	7.67	Bamboo-like structured, $R_{ct} = 0.82\ \Omega/cm^2$	2009	*(139)*
Pt	16.50	0.73	0.65	7.83			
MWCNTs	12.93	0.76	0.58	5.69	Screen printed	2010	*(145)*
Pt	12.96	0.76	0.73	7.15			
DWCNTs	10.75	0.68	0.53	6.05	Screen printed, $R_{ct} = 12.93\ \Omega/cm^2$	2010	*(146)*
Pt	10.73	0.71	0.56	6.80			
SWCNTs	14.25	0.74	0.74	7.81	Optimum film thickness was 8 μm, $R_{ct} = 0.6\ \Omega/cm^2$	2010	*(147)*
Pt	14.63	0.75	0.73	7.92			
MWCNTs	17.62	0.76	0.73	10.04	Vertically grown by CVD	2010	*(76)*
Pt	17.68	0.75	0.65	8.80			
MWCNTs	14.00	0.77	0.67	7.04	N-doped and VACNTs	2011	*(148)*
Pt	14.90	0.80	0.62	7.34			
SWCNTs	13.80	0.65	0.61	5.5	VACNT, $R_{ct} = 5-10\ \Omega/cm^2$	2011	*(149)*
Pt	13.67	0.70	0.59	5.6			
DWCNTs	15.43	0.80	0.65	8.03	$3.13\ \Omega/cm^2$	2011	*(150)*
SWCNTs	14.94	0.80	0.64	7.61	$6.72\ \Omega/cm^2$		
MWCNTs	15.25	0.80	0.56	7.06	$10.0\ \Omega/cm^2$		
Graphene	16.99	0.75	0.54	6.81	—		
Pt	16.08	0.81	0.65	8.49	$4.86\ \Omega/cm^2$		
MWCNTs	14.78	0.74	0.65	7.03	Optimum film with 12.3 μm, $R_{ct} = 4.4\ \Omega/cm^2$	2011	*(151)*
Pt	15.04	0.77	0.69	7.88			
MWCNTs	9.95	0.78	0.64	4.94	Functionalized, ball-milled, and spin-coated CNTs	2012	*(152)*
Pt	13.67	0.63	0.59	5.08			
CNTs	13.9	0.65	0.58	5.2	VACNT, $R_{ct} = 17.1\ \Omega/cm^2$	2013	*(153)*
Pt	17.00	0.61	0.64	6.6			
MWCNTs	14.01	0.73	0.64	6.5	Surface-modified CNTs, $R_{ct} = 0.33\ \Omega/cm^2$	2013	*(154)*
Pt	14.40	0.77	0.70	7.76			

Note that DSSCs are prepared with only I^-/I_3^- redox couple. In the PV results of each study, the upper data are the PV characteristic of CNT-based DSSCs and the bottom data are for the cells with Pt film.
CNTs, Carbon nanotubes; *CVD*, chemical vapor deposition; *DSSCs*, dye-sensitized solar cells; *DWCNTs*, double-walled carbon nanotubes; *FF*, fill factor; *FTO*, fluorine tin oxide; *MWCNTs*, multiwalled carbon nanotubes; *PV*, photovoltaic; *SWCNTs*, single-walled carbon nanotubes, *VACNT*, vertically aligned CNT.

made from CNTs have been used in perovskite solar cells *(176,177)* to replace the expensive metal electrode (usually Au) in the fabrication of perovskite solar cells *(178)*. In all cases, the CNT network is part of the hole collection side of the cell. The efficiency improved from 5.14% to 6.87% after replacing the Au with CNT because CNT films can provide a stronger driving force for the hole injection according to the band alignment. The sparsity of the CNT network allows the device to be illuminated from both sides, which makes the application of solar cells as a window a possibility. By adding spiro-OMeTAD (hole transporting substances), the performance of the devices was further improved to 9.9%. Although the PCE of this device was not as high as that from some other literature, it shows the potential to achieve a semitransparent and flexible perovskite solar cell without the energy-consuming process of Au deposition. Additionally, acting as the hole transporting layer with poly(3-hexylthiophene) (P3HT), poly(methylmethacrylate) (PMMA)-functionalized SWCNTs were found to improve the thermal stability significantly because of their better stability compared to the organic hole transport layer *(179)*.

A partially functionalized SWNCT film has been used as a front transparent conducting electrode in place of the ITO electrode in perovskite solar cells. The fabricated device structure was glass or PET/SWCNTs/PEDOT:PSS/$CH_3NH_3PbI_3$/$PC_{61}BM$/Al, in which SWCNT film acted as an electron blocking layer. The transparent conducting SWCNT electrode-based planar heterojunction perovskite solar cell achieved a PCE of 6.32%, which is 70% of an ITO-based device (9.05%). A flexible perovskite solar cell made using SWCNT on a PET substrate exhibited a promising PCE (5.38%) *(180)*.

The main advantages of applying CNTs in perovskite devices include cheap fabrication processes and materials, and the potential of two-sided illumination application.

8.5 CNTs IN ORGANOPHOTOVOLTAIC SOLAR CELLS

Organic materials have both conducting and semiconducting properties, which are really promising for optoelectronic devices *(181)*. For solar cells, organic materials are attractive because of their potential low cost, simple manufacturing process, and high throughput, suggesting that organic materials have the potential to make a major impact on the PV market *(182)*. The absorption coefficient of organic semiconductors is very high, which allows light to absorb within a very thin layer leading to low-cost solar cells *(183)*. In addition, the efficiency of organic solar cells increases with temperature, while most conventional inorganic solar cells lose efficiency with increasing temperature *(184)*. Different organic materials including organic molecules, conjugated polymers, and four typical carbon materials (e.g., amorphous carbon, fullerenes, CNTs, and graphene) are used for both organic and organic–inorganic hybrid solar cells *(181,185,186)*. Among these, research on CNTs indicates it is a potential material for the organic and hybrid solar cells.

Moreover, CNTs exhibit interesting optoelectronic, physical, and chemical properties, required for many viable applications *(187,188)*, although large-scale commercial production will still need the development of robust CNT sorting and handling protocols. Investigation of SWCNT—polymer solar cells has been conducted toward developing alternative, lightweight, flexible devices for space power applications.

The efficiencies of OPVs have tripled over the years *(189)*. One of the areas of development of these cells has been the addition of CNTs to the various components of the devices. CNTs are small in diameter, which is easily compatible with the properties of organic solar cells. Given that the active molecules in the systems are typically on the order of a few nanometers across, these species (nanotubes and polymer base units) can readily interact potentially leading to ready charge transfer. In OPV device applications the preferred diameter is up to 20 nm, and the typical diameters of SWCNTs and MWCNTs are in the range of 2—10 nm and 5—100 nm, respectively *(190)*. Both CNTs and conducting polymers possess conjugated π-systems and the nature of their electronic interaction is anticipated to occur via $\pi-\pi$ stacking.

Choosing the corresponding material for CNTs can also be very different including small molecules, oligomers, polymers, quantum dots, and semiconductors (bulk and nanostructure). One of the most extensively studied structures utilized in CNT-based solar cell design are CNTs/small molecules and CNTs/polymers, where CNTs act as an electron acceptor (with some exceptions) and light is absorbed through the CNT complementary element *(191—193)*. Research has been undertaken to incorporate CNTs as a hole extraction layer in active layers *(194—196)* and as charge transport layer or electrode *(190,197)*. An overview of the uses of CNTs in OPVs is presented in Fig. 8.2. CNTs have been incorporated into virtually every part of the structure via a number of approaches. In Table 8.3 the performance of different CNT/conducting polymer based solar cells has been summarized.

8.5.1 CNTs as a Hole Extraction Layer in OPVs

A great deal of research has been carried out using CNTs as a hole extraction layer in OPVs. High transparency and low sheet resistance are two of the most important characteristics for transparent electrodes, but for CNT-based electrodes a tradeoff is necessary between these two characteristics. A film of poly(3,4-ethylenedioxythiophene):poly(4-styrenesulfonate) (PEDOT:PSS) has the potential to replace the ITO layer over large areas (cm^2) *(31)*. This type of thin film has been implemented in poly(3-hexylthiophene):[6,6]-phenyl-C_{61}-butyric acid methyl ester (P3HT:PCBM)-based bulk heterojunction solar cells and achieved an efficiency of 1% *(215)*. Similar work done by blending SWCNTs with PEDOT:PSS showed an improved efficiency of 1.3—1.5% *(216,217)*. This improvement with the presence of the nanotubes is likely because of a combined effect of an enhanced hole collection by a three-dimensional

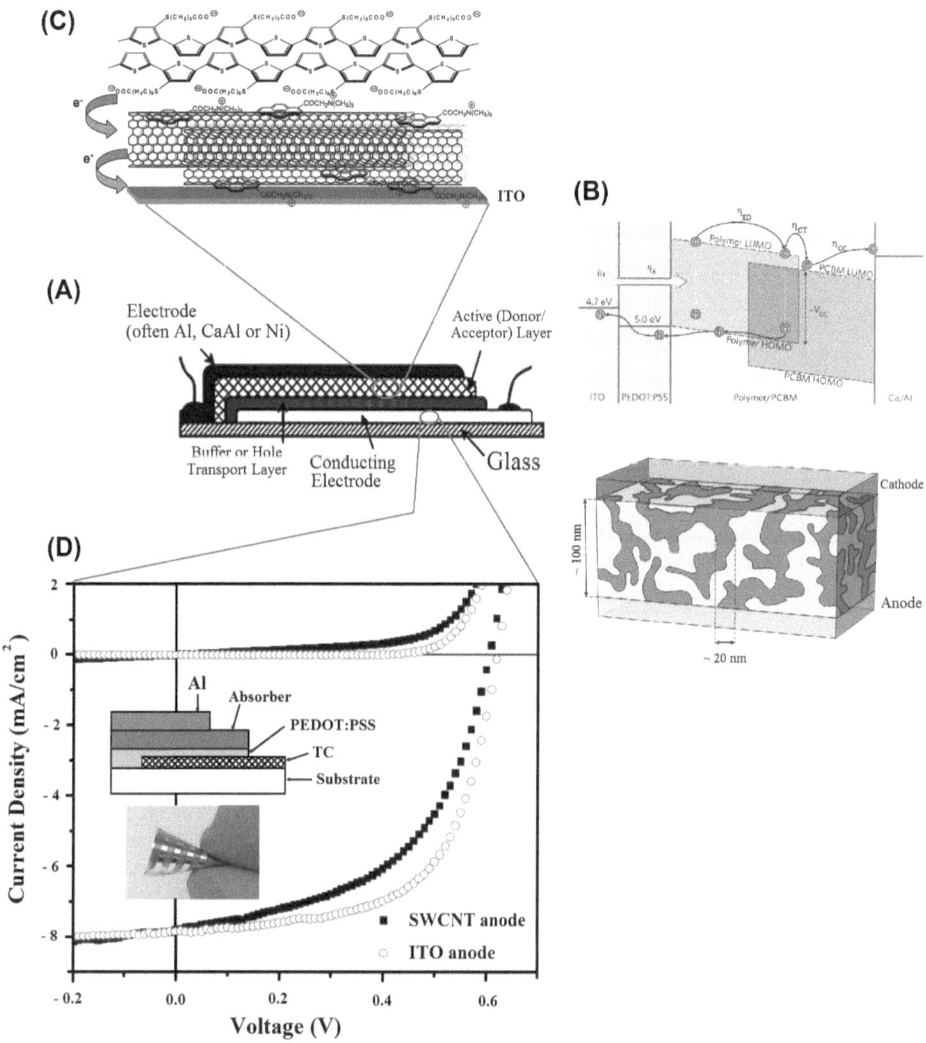

Figure 8.2 (A) Typical structure of a polymer-based solar cell. *(Adapted from Patyk, R. L.; Lomba, B. S.; Nogueira, A. F.; Furtado, C. A.; Santos, A. P.; Mello, R. M. Q.; Micaroni, L.; Hummelgen, I. A. Carbon Nanotube-Polybithiophene Photovolyaic Devices with High Open-Circuit Voltage. Physica Status Solidi (RRL)* ***2007,*** *1 R43–R45. Reproduced with permission from John Wiley and Sons.)* (B) The operating mechanism of an organophotovoltaic (OPV) with a model often presented for the network of the polymer and the acceptor. *(Adapted from Li, G.; Zhu, R.; Yang, Y. Polymer Solar Cells. Nature Photonics* ***2012,*** *6 (3), 153–161. Reproduced with permission from Nature Publishing Group.)* (C) Typical carbon nanotube (CNT) addition to the polymer layer. *(Reprinted with permission from Rahman, G. M. A.; Guldi, D. M.; Cagnoli, R.; Mucci, A.; Schenetti, L.; Vaccari, L.; Prato, M. Combining Single Wall Carbon Nanotubes and Photoactive Polymers for Photoconversion. Journal of the American Chemical Society* ***2005,*** *127 (28), 10051–10057. Copyright (2005) American Chemical Society.)* (D) Typical approach to incorporate CNT electrodes in a polymer solar cell showing the potential flexibility. *(Reproduced from Rowell, M. W.; Topinka, M. A.; McGehee, M. D.; Prall, H.-J.; Dennler, G.; Sariciftci, N. S.; Hu, L.; Gruner, G. Organic Solar Cells with Carbon Nanotube Network Electrodes. Applied Physics Letters* ***2006,*** *88 (23), 233506. With permission from AIP Publishing.) ITO*, Indium tin oxide; *PEDOT:PSS*, poly(3,4-ethylenedioxythiophene): poly(4-styrenesulfonate); *SWCNT*, single-walled carbon nanotube; *TC*, transparent conductor.

Table 8.3 Summary of performance of CNT/conducting polymer-based solar cells

Device structure	J_{sc} (mA/cm^2)	V_{oc} (V)	FF	η (%)	Spectrum (mW/cm^2)	References
Glass/ITO/P3OT/P3OT:SWCNT/Al	0.12	0.75	0.40	0.04	AM 1.5/100	*(194)*
Glass/SWCNT/PEDOT:PSS/P3HT:PCBM/Ga; In	6.50	0.50	0.30	0.99	—/100	*(190)*
Glass/SWCNT/P3HT:PCBM/Al	4.46	0.36	0.38	0.61	AM 1.5/100	*(201)*
Glass/SWCNT(H_2O:SDS)/PEDOT:PSS/P3HT:PCBM/LiF/Al	7.30	0.59	0.46	2.2	AM 1.5/100	*(41)*
PET/SWCNT/PEDOT:PSS/P3HT:PCBM/Al	7.80	0.61	0.52	2.5	AM 1.5G/100	*(200)*
PET/SWCNT/ZnO-nw/P3HT/Au	—	—	—	~0.60	AM 1.5G/100	*(202)*
Glass/ITO/PEDOT:PSS/PTEBS:MWCNT/C60/Al	1.52	0.57	0.62	0.55	AM 1.5/100	*(195)*
Glass/ITO/MWCNT/P3HT:PCBM/LiF/Al	4.00	0.50	0.47	0.93	AM 1.5/100	*(203)*
Glass/ITO/PEDOT:PSS/P3HT:PCBM:MWCNT/LiF/Al	9.33	0.57	0.38	2.00	AM 1.5/100	*(204)*
Glass/ITO/PEDOT:PSS/P3HT:C60:SWCNT/LiF/Al	2.69	0.54	0.49	0.75	AM 1.5/95	*(205)*
Glass/ITO/PEDOT:PSS/P3HT:SWCNT/Al	1.93	0.58	0.42	0.52	—/70	*(206)*
Glass/ITO/PEDOT:PSS/P3HT:PCBM:SWCNT/Al	4.95	0.55	0.52	1.40	AM 1.5/100	*(207)*
Glass/ITO/PEDOT:PSS/QTF12:PCBM:DWCNT/LiF/Al	2.37	0.56	0.37	0.50	AM 1.5/100	*(208)*
Glass/ITO/MWCNT/P3HT:PCBM/LiF/Al	7.30	0.61	0.62	2.70	AM 1.5/100	*(209)*
Glass/ITO/ODA-SWCNT:P3HT:PC70BM/BCP/Al	7.66	0.52	0.44	1.76	AM 1.5/100	*(210)*

Glass/SWCNT/PEDOT:PSS/P3HT:PCBM/ Ca/Al	13.78	0.57	0.53	4.13	AM 1.5/100	*(197)*
Glass/FTO/MWCNT/MoO_3/P3HT: PCBM/Ca/Al	8.88	0.51	0.46	2.1	AM 1.5/100	*(211)*
Glass/ITO P3HT:PCBM:SWCNTs/Al	11.46	0.57	0.46	3.02	AM 1.5/100	*(212)*
FTO/PBT/POT/SWCNT-TIOPH/Ca/Al	1.81			1.48	AM 1.5/155	*(198)*
Glass/ITO/PEDOT:PSS/*f*-MWCNT/Al	11.15	0.50	0.30	1.65	AM 1.5/100	*(213)*
Glass/ITO/HTM/PTM10:PTM21-CNT: PCBM/LiF/Al	0.45	0.88	0.38	0.15	AM 1.5/100	*(214)*

CNT, Carbon nanotube; *FTO*, fluorine tin oxide; *HTM*, hole transporting material; *ITO*, indium tin oxide; *MWCNT*, multiwalled carbon nanotube; *ODA*, octadecylamine; *PBT*, polybithiophene; *PEDOT:PSS*, poly(3,4-ethylenedioxythiophene):poly(4-styrenesulfonate); *P3HT:PCBM*, poly(3-hexylthiophene):[6,6]-phenyl-C_{61}-butyric acid methyl ester; *P3OT*, poly(3-octylthiophene); *POT*, poly(3-octylthiophene); *QTF12*, 2,7-bis-(3,3‴-didodecyl-[2,2′,5′,2″;5″,2‴] quaterthiophen-5-yl)-fluoren-9-one; *SWCNT*, single-walled carbon nanotube; *TIOPH*, 2-(2-tienyl)ethanol.

CNT network coupled with improved transport of charge carriers into the planar ITO part of the electrode *(216)*.

Other early work used MWCNTs with poly(*p*-phenylenevinylene) (PPV) as the conducting polymer instead of ITO but reported very poor efficiencies *(218)*. A simple incorporation of SWCNT at the back electrode with the PEDOT:PSS layer and P3HT: PCBM as an active layer exhibited efficiency of 0.99% on glass and 2.50% on PET substrate *(200,219,220)*. The effectiveness of the CNT electrode was attributed to the three-dimensional nature of the interface between the SWCNTs and the P3HT:PCBM nanocomposite. It has been reported that MWCNT on ITO yields a total power conversion efficiency of 0.93% *(203)* while an oxidized MWCNT can improve the conversion efficiency to 2.70% *(221)*. The oxidization process helped to increase the FF from 0.47 to 0.62 *(221)*. The increase in FF was thought to be caused by an improved matching of the work function electronic states between the polymer and nanotube mixture with the introduction of the function groups on the MWCNTs *(222)*. This provides better overlap for improved electron and hole transportation. It is also reported that the acid oxidation treatment creates an oxide layer on the CNT films and improves the conductivity as well *(223,224)*.

It has been demonstrated that a cost-effective method of fabrication is brush painting *(66)*. In this particular work a commercial ink is used and brushed on plasma-processed PET. The plasma processing helps adhesion of the CNTs. Although the type of CNT is not specified, an efficiency of 1.63% (with an FF of 0.4049 and J_{sc} of 7.12 mA/cm^2) is achieved. A report by Capasso et al. *(211)* showed that MWCNTs deposited on FTO can provide an increased efficiency of 2.1%, with improved FF of 0.46% and J_{sc} of 8.88 mA/cm^2.

Transparent MWCNT sheets as an interlayer anode electrode for tandem OPV have been reported by Tanaka et al. *(225)*. In this work, PEDOT:PSS/P3HT:PCBM on top of an ITO-coated glass substrate acts as a front cell, and the structure of the back cell is Al/ Bathocuproine (BCP)/C_{60}/CuPc:C_{60}/CuPc/PEDOT:PSS. The MWCNT layer is sandwiched between these two front cells and back cells, thus exhibiting a device efficiency of 0.31%. However, the most promising efficiency of 4.9% is obtained by incorporating a thin CNT film at the interface of the P3HT:PCBM layer *(47)*. The report also depicts that the thin CNT film can be inserted either between the ITO and PEDOT:PSS layer or between the PEDOT:PSS layer and the active layer.

8.5.2 CNTs in the Active Layer

Modification of the CNTs affords the opportunity to improve the dispersion of the nanotubes in the polymer matrix. It is important to mention that in the case of SWCNTs, covalent functionalization disrupts some of the aromatic nature of the nanotubes, leaving "holes" or "defects" in the structure of the nanotube, which significantly affect both mechanical and electrical properties. One of the earlier examples of the use of functionalization was from Pradhan et al. *(226)* who incorporated CNTs into a P3HT layer that was then used in combination with C_{60}-based solar cells but the devices exhibited very low efficiencies.

Photoexcitation of a conjugated polymer, a typical light absorber in solution-processed OPV devices, generates a bound electron–hole pair called an exciton *(227)*. This is the key feature in OPV devices that requires the use of a two component donor–acceptor film called the active layer. This layer performs the function of light absorption and photocurrent generation. Upon arrival of the photogenerated exciton at the interface, the heterojunction formed between polymer donor and acceptor helps to transfer the electron to the acceptor leaving behind a hole in the polymer.

Early research on CNTs used poly(3-octylthiophene) (P3OT) as a donor and SWCNT as an acceptor material *(194,228)*. This early introduction of CNTs used a simple blending approach to mix the two materials together. Although this report demonstrated a poor device efficiency of 0.02%, it introduced a new way of using CNTs in OPVs. Later, poly(phenylene ethynylene) as a donor layer was investigated with SWCNTs as an acceptor *(229)*. In this case the depicted efficiency is low; however, the efficiency is improved in devices yielding 0.05% efficiency *(230)*. Poly[(2-methoxy,5-octoxy)1,4-phenylenevinylene] (MO-PPV)/SWCNT systems are reported with nonstandard solar simulation and device efficiencies as low as 0.004% *(231)*. The most commonly used polymer with SWCNTs is the poly-hexylthiophene derivatives. These SWCNT/poly-hexylthiophene-based donor–acceptor solar cells exhibited an efficiency of 0.52% *(206)*. Devices using modified SWCNTs have been produced through blending with poly(3-octylthiophene). An open-circuit voltage of 1.81 V is observed, which is very encouraging although the PCE is still modest at 1.48% at higher than the normal light intensity of AM 1.5 *(198)*.

SWCNTs have been used in conjunction with C_{60} as an acceptor material for P3HT bulk heterojunction devices. Schuettfort and his group reported a highly purified nanohybrid structure consisting of an SWCNT coated with a monolayer of P3HT *(193)*. The best reported efficiency for this system is 0.57% *(205)*. Acid-functionalized SWCNTs in bulk heterojunction P3HT:PCBM devices are also yielding efficiencies as high as 1.4%, compared to reference devices of 1% efficiency *(207)*. A study of octadecylamine (ODA)-SWCNT:P3HT:$PC_{70}BM$ as an active layer shows an efficiency as high as 1.76% *(210)*. Other work has used functionalized SWCNTs in a photoactive layer with P3HT and PCBM. A maximum PCE of 3.02% was obtained with an optimum 0.3 wt% SWCNTs in a P3HT:PCBM:SWCNT device. The performance improvement was mainly attributed to the extended exciton lifetime and the rapid charge transport afforded by the nanotubes *(186)*.

DWCNTs as an accepter material in a bulk heterojunction configuration have also been investigated. Incorporation of DWCNTs with P3OT as an acceptor gives a very low efficiency of 0.001% *(232)*. However, when incorporated with *bis*-quaterthiophene, the DWCNT/*bis*-quaterthiophene donor—acceptor exhibited an improved efficiency of 0.50% *(208)*. Electrostatic attachment of a polyelectrolyte has facilitated the incorporation of N-doped functionalized nanotubes into a P3OT layer. Devices have been reported with an efficiency of 2.0% with 0.1% by weight MWCNTs in the active layer *(204)*.

Some workers have incorporated dyes into the CNT-containing layers *(233)*. Work with SWCNTs has probed the PV properties of dye, *N*-(1-pyrenyl)maleimide, functionalized SWCNT-conjugated polymer, P3OT, blend composites. Addition of dye with SWCNTs in the active layer has been found to improve performance. An open circuit voltage of 0.6—0.7 V was measured and thought to be caused by efficient transfer of holes by dye molecules to P3OT at the dye/polymer interface and the rapid transfer of the generated electrons to the SWCNTs at the dye/nanotube interface. Despite the improved short circuit current also observed, these early cells showed very poor performance *(234)*.

The structure shown in Fig. 8.2C produced a donor—acceptor nanocomposite capable of 1.2% efficiencies. More promising perhaps was the fact that multiple layers of the structure provided efficiencies of over 9%, which starts to move toward a regime where these cells could be commercially competitive *(199)*. This might also suggest that the CNTs could play a critical role in absorbing light to lead to greater efficiencies *(235)*.

8.6 CARBON NANOTUBE/SILICON OR NANOTUBE/SILICON HETEROJUNCTION SOLAR CELLS

In nanotube/silicon heterojunction (NSH) solar cells the CNTs not only transport charge carriers, but also act as a photoactive material *(236—238)*. Fig. 8.3 presents a schematic of the cells and shows representative examples of the work reported while Table 8.4 presents a summary of some of the work to date.

The first such cells were designed by Wei et al. in 2007 *(257,237,238)* using DWCNTs deposited via water expansion and aqueous film transfer of an as-grown CVD film *(237,238)*. While these cells had a photoconversion efficiency of only 1.3% (compared to commercial cells at 20—30%) many improvements have since been made to the cell design and doping methods, with 15% efficiencies reported in 2012 *(249)* and 17% efficiencies reported in 2015 *(254,258)*. Thus in less than a decade the cell efficiency was improved by a factor of 10. Such rapid improvement is unsurprising for a system based on the already mature field of silicon PV. However, in addition to the fundamental physics of the nanotube/silicon junction, interest in these solar cells is based largely on the potential to replace the costly and energy-intensive ion diffusion component of current manufacturing processes. The typical architecture for these cells is much like that of a single junction n-type crystalline silicon solar cell with a front side emitter layer replaced by a film of *p*-doped CNTs *(238)*. While the initial design

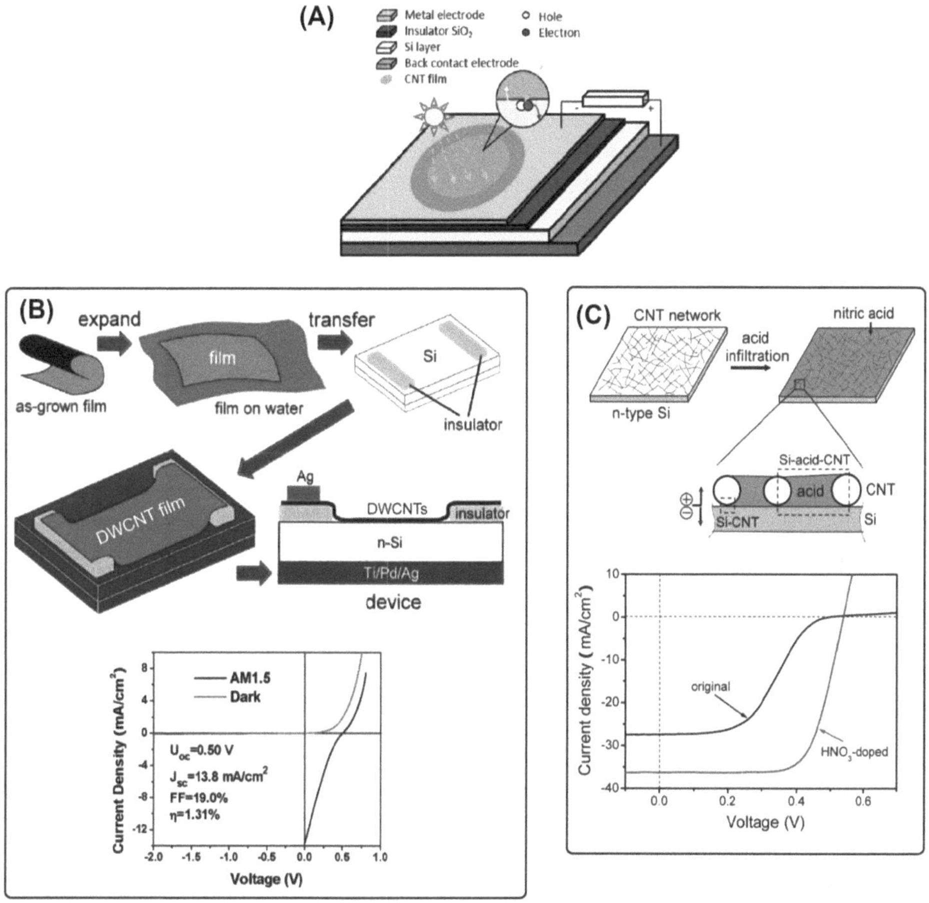

Figure 8.3 (A) Schematic of a typical nanotube/silicon heterojunction (NSH) structure. (B) Schematic illustrating the methods of production of the first NSH cell *(Adapted with permission from Wei, J.; Jia, Y.; Shu, Q.; Gu, Z.; Wang, K.; Zhuang, D.; Zhang, G.; Wang, Z.; Luo, J.; Cao, A.; Wu, D. Double-Walled Carbon Nanotube Solar Cells. Nano Letters* ***2007,*** *7 (8), 2317–2321. Copyright (2007) American Chemical Society.).* (C) Overview of treatment with nitric acid and results obtained *(Adapted with permission from Jia, Y.; Cao, A.; Bai, X.; Li, Z.; Zhang, L.; Guo, N.; Wei, J.; Wang, K.; Zhu, H.; Wu, D.; Ajayan, P.M. Achieving High Efficiency Silicon-Carbon Nanotube Heterojunction Solar Cells by Acid Doping. Nano Letters* ***2011,*** *11 (5), 1901–1905. Copyright (2011) American Chemical Society.). CNT,* Carbon nanotube; *DWCNTs,* double-walled carbon nanotubes.

used DWCNTs, most designs use SWCNTs; however, MWCNTs may also be used *(236,259)*. In all cases the CNT film acts as a component of the heterojunction setting up a built-in potential to enable charge separation, as a highly conductive network for charge collection and transport, and as a transparent electrode to allow good light illumination of, and photocurrent generation in, the silicon base *(236)*.

Table 8.4 Summary of performance of CNT/Si solar cells

Type of nanotube used	Cell active area (cm^2)	Reported efficiency (%)	Film preparation/ deposition method	References
DWCNT	0.49	1.4	CVD chamber wall/ aqueous film transfer	*(237)*
MWCNT	0.01	~3	Drop cast:SDS	*(240)*
DWCNT	0.49	7.4	CVD chamber wall/ aqueous film transfer	*(236)*
SWCNT	0.25	~1.3	Spray:DMF	*(241)*
SWCNT	0.25	4.5	Spray:DMF	*(242)*
DWCNT	0.49	4.1	CVD chamber wall/ aqueous film transfer	*(243)*
SWCNT	0.25	1.7	Spray:DCB	*(244)*
SWCNT	0.08	10.9	Filtration:MCE/MCE dissolution	*(245)*
SWCNT	0.09	13.8	CVD chamber/free transfer	*(239)*
SWCNT	0.09	10.9	CVD chamber	*(246)*
SWCNT	0.08	12.0	filtration:MCE/MCE dissolution	*(247)*
MWCNT	0.09	8.8	CVD chamber wall/ aqueous film transfer	*(248)*
SWCNT	0.15	15.1	CVD chamber wall/ aqueous film transfer	*(249)*
SWCNT	0.09	11.5	Slide casting:CSA/ aqueous film transfer	*(250)*
SWCNT	0.08	9.7	Filtration:MCE/MCE dissolution	*(251)*
DWCNT	0.13	13.1	Array drawing/free transfer	*(252)*
SWCNT	0.08	4.5	Filtration:MCE/MCE dissolution	*(253)*
SWCNT	0.008	17.0	Filtration:MCE/dry transfer	*(254)*
SWCNT	0.08	15.1	Spray, filtration:MCE/ spray, MCE dissolution	*(255)*
SWCNT	0.08	4.5	Slide casting:Na polyelectrolyte/ aqueous film transfer	*(256)*

CNT/Si, Carbon nanotube/silicon; *CSA*, chlorosulfonic acid; *CVD*, chemical vapor deposition; *DCB*, dichlorobenzene; *DMF*, dimethylformamide; *DWCNT*, double-walled carbon nanotube; *MCE*, mixed cellulose esters membrane filters; *MWCNT*, multiwalled carbon nanotube; *SDS*, sodium dodecyl sulfate; *SWCNT*, single-walled carbon nanotube.

The energy generation within PV devices is a complex process, which involves several steps including photon absorption, electron–hole diffusion, charge separation, carrier transport, and collection *(246)*. As such there are a variety of effects that must be balanced to attain higher efficiency cells. Following Wei's first report in 2007, the first significant improvement by Jia et al. *(236)*, who achieved an efficiency of 7.4% from DWCNT/Si cells, came largely by substituting the original mica window with a much thinner silicon oxide window, allowing the nanotube film to sit more smoothly on the silicon surface *(238)*.

In 2008, Li et al. *(241)* found that treatment of the SWCNT layer with the *p*-dopant thionyl chloride ($SOCl_2$) postfabrication increased the power conversion efficiency of the cells by over 45% (relative) by lowering the sheet resistance, and increasing the short circuit current density and open circuit voltage. Hall effect measurements showed that the $SOCl_2$ treatment led to an increase of carrier density from 3.1×10^{15} to 4.6×10^{17} cm^{-2} and an improvement from 0.23 to 1.02 $cm^2/V\ s$ of the effective mobilities *(238,241)*. In addition, it was found that $SOCl_2$ treatment adjusted the Fermi level and shifted the major conduction mechanism in the SWCNT layer from hopping toward tunneling *(242,253)*. This work was the first of many in the field to focus on improving performance on a variety of silicon substrates and device designs through the use of doping and other strategies applied to nanotube films to improve their conductivity/transparency relationship.

Jia et al. *(243)* performed the first comparison between SWCNTs, DWCNTs, and MWCNTs for use in CNT/Si cells in terms of the area density of the films. It was found that SWCNTs are superior to MWCNTs at low densities. It was also found that the density (and thus optical transmittance) is vastly important in the performance of these cells. Increasing film transparency (lowering CNT density) increases the efficiency of the cells by allowing more light to reach the silicon, while decreasing the transparency (increasing the CNT density) increases the efficiency by lowering the sheet resistance across the film *(238,243)*. Thus there must exist some optimal thickness to achieve maximum efficiency. This research team also found their DWCNT cells to be significantly superior to both the SWCNT and MWCNT cells. However, their DWCNT films were produced using a different method to the SWCNT and MWCNT films *(243)*. This makes it difficult to draw a good comparison, because the nanotube film properties are highly dependent on film morphology *(238)*.

A 2013 study by Tune et al. *(260)* found the optimal film thickness for the type of SWCNTs used (large diameter arc-discharge nanotubes with high carboxyl functionality). It was determined that a film transmittance of 70% gave the optimal efficiency and this was found to taper significantly for thinner films. This was because of a decrease in the amount of SWCNT/Si contacts for charge separation and an exponential increase in the sheet resistance of the film because the conduction mechanism became dominated by the limited percolation pathways in very thin nanotube films *(260)*.

Wadhwa et al. suggested that the mechanism of photocurrent generation in the CNT/Si heterojunction solar cell is consistent with the Schottky–Mott model, that is, via thermionic emission over a barrier formed at the interface of a metal and a semiconductor *(245)*. However, the performance enhancement via Fermi-level modulation could apply to both Schottky and *p–n* junctions *(238)*. Work done by Ong et al. *(244)* found that there was a possible correlation between an SWCNT absorption band [S_{11}, the first interband transition for the (7,6) and (8,6) chiralities] and a photocurrent band at around 1150 nm. This seems to imply that the nanotubes are involved in the production of the photocurrent as a light absorber, and not simply as a charge separator, transporter, and collector, which indicates that the cell architecture is that of a *p–n* junction *(238,244)*. However, as shown in a later study by Tune et al. the feature in the photocurrent spectrum could have other origins *(261)*. In 2007, Tzolov et al. *(262)* achieved a broad photoresponse in the mid-IR range separate from the expected strong silicon response at photon energies of 1.1–1.2 eV. It was claimed that this response was proof that CNTs participate in exciton generation themselves. However, in addition to questions about whether or not the tubular carbon material formed via their template method was, in fact, CNTs in the usual sense, this feature could also be caused by free carrier absorption in the silicon base *(238,263,264)*. Thus the work of neither Tzolov et al. nor Ong et al. conclusively proves that the light adsorption in the nanotubes contribute to the photocurrent and thus create a *p–n* junction *(238,261)*.

Jia et al. *(239)* experimented with doping the nanotube film using nitric acid (HNO_3) in 2011. Post acid-doped cells achieved very high power conversion efficiencies of over 13% while the undoped cells exhibited much lower efficiencies of 6.2%. The increase in efficiency was attributed to a reduction of the series resistance in the CNT network, producing J–V curves with a higher fill factor than in undoped cells. This decrease in series resistance is thought to be caused by *p*-type doping via oxidation, plus the removal of residual molecules (such as surfactant) from the network *(60,239,265,266)*. It is known that HNO_3, like $SOCl_2$, injects large hole densities into CNTs and thus lowers the Fermi level in SWCNT films *(267,268)*.

One issue with current CNT/Si heterojunction solar cells is the formation of a silicon oxide layer between the silicon substrate and the CNT film. Of course, cells with a thick insulating oxide cannot function. However, a tunneling barrier in the form of a thin passivating oxide can inhibit surface recombination increasing the minority carrier lifetime and blocking majority carrier transport, thus suppressing the dark saturation current, and ultimately leading to higher voltages, larger photocurrents, and improved fill factors *(246,248)*. The HNO_3 treatment used by Jia et al. was shown to produce an oxide on the silicon surface with a thickness of approximately 1 nm, which is optimal for a tunneling barrier *(246)*. In addition, Jia et al. *(246)* encapsulated an SWCNT/Si device with a poly(dimethylsiloxane) (PDMS) coating and found that this improved the stability over a period of 20 days. However, after this point there was a slight reduction in efficiency, which the authors attributed to the growth of an oxide layer that was too thick for tunneling *(238,246)*.

Other *p*-dopants can be used in the treatment of CNT films to increase the efficiency of the CNT/Si heterojunction cell including gold chloride ($AuCl_3$) *(269)*. It may be assumed that $AuCl_3$ functions as a *p*-dopant because the high reduction potential of the Au^{3+} ion allows it to extract valance electrons from the CNTs, resulting in the formation of gold nanoparticles throughout the film. However, it is also proposed that the Au^{3+} ions act as a precursor for Cl^- ion adsorption allowing for further electron extraction from the CNTs *(269)*. Several other groups *(250,270,271)* have performed $AuCl_3$ treatment and observed the expected increase in short-circuit current density, leading to improved efficiencies to above 10% in both cases. In addition, Li et al. found that if a large amount of gold nanoparticles were deposited on the nanotube film the cell performance would degrade because of a "shorting" effect in which the Au particles form contacts with the silicon substrate *(250)*.

The majority of research performed on nanotube/silicon cells is performed with a mixture of nanotube types to form the film, because nanotubes are grown as a mixture of types. Tune et al. *(253)* investigated the effects of fabricating cells with pure metallic or semiconducting nanotubes versus cells with an as-grown mix of types. It was discovered that, initially, the films containing purely metallic nanotubes significantly outperformed both the semiconducting and unsorted films. However, after the films were treated with $SOCl_2$ the situation became very different. The *p*-dopant greatly improved the efficiency of the semiconducting and mixed film cells, but had a much smaller effect on the metallic film cells *(253)*. This is expected since the $SOCl_2$ both lowers the contact resistance between the nanotubes and increases the Fermi energy of the semiconducting nanotubes such that the Fermi level drops from a point where there are no states (in the gap) to well within the valance band/first van Hove singularity, where the density of states is very high. However, although there is a comparable improvement in contact resistance in the metallic films, the reduction potential of $SOCl_2$ is insufficient to lower the Fermi level of the metallic nanotubes far enough to reach the first van Hove singularity. This means that the $SOCl_2$ treatment does not result in any appreciable change in the density of states at the Fermi level, and also does not increase the Fermi energy to the same extent as in the semiconducting nanotubes *(253)*. Similar observations have been made by Harris et al. in their comparison of semiconducting and metallic nanotube films on silicon *(271)*. Overall, it is evident that the metallicity of the nanotube film plays a significant role in the cell operation.

One of the potentially limiting factors in the performance of the CNT/Si heterojunctions discussed so far is the film morphology. Generally, the films formed consist of a random array of nanotubes, which can suppress some of the film properties. This is because a random alignment of tubes can lead to "charge traps" and holes within the film, which can inhibit charge carrier transport across the network. For this reason there has been much research done to produce highly ordered, aligned CNT films. Many techniques have been tried such as solution shearing from superacids *(250,272)*, controlled collapsing of vertically aligned CVD-grown CNT forests *(273–275)*, horizontal CNT growth from lithography-controlled CVD *(276)*, exfoliation of CVD-grown CNT forests *(277)*, evaporation-driven self-assembly of CNTs via surfactant stabilization, sidewall functionalization, polymer wrapping, and floating evaporative self-assembly *(278–282)*. Jung et al. *(270)* and Li et al. *(250)* have used the superacid chlorosulfonic acid, and Tune et al. have similarly used sodium nanotubide polyelectrolytes to form true solutions (rather than metastable suspensions) of SWCNTs at volume concentrations high enough to generate a nematic liquid crystal phase, allowing the nanotubes to be aligned on a surface when a droplet of the solution is sheared rapidly between two glass slides *(250,256,270)*. This ordered alignment of the SWCNTs allows a thin, highly transparent, very smooth film to be produced, which does not suffer from the drawbacks of thin films produced via other methods (such as poor contact with the silicon substrate and a poorly formed nanotube network with many electronic charge traps and physical holes in the film). These films were found to benefit the cells by allowing more light to reach the silicon while also providing morphological uniformity, greatly improving the maximum solar cell efficiency compared to disordered films on the same devices. Jung et al. *(270)* also showed that hemispherical droplets of liquid resting on the cells' active area act as lenses to give a light concentrating effect, artificially inflating the calculated photocurrent density and thus power conversion efficiency. This calls into question the validity of some earlier reports of improved performance when a droplet of acid was placed on the cells *(250)*.

Front side reflectance detracts significantly from the photocurrent production of PV devices *(252)*. Smooth silicon is a very reflective surface, with an average reflectance of 36% in the UV–Visible region *(249,252,283,284)*. This is caused by the large difference in refractive index between air and silicon *(285)*. As such, efforts to subdue the reflective properties of silicon have been of high priority in solar cell research *(286,287)*. The usual industry methods of reducing light reflectance are by a combination of refractive index matching with a coating of silicon nitride, and an alkaline etching process producing a random pyramid surface that acts as a scattering layer *(287)*. Although random pyramids have not yet been implemented in the CNT/Si heterojunction design, the use of silicon nanowires has been investigated by Petterson et al. *(255)* with significant improvements in surface reflectance and performance observed.

Another method of reducing light reflection in CNT/Si solar cells is via the use of a polymer coating over the CNT film *(258)*. In this way the polymer layer can be doubly effective, both by reducing the reflectance and by protecting the nanotube film and the nanotube/silicon interface from the environment to some degree *(258)*. In addition, this method does not damage the silicon surface, thus causing no interference between the CNT film and the silicon *(249)*. Polymeric antireflection layers can also assist in the stability of CNT/Si cells, by reducing the amount of oxygen that can pass through to the silicon surface. Atmospheric oxygen can react with the silicon surface in the regular cell design, growing a silicon oxide layer between the silicon and the CNT film, which, if it becomes thick enough, can lead to a decrease in cell performance *(258)*. As well as preventing oxygen from getting to the interface, polymer encapsulation can assist in preventing beneficial adsorbates such as Cl^- and SO_2 from getting out, dedoping the nanotubes, and again leading to a decrease in performance.

Li et al. *(252)* used a coating of PMMA as an antireflection layer for a CNT/Si junction. This film improved efficiency from 7.1% to 11.5%. This was mainly because of a large increase in the short circuit current density in the cell, which was correlated with the PMMA's antireflection properties. The external quantum efficiency increased from 60% at wavelengths of 400–900 nm to above 80% from 500 to 800 nm, indicating more photon–electron conversion was taking place *(252)*. Yu et al. *(258)* used PDMS and polystyrene in addition to PMMA as antireflection coatings. It was found that, while all the polymers performed successfully as antireflection coatings and managed to increase the efficiency of the cell design, the effect was dependent on the thickness of the coating.

In addition to using polymer layers as an antireflection strategy, research has been done on the use of a thin, conductive layer of polymer placed in between the CNT layer and the silicon substrate *(251)*. Polyaniline (PANI) has previously been studied as a transparent electrode material by spin coating a thin film on a silicon substrate *(288,289)*. These junctions suffered because of the limited lateral conductivity in the PANI film. CNT films can overcome this issue because of their good conductivity *(251)*. Tune et al. *(251)* reported a cell design combining a first layer of PANI and then a second one of SWCNTs on an *n*-silicon substrate. The SWCNT–PANI–Si junctions showed a maximum efficiency of 9.7%, three orders of magnitude higher than when only PANI was used (0.005% efficiency) and three times higher than the SWCNT/Si cells in that study (3.3%). This was attributed to the combined properties of both substances; the improved depletion region because of a more complete surface coverage provided by the polymer, and the high conductivity and current carrying capacity of the SWCNT film *(251)*.

One approach to reduce the silicon reflectance involved spin coating of a TiO_2 colloid giving a nanoparticulate layer of 50–80 nm in thickness on the surface. This method had previously been used to lower the reflectance of silicon solar cells, leading to greatly reduced reflectance (approaching 0.04%) and improved cell efficiencies of up to 44% relative when coated with thermal SiO_2 and a TiO_2 thin film *(290)*. In the case of the nanotube/silicon cells the layer subdued the light reflection significantly (decreased reflectance from 30% to 40% to below 5% in some wavelength regions) and increased the incident photon to current efficiency from 60% to 90%, while still allowing postfabrication doping treatments to reach the nanotube network. Together with nitric acid and hydrogen peroxide doping this antireflection layer then recorded high cell efficiencies of above 15% *(249)*. The authors asserted that the TiO_2 layer improved performance through better light management in the front surface, acting as an interference coating by creating destructive interference at the interface, instead of the light scattering and trapping effects of patterned silicon surfaces *(249)*.

Wang et al. investigated the application of molybdenum oxide (MoO_x) and ZnO as cheap, environmentally friendly antireflection layers *(254)*. In addition to the antireflection properties, these layers also assisted with carrier doping of the nanotube film and helped to improve the carrier transport efficiency across the junction between the nanotubes and the front side metal electrode in both the standard *p*-SWCNT/*n*-Si cell and the *n*-SWCNT/*p*-Si cell *(254)*. Using these films allowed the team to set a record efficiency for CNT/Si solar cells of 17% (albeit with a submillimeter active area and no aperture during testing) and achieve a relatively high efficiency for a cell using *n*-doped nanotubes of 4% *(254)*. The MoO_x coating improved the properties of the standard CNT/Si cell, with a higher J_{sc} and much improved diode properties.

Despite all these advances, and the clear interest in CNT/Si solar cells from a fundamental point of view, one may well question the real benefit of adding what is still an expensive commodity nanomaterial onto an already costly product. However, with current commercial silicon PV module efficiencies nearing laboratory-scale records, which are themselves already near to fundamental limits, and with the very steep cost reductions because of economies of scale involved in recent massive expansion of worldwide production capacity yielding prices that are not far above wholesale material costs, meeting the market demand for continued improvements in the levelized cost of energy to the consumer will require more than simply following the price learning curve to plateau. The cost of CNTs is only now starting to benefit from economies of scale, and this material is reasonably predicted to become very cheap as capacity ramps up and production technologies mature. As well, the quantity of material required in CNT/Si solar cells is miniscule, on the order of milligrams per square meter. Taking these factors into account, it is not difficult to envisage a scenario where CNTs replace the intrinsically costly and throughput-limiting ion diffusion processes used in current state-of-the-art silicon PVs, when our understanding of the junction physics improves alongside technological advancements in high-throughput methods of incorporating them into the fabrication sequence.

8.7 SUMMARY AND OUTLOOK

While an extensive body of work using CNTs in solar cells already exists as detailed in this chapter, there are still many avenues of investigation to be pursued. The ability to sort nanotubes both in terms of the electronic characteristics (metallic vs. semiconducting) and further according to their chirality *(291–302)* presents opportunities to tune conduction pathways and select band gaps and/or light absorption/transmission wavelengths. As with most nanomaterials, an important factor determining performance is the morphology of the CNT films and future investigations will have to pay particular attention to characteristics such as the smoothness of the nanotube film to obtain the best performance. Various schemes to align CNTs on surfaces have been demonstrated and this provides the opportunity to significantly increase the conductivity of films that already have exceptionally high transparency. The use of such aligned films in solar cells is an area worthy of further study. Hybrid structures combining CNTs with other nanomaterials present many possibilities and there is little doubt this space will be explored extensively to find optimal solutions for various components of solar cells. Chemical doping is known to be a powerful method to improve the CNTs' properties including conductivity and catalytic activity *(303–305)* but the effects are often short lived. Work to improve the long-term stability of the performance enhancement, especially in air, would be of great value.

Semiconducting nanotubes have a Fermi level in the middle of the band gap and could thus be termed "intrinsic" similarly to undoped silicon. However, when exposed to air they develop *p*-type characteristics because of the electron withdrawing nature of adsorbed oxygen molecules. Even in an ultrahigh vacuum this adsorbed oxygen is very difficult to remove and has likely skewed the results of much research carried out to date *(306)*. In addition, SWCNTs can be chemically doped *p*-type by electron-withdrawing species such as strong oxidizers or oxidizing acids *(265,268)*, or *n*-type by electron-donating species such as alkali metals *(307)*, hydrazine *(268,308)*, nitrogen-rich polymers *(309)*, and nitrogen-rich aromatic compounds *(310)*. Substitutional doping has also been demonstrated with boron (*p*-type) *(311)* and nitrogen (*n*-type) *(312,313)*. This raises the exciting prospect of all-nanotube PVs based on *p*–*n* or *p*–*i*–*n* homo- or heterojunctions, although such architectures are yet to be reported for other than single tube devices *(235)*.

There can be little doubt that CNTs will play a key role in the development of solar cells, which will provide energy at competitive costs compared to current energy production approaches. New processes are continually being reported and these will present opportunities for innovation in both research and commercial development.

ACKNOWLEDGMENTS

The authors would like to acknowledge deanship of Scientific Research, College of Science Research Center, King Saud University, Riyadh, Saudi Arabia for funding this project.

REFERENCES

1. Hoffert, M. I.; Caldeira, K.; Jain, A. K.; Haites, E. F.; Harvey, L. D. D.; Potter, S. D.; Schlesinger, M. E.; Schneider, S. H.; Watts, R. G.; Wigley, T. M. L.; Wuebbles, D. J. Energy Implications of Future Stabilization of Atmospheric CO_2 Content. *Nature* **1998,** *395* (6705), 881–884.

2. Clerici, A. *World Energy Resources 2013 Survey,* 2013; p. 468. London, ISBN 978 0 946121 29 8.
3. Díaz-González, F.; Sumper, A.; Gomis-Bellmunt, O.; Villafáfila-Robles, R. A Review of Energy Storage Technologies for Wind Power Applications. *Renewable and Sustainable Energy Reviews* **2012,** *16* (4), 2154–2171.
4. Brennan, L.; Owende, P. Biofuels from Microalgae-a Review of Technologies for Production, Processing, and Extractions of Biofuels and Co-products. *Renewable and Sustainable Energy Reviews* **2010,** *14* (2), 557–577.
5. Wenham, S. R.; Green, M. A. Silicon Solar Cells. *Progress in Photovoltaics: Research and Applications* **1996,** *4* (1), 3–33.
6. Hawkes, A.; Staffell, I.; Brett, D.; Brandon, N. Fuel Cells for Micro-combined Heat and Power Generation. *Energy & Environmental Science* **2009,** *2* (7), 729–744.
7. Fthenakis, V. Sustainability of Photovoltaics: The Case for Thin-film Solar Cells. *Renewable and Sustainable Energy Reviews* **2009,** *13* (9), 2746–2750.
8. Conibeer, G. Third-Generation Photovoltaics. *Materials Today* **2007,** *10* (11), 42–50.
9. Fritts, C. E. On a New Form of Selenium Photocell. *Proceedings of the American Association for the Advancement of Science* **1883,** *33,* 97.
10. Chapin, D. M.; Fuller, C. S.; Pearson, G. L. A New Silicon p-n Junction Photocell for Converting Solar Radiation into Electrical Power. *Journal of Applied Physics* **1954,** *25* (5), 676–677.
11. Green, M. A.; Blakers, A. W.; Osterwald, C. R. Characterization of High-Efficiency Silicon Solar Cells. *Journal of Applied Physics* **1985,** *58* (11), 4402–4408.
12. Green, M. A.; Emery, K.; Hishikawa, Y.; Dunlop, E. D. Solar Cell Efficiency Tables (Version 39). *Progress in Photovoltaics: Research and Applications* **2012,** *20,* 12–20.
13. Kabir, M. I.; Ibarahim, Z.; Sopian, K.; Amin, N. A Review on Progress of Amorphous and Microcrystalline Silicon Thin-Film Solar Cells. *Recent Patents on Electrical Engineering* **2011,** *4,* 50–62.
14. Jiang, M.; Yan, X. Cu_2ZnSnS_4 Thin Film Solar Cells: Present Status and Future Prospects. In *Solar Cells – Research and Application Perspectives;* Morales-Acevedo, A., Ed.; Intech: Croatia, 2013.
15. Bosi, M.; Pelosi, C. The Potential of III-V Semiconductors as Terrestrial Photovoltaic Devices. *Progress in Photovoltaics: Research and Applications* **2007,** *15* (1), 51–68.
16. Jackson, P.; Hariskos, D.; Lotter, E.; Paetel, S.; Wuerz, R.; Menner, R.; Wischmann, W.; Powalla, M. New World Record Efficiency for $Cu(In,Ga)Se_2$ Thin-Film Solar Cells Beyond 20%. *Progress in Photovoltaics: Research and Applications* **2011,** *19* (7), 894–897.
17. Kranz, L.; Gretener, C.; Perrenoud, J.; Schmitt, R.; Pianezzi, F.; La Mattina, F.; Blösch, P.; Cheah, E.; Chirilă, A.; Fella, C. M.; Hagendorfer, H.; Jäger, T.; Nishiwaki, S.; Uhl, A. R.; Buecheler, S.; Tiwari, A. N. Doping of Polycrystalline Cdte for High-Efficiency Solar Cells on Flexible Metal Foil. *Nature Communications* **2013,** *4,* 2306.
18. Zhou, H.; Hsu, W.-C.; Duan, H.-S.; Bob, B.; Yang, W.; Song, T.-B.; Hsu, C.-J.; Yang, Y. Czts Nanocrystals: A Promising Approach for Next Generation Thin Film Photovoltaics. *Energy & Environmental Science* **2013,** *6* (10), 2822–2838.
19. Iijima, S. Helical Microtubules of Graphitic Carbon. *Nature* **1991,** *354* (6348), 56–58.
20. Thostenson, E. T.; Ren, Z.; Chou, T.-W. Advances in the Science and Technology of Carbon Nanotubes and Their Composites: A Review. *Composites Science and Technology* **2001,** *61* (13), 1899–1912.
21. Zhang, Q.; Huang, J.-Q.; Qian, W.-Z.; Zhang, Y.-Y.; Wei, F. The Road for Nanomaterials Industry: A Review of Carbon Nanotube Production, Post-Treatment, and Bulk Applications for Composites and Energy Storage. *Small* **2013,** *9* (8), 1237–1265.
22. Prasek, J.; Drbohlavova, J.; Chomoucka, J.; Hubalek, J.; Jasek, O.; Adam, V.; Kizek, R. Methods for Carbon Nanotubes Synthesis-Review. *Journal of Materials Chemistry* **2011,** *21* (40), 15872–15884.
23. Peng, X.; Chen, J.; Misewich, J. A.; Wong, S. S. Carbon Nanotube-Nanocrystal Heterostructures. *Chemical Society Reviews* **2009,** *38* (4), 1076–1098.
24. Andersson, A.; Johansson, N.; Bröms, P.; Yu, N.; Lupo, D.; Salaneck, W. R. Fluorine Tin Oxide as an Alternative to Indium Tin Oxide in Polymer Leds. *Advanced Materials* **1998,** *10* (11), 859–863.
25. Wang, X.; Zhi, L.; Müllen, K. Transparent, Conductive Graphene Electrodes for Dye-Sensitized Solar Cells. *Nano Letters* **2008,** *8* (1), 323–327.

26. Wang, X.; Zhi, L.; Tsao, N.; Tomović, Ž.; Li, J.; Müllen, K. Transparent Carbon Films as Electrodes in Organic Solar Cells. *Angewandte Chemie International Edition* **2008,** *47* (16), 2990–2992.
27. Na, S.-I.; Kim, S.-S.; Jo, J.; Kim, D.-Y. Efficient and Flexible ITO-Free Organic Solar Cells Using Highly Conductive Polymer Anodes. *Advanced Materials* **2008,** *20* (21), 4061–4067.
28. Kang, M. G.; Guo, L. J. Nanoimprinted Semitransparent Metal Electrodes and Their Application in Organic Light-Emitting Diodes. *Advanced Materials* **2007,** *19* (10), 1391–1396.
29. Rathmell, A. R.; Bergin, S. M.; Hua, Y.-L.; Li, Z.-Y.; Wiley, B. J. The Growth Mechanism of Copper Nanowires and Their Properties in Flexible, Transparent Conducting Films. *Advanced Materials* **2010,** *22* (32), 3558–3563.
30. Eda, G.; Fanchini, G.; Chhowalla, M. Large-Area Ultrathin Films of Reduced Graphene Oxide as a Transparent and Flexible Electronic Material. *Nature Nanotechnology* **2008,** *3* (5), 270–274.
31. Wu, Z.; Chen, Z.; Du, X.; Logan, J. M.; Sippel, J.; Nikolou, M.; Kamaras, K.; Reynolds, J. R.; Tanner, D. B.; Hebard, A. F.; Rinzler, A. G. Transparent, Conductive Carbon Nanotube Films. *Science* **2004,** *305* (5688), 1273–1276.
32. Brennan, L. J.; Byrne, M. T.; Bari, M.; Gun'ko, Y. K. Carbon Nanomaterials for Dye-sensitized Solar Cell Applications: A Bright Future. *Advanced Energy Materials* **2011,** *1* (4), 472–485.
33. Ellmer, K. Past Achievements and Future Challenges in the Development of Optically Transparent Electrodes. *Nature Photonics* **2012,** *6* (12), 809–817.
34. Hecht, D. S.; Hu, L.; Irvin, G. Emerging Transparent Electrodes Based on Thin Films of Carbon Nanotubes, Graphene, and Metallic Nanostructures. *Advanced Materials* **2011,** *23* (13), 1482–1513.
35. Hu, L.; Hecht, D. S.; Grüner, G. Carbon Nanotube Thin Films: Fabrication, Properties, and Applications. *Chemical Reviews* **2010,** *110* (10), 5790–5844.
36. Kumar, A.; Zhou, C. The Race to Replace Tin-Doped Indium Oxide: Which Material Will Win? *ACS Nano* **2010,** *4* (1), 11–14.
37. Choi, G. S.; Cho, Y. S.; Son, K. H.; Kim, D. J. Mass Production of Carbon Nanotubes Using Spin-Coating of Nanoparticles. *Microelectronic Engineering* **2003,** *66* (1–4), 77–82.
38. Dan, B.; Irvin, G. C.; Pasquali, M. Continuous and Scalable Fabrication of Transparent Conducting Carbon Nanotube Films. *ACS Nano* **2009,** *3* (4), 835–843.
39. De, S.; Lyons, P. E.; Sorel, S.; Doherty, E. M.; King, P. J.; Blau, W. J.; Nirmalraj, P. N.; Boland, J. J.; Scardaci, V.; Joimel, J.; Coleman, J. N. Transparent, Flexible, and Highly Conductive Thin Films Based on Polymer–Nanotube Composites. *ACS Nano* **2009,** *3* (3), 714–720.
40. Jo, J. W.; Jung, J. W.; Lee, J. U.; Jo, W. H. Fabrication of Highly Conductive and Transparent Thin Films from Single-Walled Carbon Nanotubes Using a New Non-Ionic Surfactant Via Spin Coating. *ACS Nano* **2010,** *4* (9), 5382–5388.
41. Kim, S.; Yim, J.; Wang, X.; Bradley, D. D. C.; Lee, S.; deMello, J. C. Spin- and Spray-Deposited Single-Walled Carbon-Nanotube Electrodes for Organic Solar Cells. *Advanced Functional Materials* **2010,** *20* (14), 2310–2316.
42. Schindler, A.; Brill, J.; Fruehauf, N.; Novak, J. P.; Yaniv, Z. Solution-Deposited Carbon Nanotube Layers for Flexible Display Applications. *Physica E: Low-Dimensional Systems and Nanostructures* **2007,** *37* (1–2), 119–123.
43. Song, Y. I.; Yang, C. M.; Kim, D. Y.; Kanoh, H.; Kaneko, K. Flexible Transparent Conducting Single-Wall Carbon Nanotube Film with Network Bridging Method. *Journal of Colloid and Interface Science* **2008,** *318* (2), 365–371.
44. Zhang, D.; Ryu, K.; Liu, X.; Polikarpov, E.; Ly, J.; Tompson, M. E.; Zhou, C. Transparent, Conductive, and Flexible Carbon Nanotube Films and Their Application in Organic Light-Emitting Diodes. *Nano Letters* **2006,** *6* (9), 1880–1886.
45. Zhou, Y.; Hu, L.; Grüner, G. A Method of Printing Carbon Nanotube Thin Films. *Applied Physics Letters* **2006,** *88* (12), 123109.
46. Ulbricht, R.; Jiang, X.; Lee, S.; Inoue, K.; Zhang, M.; Fang, S.; Baughman, R.; Zakhidov, A. Polymeric Solar Cells with Oriented and Strong Transparent Carbon Nanotube Anode. *Physica Status Solidi (B)* **2006,** *243* (13), 3528–3532.

47. Chaudhary, S.; Lu, H.; Müller, A. M.; Bardeen, C. J.; Ozkan, M. Hierarchical Placement and Associated Optoelectronic Impact of Carbon Nanotubes in Polymer-Fullerene Solar Cells. *Nano Letters* **2007,** 7 (7), 1973–1979.
48. Gabor, N. M.; Zhong, Z.; Bosnick, K.; Park, J.; McEuen, P. L. Extremely Efficient Multiple Electron-Hole Pair Generation in Carbon Nanotube Photodiodes. *Science* **2009,** *325* (5946), 1367–1371.
49. Batmunkh, M.; Biggs, M. J.; Shapter, J. G. Carbon Nanotubes for Dye-Sensitized Solar Cells. *Small* **2015,** *11* (25), 2963–2989.
50. Cheng, H. M.; Li, F.; Sun, X.; Brown, S. D. M.; Pimenta, M. A.; Marucci, A.; Dresselhaus, G.; Dresselhaus, M. S. Bulk Morphology and Diameter Distribution of Single-Walled Carbon Nanotubes Synthesized by Catalytic Decomposition of Hydrocarbons. *Chemical Physics Letters* **1998,** *289* (5–6), 602–610.
51. Kaskela, A.; Nasibulin, A. G.; Timmermans, M. Y.; Aitchison, B.; Papadimitratos, A.; Tian, Y.; Zhu, Z.; Jiang, H.; Brown, D. P.; Zakhidov, A.; Kauppinen, E. I. Aerosol-Synthesized Swcnt Networks with Tunable Conductivity and Transparency by a Dry Transfer Technique. *Nano Letters* **2010,** *10* (11), 4349–4355.
52. Ma, W.; Song, L.; Yang, R.; Zhang, T.; Zhao, Y.; Sun, L.; Ren, Y.; Liu, D.; Liu, L.; Shen, J.; Zhang, Z.; Xiang, Y.; Zhou, W.; Xie, S. Directly Synthesized Strong, Highly Conducting, Transparent Single-Walled Carbon Nanotube Films. *Nano Letters* **2007,** 7 (8), 2307–2311.
53. Nasibulin, A. G.; Kaskela, A.; Mustonen, K.; Anisimov, A. S.; Ruiz, V.; Kivistö, S.; Rackauskas, S.; Timmermans, M. Y.; Pudas, M.; Aitchison, B.; Kauppinen, M.; Brown, D. P.; Okhotnikov, O. G.; Kauppinen, E. I. Multifunctional Free-Standing Single-Walled Carbon Nanotube Films. *ACS Nano* **2011,** *5* (4), 3214–3221.
54. Feng, C.; Liu, K.; Wu, J. S.; Liu, L.; Cheng, J. S.; Zhang, Y.; Sun, Y.; Li, Q.; Fan, S.; Jiang, K. Flexible, Stretchable, Transparent Conducting Films Made from Superaligned Carbon Nanotubes. *Advanced Functional Materials* **2010,** *20* (6), 885–891.
55. Jiang, K.; Li, Q.; Fan, S. Spinning Continuous Carbon Nanotube Yarns. *Nature* **2002,** *419* (6909), 801.
56. Zhang, M.; Fang, S.; Zakhidov, A. A.; Lee, S. B.; Aliev, A. E.; Williams, C. D.; Atkinson, K. R.; Baughman, R. H. Materials Science: Strong, Transparent, Multifunctional, Carbon Nanotube Sheets. *Science* **2005,** *309* (5738), 1215–1219.
57. Saran, N.; Parikh, K.; Suh, D. S.; Muñoz, E.; Kolla, H.; Manohar, S. K. Fabrication and Characterization of Thin Films of Single-Walled Carbon Nanotube Bundles on Flexible Plastic Substrates. *Journal of the American Chemical Society* **2004,** *126* (14), 4462–4463.
58. Hu, L.; Hecht, D. S.; Gruner, G. Percolation in Transparent and Conducting Carbon Nanotube Networks. *Nano Letters* **2004,** *4* (12), 2513–2517.
59. Mirri, F.; Ma, A. W. K.; Hsu, T. T.; Behabtu, N.; Eichmann, S. L.; Young, C. C.; Tsentalovich, D. E.; Pasquali, M. High-Performance Carbon Nanotube Transparent Conductive Films by Scalable Dip Coating. *ACS Nano* **2012,** *6* (11), 9737–9744.
60. Geng, H. Z.; Ki, K. K.; Kang, P. S.; Young, S. L.; Chang, Y.; Young, H. L. Effect of Acid Treatment on Carbon Nanotube-Based Flexible Transparent Conducting Films. *Journal of the American Chemical Society* **2007,** *129* (25), 7758–7759.
61. Maeda, Y.; Hashimoto, M.; Kaneko, S.; Kanda, M.; Hasegawa, T.; Tsuchiya, T.; Akasaka, T.; Naitoh, Y.; Shimizu, T.; Tokumoto, H.; Lu, J.; Nagase, S. Preparation of Transparent and Conductive Thin Films of Metallic Single-Walled Carbon Nanotubes. *Journal of Materials Chemistry* **2008,** *18* (35), 4189–4192.
62. Lima, M. D.; De Andrade, M. J.; Bergmann, C. P.; Roth, S. Thin, Conductive, Carbon Nanotube Networks over Transparent Substrates by Electrophoretic Deposition. *Journal of Materials Chemistry* **2008,** *18* (7), 776–779.
63. Pei, S.; Du, J.; Zeng, Y.; Liu, C.; Cheng, H. M. The Fabrication of a Carbon Nanotube Transparent Conductive Film by Electrophoretic Deposition and Hot-Pressing Transfer. *Nanotechnology* **2009,** *20,* 235707.
64. Kordás, K.; Mustonen, T.; Tóth, G.; Jantunen, H.; Lajunen, M.; Soldano, C.; Talapatra, S.; Kar, S.; Vajtai, R.; Ajayan, P. M. Inkjet Printing of Electrically Conductive Patterns of Carbon Nanotubes. *Small* **2006,** *2* (8–9), 1021–1025.

65. Lau, P. H.; Takei, K.; Wang, C.; Ju, Y.; Kim, J.; Yu, Z.; Takahashi, T.; Cho, G.; Javey, A. Fully Printed, High Performance Carbon Nanotube Thin-Film Transistors on Flexible Substrates. *Nano Letters* **2013,** *13* (8), 3864–3869.
66. Cho, D.-Y.; Eun, K.; Choa, S.-H.; Kim, H.-K. Highly Flexible and Stretchable Carbon Nanotube Network Electrodes Prepared by Simple Brush Painting for Cost-Effective Flexible Organic Solar Cells. *Carbon* **2014,** *66,* 530–538.
67. Du, J.; Pei, S.; Ma, L.; Cheng, H. M. 25th Anniversary Article: Carbon Nanotube- and Graphene-Based Transparent Conductive Films for Optoelectronic Devices. *Advanced Materials* **2014,** *26* (13), 1958–1991.
68. Park, S.; Vosguerichian, M.; Bao, Z. A Review of Fabrication and Applications of Carbon Nanotube Film-Based Flexible Electronics. *Nanoscale* **2013,** *5* (5), 1727–1752.
69. Mathew, S.; Yella, A.; Gao, P.; Humphry-Baker, R.; Curchod, B. F. E.; Ashari-Astani, N.; Tavernelli, I.; Rothlisberger, U.; Nazeeruddin, M. K.; Grätzel, M. Dye-Sensitized Solar Cells with 13% Efficiency Achieved through the Molecular Engineering of Porphyrin Sensitizers. *Nature Chemistry* **2014,** *6* (3), 242–247.
70. Kakiage, K.; Aoyama, Y.; Yano, T.; Oya, K.; Fujisawa, J-i.; Hanaya, M. Highly-Efficient Dye-Sensitized Solar Cells with Collaborative Sensitization by Silyl-Anchor and Carboxy-Anchor Dyes. *Chemical Communications* **2015,** *51* (88), 15894–15897.
71. Scharber, M. C.; Sariciftci, N. S. Efficiency of Bulk-Heterojunction Organic Solar Cells. *Progress in Polymer Science* **2013,** *38* (12), 1929–1940.
72. O'Regan, B.; Grätzel, M. A Low-Cost, High-Efficiency Solar Cell Based on Dye-Sensitized Colloidal TiO_2 Films. *Nature* **1991,** *353* (6346), 737–740.
73. Grätzel, M. Dye-Sensitized Solar Cells. *Journal of Photochemistry and Photobiology C: Photochemistry Reviews* **2003,** *4* (2), 145–153.
74. Hamann, T. W.; Jensen, R. A.; Martinson, A. B. F.; Van Ryswyk, H.; Hupp, J. T. Advancing Beyond Current Generation Dye-Sensitized Solar Cells. *Energy & Environmental Science* **2008,** *1* (1), 66–78.
75. Chittibabu, K. G. Titanium Oxide Nano Colloids for Dye Sensitized Solar Cells and Other Applications. *G24 Power Limited: United Kingdom* **2014**.
76. Nam, J. G.; Park, Y. J.; Kim, B. S.; Lee, J. S. Enhancement of the Efficiency of Dye-Sensitized Solar Cell by Utilizing Carbon Nanotube Counter Electrode. *Scripta Materialia* **2010,** *62* (3), 148–150.
77. Aitola, K.; Kaskela, A.; Halme, J.; Ruiz, V.; Nasibulin, A. G.; Kauppinen, E. I.; Lund, P. D. Single-Walled Carbon Nanotube Thin-Film Counter Electrodes for Indium Tin Oxide-Free Plastic Dye Solar Cells. *Journal of the Electrochemical Society* **2010,** *157* (12), B1831–B1837.
78. Liu, C. T.; Wang, Y. C.; Dong, R. X.; Wang, C. C.; Huang, K. C.; Vittal, R.; Ho, K. C.; Lin, J. J. A Dual-Functional Pt/CNT TCO-Free Counter Electrode for Dye-Sensitized Solar Cell. *Journal of Materials Chemistry* **2012,** *22* (48), 25311–25315.
79. Malara, F.; Manca, M.; De Marco, L.; Pareo, P.; Gigli, G. Flexible Carbon Nanotube-Based Composite Plates as Efficient Monolithic Counter Electrodes for Dye Solar Cells. *ACS Applied Materials and Interfaces* **2011,** *3* (9), 3625–3632.
80. Tantang, H.; Kyaw, A. K. K.; Zhao, Y.; Chan-Park, M. B.; Tok, A. I. Y.; Hu, Z.; Li, L. J.; Sun, X. W.; Zhang, Q. Nitrogen-Doped Carbon Nanotube-Based Bilayer Thin Film as Transparent Counter Electrode for Dye-Sensitized Solar Cells (DSSCs). *Chemistry – An Asian Journal* **2012,** 7 (3), 541–545.
81. Du, J.; Bittner, F.; Hecht, D. S.; Ladous, C.; Ellinger, J.; Oekermann, T.; Wark, M. A Carbon Nanotube-Based Transparent Conductive Substrate for Flexible ZnO Dye-Sensitized Solar Cells. *Thin Solid Films* **2013,** *531,* 391–397.
82. Wei, D.; Unalan, H. E.; Han, D.; Zhang, Q.; Niu, L.; Amaratunga, G.; Ryhanen, T. A Solid-State Dye-Sensitized Solar Cell Based on a Novel Ionic Liquid Gel and ZnO Nanoparticles on a Flexible Polymer Substrate. *Nanotechnology* **2008,** *19,* 424006.
83. Kyaw, A. K. K.; Tantang, H.; Wu, T.; Ke, L.; Peh, C.; Huang, Z. H.; Zeng, X. T.; Demir, H. V.; Zhang, Q.; Sun, X. W. Dye-Sensitized Solar Cell with a Titanium-Oxide-Modified Carbon Nanotube Transparent Electrode. *Applied Physics Letters* **2011,** *99,* 021107.

84. Rensmo, H.; Keis, K.; Lindström, H.; Södergren, S.; Solbrand, A.; Hagfeldt, A.; Lindquist, S. E.; Wang, L. N.; Muhammed, M. High Light-to-Energy Conversion Efficiencies for Solar Cells Based on Nanostructured ZnO Electrodes. *Journal of Physical Chemistry B* **1997,** *101* (14), 2598–2601.
85. Zhang, Q.; Chou, T. P.; Russo, B.; Jenekhe, S. A.; Cao, G. Aggregation of ZnO Nanocrystallites for High Conversion Efficiency in Dye-Sensitized Solar Cells. *Angewandte Chemie – International Edition* **2008,** *47* (13), 2402–2406.
86. Ferrere, S.; Zaban, A.; Gregg, B. A. Dye Sensitization of Nanocrystalline Tin Oxide by Perylene Derivatives. *Journal of Physical Chemistry B* **1997,** *101* (23), 4490–4493.
87. Kay, A.; Grätzel, M. Dye-Sensitized Core-Shell Nanocrystals: Improved Efficiency of Mesoporous Tin Oxide Electrodes Coated with a Thin Layer of an Insulating Oxide. *Chemistry of Materials* **2002,** *14* (7), 2930–2935.
88. Chen, S. G.; Chappel, S.; Diamant, Y.; Zaban, A. Preparation of Nb_2O_5 Coated TiO_2 Nanoporous Electrodes and Their Application in Dye-Sensitized Solar Cells. *Chemistry of Materials* **2001,** *13* (12), 4629–4634.
89. Yang, S.; Kou, H.; Wang, J.; Xue, H.; Han, H. Tunability of the Band Energetics of Nanostructured $SrTiO_3$ Electrodes for Dye-Sensitized Solar Cells. *Journal of Physical Chemistry C* **2010,** *114* (9), 4245–4249.
90. Shin, S. S.; Kim, J. S.; Suk, J. H.; Lee, K. D.; Kim, D. W.; Park, J. H.; Cho, I. S.; Hong, K. S.; Kim, J. Y. Improved Quantum Efficiency of Highly Efficient Perovskite $BaSnO_3$-Based Dye-Sensitized Solar Cells. *ACS Nano* **2013,** *7* (2), 1027–1035.
91. Chen, H. Y.; Kuang, D. B.; Su, C. Y. Hierarchically Micro/Nanostructured Photoanode Materials for Dye-Sensitized Solar Cells. *Journal of Materials Chemistry* **2012,** *22* (31), 15475–15489.
92. Maçaira, J.; Andrade, L.; Mendes, A. Review on Nanostructured Photoelectrodes for Next Generation Dye-Sensitized Solar Cells. *Renewable and Sustainable Energy Reviews* **2013,** *27,* 334–349.
93. Tétreault, N.; Grätzel, M. Novel Nanostructures for Next Generation Dye-Sensitized Solar Cells. *Energy & Environmental Science* **2012,** *5* (9), 8506–8516.
94. Zhang, Q.; Dandeneau, C. S.; Zhou, X.; Cao, C. ZnO Nanostructures for Dye-Sensitized Solar Cells. *Advanced Materials* **2009,** *21* (41), 4087–4108.
95. Chandiran, A. K.; Abdi-Jalebi, M.; Nazeeruddin, M. K.; Grätzel, M. Analysis of Electron Transfer Properties of ZnO and TiO_2 Photoanodes for Dye-Sensitized Solar Cells. *ACS Nano* **2014,** *8* (3), 2261–2268.
96. Hagfeldt, A.; Boschloo, G.; Sun, L.; Kloo, L.; Pettersson, H. Dye-Sensitized Solar Cells. *Chemical Reviews* **2010,** *110* (11), 6595–6663.
97. Batmunkh, M.; Biggs, M. J.; Shapter, J. G. Carbonaceous Dye-sensitised Solar Cell Photoelecrodes. *Advanced Science* **2015,** *2,* 1400025.
98. Jang, S. R.; Vittal, R.; Kim, K. J. Incorporation of Functionalized Single-Wall Carbon Nanotubes in Dye-Sensitized TiO_2 Solar Cells. *Langmuir* **2004,** *20* (22), 9807–9810.
99. Brown, P.; Takechi, K.; Kamat, P. V. Single-Walled Carbon Nanotube Scaffolds for Dye-Sensitized Solar Cells. *Journal of Physical Chemistry C* **2008,** *112* (12), 4776–4782.
100. Kongkanand, A.; Domínguez, R. M.; Kamat, P. V. Single Wall Carbon Nanotube Scaffolds for Photoelectrochemical Solar Cells. Capture and Transport of Photogenerated Electrons. *Nano Letters* **2007,** *7* (3), 676–680.
101. Lee, T. Y.; Alegaonkar, P. S.; Yoo, J. B. Fabrication of Dye Sensitized Solar Cell Using TiO_2 Coated Carbon Nanotubes. *Thin Solid Films* **2007,** *515* (12), 5131–5135.
102. Lee, W.; Lee, J.; Lee, S. H.; Chang, J.; Yi, W.; Han, S. H. Improved Photocurrent in Ru(2,2′-bipyridine-4,4′-dicarboxylic Acid)$_2$(NCS)$_2$/Di(3-aminopropyl)Viologen/Single-Walled Carbon Nanotubes/Indium Tin Oxide System: Suppression of Recombination Reaction by Use of Single-Walled Carbon Nanotubes. *Journal of Physical Chemistry C* **2007,** *111* (26), 9110–9115.
103. Guai, G. H.; Li, Y.; Ng, C. M.; Li, C. M.; Chan-Park, M. B. TiO_2 Composing with Pristine, Metallic or Semiconducting Single-Walled Carbon Nanotubes: Which Gives the Best Performance for a Dye-Sensitized Solar Cell. *ChemPhysChem* **2012,** *13* (10), 2566–2572.

104. Khamwannah, J.; Noh, S. Y.; Frandsen, C.; Zhang, Y.; Kim, H.; Kong, S. D.; Jin, S. Nanocomposites of TiO_2 and Double-Walled Carbon Nanotubes for Improved Dye-Sensitized Solar Cells. *Journal of Renewable and Sustainable Energy* **2012,** *4* (2).
105. Bakhshayesh, A. M.; Mohammadi, M. R.; Masihi, N.; Akhlaghi, M. H. Improved Electron Transportation of Dye-Sensitized Solar Cells Using Uniform Mixed CNTs-TiO_2 Photoanode Prepared by a New Polymeric Gel Process. *Journal of Nanoparticle Research* **2013,** *15* (9).
106. Song, C. B.; Qiang, Y. H.; Zhao, Y. L.; Gu, X. Q.; Song, D. M.; Zhu, L. Adhesion of TiO_2 Nanotube Arrays on Transparent Conducting Substrates Using CNT-TiO_2 Composite Pastes. *Applied Surface Science* **2014,** *305,* 792–796.
107. Zhang, S.; Niu, H.; Lan, Y.; Cheng, C.; Xu, J.; Wang, X. Synthesis of TiO_2 Nanoparticles on Plasma-Treated Carbon Nanotubes and Its Application in Photoanodes of Dye-Sensitized Solar Cells. *Journal of Physical Chemistry C* **2011,** *115* (44), 22025–22034.
108. Cheng, C.; Wu, J.; Xiao, Y.; Chen, Y.; Yu, H.; Tang, Z.; Lin, J.; Huang, M. Preparation of Titanium Dioxide-Double-Walled Carbon Nanotubes and Its Application in Flexible Dye-Sensitized Solar Cells. *Frontiers of Optoelectronics* **2012,** *5* (2), 224–230.
109. Meng, L.; Fu, C.; Lu, Q. Advanced Technology for Functionalization of Carbon Nanotubes. *Progress in Natural Science* **2009,** *19* (7), 801–810.
110. Sun, S.; Gao, L.; Liu, Y. Optimization of the Cutting Process of Multi-Wall Carbon Nanotubes for Enhanced Dye-Sensitized Solar Cells. *Thin Solid Films* **2011,** *519* (7), 2273–2279.
111. Xu, J.; Wang, L.; Zhu, Y. Decontamination of Bisphenol a from Aqueous Solution by Graphene Adsorption. *Langmuir* **2012,** *28* (22), 8418–8425.
112. Lee, K. M.; Hu, C. W.; Chen, H. W.; Ho, K. C. Incorporating Carbon Nanotube in a Low-Temperature Fabrication Process for Dye-Sensitized TiO_2 Solar Cells. *Solar Energy Materials and Solar Cells* **2008,** *92* (12), 1628–1633.
113. Sawatsuk, T.; Chindaduang, A.; Sae-kung, C.; Pratontep, S.; Tumcharern, G. Dye-Sensitized Solar Cells Based on TiO_2-MWCNTs Composite Electrodes: Performance Improvement and Their Mechanisms. *Diamond and Related Materials* **2009,** *18* (2–3), 524–527.
114. Yen, C. Y.; Lin, Y. F.; Liao, S. H.; Weng, C. C.; Huang, C. C.; Hsiao, Y. H.; Ma, C. C. M.; Chang, M. C.; Shao, H.; Tsai, M. C.; Hsieh, C. K.; Tsai, C. H.; Weng, F. B. Preparation and Properties of a Carbon Nanotube-Based Nanocomposite Photoanode for Dye-Sensitized Solar Cells. *Nanotechnology* **2008,** *19,* 375305.
115. Chan, Y. F.; Wang, C. C.; Chen, B. H.; Chen, C. Y. Dye-Sensitized TiO_2 Solar Cells Based on Nanocomposite Photoanode Containing Plasma-Modified Multi-Walled Carbon Nanotubes. *Progress in Photovoltaics: Research and Applications* **2013,** *21* (1), 47–57.
116. Chen, J.; Li, B.; Zheng, J.; Zhao, J.; Zhu, Z. Role of Carbon Nanotubes in Dye-Sensitized TiO_2-Based Solar Cells. *Journal of Physical Chemistry C* **2012,** *116* (28), 14848–14856.
117. Dang, X.; Yi, H.; Ham, M. H.; Qi, J.; Yun, D. S.; Ladewski, R.; Strano, M. S.; Hammond, P. T.; Belcher, A. M. Virus-Templated Self-Assembled Single-Walled Carbon Nanotubes for Highly Efficient Electron Collection in Photovoltaic Devices. *Nature Nanotechnology* **2011,** *6* (6), 377–384.
118. Dembele, K. T.; Nechache, R.; Nikolova, L.; Vomiero, A.; Santato, C.; Licoccia, S.; Rosei, F. Effect of Multi-Walled Carbon Nanotubes on the Stability of Dye Sensitized Solar Cells. *Journal of Power Sources* **2013,** *233,* 93–97.
119. Dembele, K. T.; Selopal, G. S.; Soldano, C.; Nechache, R.; Rimada, J. C.; Concina, I.; Sberveglieri, G.; Rosei, F.; Vomiero, A. Hybrid Carbon Nanotubes-TiO_2 Photoanodes for High Efficiency Dye-Sensitized Solar Cells. *Journal of Physical Chemistry C* **2013,** *117* (28), 14510–14517.
120. Yu, J.; Fan, J.; Cheng, B. Dye-Sensitized Solar Cells Based on Anatase TiO_2 Hollow Spheres/Carbon Nanotube Composite Films. *Journal of Power Sources* **2011,** *196* (18), 7891–7898.
121. Zhang, X.; Liu, J.; Li, S.; Tan, X.; Zhang, J.; Yu, M.; Zhao, M. DNA Assembled Single-Walled Carbon Nanotube Nanocomposites for High Efficiency Dye-Sensitized Solar Cells. *Journal of Materials Chemistry A* **2013,** *1* (36), 11070–11077.
122. Shearer, C. J.; Cherevan, A.; Eder, D. Application and Future Challenges of Functional Nanocarbon Hybrids. *Advanced Materials* **2014,** *26* (15), 2295–2318.

123. Papageorgiou, N.; Maier, W. F.; Grätzel, M. An Iodine/Triiodide Reduction Electrocatalyst for Aqueous and Organic Media. *Journal of the Electrochemical Society* **1997,** *144* (3), 876–884.
124. Murakami, T. N.; Grätzel, M. Counter Electrodes for DSC: Application of Functional Materials as Catalysts. *Inorganica Chimica Acta* **2008,** *361* (3), 572–580.
125. Kim, J.; Kang, J.; Jeong, U.; Kim, H.; Lee, H. Catalytic, Conductive, and Transparent Platinum Nanofiber Webs for Fto-Free Dye-Sensitized Solar Cells. *ACS Applied Materials and Interfaces* **2013,** *5* (8), 3176–3181.
126. Wu, J.; Tang, Z.; Huang, Y.; Huang, M.; Yu, H.; Lin, J. A Dye-Sensitized Solar Cell Based on Platinum Nanotube Counter Electrode with Efficiency of 9.05%. *Journal of Power Sources* **2014,** *257,* 84–89.
127. Jeong, H.; Pak, Y.; Hwang, Y.; Song, H.; Lee, K. H.; Ko, H. C.; Jung, G. Y. Enhancing the Charge Transfer of the Counter Electrode in Dye-Sensitized Solar Cells Using Periodically Aligned Platinum Nanocups. *Small* **2012,** *8* (24), 3757–3761.
128. Hsieh, T. L.; Chen, H. W.; Kung, C. W.; Wang, C. C.; Vittal, R.; Ho, K. C. A Highly Efficient Dye-Sensitized Solar Cell with a Platinum Nanoflowers Counter Electrode. *Journal of Materials Chemistry* **2012,** *22* (12), 5550–5559.
129. Zhang, B.; Wang, D.; Hou, Y.; Yang, S.; Yang, X. H.; Zhong, J. H.; Liu, J.; Wang, H. F.; Hu, P.; Zhao, H. J.; Yang, H. G. Facet-Dependent Catalytic Activity of Platinum Nanocrystals for Triiodide Reduction in Dye-Sensitized Solar Cells. *Scientific Reports* **2013,** *3,* 1836.
130. Olsen, E.; Hagen, G.; Eric Lindquist, S. Dissolution of Platinum in Methoxy Propionitrile Containing LiI/I_2. *Solar Energy Materials and Solar Cells* **2000,** *63* (3), 267–273.
131. Kay, A.; Grätzel, M. Low Cost Photovoltaic Modules Based on Dye Sensitized Nanocrystalline Titanium Dioxide and Carbon Powder. *Solar Energy Materials and Solar Cells* **1996,** *44* (1), 99–117.
132. Thomas, S.; Deepak, T. G.; Anjusree, G. S.; Arun, T. A.; Nair, S. V.; Nair, A. S. A Review on Counter Electrode Materials in Dye-Sensitized Solar Cells. *Journal of Materials Chemistry A* **2014,** *2* (13), 4474–4490.
133. Poudel, P.; Qiao, Q. Carbon Nanostructure Counter Electrodes for Low Cost and Stable Dye-Sensitized Solar Cells. *Nano Energy* **2014,** *4,* 157–175.
134. Hwang, S.; Batmunkh, M.; Nine, M. J.; Chung, H.; Jeong, H. Dye-Sensitized Solar Cell Counter Electrodes Based on Carbon Nanotubes. *ChemPhysChem* **2015,** *16* (1), 53–65.
135. Costa, R. D.; Lodermeyer, F.; Casillas, R.; Guldi, D. M. Recent Advances in Multifunctional Nanocarbons Used in Dye-Sensitized Solar Cells. *Energy & Environmental Science* **2014,** 7 (4), 1281–1296.
136. Wu, M.; Ma, T. Recent Progress of Counter Electrode Catalysts in Dye-Sensitized Solar Cells. *Journal of Physical Chemistry C* **2014,** *118* (30), 16727–16742.
137. Yun, S.; Hagfeldt, A.; Ma, T. Pt-Free Counter Electrode for Dye-Sensitized Solar Cells with High Efficiency. *Advanced Materials* **2014,** *26,* 6210–6237.
138. Suzuki, K.; Yamaguchi, M.; Kumagai, M.; Yanagida, S. Application of Carbon Nanotubes to Counter Electrodes of Dye-Sensitized Solar Cells. *Chemistry Letters* **2003,** *32* (1), 28–29.
139. Lee, W. J.; Ramasamy, E.; Lee, D. Y.; Song, J. S. Efficient Dye-Sensitized Solar Cells with Catalytic Multiwall Carbon Nanotube Counter Electrodes. *ACS Applied Materials and Interfaces* **2009,** *1* (6), 1145–1149.
140. Ramasamy, E.; Lee, W. J.; Lee, D. Y.; Song, J. S. Spray Coated Multi-Wall Carbon Nanotube Counter Electrode for Tri-Iodide (I_3^-) Reduction in Dye-Sensitized Solar Cells. *Electrochemistry Communications* **2008,** *10* (7), 1087–1089.
141. Roy-Mayhew, J. D.; Bozym, D. J.; Punckt, C.; Aksay, I. A. Functionalized Graphene as a Catalytic Counter Electrode in Dye-Sensitized Solar Cells. *ACS Nano* **2010,** *4* (10), 6203–6211.
142. Wang, H.; Hu, Y. H. Graphene as a Counter Electrode Material for Dye-Sensitized Solar Cells. *Energy & Environmental Science* **2012,** *5* (8), 8182–8188.
143. Roy-Mayhew, J. D.; Aksay, I. A. Graphene Materials and Their Use in Dye-Sensitized Solar Cells. *Chemical Reviews* **2014,** *114* (12), 6323–6348.
144. Trancik, J. E.; Barton, S. C.; Hone, J. Transparent and Catalytic Carbon Nanotube Films. *Nano Letters* **2008,** *8* (4), 982–987.

145. Choi, H. J.; Shin, J. E.; Lee, G. W.; Park, N. G.; Kim, K.; Hong, S. C. Effect of Surface Modification of Multi-Walled Carbon Nanotubes on the Fabrication and Performance of Carbon Nanotube Based Counter Electrodes for Dye-Sensitized Solar Cells. *Current Applied Physics* **2010,** *10* (2 Suppl.), S165−S167.
146. Zhang, D. W.; Li, X. D.; Chen, S.; Tao, F.; Sun, Z.; Yin, X. J.; Huang, S. M. Fabrication of Double-Walled Carbon Nanotube Counter Electrodes for Dye-Sensitized Solar Sells. *Journal of Solid State Electrochemistry* **2010,** *14* (9), 1541−1546.
147. Mei, X.; Cho, S. J.; Fan, B.; Ouyang, J. High-Performance Dye-Sensitized Solar Cells with Gel-Coated Binder-Free Carbon Nanotube Films as Counter Electrode. *Nanotechnology* **2010,** *21,* 395202.
148. Lee, K. S.; Lee, W. J.; Park, N. G.; Kim, S. O.; Park, J. H. Transferred Vertically Aligned n-doped Carbon Nanotube Arrays: Use in Dye-Sensitized Solar Cells as Counter Electrodes. *Chemical Communications* **2011,** *47* (14), 4264−4266.
149. Dong, P.; Pint, C. L.; Hainey, M.; Mirri, F.; Zhan, Y.; Zhang, J.; Pasquali, M.; Hauge, R. H.; Verduzco, R.; Jiang, M.; Lin, H.; Lou, J. Vertically Aligned Single-Walled Carbon Nanotubes as Low-Cost and High Electrocatalytic Counter Electrode for Dye-Sensitized Solar Cells. *ACS Applied Materials and Interfaces* **2011,** *3* (8), 3157−3161.
150. Zhang, D.; Li, X.; Chen, S.; Sun, Z.; Yin, X. J.; Huang, S. Performance of Dye-Sensitized Solar Cells with Various Carbon Nanotube Counter Electrodes. *Microchimica Acta* **2011,** *174* (1), 73−79.
151. Zhu, G.; Pan, L.; Lu, T.; Liu, X.; Lv, T.; Xu, T.; Sun, Z. Electrophoretic Deposition of Carbon Nanotubes Films as Counter Electrodes of Dye-Sensitized Solar Cells. *Electrochimica Acta* **2011,** *56* (27), 10288−10291.
152. Munkhbayar, B.; Hwang, S.; Kim, J.; Bae, K.; Ji, M.; Chung, H.; Jeong, H. Photovoltaic Performance of Dye-Sensitized Solar Cells with Various Mwcnt Counter Electrode Structures Produced by Different Coating Methods. *Electrochimica Acta* **2012,** *80,* 100−107.
153. Anwar, H.; George, A. E.; Hill, I. G. Vertically-Aligned Carbon Nanotube Counter Electrodes for Dye-Sensitized Solar Cells. *Solar Energy* **2013,** *88,* 129−136.
154. Choi, H. J.; Gong, H. H.; Park, J. Y.; Hong, S. C. Characteristics of Dye-Sensitized Solar Cells with Surface-Modified Multi-Walled Carbon Nanotubes as Counter Electrodes. *Journal of Materials Science* **2013,** *48* (2), 906−912.
155. Qiu, L.; Wu, Q.; Yang, Z.; Sun, X.; Zhang, Y.; Peng, H. Freestanding Aligned Carbon Nanotube Array Grown on a Large-Area Single-Layered Graphene Sheet for Efficient Dye-Sensitized Solar Cell. *Small* **2014,** *11* (9−10), 1150−1155.
156. Sun, X.; Chen, T.; Yang, Z.; Peng, H. The Alignment of Carbon Nanotubes: An Effective Route to Extend Their Excellent Properties to Macroscopic Scale. *Accounts of Chemical Research* **2013,** *46* (2), 539−549.
157. Yang, Z.; Chen, T.; He, R.; Li, H.; Lin, H.; Li, L.; Zou, G.; Jia, Q.; Peng, H. A Novel Carbon Nanotube/Polymer Composite Film for Counter Electrodes of Dye-Sensitized Solar Cells. *Polymer Chemistry* **2013,** *4,* 1680−1684.
158. Guan, G.; Yang, Z.; Qiu, L.; Sun, X.; Zhang, Z.; Ren, J.; Peng, H. Oriented PEDOT: Pss on Aligned Carbon Nanotubes for Efficient Dye-Sensitized Solar Cells. *Journal of Materials Chemistry A* **2013,** *1* (42), 13268−13273.
159. Yang, Z.; Li, L.; Luo, Y.; He, R.; Qiu, L.; Lin, H.; Peng, H. An Integrated Device for Both Photoelectric Conversion and Energy Storage Based on Free-Standing and Aligned Carbon Nanotube Film. *Journal of Materials Chemistry A* **2013,** *1* (3), 954−958.
160. Yang, Z.; Chen, T.; He, R.; Guan, G.; Li, H.; Qiu, L.; Peng, H. Aligned Carbon Nanotube Sheets for the Electrodes of Organic Solar Cells. *Advanced Materials* **2011,** *23,* 5436−5439.
161. Huang, S.; Li, L.; Yang, Z.; Zhang, L.; Saiyin, H.; Chen, T.; Peng, H. A New and General Fabrication of an Aligned Carbon Nanotube/Polymer Film for Electrode Applications. *Advanced Materials* **2011,** *23* (40), 4707−4710.
162. Chen, T.; Wang, S.; Yang, Z.; Feng, Q.; Sun, X.; Li, L.; Wang, Z. S.; Peng, H. Flexible, Light-Weight, Ultrastrong, and Semiconductive Carbon Nanotube Fibers for a Highly Efficient Solar Cell. *Angewandte Chemie - International Edition* **2011,** *50* (8), 1815−1819.

163. Zhang, M.; Atkinson, K. R.; Baughman, R. H. Multifunctional Carbon Nanotube Yarns by Downsizing an Ancient Technology. *Science* **2004,** *306* (5700), 1358−1361.
164. Fang, X.; Yang, Z.; Qiu, L.; Sun, H.; Pan, S.; Deng, J.; Luo, Y.; Peng, H. Core-Sheath Carbon Nanostructured Fibers for Efficient Wire-Shaped Dye-Sensitized Solar Cells. *Advanced Materials* **2014,** *26* (11), 1694−1698.
165. Sun, H.; Li, H.; You, X.; Yang, Z.; Deng, J.; Qiu, L.; Peng, H. Quasi-Solid-State, Coaxial, Fiber-Shaped Dye-Sensitized Solar Cells. *Journal of Materials Chemistry A* **2014,** *2* (2), 345−349.
166. Sun, H.; Yang, Z.; Chen, X.; Qiu, L.; You, X.; Chen, P.; Peng, H. Photovoltaic Wire with High Efficiency Attached onto and Detached from a Substrate Using a Magnetic Field. *Angewandte Chemie International Edition* **2013,** *52,* 8276−8280.
167. Pan, S.; Yang, Z.; Li, H.; Qiu, L.; Sun, H.; Peng, H. Efficient Dye-Sensitized Photovoltaic Wires Based on an Organic Redox Electrolyte. *Journal of the American Chemical Society* **2013,** *135* (29), 10622−10625.
168. Chen, T.; Qiu, L.; Yang, Z.; Peng, H. Novel Solar Cells in a Wire Format. *Chemical Society Reviews* **2013,** *42* (12), 5031−5041.
169. Sun, H.; You, X.; Deng, J.; Chen, X.; Yang, Z.; Ren, J.; Peng, H. Novel Graphene/Carbon Nanotube Composite Fibers for Efficient Wire-Shaped Miniature Energy Devices. *Advanced Materials* **2014,** *26* (18), 2868−2873.
170. Yang, Z.; Deng, J.; Sun, X.; Li, H.; Peng, H. Stretchable, Wearable Dye-Sensitized Solar Cells. *Advanced Materials* **2014,** *26* (17), 2643−2647.
171. Chen, X.; Sun, H.; Yang, Z.; Guan, G.; Zhang, Z.; Qiu, L.; Peng, H. A Novel "Energy Fiber" by Coaxially Integrating Dye-Sensitized Solar Cell and Electrochemical Capacitor. *Journal of Materials Chemistry A* **2014,** *2* (6), 1897−1902.
172. Li, H.; Yang, Z.; Qiu, L.; Fang, X.; Sun, H.; Chen, P.; Pan, S.; Peng, H. Stable Wire-Shaped Dye-Sensitized Solar Cells Based on Eutectic Melts. *Journal of Materials Chemistry A* **2014,** *2* (11), 3841−3846.
173. Chen, T.; Qiu, L.; Yang, Z.; Cai, Z.; Ren, J.; Li, H.; Lin, H.; Sun, X.; Peng, H. An Integrated "Energy Wire" for Both Photoelectric Conversion and Energy Storage. *Angewandte Chemie − International Edition* **2012,** *51* (48), 11977−11980.
174. Kojima, A.; Teshima, K.; Shirai, Y.; Miyasaka, T. Organometal Halide Perovskites as Visible-Light Sensitizers for Photovoltaic Cells. *Journal of the American Chemical Society* **2009,** *131* (17), 6050−6051.
175. Batmunkh, M.; Shearer, C. J.; Biggs, M. J.; Shapter, J. G. Nanocarbons for Mesoscopic Perovskite Solar Cells. *Journal of Materials Chemistry A* **2015,** *3* (17), 9020−9031.
176. Cai, M.; Tiong, V. T., Hreid, T.; Bell, J.; Wang, H. An Efficient Hole Transport Material Composite Based on Poly(3-hexylthiophene) and Bamboo-Structured Carbon Nanotubes for High Performance Perovskite Solar Cells. *Journal of Materials Chemistry A* **2015,** *3* (6), 2784−2793.
177. Habisreutinger, S. N.; Leijtens, T.; Eperon, G. E.; Stranks, S. D.; Nicholas, R. J.; Snaith, H. J. Enhanced Hole Extraction in Perovskite Solar Cells through Carbon Nanotubes. *The Journal of Physical Chemistry Letters* **2014,** *5* (23), 4207−4212.
178. Li, Z.; Kulkarni, S. A.; Boix, P. P.; Shi, E.; Cao, A.; Fu, K.; Batabyal, S. K.; Zhang, J.; Xiong, Q.; Wong, L. H.; Mathews, N.; Mhaisalkar, S. G. Laminated Carbon Nanotube Networks for Metal Electrode-Free Efficient Perovskite Solar Cells. *ACS Nano* **2014,** *8* (7), 6797−6804.
179. Habisreutinger, S. N.; Leijtens, T.; Eperon, G. E.; Stranks, S. D.; Nicholas, R. J.; Snaith, H. J. Carbon Nanotube/Polymer Composites as a Highly Stable Hole Collection Layer in Perovskite Solar Cells. *Nano Letters* **2014,** *14* (10), 5561−5568.
180. Jeon, I.; Chiba, T.; Delacou, C.; Guo, Y.; Kaskela, A.; Reynaud, O.; Kauppinen, E. I.; Maruyama, S.; Matsuo, Y. Single-Walled Carbon Nanotube Film as Electrode in Indium-Free Planar Heterojunction Perovskite Solar Cells: Investigation of Electron-Blocking Layers and Dopants. *Nano Letters* **2015,** *15* (10), 6665−6671.
181. Mutolo, K. L.; Mayo, E. I.; Rand, B. P.; Forrest, S. R.; Thompson, M. E. Enhanced Open-Circuit Voltage in Subphthalocyanine/C60 Organic Photovoltaic Cells. *Journal of the American Chemical Society* **2006,** *128* (25), 8108−8109.
182. Wöhrle, D.; Meissner, D. Organic Solar Cells. *Advanced Materials* **1991,** *3* (3), 129−138.

183. Zhokhavets, U.; Erb, T.; Gobsch, G.; Al-Ibrahim, M.; Ambacher, O. Relation Between Absorption and Crystallinity of Poly(3-hexylthiophene)/Fullerene Films for Plastic Solar Cells. *Chemical Physics Letters* **2006,** *418* (4–6), 347–350.
184. Katz, E. A.; Faiman, D.; Tuladhar, S. M.; Kroon, J. M.; Wienk, M. M.; Fromherz, T.; Padinger, F.; Brabec, C. J.; Sariciftci, N. S. Temperature Dependence for the Photovoltaic Device Parameters of Polymer-Fullerene Solar Cells under Operating Conditions. *Journal of Applied Physics* **2001,** *90* (10), 5343–5350.
185. Wright, M.; Uddin, A. Organic—Inorganic Hybrid Solar Cells: A Comparative Review. *Solar Energy Materials and Solar Cells* **2012,** *107,* 87–111.
186. Yan, J.; Uddin, M. J.; Dickens, T. J.; Okoli, O. I. Carbon Nanotubes (CNTs) Enrich the Solar Cells. *Solar Energy* **2013,** *96,* 239–252.
187. Candelaria, S. L.; Shao, Y.; Zhou, W.; Li, X.; Xiao, J.; Zhang, J.-G.; Wang, Y.; Liu, J.; Li, J.; Cao, G. Nanostructured Carbon for Energy Storage and Conversion. *Nano Energy* **2012,** *1* (2), 195–220.
188. Wang, C.; Guo, Z.-X.; Fu, S.; Wu, W.; Zhu, D. Polymers Containing Fullerene or Carbon Nanotube Structures. *Progress in Polymer Science* **2004,** *29* (11), 1079–1141.
189. Li, G.; Zhu, R.; Yang, Y. Polymer Solar Cells. *Nature Photonics* **2012,** *6* (3), 153–161.
190. Hatton, R. A.; Miller, A. J.; Silva, S. R. P. Carbon Nanotubes: A Multi-Functional Material for Organic Optoelectronics. *Journal of Materials Chemistry* **2008,** *18* (11), 1183–1192.
191. Capasso, A.; Salamandra, L.; Carlo, A. D.; Bell, J. M.; Motta, N. Low-Temperature Synthesis of Carbon Nanotubes on Indium Tin Oxide Electrodes for Organic Solar Cells. *Beilstein Journal of Nanotechnology* **2012,** *3,* 524–532.
192. Ren, S.; Bernardi, M.; Lunt, R. R.; Bulovic, V.; Grossman, J. C.; Gradečak, S. Toward Efficient Carbon Nanotube/P3HT Solar Cells: Active Layer Morphology, Electrical, and Optical Properties. *Nano Letters* **2011,** *11* (12), 5316–5321.
193. Schuettfort, T.; Snaith, H. J.; Nish, A.; Nicholas, R. J. Synthesis and Spectroscopic Characterization of Solution Processable Highly Ordered Polythiophene–Carbon Nanotube Nanohybrid Structures. *Nanotechnology* **2010,** *21* (2), 025201.
194. Kymakis, E.; Amaratunga, G. A. J. Single-Wall Carbon Nanotube/Conjugated Polymer Photovoltaic Devices. *Applied Physics Letters* **2002,** *80* (1), 112–114.
195. Miller, A. J.; Hatton, R. A.; Silva, S. R. P. Water-Soluble Multiwall-Carbon-Nanotube-Polythiophene Composite for Bilayer Photovoltaics. *Applied Physics Letters* **2006,** *89* (12), 123115.
196. Landi, B. J.; Raffaelle, R. P.; Castro, S. L.; Bailey, S. G. Single-Wall Carbon Nanotube-Polymer Solar Cells. *Progress in Photovoltaics: Research and Applications* **2005,** *13* (2), 165–172.
197. Barnes, T. M.; Bergeson, J. D.; Tenent, R. C.; Larsen, B. A.; Teeter, G.; Jones, K. M.; Blackburn, J. L.; van de Lagemaat, J. Carbon Nanotube Network Electrodes Enabling Efficient Organic Solar Cells without a Hole Transport Layer. *Applied Physics Letters* **2010,** *96* (24), 243309.
198. Patyk, R. L.; Lomba, B. S.; Nogueira, A. F.; Furtado, C. A.; Santos, A. P.; Mello, R. M. Q.; Micaroni, L.; Hummelgen, I. A. Carbon Nanotube-Polybithiophene Photovolyaic Devices with High Open-Circuit Voltage. *Physica Status Solidi (RRL)* **2007,** *1,* R43–R45.
199. Rahman, G. M. A.; Guldi, D. M.; Cagnoli, R.; Mucci, A.; Schenetti, L.; Vaccari, L.; Prato, M. Combining Single Wall Carbon Nanotubes and Photoactive Polymers for Photoconversion. *Journal of the American Chemical Society* **2005,** *127* (28), 10051–10057.
200. Rowell, M. W.; Topinka, M. A.; McGehee, M. D.; Prall, H.-J.; Dennler, G.; Sariciftci, N. S.; Hu, L.; Gruner, G. Organic Solar Cells with Carbon Nanotube Network Electrodes. *Applied Physics Letters* **2006,** *88* (23), 233506.
201. Feng, Y.; Ju, X.; Feng, W.; Zhang, H.; Cheng, Y.; Liu, J.; Fujii, A.; Ozaki, M.; Yoshino, K. Organic Solar Cells Using Few-Walled Carbon Nanotubes Electrode Controlled by the Balance between Sheet Resistance and the Transparency. *Applied Physics Letters* **2009,** *94* (12), 123302.
202. Unalan, H. E.; Hiralal, P.; Kuo, D.; Parekh, B.; Amaratunga, G.; Chhowalla, M. Flexible Organic Photovoltaics from Zinc Oxide Nanowires Grown on Transparent and Conducting Single Walled Carbon Nanotube Thin Films. *Journal of Materials Chemistry* **2008,** *18* (48), 5909–5912.
203. Miller, A. J.; Hatton, R. A.; Chen, G. Y.; Silva, S. R. P. Carbon Nanotubes Grown on In_2O_3:Sn Glass as Large Area Electrodes for Organic Photovoltaics. *Applied Physics Letters* **2007,** *90* (2), 023105.

204. Berson, S.; de Bettignies, R.; Bailly, S.; Guillerez, S.; Jousselme, B. Elaboration of P3HT/CNT/PCBM Composites for Organic Photovoltaic Cells. *Advanced Functional Materials* **2007,** *17* (16), 3363–3370.
205. Li, C.; Mitra, S. Processing of Fullerene-Single Wall Carbon Nanotube Complex for Bulk Heterojunction Photovoltaic Cells. *Applied Physics Letters* **2007,** *91* (25), 253112.
206. Lanzi, M.; Paganin, L.; Caretti, D. New Photoactive Oligo- and Poly-Alkylthiophenes. *Polymer* **2008,** *49* (23), 4942–4948.
207. Kymakis, E.; Kornilios, N.; Koudoumas, E. Carbon Nanotube Doping of P3HT: PCBM Photovoltaic Devices. *Journal of Physics D: Applied Physics* **2008,** *41* (16), 165110.
208. Picard, L.; Lincker, F.; Kervella, Y.; Zagorska, M.; DeBettignies, R.; Peigney, A.; Flahaut, E.; Louarn, G.; Lefrant, S.; Demadrille, R.; Pron, A. Composites of Double-Walled Carbon Nanotubes with Bis-Quaterthiophene-Fluorenone Conjugated Oligomer: Spectroelectrochemical and Photovoltaic Properties. *The Journal of Physical Chemistry C* **2009,** *113* (40), 17347–17354.
209. Hatton, R. A.; Blanchard, N. P.; Tan, L. W.; Latini, G.; Cacialli, F.; Silva, S. R. P. Oxidised Carbon Nanotubes as Solution Processable, High Work Function Hole-Extraction Layers for Organic Solar Cells. *Organic Electronics* **2009,** *10* (3), 388–395.
210. Adikaari, A. A. D. T.; Dissanayake, D. M. M. N.; Silva, S. R. P. Organic-Inorganic Solar Cells: Recent Developments and Outlook. *IEEE Journal of Selected Topics in Quantum Electronics* **2010,** *16* (6), 1595–1606.
211. Capasso, A.; Salamandra, L.; Chou, A.; Di Carlo, A.; Motta, N. Multi-Wall Carbon Nanotube Coating of Fluorine-Doped Tin Oxide as an Electrode Surface Modifier for Polymer Solar Cells. *Solar Energy Materials and Solar Cells* **2014,** *122* (0), 297–302.
212. Yan, J.; Ni, T.; Zou, F.; Zhang, L.; Yang, D.; Yang, S.; Zou, B. Towards Optimization of Functionalized Single-Walled Carbon Nanotubes Adhering with Poly(3-hexylthiophene) for Highly Efficient Polymer Solar Cells. *Diamond and Related Materials* **2014,** *41,* 79–83.
213. Lee, I.; Lee, S.; Kim, H.; Lee, H.; Kim1, Y. Polymer Solar Cells with Polymer/Carbon Nanotube Composite Hole-Collecting Buffer Layers. *The Open Physical Chemistry Journal* **2010,** *4,* 1–3.
214. Lee, R.-H.; Lee, L.-Y.; Huang, J.-L.; Huang, C.-C.; Hwang, J.-C. Conjugated Polymer-Functionalized Carbon Nanotubes Enhance the Photovoltaic Properties of Polymer Solar Cells. *Colloid and Polymer Science* **2011,** *289* (15–16), 1633–1641.
215. Kymakis, E.; Klapsis, G.; Koudoumas, E.; Stratakis, E.; Kornilios, N.; Vidakis, N.; Franghiadakis, Y. Carbon Nanotube/PEDOT: PSS Ectrodes for Organic Photovoltaics. *The European Physical Journal – Applied Physics* **2006,** *36* (03), 257–259.
216. Ulbricht, R.; Lee, S. B.; Jiang, X.; Inoue, K.; Zhang, M.; Fang, S.; Baughman, R. H.; Zakhidov, A. A. Transparent Carbon Nanotube Sheets as 3-D Charge Collectors in Organic Solar Cells. *Solar Energy Materials and Solar Cells* **2007,** *91* (5), 416–419.
217. van de Lagemaat, J.; Barnes, T. M.; Rumbles, G.; Shaheen, S. E.; Coutts, T. J.; Weeks, C.; Levitsky, I.; Peltola, J.; Glatkowski, P. Organic Solar Cells with Carbon Nanotubes Replacing In_2O_3:Sn as the Transparent Electrode. *Applied Physics Letters* **2006,** *88* (23), 233503.
218. Ago, H.; Petritsch, K.; Shaffer, M. S. P.; Windle, A. H.; Friend, R. H. Composites of Carbon Nanotubes and Conjugated Polymers for Photovoltaic Devices. *Advanced Materials* **1999,** *11* (15), 1281–1285.
219. Pasquier, A. D.; Unalan, H. E.; Kanwal, A.; Miller, S.; Chhowalla, M. Conducting and Transparent Single-Wall Carbon Nanotube Electrodes for Polymer-Fullerene Solar Cells. *Applied Physics Letters* **2005,** *87* (20), 203511.
220. Kymakis, E.; Stratakis, E.; Koudoumas, E. Integration of Carbon Nanotubes as Hole Transport Electrode in Polymer/Fullerene Bulk Heterojunction Solar Cells. *Thin Solid Films* **2007,** *515* (24), 8598–8600.
221. Khatri, I.; Adhikari, S.; Aryal, H. R.; Soga, T.; Jimbo, T.; Umeno, M. Improving Photovoltaic Properties by Incorporating Both Single Walled Carbon Nanotubes and Functionalized Multiwalled Carbon Nanotubes. *Applied Physics Letters* **2009,** *94* (9), 093509.
222. Miller, A. J.; Hatton, R. A.; Silva, S. R. P. Interpenetrating Multiwall Carbon Nanotube Electrodes for Organic Solar Cells. *Applied Physics Letters* **2006,** *89* (13), 133117.

223. Jackson, R.; Domercq, B.; Jain, R.; Kippelen, B.; Graham, S. Stability of Doped Transparent Carbon Nanotube Electrodes. *Advanced Functional Materials* **2008,** *18* (17), 2548–2554.
224. Parekh, B. B.; Fanchini, G.; Eda, G.; Chhowalla, M. Improved Conductivity of Transparent Single-Wall Carbon Nanotube Thin Films Via Stable Postdeposition Functionalization. *Applied Physics Letters* **2007,** *90,* 121913.
225. Tanaka, S.; Mielczarek, K.; Ovalle-Robles, R.; Wang, B.; Hsu, D.; Zakhidov, A. A. Monolithic Parallel Tandem Organic Photovoltaic Cell with Transparent Carbon Nanotube Interlayer. *Applied Physics Letters* **2009,** *94* (11), 113506.
226. Pradhan, B.; Batabyal, S. K.; Pal, A. J. Functionalized Carbon Nanotubes in Donor/Acceptor-type Photovoltaic Devices. *Applied Physics Letters* **2006,** *88* (9), 093106.
227. Gregg, B. A. Excitonic Solar Cells. *The Journal of Physical Chemistry B* **2003,** *107* (20), 4688–4698.
228. Kymakis, E.; Alexandrou, I.; Amaratunga, G. A. J. High Open-Circuit Voltage Photovoltaic Devices from Carbon-Nanotube-Polymer Composites. *Journal of Applied Physics* **2003,** *93* (3), 1764–1768.
229. Mao, J.; Liu, Q.; Lv, X.; Liu, Z.; Huang, Y.; Ma, Y.; Chen, Y.; Yin, S. A Water-Soluble Hybrid Material of Single-Walled Carbon Nanotubes with an Amphiphilic Poly(Phenyleneethynylene): Preparation, Characterization, and Photovoltaic Properties. *Journal of Nanoscience and Nanotechnology* **2007,** *7* (8), 2709–2718.
230. Liu, Q.; Mao, J.; Liu, Z.; Zhang, N.; Wang, Y.; Yang, L.; Yin, S.; Chen, Y. A Photovoltaic Device Based on a poly(phenyleneethynylene)/SWNT Composite Active Layer. *Nanotechnology* **2008,** *19* (11), 115601.
231. Yun, D.; Feng, W.; Wu, H.; Li, B.; Liu, X.; Yi, W.; Qiang, J.; Gao, S.; Yan, S. Controllable Functionalization of Single-Wall Carbon Nanotubes by in Situ Polymerization Method for Organic Photovoltaic Devices. *Synthetic Metals* **2008,** *158* (21–24), 977–983.
232. Somani, S. P.; Somani, P. R.; Flahaut, E.; Kalita, G.; Umeno, M. Double-Walled Carbon Nanotubes-Incorporated Donor–Acceptor-Type Organic Photovoltaic Devices Using Poly(3-octylthiophene) and C_{60}. *Japanese Journal of Applied Physics* **2008,** *47* (2S), 1219.
233. Kymakis, E.; Amaratunga, G. A. J. Photovoltaic Cells Based on Dye-Sensitisation of Single-Wall Carbon Nanotubes in a Polymer Matrix. *Solar Energy Materials and Solar Cells* **2003,** *80* (4), 465–472.
234. Bhattacharyya, S.; Kymakis, E.; Amaratunga, G. A. J. Photovoltaic Properties of Dye Functionalized Single-Wall Carbon Nanotube/Conjugated Polymer Devices. *Chemistry of Materials* **2004,** *16* (23), 4819–4823.
235. Tune, D. D.; Shapter, J. G. The Potential Sunlight Harvesting Efficiency of Carbon Nanotube Solar Cells. *Energy & Environmental Science* **2013,** *6* (9), 2572–2577.
236. Jia, Y.; Wei, J.; Wang, K.; Cao, A.; Shu, Q.; Gui, X.; Zhu, Y.; Zhuang, D.; Zhang, G.; Ma, B.; Wang, L.; Liu, W.; Wang, Z.; Luo, J.; Wu, D. Nanotube-Silicon Heterojunction Solar Cells. *Advanced Materials* **2008,** *20* (23), 4594–4598.
237. Wei, J.; Jia, Y.; Shu, Q.; Gu, Z.; Wang, K.; Zhuang, D.; Zhang, G.; Wang, Z.; Luo, J.; Cao, A.; Wu, D. Double-Walled Carbon Nanotube Solar Cells. *Nano Letters* **2007,** *7* (8), 2317–2321.
238. Tune, D. D.; Flavel, B. S.; Krupke, R.; Shapter, J. G. Carbon Nanotube-Silicon Solar Cells. *Advanced Energy Materials* **2012,** *2* (9), 1043–1055.
239. Jia, Y.; Cao, A.; Bai, X.; Li, Z.; Zhang, L.; Guo, N.; Wei, J.; Wang, K.; Zhu, H.; Wu, D.; Ajayan, P. M. Achieving High Efficiency Silicon-Carbon Nanotube Heterojunction Solar Cells by Acid Doping. *Nano Letters* **2011,** *11* (5), 1901–1905.
240. Arena, A.; Donato, N.; Saitta, G.; Galvagno, S.; Milone, C.; Pistone, A. Photovoltaic Properties of Multi-Walled Carbon Nanotubes Deposited on n-doped Silicon. *Microelectronics Journal* **2008,** *39* (12), 1659–1662.
241. Li, Z.; Kunets, V. P.; Saini, V.; Xu, Y.; Dervishi, E.; Salamo, G. J.; Biris, A. R.; Biris, A. S. $SOCl_2$ Enhanced Photovoltaic Conversion of Single Wall Carbon Nanotube/n-silicon Heterojunctions. *Applied Physics Letters* **2008,** *93* (24), 3117.
242. Li, Z.; Kunets, V. P.; Saini, V.; Xu, Y.; Dervishi, E.; Salamo, G. J.; Biris, A. R.; Biris, A. S. Light-Harvesting Using High Density p-type Single Wall Carbon Nanotube/n-type Silicon Heterojunctions. *ACS Nano* **2009,** *3* (6), 1407–1414.

243. Jia, Y.; Li, P.; Wei, J.; Cao, A.; Wang, K.; Li, C.; Zhuang, D.; Zhu, H.; Wu, D. Carbon Nanotube Films by Filtration for Nanotube-Silicon Heterojunction Solar Cells. *Materials Research Bulletin* **2010,** *45* (10), 1401–1405.
244. Ong, P. L.; Euler, W. B.; Levitsky, I. A. Hybrid Solar Cells Based on Single-Walled Carbon Nanotubes/Si Heterojunctions. *Nanotechnology* **2010,** *21,* 105203.
245. Wadhwa, P.; Liu, B.; McCarthy, M. A.; Wu, Z.; Rinzler, A. G. Electronic Junction Control in a Nanotube-Semiconductor Schottky Junction Solar Cell. *Nano Letters* **2010,** *10* (12), 5001–5005.
246. Jia, Y.; Li, P.; Gui, X.; Wei, J.; Wang, K.; Zhu, H.; Wu, D.; Zhang, L.; Cao, A.; Xu, Y. Encapsulated Carbon Nanotube-Oxide-Silicon Solar Cells with Stable 10% Efficiency. *Applied Physics Letters* **2011,** *98* (13), 133115.
247. Wadhwa, P.; Seol, G.; Petterson, M. K.; Guo, J.; Rinzler, A. G. Electrolyte-Induced Inversion Layer Schottky Junction Solar Cells. *Nano Letters* **2011,** *11* (6), 2419–2423.
248. Jia, Y.; Cao, A.; Kang, F.; Li, P.; Gui, X.; Zhang, L.; Shi, E.; Wei, J.; Wang, K.; Zhu, H.; Wu, D. Strong and Reversible Modulation of Carbon Nanotube-Silicon Heterojunction Solar Cells by an Interfacial Oxide Layer. *Physical Chemistry Chemical Physics* **2012,** *14* (23), 8391–8396.
249. Shi, E.; Zhang, L.; Li, Z.; Li, P.; Shang, Y.; Jia, Y.; Wei, J.; Wang, K.; Zhu, H.; Wu, D.; Zhang, S.; Cao, A. TiO_2-Coated Carbon Nanotube-Silicon Solar Cells with Efficiency of 15%. *Scientific Reports* **2012,** *2,* 884.
250. Li, X.; Jung, Y.; Sakimoto, K.; Goh, T.-H.; Reed, M. A.; Taylor, A. D. Improved Efficiency of Smooth and Aligned Single Walled Carbon Nanotube/Silicon Hybrid Solar Cells. *Energy & Environmental Science* **2013,** *6* (3), 879–887.
251. Tune, D. D.; Flavel, B. S.; Quinton, J. S.; Ellis, A. V.; Shapter, J. G. Single-Walled Carbon Nanotube/Polyaniline/n-Silicon Solar Cells: Fabrication, Characterization, and Performance Measurements. *ChemSusChem* **2013,** *6* (2), 320–327.
252. Li, R.; Di, J.; Yong, Z.; Sun, B.; Li, Q. Polymethylmethacrylate Coating on Aligned Carbon Nanotube-Silicon Solar Cells for Performance Improvement. *Journal of Materials Chemistry A* **2014,** *2* (12), 4140–4143.
253. Tune, D. D.; Blanch, A. J.; Krupke, R.; Flavel, B. S.; Shapter, J. G. Nanotube Film Metallicity and Its Effect on the Performance of Carbon Nanotube–Silicon Solar Cells. *Physica Status Solidi (A)* **2014,** *211* (7), 1479–1487.
254. Wang, F.; Kozawa, D.; Miyauchi, Y.; Hiraoka, K.; Mouri, S.; Ohno, Y.; Matsuda, K. Considerably Improved Photovoltaic Performance of Carbon Nanotube-Based Solar Cells Using Metal Oxide Layers. *Nature Communications* **2015,** *6,* 6305.
255. Petterson, M. K.; Lemaitre, M. G.; Shen, Y.; Wadhwa, P.; Hou, J.; Vasilyeva, S. V.; Kravchenko, I. I.; Rinzler, A. G. On Field-Effect Photovoltaics: Gate Enhancement of the Power Conversion Efficiency in a Nanotube/Silicon-Nanowire Solar Cell. *ACS Applied Materials & Interfaces* **2015,** *7* (38), 21182–21187.
256. Tune, D. D.; Blanch, A. J.; Shearer, C. J.; Moore, K. E.; Pfohl, M.; Shapter, J. G.; Flavel, B. S. Aligned Carbon Nanotube Thin Films from Liquid Crystal Polyelectrolyte Inks. *ACS Applied Materials & Interfaces* **2015,** *7* (46), 25857–25864.
257. Castrucci, P. Carbon Nanotube/Silicon Hybrid Heterojunctions for Photovoltaic Devices. *Advances in Nano Research* **2014,** *2* (1), 23–56.
258. Yu, L.; Tune, D. D.; Shearer, C. J.; Shapter, J. G. Implementation of Antireflection Layers for Improved Efficiency of Carbon Nanotube–Silicon Heterojunction Solar Cells. *Solar Energy* **2015,** *118,* 592–599.
259. Castrucci, P.; Scilletta, C.; Del Gobbo, S.; Scarselli, M.; Camilli, L.; Simeoni, M.; Delley, B.; Continenza, A.; De Crescenzi, M. Light Harvesting with Multiwall Carbon Nanotube/Silicon Heterojunctions. *Nanotechnology* **2011,** *22,* 115701.
260. Tune, D.; Shapter, J. Effect of Nanotube Film Thickness on the Performance of Nanotube-Silicon Hybrid Solar Cells. *Nanomaterials* **2013,** *3* (4), 655–673.
261. Tune, D. D.; Hennrich, F.; Dehm, S.; Klein, M. F. G.; Glaser, K.; Colsmann, A.; Shapter, J. G.; Lemmer, U.; Kappes, M. M.; Krupke, R.; Flavel, B. S. The Role of Nanotubes in Carbon Nanotube–Silicon Solar Cells. *Advanced Energy Materials* **2013,** *3* (8), 1091–1097.

262. Tzolov, M. B.; Kuo, T. F.; Straus, D. A.; Yin, A.; Xu, J. Carbon Nanotube-Silicon Heterojunction Arrays and Infrared Photocurrent Responses. *Journal of Physical Chemistry C* **2007,** *111* (15), 5800–5804.
263. Schroder, D. K.; Thomas, R. N.; Swartz, J. C. Free Carrier Absorption in Silicon. *IEEE Journal of Solid-State Circuits* **1978,** *13* (1), 180–187.
264. Spitzer, W.; Fan, H. Y. Infrared Absorption in n-type Silicon. *Physical Review* **1957,** *108* (2), 268–271.
265. Tantang, H.; Ong, J. Y.; Loh, C. L.; Dong, X.; Chen, P.; Chen, Y.; Hu, X.; Tan, L. P.; Li, L. J. Using Oxidation to Increase the Electrical Conductivity of Carbon Nanotube Electrodes. *Carbon* **2009,** *47* (7), 1867–1870.
266. Pintossi, C.; Pagliara, S.; Drera, G.; De Nicola, F.; Castrucci, P.; De Crescenzi, M.; Crivellari, M.; Boscardin, M.; Sangaletti, L. Steering the Efficiency of Carbon Nanotube–Silicon Photovoltaic Cells by Acid Vapor Exposure: A Real-Time Spectroscopic Tracking. *ACS Applied Materials & Interfaces* **2015,** 7 (18), 9436–9444.
267. Blackburn, J. L.; Barnes, T. M.; Beard, M. C.; Kim, Y. H.; Tenent, R. C.; McDonald, T. J.; To, B.; Coutts, T. J.; Heben, M. J. Transparent Conductive Single-Walled Carbon Nanotube Networks with Precisely Tunable Ratios of Semiconducting and Metallic Nanotubes. *ACS Nano* **2008,** *2* (6), 1266–1274.
268. Barnes, T. M.; Blackburn, J. L.; van de Lagemaat, J.; Coutts, T. J.; Heben, M. J. Reversibility, Dopant Desorption, and Tunneling in the Temperature-Dependent Conductivity of Type-Separated, Conductive Carbon Nanotube Networks. *ACS Nano* **2008,** *2* (9), 1968–1976.
269. Kim, S. M.; Kim, K. K.; Jo, Y. W.; Park, M. H.; Chae, S. J.; Duong, D. L.; Yang, C. W.; Kong, J.; Lee, Y. H. Role of Anions in the $AuCl_3$-Doping of Carbon Nanotubes. *ACS Nano* **2011,** *5* (2), 1236–1242.
270. Jung, Y.; Li, X.; Rajan, N. K.; Taylor, A. D.; Reed, M. A. Record High Efficiency Single-Walled Carbon Nanotube/Silicon p-n Junction Solar Cells. *Nano Letters* **2013,** *13* (1), 95–99.
271. Harris, J. M.; Semler, M. R.; May, S.; Fagan, J. A.; Hobbie, E. K. Nature of Record Efficiency Fluid-Processed Nanotube–Silicon Heterojunctions. *The Journal of Physical Chemistry C* **2015,** *119* (19), 10295–10303.
272. Park, S.; Pitner, G.; Giri, G.; Koo, J. H.; Park, J.; Kim, K.; Wang, H.; Sinclair, R.; Wong, H. S. P.; Bao, Z. Large-aArea Assembly of Densely Aligned Single-Walled Carbon Nanotubes Using Solution Shearing and Their Application to Field-Effect Transistors. *Advanced Materials* **2015,** *27* (16), 2656–2662.
273. Bradford, P. D.; Wang, X.; Zhao, H.; Maria, J.-P.; Jia, Q.; Zhu, Y. T. A Novel Approach to Fabricate High Volume Fraction Nanocomposites with Long Aligned Carbon Nanotubes. *Composites Science and Technology* **2010,** *70* (13), 1980–1985.
274. Ding, W.; Pengcheng, S.; Changhong, L.; Wei, W.; Shoushan, F. Highly Oriented Carbon Nanotube Papers Made of Aligned Carbon Nanotubes. *Nanotechnology* **2008,** *19* (7), 075609.
275. deHeer, W. A.; Bacsa, W. S.; Châtelain, A.; Gerfin, T.; Humphrey-Baker, R.; Forro, L.; Ugarte, D. Aligned Carbon Nanotube Films: Production and Optical and Electronic Properties. *Science* **1995,** *268* (5212), 845–847.
276. Ren, L.; Pint, C. L.; Booshehri, L. G.; Rice, W. D.; Wang, X.; Hilton, D. J.; Takeya, K.; Kawayama, I.; Tonouchi, M.; Hauge, R. H.; Kono, J. Carbon Nanotube Terahertz Polarizer. *Nano Letters* **2009,** *9* (7), 2610–2613.
277. Wang, K.; Luo, S.; Wu, Y.; He, X.; Zhao, F.; Wang, J.; Jiang, K.; Fan, S. Super-Aligned Carbon Nanotube Films as Current Collectors for Lightweight and Flexible Lithium Ion Batteries. *Advanced Functional Materials* **2013,** *23* (7), 846–853.
278. Shimoda, H.; Oh, S. J.; Geng, H. Z.; Walker, R. J.; Zhang, X. B.; McNeil, L. E.; Zhou, O. Self-Assembly of Carbon Nanotubes. *Advanced Materials* **2002,** *14* (12), 899–901.
279. Ko, H.; Tsukruk, V. V. Liquid-Crystalline Processing of Highly Oriented Carbon Nanotube Arrays for Thin-Film Transistors. *Nano Letters* **2006,** *6* (7), 1443–1448.
280. Engel, M.; Small, J. P.; Steiner, M.; Freitag, M.; Green, A. A.; Hersam, M. C.; Avouris, P. Thin Film Nanotube Transistors Based on Self-Assembled, Aligned, Semiconducting Carbon Nanotube Arrays. *ACS Nano* **2008,** *2* (12), 2445–2452.

281. Joo, Y.; Brady, G. J.; Arnold, M. S.; Gopalan, P. Dose-Controlled, Floating Evaporative Self-Assembly and Alignment of Semiconducting Carbon Nanotubes from Organic Solvents. *Langmuir* **2014,** *30* (12), 3460–3466.
282. Olivier, J.-H.; Deria, P.; Park, J.; Kumbhar, A.; Andrian-Albescu, M.; Therien, M. J. Ionic Self-Assembly Provides Dense Arrays of Individualized, Aligned Single-Walled Carbon Nanotubes. *Angewandte Chemie* **2013,** *125* (49), 13318–13323.
283. Chattopadhyay, S.; Huang, Y. F.; Jen, Y. J.; Ganguly, A.; Chen, K. H.; Chen, L. C. Anti-Reflecting and Photonic Nanostructures. *Materials Science and Engineering: R: Reports* **2010,** *69* (1–3), 1–35.
284. Li, X.; Yu, X.; Han, Y. Polymer Thin Films for Antireflection Coatings. *Journal of Materials Chemistry C* **2013,** *1* (12), 2266–2285.
285. Zhong, S.; Liu, B.; Xia, Y.; Liu, J.; Liu, J.; Shen, Z.; Xu, Z.; Li, C. Influence of the Texturing Structure on the Properties of Black Silicon Solar Cell. *Solar Energy Materials and Solar Cells* **2013,** *108,* 200–204.
286. Xia, Y.; Liu, B.; Liu, J.; Shen, Z.; Li, C. A Novel Method to Produce Black Silicon for Solar Cells. *Solar Energy* **2011,** *85* (7), 1574–1578.
287. Liu, Y.; Das, A.; Lin, Z.; Cooper, I. B.; Rohatgi, A.; Wong, C. P. Hierarchical Robust Textured Structures for Large Scale Self-Cleaning Black Silicon Solar Cells. *Nano Energy* **2014,** *3,* 127–133.
288. Wang, W.; Schiff, E. A. Polyaniline on Crystalline Silicon Heterojunction Solar Cells. *Applied Physics Letters* **2007,** *91* (13), 133504.
289. Cárdenas, J. R.; de Vasconcelos, E. A.; de Azevedo, W. M.; da Silva, E. F.; Pepe, I.; da Silva, A. F.; Ribeiro, S. S.; Silva, K. A. A Conducting Polymer–Silicon Heterojunction as a New Ultraviolet Photodetector. *Applied Surface Science* **2008,** *255* (3), 688–690.
290. Pettit, R. B.; Brinker, C. J.; Ashley, C. S. Sol-Gel Double-Layer Antireflection Coatings for Silicon Solar Cells. *Solar Cells* **1985,** *15* (3), 267–278.
291. Hersam, M. C. Progress Towards Monodisperse Single-Walled Carbon Nanotubes. *Nature Nanotechnology* **2008,** *3* (7), 387–394.
292. Liu, J.; Hersam, M. C. Recent Developments in Carbon Nanotube Sorting and Selective Growth. *MRS Bulletin* **2010,** *35* (4), 315–321.
293. Liu, H.; Nishide, D.; Tanaka, T.; Kataura, H. Large-Scale Single-Chirality Separation of Single-Wall Carbon Nanotubes by Simple Gel Chromatography. *Nature Communications* **2011,** *2* (1), 309.
294. Tanaka, T.; Liu, H.; Fujii, S.; Kataura, H. From Metal/Semiconductor Separation to Single-Chirality Separation of Single-Wall Carbon Nanotubes Using Gel. *Physica Status Solidi – Rapid Research Letters* **2011,** *5* (9), 301–306.
295. Chen, Y.; Zhang, Y.; Hu, Y.; Kang, L.; Zhang, S.; Xie, H.; Liu, D.; Zhao, Q.; Li, Q.; Zhang, J. State of the Art of Single-Walled Carbon Nanotube Synthesis on Surfaces. *Advanced Materials* **2014,** *26,* 5898–5922.
296. Krupke, R.; Linden, S.; Rapp, M.; Hennrich, F. Thin Films of Metallic Carbon Nanotubes Prepared by Dielectrophoresis. *Advanced Materials* **2006,** *18* (11), 1468–1470.
297. Arnold, M. S.; Green, A. A.; Hulvat, J. F.; Stupp, S. I.; Hersam, M. C. Sorting Carbon Nanotubes by Electronic Structure Using Density Differentiation. *Nature Nanotechnology* **2006,** *1* (1), 60–65.
298. Hwang, J. Y.; Nish, A.; Doig, J.; Douven, S.; Chen, C. W.; Chen, L. C.; Nicholas, R. J. Polymer Structure and Solvent Effects on the Selective Dispersion of Single-Walled Carbon Nanotubes. *Journal of the American Chemical Society* **2008,** *130* (11), 3543–3553.
299. Hennrich, F.; Lebedkin, S.; Kappes, M. M. Improving Separation Techniques for Single-Walled Carbon Nanotubes: Towards Monodisperse Samples. *Physica Status Solidi (B)* **2008,** *245* (10), 1951–1953.
300. Moshammer, K.; Hennrich, F.; Kappes, M. M. Selective Suspension in Aqueous Sodium Dodecyl Sulfate According to Electronic Structure Type Allows Simple Separation of Metallic from Semiconducting Single-Walled Carbon Nanotubes. *Nano Research* **2009,** *2* (8), 599–606.
301. Stürzl, N.; Hennrich, F.; Lebedkin, S.; Kappes, M. M. Near Monochiral Single-Walled Carbon Nanotube Dispersions in Organic Solvents. *Journal of Physical Chemistry C* **2009,** *113* (33), 14628–14632.
302. Tu, X.; Manohar, S.; Jagota, A.; Zheng, M. DNA Sequence Motifs for Structure-Specific Recognition and Separation of Carbon Nanotubes. *Nature* **2009,** *460* (7252), 250–253.

303. Sun, D. M.; Liu, C.; Ren, W. C.; Cheng, H. M. A Review of Carbon Nanotube- and Graphene-Based Flexible Thin-Film Transistors. *Small* **2013,** *9* (8), 1188–1205.
304. Lee, W. J.; Maiti, U. N.; Lee, J. M.; Lim, J.; Han, T. H.; Kim, S. O. Nitrogen-Doped Carbon Nanotubes and Graphene Composite Structures for Energy and Catalytic Applications. *Chemical Communications* **2014,** *50* (52), 6818–6830.
305. Maiti, U. N.; Lee, W. J.; Lee, J. M.; Oh, Y.; Kim, J. Y.; Kim, J. E.; Shim, J.; Han, T. H.; Kim, S. O. 25th Anniversary Article: Chemically Modified/Doped Carbon Nanotubes & Graphene for Optimized Nanostructures & Nanodevices. *Advanced Materials* **2014,** *26* (1), 40–67.
306. Collins, P. G.; Bradley, K.; Ishigami, M.; Zettl, A. Extreme Oxygen Sensitivity of Electronic Properties of Carbon Nanotubes. *Science* **2000,** *287* (5459), 1801–1804.
307. Zhou, C.; Kong, J.; Yenilmez, E.; Dai, H. Modulated Chemical Doping of Individual Carbon Nanotubes. *Science* **2000,** *290* (5496), 1552–1555.
308. Klinke, C.; Chen, J.; Afzali, A.; Avouris, P. Charge Transfer Induced Polarity Switching in Carbon Nanotube Transistors. *Nano Letters* **2005,** *5* (3), 555–558.
309. Li, Z.; Saini, V.; Dervishi, E.; Kunets, V. P.; Zhang, J.; Xu, Y.; Biris, A. R.; Salamo, G. J.; Biris, A. S. Polymer Functionalized n-type Single Wall Carbon Nanotube Photovoltaic Devices. *Applied Physics Letters* **2010,** *96* (3), 033110.
310. Biswas, C.; Lee, S. Y.; Ly, T. H.; Ghosh, A.; Dang, Q. N.; Lee, Y. H. Chemically Doped Random Network Carbon Nanotube p-n Junction Diode for Rectifier. *ACS Nano* **2011,** *5* (12), 9817–9823.
311. Redlich, P.; Loeffler, J.; Ajayan, P. M.; Bill, J.; Aldinger, F.; Rühle, M. B-C-N Nanotubes and Boron Doping of Carbon Nanotubes. *Chemical Physics Letters* **1996,** *260* (3–4), 465–470.
312. Ayala, P.; Arenal, R.; Rümmeli, M.; Rubio, A.; Pichler, T. The Doping of Carbon Nanotubes with Nitrogen and Their Potential Applications. *Carbon* **2010,** *48* (3), 575–586.
313. Pint, C. L.; Sun, Z.; Moghazy, S.; Xu, Y. Q.; Tour, J. M.; Hauge, R. H. Supergrowth of Nitrogen-Doped Single-Walled Carbon Nanotube Arrays: Active Species, Dopant Characterization, and Doped/Undoped Heterojunctions. *ACS Nano* **2011,** *5* (9), 6925–6934.

CHAPTER 9

Application of Carbon Nanotubes in Lithium-Ion Batteries

S. Fang, L. Shen, X. Zhang
Nanjing University of Aeronautics and Astronautics, Nanjing, China

Contents

9.1 INTRODUCTION

The need to have better energy storage for technological applications like consumer electronics, hybrid electric vehicles, and remote sensing applications is propelling electrochemical devices to the forefront of research goals. Lithium-ion batteries (LIBs) are noted for their impressively high energy density and their high open circuit voltage attracts a lot of attention.

LIBs typically consist of negative electrode (anode), positive electrode (cathode), separator, and conducting electrolyte, storing electrical energy in the two electrodes in the form of Li-intercalation compounds. During the charge of the LIBs, lithium ions released from the cathode move through the electrolyte and insert into the anode. Upon discharging, lithium ions are extracted from the anode and move back to the cathode. Since electrolytes are not responsible for the conduction of free electrons, electrons that complete the half reaction will move through an extra external wire.

The performance of LIBs is mainly determined by the intrinsic properties of electrode materials. Carbonaceous materials used as anodes for LIBs exhibit significant advantages. As a new member in the carbonaceous material family, the carbon

Industrial Applications of Carbon Nanotubes
ISBN 978-0-323-41481-4

nanotube (CNT) is distinguished at improving the performance of current electrode materials. CNTs, an allotrope of graphite, have been reported to show much improved lithium storage capacity compared to graphite, because of their unique structures and properties. CNTs have been reported to display conductivities as high as 10^6 S/m and 10^5 S/m for single-walled carbon nanotubes (SWCNTs) and multi-walled carbon nanotubes (MWCNTs), respectively, and high tensile strength up to 60 GPa. Recently, many investigations have focused on CNT-based anodes for LIBs with varying success, depending on the treatments employed. This chapter provides an overview of the recent research of CNTs in LIB anode applications. We discuss the influence of different morphology structures of CNTs on electrochemical performance, like various diameters and lengths of the CNTs; defective CNTs were also included. At the end of this chapter, we discuss the CNTs as a role for the framework and conductive additives when used in LIBs.

9.2 MECHANISM OF LITHIUM ION INTERCALATION AND ADSORPTION IN CNTs

As previously explained, lithium ions can be stored via two fundamentally different mechanisms: intercalation and alloying. The former is employed by carbonaceous materials such as graphite and CNTs, while the latter is employed by metals. We briefly describe the intercalation and diffusion of Li at different sites on CNTs.

To investigate the mechanisms through which lithium ions are stored in CNTs, many ab initio studies have been performed *(1–5)*. Nishidate and Hasegawa used molecular dynamics simulations to test the diffusion characteristics of lithium entering SWCNTs and quantified the effect of defects on lithium insertion in CNTs *(1)*. In Figs. 9.1 and 9.2 we can observe the effects that defects have on the morphology of the CNTs. As carbon atoms begin to be removed, a hole begins to occur in the wall of the CNT as each of the carbon atoms attempts to remain bonded with its neighbors. The results showed that while lithium rarely diffused into defect-free, $n = 7$ and $n = 8$-defected CNTs, it was able to diffuse readily into CNTs with $n = 9$ defect. Once inside, the lithium is able to move through the interior of the CNTs and can be absorbed successfully, which indicates that lithium ions can be accumulated in the interior of CNTs in addition to the exterior.

While Nishidate et al. focused on defects introduced through the sidewall of the CNTs, it is important to note that lithium can also enter CNTs in other places. Meunier et al. determined through ab initio simulations that lithium could also come in through the ends of open-ended nanotubes (as opposed to closed nanotubes that had a buckyball-like cap at the ends) *(6)*. Meunier et al. noted that to be effective, however, open-ended CNTs would need to be relatively short to allow ions to freely enter and exit. Experimental evidence to support this can be found whenever ball milling open-ended

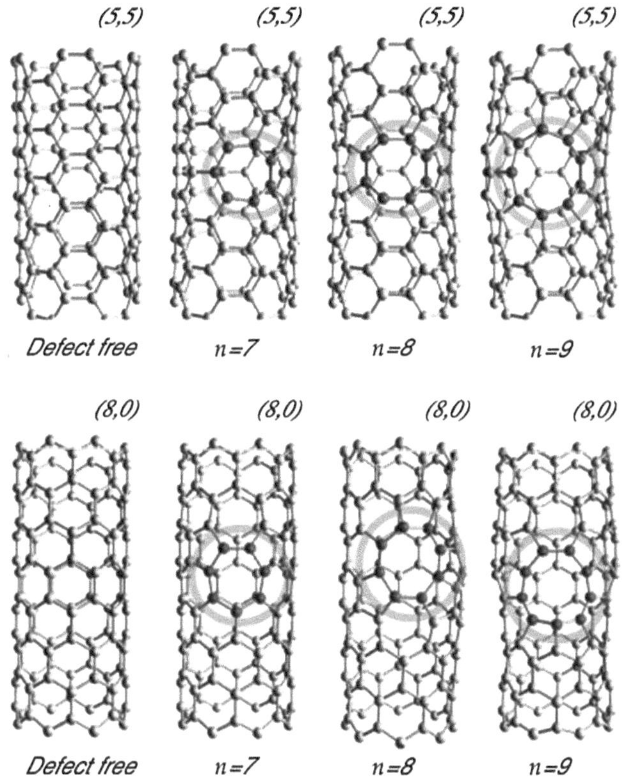

Figure 9.1 Types of topological defects in a (5, 5) single-walled carbon nanotube (SWCNT). Here $n = 7$ means 7-carbon ring defect, $n = 8$ means 8-carbon ring defect, and $n = 9$ means 9-carbon ring defect. *(Reprinted with permission from Nishidate, K.; Hasegawa, M. Energetics of Lithium Ion Adsorption on Defective Carbon Nanotubes.* Physical Review B ***2005,*** *71 (24). Copyright 2005 American Physical Society.)*

CNT samples, which immediately results in higher capacity. Kumar et al., for example, determined the difference between the capacities of opened and unopened CNTs to be almost 120 mAh/g *(7)*. Meunier et al. concluded that the key factor in determining the diffusion coefficient of the CNTs is the ability for the Li ions to reach either end of the open CNTs or a nine-sided (9-carbon ring) defect in between 6. This is because once inserted, the lithium ions undergo a one-dimensional random walk inside the tube, and if the tube is too long, the effective diffusion will decrease as lithium ions are able to enter but never exit.

9.3 SWCNTs AND MWCNTs AS ANODE FOR LIBs

CNTs have generated great interest among researchers as a unique material since their first discovery by Iijima in 1991 using an arc-charge method *(8)*. Structurally, CNTs

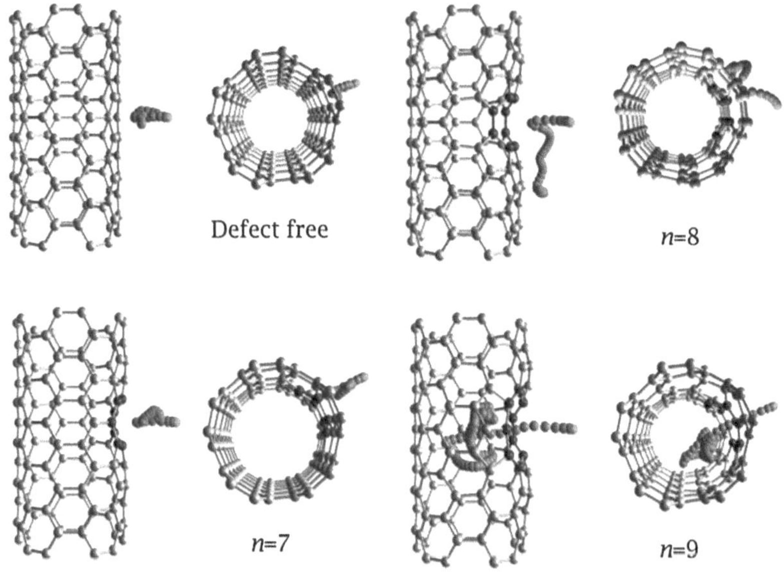

Figure 9.2 Effect of defects on Li insertion in (5, 5) single-walled carbon nanotube (SWCNT). Note: For each nanotube, both the side view and the top view are presented. *(Reprinted with permission from Nishidate, K.; Hasegawa, M. Energetics of Lithium Ion Adsorption on Defective Carbon Nanotubes.* Physical Review B ***2005,*** *71 (24). Copyright 2005 American Physical Society.)*

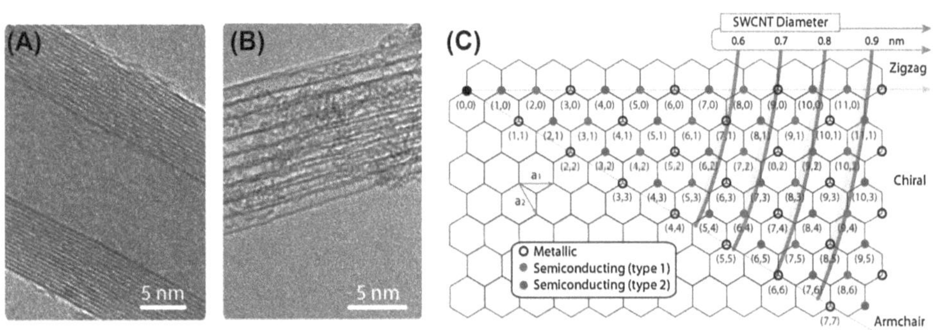

Figure 9.3 Transmission electron microscopy (TEM) images of an as-produced (A) multiwalled carbon nanotube (MWCNT) and (B) single-walled carbon nanotube (SWCNT) bundle. (C) Chirality chart illustrating the assignment of (*n*, *m*) SWCNT structures using a role-up vector from an origin point on a graphene sheet and the nanotube diameter. *(Reprinted with permission from Landi, B. J.; Ganter, M. J.; Cress, C. D.; DiLeo, R. A.; Raffaelle, R. P. Carbon Nanotubes for Lithium Ion Batteries. Energy & Environmental Science* ***2009,*** *2, 638–654. Copyright 2002 Elsevier Science BV.)*

can be envisioned as a rolled-up graphene sheet into a seamless cylinder with fullerene caps for single wall (SWCNTs) or multiple sheets as in the case of MWCNTs where concentric layers are spaced 0.34 nm apart, as illustrated by the transmission electron microscopy (TEM) images in Fig. 9.3A *(9)*. In the case of SWCNTs, there are prominent van der Waals interactions between sidewalls that lead to hexagonally close-packed bundles *(9)*. The bundles are an important physical property; the corresponding TEM image from as-produced SWCNTs is shown in Fig. 9.3B. The electronic properties of SWCNTs are determined by the chirality of the SWCNT, which represents the diameter and helicity of the structure. The chirality determines whether the nanotube will be metallic or semiconducting in nature. Fig. 9.3C demonstrates the basis for assigning SWCNT chirality with a roll-up vector ($na1 + ma2$) from a point of origin on a graphene sheet. The chiral angle ($\leq$30 degrees) is the angle between the zigzag axis and the roll-up vector. The result from this exercise is a series of chiral indices with integer values of n and m, which are used to assign the uniqueness of each SWCNT. Based upon quantum confinement effects, the optoelectronic properties are unique to each (n, m) SWCNT and show a chirality-dependent relationship *(10)*.

The superior mechanical properties of CNTs, with a Young's modulus and tensile strength of 1.2 TPa and 50–200 GPa, respectively, make CNTs one of the strongest and stiffest materials *(11)*. The extraordinary properties and unique structure of both SWCNTs and MWCNTs make them promising materials for a variety of applications. CNTs are especially interesting as one-dimensional hosts for the intercalation of Li, and are regarded as one of the most promising candidates for the anode materials of LIBs.

9.3.1 SWCNTs as Anode for LIBs

The initial evaluation of free-standing SWCNT electrodes showed that the lithium-ion capacity is generally between 400 and 460 mAh/g for purified materials *(12–14)*. The capacity values can approach 1000 mAh/g when sidewall defects are introduced, which shorten the SWCNTs *(15,16)*. In all cases to date, there is a large irreversibility capacity in the first cycle of SWCNTs, which is attributed to solid electrolyte interface (SEI) formation at around 0.9 V (exact value is dependent on electrolyte and constant current). This effect has been attributed to the high mesoporous volume of SWCNTs, which affects the extent of solvent decomposition leading to SEI formation *(12)*. However, because of the varying roles of additives and cosolvents on this process *(17)*, a more thorough examination is warranted for SWCNTs in differing electrolytes [cosolvent mixtures with ethylene carbonate have strictly been used to date *(12–14, 18,19)*].

Another important consideration when evaluating the early measurements on SWCNTs is that they were performed for only a few cycles using materials without a standard of purity. Given the developments on purity standards for SWCNTs *(20,21)*, it is now possible to investigate whether purity dependence exists for lithium-ion capacity in SWCNT electrodes. Raw SWCNTs of sufficient quality can also be fabricated into free-standing papers and would be an attractive material since the purification costs would be eliminated.

9.3.2 MWCNTs as Anode for LIBs

MWCNTs, with concentric graphene layers spaced 0.34 nm apart, display diameters from 10 to 20 nm and lengths of hundreds of microns *(22)*. MWCNTs contain multiple layers of graphene sheets and have more complicated electronic properties. MWCNTs are a promising candidate as an anode material for LIBs, since the Li ions inserted in the multiple rolled layers are similar to graphite. Not only the interplanar spacing of graphitic sheets that allows for the intercalation/deintercalation of lithium ions, but also the hollow cores of MWCNTs are available for lithium-ion intercalation, therefore the unique structures of MWCNTs result in a higher capacity than those of graphite. Much work has been done to demonstrate the excellent electrochemical performance of MWCNT-based anode materials in LIB applications *(23–27)*.

Eom et al. *(28)* reported the chemical processing of Li insertion into purified and chemically etched MWCNTs. The reversible capacity increased along with increase in the etching time from 351 mAh/g ($Li_{0.9}C_6$) for the purified MWCNTs to 681 mAh/g ($Li_{1.8}C_6$) for the etched MWCNTs. The irreversible capacity also increased with increase in the etching time from 1012 mAh/g ($Li_{2.7}C_6$) for the purified MWCNTs to 1229 mAh/g ($Li_{3.3}C_6$) for the etched MWCNTs. The structural and chemical modifications in the etched MWCNTs facilitated an insertion of Li ions into the etched MWCNTs and enhanced a reversible capacity, but the large surface area of the etched MWCNTs induced a large irreversible capacity. The extraction of Li ions from the purified MWCNTs and the etched MWCNTs was a great hindrance.

Huang's *(29)* group improved electrochemical storage of lithium into MWCNTs by xenon light irradiation. With xenon light irradiation on the surface of the electrode, the reversible capacity of the MWCNTs increases remarkably. This is because the electrode irradiated by light has good Li ion reaction kinetics, which facilitates the insertion/extraction of Li ions into/from CNTs and thus enhances the reversible capacity.

Choi et al. *(30)* present a novel binder-free MWCNT structure as an anode in LIBs. The structure was synthesized through a two-step process of catalyst deposition and chemical vapor deposition (CVD), directly grown on a copper current collector, which showed very high specific capacity, excellent rate capability, and no capacity degradation up to 50 cycles.

Researchers have reported that CNT arrays could allow for better contact with the current collector and increase ion diffusivity to significantly improve bulk electron transport properties thereby allowing for improved rate capabilities *(31–33)*. Welna et al. *(34)* have incorporated aligned CNTs into LIB electrodes. The vertically aligned MWCNT electrodes can lead to higher storage capacities and significantly excellent rate capabilities compared to graphite. These structural features increased the effective surface area of the electrode and presumably provided defect sites with which the lithium ions interact. Additionally, the intimate contact between each of the vertically aligned nanotubes and the nickel metal current collector provides a low-resistance pathway allowing for effective connectivity throughout the electrode. Furthermore, if the material density can be increased and the irreversibly specific capacity decreased, the large reversible specific capacity of the aligned CNTs and their ability to maintain moderate specific capacities at very high discharge rates make this concept attractive for high-power LIBs.

Wei et al. *(35)* demonstrated a long-cycle performance of aligned MWCNTs directly synthesized on stainless-steel foil as an anode material in lithium batteries. The electrode shows an increase in the specific capacity with an increase in the cycle number. Starting at a value of 132 mAh/g in the first cycle at a current rate of 1 C to a value of 460 mAh/g after 1200 cycles, the specific capacity increased by about 250%. This is because the continuous charge–discharge process broke the nanotubes preferably at the structural and surface defects along the sidewalls and created more active sites for Li storage in every subsequent cycle. The nanotube disruption gradually exposed the stainless-steel substrate and allowed it to contribute to the Li intercalation.

9.4 CNTs WITH DIFFERENT MORPHOLOGIES AS ANODE FOR LIBs

Morphology of CNTs includes the diameter, diameter distribution, and length of individual tube or bundle. CNTs with controlled structure were essential to exhibit a uniform property in microscopic or macroscopic scale for applications. Chemical etching and ball milling are two widely applied methods to modify the morphologies of CNTs. These treatments are reported to result in structural changes and the formation of surface functional groups on CNTs. The structural changes, including the lateral defects on the surface of CNTs and the shortening of the length of CNTs, can increase the Li insertion capacity. The diameter of the CNTs can be controlled by adjusting the synthesizing conditions such as temperature, catalyst, pressure, and so on.

9.4.1 CNTs With Different Diameters as Anode for LIBs

The diameter is an important parameter of the structure of CNTs and can affect lithium adsorption and diffusion, both inside and outside the CNTs. Lithium ions intercalated into CNTs will interact with CNTs, and the charge transfer between CNTs and lithium ions will further enhance these interactions *(36)*. The interaction is related to the curvature of the tubes and will result in different lithium capacities in CNTs. In addition, exfoliation from MWCNTs during electrochemical insertion of lithium is also dependent on the inner diameter of the tubes and will consequently influence lithium capacities as well *(23)*.

Udomvech et al. *(37)* performed first principle calculations of Li and Li^+ adsorbed on the sidewall of CNTs. Dependence of the atomic net charge of Li/Li^+ on tubule diameter and length has been investigated based on the natural bond orbital (NBO) and Mulliken population analysis, as illustrated in Fig. 9.4. It was found that the

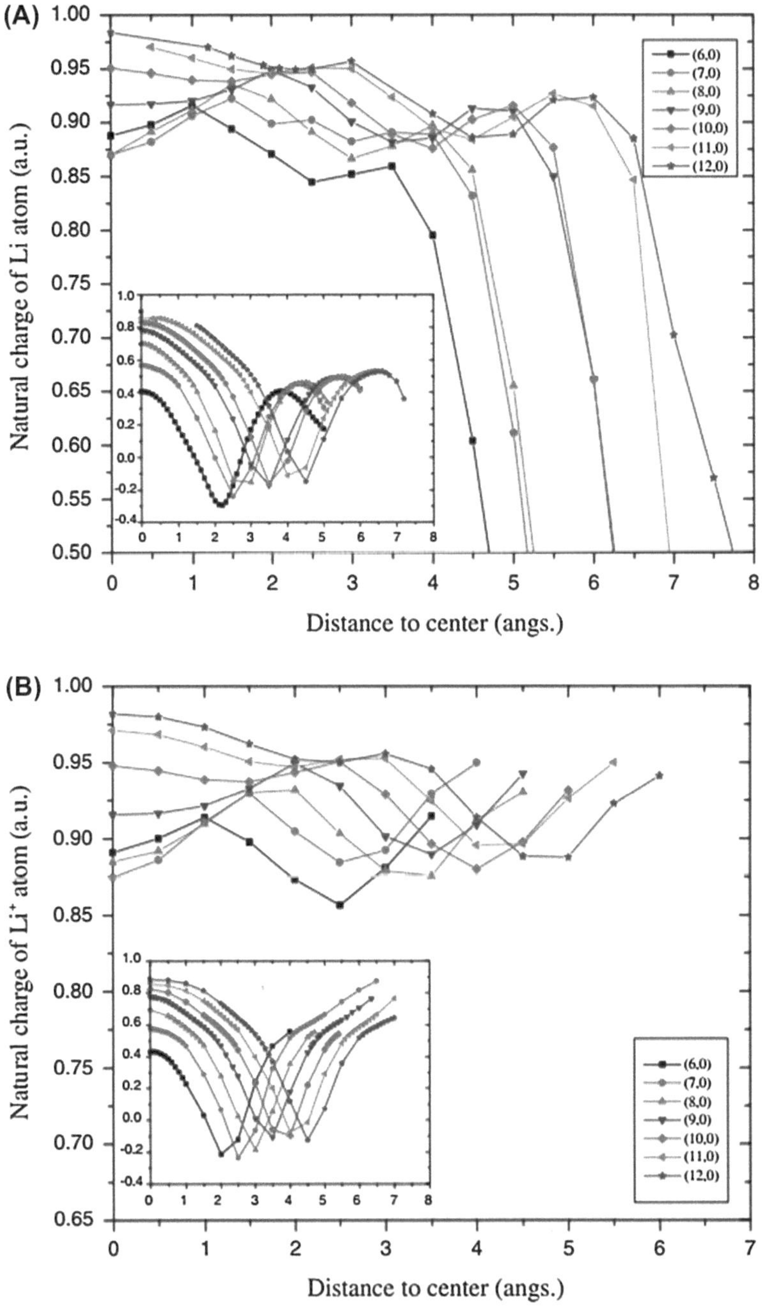

Figure 9.4 Dependence of natural bond orbital (NBO) atomic net charge of (A) Li and (B) Li^+ on the tubule diameter. Mulliken charges are also provided for comparison in the inset. *(Reprinted with permission from Udomvech, A.; Kerdcharoen, T.; Osotchan, T. First Principles Study of Li and Li^+ Adsorbed on Carbon Nanotube: Variation of Tubule Diameter and Length.* Chemical Physics Letters ***2005,*** *406, 161–166. Copyright 2005 Elsevier Science BV.)*

intercalated Li/Li^+ prefers to localize near the CNT sidewall, instead of being located at the tubule center. Therefore larger CNTs, which have useless space inside, will be at a disadvantage compared with the small diameter ones when used as batteries.

More directly, the research performed by Zhao et al. *(38)* showed a clear relationship between the Li/C ratio and tube diameter. With an increase in tube diameter, the intercalated lithium atoms tended to form a multishell structure (Fig. 9.5) at the equilibrium state, composed of coaxial tubes with a linear chain in the axis, which will improve lithium capacity. Additionally, Garau et al. *(39)* have shown that the interaction potential at the central region is dependent on the diameter of the nanotubes, and CNTs with a diameter of 4.68 Å have greater interaction energy, thus making them the best candidate for lithium storage as an anode material of LIBs.

Zhang et al. *(40)* employed different diameter range MWCNTs as the anode materials in LIBs and reported that MWCNTs with a diameter of 40–60 nm display the highest specific capacity (187.4 mAh/g at a charge–discharge rate of 50 mA/g) and good cyclability; coulombic efficiency is up to 101.9% after 50 cycles. The excellent electrochemical diameter range of 40–60 nm can preserve the highest conductivity as the electrode (Fig. 9.6), and electron transport is greatly enhanced during the lithium insertion/extraction process, which results in significant advantages in the electrochemical performances of CNTs.

9.4.2 CNTs With Different Lengths for LIB Anode Materials

The length of CNTs is another important influential morphological factor that is able to influence the lithium diffusion of CNTs. This is because short MWCNTs provide short

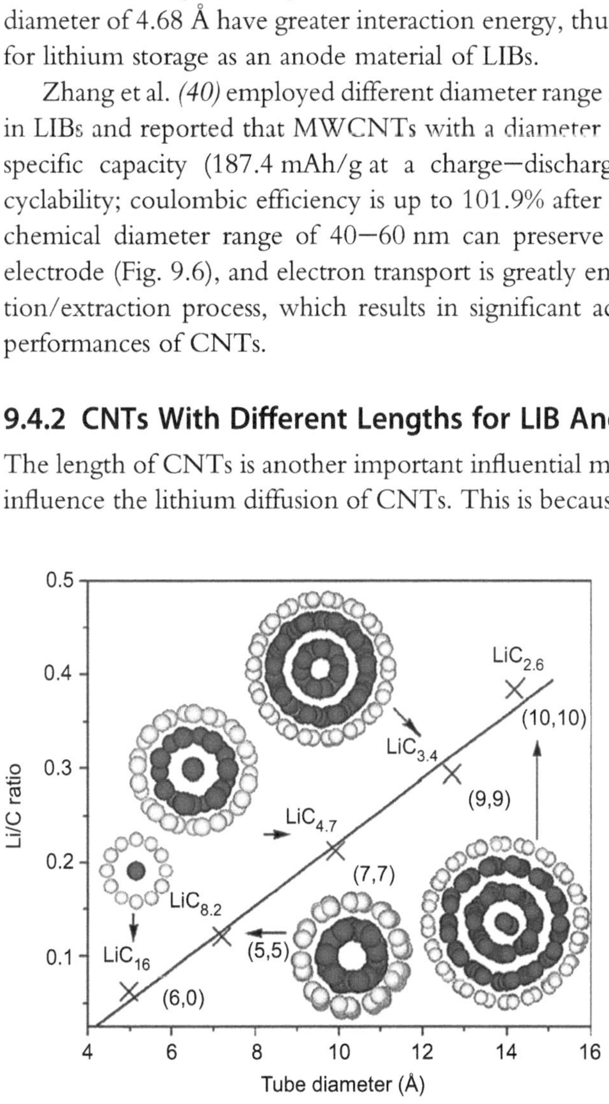

Figure 9.5 The variation of Li/C ratio as a function of tube diameter. The equilibrium configurations single-walled carbon nanotubes (SWCNTs) filled with Li atoms are shown in the insets of this figure. *(Reprinted with permission from Zhao, M. W.; Xia, Y. Y.; Liu, X. D.; Tan, Z. Y.; Huang, B. D.; Li, F.; Ji, Y. J.; Song, C. Curvature-Induced Condensation of Lithium Confined Inside Single-Walled Carbon Nanotubes: First-Principles Calculations.* Physics Letters A ***2005,*** *340, 434–439. Copyright 2004 Elsevier Science BV.)*

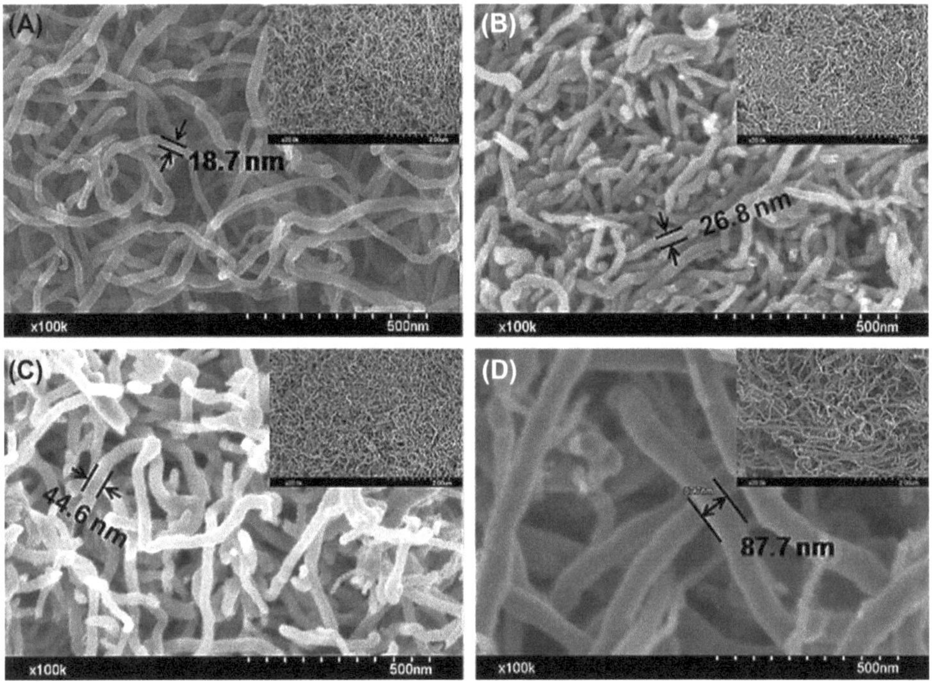

Figure 9.6 Field emission scanning electron microscopy images of multiwalled carbon nanotubes (MWCNTs) with diameter ranges of (A) 10–20 nm, (B) 20–40 nm, (C) 40–60 nm, and (D) 60–100 nm. *(Reprinted with permission from Zhang, Y. P.; Chen, T. Q.; Wang, J. H.; Min, G. Q.; Pan, L. K.; Song, Z. T.; Sun, Z.; Zhou, W. M.; Zhang, J. The Study of Multi-Walled Carbon Nanotubes With Different Diameter as Anodes for Lithium-Ion Batteries.* Applied Surface Science ***2012,*** *258, 4729–4732. Copyright 2012 Elsevier Science BV.)*

reversible routes for Li ions and additional intercalation sites at abundant open ends for Li ion storage. Commonly used chemical etching and ball milling treatments are employed to increase the reversible capacity, but the etching method usually reduces the coulombic efficiency because the introduction of large structure defects *(41)* and ball milling treatment will result in alarge voltage hysteresis *(42)* because of the amount of surface functional groups formed on CNTs *(43)*. Gao et al. *(42)* believed that it was at least partially related to the kinetics of lithium diffusion into the inner cores of CNTs, and indicated that the hysteresis could be reduced by cutting the nanotubes to shorter segments from their preliminary experiments.

Yang et al. *(44)* comparatively investigated two kinds of CNTs, short carbon nanotubes (CNTs-1) synthesized by the copyrolysis method and long carbon nanotubes (CNTs-2) produced using common CVD as anode materials for LIBs. As shown in Fig. 9.7A and B, the CNTs-1 is very straight with an outer diameter of 35–65 nm, inner diameter of 15–40 nm, and length of 150–400 nm, while the CNTs-2 entangles together with an outer diameter of 15–30 nm, an inner diameter of 5–10 nm, and a

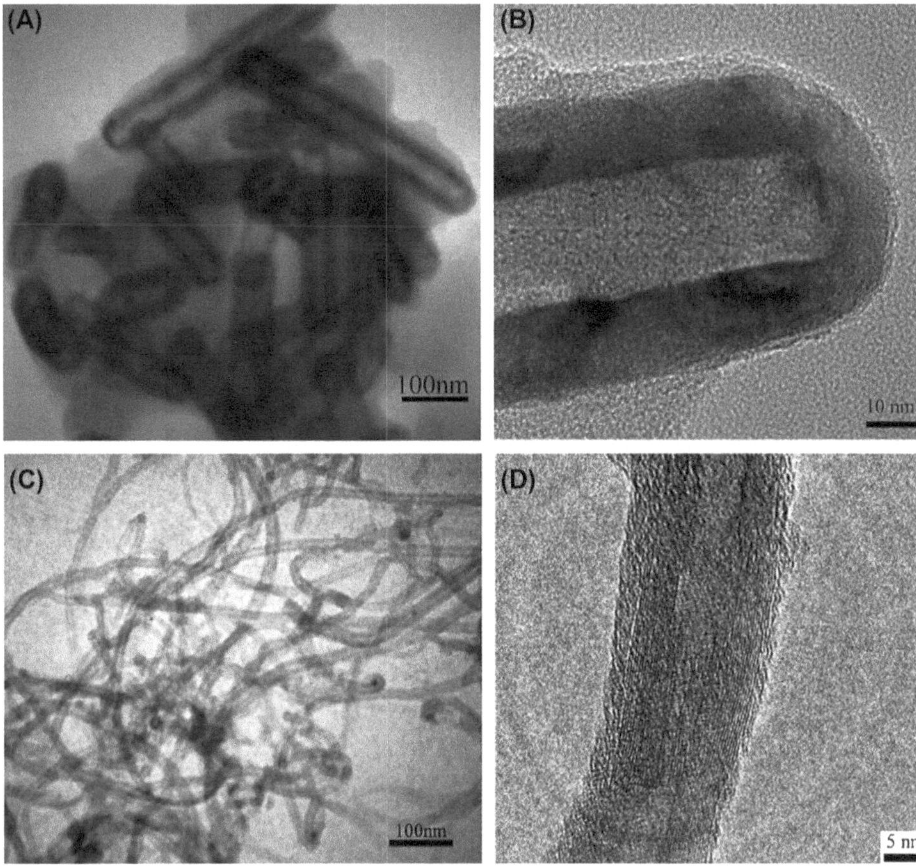

Figure 9.7 Transmission electron microscopy (TEM) and high-resolution (HR)TEM images of CNTs-1 (A and B) and CNTs-2 (C and D), respectively. *(Reprinted with permission from Yang, S. B.; Huo, J. P.; Song, H. H.; Chen, X. H. A Comparative Study of Electrochemical Properties of Two Kinds of Carbon Nanotubes as Anode Materials for Lithium Ion Batteries.* Electrochimica Acta ***2008,*** *53, 2238–2244. Copyright 2008 Elsevier Science BV.)*

length of more than several micrometers. The electrochemical properties of the two kinds of CNTs, CNTs-1 and CNTs-2 as anode materials for LIBs, were systemically investigated. It was found that the reversible capacities of the CNTs-1 electrode were 266 and 170 mAh/g at current densities of 0.2 and 0.8 mA/cm^2, respectively, which were almost twice those of the CNTs-2 electrode. Moreover, the storage mechanism of lithium ions and voltage hysteresis in CNTs-1 and CNTs-2 electrodes were different. The larger voltage hysteresis in the CNTs-2 electrode was related not only to the surface functional groups in CNTs, but also to the surface resistance of CNTs.

Wang et al. *(45)* developed a new approach for cutting conventional micrometer-long entangled CNTs too short to improve the reversible capacities and decrease the irreversible capacities of CNT-based LIBs. Such CNTs with typical lengths of ca. 200 nm were obtained from conventional micrometer-long CNTs by a novel solid-state cutting method. This method involves the deposition of NiO nanoparticles (NPs) on the CNTs-2, followed by reaction at high temperatures (Fig. 9.8). The electrochemical performance

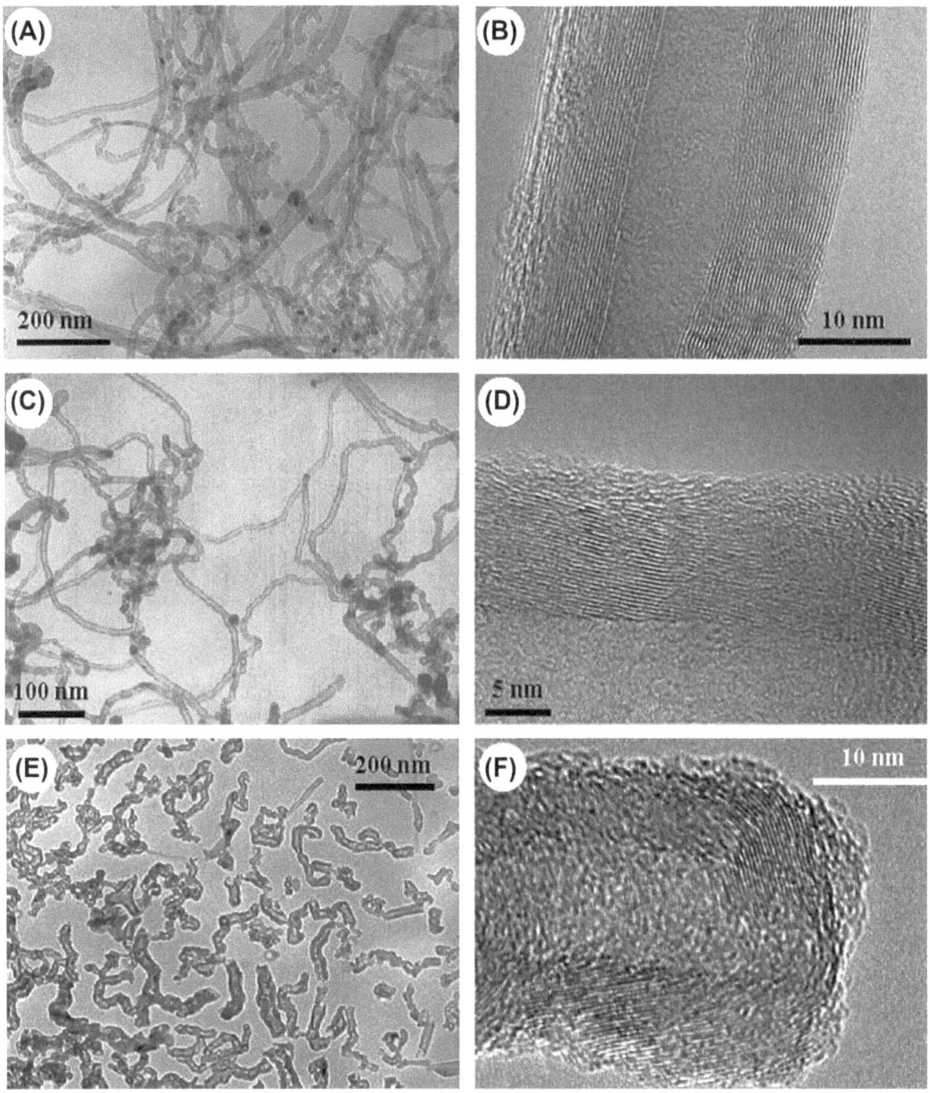

Figure 9.8 Transmission electron microscopy (TEM) and high-resolution (HR)TEM images of (A, B) very long, (C, D) long, and (E, F) short carbon nanotubes (CNTs). *(Reprinted with permission from Wang, X. X.; Wang, J. N.; Chang, H.; Zhang, Y. F. Preparation of Short Carbon Nanotubes and Application as an Electrode Material in Li-ion Batteries.* Advanced Functional Materials ***2007,*** *17, 3613–3618. Copyright 2007, Wiley-VCH.)*

of the obtained CNTs-1 was far superior to that of CNTs-2. Specifically, the reversible capacity increases and the irreversible capacity decreases upon reducing the length of the CNTs. Upon increasing the current density from 25 to 60 mA/g, the decrease in reversible capacity for CNTs-1 is smaller than that for CNTs-2, indicating a better rate performance for CNTs-1. The reason for this may be that the extraction of Li ions from CNTs-1 is facilitated by the shortened length of the tubes. Moreover, the cutting process may also increase the number of edges in the graphite layers and induce the formation of lateral defects on the CNT surface. Additionally, the cutting process also opens the ends of the MWCNTs. These factors favor the enhanced insertion of Li ions into the cut CNTs and thus lead to increased reversible capacity values.

Fig. 9.9 shows the charge—discharge curves for CNTs-1 and CNTs-2 at current densities of 25 and 60 mA/g from the fourth cycle. The clear decrease in reversible capacity at higher current densities for CNTs-2 is about 69% retention of the charging capacity. In contrast, the decrease at higher current densities is very small for CNTs-1, corresponding to about 96% retention of the charging capacity. This result indicates that CNTs-1 exhibit better performance at higher cycling rates, which can be attributed to the rapid intercalation/deintercalation of Li ions in CNTs-1.

In subsequent work, the same group reported for the first time on the use of the Fe compound FeS as a catalyst to produce CNTs-1 directly *(46)*. FeS formed during the process promoted the growth of CNTs and limited their length. The directly grown CNTs-1 were about 200—500 nm in length and 20—30 nm in diameter. These directly produced CNTs-1, as well as CNTs-1 from solid-state cutting and conventional CNTs-2, were used in electrochemical tests as anodes in LIB devices. The result showed that the

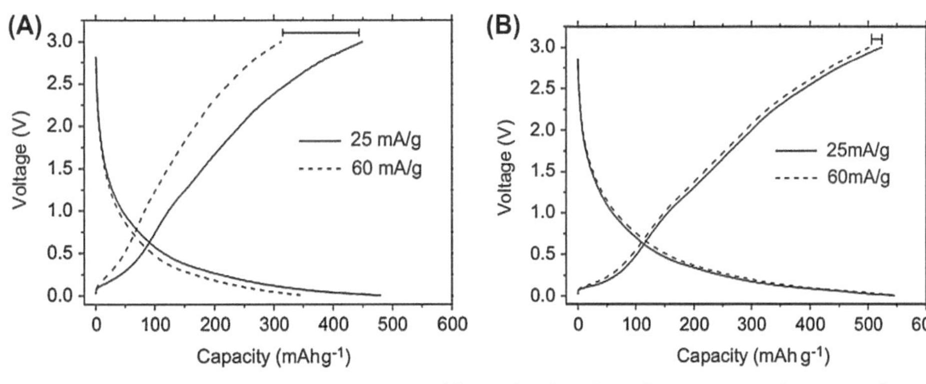

Figure 9.9 Discharge—charge curves obtained from the fourth cycle at current densities of 25 and 60 mA/g for (A) long carbon nanotubes (CNTs-2) and (B) short carbon nanotubes (CNTs-1). Note the decrease in Crev as the current density is increased from 25 to 60 mA/g. *(Reprinted with permission from Wang, X. X.; Wang, J. N.; Chang, H.; Zhang, Y. F. Preparation of Short Carbon Nanotubes and Application as an Electrode Material in Li-ion Batteries.* Advanced Functional Materials ***2007,*** *17, 3613—3618. Copyright 2007, Wiley-VCH.)*

reversible capacity (Crev) for the CNTs-2 was 188 mAh/g, while for the directly grown CNTs-1 and the cut ones, the values were dramatically increased to 502 and 577 mAh/g, respectively (Fig. 9.8). The stable charge capacities of 230 mAh/g and 142 mAh/g after 20 cycles for CNTs-1 and CNTs-2, respectively, indicate a much better cyclic performance for the CNTs-1. The synthetic technique they presented can be used to produce CNTs-1 in large amounts and facilitate a wide range of possible applications.

9.4.3 Doped Carbon Nanotubes as Anode for LIB

It is well known that heteroatom doping in carbon materials is an effective approach to tailor both the chemical and physical nature of carbon-based anode materials. For example, nitrogen-doped carbon nanofibers are shown to be metallic because of the pyridine-like nitrogen and bridgehead-nitrogen type *(47)*. Doping heteroatoms into carbon anodes has an important influence on Li^+ storage performance for LIBs, for instance, the polymerized carbon nitride (10 at.%) nanobells improved Li^+ intercalation/deintercalation performance with a high reversible capacity of 480 mAh/g *(48)*. Similarly, the nitrogen concentration in a CNT structure could effectively impact Li^+ storage properties; the higher concentration of nitrogen doping, the greater the influence of Li^+ storage performance on the CNTs *(49)*.

Ren et al. detected the performance of CNTs with different nitrogen contents as anodes in LIBs. CNTs with different nitrogen contents were synthesized by a floating catalyst CVD method. Melamine was employed as nitrogen precursor to effectively monitor nitrogen content in the nitrogen-doped carbon nanotubes (CNx-NTs) and to modify their structures. The study showed that with the increase in the content of melamine, the growth rate of the tubes was enhanced and the diameter became larger. The electrochemical properties showed that the discharge capacity of CNx-NT anodes improved as their nitrogen concentration increased. Also higher reversible capacities and better rate capabilities were observed. The superior electrochemical performance could be related to the high electrical conductivity and the larger number of defect sites. These performances revealed that the electrochemical activities of CNTs have been enhanced by the incorporation of nitrogen atoms.

Li et al. *(50)* reported the different electrochemical performances of Li^+ intercalation/deintercalation in 16.4 at.% CNx-NTs synthesized by using a floating catalyst CVD approach in comparison to the pristine CNTs as anodes for LIBs. It has been found that high nitrogen concentration-doped carbon nanotubes (HN-CNTs) can exhibit a much higher reversible capacity and better rate capability compared with nondoped CNTs for LIBs. Fig. 9.10 shows the morphology of CNTs and HN-CNTs, respectively. The aligned CNTs vary from 10 to 80 nm in diameter and because the C—N bond length is shorter than C—C, the result is the curving of basal planes forming the bamboo-like structure after nitrogen doping in the CNTs. The impedance value of

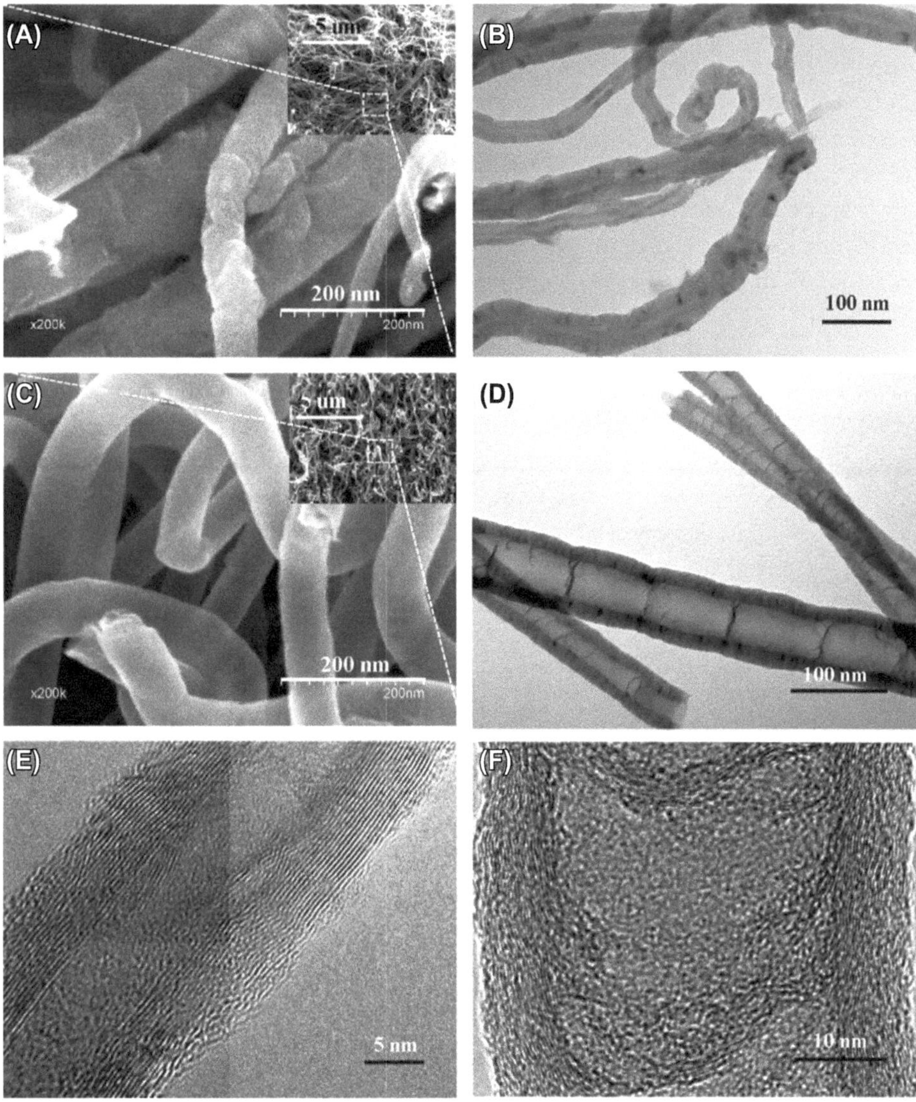

Figure 9.10 Scanning electron microscope (SEM), transmission electron microscopy (TEM), and high-resolution (HR)TEM images of (A, B, and E) carbon nanotubes (CNTs) and (C, D, and F) high nitrogen concentration-doped carbon nanotubes (HN-CNTs). *(Reprinted with permission from Li, X. F.; Liu, J.; Zhang, Y.; Li, Y. L.; Liu, H.; Meng, X. B.; Yang, J. L.; Geng, D. S.; Wang, D. N.; Li, R. Y.; Sun, X. L. High Concentration Nitrogen Doped Carbon Nanotube Anodes with Superior Li^+ Storage Performance for Lithium Rechargeable Battery Application.* Journal of Power Sources ***2012,*** *197, 238–245. Copyright 2012 Elsevier Science BV.)*

CNTs is higher than that of HN-CNTs, as shown in Fig. 9.11. After simulating by the equivalent circuit, it is found that the charge transfer resistance of CNTs and HN-CNTs is 58 Ω and 38 Ω, respectively. The nitrogen atom has additional lone pairs of electrons, contributes additional electrons, and provides electron carriers for the conduction band *(51,52)*. Thereby HN-CNTs show higher electrical conductivity than CNTs. As a consequence, the higher electrical conductivity of HN-CNTs decreases the charge transfer resistance of anode materials *(53–55)*. This is beneficial for Li^+ intercalation/deintercalation into the anodes. The results demonstrated that the HN-CNT anode shows a much higher specific capacity of 494 mAh/g than CNTs, and still delivers the discharge capacity of 397 mAh/g in the 100th cycle (only 266 mAh/g for CNTs). Moreover, HN-CNTs display better rate capability than CNTs *(30)*.

Doping of B in MWCNTs (B-MWCNTs) has been found to increase the overall 3D ordering and also the electric conductivity *(56–58)*. Carrol et al. showed that doping of B in MWCNTs induced a semimetallic character via the formation of ordered BC_3 nanodomains with a vanishing or narrow bandgap. Moreover, the local in-plane hexagonal lattice symmetry is broken by the introduction of boron into the MWCNT matrix *(56,58)*, enhancing the number of defect sites. Mukhopadhyay et al. *(59)* investigated the effect of boron doping on electrochemical Li insertion capability of MWCNTs, prepared by the graphite arc technique, in aprotic medium. The galvanostatic discharge–

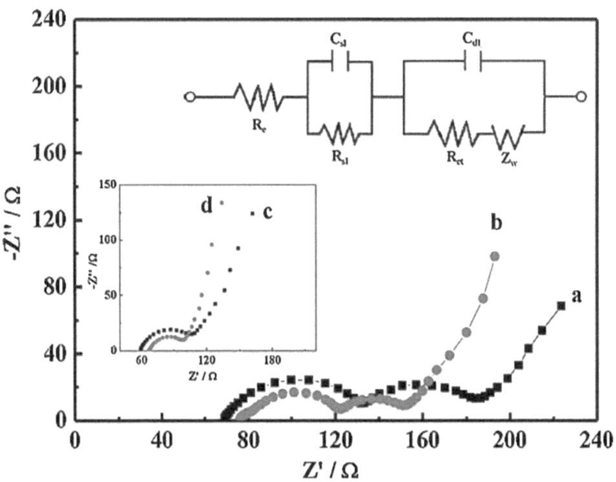

Figure 9.11 Electrochemical impedance spectra of (A) carbon nanotubes (CNTs) and (B) high nitrogen concentration-doped carbon nanotubes (HN-CNTs) at 0.01 V in the first discharge; *inset*: the electrochemical impedance spectra of (C) CNTs and (D) HN-CNTs at the open circuit voltage and the equivalent circuit to fit the electrochemical impedance spectroscopy. *(Reprinted with permission from Li, X. F.; Liu, J.; Zhang, Y.; Li, Y. L.; Liu, H.; Meng, X. B.; Yang, J. L.; Geng, D. S.; Wang, D. N.; Li, R. Y.; Sun, X. L. High Concentration Nitrogen Doped Carbon Nanotube Anodes With Superior* Li^+ *Storage Performance for Lithium Rechargeable Battery Application.* Journal of Power Sources **2012,** *197, 238–245. Copyright 2012 Elsevier Science BV.)*

charge measurements revealed that the reversible capacity was 156 mAh/g for undoped and 180 mAh/g for B-doped nanotubes in the first cycle with almost equal coulombic efficiencies of 55–58%. The coulombic efficiency increased to more than 90% after the second cycle.

9.4.4 Free-Standing CNT "Papers" for LIB Anodes

The investigation and development of flexible power sources has motivated the development of flexible, lightweight, binder-free, and current-collector-free electrodes for LIBs *(18,30,60–63)*. In conventional methods for fabricating LIB electrodes, binders are introduced to inhibit the collapse of the active materials from metal current collectors, while current collectors are used to maintain the electrode conductivity. The employment of binders and current collectors contributes nothing but dead weight to lithium storage, which decreases the energy density of LIBs. A free-standing CNT film can be applied directly as a flexible electrode material in a test cell without the need for a current collector and a binder. Free-standing film electrodes are attractive for basic electrochemical characterization as model electrodes because of their ease of handling, favorable mechanical properties, and, most importantly, the absence of the binder.

Chew et al. *(26)* preserve the benefits of electrochemistry at the nanoscale by introducing a simple filtration technique to prepare flexible CNT films. They compare the properties of SWCNTs, double-walled carbon nanotubes (DWCNTs), and MWCNTs, all as free-standing films. When used as flexible anodes, the investigated films possess the functions of both charge storage and of a current collector. For the first reduction half-cycle (Fig. 9.12A), both SWCNT (2390 mAh/g) and DWCNT (2110 mAh/g) films

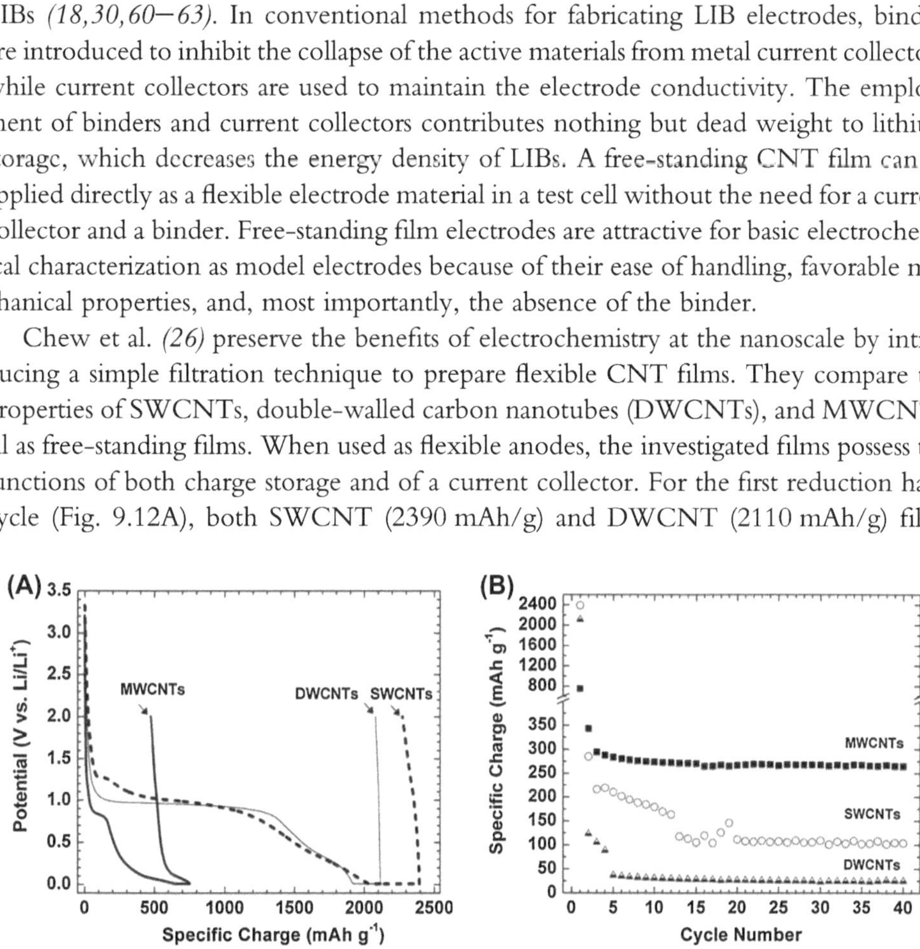

Figure 9.12 The electrochemical performance of the freestanding carbon nanotube (CNT) films. (A) Galvanostatic charge–discharge profiles in the first cycle and (B) cycling behavior between 0.01 and 2.00 V versus Li/Li$^+$ at a specific current of 25 mA/g. *(Reprinted with permission from Chew, S. Y.; Ng, S. H.; Wang, J. Z.; Novak, P.; Krumeich, F.; Chou, S. L.; Chen, J.; Liu, H. K. Flexible Free-Standing Carbon Nanotube Films for Model Lithium-Ion Batteries.* Carbon *2009, 47, 2976–2983. Copyright 2009 Elsevier Science BV.)*

exhibit much higher specific charge compared to the MWCNT film (750 mAh/g). The MWCNT film exhibits a reversible charge of approximately 300 mAh/g with a stable cycling behavior. However, the SWCNT and DWCNT films show strong capacity fading. The coulombic efficiency of the MWCNT film is 80% in the second cycle, much higher than those of the SWCNT film (59%) and the DWCNT film (55%), respectively. Taking into account the relatively low price of the MWCNT feedstock, the straightforward vacuum filtration preparation process, and the electrochemical performance, free-standing MWCNT flexible films deserve more detailed investigation as a possible candidate anode for flexible or bendable next-generation LIBs.

Lee et al. *(64)* prepared CNT films through a direct spinning process *(65)*. The CNT films were posttreated by two treatment methods [heat treatment (HT) and acid-heat treatment (AHT)]. It is known that the HT of CNTs removes amorphous carbon from them and functional groups on their surface, while the AHT of CNTs removes metallic catalysts and introduces functional groups into them *(66)*. Fig. 9.13B illustrates the cycle performance of the discharge capacity. The discharge capacities at the 10th cycle with a 0.5 C rate resulted in 163 for the Raw cell and around 250 mAh/g for the HT cell and the AHT cell, respectively. The HT cell exhibited higher capacities than the Raw cell. As a result, the HT sample enhanced crystalline perfection by removing the amorphous carbon. Eventually, it could happen that the crystalline perfection of CNT films would affect the charge–discharge capacity of LIBs. Otherwise, the AHT cell showed a similar capacity to the HT cell. It is suggested that the defects in the CNTs were generated during the AHT so that an increased number of Li ions might be easier to store at the largely formed defects.

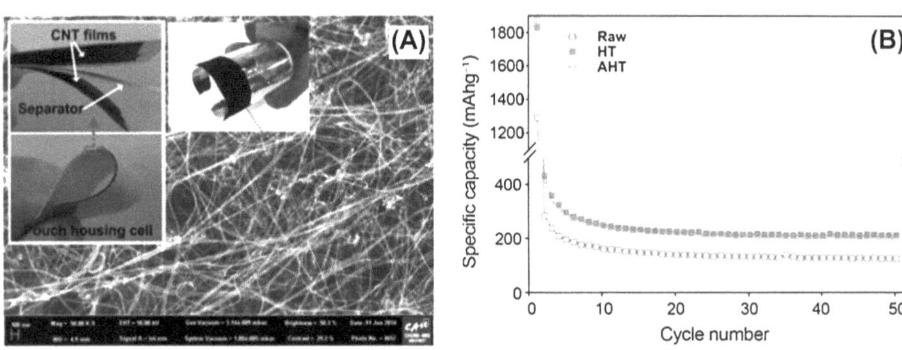

Figure 9.13 (A) Scanning electron microscope (SEM) image of a carbon nanotube (CNT) film [*inset*: cross-section photo image (*up*) of pouch housing cell (*down*) including anode/cathode CNT films and a separator, and photo image (*middle*) of wound CNT film on a vial]. (B) Discharge capacity of cells versus cycle at 0.5 C rate. *(Reprinted with permission from Lee, S.; Cho, D.; Jeong, Y. Effects of Post-treated Carbon Nanotube Films on Electrochemical Performance for Bendable Lithium-Ion Batteries.* Fiber Polymers ***2015,*** *16, 1600–1604. Copyright 2015 Springer.)*

Compared to conventional processes for forming LIB electrodes, the CNT papers can fabricate current-collector-free electrodes without using binder polymers and carbon particles, which can simplify the electrode formation process. This has a high potential for use as an anode electrode since good results could be achieved in terms of retention rates and stable rate capabilities.

9.5 CNT-BASED COMPOSITE ELECTRODES

An immediate solution to improve performance of CNT-based electrodes, as envisioned, was the use of composite electrodes and many research efforts were concentrated on developing new composite structures. Different materials were used/mixed with CNTs to prepare composite electrodes. In past decades, to improve the performance of CNT electrodes, materials with high specific capacity such as metallic oxide and Si were combined with CNTs in many studies; this had the advantage of a huge specific surface area and the role of supporting and conducting the frame. MWCNT composite material coated with a layer of even SnO_2 was produced by Wang et al. *(67)* through a thioglycolic acid-assisted hydrothermal method and the thickness of SnO_2 was controlled by reaction conditions. After the SEI was formed in the first cycle, the specific capacity improved and the stability was quite good. After 50 cycles of complete charge and discharge at a 0.2 C rate, the specific capacity stood at 435 mAh/g, which was higher than that of graphite. Zhang et al. *(68)* have produced a CNT array with CVD before drawing the array into a CNT film on which nano-SnO_2 is produced. The specific capacity is as high as 850 mAh/g, and it retains 100% after 65 cycles at the voltage range of 0.01−3 V. Park et al. *(69)* fabricated vertically aligned mesoporous carbon nanotubes (MCTs) by a dual template method: a hard template-assisted sol−gel process. This kind of CNT was filled with Co_3O_4 NPs. As an anode material, it shows a high reversible capacity of about 627 mAh/g after the 50th discharge. The small size of Co_3O_4 NPs improved electrochemical performance of Co_3O_4/MCT and the rapid diffusion of Li ions was induced by the 3D structure of the MCT. Therefore the 3D nanostructure can be widely applied to other promising conversion reaction electrodes to improve LIB performance.

Si and lithium can produce an Si−Li alloy phase of $Li_{4.4}Si$ whose specific capacity can reach 4200 mAh/g, having great potential of replacing graphite; thus the study of silicon nanowire, porous silicon, and carbon fiber covering amorphous silicon has been accelerated. One-dimensional composite material has been produced by Wang and Kumta *(70)* through depositing nanocrystalline silicon on a CNT array with a simple two-step pouring liquor into CVD. Nanosilicon clusters connect tightly with CNTs, which minimize the loss when the silicon connects with and desorbs from Li^+. This novel structure has good rate capability, high reversible capacity (2050 mAh/g), and good stability (after 25 cycles, only 20% has been reduced). The capacity loss results mainly from the damage on the boundary between the silicon and the CNTs as well as the hindering of Li^+ transport by SEI weakening. Weng et al. *(71)* used a new type of anode that was fabricated by three-dimensionally aligned CNT/Si hybrids to develop an LIB. The LIBs have high specific capacity and cyclic stability, which were retained at 1055 mAh/g after 1000 cycles at 5 A/g.

Cathode material is the provider of Li^+. Currently, the materials in LIBs are mainly inorganic salts including $LiCoO_2$ of hexagonal layered structure, $LiMn_2O_4$ of spinel structure, and $LiFePO_4$ of olivine structure. In the circle of research, the conducting polymer used as the cathode material of LIBs has received increasing attention. Hosono et al. *(72)* have produced a spinning $LiFePO_4$ whose core is MWCNTs and a complex shell of $LiFePO_4$ and amorphous carbon, and this nanowire is used as the cathode of LIBs through electrostatic spinning. CNTs play a leading role in the electron transportation and inhibition of oxidization of Fe^{2+}. At 0.1 A/g, the specific capacity is 130 mAh/g, and the rate capability and cycle properties are excellent. $LiMn_2O_4$–MWCNT composite material has been produced by Liu et al. *(73)* with the sol–gel method. As the 1D conductor in nanocomposite material, CNTs facilitate the transport of electrons. The resistance of composite material is far lower than that of $LiMn_2O_4$ of spinel structure and displays high cycle stability after 20 cycles; capacity retention is 99% while that of pure $LiMn_2O_4$ is only 9%. Porous $LiFePO_4$–CNT composite material has been synthesized by Zhou et al. with the sol–gel method *(74)*. The oxidation peak and reduction peak appear at 3.51 V and 3.38 V, respectively, and the potential difference is 0.13 V, which improves the irreversibility and reaction activity.

Because of the innovative work of Tarascon, which revealed a new strategy for high-capacity electrodes by exhibiting that metal oxides can store more than one Li ion per transition metal atom, many researchers have investigated various conversion reaction compounds, such as metal nitrides, sulfides, fluorides, and oxides. As a kind of novel cathode material, vanadium pentoxide (V_2O_5) has the characteristics of low cost and being earth rich, and shows much higher energy density than traditional $LiMn_2O_4$, $LiCoO_2$, and $LiFePO_4$ materials *(75)*. But the multiphase transitions of V_2O_5 in the charge–discharge process correspond to the multiple voltage plateaus, which cause repeated damage to the structure of V_2O_5. To improve the structural stability of V_2O_5, Zhou et al. *(76)* synthesized a vanadium oxide nanosheet–MWCNT composite via the sol–gel method and subsequent hydrothermal treatment process. This nanosheet–MWCNT composite with a distinct single-phase transition feature displays high specific capacity and good cycling stability because of its sheet-like nanostructure and uniform adding of MWCNTs, which makes this novel composite quite suitable. Cao and Wei *(77)* prepared V_2O_5/SWCNT hybrid films with mesoporous structure by a simple floating CVD method. Li^+ diffusivity can be promoted by two–four orders because of SWCNTs. The hybrid films as cathodes for LIBs demonstrate a high rate capacity of 548 mAh/g at the discharge rate of 1 C. Kim et al. *(78)* fabricated nanoarchitectures composed of

FeF_3 nanoflowers on CNT branches by functionalizing the surfaces of CNT with FeF_3. Electron transport made the nanostructured cathode deliver more than 210 mAh/g at a current rate of 20 mA/g with the specific energy about 650 Wh/kg under conditions of average discharge voltage of about 3.1 V in the voltage range of 2.0–4.5 V, which was higher than conventional cathode materials.

9.6 CNTs AS CONDUCTIVE ADDITIVES

Many electrode materials are oxides with poor conductivity. The performance of such electrode materials highly relies on the adequate addition of carbon conductive additives. The high electrical conductivity, large aspect ratio, excellent mechanical properties, and high chemical stability suggest that CNTs are a distinctive choice for improving electrochemical properties of both cathodes and anodes. The high aspect ratio of CNTs (ratio of length to diameter, which is >10,000 for typical materials) compared to the other carbon additives allows for lower weight doping levels to achieve a comparable percolation threshold. Accordingly, the conductivity and electrochemical properties of the electrodes can be improved at a smaller amount of CNT, leading to a higher gravimetric capacity and power density for the entire electrode.

Many works have investigated the addition of CNTs as a conductive additive for both positive and negative electrodes, using several active materials such as $LiCoO_2$ *(79)*, $LiNi_{0.7}Co_{0.3}O_2$ *(80)*, MnO_2 *(81)*, $LiMn_2O_4$ *(82,83)*, $LiFePO_4$ *(84–86)*, and $LiMnPO_4$ *(86)*. According to these studies, CNTs can not only provide higher conductivity and thus better capability at high C rates, but they could also improve the capacity retention upon cycling by avoiding contact loss between the active particles. In terms of anode improvement, work investigating the enhancement in capacity using MWCNTs with nanobrookite TiO_2 has shown a threefold improvement in capacity over comparable composites using carbon black *(87)*.

While the ability to achieve electrical percolation at low doping levels is a crucial characteristic, there are other potential advantages of using CNTs in the composite blend. Because of the nature of π-orbital overlap in metallic CNT chiralities, electron conduction can occur via ballistic transport (i.e., electrons can transfer with mean free paths on the order of microns along the length of the nanotube unless scattered by a defect). This type of property has the potential for increased C-rate performance when used as an additive, particularly in conjunction with the poor transport inherent in cathode materials.

CNTs also have remarkable thermal conductivity properties that can promote effective heat dissipation within a composite *(88)*, potentially enhancing safety over composite electrodes with inferior carbon additives. In addition, the mechanical properties of CNTs exhibit both strength and flexibility: attractive features to prevent cracking during operation or in vibration environments. The ability to entangle anode and cathode particles using CNTs can obviate issues of active particle separation with extended cycling. All of the recent reports using CNT additives in the composite electrode formulation showed equivalent or improved cycling behavior with CNTs over other carbon additives. Thus the ability to use CNTs as an additive in conventional electrode composites shows tremendous promise to enable higher reversible capacities and rate capability while improving battery cyclability.

9.7 SUMMARY AND OUTLOOK

LIBs have emerged as a viable candidate because of their higher energy density and cycling performance compared to other rechargeable technologies. At the same time, engineering developments warrant the pursuit of novel nanostructures like CNTs to improve the performance of conventional materials. At present, there is a clear advantage for using CNTs in LIBs as an additive within composite electrodes to increase the reversible capacity, enhance the rate capability, and improve cyclability. In addition, the ability to fabricate free-standing electrodes using CNTs (without binder or metal substrate) represents a novel approach to store lithium intrinsically or as a support for ultrahigh capacity materials. As an electrode material for LIBs, CNTs have promising prospects for application. Although the study on CNTs has made much progress, there is great potential. In future studies and research, more effort and energy should be invested in the following aspects. First, it is necessary to further study the lithium storage mechanism of CNTs. The influence of the tubes' length, diameter, number of walls, and defects on the wall on the performance of lithium storage should be analyzed and tested both theoretically and empirically to minimize the loss of specific capacity because of the SEI formation and to improve stability and cycle performance. Second, as a surface template and conductive frame with high specific capacity, CNTs can greatly enhance lithium storage capacity of the composites electrode. Hence, the CNT-based composites electrode needs to be expanded. Third, cost should be reduced, technology should be simplified, and performance should be stabilized in the real application of CNTs in electrode material.

REFERENCES

1. Nishidate, K.; Hasegawa, M. Energetics of Lithium Ion Adsorption on Defective Carbon Nanotubes. *Physical Review B* **2005,** *71* (24).
2. Garau, C.; Frontera, A.; Quinonero, D.; Costa, A.; Ballester, P.; Deya, P. M. Lithium Diffusion in Single-Walled Carbon Nanotubes: A Theoretical Study. *Chemical Physics Letters* **2003,** *374,* 548–555.
3. Garau, C.; Frontera, A.; Quinonero, D.; Ballester, P.; Costa, A.; Deya, P. M. Cation-pi Versus Anion-pi Interactions: A Comparative Ab Initio Study Based on Energetic, Electron Charge Density and Aromatic Features. *Chemical Physics Letters* **2004,** *392,* 85–89.
4. Fagan, S. B.; Guerini, S.; Mendes, J.; Lemos, V. Lithium Intercalation into Single-Wall Carbon Nanotube Bundles. *Microelectronics Journal* **2005,** *36,* 499–501.
5. De las Casas, C.; Li, W. Z. A Review of Application of Carbon Nanotubes for Lithium Ion Battery Anode Material. *Journal of Power Sources* **2012,** *208,* 74–85.

6. Meunier, V.; Kephart, J.; Roland, C.; Bernholc, J. Ab Initio Investigations of Lithium Diffusion in Carbon Nanotube Systems. *Physical Review Letters* **2002,** *88,* 075506.
7. Kumar, T. P.; Ramesh, R.; Lin, Y. Y.; Fey, G. T. K. Tin-Filled Carbon Nanotubes as Insertion Anode Materials for Lithium-Ion Batteries. *Electrochemistry Communications* **2004,** *6,* 520–525.
8. Iijima, S. Helical Microtubules of Graphitic Carbon. *Nature* **1991,** *354,* 56–58.
9. Dai, H. J. Carbon Nanotubes: Opportunities and Challenges. *Surface Science* **2002,** *500,* 218–241.
10. Landi, B. J.; Ganter, M. J.; Cress, C. D.; DiLeo, R. A.; Raffaelle, R. P. Carbon Nanotubes for Lithium Ion Batteries. *Energy & Environmental Science* **2009,** *2,* 638–654.
11. Xiong, Z.; Yun, Y. S.; Jin, H. J. Applications of Carbon Nanotubes for Lithium Ion Battery Anodes. *Materials* **2013,** *6,* 1138–1158.
12. Frackowiak, E.; Beguin, F. Electrochemical Storage of Energy in Carbon Nanotubes and Nanostructured Carbons. *Carbon* **2002,** *40,* 1775–1787.
13. Claye, A. S.; Fischer, J. E.; Huffman, C. B.; Rinzler, A. G.; Smalley, R. E. Solid-State Electrochemistry of the Li Single Wall Carbon Nanotube System. *Journal of the Electrochemical Society* **2000,** *147,* 2845–2852.
14. Mukhopadhyay, I.; Kawasaki, S.; Okino, F.; Govindaraj, A.; Rao, C. N. R.; Touhara, H. Electrochemical Li Insertion into Single-Walled Carbon Nanotubes Prepared by Graphite Arc-Discharge Method. *Physica B* **2002,** *323,* 130–132.
15. Gao, B.; Yue, G. Z.; Qiu, Q.; Cheng, Y.; Shimoda, H.; Fleming, L.; Zhou, O. Fabrication and Electron Field Emission Properties of Carbon Nanotube Films by Electrophoretic Deposition. *Advanced Materials* **2001,** *13,* 1770–1773.
16. Shimoda, H.; Gao, B.; Tang, X. P.; Kleinhammes, A.; Fleming, L.; Wu, Y.; Zhou, O. Lithium Intercalation into Opened Single-Wall Carbon Nanotubes: Storage Capacity and Electronic Properties. *Physical Review Letters* **2002,** *88,* 015502.
17. Jeong, S. K.; Inaba, M.; Mogi, R.; Iriyama, Y.; Abe, T.; Ogumi, Z. Surface Film Formation on a Graphite Negative Electrode in Lithium-Ion Batteries: Atomic Force Microscopy Study on the Effects of Film-Forming Additives in Propylene Carbonate Solutions. *Langmuir* **2001,** *17,* 8281–8286.
18. Ng, S. H.; Wang, J.; Guo, Z. P.; Wang, G. X.; Liu, H. K. Single Wall Carbon Nanotube Paper as Anode for Lithium-Ion Battery. *Electrochimica Acta* **2005,** *51* (1), 23–28.
19. Gao, B.; Kleinhammes, A.; Tang, X. P.; Bower, C.; Fleming, L.; Wu, Y.; Zhou, O. Electrochemical Intercalation of Single-Walled Carbon Nanotubes with Lithium. *Chemical Physics Letters* **1999,** *307,* 153–157.
20. Landi, B. J.; Ruf, H. J.; Evans, C. M.; Cress, C. D.; Raffaelle, R. P. Purity Assessment of Single-Wall Carbon Nanotubes, Using Optical Absorption Spectroscopy. *Journal of Physical Chemistry B* **2005,** *109,* 9952–9965.
21. Landi, B. J.; Cress, C. D.; Evans, C. M.; Raffaelle, R. P. Thermal oxidation Profiling of Single-Walled Carbon Nanotubes. *Chemistry of Materials* **2005,** *17,* 6819–6834.
22. Dai, H. J. Carbon Nanotubes: Synthesis, Integration, and Properties. *Account of Chemical Research* **2002,** *35* (12), 1035–1044.
23. Zhang, D. W.; Zhao, Y. B.; Goodenough, J. B.; Lu, Y. H.; Chen, C. H.; Wang, L.; Liu, J. W. Exfoliation from Carbon Nanotubes versus Tube Size on Lithium Insertion. *Electrochemistry Communications* **2011,** *13,* 125–128.
24. Kang, C.; Lahiri, I.; Baskaran, R.; Kim, W. G.; Sun, Y. K.; Choi, W. 3-Dimensional Carbon Nanotube for Li-Ion Battery Anode. *Journal of Power Sources* **2012,** *219,* 364–370.
25. Varzi, A.; Taubert, C.; Wohlfahrt-Mehrens, M.; Kreis, M.; Schutz, W. Study of Multi-walled Carbon Nanotubes for Lithium-Ion Battery Electrodes. *Journal of Power Sources* **2011,** *196,* 3303–3309.
26. Chew, S. Y.; Ng, S. H.; Wang, J. Z.; Novak, P.; Krumeich, F.; Chou, S. L.; Chen, J.; Liu, H. K. Flexible Free-Standing Carbon Nanotube Films for Model Lithium-Ion Batteries. *Carbon* **2009,** *47,* 2976–2983.
27. Sheem, K.; Lee, Y. H.; Lim, H. S. High-Density Positive Electrodes Containing Carbon Nanotubes for Use in Li-ion Cells. *Journal of Power Sources* **2006,** *158,* 1425–1430.
28. Eom, J. Y.; Kwon, H. S.; Liu, J.; Zhou, O. Lithium Insertion into Purified and Etched Multi-walled Carbon Nanotubes Synthesized on Supported Catalysts by Thermal CVD. *Carbon* **2004,** *42,* 2589–2596.

29. Huang, H.; Zhang, W. K.; Gan, X. P. Improved Electrochemical Storage of Lithium into Multi-walled Carbon Nanotubes by Light Irradiation. *Materials Chemistry and Physics* **2007,** *104,* 271–275.
30. Lahiri, I.; Oh, S. W.; Hwang, J. Y.; Cho, S.; Sun, Y. K.; Banerjee, R.; Choi, W. High Capacity and Excellent Stability of Lithium Ion Battery Anode Using Interface-Controlled Binder-Free Multiwall Carbon Nanotubes Grown on Copper. *ACS Nano* **2010,** *4,* 3440–3446.
31. Wang, G. X.; Yao, J.; Liu, H. K.; Dou, S. X.; Ahn, J. H. Growth and Lithium Storage Properties of Vertically Aligned Carbon Nanotubes. *Metals and Materials International* **2006,** *12,* 413–416.
32. Evanoff, K.; Khan, J.; Balandin, A. A.; Magasinski, A.; Ready, W. J.; Fuller, T. F.; Yushin, G. Towards Ultrathick Battery Electrodes: Aligned Carbon Nanotube – Enabled Architecture. *Advanced Materials* **2012,** *24,* 533–537.
33. Zhang, H.; Cao, G. P.; Wang, Z. Y.; Yang, Y. S.; Shi, Z. J.; Gu, Z. A. Carbon Nanotube Array Anodes for High-Rate Li-ion Batteries. *Electrochimica Acta* **2010,** *55,* 2873–2877.
34. Welna, D. T.; Qu, L. T.; Taylor, B. E.; Dai, L. M.; Durstock, M. F. Vertically Aligned Carbon Nanotube Electrodes for Lithium-ion Batteries. *Journal of Power Sources* **2011,** *196,* 1455–1460.
35. Masarapu, C.; Subramanian, V.; Zhu, H. W.; Wei, B. Q. Long-Cycle Electrochemical Behavior of Multiwall Carbon Nanotubes Synthesized on Stainless Steel in Li Ion Batteries. *Advanced Functional Materials* **2009,** *19,* 1008–1014.
36. Senami, M.; Ikeda, Y.; Fukushima, A.; Tachibana, A. Theoretical Study of Adsorption of Lithium Atom on Carbon Nanotube. *AIP Advanves* **2011,** *1,* 042106.
37. Udomvech, A.; Kerdcharoen, T.; Osotchan, T. First Principles Study of Li and Li^+ Adsorbed on Carbon Nanotube: Variation of Tubule Diameter and Length. *Chemical Physics Letters* **2005,** *406,* 161–166.
38. Zhao, M. W.; Xia, Y. Y.; Liu, X. D.; Tan, Z. Y.; Huang, B. D.; Li, F.; Ji, Y. J.; Song, C. Curvature-Induced Condensation of Lithium Confined inside Single-Walled Carbon Nanotubes: First-Principles Calculations. *Physics Letters A* **2005,** *340,* 434–439.
39. Garau, C.; Frontera, A.; Quinonero, D.; Costa, A.; Ballester, P.; Deya, P. M. Ab Initio Investigations of Lithium Diffusion in Single-Walled Carbon Nanotubes. *Chemical Physics Letters* **2004,** *297,* 85–91.
40. Zhang, Y. P.; Chen, T. Q.; Wang, J. H.; Min, G. Q.; Pan, L. K.; Song, Z. T.; Sun, Z.; Zhou, W. M.; Zhang, J. The Study of Multi-walled Carbon Nanotubes With Different Diameter as Anodes for Lithium-Ion Batteries. *Applied Surface Science* **2012,** *258,* 4729–4732.
41. Mi, C. H.; Cao, G. S.; Zhao, X. B. A Non-GIC Mechanism of Lithium Storage in Chemical Etched MWNTs. *Journal of Electroanalytical Chemistry* **2004,** *562,* 217–221.
42. Gao, B.; Bower, C.; Lorentzen, J. D.; Fleming, L.; Kleinhammes, A.; Tang, X. P.; McNeil, L. E.; Wu, Y.; Zhou, O. Enhanced Saturation Lithium Composition in Ball-Milled Single-Walled Carbon Nanotubes. *Chemical Physics Letters* **2000,** *327,* 69–75.
43. Eom, J.; Kim, D.; Kwon, H. S. Effects of Ball-Milling on Lithium Insertion into Multi-walled Carbon Nanotubes Synthesized by Thermal Chemical Vapour Deposition. *Journal of Power Sources* **2006,** *157,* 507–514.
44. Yang, S. B.; Huo, J. P.; Song, H. H.; Chen, X. H. A Comparative Study of Electrochemical Properties of Two Kinds of Carbon Nanotubes as Anode Materials for Lithium Ion Batteries. *Electrochimica Acta* **2008,** *53,* 2238–2244.
45. Wang, X. X.; Wang, J. N.; Chang, H.; Zhang, Y. F. Preparation of Short Carbon Nanotubes and Application as an Electrode Material in Li-ion Batteries. *Advanced Functional Materials* **2007,** *17,* 3613–3618.
46. Wang, X. X.; Wang, J. N.; Su, L. F. Preparation and Electrochemical Performance of Ultra-Short Carbon Nanotubes. *Journal of Power Sources* **2009,** *186,* 194–200.
47. Terrones, M.; Redlich, P.; Grobert, N.; Trasobares, S.; Hsu, W. K.; Terrones, H.; Zhu, Y. Q.; Hare, J. P.; Reeves, C. L.; Cheetham, A. K.; et al. Carbon Nitride Nanocomposites: Formation of Aligned CxNy Nanofibers. *Advanced Materials* **1999,** *11,* 655–658.
48. Zhong, D. Y.; Zhang, G. Y.; Liu, S.; Wang, E. G.; Wang, Q.; Li, H.; Huang, X. J. Lithium Storage in Polymerized Carbon Nitride Nanobells. *Applied Physics Letters* **2001,** *79,* 3500–3502.
49. Wu, Y. P.; Fang, S. B.; Jiang, Y. Y. Effects of Nitrogen on the Carbon Anode of a Lithium Secondary Battery. *Solid State Ionics* **1999,** *120,* 117–123.
50. Li, X. F.; Liu, J.; Zhang, Y.; Li, Y. L.; Liu, H.; Meng, X. B.; Yang, J. L.; Geng, D. S.; Wang, D. N.; Li, R. Y.; et al. High Concentration Nitrogen Doped Carbon Nanotube Anodes with Superior Li^+

Storage Performance for Lithium Rechargeable Battery Application. *Journal of Power Sources* **2012,** *197,* 238–245.

51. Liu, H.; Zhang, Y.; Li, R. Y.; Sun, X. L.; Desilets, S.; Abou-Rachid, H.; Jaidann, M.; Lussier, L. S. Structural and Morphological Control of Aligned Nitrogen-Doped Carbon Nanotubes. *Carbon* **2010,** *48,* 1498–1507.
52. Terrones, M.; Ajayan, P. M.; Banhart, F.; Blase, X.; Carroll, D. L.; Charlier, J. C.; Czerw, R.; Foley, B.; Grobert, N.; Kamalakaran, R.; et al. N-Doping and Coalescence of Carbon Nanotubes: Synthesis and Electronic Properties. *Applied Physics A* **2002,** *74,* 355–361.
53. Liu, G.; Zheng, H.; Simens, A. S.; Minor, A. M.; Song, X.; Battaglia, V. S. Optimization of Acetylene Black Conductive Additive and PVDF Composition for High-Power Rechargeable Lithium-ion Cells. *Journal of The Electrochemical Society* **2007,** *154,* A1129–A1134.
54. Kim, H. J.; Kim, J. M.; Kim, W. S.; Koo, H. J.; Bae, D. S.; Kim, H. S. Synthesis of $LiFePO_4$/C Cathode Materials through an Ultrasonic-Assisted Rheological Phase Method. *Journal of Alloys and Compounds* **2011,** *509* (18), 5662–5666.
55. Idris, N. H.; Rahman, M. M.; Wang, J. Z.; Chen, Z. X.; Liu, H. K. Synthesis and Electrochemical Performance of LiV_3O_8/carbon Nanosheet Composite as Cathode Material for Lithium-Ion Batteries. *Composites Science and Technology* **2011,** *71,* 343–349.
56. Terrones, M.; Hsu, W. K.; Ramos, S.; Castillo, R.; Terrones, H. The Role of Boron Nitride in Graphite Plasma Arcs. *Fullerene Science and Technology* **1998,** *6,* 787–800.
57. Roth, S.; Blumentritt, S.; Burghard, M.; Jaschinski, O.; Liu, K.; Muster, J.; Philipp, G.; Zha, F.; Redlich, P.; Carrol, D. L.; et al. Density of States and Tunneling Spectroscopy on Molecular Nanostructures. *Thin Solid Films* **1998,** *331,* 45–50.
58. Hsu, W. K.; Firth, S.; Redlich, P.; Terrones, M.; Terrones, H.; Zhu, Y. Q.; Grobert, N.; Schilder, A.; Clark, R. J. H.; Kroto, H. W.; et al. Boron-Doping Effects in Carbon Nanotubes. *Journal of Materials Chemistry* **2000,** *10,* 1425–1429.
59. Mukhopadhyay, I.; Hoshino, N.; Kawasaki, S.; Okino, F.; Hsu, W. K.; Touhara, H. Electrochemical Li Insertion in B-doped Multiwall Carbon Nanotubes. *Journal of The Electrochemical Society* **2002,** *149,* A39–A44.
60. Cui, L. F.; Hu, L. B.; Choi, J. W.; Cui, Y. Light-Weight Free-Standing Carbon Nanotube-Silicon Films for Anodes of Lithium Ion Batteries. *ACS Nano* **2010,** *4,* 3671–3678.
61. Morris, R. S.; Dixon, B. G.; Gennett, T.; Raffaelle, R.; Heben, M. J. High-Energy, Rechargeable Li-ion Battery Based on Carbon Nanotube Technology. *Journal of Power Sources* **2004,** *138,* 277–280.
62. Li, X. F.; Yang, J. L.; Hu, Y. H.; Wang, J. J.; Li, Y. L.; Cai, M.; Li, R. Y.; Sun, X. L. Novel Approach toward a Binder-Free and Current Collector-Free Anode Configuration: Highly Flexible Nanoporous Carbon Nanotube Electrodes with Strong Mechanical Strength Harvesting Improved Lithium Storage. *Journal of Materials Chemistry* **2012,** *22,* 18847–18853.
63. Landi, B. J.; Ganter, M. J.; Schauerman, C. M.; Cress, C. D.; Raffaelle, R. P. Lithium Ion Capacity of Single Wall Carbon Nanotube Paper Electrodes. *Journal of Physical Chemistry C* **2008,** *112,* 7509–7515.
64. Lee, S.; Cho, D.; Jeong, Y. Effects of Post-treated Carbon Nanotube Films on Electrochemical Performance for Bendable Lithium-Ion Batteries. *Fiber Polym* **2015,** *16,* 1600–1604.
65. Lee, J.; Jung, Y.; Song, J.; Kim, J. S.; Lee, G. W.; Jeong, H. J.; Jeong, Y. High-Performance Field Emission from a Carbon Nanotube Carpet. *Carbon* **2012,** *50,* 3889–3896.
66. Hou, P. X.; Bai, S.; Yang, Q. H.; Liu, C.; Cheng, H. M. Multi-step Purification of Carbon Nanotubes. *Carbon* **2002,** *40,* 81–85.
67. Wang, Z. Y.; Chen, G.; Xia, D. G. Coating of Multi-walled Carbon Nanotube with SnO_2 Films of Controlled Thickness and its Application for Li-ion Battery. *Journal of Power Sources* **2008,** *184,* 432–436.
68. Zhang, H. X.; Feng, C.; Zhai, Y. C.; Jiang, K. L.; Li, Q. Q.; Fan, S. S. Cross-Stacked Carbon Nanotube Sheets Uniformly Loaded with SnO_2 Nanoparticles: A Novel Binder-Free and High-Capacity Anode Material for Lithium-Ion Batteries. *Advanced Materials* **2009,** *21,* 2299–2304.
69. Park, J.; Moon, W. G.; Kim, G. P.; Nam, I.; Park, S.; Kim, Y.; Yi, J. Three-Dimensional Aligned Mesoporous Carbon Nanotubes Filled with Co_3O_4 Nanoparticles for Li-ion Battery Anode Applications. *Electrochimica Acta* **2013,** *105,* 110–114.

70. Wang, W.; Kumta, P. N. Nanostructured Hybrid Silicon/Carbon Nanotube Heterostructures: Reversible High-Capacity Lithium-Ion Anodes. *ACS Nano* **2010,** *4,* 2233–2241.
71. Weng, W.; Lin, H. J.; Chen, X. L.; Ren, J.; Zhang, Z. T.; Qiu, L. B.; Guan, G. Z.; Peng, H. S. Flexible and Stable Lithium Ion Batteries Based on Three-Dimensional Aligned Carbon Nanotube/silicon Hybrid Electrodes. *Journal of Materials Chemistry A* **2014,** *2,* 9306–9312.
72. Hosono, E.; Wang, Y. G.; Kida, N.; Enomoto, M.; Kojima, N.; Okubo, M.; Matsuda, H.; Saito, Y.; Kudo, T.; Honma, I.; et al. Synthesis of Triaxial $LiFePO_4$ Nanowire with a VGCF Core Column and a Carbon Shell through the Electrospinning Method. *ACS Applied Materials & Interfaces* **2010,** *2,* 212–218.
73. Liu, X. M.; Huang, Z. D.; Oh, S.; Ma, P. C.; Chan, P. C. H.; Vedam, G. K.; Kang, K.; Kim, J. K. Sol-gel Synthesis of Multiwalled Carbon Nanotube-$LiMn_2O_4$ Nanocomposites as Cathode Materials for Li-ion Batteries. *Journal of Power Sources* **2010,** *195,* 4290–4296.
74. Zhou, Y. K.; Wang, J.; Hu, Y. Y.; O'Hayre, R.; Shao, Z. P. A Porous $LiFePO_4$ and Carbon Nanotube Composite. *Chemical Communications* **2010,** *46,* 7151–7153.
75. Winter, M.; Besenhard, J. O.; Spahr, M. E.; Novak, P. Insertion Electrode Materials for Rechargeable Lithium Batteries. *Advanced Materials* **1998,** *10,* 725–763.
76. Zhou, X. W.; Wu, G. M.; Gao, G. H.; Cui, C. J.; Yang, H. Y.; Shen, J.; Zhou, B.; Zhang, Z. H. The Synthesis, Characterization and Electrochemical Properties of Multi-wall Carbon Nanotube-Induced Vanadium Oxide Nanosheet Composite as a Novel Cathode Material for Lithium Ion Batteries. *Electrochimica Acta* **2012,** *74,* 32–38.
77. Cao, Z. Y.; Wei, B. Q. V_2O_5/single-Walled Carbon Nanotube Hybrid Mesoporous Films as Cathodes with High-Rate Capacities for Rechargeable Lithium Ion Batteries. *Nano Energy* **2013,** *2,* 481–490.
78. Kim, S. W.; Seo, D. H.; Gwon, H.; Kim, J.; Kang, K. Fabrication of FeF_3 Nanoflowers on CNT Branches and Their Application to High Power Lithium Rechargeable Batteries. *Advanced Materials* **2010,** *22,* 5260–5264.
79. Wang, G. P.; Zhang, Q. T.; Yu, Z. L.; Qu, M. Z. The Effect of Different Kinds of Nano-Carbon Conductive Additives in Lithium Ion Batteries on the Resistance and Electrochemical Behavior of the $LiCoO_2$ Composite Cathodes. *Solid State Ionics* **2008,** *179,* 263–268.
80. Li, X. L.; Kang, F. Y.; Shen, W. C. A Comparative Investigation on Multiwalled Carbon Nanotubes and Carbon Black as Conducting Additive in $LiNi_{0.7}Co_{0.3}O_2$. *Electrochemical and Solid-State Letters* **2006,** *9,* A126–A129.
81. Yue, H. J.; Huang, X. K.; Yang, Y. Preparation and Electrochemical Performance of Manganese Oxide/carbon Nanotubes Composite as a Cathode for Rechargeable Lithium Battery with High Power Density. *Materials Letters* **2008,** *62,* 3388–3390.
82. Chen, S. Y.; Mi, C. H.; Su, L. H.; Gao, B.; Fu, Q. B.; Zhang, X. G. Improved Performances of Mechanical-Activated $LiMn_2O_4$/MWNTs Cathode for Aqueous Rechargeable Lithium Batteries. *Journal of Applied Electrochemistry* **2009,** *39,* 1943–1948.
83. Ma, S. B.; Nam, K. W.; Yoon, W. S.; Bak, S. M.; Yang, X. Q.; Cho, B. W.; Kim, K. B. Nano-Sized Lithium Manganese Oxide Dispersed on Carbon Nanotubes for Energy Storage Applications. *Electrochemistry Communications* **2009,** *11,* 1575–1578.
84. Li, X. L.; Kang, F. Y.; Bai, X. D.; Shen, W. Energetics of Lithium Ion Adsorption on Defective Carbon Nanotubes. *Electrochemistry Communications* **2007,** *9,* 663–666.
85. Liu, Y. J.; Li, X. H.; Guo, H. J.; Wang, Z. X.; Peng, W. J.; Yang, Y.; Liang, R. F. Effect of Carbon Nanotube on the Electrochemical Performance of C-lifepo4/graphite Battery. *Journal of Power Sources* **2008,** *184,* 522–526.
86. Kavan, L.; Bacsa, R.; Tunckol, M.; Serp, P.; Zakeeruddin, S. M.; Le Formal, F.; Zukalova, M.; Graetzel, M. Multi-walled Carbon Nanotubes Functionalized by Carboxylic Groups: Activation of TiO_2 (Anatase) and Phosphate Olivines ($LiMnPO_4$; $LiFePO_4$) for Electrochemical Li-storage. *Jouranl of Power Sources* **2010,** *195,* 5360–5369.
87. Lee, D. H.; Kim, D. W.; Park, J. G. Enhanced Rate Capabilities of Nanobrookite with Electronically Conducting MWCNT Networks. *Crystal Growth & Design* **2008,** *8,* 4506–4510.
88. Berber, S.; Kwon, Y. K.; Tomanek, D. Microscopic Formation Mechanism of Nanotube Peapods. *Physical Review Letters* **2002,** *88,* 185502.

CHAPTER 10

Carbon Nanotubes for Electrochemical Capacitors

Y. Zhang[1], H. Wu[1], L. Cheng[1], L. Zheng[2], W. Yuan[1]
[1]Northwestern Polytechnical University, Xi'an, China; [2]Khalifa University of Science, Technology and Research, Abu Dhabi, United Arab Emirates

Contents

10.1 INTRODUCTION

Because of the tremendous growth of the global economy and the explosion of the world population, the demand for a safe, green, and stable energy supply has been unprecedentedly accelerated *(1—3)*. Nevertheless, with the depletion of traditional fossil fuels (coal, petroleum, natural gas, etc.), the energy crisis has thus increasingly become an urgent issue. In addition, severe environmental pollution as well as global warming caused by the burning of fossil fuels remains another great challenge. To solve the stated problems, technologies of utilizing new energy resources including wind, solar, nuclear, water, etc.

Industrial Applications of Carbon Nanotubes
ISBN 978-0-323-41481-4

have been developed, and various advanced, efficient, and environmentally friendly energy storage systems (ESSs) have also been invented *(4,5)*.

It is commonly believed that an ideal ESS should simultaneously possess high energy density and high power density, good charging/discharging rate stability, excellent recyclability, light weight, reliable performance safety, low cost, and environmental friendliness *(3–5)*. However, batteries usually possess high energy density (up to 180 Wh/kg) but relatively low power density (~2 kW/kg), which significantly inhibits their application as energy storage devices that need a rapid charge/discharge speed. However, for traditional electrostatic capacitors, a much higher power density (~40 kW/kg) and low energy density (~0.03 Wh/kg) are provided *(6,7)*. Neither of the two important ESSs could meet the need of both high energy density and high power density for future electrochemical capacitors (ECs). Interestingly, the emergence of ECs has exactly bridged the gap of electrochemical performance between batteries and traditional electrostatic capacitors, as illustrated in the Ragone plot (Fig. 10.1) *(7)*.

ECs, which are also called supercapacitors or ultracapacitors, have attracted great attention in the fields of academia and industry since they were patented in 1957 by H.I. Becker because of their (1) high power capability (60–120 s is typical), (2) relatively high energy density (compared with traditional electrostatic capacitors), (3) long cycle life

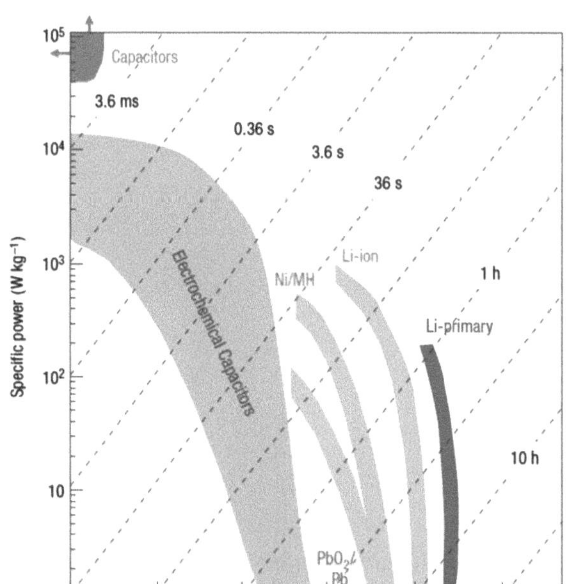

Figure 10.1 Ragone plot showing the power density against energy density for major energy storage system (ESS) devices. *(Reproduced from Simon, P.; Gogotsi, Y. Materials for Electrochemical Capacitors.* Nature Materials ***2008,*** *7 (11), 845–854. Copyright 2008 Nature Publishing Group.)*

(>10^5 cycles), and (4) excellent reversibility (90–95% or higher) *(6,8)*. ECs also have been equipped or have been considered for application in a variety of devices such as hybrid electric vehicles, electric vehicles, cranes, uninterruptible power supplies, portable consumer devices, and emergency system devices, for all of which high power density is needed. Although the energy density of ECs is much higher than traditional electrostatic capacitors, it is still insufficient especially when compared with batteries. In recent years, plenty of research has been performed to enhance the energy density of ECs while maintaining their high power density to meet the growing needs of high-performance ESSs in modern society.

The ECs can be divided into two main types according to different electric energy storage mechanisms, that is, electrical chemical double-layer capacitors (EDLCs) and pseudocapacitors. EDLCs are similar to traditional capacitors, they store electric charges electrostatically on electrode/electrolyte interfaces of both negative and positive electrodes to form two electric double layers (Fig. 10.2A) *(9)*. The mechanism of surface electrode charge generation includes surface dissociation, ion adsorption from the electrolyte, and crystal lattice defects. A theoretical model of double-layer capacitance was first established by Helmholtz, which was further improved to be more accurate by

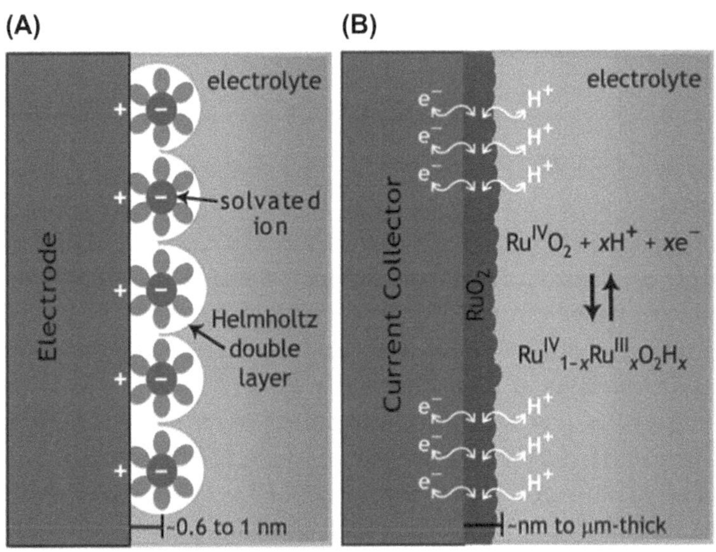

Figure 10.2 Schematic of two different charge storage mechanisms via (A) electrochemical double-layer capacitance (EDLC) or (B) redox reaction-based pseudocapacitance. *(Reproduced from Long, J. W.; Bélanger, D.; Brousse, T.; Sugimoto, W.; Sassin, M. B.; Crosnier, O. Asymmetric Electrochemical Capacitors-Stretching the Limits of Aqueous Electrolytes.* Mrs Bulletin ***2011,*** *36 (07), 513–522. Copyright 2011 Cambridge University Press.)*

Gouy, Chapman, and Stern *(10)*. The capacitance of an EDLC is generally described according to Eq. (10.1):

$$C = \varepsilon_r \varepsilon_0 A/d \quad (10.1)$$

where C is the double-layer capacitance (F), ε_r is the relative dielectric constant of electrolyte, ε_0 is the dielectric constant of the vacuum, d is the effective thickness (the charge separation distance), and A is the specific surface area (SSA) of the electrode.

Different to EDLCs, pseudocapacitors store charges mainly through fast and reversible redox reactions on the surfaces of pseudocapacitance materials (Fig. 10.2B) *(9)*. Obviously, a high surface area of an electrode is still beneficial for pseudocapacitors. The most commonly used pseudocapacitance materials include a variety of transition metal oxides such as RuO_2, Fe_3O_4, Co_3O_4, NiO, and MnO_2, conducting polymers (CPs) such as polyanilines (PANI), polypyrroles (PPy), and polythiophenes (PTh), as well as O- or N-containing surface functional groups. Compared with EDLCs the specific capacitance (SC) of which merely rely on double-layer capacitance (typically 100–250 F/g), pseudocapacitors possess much higher SC (typically 300–1000 F/g).

In addition to SC, energy density and power density are two other fateful electrochemical properties of ECs, the maximum value of which is described in the following equations, respectively *(2,11)*:

$$E = \frac{1}{2}CV^2 \quad (10.2)$$

$$P = V^2/4R \quad (10.3)$$

where C is the SC of the EC, V is the operating voltage, which is strongly limited by the stability of electrolytes (1.2 V for water, 2.5–2.7 V for organic electrolytes, 3–6 V for ionic liquids) *(6)*, and R represents the equivalent series resistance (ESR) including the resistance of the electrode materials, the diffusion resistance of electrolyte ions in the electrodes, and the resistance of electrolytes. Typically, the ESR of pseudocapacitors is much higher than that of EDLCs because charges have to transfer across the interface of the electrolyte/electrode in a pseudocapacitor *(12)*.

Based on Eq. (10.1), it is evident that the capacitive capability of an EDLC strongly depends on the surface area of the electrode. As a result, carbon-based active materials such as activated carbons (ACs), mesoporous carbon, carbide-derived carbons (CDCs), carbon nanotubes (CNTs), and graphene with high SSA, low weight, low electrical resistance, excellent thermal and chemical stability have been selected as electrode materials for EDLCs. Among these carbon materials, ACs are the most successful commercial electrode materials applied in EDLCs because of their many merits including high SSA, easy processability, good accessibility, relatively low cost, high chemical stability, and relatively good electrical conductivity. Unfortunately, in spite of the SSA of commercial

ACs, which can reach as high as 1000–3000 m^2/g, the SC of AC-based EDLCs does not increase linearly with its SSA when it exceeds 1500 m^2/g.

Porous carbon materials typically contain three levels of pores: macropores (>50 nm), mesopores (2–50 nm), and micropores (<2 nm). When the porous carbon is employed as an electrode of ECs, all levels of pores possess their particular function during the charge storage process. For example, macropores could serve as a buffer zone for electrolyte ions, mesopores are beneficial for the rapid diffusion of electrolyte ions, and micropores with the highest SSA could optimize the charge storage in ECs *(13)*. It is believed that the relative content of macropores, mesopores, and micropores as well as their connection form of most ACs is not optimal or reasonable because the control on pore size distribution (PSD) during activation is relatively difficult. As a result, abundant micropores in ACs are inaccessible for electrolyte ions, especially for large electrolyte ions of organic electrolytes and ionic liquids, which could provide a higher operating voltage than aqueous electrolytes *(14,15)*. Ironically, the high SSA of ACs is mainly contributed by micropores, which make ACs the most successful commercial electrode material; however, this is simultaneously the major drawback of ACs because abundant micropores severely decrease the rate capability of AC-based ECs. Together with other inherent drawbacks of ACs such as low electrical conductivity, disordered texture, and impurities inherited from precursors, their wider application in ECs has been seriously limited *(16)*. Thus to fabricate next-generation ECs with high performance, it is highly desired to select a suitable electrode material that should simultaneously possess high electrical conductivity, high effective SSA, tunable PSD, multiple and short diffusion paths, and high mechanical and chemical stability *(17)*. As a consequence, nanostructured carbons such as CNTs, graphene, mesoporous carbon, CDCs, and carbon aerogels have been developed *(5)*.

Among all of the nanostructured carbon materials, CNTs including single-walled carbon nanotubes (SWCNTs), double-walled carbon nanotubes (DWCNTs), and multiwalled carbon nanotubes (MWCNTs) are undoubtedly one of the most important and attractive carbon materials for electrodes of ECs because of their excellent comprehensive performance *(18)*. Different to ACs with a low electrical conductivity ($10-10^2$ S/m), CNTs possess an ultrahigh electrical conductivity up to 10^7 S/m because of their unique characteristic of superior ballistic electronic conduction *(19–22)*. Besides, unlike ACs whose high SSA cannot be fully utilized because many micropores are inaccessible for electrolyte ions, the high SSA (typically over 1500 m^2/g) of CNTs would be easy to access if the CNT-based electrode had been designed and fabricated with a proper structure. The high electrical conductivity and high SSA as well as combined high mechanical strength, low mass density, outstanding chemical stability, and electrocatalytic activity are keys to high-performance ECs by providing a high SC and a low ESR.

10.2 ARCHITECTURE DEPENDENCE OF CNTs FOR CAPACITANCE

10.2.1 CNT Architecture Variety for Electrodes

CNTs are 1D hollow tubular graphite with ultrahigh length-diameter ratio. CNTs including SWCNTs, DWCNTs, and MWCNTs can be prepared through various methods such as chemical vapor decomposition (CVD), laser ablation, arc discharge, etc. *(23)*. Like other carbon materials, electric charges are stored electrostatically in the electric double layers on the huge surface of individual CNTs. However, by simply mixing pure CNTs with carbon black and polymer binders to form an electrode, typically low SC (varied from 20 to 80 F/g) as well as energy density (~7 Wh/kg) and power densities (~20 kW/kg) of the as-prepared ECs are exhibited *(24–26)*. To make full use of the many merits of individual CNTs at a macroscopic level in energy storage devices including ECs, it is desirable to assemble CNTs into various macroscopic CNT-based materials with huge SSA and proper PSD *(6,27)*. Besides, the effective SSA and electrochemical performance of CNT-based electrodes are closely related to their architecture, which largely determines the transportation paths of both ions and electrons.

Macroscale CNT assemblies can be strategically divided into two groups, unaligned and aligned. The unaligned CNT assemblies means that most CNTs are arranged or grown randomly on the base; agglomeration of large amounts of CNTs, which are detrimental for effective SSA and ion transportation, can be observed anywhere in the unaligned CNT films and networks *(28,29)*. The aligned CNT assemblies with highly organized alignments include 1D-CNT fibers/yarns, 2D-CNT films/papers, and 3D-CNT arrays/solids; they are usually produced based on initially elaborate prototype designing. Thus far, plenty of work has been carried out to fabricate regular macroscopic CNT assemblies. Compared with the unaligned ones, aligned CNT assemblies are more easily handled and applied under various conditions in addition to larger values of both SC and capacitance retention because of the following advantages *(28,30,31)*:

1. Aligned designs can provide proper PSD and ordered arrays with highly organized transportation paths.
2. Appropriate distance between CNTs, fibers, films, and arrays can be adjusted in the aligned architecture.
3. Aligned architecture is considered to have direct and firm electrical contact with current collectors, similar to a metallic plate or foamed network.

Continuous and ordered CNT fibers with diameters ranging from micrometers to millimeters are 1D-CNT assemblies. During the past several years, there have been numerous reports in the literature of the fabrication of neat CNT fibers through a wide range of technologies *(32)*. Typically, CNT fibers can be prepared through two kinds of spinning methods: solution-state spinning and solid-state spinning, which consists of spinning from a vertically aligned CNT array, spinning from CNT aerogel, and twisting/rolling from a CNT film *(33–35)*. Similar to carbon fibers, the CNT fibers are also flexible, integratable, wearable, and lightweight. Supercapacitors based on CNT fibers can be easily fabricated through directly twisting two fiber electrodes or wrapping two electrodes on fiber substrate to form a coaxial structure in the air under mild conditions; they can be also more stably operated with a long lifetime during application *(35)*. Ever since a 30 cm-long CNT fiber was fabricated in 2002, plenty of research has been focused on the electrochemical performance of CNT fiber-based ECs.

As another macroscale CNT assembly, 2D-CNT films with thicknesses ranging from tens of nanometers to micrometers can be prepared through direct growth using the CVD method, pulling out a CNT bundle with tweezers from CNT arrays, posttreating CNT suspensions, and dry drawing from CNT arrays *(36,37)*. CNT films not only possess all the merits of CNTs, their interconnected structure makes them a suitable self standing electrode material with better electrical conductivity for supercapacitors. Besides, CNT films are highly stretchable and offer a high fault tolerance because they possess many current pathways; CNT films can still transport electrical charges even when certain links are disconnected. Besides, highly stretchable CNT films could offer a high fault tolerance because of the existing abundant current pathways. CNT films can still transport electrical charges even when some links are destroyed during stretching *(11,38)*.

The third kind of macroscale CNT assembly is 3D-CNT arrays/solids, which refer to aligned CNTs vertically grown on certain plane substrates with heights up to the millimeter scale. Typically, aligned CNT arrays can be prepared by the CVD method on nonconductive substrates such as silicon wafers and silicon dioxide films, which were previously deposited with catalysts; the grown CNT arrays could be further transferred to another substrate according to their final application *(39,40)*. For application in ECs as free-standing electrodes, CNT arrays are often transferred to conductive substrates such as metal foils, which could simultaneously serve as current collectors. In fact, because of their high electrical conductivity, ordered porous structures with multiple paths for diffusion of ions and electrons, excellent mechanical properties, highly effective SSA, and tunable PSD, aligned CNT arrays are very promising for electrode materials of ECs *(5,41,42)*. Abundant research has been carried out to explore the electrochemical potential of CNT arrays ever since they were first reported in 1996 *(39,43–46)*.

In addition to maintaining the unique intrinsic properties of individual CNTs, regular macroscale CNT assemblies have been endowed with a number of new properties that are very appealing in novel ECs. For example, highly flexible and stretchable 1D-CNT fibers with high electrical conductivity and tensile strength are very promising materials for wearable ECs. As a basic building block, CNT fibers can be fabricated into fabrics and any other designed preforms through a simple knitting process. Similar to 1D-CNT fibers, the 2D-CNT films also possess excellent mechanical flexibility and stretchability. Moreover, CNT films with optical transparency could be developed into stretchable thin film-based transparent electronics. With the help of the zipping effect of liquids, the 3D-CNT arrays can be drawn into superaligned CNT arrays of different shapes with higher density and more regular structure, while the intrinsic properties of pristine CNT arrays such as high SSA and high electrical conductivity could be maintained *(47)*. With decreased average tube separation (~0.9 nm), which is still large enough for the access of electrolyte ions, a higher volumetric capacitance of the CNT array-based electrode is thus expected *(47,48)*.

10.2.2 Electrochemical Performance of CNT-Based Capacitors

As a very promising electrode material for ECs, CNTs with high SSA and high electrical conductivity have been extensively studied in both aqueous and nonaqueous electrolytes, exhibiting an SC of 5–200 F/g based on SSA ranging from 120 to 500 m^2/g. Because of an efficient diffusion within the unique porous structure for the electrolyte ions, CNT-based ECs could provide a considerable SC even at an ultrahigh current density, showing a much higher rate capability than traditional ACs. Remarkably, the SC of CNT-based ECs is highly dependent on the SSA and PSD of the used electrodes. As a result, the most direct way to boost the SC of CNT-based ECs is to increase their accessible SSA.

Pioneering work has been reported by Niu et al. who prepared an EC from catalytically grown MWCNTs with high purity and a diameter of ~8 nm (Fig. 10.3) *(49)*. After being treated with nitric acid, the MWCNTs were fabricated into free-standing mats with open porosity and SSA of 430 m^2/g, which is much higher than the pristine nanotubes (~250 m^2/g). Different to ACs, which contain abundant micropores, the measured micropore volume of such CNT mats is negligible, with an average pore diameter of 9.2 nm, which is easily accessible for most aqueous and nonaqueous electrolyte ions. As a result, the as-prepared ECs employing an electrolyte of 38 wt% H_2SO_4 solution exhibited an SC of 102 F/g at 1 Hz, corresponding to areal capacitance of 24.2 F/m^2. In addition, because of the low electrical resistance of the MWCNTs, the same EC could easily provide a power density of >8 kW/kg. In another work carried out by Fang et al., the MWCNTs directly grown on Ti foil were treated with nitric acid *(50)*. Because of a faster charge transfer through the graphite edge planes of MWCNTs with open tips after HNO_3 treating, more stable cyclic voltammograms of acid-treated MWCNTs were shown at the scan rate >80 mV/s. An et al. submitted CNTs to heat treating at high temperature to obtain a higher SSA, a higher content of mesopores in the range of 3–5 nm, and a higher electrical conductivity *(51)*. The supercapacitor based on the treated CNTs showed a maximum SC of 180 F/g as well as a high power density (20 kW/kg at an energy density of 6.5 Wh/kg) because of the low ESR.

Besides, an activation process has also been employed by many researchers to further enhance the SSA and SC of CNT-based ECs. Unlike traditional AC powders, activated CNTs possess shorter paths for ion diffusion because of their regular nanotubular

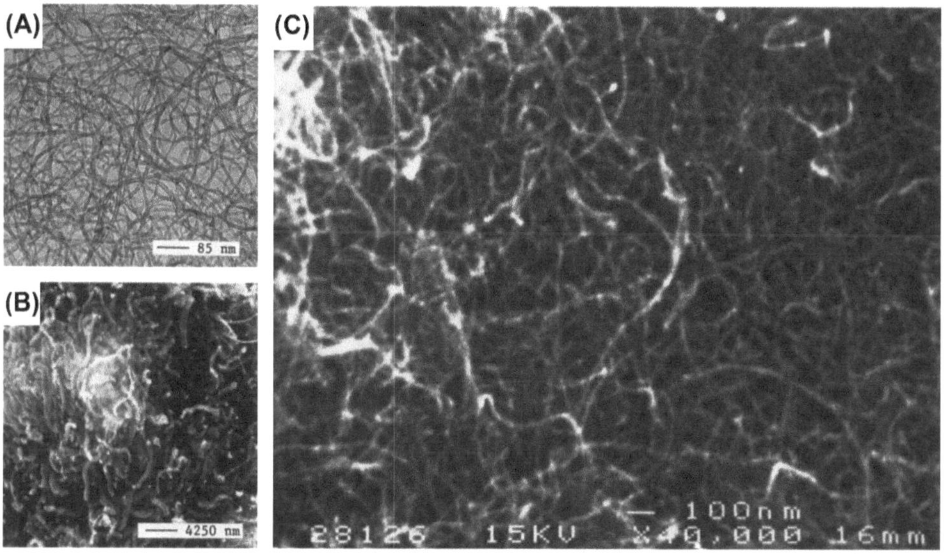

Figure 10.3 (A) Transmission electron microscopy (TEM) micrograph of highly dispersed catalytically grown carbon nanotubes. The average diameter is ~80 Å; (B) Scanning electron microscope (SEM) micrograph of as-made catalytically grown carbon nanotubes (CNTs). The CNTs are formed as bundles with individual nanotubes arranged semiparallel to each other. The diameter of the bundles is ~2 mm with length of ~20 mm; and (C) SEM micrograph of a nanotube electrode. The electrode consists of randomly entangled and cross-linked CNTs that have uniform diameters of ~80 Å. *(Reproduced from Niu, C.; Sichel, E.K.; Hoch, R.; Moy, D.; Tennent, H. High Power Electrochemical Capacitors Based on Carbon Nanotube Electrodes.* Applied Physics Letters ***1997,*** *70, (11), 1480–1482. Copyright 1997 AIP Publishing LLC.)*

morphology (Fig. 10.4) *(52)*. Jiang et al. prepared ECs employing chemically activated CNTs by KOH as electroactive material *(52)*. With an increased Brunauer–Emmett–Teller SSA of ~510.5 m^2/g (three times than pristine CNTs), the activated CNT-based EC exhibited a doubled SC of 50 F/g. In another work, Frackowiak et al. prepared activated MWCNTs with a much higher SSA (1050 m^2/g) at a KOH/C ratio of 4:1 *(53)*. With a well-kept noticeable mesoporosity as well as abundant micropores on the surface of MWCNTs, the ECs based on such activated MWCNTs provided an SC of 90 F/g in aqueous electrolyte of 6 M KOH. However, although a higher surface area of CNTs could be obtained through corroding their outer surface, the electrical conductivity of CNTs is simultaneously decreasing. Therefore finding a proper balance between the conductivity and the SSA is of great importance to obtain both high SC and high rate capability *(54)*.

Another effective strategy to improve the SC of CNT-based ECs is to introduce pseudocapacitance through functionalizing CNTs. It has been reported that carboxyl groups could be introduced on the surfaces of MWCNTs through oxidation reaction

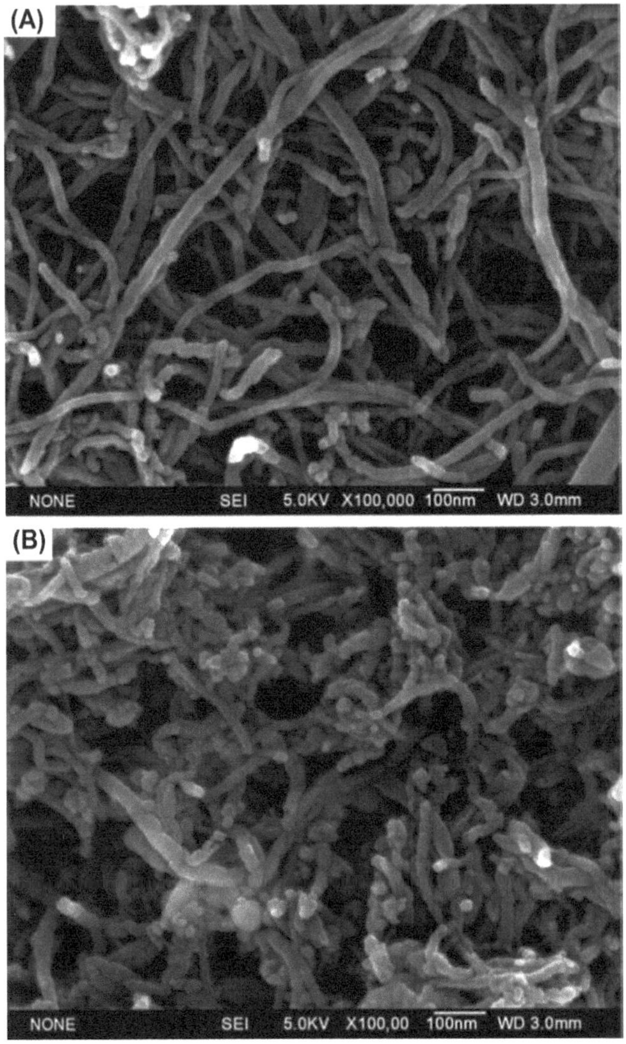

Figure 10.4 Scanning electron microscope (SEM) images of the normal carbon nanotubes (CNTs) (A) and the activated CNTs (B). *(Reproduced from Jiang, Q.; Qu, M.; Zhou, G.; Zhang, B.; Yu, Z. A Study of Activated Carbon Nanotubes as Electrochemical Super Capacitors Electrode Materials.* Materials Letters ***2002,*** *57 (4), 988–991. Copyright 2002 Elsevier.)*

in an extremely aggressive oxidizing agent such as mixed acid H_2SO_4/HNO_3 *(50,55)*. In addition to the introduced pseudocapacitance, the good hydrophilicity of the carboxyl groups significantly improved the wettability of MWCNTs in aqueous electrolyte, which in fact increased the effective surface area of MWCNTs. As a result, with the increasing oxidation time that resulted in a better hydrophilicity of MWCNTs, an increasing SC was meanwhile exhibited. However, the excessive oxidation process also severely damaged the surface structure of MWCNTs and significantly decreased their electrical conductivities. Therefore a proper balance between hydrophilicity and conductivity was required.

Wang et al. reported a kind of functionalized MWCNTs prepared by heating MWCNTs in NaOH solution followed by treating in mixed acid H_2SO_4/HNO_3 under ultrasonic conditions *(56)*. Benefiting from the increased SSA and appropriate PSD, the functionalized MWCNTs showed a much improved SC of 85 F/g in H_2SO_4 electrolyte, which was about three times the raw MWCNTs. By further introducing hydroquinone with redox activity into the H_2SO_4 electrolyte, the electrodes based on modified MWCNTs could present an ultrahigh SC of 3199 F/g at 5 mV/s because of the extra pseudocapacitance provided by hydroquinone. After 350 cycles of charging/discharging, the electrodes could still maintain a stable SC of ~2250 F/g in H_2SO_4 electrolyte containing 0.075 M hydroquinone.

By surface treating CNTs, which were directly grown on the bulk Ni substrates through the plasma enhanced CVD method using an ammonia plasma, the SC of the as-prepared electrodes was significantly increased from 38.7 F/g to 207.3 F/g by Yoon and coworkers (Fig. 10.5) *(57)*. Jurewicz et al. modified catalytic MWCNTs through KOH activation at 800°C and/or ammoxidation at 350°C, and investigated the changes of both texture and chemical composition of MWCNTs after ammoxidation *(58)*. It has been proved that the introduced oxygenated functionality and nitrogen-containing groups were both beneficial for the energy storage capability of the modified MWCNTs. Particularly, the nitrogen-containing groups improved the SC of the negative electrode in the alkaline electrolyte. As a result, a few-fold increase of SC was observed in the modified MWCNT-based electrodes. In addition, because of the presence of nitrogen in MWCNTs, the leakage current and the self-discharge of supercapacitors have been effectively limited.

10.3 SUPERCAPACITORS BASED ON ALIGNED CNTs

10.3.1 Supercapacitor Electrode Based on Aligned SWCNT

Compared with DWCNTs and MWCNTs, SWCNTs possess the highest theoretical SSA (1300 m^2/g). As a result, SWCNT-based macroscale assemblies are expected to provide the highest SC. Kozlov and his coworkers prepared purified SWCNT fibers with different morphology (hollow, folded ribbon, and solid) through spinning of SWCNT solution followed by annealing *(27)*. The achieved hollow-fiber structures are unprecedented for directly spun polymer-free fibers, and suggest applications such as energy storage devices, materials separation, materials absorption, and sensing. With a degree of SWCNT alignment, which is beneficial for electron transportation and ion diffusion, the purified SWCNT fibers without polymer binder exhibited a high SC of 100 F/g and a high electrical conductivity of 140 S/cm at room temperature.

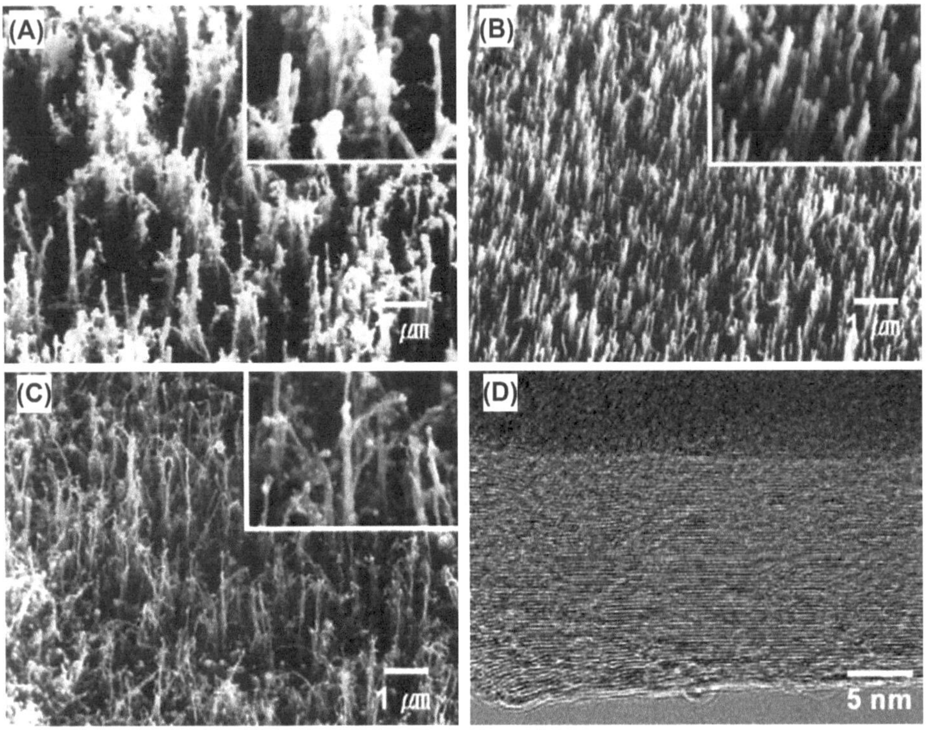

Figure 10.5 Electron microscope images of the carbon nanotubes (CNTs) directly grown on Ni substrates. (A)–(C) are the scanning electron microscope (SEM) images of CNTs, which are grown on the Ni substrates etched by NH_3 plasma for (A) 0 (bare Ni foil), (B) 5, and (C) 10 min, respectively. (D) High-resolution transmission electron microscopy (TEM) image of the CNT shown in (B). *(Reproduced from Yoon, B. J.; Jeong, S. H.; Lee, K. H.; Kim, H. S.; Park, C. G.; Han, J. H. Electrical Properties of Electrical Double Layer Capacitors With Integrated Carbon Nanotube Electrodes.* Chemical Physics Letters ***2004,*** *388 (1), 170–174. Copyright 2004 Elsevier.)*

2D SWCNT films have been prepared by Izadi-Najafabadi et al. through shearing water-assisted CVD-grown SWCNT arrays between two glass slides *(59)*. With high carbon purity (99.98%), exclusive SWCNT selectivity (>99%), negligible carbonaceous impurity (<2% amorphous carbon), and near-ideal SSA (1300 m^2/g), the water-assisted CVD-grown SWCNTs satisfy all the fundamental requirements for supercapacitor electrodes. As a result, electrodes based on such SWCNT films possessed mesoporous structure in addition to a high SSA. While the ECs were operated at a higher voltage (4 V), a high SC of 160 F/g with durable full charge/discharge cyclability could be provided. In addition to a high-energy density of 94 Wh/kg (47 Wh/L) and power density of 210 kW/kg (105 kW/L), far exceeding those of AC both gravimetrically (33 Wh/kg, 60 kW/kg) and volumetrically (19.8 Wh/L, 36 kW/L). Compared with postdeposited SWCNT films, directly grown SWCNT films through the CVD method usually possess much higher electrical conductivity of ~2000 S/cm as well as higher strain tolerance because of their regular interconnected architecture with multiple current paths. Ma et al. prepared continuous reticulate SWCNT films through improving the floating catalyst chemical vapor deposition (FCCVD) technique *(36)*. By combining directly grown SWCNT films with continuous reticulate architecture with polydimethylsiloxane (PDMS) with enhanced prestrain, Niu et al. successfully fabricated highly stretchable buckled SWCNT films, which can be stretched under a strain of 140% without a significant change of the resistance *(37)*. Employing these buckled SWCNT films as electrodes, highly stretchable supercapacitors with H_2SO_4-polyvinyl alcohol (PVA) gel as both electrolyte and separator were fabricated. Benefiting from the particular buckled structure, the ECs showed an SC of 53 F/g and a power density of 32 kW/kg, both of which kept nearly unchanged even when the EC was stretched under a strain of 120%, indicating that such SWCNT films are a promising candidate for stretchable wearable electronics and supercapacitors.

Catalyst-free SWCNT arrays with little bundling have been prepared by Kimizuka et al. through a highly effective water-assisted CVD method, showing an excellent capacitor performance attributable to the intrinsic nature of the SWCNTs such as high carbon purity (>99.9%) and high SSA (1100 m^2/g) *(29,60)*. While serving as electrodes of ECs, an unusual electrochemical property characterized by a butterfly-shaped cyclic voltammogram was observed because of the effect of electrochemical doping in existing semiconductive SWCNTs, showing a higher capacitance at the higher charging potentials (18 F/g at a charging potential of 2 V and 25 F/g at 3 V) *(29)*. Futaba et al. reported a rational and general method to fabricate high densely packed SWCNT arrays with a more regular structure by using the zipping effect of liquids to draw CVD-grown aligned SWCNT array tubes together *(47)*. This bulk superaligned SWCNT array could retain the intrinsic properties of individual SWCNTs such as high surface area (1000 m^2/g) and electrical conductivity. Through controlling the fabrication process, this new "SWCNT solid" can be fabricated into a wide range of solids with different shapes and structures. Because of the regular structure with average tube separation (~0.9 nm) large enough for electrolyte ions and the abundant "high-speed" channels for both electrons and ions, the as-prepared ECs exhibited excellent electrochemical performance, including a high volumetric capacitance of ~12 F/cm^3 (80 F/g), high energy density (69.4 Wh/kg), and high power density (43.3 kW/kg) *(47,48)*.

10.3.2 CNT Fiber-Based Electrode for Capacitance

1D-CNT fibers/yarns are continuous and ordered CNT assemblies with fiber diameters ranging from micrometers to millimeters. CNT fibers can be prepared through different methods including wet spinning from CNT solution, dry drawing from spinnable CNT array, spinning from CNT aerogels, and direct synthesis *(33–35)*. Since 30 cm-long

CNT fibers were spun from a 100 μm-high CNT array in 2002, abundant research has been carried out to study the properties and applications of CNT fibers *(61–63)*. Highly flexible CNT fibers with low density and high tensile strength can be woven into fabrics or other desired preforms, which also make them very appealing in wearable electronics including wearable ECs *(64)*.

It is well known that the capacitance of carbon materials is largely dependent on their accessible SSA. Therefore the SC of CNT fiber-based ECs varies according to the type of used CNTs (SWCNTs, MWCNTs) because of their different theoretical SSA. Baughman et al. prepared SWCNT fibers through spinning of SWCNT solution *(27)*. After the following annealing process, the polymer binders were thoroughly removed. The purified SWCNT fibers with a degree of CNT alignment (Fig. 10.6A), which is

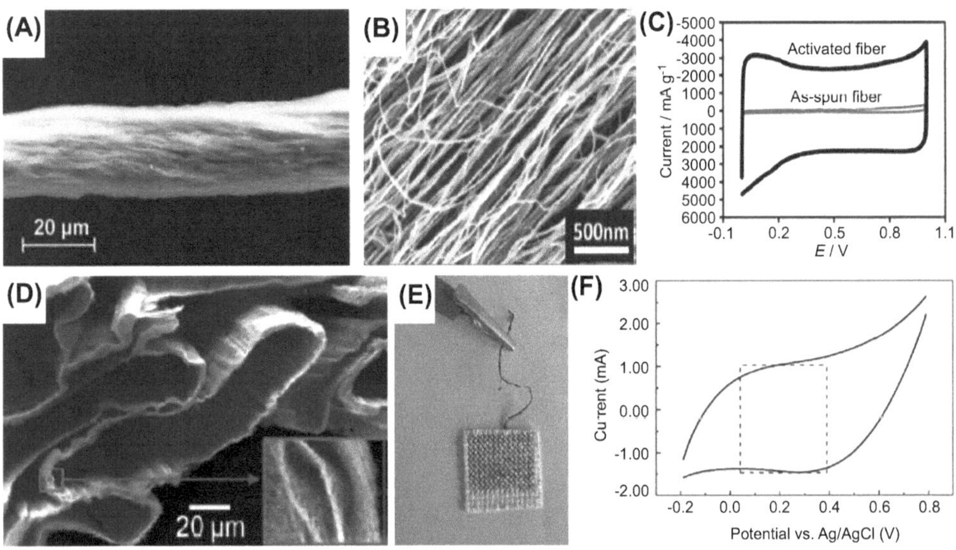

Figure 10.6 (A) Scanning electron microscope (SEM) image of single-walled carbon nanotube (SWCNT) fibers showing a degree of alignment. (*Reproduced from Kozlov, M. E.; Capps, R. C.; Sampson, W. M.; Ebron, V. H.; Ferraris, J. P.; Baughman, R. H. Spinning Solid and Hollow Polymer-Free Carbon Nanotube Fibers.* Advanced Materials ***2005,*** *17 (5), 614–617. Copyright 2005 John Wiley and Sons.*) (B) Aligned multiwalled carbon nanotube (MWCNT) fibers drawn from aligned MWCNT array. *(Reproduced from Ren, J.; Li, L.; Chen, C.; Chen, X.; Cai, Z.; Qiu, L.; et al. Twisting Carbon Nanotube Fibers for Both Wire-Shaped Micro-supercapacitor and Micro-battery.* Advanced Materials ***2013,*** *25 (8), 1155–1159. Copyright 2012 John Wiley and Sons.)* (C) Cyclic voltammogram (CV) results of as-spun and activated carbon nanotube (CNT) fibers. *(Reproduced from Sun, G.; Zhou, J.; Yu, F.; Zhang, Y.; Pang, J. H. L.; Zheng, L. Electrochemical Capacitive Properties of CNT Fibers Spun From Vertically Aligned CNT Arrays.* Journal of Solid State Electrochemistry ***2012,*** *16 (5), 1775–1780. Copyright 2011 Springer.)* (D) Cross-sections of as-spun yarns, showing the hollow, multilayer microstructure (*inset*), (E) a woven CNT fabric, and (F) a CV curve. *(Reproduced from Zhong, X. H.; Li, Y. L.; Liu, Y. K.; Qiao, X. H.; Feng, Y.; Liang, J.; et al. Continuous Multilayered Carbon Nanotube Yarns.* Advanced Materials ***2010,*** *22 (6), 692–696. Copyright 2009 John Wiley and Sons.)*

beneficial for electron transportation and ion diffusion, thus exhibited a high electrical conductivity (140 S/cm) at room temperature and a high SC of 100 F/g *(27)*. In another work, Peng et al. prepared highly aligned MWCNT fibers (Fig 10.6B) by dry drawing aligned MWCNT arrays, and two of the as-drawn fibers were further fabricated into micro-EC by simply twisting them together *(65)*. Although the prepared EC based on MWCNTs possess low SC (13.31 F/g measured at 1.67 A/g) because of their relatively low theoretical SSA when compared with SWCNTs, this method provided an effective and facile way to prepare flexible fiber-shaped ECs whose capacitance could be further improved by incorporation of other electroactive materials.

As discussed, the measured SC of neither SWCNT fibers nor MWCNT fibers matches well with their theoretical SSA. In addition to the intrinsic low effective SSA of CNT fibers related to architecture feature (tiny distance between adjacent CNTs inaccessible for electrolyte), the wetting problem has been proved by Zheng et al. to be another reason for the low SC *(66)*. After cycles of electrochemical activation in a wide potential window, the SC of aligned dry-drawn MWCNT fibers has significantly increased from 2.6 F/g to 52 F/g (or 95 F/g by adding more functionalities), demonstrating that activation could greatly improve the wettability of CNT fibers to allow electrolytes access to their inner part (Fig. 10.6C) *(66)*. In other words, the accessible SSA have been increased.

As a basic structure unit for flexible wearable electronics, the capacitance potential of CNT fabric has been studied. Li et al. successfully wove aligned DWNT fibers prepared by direct CVD spinning into a jersey fabric (18 mm × 23 mm) *(67)*. Because the CNT fibers possess a unique multiple-layered structure for effective contact with electrolyte, the jersey fabric exhibited a high SC of 79.8 F/g (Fig. 10.6D−F).

To increase the stretchability of CNT fibers to meet the demand for wider application in wearable electronics, elastic polymers have always been employed as substrate *(68)*. However, these polymers contribute no capacitance and are detrimental for electrical conductivity. Recently, binder-free spring-like MWCNT fibers were prepared by Peng et al. through overtwisting several small CNT fibers dry drawn from an aligned MWCNT array *(69)*. Benefiting from the hierarchical helical structure, the stretchability and elongation (over 300%) of aligned CNT fibers have significantly increased. Meanwhile, the fibers showed a high electrical conductivity (10^2-10^3 S/cm) and a high SC of 12.25−18.12 F/cm^3 both of which remain stable during 1000 cycles of stretching (Fig. 10.7A−D).

10.3.3 Electrochemical Properties of Aligned CNT Film and Papers

2D-CNT films/papers are another macroscopic structure of CNT assemblies with thicknesses ranging from tens of nanometers to micrometers, which can be prepared by posttreating CNT suspensions, dry drawing from CNT arrays, and direct growth

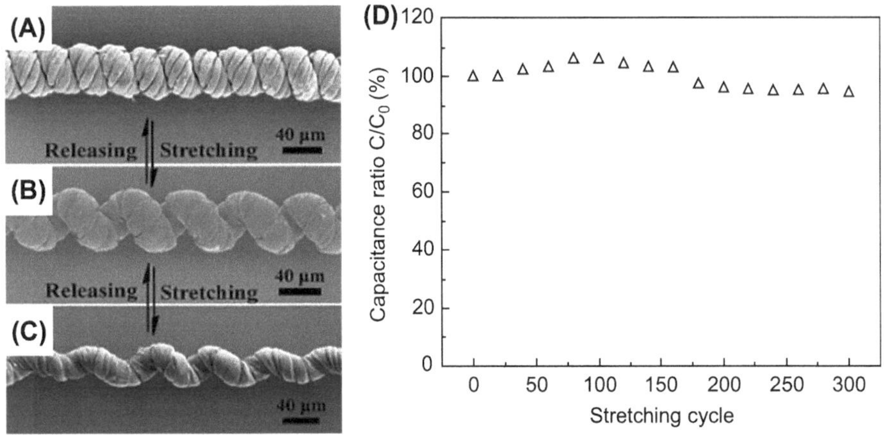

Figure 10.7 (A–C) Scanning electron microscope (SEM) images of a fiber at different strains: 0% (A), 50% (B), and 100% (C). (D) Evolution of specific capacitance over number of stretching cycles at a strain of 100%. *(Reproduced from Ye, Z.; Wenyu, B.; Xunliang, C.; Jing, R.; Wei, W.; Peining, C.; et al. Flexible and Stretchable Lithium-Ion Batteries and Supercapacitors Based on Electrically Conducting Carbon Nanotube Fiber Springs.* Angewandte Chemie ***2014,*** *53 (52), 14564–14568. Copyright 2014 John Wiley and Sons.)*

(36,37). This unique 2D material possesses high tensile strength, excellent electrical conductivity, and high chemical stability, all of which make them very appealing as self-standing electrode materials for ECs *(39,70)*. Besides, flexible CNT films are highly stretchable, can offer a high fault tolerance as they possess multiple current pathways, and can still transport electric charges even when some links are disconnected in wearable electronics *(11,38)*.

In the past decade, much research has been carried out to study the electrochemical performance of CNT films. Pan et al. employed a simple deposition method to prepare MWCNT films from highly concentrated colloidal MWCNT suspensions *(71)*. Because of the high packing density and some local alignments, the film-based EC exhibited an SC of 20 F/g at 1000 mV/s and a high power density of 30 kW/kg. In another work, a simple and effective method similar to "domino pushing" was presented by Fan et al. to prepare aligned paper-like MWCNT films from aligned CNT arrays *(72)*. The as-prepared aligned MWCNT films in which most CNTs maintain their straight morphology and regular interspace showed a high electrical conductivity (200 S/cm) and SC of 81 F/g. In the same work, random MWCNT films have been simultaneously prepared from the same arrays for comparison. Although the electrical conductivity of this random sample is relatively low (150 S/cm), the SC is higher (87.5 F/g) because of the porous structure and larger effective SSA.

To study the full potential of SWCNTs, film-based electrodes have been prepared from vertically aligned SWCNT arrays with ultrahigh carbon purity (99.98%) by shearing them between two glass slides (Fig. 10.8A) *(59)*. On account of the mesoporous structure

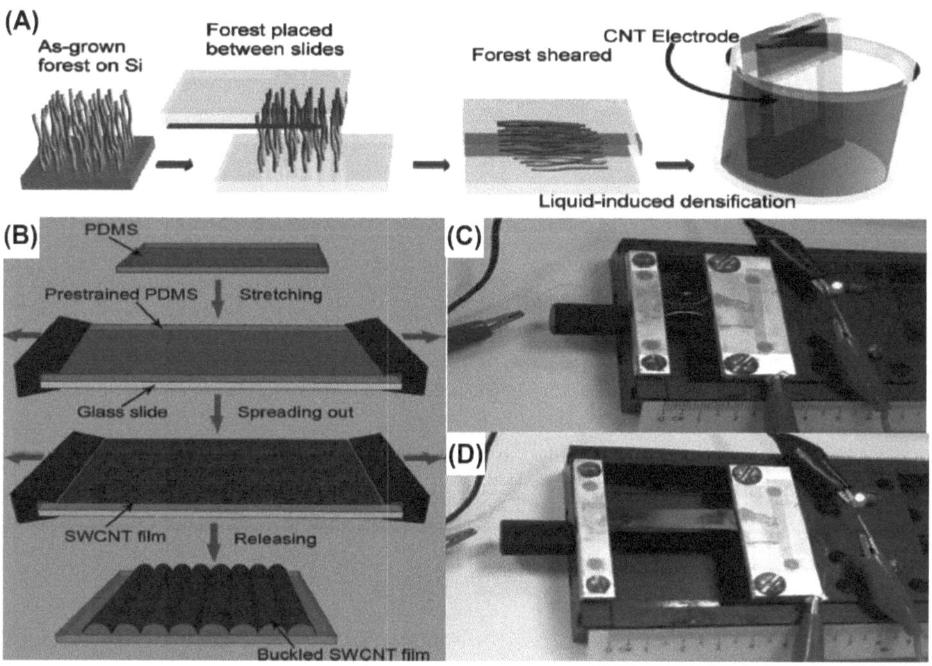

Figure 10.8 (A) Single-walled carbon nanotube (SWCNT) electrode fabrication. *(Reproduced from Izadi-Najafabadi, A.; Yasuda, S.; Kobashi, K.; Yamada, T.; Futaba, D. N.; Hatori, H.; et al. Extracting the Full Potential of Single-Walled Carbon Nanotubes as Durable Supercapacitor Electrodes Operable at 4V with High Power and Energy Density.* Advanced Materials ***2010,*** *22 (35), E235–E41. Copyright 2010 John Wiley and Sons.)* (B) Schematic of preparing buckled SWCNT film on polydimethylsiloxane (PDMS). Optical images of an LED illuminated by using a buckled SWCNT film as the connecting wire under (C) no strain and (D) about 140% strain. *(Reproduced from Niu, Z.; Dong, H.; Zhu, B.; Li, J.; Hng, H. H.; Zhou, W.; et al. Highly Stretchable, Integrated Supercapacitors Based on Single-Walled Carbon Nanotube Films with Continuous Reticulate Architecture.* Advanced Materials ***2013,*** *25 (7), 1058–1064. Copyright 2012 John Wiley and Sons.)*

and near-ideal SSA (1250 m^2/g) of the as-prepared electrodes, they can be operated at a higher voltage (4 V) with a high SC of 160 F/g, and the resulting ECs possess durable full charge/discharge cyclability with an energy density of 94 Wh/kg and a power density of 210 kW/kg.

Continuous reticulate SWCNT films have been prepared by Xie et al. through improving the FCCVD technique *(36)*. Compared with postdeposited CNT films, the directly grown CNT films possess much higher electrical conductivity (~2000 S/cm) and strain tolerance because of the regular interconnected architecture of SWCNTs and the provided multiple current paths. Niu et al. further employed these films to fabricate highly stretchable buckled SWCNT films with the help of prestrained PDMS substrates (Fig. 10.8B) *(37)*. Benefiting from the particular buckled structure, the SC (53 F/g) and power density (32 kW/kg) of the film-based EC remained nearly unchanged even when it was stretched under a strain of 120%, indicating this SWCNT film with excellent flexibility is thus a promising candidate for future wearable electronics (Fig. 10.8C and D).

10.3.4 Supercapacitors Utilizing Vertically Aligned CNT Arrays

3D-CNT arrays/solids that typically refer to vertically aligned CNTs with heights up to the millimeter scale are the third kind of CNT macroscopic structure. CNT arrays are usually prepared through the CVD method on a smooth catalytic plate predeposited on nonconductive substrates such as silicon wafers and could be further transferred to other conducting substrates *(39,40)*.

Since vertically aligned CNT arrays were first reported by Li et al. in 1996, plenty of work has been done to scale up the synthesis and to explore their commercial potential *(39,43–46)*. The aligned CNT arrays with high electrical conductivity and excellent mechanical properties are quite suitable for EC electrodes because of their ordered porous structures and conductive paths, which respectively provide higher accessible SSA and faster electron/ion transportation *(5,41,42)*.

Tanaike et al. prepared SWCNT arrays through a highly effective water-assisted CVD method; this catalyst-free SWCNT array is a potential material for next-generation EC because of its high carbon purity (>99.9%) and high SSA (1100 m^2/g) *(29,60)*. As an electrode, the measured SC increased from 18 F/g at charging potential of 2 V to 25 F/g at 3 V because of the effect of electrochemical doping in semiconductive SWCNTs *(29)*. In another work, to overcome the disadvantages of semiconductive SWCNT arrays (relatively low SC and electrical conductivity), Ishikawa et al. employed DWCNT arrays (Fig. 10.9A and B) to fabricate EC *(73)*. By transferring CVD-grown arrays from silicon substrate to an aluminum sheet, which serves as current collector,

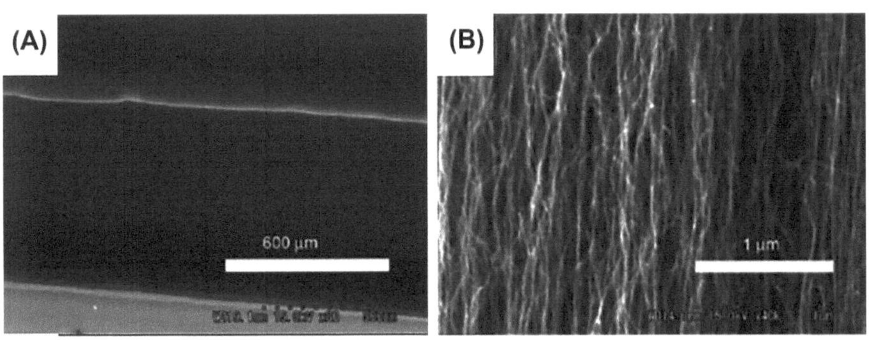

Figure 10.9 Scanning electron microscope (SEM) images of as-grown double-walled carbon nanotubes (DWCNTs) at (A) low and (B) high magnifications. *(Reproduced from Honda, Y.; Takeshige, M.; Shiozaki, H.; Kitamura, T.; Yoshikawa, K.; Chakrabarti, S.; et al. Vertically Aligned Double-Walled Carbon Nanotube Electrode Prepared by Transfer Methodology for Electric Double Layer Capacitor.* Journal of Power Sources ***2008,*** *185 (2), 1580–1584. Copyright 2008 Elsevier.)*

the prepared EC exhibited an SC of 47 F/g at 10 mV/s and could almost maintain its intrinsic SC even at a high current density of 20 A/g. In addition, ECs based on MWCNT arrays have also been prepared by utilizing the "transfer methodology" *(42,74)*. Because of the suitable paths for both ion adsorption/desorption and electron transportation provided by the aligned architecture, a discharge capacity of 10−15 F/g as well as a high power density of 125 kW/kg was obtained even at an ultrahigh current density of 200 A/g.

To get superaligned CNT arrays with higher density and a more regular structure, high densely packed SWCNT arrays have been prepared by using the zipping effect of liquids to draw CVD-grown aligned SWCNT arrays together (Fig. 10.10A−C) *(47)*.

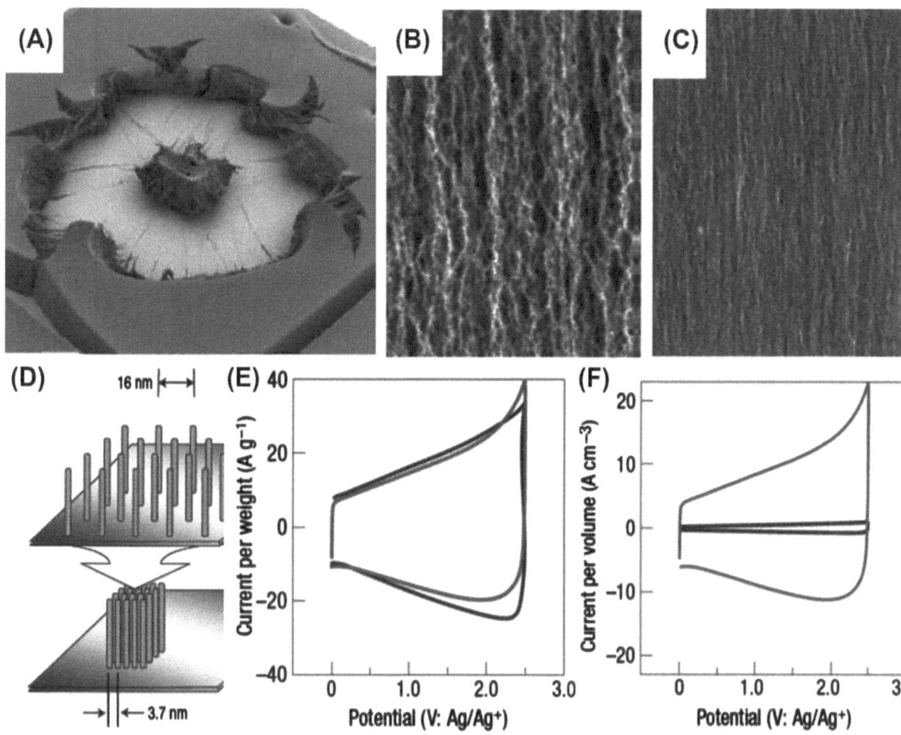

Figure 10.10 (A) Scanning electron microscope (SEM) image of single-walled carbon nanotube (SWCNT) forest structural collapse from a single drop of liquid. (B and C) SEM images of the as-grown forest (B) and solid (C). (D) Schematic diagram of the collapse of the aligned low-density as-grown forest (*above*) to the highly densely packed SWCNT solid (*below*). (E and F) Cyclic voltammograms of the electrical chemical double-layer capacitor (EDLC) using the SWCNT solid sheet (*red*) and as-grown forest (*black*) as electrodes comparing the capacitance per weight (E) and capacitance per volume (F). *(Reproduced from Futaba, D. N.; Hata, K.; Yamada, T.; Hiraoka, T.; Hayamizu, Y.; Kakudate, Y.; et al. Shape-engineerable and Highly Densely Packed Single-Walled Carbon Nanotubes and Their Application as Super-Capacitor Electrodes.* Nature Materials ***2006,*** *5 (12), 987–994. Copyright 2006 Nature Publishing Group.)*

In addition, to maintain the intrinsic properties of SWCNT arrays such as high SSA ($1300\ m^2/g$) and electrical conductivity, this new "SWCNT solid" can be fabricated into different shapes through process controlling. In spite of the absence of mesopores in the "solid," the average tube separation (~0.9 nm) is large enough for the access of electrolyte ions, the as-prepared EC thus exhibiting a high volumetric capacitance of $\sim 12\ F/cm^3$ (80 F/g) and a high energy density (69.4 Wh/kg) (Fig. 10.10D–F) *(47,48)*. Besides, a high power density (43.3 kW/kg) was obtained because of the regular structure, which provides a "high-speed" channel for ions.

As CNT arrays cannot be directly used as current collectors because of their low electrical conductivity along the radial direction, they are usually bonded to another metal or carbon collector *(75)*. However, the huge internal resistance between CNT arrays and current collectors seriously limits the EC's power density *(76)*. To reduce the contact resistance, many attempts have been carried out to eliminate the additive binders by growing CNT arrays directly on current collectors *(77–80)*. Meanwhile, the electrochemical properties of such arrays have been reported.

Aligned MWCNT arrays with lengths of about 200 μm were prepared on an Inconel 600 substrate by Talapatra et al. through a novel vapor-phase catalyst delivery CVD method (Fig. 10.11A–C) *(40)*. Because of the strong adhesion and low contact resistance between the arrays and the metal substrate, the as-prepared EC showed a well-satisfied SC (18 F/g) and power density (7 kW/kg) at 1 V/s scan rate when compared with other ECs, which employ traditional CVD-grown MWCNT arrays as electrodes (Fig. 10.11D and E).

Based on these results and discussion, it is evident that pure macroscopic CNT assemblies (fibers, films, and arrays) possess great potential as next-generation electrode materials because of their excellent comprehensive performance, which includes high mechanical properties, high electrical conductivity, reasonable SSA, and tunable PSD. In addition, highly flexible and stretchable CNT fibers/films, which could be fabricated into desired shapes, are very promising candidates for wearable electronics. Nevertheless, pure double-layer capacitance dependent on the surface area of CNTs cannot meet the growing demands for higher SC and energy density, which seriously restricted their commercial application. Thus in recent years, incorporation of pseudocapacitance materials (metal oxides, CPs) into CNTs to improve their energy storage capability have attracted great attention, which will be discussed later.

10.3.5 Current Challenges in CNT Electrodes

Nevertheless, despite all these merits and developing trends, some shortcomings are still evident. For developing CNT-based electrodes, the major challenge is their relatively low pure double-layer capacitance dependent on the effective surface area of CNTs in practical application. Even through elaborate designing, the aligned macroscale

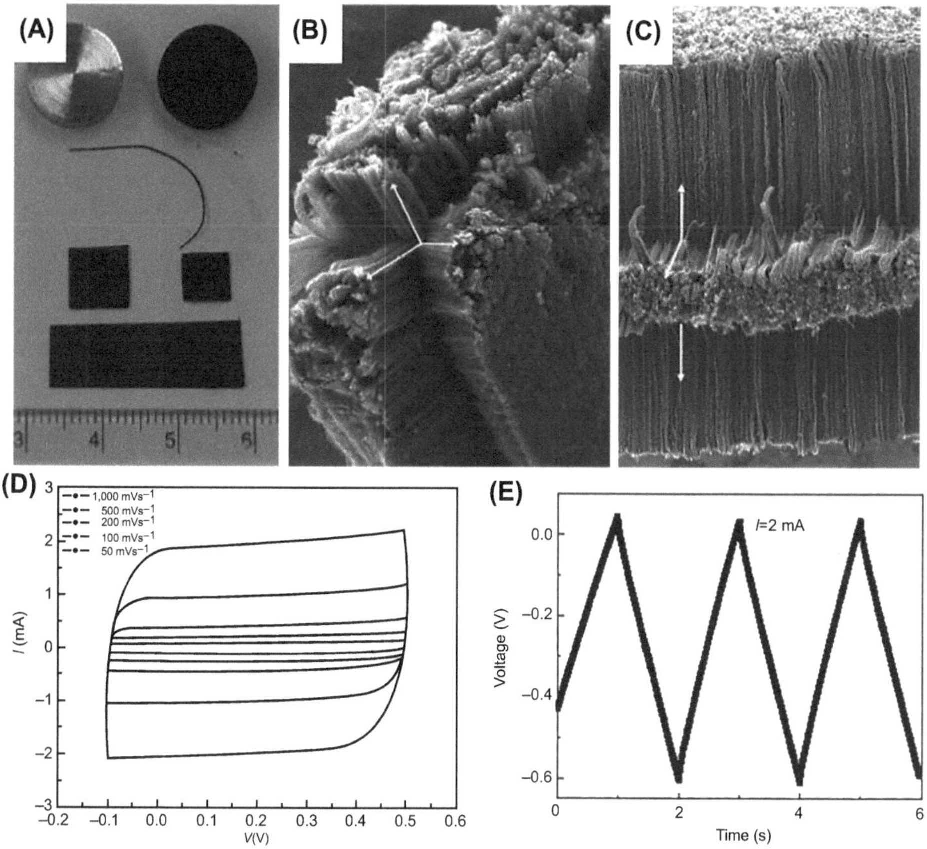

Figure 10.11 (A) Optical image of differently shaped and sized Inconel substrates used for carbon nanotube (CNT) growth. The *top-left* and *top-right* examples are the as-received Inconel rod and a similar substrate with uniform CNTs grown on it. An Inconel wire and sheets are also shown. (B) Scanning electron microscope (SEM) image showing the corner of a CNT-grown Inconel sheet. Note the highly directional nature of the CNT growth, with the orientation of the nanotubes being perpendicular to the substrate surface. The *arrows* indicate the direction of CNT growth. (C) SEM image of one side of an Inconel sheet (0.075 mm thick) showing well-aligned CNTs on the top, side, and bottom of the substrate. The *arrows* indicate the direction of CNT growth. (D) Cyclic voltammograms (CVs) measured for a particular double-layer capacitor (DLC) device at scan rates ranging from 50 mV/s to 1000 mV/s. The rectangular shape of the CVs are maintained, even at very high scan rates. The CV data obtained at 1000 mV/s were used for calculating the specific power density of the device (reported in the main text). (E) Charge/discharge cycles taken at a constant current of 2 mA. The discharge slope was used to calculate the specific capacitance of the device. *(Reproduced from Talapatra, S.; Kar, S.; Pal, S.; Vajtai, R.; Ci, L.; Victor, P.; et al. Direct Growth of Aligned Carbon Nanotubes on Bulk Metals.* Nature Nanotechnology ***2006,*** *1 (2), 112–116. Copyright 2006 Nature Publishing Group.)*

architecture/structure including 1D-CNT fibers/yarns, 2D-CNT films/papers, and 3D-CNT arrays/solids can be realized to make full use of the outside surface of CNTs; however, the stored energy is still not high enough, usually below 100 F/g *(40,81)*. High cost is another retarding factor for the application of CNTs in ECs because of their complicated fabrication processes. The current market price of CNTs (purity: $\geq$95 wt%) is usually above $15/g.

To partially solve the stated issues, nanohybrid electrodes based on CNTs with well-defined hierarchical nanostructures have been developed. By incorporating complementary materials (mainly pseudocapacitance materials) into the system, they first function as additives to improve the electrochemical properties according to the rule of mixtures *(82)*. In addition, hierarchical nanostructures could induce a synergistic effect between components with different length scales. In a nanohybrid, the smaller component could shorten the path of ion diffusion and transportation in addition to increase the effective SSA, which is beneficial for both EDLC and pseudocapacitance, while the larger one could increase the internal electrical conductivity and provide efficient current pathway for electrons *(83)*. Furthermore, nanohybrid electrodes may be endowed with some new features such as self-supporting, binder free, flexibility, and transparency.

10.4 HIGH-PERFORMANCE CNT HYBRIDS FOR CAPACITANCE

10.4.1 CNT—Polymer Composite for Supercapacitors

As discussed previously, a major obstacle to commercial application of pure CNTs in ECs is their relatively low SC and energy density. To effectively solve this problem, it is necessary to incorporate other functional materials into the system. CPs such as PPy, PANI, and PTh store electric charges through rapid and reversible redox and doping/dedoping reactions on π-conjugated polymer chains *(84,85)*. Compared with carbon-based electrodes, they can provide a higher SC because of their pseudocapacitance characteristics. Unfortunately, CPs often suffer from a poor cyclability and fast decayed capacitance caused by volume changes during the intercalating/deintercalating process *(84)*. Besides, their poor electrical conductivity also seriously limits rapid transportation of electrons for a higher power density. Thus to fully utilize the respective electrochemical advantages of nanocarbons (high conductivity and high SSA) and CPs, they have been composited as hybrid electrodes with controllable architecture.

10.4.1.1 Fiber-Shaped CNT/CP Electrodes

To increase the SC and energy density of CNT fibers, Peng et al. deposited PANI on textiles composed of aligned drawn MWCNT fiber and studied the effect of PANI content on their electrochemical performance (Fig. 10.12A and B) *(86)*. It was found that when the added PANI had an equal weight with MWCNT fibers, the highest measured SC (272.7 F/g) could be achieved, which was about 35 times that of pure MWCNT

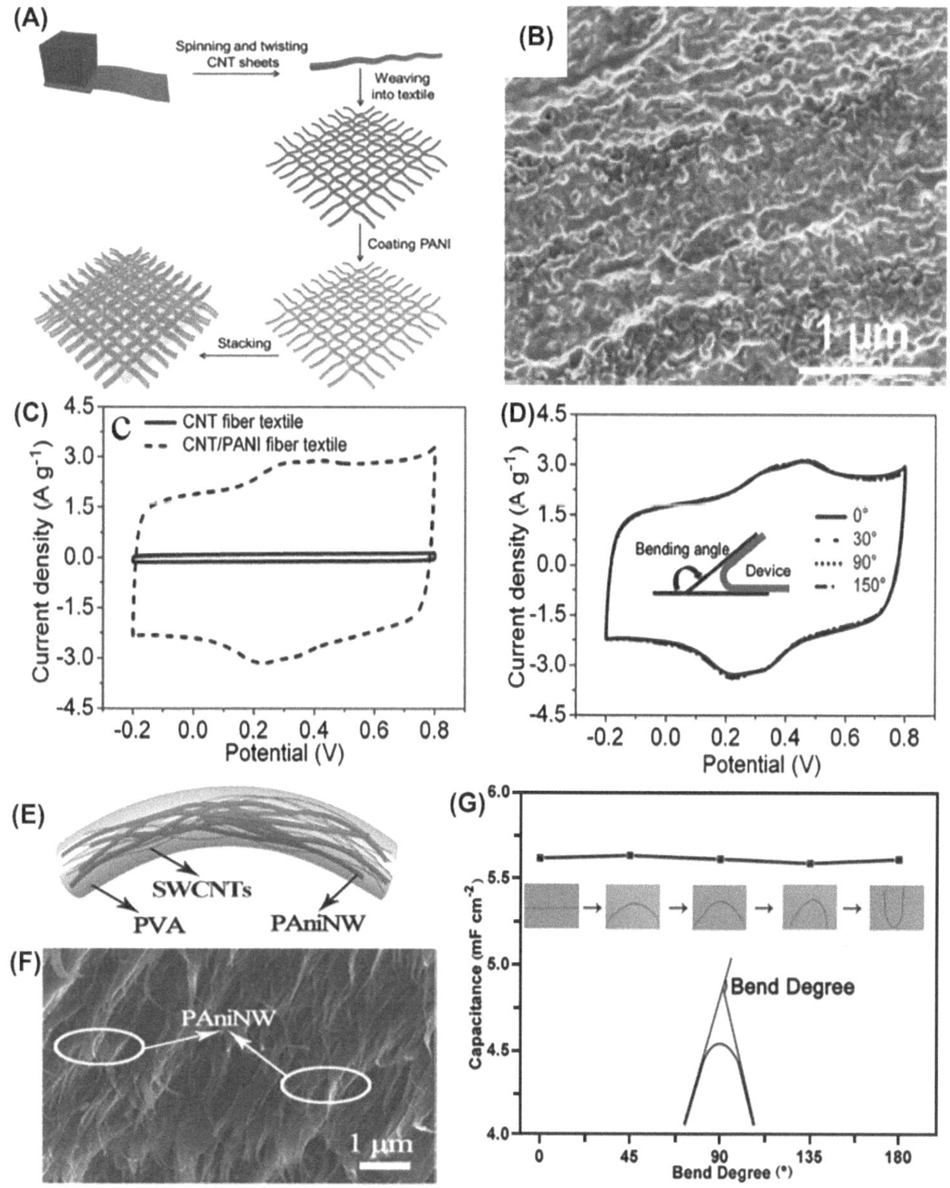

Figure 10.12 (A) Schematic illustration of fabrication of a supercapacitor textile based on the carbon nanotube/polyaniline (CNT/PANI) composite fiber. (B) Scanning electron microscope (SEM) image of aligned CNT/PANI composite fibers with PANI weight percentages of 50%. (C) Cyclic voltammogram (CV) curves of bare CNT fiber-based textile and CNT/PANI composite fiber-based textile with PANI weight percentage of 50% as electrodes at a scan rate of 20 mV/s and potential range from −0.2 to 0.8 V. (d) CV curves of a supercapacitor textile being bent into different angles at a scan rate of 20 mV/s *(Reproduced from Pan, S.; Lin, H.; Deng, J.; Chen, P.; Chen, X.; Yang, Z.; et al. Novel Wearable Energy Devices Based on Aligned Carbon Nanotube Fiber Textiles.* Advanced Energy Materials ***2015,*** *5 (4), 1401438. Copyright 2014 John Wiley and Sons.)* (E) Illustration of single-walled carbon nanotube/polyaniline nanowire (SWCNTs/PANINW) yarn and (F) SEM image of cross-section of SWCNT/PANINW yarns. (G) Capacitance retention under different bending states. *(Reproduced from Meng, Q.; Wang, K.; Guo, W.; Fang, J.; Wei, Z.; She, X. Thread-Like Supercapacitors Based on One-Step Spun Nanocomposite Yarns.* Small ***2014****, 10 (15), 3187–3193. Copyright 2014 John Wiley and Sons.)*

fibers, indicating a more obvious synergistic effect (Fig. 10.12C). What is more, this flexible and lightweight textile possessed good cycling stability under bending conditions and showed huge potential in future wearable electronics (Fig. 10.12D). Very recently, the same group reported a one-step method to prepare MWCNT fiber-based ECs within several minutes *(87)*. The continuous process consisted of growth of MWCNT arrays, dry drawing of MWCNT fibers, electrochemical deposition of various active materials (including PANI and PPy) on MWCNT fibers, immersing of hybrid fibers in electrolytes, and finally twisting of two fibers into an EC. In addition to the obtained remarkable energy-storage capability, this method presented a general and efficient strategy for developing wearable ECs on a large scale. Compared with solid state spinning, solution state spinning is a relatively facile way to functionalize CNT fibers *(88)*. Thread-like and flexible yarns made of SWCNTs and polyaniline nanowires (PANINWs) were prepared by Meng et al. through a one-step spinning process (Fig. 10.12E–G) *(89)*. In addition to high rate capability and good stability, the areal capacitance of ECs based on this hybrid electrode can reach 6.23 mF/cm^2, which was 18 times that of pure SWCNT yarn-based ECs.

In another work, core–shell structured CNT/PPy fibers were spun from CNT films (mainly composed of SWCNTs and DWCNTs) by Wei et al. with PPy electrochemically deposited on the surface; this unique structure made charge transfer easily between PPy and CNTs *(90)*. As a result, a very low ESR and high SC of 350 F/g were obtained. Besides, the same group further fabricated fiber-shaped all-solid-state ECs by employing the same CNT/PPy fibers as electrodes and PVA/H_2SO_4/hydroquinone gel as electrolyte (Fig. 10.13) *(91)*. Because of the unique core–shell structure and synergistic effect between the two redox additives of PPy and hydroquinone, the ECs exhibited a high SC (56 F/g) in addition to high stability and good flexibility.

10.4.1.2 CNT/CP Film-Based Electrodes

CNT films could be simply prepared through filtration of CNT suspensions under vacuum; Chen et al. further employed them as substrate and fabricated transparent and flexible SWCNT/PANI films through electrochemical polymerization (Fig. 10.14) *(92)*. Because of the introduced pseudocapacitance, the SWCNT/PANI film exhibited an SC of 234 F/g (55 F/g for an EC) at a current density of 2.6 A/g, showing its possibility for wearable electronics. The same method has been employed by Gao and coworkers; after electrodegradation process, which brought more available current pathways and increased polycrystalline PANI regions, the SC of the SWCNT/PANI film can be significantly increased to 706.7 F/g *(93)*.

In another work, dry-drawn MWCNT films from aligned spinnable arrays have been electrochemically deposited with PANI and PPy to get transparent and flexible hybrid films for ECs *(94)*. Unlike the smooth surface of PPy, PANI on the aligned CNT films showed a rough surface, which could provide a higher SSA. As a result, the ECs based on PANI/MWCNT films exhibited a high SC of 233 F/g, which was 36 times that of bare

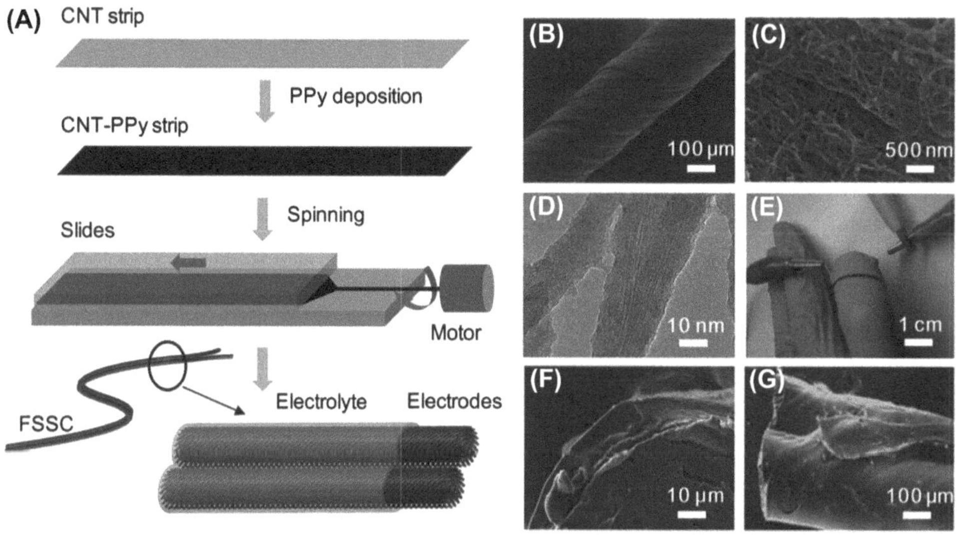

Figure 10.13 (A) Schematic illustration of the fabrication process of fiber-shaped supercapacitors (FSSCs). (B) Low magnification scanning electron microscope (SEM) image of a carbon nanotube-polypyrrole (CNT-PPy) fiber. (C) High-magnification SEM image of a CNT-PPy fiber. (D) Transmission electron microscopy (TEM) image of the CNT-PPy nanocomposite. (E) Optical image of an FSSC wrapping on a finger. (F) Cross-sectional SEM image of an FSSC. (G) Side-viewed SEM image of an FSSC. *(Reproduced from Xu, R.; Guo, F.; Cui, X.; Zhang, L.; Wang, K.; Wei, J. High Performance Carbon Nanotube Based Fiber-Shaped Supercapacitors Using Redox Additives of Polypyrrole and Hydroquinone.* Journal of Materials Chemistry A ***2015,*** *3 (44), 22353–22360. Copyright 2015 Royal Society of Chemistry.)*

MWCNT films and 23 times that of pure PANI. When employed as electrodes, the thickness of the films significantly affected their electrochemical performance. It was found that the SC as well as power density and energy density decreased with the increasing electrode thickness mainly because of the gradient potential distribution across the electrode *(95)*. For the MWCNT/PANI films prepared by Liu et al. through in situ polymerization on dry-drawn MWCNT films, the optimal thickness (~5 μm or less) for achieving the best electrochemical performance was obtained *(95)*.

Compared with CNT films prepared through dry drawing and filtration, directly grown films through the floating CVD method a possessed continuous reticulate structure rather than overlapped bundles for which electrons must transfer across overlapped connection points with high contact resistance *(96)*. After incorporation of PANI on the highly conductive continuous reticulate SWCNT films by Zhou et al., this "skeleton/skin" film exhibited a high power density of 62.5 kW/kg in addition to high SC (236 F/g) and high energy density (131 Wh/kg) (Fig. 10.15) *(96)*. However, although the introduced CPs could largely increase the SC of CNT films, their content should be controlled within a certain range *(94,96)*. It was found that excessive CPs could fill the pores on the hybrid films and thus resulted in a low SC because of the decreased accessible SSA.

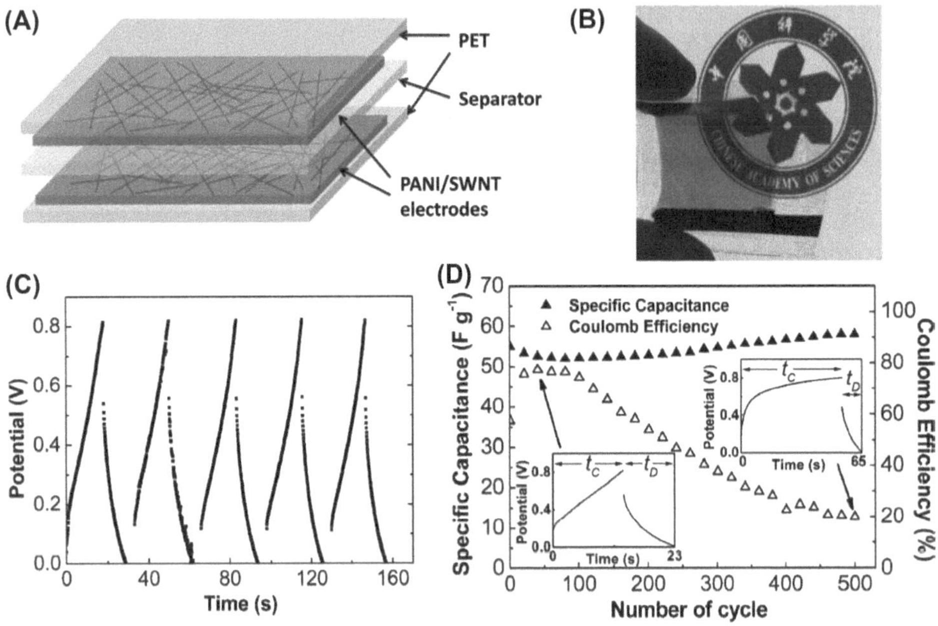

Figure 10.14 (A) Schematic illustration of the sandwich structure for transparent and flexible supercapacitors. (B) Photograph of a transparent and flexible supercapacitor. (C) Cycles of galvanostatic charge/discharge measurement of the as-prepared supercapacitor in (B). (D) Specific capacitance and coulomb efficiency of the supercapacitor as functions of cycle number, insets are the particular galvanostatic charge/discharge curves of cycle 41 and 500. *(Reproduced from Ge, J.; Cheng, G.; Chen, L. Transparent and Flexible Electrodes and Supercapacitors Using Polyaniline/Single-Walled Carbon Nanotube Composite Thin Films.* Nanoscale ***2011,*** *3 (8), 3084–3088. Copyright 2011 Royal Society of Chemistry.)*

10.4.1.3 Hybrid Electrodes of CNT/CP Arrays

Pioneering work has been conducted by Windle et al. through electrochemically depositing PPy on the CVD-grown aligned MWCNT arrays *(97)*. The superior performance (SC of 2.55 F/cm^2) of the hybrid array was closely related to the synergistic effect, which provided high electrolyte accessibility, high conductivity of the hybrid, and shorter diffusion paths for ions. In addition, Windle further pointed out that the thickness of coated PPy and the size of the intertubular pores were two key factors affecting the overall performance of the hybrid since they respectively determined the added amount of PPy and PSD of the whole electrode. As one of the most popular CPs, poly(3,4-ethylenedioxythiophene) (PEDOT) was highly conductive and operable in both aqueous (acidic and alkaline) and organic electrolyte with a higher electrochemical window, which made them very appealing for practical applications *(98–100)*. Zhang et al. have deposited PEDOT on CNT arrays and studied their electrochemical performance (Fig. 10.16) *(98)*. With conformal PEDOT coated on aligned MWCNT arrays, an improved cycle life as well as enhanced SC of 80–110 F/cm^3 was obtained. Compared

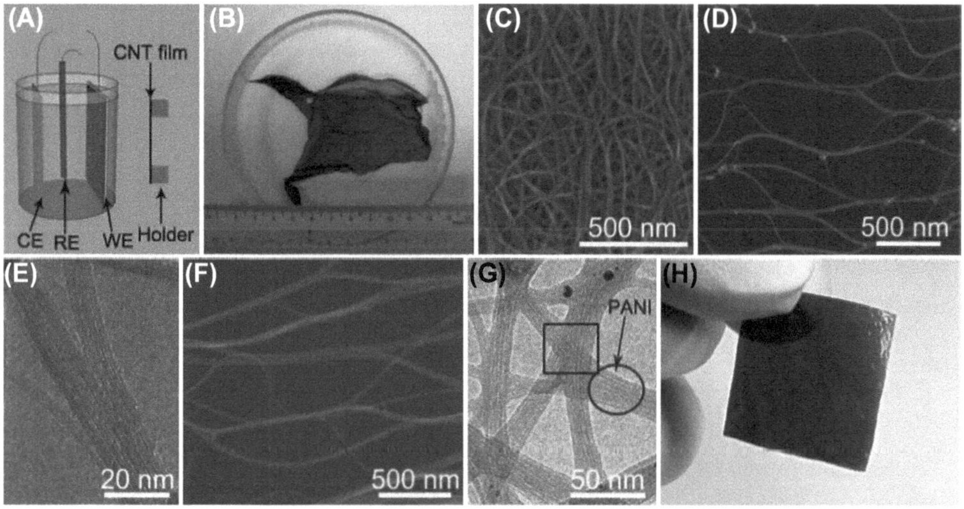

Figure 10.15 (A) Sketch of polyaniline (PANI) electrodeposition using the directly grown single-walled carbon nanotube (SWCNT) film as template. (B) Optical image of a directly grown SWCNT film with about 200 nm thickness. (C) Scanning electron microscope (SEM) image of a directly grown SWCNT film. (D) SEM image of a single layer of SWCNT bundles peeled off from a thick film. (E) Transmission electron microscopy (TEM) image of an SWCNT junction in the directly grown film. (F) SEM image of a single layer of single-walled carbon nanotube/polyaniline nanowire (SWCNT/PANI) bundles peeled off from a thick SWCNT/PANI film. (G) TEM image of SWCNT/PANI in the hybrid film. (H) Optical image of SWCNT/PANI hybrid film (size: 2 cm × 2 cm, thickness: 240 nm, deposition time: 30 s). *CE*, Counter electrode; *RE*, reference electrode; *WE*, working electrodes. *(Reproduced from Niu, Z.; Luan, P.; Shao, Q.; Dong, H.; Li, J.; Chen, J.; et al. A "skeleton/skin" Strategy for Preparing Ultrathin Free-Standing Single-Walled Carbon Nanotube/Polyaniline Films for High Performance Supercapacitor Electrodes.* Energy & Environmental Science ***2012,*** *5 (9), 8726–8733. Copyright 2012 Royal Society of Chemistry.)*

with CNT/PPy and CNT/PEDOT arrays, more research has been focused on CNT/PANI arrays because of their better electrochemical performance. Through controlling the content of introduced PANI, a hierarchically porous structure with higher SSA was obtained by Cao et al., which was highly beneficial for ion diffusion and energy storage. Typically, aligned MWCNT arrays with a certain amount of coated PANI could provide a high SC up to ~1000 F/g as well as high stability and high rate capability (<5% capacitance loss at 118 A/g) *(101,102)*.

Recently, directly growing CNT arrays on metal substrate, which could also serve as current collector of ECs, has attracted great attention. Because of a lower contact resistance between the end of CNTs and current collector, a higher power density was expected. Chen et al. directly prepared aligned MWCNT arrays on Ti foils and electrochemically deposited PANI on them with different contents (Fig. 10.17) *(103)*. After fabricating the MWCNT/PANI/Ti into ECs, they exhibited a high SC of ~540 F/g and high power density of ~270 W/kg. In another work carried out by the same group,

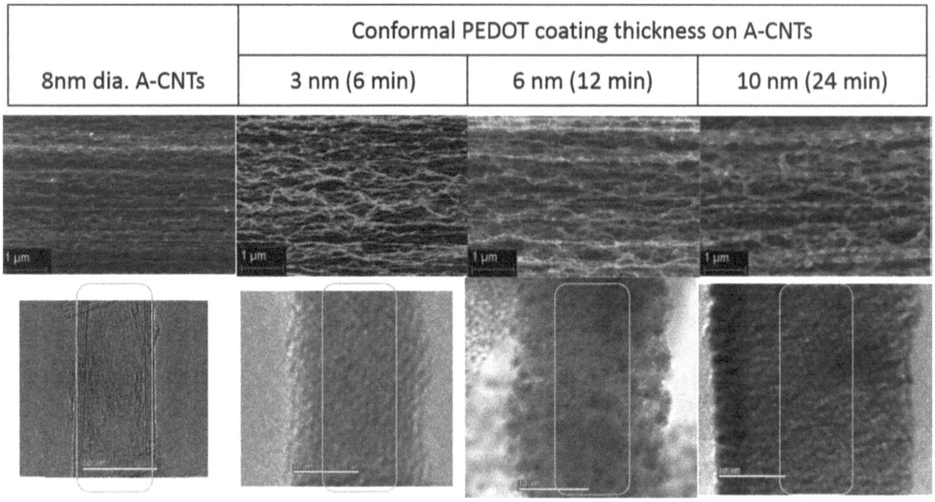

Figure 10.16 Scanning electron microscope (SEM) (*top*) and transmission electron microscopy (TEM) (*bottom*) images of carbon nanotubes (CNTs) with varying conducting polymer (CP) thickness. All the scale bars in the TEM images are 10 nm and the size and location of the CNT outer wall is indicated in the CP-coated images. Note that the alignment is maintained after coating, and that the coating is uniform. *(Reproduced from Lachman, N.; Xu, H.; Zhou, Y.; Ghaffari, M.; Lin, M.; Bhattacharyya, D.; et al. Tailoring Thickness of Conformal Conducting Polymer Decorated Aligned Carbon Nanotube Electrodes for Energy Storage.* Advanced Materials Interfaces ***2014,*** *1 (7), 1400076101. Copyright 2014 John Wiley and Sons.)*

aligned MWCNT arrays were prepared on household Al foil *(104)*. With the thin Al foil as current collector, the flexible MWCNT/PANI/Al array possessed a low ESR of 1.3 Ω, which resulted in a high power density (11.3 kW/kg), in addition to a high SC (705.8 F/g) and high energy density (18.9 Wh/kg).

10.4.2 Electrochemical Performance of CNT/Metal Oxide Hybrids

A variety of metal oxides including RuO_2, MnO_2, Fe_3O_4, Co_3O_4, NiO, CeO_2, etc. have been used to offer a significant rise on both SC and energy density of carbon-based ECs because of their rapid faradic reactions *(105,106)*. Among these metal oxides, RuO_2 is the most extensively studied material because it has the highest theoretical SC (2000 F/g), good electrical conductivity, and chemical stability *(107)*. However, their relatively high cost and toxicity significantly restricted their practical applications in ECs. As a result, other oxides such as MnO_2, NiO, and Co_3O_4 have been developed and extensively investigated because of their low cost and reasonable theoretical SC *(108–110)*. To gain the maximum potential in EC application, nanostructured metal oxides are usually prepared, which poses the challenge of charge transport. Therefore a conductive template is highly required for facilitating the charge transport as well as the metal oxide loading and immobilization. CNTs are usually used as such templates because of their good conductivity and large SSA.

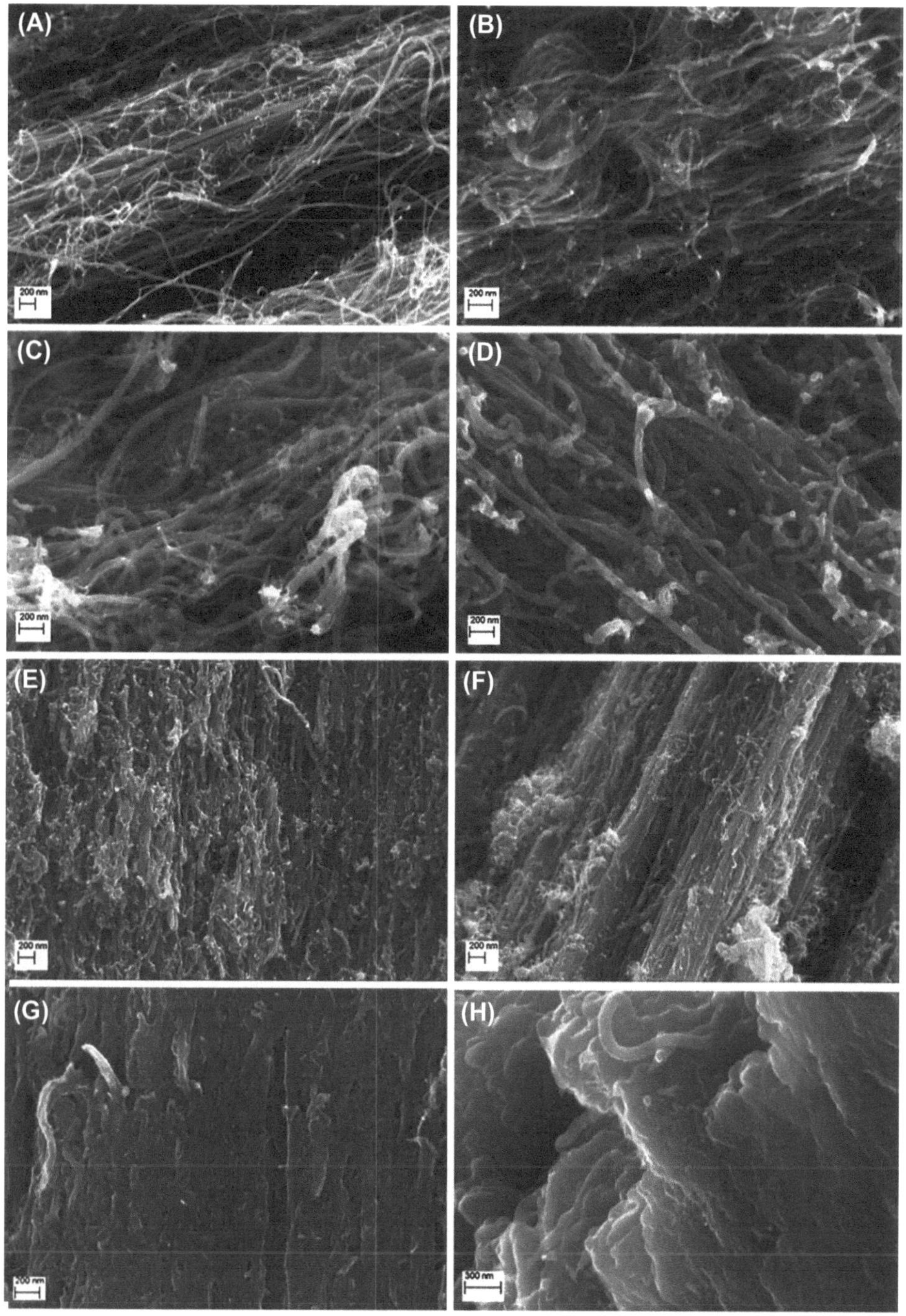

Figure 10.17 Scanning electron microscope (SEM) images of polyaniline/activated carbon nanotube (PANI/A-CNT) composites synthesized from 10 to 150 cyclic voltammogram (CV) cycles. (A) A-CNTs. (B) 10 cycles, 15.4% PANI. (C) 30 cycles, 21.4% PANI. (D) 50 cycles, 31.3% PANI. (E) 70 cycles, 52.3% PANI. (F) 100 cycles, 63.3% PANI. (G) 120 cycles, 67.6% PANI. (H) 150 cycles, 71.8% PANI. *(Reproduced from Huang, F.; Chen, D. Towards the Upper Bound of Electrochemical Performance of ACNT@polyaniline Arrays as Supercapacitors.* Energy & Environmental Science ***2012,*** *5 (2), 5833–5841. Copyright 2011 Royal Society of Chemistry.)*

10.4.2.1 Fiber-Shaped CNT/Metal Oxide Electrodes

Among various metal oxides, MnO_2 possess relatively low cost and high SC and have been most reported in fiber-shaped CNT/metal oxide electrodes. Peng et al. employed dry-drawn MWCNT fibers as substrates to electrodeposit MnO_2 nanoparticles with different weight percentages *(65)*. With a 30 wt% addition of MnO_2, the areal capacitance increased by 23% from 3.010 mF/cm^2 (for bare MWCNT fibers) to 3.707 mF/cm^2 (for hybrid fibers). In another work also reported by Peng, flower-like MnO_2 were electrochemically deposited on MWCNT fibers in a one-step process *(87)*. Benefiting from the highly conductive CNT core and high pseudocapacitance provided by the flower-like MnO_2 sheath with high surface area, the SC of the hybrid fibers increased by 400% up to 20.2 F/cm^3. In addition to dry drawing from arrays, Kim et al. prepared MWCNT fibers through twisting MWCNT films and electrodeposited MnO_2 on them (Fig. 10.18) *(111)*. Because of the abundant nanopores inside the MWCNT fibers, nanosized MnO_2 with high SSA and short diffusion paths were well trapped in the pores. As a result, a high SC (25.4 F/cm^3) as well as high energy density (3.52 mWh/cm^3) and power density (127 mW/cm^3) were achieved in the as-prepared all-solid-state yarn EC.

Su et al. electrodeposited Co_3O_4 and NiO, respectively, on MWCNT fibers and studied their electrochemical performance (Fig. 10.19) *(112)*. Because of the unique core—shell structure and uniformly distributed Co_3O_4 on the surface of MWCNT fibers, the EC based on two MWCNT/Co_3O_4 fibers exhibited a high SC of 87.6 F/g, while a relatively low SC of 25.9 F/g for MWCNT/NiO-based EC was attributed to a certain content of NiO aggregation.

10.4.2.2 CNT/Metal Oxide Film-Based Electrodes

As a facile and effective method, vacuum filtering was employed by Mai et al. to prepare free-standing CNT/MnO_2 films from suspensions containing CNTs and as-prepared MnO_2 nanotubes (Fig. 10.20) *(113)*. Benefiting from the hierarchical porous structure and synergistic effects between CNTs and MnO_2 nanotubes, a high SC of 5.1 F/cm^3 and a high energy density of 0.45 mWh/cm^3 were obtained. The same method was employed by Liu et al. to prepare two hybrid films through combining CNTs with MnO_2 and MoO_3 nanowires *(114)*. By respectively employing the two films as anode and cathode for asymmetric EC, the device with a high potential window of 4.0 V could provide a high energy density (28.6 mWh/cm^3) as well as high SC. Besides, CNT films prepared through vacuum filtration could also serve as good substrates to deposit other metal oxides such as In_2O_3 and WO_3 on the surface *(115,116)*. These hybrid films with good electrochemical performance showed a promising strategy for fabricating CNT/metal oxide film-based ECs on a large scale.

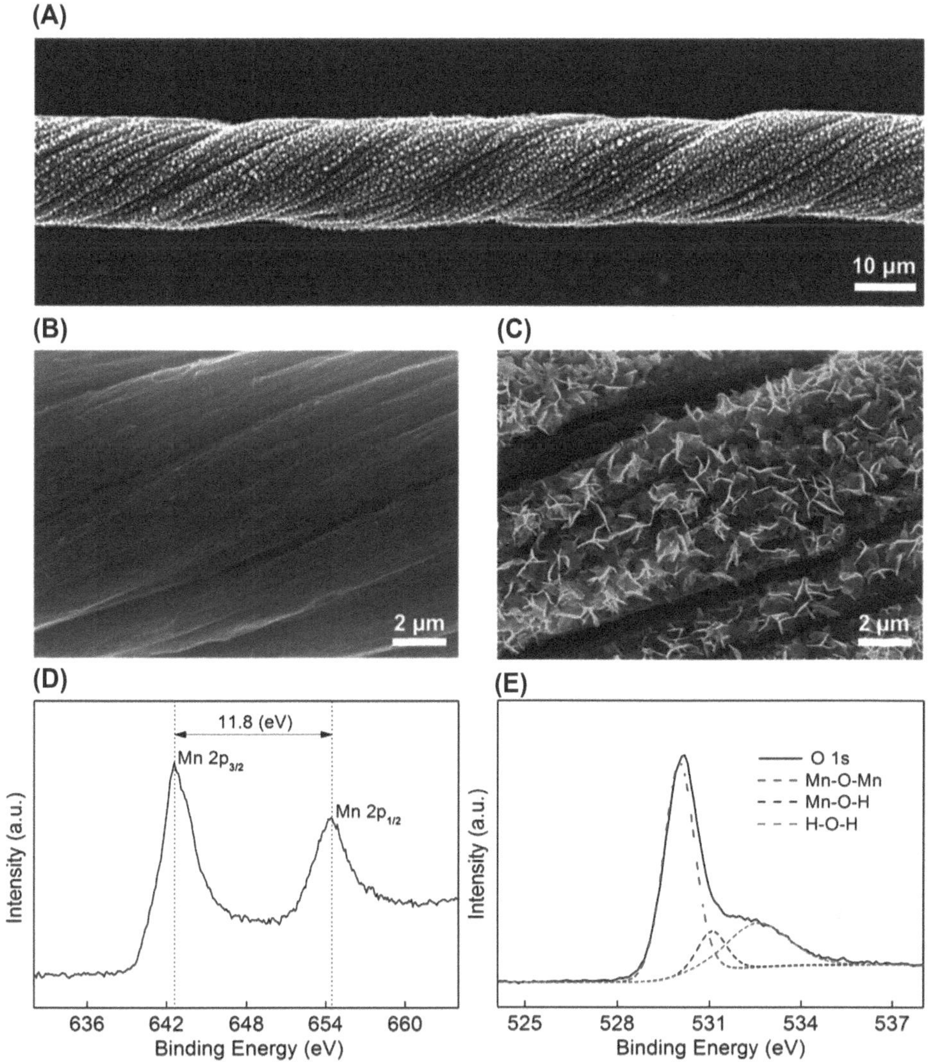

Figure 10.18 (A) Overview scanning electron microscope (SEM) image of CNT/MnO_2 composite yarn (CMY). The as-calculated volume is about 2.1×10^{-6} cm^3, assuming that the yarn is cylindrical. (B and C) Magnified surface SEM images of bare yarn (B) and CMY (C). Well-coated flower-shaped MnO_2 particles can be seen along the carbon nanotube (CNT) bundles. (D and E) Results of X-ray photoelectron spectroscopy (XPS) analysis of MnO_2 coating layer for oxidation state determination. Binding energies of Mn 2p (D) and O1s (E) components are shown. *(Reproduced from Choi, C.; Lee, J. A.; Choi, A. Y.; Kim, Y. T.; Lepró, X.; Lima, M. D.; et al. Flexible Supercapacitor Made of Carbon Nanotube Yarn With Internal Pores.* Advanced Materials ***2014,*** *26 (13), 2059–2065. Copyright 2013 John Wiley and Sons.)*

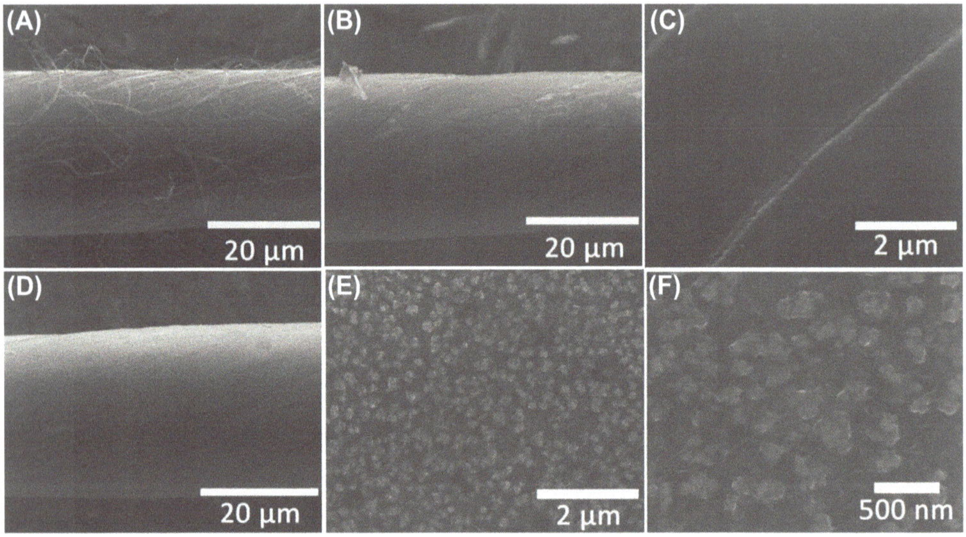

Figure 10.19 (A) Field emission scanning electron microscope (FESEM) image of a pure carbon nanotube (CNT) yarn. (B and C) FESEM images of CNT@NiO composite yarn. (D and E) FESEM images of CNT@Co_3O_4 composite yarn. *(Reproduced from Su, F.; Lv, X.; Miao, M. High-Performance Two-ply Yarn Supercapacitors Based on Carbon Nanotube Yarns Dotted with Co_3O_4 and NiO Nanoparticles.* Small ***2015,*** *11 (7), 854–861. Copyright 2014 John Wiley and Sons.)*

In addition to vacuum filtration, depositing metal oxides on dry-drawn MWCNT films has also been extensively studied. Jiang et al. employed MWCNT films dry drawn from arrays as a template to hold the nitrate precursors that were further transformed into metal oxide (NiO, Co_3O_4, or Mn_2O_3) nanoparticles after thermal decomposition *(70)*. These as-prepared hybrid electrodes presented excellent electrochemical performance including high average SC (336 F/g, 302 F/g, and 508 F/g for CNT/NiO, CNT/Co_3O_4, and CNT/Mn_2O_3, respectively), high cycling stability (<4.5% in 2500 cycles), and high rate capability (245 F/g at 155 A/g). To further increase the surface area of metal oxides, Lee et al. directly deposited flower-shaped MnO_x nanoparticles on aligned MWCNT films with well-ordered conductive paths (Fig. 10.21) *(117)*, which ensured highly accessible SSA and multiple paths for rapid electron transportation and ion diffusion, and obtained a high SC of 500–1250 F/g and high rate capability according to the sizes of MnO_x nanoparticles.

Continuous reticulate CNT films directly grown through the CVD method have been employed as templates to deposit metal oxides because of their high electrical conductivity and relatively high SSA mainly contributed by mesopores *(118,119)*. Hou et al. chemically deposited $NiCo_2O_4$ on such MWCNT films and studied their performance (Fig. 10.22) *(119)*. As the mesoporous structure of MWCNT/$NiCo_2O_4$ provided highly accessible SSA and short paths for ion diffusion, the hybrid films exhibited excellent

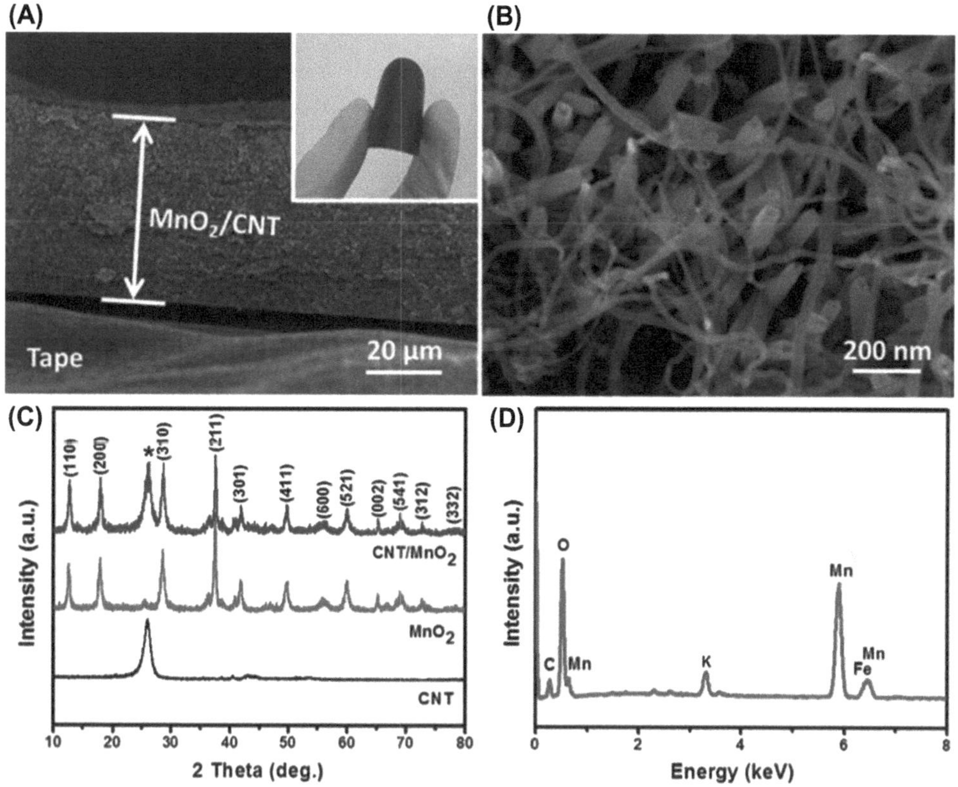

Figure 10.20 (A) Cross-sectional scanning electron microscope (SEM) image of the free-standing carbon nanotube (CNT)/MnO_2 nanotube hybrid film, and the *inset* is a digital image of the film showing the high flexibility. (B) Magnified cross-sectional SEM image of the free-standing CNT/MnO_2 nanotube hybrid film. (C) X-ray powder diffraction (XRD) patterns of the CNT film, MnO_2 nanotubes, and CNT/MnO_2 nanotube hybrid film. (D) Energy dispersive X-ray spectroscopy (EDS) spectrum of CNT/MnO_2 nanotube hybrid film. *(Reproduced from Du, L.; Yang, P.; Yu, X.; Liu, P.; Song, J.; Mai, W. Flexible Supercapacitors Based on Carbon Nanotube/MnO_2 Nanotube Hybrid Porous Films for Wearable Electronic Devices.* Journal of Materials Chemistry A ***2014,*** *2 (41), 17561–17567. Copyright 2014 Royal Society of Chemistry.)*

performance including high flexibility, high SC (828 F/g at 1 A/g), high rate capability, and cycling stability.

10.4.2.3 Hybrid Electrodes of CNT/Metal Oxide Arrays

Vertically aligned CNT arrays are quite suitable substrates for depositing metal oxides mainly because of their high electrical conductivity, ordered porous structures, and multiple conductive paths. Through incorporating metal oxides into CNT arrays, not only a better mechanical strength and electrochemical performance of the hybrids could be achieved, the induced synergistic effect could make better utilization of various active materials *(120)*. In recent years, much work has been carried out to introduce metal

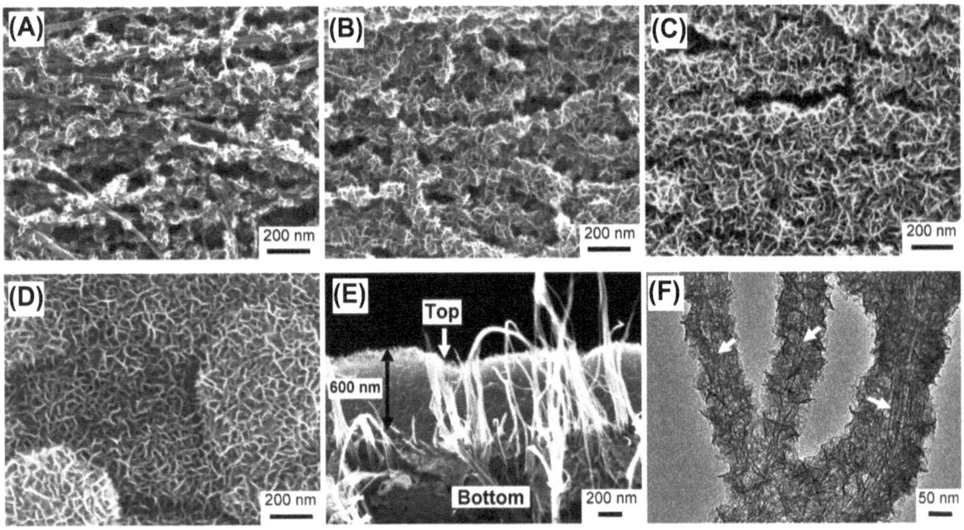

Figure 10.21 Morphology and microstructures of carbon nanotube (CNT) sheet/MnO_x composites. Scanning electron microscope (SEM) images of the MnO_x composites deposited for 5 (A), 15 (B), 30 (C), and 180 s (D) along with a cross-sectional view of the 180 s-deposited composite (E). The CNTs are shown as bright white lightning wires in panel E. A transmission electron microscopy (TEM) image of the 15 s-deposited composite (F). *White arrows* in panel F indicate CNTs covered by the MnO_x. *(Reproduced from Kim, J. H.; Lee, K. H.; Overzet, L. J.; Lee, G. S. Synthesis and Electrochemical Properties of Spin-Capable Carbon Nanotube Sheet/MnOx Composites for High-Performance Energy Storage Devices.* Nano Letters ***2011,*** *11 (7), 2611–2617. Copyright 2011 American Chemical Society.)*

oxides into CNT arrays and study their performance. Chen et al. prepared hybrid arrays through dip casting Mn_3O_4 nanoparticles into CVD-grown aligned MWCNT arrays previously removed from nonconductive substrates (Fig. 10.23) *(121)*. Compared with pure as-grown MWCNT arrays with an SC of only 1.84 F/g, this hybrid array with Mn_3O_4 loading of 84% exhibited a high SC (up to 143 F/g at a scan rate of 2 mV/s) and high rate capability because of its excellent nanostructured scaffold with uniformly distributed Mn_3O_4. However, introducing metal oxides in aligned CNT arrays often meets with problems such as limited intertube spacing, intrinsic hydrophobicity preventing the access of aqueous solutions, and the transferring step from the growth substrate to another substrate, which is detrimental for the quality of CNT arrays *(122)*. To solve these problems, Yang et al. developed a simple and novel method involving nebulized ethanol-assisted infiltration to prepare CNT/metal oxides with high quality *(122)*. Through this method, various metal oxides (including MnO_2, ZnO, NiO, Fe_3O_4, and TiO_2) have been effectively and uniformly deposited on CNT arrays, showing great potential in the energy storage field.

When serving as electrodes for ECs, CNT arrays directly grown on conductive substrates could provide higher power density because of the low contact resistance. As a

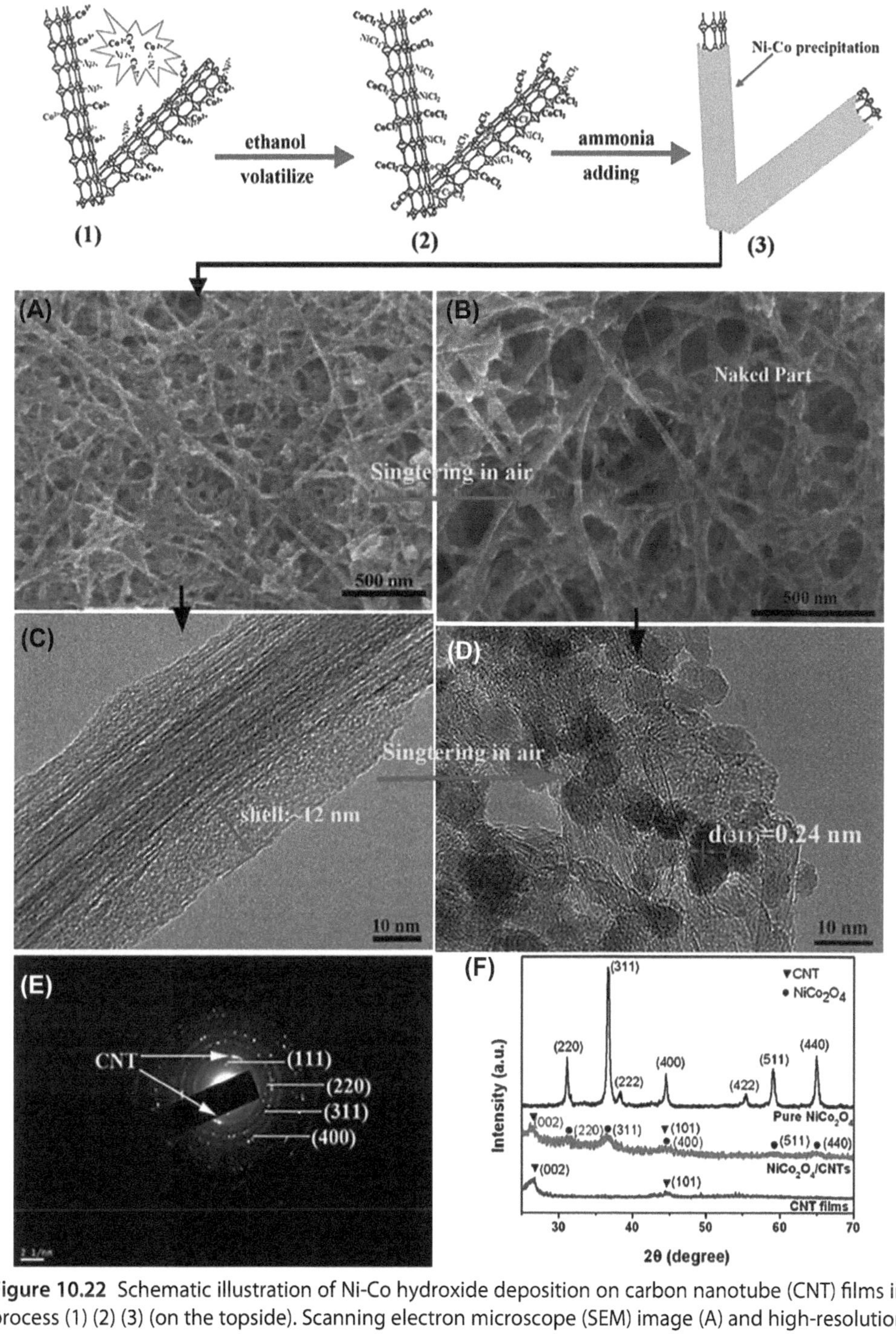

Figure 10.22 Schematic illustration of Ni-Co hydroxide deposition on carbon nanotube (CNT) films in process (1) (2) (3) (on the topside). Scanning electron microscope (SEM) image (A) and high-resolution transmission electron microscopy (HRTEM) image (C) of the corresponding Ni-Co precursor; SEM image (B) and HRTEM image (D) and the selected area electron diffraction (SAED) pattern (E) and X-ray powder diffraction (XRD) patterns (F) of flexible $NiCo_2O_4$/CNTs-21% composite film. *(Reproduced from Xu, S.; Yang, D.; Zhang, F.; Liu, J.; Guo, A.; Hou, F. Fabrication of $NiCo_2O_4$ and Carbon Nanotube Nanocomposite Films as a High-Performance Flexible Electrode of Supercapacitors.* RSC Advances ***2015,*** *5 (90), 74032–74039. Copyright 2015 Royal Society of Chemistry.)*

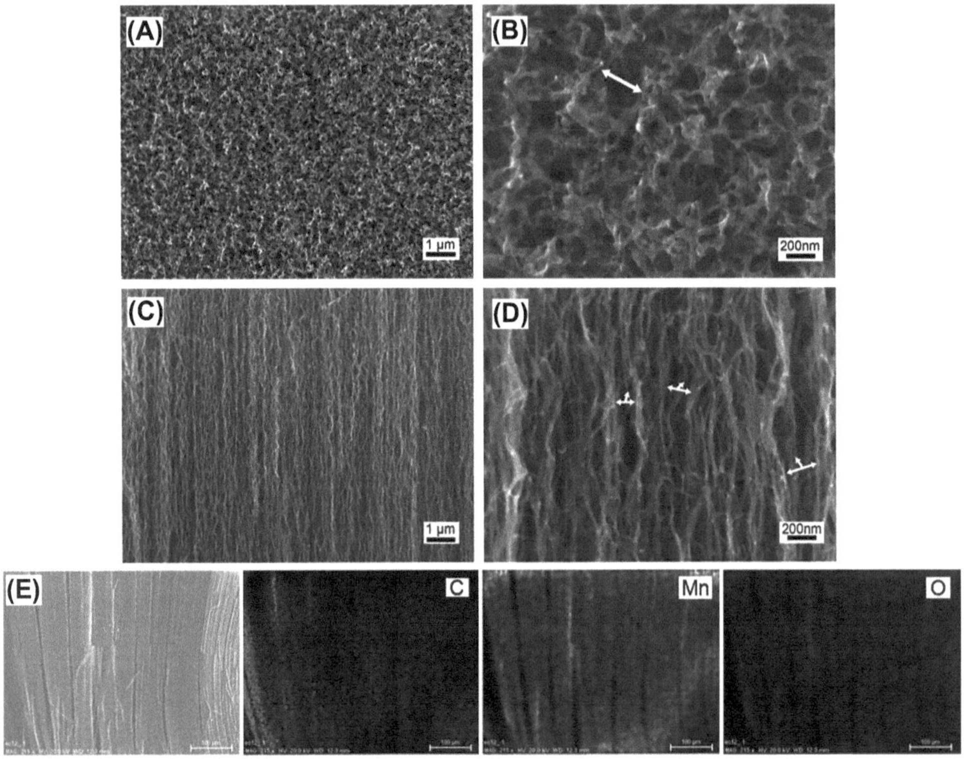

Figure 10.23 (A and B) Surface morphology and (C and D) cross-section field emission scanning electron microscope (FESEM) images of Mn_3O_4/carbon nanotube array (CNTA) composites at different magnifications. (E) Cross-section scanning electron microscope (SEM) image of Mn_3O_4/CNTA composite and its corresponding energy dispersive X-ray spectroscopy (EDS) mapping patterns for C, Mn, and O. *(Reproduced from Cui, X.; Hu, F.; Wei, W.; Chen, W. Dense and Long Carbon Nanotube Arrays Decorated With Mn_3O_4 Nanoparticles for Electrodes of Electrochemical Supercapacitors.* Carbon ***2011,*** *49 (4), 1225–1234. Copyright 2011 Elsevier.)*

result, depositing metal oxides on such CNT arrays is very appealing *(123–126)*. Tay et al. deposited $NiCo_2O_4$ nanoparticles uniformly on aligned MWCNT arrays grown on stainless steel and studied their electrochemical properties (Fig. 10.24) *(123)*. With unique 3D core–shell structure and low ESR, the hybrid array exhibited a high SC (695 F/g at 1 A/g) as well as a high power density (about 4 kW/kg). In another work, aligned MWCNT arrays have been grown on aluminum thin foils through a low-pressure CVD method by Ready and coworkers *(125)*. After deposition of conformal and uniform TiO_2 through atomic layer deposition, hierarchical 3D VACNTs@TiO_2/Al electrodes were prepared. Because of the firm connection between CNTs and Al foil with low ESR (1.21 Ω), the electrodes exhibited a high power density up to 24.8 ± 4.8 kW/kg and an SC of 73 F/g. In addition to thin metal foil, aligned

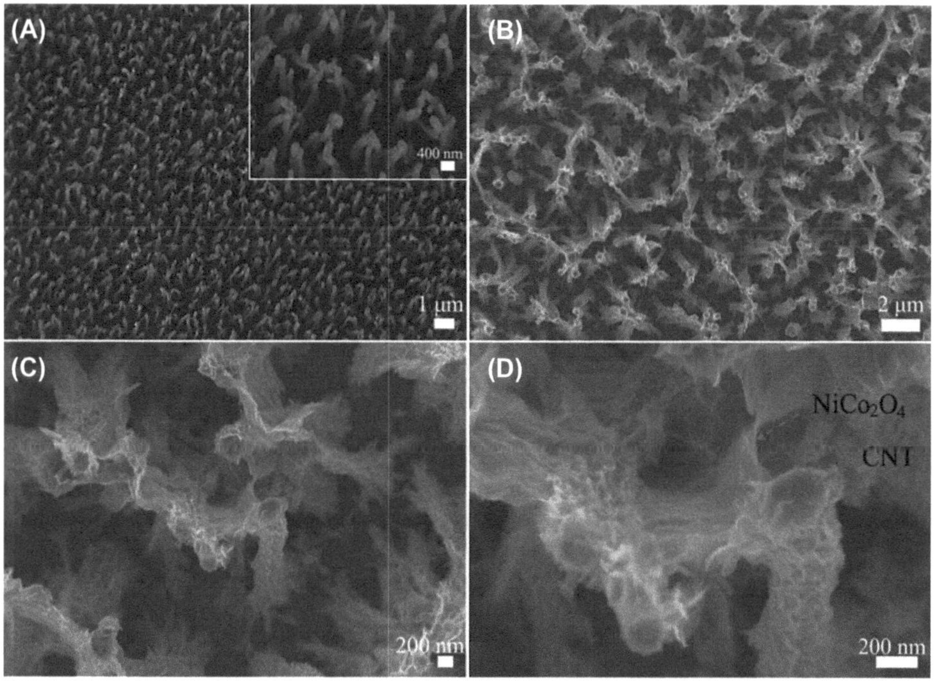

Figure 10.24 (a) Field emission scanning electron microscope (FESEM) images of bare 3D carbon nanotube (CNT) array and (B–D) FESEM images with different magnifications for the 3D vertically aligned CNT/$NiCo_2O_4$ core–shell structure. The inset in panel (A) is an enlargement of the part indicated by the circle in panel (A). The image in (C) is taken from the region marked with the rectangle in panel (B). *(Reproduced from Liu, W. W.; Lu, C.; Liang, K.; Tay, B. K. A Three Dimensional Vertically Aligned Multiwall Carbon Nanotube/NiCo*$_2$*O*$_4$ *Core/shell Structure for Novel High-Performance Supercapacitors.* Journal of Materials Chemistry A ***2014,*** *2 (14), 5100–5107. Copyright 2014 Royal Society of Chemistry.)*

CNT arrays could also be prepared on flexible carbon cloth *(124)*. With the assistance of pseudocapacitance provided by MnO_2, the hybrid array with low ESR (3.3 Ω) exhibited a remarkable SC of 235 F/g.

10.5 SUMMARY AND OUTLOOK

10.5.1 Current Challenges

Since traditional ESSs such as batteries and electrostatic capacitors cannot simultaneously meet the requirements for both high energy density and power density, ECs that partially solve this problem have been developed. In the past decades, much effort has been made to further increase the capacitance and energy density of ECs without sacrificing their power density and cycle life.

Because the electrochemical performance of an EC is largely determined by electrodes, the development of electrode material has always been the central issue. For now, most commercial ECs employ carbon materials such as ACs as electrode materials because of their high SSA and low cost. However, ACs encounter fatal weaknesses such as low electrical conductivity and unmanageable PSD, both of which will result in low rate capability of ECs.

With the rapid development of nanostructured carbons, especially CNTs and graphene, many new properties including high electric conductivity, large accessible SSA, high strength, and chemical stability stand a great chance of being endowed to the electrodes. To achieve this goal, macroscale 1D-, 2D-, and 3D-CNT assemblies with designed architecture have been developed. However, although CNTs possess excellent comprehensive performance and show great potential in ECs, they still face practical problems such as relatively low effective SSA of CNTs, high cost, and complicated preparation technology. Very recently, nanohybrid materials for electrodes based on CNTs have been extensively studied. By introducing other electroactive materials including conducting polymers and metal oxides, which simultaneously induce a synergistic effect, the electrochemical performance has been significantly improved.

10.5.2 Future Potentials

Finally, based on the previous discussion and comprehension of ECs, an ideal electrode should be designed according to the following principles:

1. Skeleton with high electrical conductivity. The process of energy storage in ECs consists of the migration of electrons and the diffusion of electrolyte ions, they occur simultaneously and are mutual restraint. To achieve a high power density and rate capability, the skeleton of the electrode must be highly conductive and lightweight. For now, undoubtedly, CNTs and graphene are the best candidates among various carbon materials. Moreover, to minimize the contact resistance between electrode and current collector, macroscale CNT assemblies can directly serve as current collector or be grown on the current collector.
2. Aligned 3D architecture with highly effective SSA and reasonable PSD. It is well known that a higher accessible SSA is beneficial for SC and energy density, while the aligned 3D architecture of the skeleton maximizes the exposed surfaces of CNTs *(127)*. In addition, the aligned architecture can provide reasonable PSD and ordered arrays. A highly organized path, which will both facilitate the rapid transportation of electrons and diffusion of electrolyte ions, is thus achieved and will improve the power density and the response time of ECs.
3. Nanohybrid electrodes with nanosized pseudocapacitance materials on the conductive skeleton. ECs with pure physical double-layer capacitance will no longer meet the growing demands for energy density. Future work should focus on developing nanohybrid electrodes with proper nanosized pseudocapacitance materials on the

conductive skeleton. In addition to introducing pseudocapacitance with better energy storage capability, an expected synergistic effect between different materials would be induced.

4. Low-cost, renewable, environmentally friendly raw electrode materials as well as simple and facile processes. Because of the complicated fabrication processes, high cost of CNTs (>$15/g) has always been the greatest challenge for their application in ECs. For most commercial applications, CNTs have been merely employed as additives whose content is usually low. Only when the market price of CNTs is decreased to below $1/g, would they be abundantly employed (>50 wt%) and fundamentally improve the electrochemical performance of ECs.

REFERENCES

1. Chu, S.; Majumdar, A. Opportunities and Challenges for a Sustainable Energy Future. *Nature* **2012,** *488* (7411), 294–303.
2. Wang, G.; Zhang, L.; Zhang, J. A Review of Electrode Materials for Electrochemical Supercapacitors. *Chemical Society Reviews* **2012,** *41* (2), 797–828.
3. Holdren, J. P. Energy and Sustainability. *Science* **2007,** *315* (5813), 737.
4. Winter, M.; Brodd, R. J. What Are Batteries, Fuel Cells, and Supercapacitors? *Chemical Reviews* **2004,** *104* (10), 4245–4270.
5. Yan, J.; Wang, Q.; Wei, T.; Fan, Z. Recent Advances in Design and Fabrication of Electrochemical Supercapacitors with High Energy Densities. *Advanced Energy Materials* **2014,** *4* (4), 1300816.
6. Lu, M.; Beguin, F.; Frackowiak, E. *Supercapacitors: Materials, Systems and Applications;* John Wiley & Sons, 2013.
7. Simon, P.; Gogotsi, Y. Materials for Electrochemical Capacitors. *Nature Materials* **2008,** *7* (11), 845–854.
8. Zhang, Y.; Feng, H.; Wu, X.; Wang, L.; Zhang, A.; Xia, T.; et al. Progress of Electrochemical Capacitor Electrode Materials: A Review. *International Journal of Hydrogen Energy* **2009,** *34* (11), 4889–4899.
9. Long, J. W.; Bélanger, D.; Brousse, T.; Sugimoto, W.; Sassin, M. B.; Crosnier, O. Asymmetric Electrochemical Capacitors-Stretching the Limits of Aqueous Electrolytes. *Mrs Bulletin* **2011,** *36* (07), 513–522.
10. Zhang, S.; Pan, N. Supercapacitors Performance Evaluation. *Advanced Energy Materials* **2015,** *5* (6), 1401401.
11. Zhang, L. L.; Zhao, X. S. Carbon-Based Materials as Supercapacitor Electrodes. *Chemical Society Reviews* **2009,** *38* (9), 2520–2531.
12. Danaee, I.; Jafarian, M.; Forouzandeh, F.; Gobal, F.; Mahjani, M. G. Electrochemical Impedance Studies of Methanol Oxidation on GC/Ni and GC/NiCu Electrode. *International Journal of Hydrogen Energy* **2009,** *34* (2), 859–869.
13. Wang, D. W.; Li, F.; Liu, M.; Lu, G. Q.; Cheng, H. M. 3D Aperiodic Hierarchical Porous Graphitic Carbon Material for High-Rate Electrochemical Capacitive Energy Storage. *Angewandte Chemie International Edition* **2008,** *120* (2), 379–382.
14. Azaıs, P. In *Manufacturing of Industrial Supercapacitors;* Béguin, F., Fr sackowiak, E., Eds.; 2013.
15. Zhang, L.; Zhang, F.; Yang, X.; Leng, K.; Huang, Y.; Chen, Y. High-Performance Supercapacitor Electrode Materials Prepared from Various Pollens. *Small* **2013,** *9* (8), 1342–1347.
16. Xu, B.; Wu, F.; Chen, S.; Zhang, C.; Cao, G.; Yang, Y. Activated Carbon Fiber Cloths as Electrodes for High Performance Electric Double Layer Capacitors. *Electrochimica Acta* **2007,** *52* (13), 4595–4598.
17. Zhang, L. L.; Zhou, R.; Zhao, X. Graphene-Based Materials as Supercapacitor Electrodes. *Journal of Materials Chemistry A* **2010,** *20* (29), 5983–5992.

18. Yu, G.; Xie, X.; Pan, L.; Bao, Z.; Cui, Y. Hybrid Nanostructured Materials for High-Performance Electrochemical Capacitors. *Nano Energy* **2013,** *2* (2), 213–234.
19. Kreup, F.; Graham, A. P.; Liebau, M.; Duesberg, G. S.; Seidel, R.; Unger, E. Carbon Nanotubes for Interconnect Applications. *Electron Devices Meeting, 2004 IEDM Technical Digest IEEE International* **2004,** 683–686.
20. Ebbesen, T.; Lezec, H.; Hiura, H.; Bennett, J.; Ghaemi, H.; Thio, T. Electrical Conductivity of Individual Carbon Nanotubes. *Nature* **1996,** *382.*
21. Yao, Z.; Kane, C. L.; Dekker, C. High-Field Electrical Transport in Single-Wall Carbon Nanotubes. *Physical Review Letters* **2000,** *84* (13), 2941.
22. Radosavljević, M.; Lefebvre, J.; Johnson, A. High-Field Electrical Transport and Breakdown in Bundles of Single-Wall Carbon Nanotubes. *Physical Review B* **2001,** *64* (24), 241307.
23. Zhang, Y.; Zheng, L. Towards Chirality-Pure Carbon Nanotubes. *Nanoscale* **2010,** *2* (10), 1919–1929.
24. Pandolfo, A.; Hollenkamp, A. Carbon Properties and Their Role in Supercapacitors. *Journal of Power Sources* **2006,** *157* (1), 11–27.
25. Hahn, M.; Baertschi, M.; Barbieri, O.; Sauter, J.-C.; Kötz, R.; Gallay, R. Interfacial Capacitance and Electronic Conductance of Activated Carbon Double-Layer Electrodes. *Electrochemical and Solid-State Letters* **2004,** 7 (2), A33–A36.
26. Kh, A.; Jeon, K. K.; Kim, W. S.; Park, Y. S.; Lim, S. C.; Bae, D. J. Characterization of Supercapacitors Using Singlewalled Carbon Nanotube Electrodes. *Journal-Korean Physical Society* **2001,** *39,* S511–S517.
27. Kozlov, M. E.; Capps, R. C.; Sampson, W. M.; Ebron, V. H.; Ferraris, J. P.; Baughman, R. H. Spinning Solid and Hollow Polymer-Free Carbon Nanotube Fibers. *Advanced Materials* **2005,** *17* (5), 614–617.
28. Yuan, Y.; Hong, L. Hierarchical Micro-Architectures of Electrodes for Energy Storage. *Journal of Power Sources* **2015,** *284,* 435–445.
29. Kimizuka, O.; Tanaike, O.; Yamashita, J.; Hiraoka, T.; Futaba, D. N.; Hata, K.; et al. Electrochemical Doping of Pure Single-Walled Carbon Nanotubes Used as Supercapacitor Electrodes. *Carbon* **2008,** *46* (14), 1999–2001.
30. Zhou, W.; Sun, F.; Pan, K.; Tian, G.; Jiang, B.; Ren, Z.; et al. Well-Ordered Large-Pore Mesoporous Anatase TiO_2 with Remarkably High Thermal Stability and Improved Crystallinity: Preparation, Characterization, and Photocatalytic Performance. *Advanced Functional Materials* **2011,** *21* (10), 1922–1930.
31. Liu, R.; Lee, S. B. MnO_2/poly (3,4-ethylenedioxythiophene) Coaxial Nanowires by One-Step Coelectrodeposition for Electrochemical Energy Storage. *Journal of the American Chemical Society* **2008,** *130* (10), 2942–2943.
32. Liu, L.; Ma, W.; Zhang, Z. Macroscopic Carbon Nanotube Assemblies: Preparation, Properties, and Potential Applications. *Small* **2011,** 7 (11), 1504–1520.
33. Lu, W.; Zu, M.; Byun, J. H.; Kim, B. S.; Chou, T. W. State of the Art of Carbon Nanotube Fibers: Opportunities and Challenges. *Advanced Materials* **2012,** *24* (14), 1805–1833.
34. Chou, T. W.; Gao, L.; Thostenson, E. T.; Zhang, Z.; Byun, J. H. An Assessment of the Science and Technology of Carbon Nanotube-Based Fibers and Composites. *Composites Science & Technology* **2010,** *70* (1), 1–19.
35. Yang, Z.; Ren, J.; Zhang, Z.; Chen, X.; Guan, G.; Qiu, L.; et al. Recent Advancement of Nanostructured Carbon for Energy Applications. *Chemical Reviews* **2015,** *115* (11), 5159–5223.
36. Ma, W.; Li, S.; Rong, Y.; Zhang, T.; Zhao, Y.; Sun, L.; et al. Directly Synthesized Strong, Highly Conducting, Transparent Single-Walled Carbon Nanotube Films. *Nano Letters* **2007,** 7 (8), 2307–2311.
37. Niu, Z.; Dong, H.; Zhu, B.; Li, J.; Hng, H. H.; Zhou, W.; et al. Highly Stretchable, Integrated Supercapacitors Based on Single-Walled Carbon Nanotube Films with Continuous Reticulate Architecture. *Advanced Materials* **2013,** *25* (7), 1058–1064.
38. Niu, Z.; Zhou, W.; Chen, J.; Feng, G.; Li, H.; Ma, W.; et al. Compact-Designed Supercapacitors Using Free-Standing Single-Walled Carbon Nanotube Films. *Energy & Environ Science* **2011,** *4* (4), 1440–1446.

39. Jiang, K.; Wang, J.; Li, Q.; Liu, L.; Liu, C.; Fan, S. Superaligned Carbon Nanotube Arrays, Films, and Yarns: A Road to Applications. *Advanced Materials* **2011,** *23* (9), 1154–1161.
40. Talapatra, S.; Kar, S.; Pal, S.; Vajtai, R.; Ci, L.; Victor, P.; et al. Direct Growth of Aligned Carbon Nanotubes on Bulk Metals. *Nature Nanotechnology* **2006,** *1* (2), 112–116.
41. Inagaki, M.; Konno, H.; Tanaike, O. Carbon Materials for Electrochemical Capacitors. *Journal of Power Sources* **2010,** *195* (24), 7880–7903.
42. Zhang, H.; Cao, G.; Yang, Y.; Gu, Z. Comparison Between Electrochemical Properties of Aligned Carbon Nanotube Array and Entangled Carbon Nanotube Electrodes. *Journal of the Electrochemical Society* **2008,** *155* (2), K19–K22.
43. Zhang, M.; Fang, S.; Zakhidov, A. A.; Lee, S. B.; Aliev, A. E.; Williams, C. D.; et al. Strong, Transparent, Multifunctional, Carbon Nanotube Sheets. *Science* **2005,** *309* (5738), 1215–1219.
44. Li, Q. W.; Zhang, X. F.; Depaula, R. F.; Zheng, L. X.; Zhao, Y. H.; Stan, L.; et al. Sustained Growth of Ultralong Carbon Nanotube Arrays for Fiber Spinning. *Advanced Materials* **2006,** *18* (23), 3160–3163.
45. Zhang, H. X.; Feng, C.; Zhai, Y. C.; Jiang, K. L.; Li, Q. Q.; Fan, S. S. Cross-Stacked Carbon Nanotube Sheets Uniformly Loaded with SnO_2 Nanoparticles: A Novel Binder-Free and High-Capacity Anode Material for Lithium-Ion Batteries. *Advanced Materials* **2009,** *21* (22), 2299–2304.
46. Li, W.; Xie, S.; Qian, L.; Chang, B.; Zou, B. S.; Zhou, W. Y.; Zhao, R. A.; Wang, G. *Science* **1996,** *274* (1701), 19.
47. Futaba, D. N.; Hata, K.; Yamada, T.; Hiraoka, T.; Hayamizu, Y.; Kakudate, Y.; et al. Shape-engineerable and Highly Densely Packed Single-Walled Carbon Nanotubes and Their Application as Super-Capacitor Electrodes. *Nature Materials* **2006,** *5* (12), 987–994.
48. Largeot, C.; Portet, C.; Chmiola, J.; Taberna, P.-L.; Gogotsi, Y.; Simon, P. Relation Between the Ion Size and Pore Size for an Electric Double-Layer Capacitor. *Journal of the American Chemical Society* **2008,** *130* (9), 2730–2731.
49. Niu, C.; Sichel, E. K.; Hoch, R.; Moy, D.; Tennent, H. High Power Electrochemical Capacitors Based on Carbon Nanotube Electrodes. *Applied Physics Letters* **1997,** *70* (11), 1480–1482.
50. Fang, W. C.; Huang, J. H.; Chen, L. C.; Su, Y. O.; Chen, K. H.; Sun, C. L. Carbon Nanotubes Grown Directly on Ti Electrodes and Enhancement of Their Electrochemical Properties by Nitric Acid Treatment. *Electrochemical and Solid-State Letters* **2006,** *9* (1), A5–A8.
51. An, K. H.; Kim, W. S.; Park, Y. S.; Choi, Y. C.; Lee, S. M.; Chung, D. C.; et al. Supercapacitors Using Single-Walled Carbon Nanotube Electrodes. *Advanced Materials* **2001,** *13* (7), 497–500.
52. Jiang, Q.; Qu, M.; Zhou, G.; Zhang, B.; Yu, Z. A Study of Activated Carbon Nanotubes as Electrochemical Super Capacitors Electrode Materials. *Materials Letters* **2002,** *57* (4), 988–991.
53. Frackowiak, E.; Delpeux, S.; Jurewicz, K.; Szostak, K.; Cazorla-Amoros, D.; Beguin, F. Enhanced Capacitance of Carbon Nanotubes through Chemical Activation. *Chemical Physics Letters* **2002,** *361* (1), 35–41.
54. Zhang, L. L.; Zhao, X. Carbon-Based Materials as Supercapacitor Electrodes. *Chemical Society Reviews* **2009,** *38* (9), 2520–2531.
55. Obreja, V. V. On the Performance of Supercapacitors With Electrodes Based on Carbon Nanotubes and Carbon Activated Material – A Review. *Physica E: Low-Dimensional Systems and Nanostructures* **2008,** *40* (7), 2596–2605.
56. Wang, G.; Liang, R.; Liu, L.; Zhong, B. Improving the Specific Capacitance of Carbon Nanotubes-Based Supercapacitors by Combining Introducing Functional Groups on Carbon Nanotubes With Using Redox-Active Electrolyte. *Electrochimica Acta* **2014,** *115,* 183–188.
57. Yoon, B. J.; Jeong, S. H.; Lee, K. H.; Kim, H. S.; Park, C. G.; Han, J. H. Electrical Properties of Electrical Double Layer Capacitors With Integrated Carbon Nanotube Electrodes. *Chemical Physics Letters* **2004,** *388* (1), 170–174.
58. Jurewicz, K.; Babeł, K.; Pietrzak, R.; Delpeux, S.; Wachowska, H. Capacitance Properties of Multi-Walled Carbon Nanotubes Modified by Activation and Ammoxidation. *Carbon* **2006,** *44* (12), 2368–2375.
59. Izadi-Najafabadi, A.; Yasuda, S.; Kobashi, K.; Yamada, T.; Futaba, D. N.; Hatori, H.; et al. Extracting the Full Potential of Single-Walled Carbon Nanotubes as Durable Supercapacitor Electrodes Operable at 4V with High Power and Energy Density. *Advanced Materials* **2010,** *22* (35), E235–E241.

60. Hata, K.; Futaba, D. N.; Mizuno, K.; Namai, T.; Yumura, M.; Iijima, S. Water-Assisted Highly Efficient Synthesis of Impurity-Free Single-Walled Carbon Nanotubes. *Science* **2004,** *306* (5700), 1362–1364.
61. Kaili, J.; Qunqing, L.; Shoushan, F. Nanotechnology: Spinning Continuous Carbon Nanotube Yarns. *Nature International Weekly Journal of Science* **2002,** *419* (19), 801.
62. Zhan, Z. Y.; Zhang, Y. N.; Sun, G. Z.; Zheng, L. X.; Liao, K. The Effects of Catalyst Treatment on Fast Growth of Millimeter-Long Multi-walled Carbon Nanotube Arrays. *Applied Surface Science* **2011,** *257* (17), 7704–7708.
63. Zhang, Y. N.; Sun, G. Z.; Zheng, L. X. Fabrication and Properties of Macroscopic Carbon Nanotube Assemblies Transforming From Aligned Nanotubes. *Journal of Nanomaterials* **2015,** *2015,* 1–8.
64. Yu, D.; Qian, Q.; Wei, L.; Jiang, W.; Goh, K.; Wei, J.; et al. Emergence of Fiber Supercapacitors. *Chemical Society Reviews* **2015,** *44* (3), 647–662.
65. Ren, J.; Li, L.; Chen, C.; Chen, X.; Cai, Z.; Qiu, L.; et al. Twisting Carbon Nanotube Fibers for Both Wire-Shaped Micro-supercapacitor and Micro-battery. *Advanced Materials* **2013,** *25* (8), 1155–1159.
66. Sun, G.; Zhou, J.; Yu, F.; Zhang, Y.; Pang, J. H. L.; Zheng, L. Electrochemical Capacitive Properties of CNT Fibers Spun From Vertically Aligned CNT Arrays. *Journal of Solid State Electrochemistry* **2012,** *16* (5), 1775–1780.
67. Zhong, X. H.; Li, Y. L.; Liu, Y. K.; Qiao, X. H.; Feng, Y.; Liang, J.; et al. Continuous Multilayered Carbon Nanotube Yarns. *Advanced Materials* **2010,** *22* (6), 692–696.
68. Zhang, Y.; Bai, W.; Ren, J.; Weng, W.; Lin, H.; Zhang, Z.; et al. Super-Stretchy Lithium-Ion Battery Based on Carbon Nanotube Fiber. *Journal of Materials Chemistry A* **2014,** *2* (29), 11054–11059.
69. Ye, Z.; Wenyu, B.; Xunliang, C.; Jing, R.; Wei, W.; Peining, C.; et al. Flexible and Stretchable Lithium-Ion Batteries and Supercapacitors Based on Electrically Conducting Carbon Nanotube Fiber Springs. *Angewandte Chemie* **2014,** *53* (52), 14564–14568.
70. Zhou, R.; et al. High-Performance Supercapacitors Using a Nanoporous Current Collector Made from Super-Aligned Carbon Nanotubes. *Nanotechnology* **2010,** *21* (34), 345701–345707 (7).
71. Du, C.; Yeh, J.; Pan, N. High Power Density Supercapacitors Using Locally Aligned Carbon Nanotube Electrodes. *Nanotechnology* **2005,** *16* (4), 350.
72. Wang, D.; Song, P. C.; Liu, C. H.; Wu, W.; Fan, S. S. Highly Oriented Carbon Nanotube Papers Made of Aligned Carbon Nanotubes. *Nanotechnology* **2008,** *19* (7), 369–381.
73. Honda, Y.; Takeshige, M.; Shiozaki, H.; Kitamura, T.; Yoshikawa, K.; Chakrabarti, S.; et al. Vertically Aligned Double-Walled Carbon Nanotube Electrode Prepared by Transfer Methodology for Electric Double Layer Capacitor. *Journal of Power Sources* **2008,** *185* (2), 1580–1584.
74. Honda, Y.; Haramoto, T.; Takeshige, M.; Shiozaki, H.; Kitamura, T.; Ishikawa, M. Aligned MWCNT Sheet Electrodes Prepared by Transfer Methodology Providing High-Power Capacitor Performance. *Electrochemical and Solid-State Letters* **2007,** *10* (4), A106–A110.
75. Gao, L.; Peng, A.; Wang, Z. Y.; Zhang, H.; Shi, Z.; Gu, Z.; et al. Growth of Aligned Carbon Nanotube Arrays on Metallic Substrate and its Application to Supercapacitors. *Solid State Communications* **2008,** *146* (9), 380–383.
76. Zhang, H.; Cao, G.; Yang, Y. Using a Cut-Paste method to Prepare a Carbon Nanotube Fur Electrode. *Nanotechnology* **2007,** *18* (19), 195607.
77. Parthangal, P. M.; Cavicchi, R. E.; Zachariah, M. R. A Generic Process of Growing Aligned Carbon Nanotube Arrays on Metals and Metal Alloys. *Nanotechnology* **2007,** *18* (18), 185605.
78. Lin, C. L.; Chen, C. F.; Shi, S. C. Field Emission Properties of Aligned Carbon Nanotubes Grown on Stainless Steel Using CH_4/CO_2 Reactant Gas. *Diamond and Related Materials* **2004,** *13* (4), 1026–1031.
79. Xu, F.; Liu, X.; Stephen, D. T. Synthesis of Carbon Nanotubes on Metal Alloy Substrates with Voltage Bias in Methane Inverse Diffusion Flames. *Carbon* **2006,** *44* (3), 570–577.
80. Hiraoka, T.; Yamada, T.; Hata, K.; Futaba, D. N.; Kurachi, H.; Uemura, S.; et al. Synthesis of Single- and Double-Walled Carbon Nanotube Forests on Conducting Metal Foils. *Journal of the American Chemical Society* **2006,** *128* (41), 13338–13339.
81. Frackowiak, E.; Jurewicz, K.; Szostak, K.; Delpeux, S.; Beguin, F. Nanotubular Materials as Electrodes for Supercapacitors. *Fuel Processing Technology* **2002,** *77,* 213–219.

82. Jha, N.; Ramesh, P.; Bekyarova, E.; Itkis, M. E.; Haddon, R. C. High Energy Density Supercapacitor Based on a Hybrid Carbon Nanotube-Reduced Graphite Oxide Architecture. *Advanced Energy Materials* **2012,** *2* (4), 438–444.
83. Peng, S.; Li, L.; Wu, H. B.; Madhavi, S.; Lou, X. W. Controlled Growth of $NiMoO_4$ Nanosheet and Nanorod Arrays on Various Conductive Substrates as Advanced Electrodes for Asymmetric Supercapacitors. *Advanced Energy Materials* **2014,** *5* (2), 1401172.
84. Pan, L.; Qiu, H.; Dou, C.; Li, Y.; Lin, P.; Xu, J.; et al. Conducting Polymer Nanostructures: Template Synthesis and Applications in Energy Storage. *International Journal of Molecular Sciences* **2010,** *11* (7), 2636–2657.
85. Woo, S. W.; Dokko, K.; Kanamura, K. Composite Electrode Composed of Bimodal Porous Carbon and Polypyrrole for Electrochemical Capacitors. *Journal of Power Sources* **2008,** *185* (2), 1589–1593.
86. Pan, S.; Lin, H.; Deng, J.; Chen, P.; Chen, X.; Yang, Z.; et al. Novel Wearable Energy Devices Based on Aligned Carbon Nanotube Fiber Textiles. *Advanced Energy Materials* **2015,** *5* (4), 1401438.
87. Wang, B.; Fang, X.; Sun, H.; He, S.; Ren, J.; Zhang, Y.; et al. Fabricating Continuous Supercapacitor Fibers with High Performances by Integrating All Building Materials and Steps into One Process. *Advanced Materials* **2015,** *27* (47), 7854–7860.
88. Dalton, A. B.; Collins, S.; Muñoz, E.; Razal, J. M.; Ebron, V. H.; Ferraris, J. P.; et al. Super-Tough Carbon-Nanotube Fibres. *Nature* **2003,** *423* (6941), 703.
89. Meng, Q.; Wang, K.; Guo, W.; Fang, J.; Wei, Z.; She, X. Thread-Like Supercapacitors Based on One-Step Spun Nanocomposite Yarns. *Small* **2014,** *10* (15), 3187–3193.
90. Xu, R.; Wei, J.; Guo, F.; Cui, X.; Zhang, T.; Zhu, H.; et al. Highly Conductive, Twistable and Bendable Polypyrrole-Carbon Nanotube Fiber for Efficient Supercapacitor Electrodes. *RSC Advances* **2015,** *5* (28), 22015–22021.
91. Xu, R.; Guo, F.; Cui, X.; Zhang, L.; Wang, K.; Wei, J. High Performance Carbon Nanotube Based Fiber-Shaped Supercapacitors Using Redox Additives of Polypyrrole and Hydroquinone. *Journal of Materials Chemistry A* **2015,** *3* (44), 22353–22360.
92. Ge, J.; Cheng, G.; Chen, L. Transparent and Flexible Electrodes and Supercapacitors Using Polyaniline/Single-Walled Carbon Nanotube Composite Thin Films. *Nanoscale* **2011,** *3* (8), 3084–3088.
93. Liu, J.; Sun, J.; Gao, L. A Promising Way to Enhance the Electrochemical Behavior of Flexible Single-Walled Carbon Nanotube/Polyaniline Composite Films. *Journal of Physical Chemistry C* **2010,** *114* (46), 19614–19620.
94. Lin, H.; Li, L.; Ren, J.; Cai, Z.; Qiu, L.; Yang, Z.; et al. Conducting Polymer Composite Film Incorporated With Aligned Carbon Nanotubes for Transparent, Flexible and Efficient Supercapacitor. *Scientific Reports* **2013,** *3,* 1353.
95. Yin, Y.; Liu, C.; Fan, S. Well-Constructed CNT Mesh/PANI Nanoporous Electrode and its Thickness Effect on the Supercapacitor Properties. *The Journal of Physical Chemistry C* **2012,** *116* (50), 26185–26189.
96. Niu, Z.; Luan, P.; Shao, Q.; Dong, H.; Li, J.; Chen, J.; et al. A "skeleton/skin" Strategy for Preparing Ultrathin Free-Standing Single-Walled Carbon Nanotube/Polyaniline Films for High Performance Supercapacitor Electrodes. *Energy & Environmental Science* **2012,** *5* (9), 8726–8733.
97. Hughes, M.; Shaffer, M. S.; Renouf, A. C.; Singh, C.; Chen, G. Z.; Fray, D. J.; et al. Electrochemical Capacitance of Nanocomposite Films Formed by Coating Aligned Arrays of Carbon Nanotubes With Polypyrrole. *Advanced Materials* **2002,** *14* (5), 382–385.
98. Lachman, N.; Xu, H.; Zhou, Y.; Ghaffari, M.; Lin, M.; Bhattacharyya, D.; et al. Tailoring Thickness of Conformal Conducting Polymer Decorated Aligned Carbon Nanotube Electrodes for Energy Storage. *Advanced Materials Interfaces* **2014,** *1* (7), 1400076101.
99. Ghaffari, M.; Kosolwattana, S.; Zhou, Y.; Lachman, N.; Lin, M.; Bhattacharya, D.; et al. Hybrid Supercapacitor Materials from Poly (3,4-ethylenedioxythiophene) Conformally Coated Aligned Carbon Nanotubes. *Electrochimica Acta* **2013,** *112,* 522–528.
100. Zhou, Y.; Xu, H.; Lachman, N.; Ghaffari, M.; Wu, S.; Liu, Y.; et al. Advanced Asymmetric Supercapacitor Based on Conducting Polymer and Aligned Carbon Nanotubes with Controlled Nanomorphology. *Nano Energy* **2014,** *9,* 176–185.

101. Zhang, H.; Cao, G.; Wang, Z.; Yang, Y.; Shi, Z.; Gu, Z. Tube-Covering-Tube Nanostructured Polyaniline/carbon Nanotube Array Composite Electrode with High Capacitance and Superior Rate Performance as Well as Good Cycling Stability. *Electrochemistry Communications* **2008,** *10* (7), 1056–1059.
102. Zhang, H.; Cao, G.; Wang, W.; Yuan, K.; Xu, B. Influence of Microstructure on the Capacitive Performance of Polyaniline/Carbon Nanotube Array Composite Electrodes. *Electrochimica Acta* **2009,** *54* (4), 1153–1159.
103. Huang, F.; Chen, D. Towards the Upper Bound of Electrochemical Performance of ACNT@polyaniline Arrays as Supercapacitors. *Energy & Environmental Science* **2012,** *5* (2), 5833–5841.
104. Huang, F.; Lou, F.; Chen, D. Exploring Aligned-Carbon- Nanotubes@Polyaniline Arrays on Household Al as Supercapacitors. *ChemSusChem* **2012,** *5* (5), 888–895.
105. Soin, N.; Roy, S. S.; Mitra, S. K.; Thundat, T.; Mclaughlin, J. A. Nanocrystalline Ruthenium Oxide Dispersed Few Layered Graphene (FLG) Nanoflakes as Supercapacitor Electrodes. *Journal of Material Chemistry* **2012,** *22* (30), 14944–14950.
106. Wang, B.; Chen, J. S.; Wang, Z.; Madhavi, S.; Lou, X. W. D. Green Synthesis of NiO Nanobelts With Exceptional Pseudo-Capacitive Properties. *Advanced Energy Materials* **2012,** *2* (10), 1188–1192.
107. Ye, J. S.; Cui, H. F.; Liu, X.; Lim, T. M.; Zhang, W. D.; Sheu, F. S. Preparation and Characterization of Aligned Carbon Nanotube-Ruthenium Oxide Nanocomposites for Supercapacitors. *Small* **2005,** *1* (5), 560–565.
108. Wu, Z. S.; Zhou, G.; Yin, L. C.; Ren, W.; Li, F.; Cheng, H. M. Graphene/Metal Oxide Composite Electrode Materials for Energy Storage. *Nano Energy* **2012,** *1* (1), 107–131.
109. Zhang, J.; Jiang, J.; Zhao, X. S. Synthesis and Capacitive Properties of Manganese Oxide Nanosheets Dispersed on Functionalized Graphene Sheets. *Journal of Physics Chemistry C* **2011,** *115* (14), 6448–6454.
110. Wen, J.; Ruan, X.; Zhou, Z. Preparation and Electrochemical Performance of Novel Ruthenium-Manganese Oxide Electrode Materials for Electrochemical Capacitors. *Journal of Physics & Chemistry of Solids* **2009,** *70,* 816–820.
111. Choi, C.; Lee, J. A.; Choi, A. Y.; Kim, Y. T.; Lepró, X.; Lima, M. D.; et al. Flexible Supercapacitor Made of Carbon Nanotube Yarn With Internal Pores. *Advanced Materials* **2014,** *26* (13), 2059–2065.
112. Su, F.; Lv, X.; Miao, M. High-Performance Two-ply Yarn Supercapacitors Based on Carbon Nanotube Yarns Dotted with Co_3O_4 and NiO Nanoparticles. *Small* **2015,** *11* (7), 854–861.
113. Du, L.; Yang, P.; Yu, X.; Liu, P.; Song, J.; Mai, W. Flexible Supercapacitors Based on Carbon Nanotube/MnO_2 Nanotube Hybrid Porous Films for Wearable Electronic Devices. *Journal of Materials Chemistry A* **2014,** *2* (41), 17561–17567.
114. Tao, J.; Liu, N.; Rao, J.; Ding, L.; Bahrani, M. R. A.; Li, L.; et al. Series Asymmetric Supercapacitors Based on Free-Standing Inner-Connection Electrodes for High Energy Density and High Output Voltage. *Nanoscale* **2014,** *6* (24), 15073–15079.
115. Chen, P. C.; Shen, G.; Sukcharoenchoke, S.; Zhou, C. Flexible and Transparent Supercapacitor Based on In_2O_3 Nanowire/Carbon Nanotube Heterogeneous Films. *Applied Physics Letters* **2009,** *94* (4), 043113.
116. Sun, P.; Deng, Z.; Yang, P.; Yu, X.; Chen, Y.; Liang, Z.; et al. Freestanding CNT-wo_3 Hybrid Electrodes for Flexible Asymmetric Supercapacitors. *Journal of Materials Chemistry A* **2015,** *3* (22), 12076–12080.
117. Kim, J. H.; Lee, K. H.; Overzet, L. J.; Lee, G. S. Synthesis and Electrochemical Properties of Spin-Capable Carbon Nanotube Sheet/MnOx Composites for High-Performance Energy Storage Devices. *Nano Letters* **2011,** *11* (7), 2611–2617.
118. Guo, F.; Xu, R.; Cui, X.; Zang, X.; Zhang, L.; Chen, Q.; et al. Highly Flexible, Tailorable and All-Solid-State Supercapacitors From Carbon Nanotube-MnO_x Composite Films. *RSC Advances* **2015,** *5* (108), 89188–89194.
119. Xu, S.; Yang, D.; Zhang, F.; Liu, J.; Guo, A.; Hou, F. Fabrication of $NiCo_2O_4$ and Carbon Nanotube Nanocomposite Films as a High-Performance Flexible Electrode of Supercapacitors. *RSC Advances* **2015,** *5* (90), 74032–74039.

120. Jiang, H.; Ma, J.; Li, C. Mesoporous Carbon Incorporated Metal Oxide Nanomaterials as Supercapacitor Electrodes. *Advanced Materials* **2012,** *24* (30), 4197–4202.
121. Cui, X.; Hu, F.; Wei, W.; Chen, W. Dense and Long Carbon Nanotube Arrays Decorated With Mn_3O_4 Nanoparticles for Electrodes of Electrochemical Supercapacitors. *Carbon* **2011,** *49* (4), 1225–1234.
122. Yang, Z.; Zhou, X.; Jin, Z.; Liu, Z.; Nie, H.; Xa, C.; et al. A Facile and General Approach for the Direct Fabrication of 3D, Vertically Aligned Carbon Nanotube Array/Transition Metal Oxide Composites as Non-pt Catalysts for Oxygen Reduction Reactions. *Advanced Materials* **2014,** *26* (19), 3156–3161.
123. Liu, W. W.; Lu, C.; Liang, K.; Tay, B. K. A Three Dimensional Vertically Aligned Multiwall Carbon Nanotube/$NiCo_2O_4$ Core/shell Structure for Novel High-Performance Supercapacitors. *Journal of Materials Chemistry A* **2014,** *2* (14), 5100–5107.
124. Li, X. J.; Zhao, Y.; Chu, W. G.; Wang, Y.; Li, Z. J.; Jiang, P.; et al. Vertically Aligned Carbon Nanotube@MnO_2 Nanosheet Arrays Grown on Carbon Cloth for High Performance Flexible Electrodes of Supercapacitors. *RSC Advances* **2015,** *5* (94), 77437–77442.
125. Fisher, R. A.; Watt, M. R.; Konjeti, R.; Ready, W. J. Atomic Layer Deposition of Titanium Oxide for Pseudocapacitive Functionalization of Vertically-Aligned Carbon Nanotube Supercapacitor Electrodes. *ECS Journal of Solid State Science and Technology* **2015,** *4* (2), M1–M5.
126. Jiang, Y.; Wang, P.; Zang, X.; Yang, Y.; Kozinda, A.; Lin, L. Uniformly Embedded Metal Oxide Nanoparticles in Vertically Aligned Carbon Nanotube Forests as Pseudocapacitor Electrodes for Enhanced Energy Storage. *Nano Letters* **2013,** *13* (8), 3524–3530.
127. Wang, L. J.; El-Kady, M. F.; Dubin, S.; Hwang, J. Y.; Shao, Y.; Marsh, K.; et al. Flash Converted Graphene for Ultra-High Power Supercapacitors. *Advanced Energy Materials* **2015,** *5* (18), 1500786.

CHAPTER 11

Carbon Nanotubes for Biomedical Applications

Y. Xue
Wenzhou Medical University, Wenzhou, China

Contents

11.1 INTRODUCTION

Carbon nanotubes (CNTs) are tube-like hollow cylindrical nanostructures made of carbon atoms, which can be considered as rolls of graphene sheets *(1,2)*. According to the thickness of the concentric graphene sheets, CNTs can be divided into single-walled carbon nanotubes (SWNTs), double-walled carbon nanotubes (DWNTs), and multiwalled carbon nanotubes (MWNTs). The diameter of the CNTs is usually from several angstroms to tens of nanometers, but the length of CNTs can be as long as several centimeters and even up to half a meter *(3)*. Because of their unique structure, extremely high specific surface area, and excellent electrical and mechanical properties, CNTs have attracted considerable interest since their first discovery in 1991 *(4)*. CNTs have been intensively studied in the past decades because of their potential applications in various fields that might affect our lives. One of the greatest values of carbon nanomaterials will be in the development of novel and effective biomedical treatments.

It is believed that the emergence of nanomaterials and nanotechnology has greatly promoted the development of biomedicine and biotechnology, which provides efficient

Industrial Applications of Carbon Nanotubes
ISBN 978-0-323-41481-4

strategies as promising alternatives for disease diagnosis and therapy *(5)*. As a kind of unique one-dimensional nanomaterial, CNTs were explored for biomedical applications soon after they were discovered. The hollow structure, high specific surface area, and specific physicochemical properties of CNTs have made them a very promising platform for biomedical applications, including drug delivery, therapy, tissue engineering, biosensing, and bioimaging. Although pure CNTs were usually not soluble and tended to aggregate in aqueous media, surface modification of CNTs has played a significant role in biomedical applications. In the past decades, both covalent and noncovalent surface modifications have been developed to improve their solubility *(6, 7)*.

In this chapter, we review the recent development of CNTs for biomedical applications, including toxicity of CNTs, drug delivery, therapy, tissue engineering, biosensing, and bioimaging, by discussing some selected examples from many research groups.

11.2 TOXICITY OF CARBON NANOTUBES

It is necessary to understand the extensive toxicity of CNTs before developing their biomedical applications. CNTs may be toxic to the lungs, skin, or heart. The toxicity of CNTs depends on their structures, morphology, surface functional groups, and dose *(8)*.

Because CNTs are very light and small, they could become airborne and be inhaled. Ryman-Rasmussen et al. *(9)* studied the inhaled CNTs in mice. They found that MWNTs could reach the subpleural tissue in mice and be embedded in the subpleural wall as well as within subpleural macrophages after an inhalation exposure of 30 mg/m^3 for 6 h. After 1 day, mononuclear cell aggregates on the pleural surface increased in both in number and size. Subpleural fibrosis increased after 2 and 6 weeks following inhalation. However, they did not find these effects in mice when a lower dose of CNTs was used, which suggested that minimizing inhalation of nanotubes during handling was prudent *(9)*. Lam et al. *(10)* studied the pulmonary toxicity of SWNTs in mice 7 and 90 days after intratracheal instillation. Mitchell et al. *(11)* found that the inhaled MWNTs caused few pulmonary effects but suppressed systemic immune function. Dose also played a vital role. Three different doses were used in their experiments, but only the dose of 1 mg/m^3 presented suppressed immune function. They also studied the mechanisms for how inhaled CNTs suppressed systemic immune function in mice and found that cyclooxygenase enzymes in the spleen, which responded to the signal from the lungs, played a very important role in suppression. Those knockout mice without cyclooxygenase enzymes were not affected when exposed to the same dose of CNTs. Their findings suggested signals from the lung activated signals in the spleen to suppress systemic immune function of exposed mice *(11)*.

CNTs may also increase the incidence of other diseases, such as skin diseases. Shvedova et al. *(12)* found that dermal toxicity may be caused after dermal exposure to

SWNTs because of accelerated oxidative stress in the skin. Bai et al. discovered that the repeated administration of CNTs in male mice can cause reversible testis damage. They found that CNTs accumulated in the testes and decreased the thickness of the seminiferous epithelium after 15 days, but the tissue damage was repaired at 60 and 90 days *(13)*.

Zhu et al. *(14)*, for the first time, assessed the DNA damage response to MWNTs in mouse embryonic stem (ES) cells. They found that MWNTs accumulated and induced apoptosis in mouse ES cells. Fig. 11.1A shows the MWNT uptake, induced differentiation, and apoptosis in mouse ES cells, and indicates that the amount of MWNTs

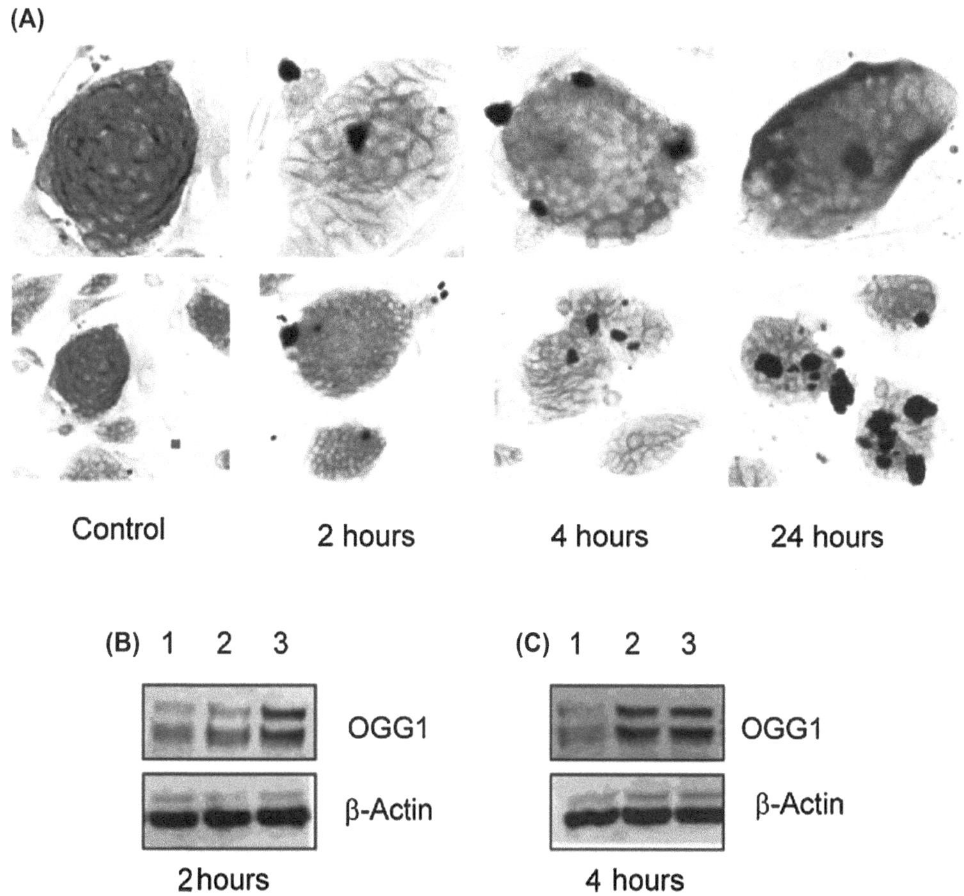

Figure 11.1 (A) Images of multiwalled carbon nanotube (MWNT) uptake and induced differentiation and apoptosis in mouse embryonic stem (ES) cells following MWNT treatment for 2, 4, and 24 h. (B and C) Upregulation of two isoforms of the key base excision repair protein 8-oxoguanine-DNA glycosylase 1 (OGG1) in response to the MWNT treatment. The ES cells were lysed with RIPA buffer, and cell extracts subjected to Western blots with OGG1 polyclonal antibody. Lane 1, untreated; Lanes 2 and 3, subjected to 5 and 100 μg/mL MWNT treatment, respectively *(14)*.

accumulated in the ES cells was gradually increased as time goes by. Fig. 11.1B and C shows two isoforms of the key base excision repair protein 8-oxoguanine-DNA glycosylase 1 (OGG1) in response to MWNT treatment by Western blot analysis. They found that the expression of both isoforms of OGG1 was elevated by the MWNT treatment, which suggested the occurrence of both nuclear and mitochondrial DNA damage through a mutagenic guanine base lesion *(14)*.

Surface modification is regarded as an efficient way to change the toxicity of CNTs. Some literature has reported that the toxicity of CNTs could be reduced by surface functionalization. For example, Schipper et al. *(15)* found that functionalized SWNTs have no toxicity over 4 months. SWNTs persisted within liver and spleen macrophages without apparent toxicity for 4 months. Jain et al. *(16)* studied the relationships of toxicity and functionalization density. They found that the toxicity of CNTs systematically decreased by increasing the functionalization density. The functionalization methods also affected the toxicity of CNTs. Sayes et al. *(17)* discovered that the chemically functionalized SWNT samples were substantially less cytotoxic than surfactant-stabilized SWNTs. We also did some work on the toxicity of SWNTs with different functional groups. The cytotoxicity of SWNTs, SWNTs with acid groups (SWNTs-COOH), and polyhedral oligomeric silsesquioxane (POSS)-modified SWNTs (SWNTs-POSS) on human retinal pigment epithelial (RPE) cells was assessed by using CCK-8 methods. As shown in Fig. 11.2, SWNTs-POSS had extremely low cytotoxicity to RPE cells compared with pure SWNTs and SWNTs-COOH *(18)*. The mechanism of the toxicity change induced by surface functionalization is very complicated, and depends on the functionalized methods, the functionalization density, and different functional groups. Generally speaking, surface functionalization increases the dispersion of CNTs. Chemical functionalization can even shorten the length of CNTs and break their graphitic structure, which makes the CNTs much more degradable than pristine CNTs. Surface modification may also induce abundant biocompatible groups onto CNTs, which reduces toxicity.

11.3 CARBON NANOTUBES FOR DELIVERY SYSTEMS

11.3.1 Carbon Nanotubes for Drug Delivery

As is discussed earlier, pristine CNTs have very low solubility and tend to aggregate in aqueous solution, which restricts their applications both in vitro and in vivo. Surface modification of CNTs is an efficient way to explore CNTs as drug carries. They can be functionalized with proteins *(19)*, DNA, and polymers, such as poly(ethylene glycol) (PEG). A worthwhile goal for the drug delivery system is to deliver the drug molecules precisely to the tumor target without effects on normal cells. As drug carries, functionalized CNTs were also designed to be tumor targetable.

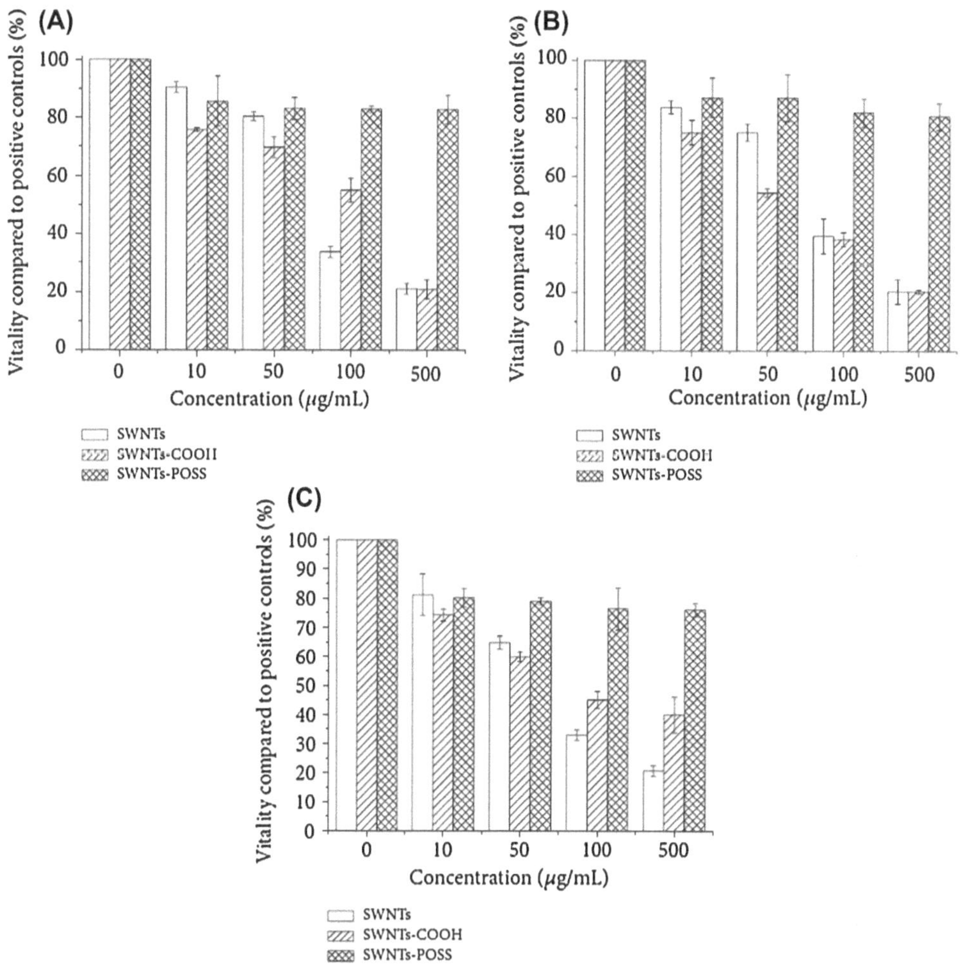

Figure 11.2 Vitality of a retinal pigment epithelial (RPE) cell incubated with single-walled carbon nanotubes (SWNTs), SWNTs with acid groups (SWNTs-COOH), and polyhedral oligomeric silsesquioxane (POSS)-modified SWNTs (SWNTs-POSS) at (A) 24 h, (B) 48 h, and (C) 72 h *(18)*.

Liu et al. *(20)* reported water-soluble SWNTs functionalized with PEG for drug loading and delivery. The PEG-functionalized SWNTs had an ultrahigh loading capacity for a cancer drug (doxorubicin, DOX) by $\pi-\pi$ stacking of aromatic molecules onto the surface of SWNTs (Fig. 11.3A). They found that the amount of DOX loaded on PEG-SWNTs was pH dependent. As shown in Fig. 11.3C, the loaded DOX-SWNT weight ratio decreased from 4 to 2 and 0.5 when pH was reduced from 9 to 7 and 5, respectively. As for drug release, they found that DOX loaded on SWNTs remained stably bound in basic buffer solutions at pH 7.4 at room temperature (Fig. 11.3D, green line). However, when the pH value decreased to 5.5, the amount of DOX released

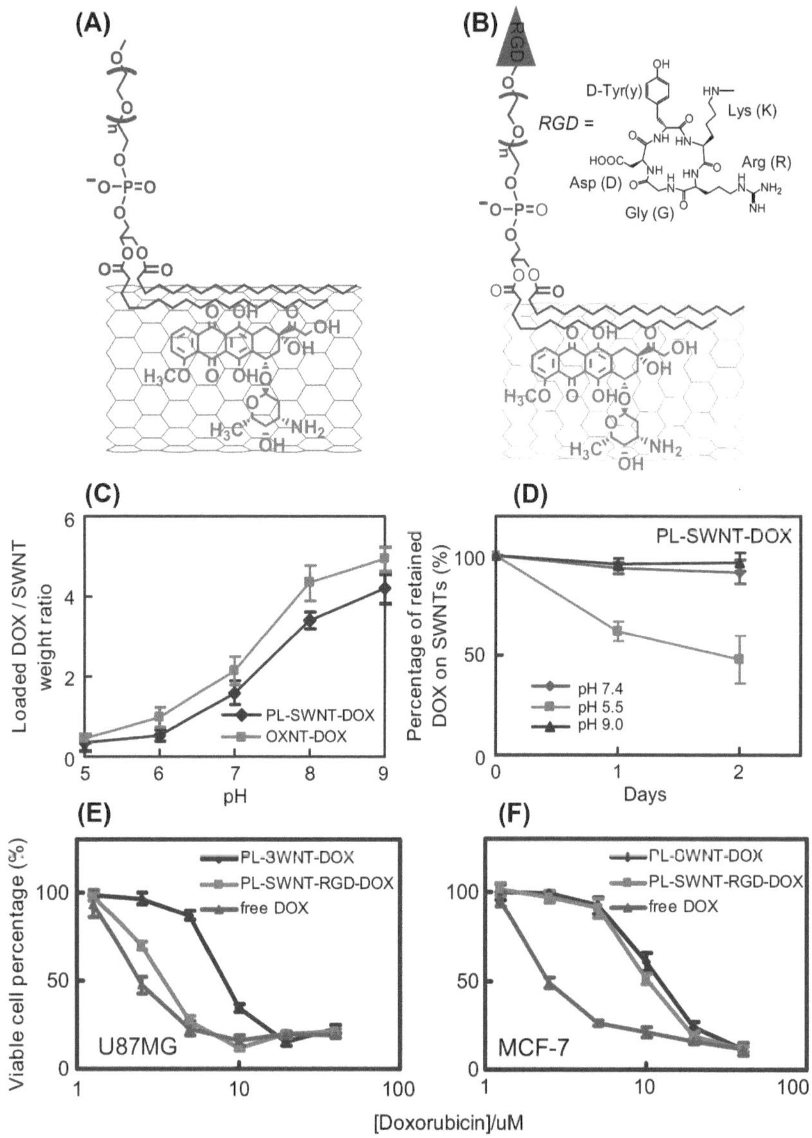

Figure 11.3 (A) Schematic drawings of doxorubicin π-stacking onto a nanotube prefunctionalized noncovalently by phospholipid-polyethylene glycol (PL-PEG). (B) Schematic structure of phospholipid-single-walled carbon nanotubes-Arg-Gly-Asp-doxorubicin (PL-SWNT-RGD-DOX), that is, SWNTs functionalized with RGD at the termini of PEG and loaded with DOX on the sidewall by π-stacking. (C) DOX loading efficiency at various pH values for the two types of SWNTs. (D) Doxorubicin retained on PL-PEG-SWNTs over time in buffers at the three pH values indicated. (E and F) Concentration-dependent survival curves of U87MG cells (E) and MCF-7 cells (F) treated by various samples, as indicated *(20)*.

from DOX-SWNT was 40% over 1 day, because this pH increased the hydrophilicity and solubility of DOX *(20)*.

They further induced a cyclic Arg-Gly-Asp (RGD) peptide at the terminal groups of PEG on SWNTs to demonstrate the targeted delivery of DOX by SWNTs (Fig. 11.3B). Because of the RGD-integrin recognition, DOX-loaded RGD-functionalized nanotubes [phospholipid (PL)-SWNT-RGD-DOX] showed a much higher cell-killing effect toward U87MG human glioblastoma cancer cells than that found prior to RGD conjugation (Fig. 11.3E). However, this was not found for MCF-7 cells (Fig. 11.3F).

MWNTs could also be used for drug delivery. Ran et al. *(21)* used the PEGylated oxidized MWNTs for targeted delivery of anticancer drugs to brain glioma.

CNTs can also be functionalized with other polymers for drug delivery. Wu et al. *(22)* functionalized the short MWNTs with polyethylenimine (PEI) and further conjugated with fluorescein isothiocyanate (FITC) and prostate stem cell antigen (PSCA) monoclonal antibody (mAb), and finally formed a composite as CNT-PEI(FITC)-mAb (Fig. 11.4A). The resulting composite showed very good biocompatibility and can specifically target the cancer cells. Drug loading and release performance of CNT-PEI(FITC)-mAb were studied by incubating the CNT-PEI(FITC)-mAb with different concentrations of DOX. As shown in Fig. 11.4C, the loading amount of DOX on CNT-PEI(FITC)-mAb increased up to 300 mg/g as the concentration increased to 2000 μg/mL. The release performance was pH sensitive. The amount of DOX released from CNT-PEI(FITC)-mAb in an acid environment (pH 5.2) was much higher than that at pH 7.4 and pH 6.3 (Fig. 11.4C)

Functionalized CNTs for delivery of drugs for both cancer and anticancer indications have been widely investigated in vivo or in vitro *(23)*. The kind of loading drugs on CNTs were varied. Drugs can either be physically adsorbed on the surface of CNTs by $\pi-\pi$ stacking or chemically bonded on the CNTs; they can even be loaded into the pores of CNTs.

11.3.2 Carbon Nanotubes for Gene Delivery

Besides drug delivery, CNTs were also widely explored for gene delivery. Pantarotto et al. *(24)* used the ammonium-functionalized CNTs (f-CNTs) to adsorb plasmid DNA through electrostatic interactions. F-CNTs associated with DNA can penetrate the cell membranes and enter the cells. Then, f-CNT-associated plasmid DNA was delivered to cells efficiently and gene expression leveled up to 10 times higher than those achieved with free DNA without CNTs. Tonelli et al. *(25)* complexed carboxi-functionalized multiwalled carbon nanotubes (f-MWNTs) with DNA. They found that f-MWNTs complexed with DNA can enhance gene delivery to Nile tilapia spermatogonial stem cells and their transfection efficiency was much higher than cationic lipids or electroporation, resulting in less cell death.

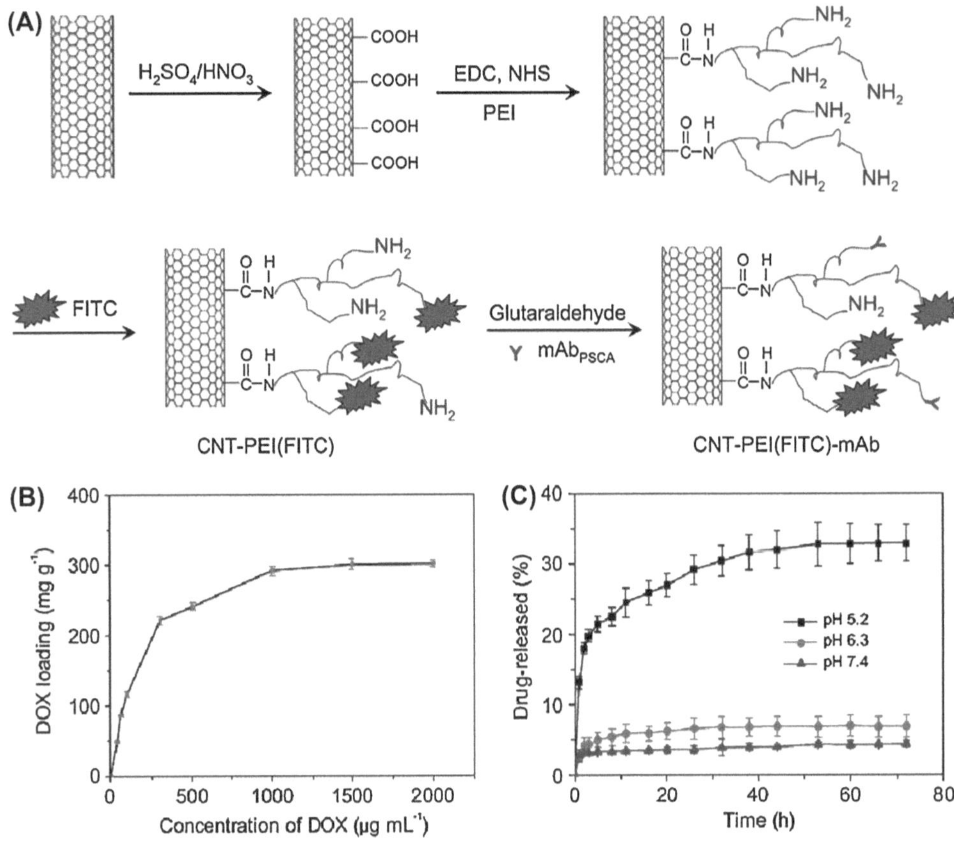

Figure 11.4 (A) Schematic diagram of the synthesis process for carbon nanotube-polyethylenimine-(fluorescein isothiocyanate)-monoclonal antibody [CNT-PEI(FITC)-mAb]. (B) Plot of doxorubicin (DOX) loading amount for CNT-PEI(FITC)-mAb versus the drug concentration. (C) In vitro release profiles of DOX from CNT-PEI(FITC)-mAb under different pH conditions *(22)*.

11.4 CARBON NANOTUBES FOR THERAPY

Great efforts have been made to cure diseases, especially cancers. Usually, radical resection of the tumor at its early stage is the only way to tackle cancers. However, the development of nanotechnology and biotechnology has provided a new way to treat diseases. Recently, CNTs have been developed for disease therapy by many groups.

11.4.1 Drug Therapy

Liu et al. *(26)* stacked DOX on CNTs for in vivo cancer therapy. The in vivo therapeutic efficacy of the SWNT-DOX, free DOX, and liposomes (Doxil) was investigated. Mice bearing raji lymphoma xenograft tumors were treated weekly with free DOX, Doxil, or SWNT-DOX. The relative tumor volume of the mice treated with SWNT-DOX was much smaller than those treated with free DOX, which indicated that SWNT-DOX had much higher inhibition efficiency of tumor growth than free DOX (Fig. 11.5A). However, the weight of the mice treated with SWNT-DOX is much greater than the mice treated with free DOX or Doxil, indicating the toxicity of SWNT-DOX was much smaller than free DOX or Doxil (Fig. 11.5B and C). As shown in Fig. 11.5D, when

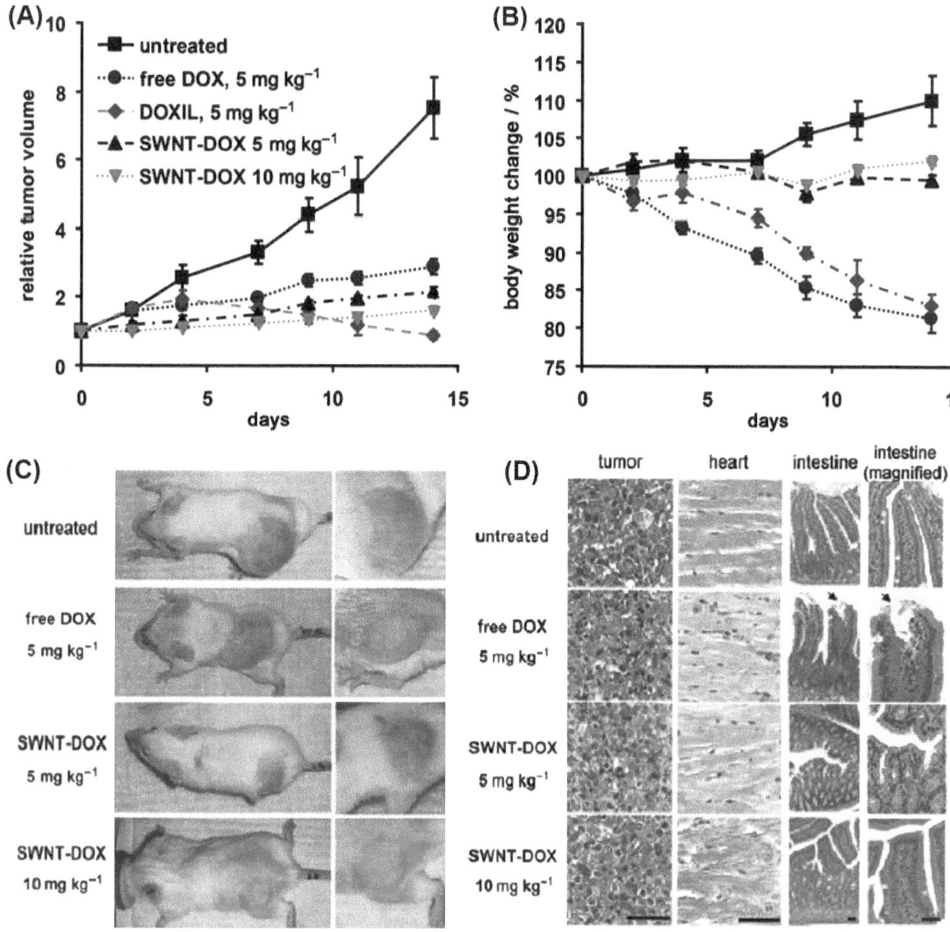

Figure 11.5 Supramolecular packing of doxorubicin (DOX) to single-walled carbon nanotubes (SWNTs) increased clinical efficacy in vivo. (A) Tumor sizes and (B) body weights of the raji tumor-bearing SCID mice, treated with different DOX formulations once per week at day 0 and day 7, were recorded three times per week. (C) Representative photos of mice from different groups were taken at the end of treatment. Mice became visibly thinner after treatment with free DOX. (D) Gastrointestinal toxicity was observed in mice treated with free DOX but not in mice treated with SWNT-DOX. Histological sections of intestinal epithelium showed damage of the intestinal epithelium in the free DOX-treated group. The *arrows* show the area of loss of columnar epithelial cells in tips of villi. Scale bars: 100 μm *(26)*.

the mice were treated with free DOX, disruption of the intestinal lining consistent with gastrointestinal mucositis and complete loss of columnar epithelial cells were observed. However, mouse tissues of those mice treated with SWNT-DOX did not exhibit gross gastrointestinal, hepatic, cardiac, or renal toxicity.

Wu et al. *(22)* investigated in vivo therapy effects of CNT-PEI(FITC)-mAb/DOX in nude mice bearing PC-3 tumor. The relative tumor volume of those mice treated with CNT-PEI(FITC)-mAb/DOX was much smaller than those treated with free DOX and CNT-PEI(FITC)-mAb or CNT-PEI(FITC)/DOX. But the relative body weight of the mice treated with CNT-PEI(FITC)-mAb/DOX was higher than those treated with others (Fig. 11.6). These results indicated that CNT-PEI(FITC)-mAb/DOX has much better therapeutic effect and less toxicity than free DOX.

11.4.2 Photoacoustic Therapy

Kang et al. *(27)* developed a unique approach using the large photoacoustic effect of SWNTs for targeting and selective destruction of cancer cells. They have synthesized folate acid functionalized SWNTs (FA-SWNTs) with the ability to selectively bind to cancer cells. When irradiated with a 1064-nm Q-switched millisecond pulsed laser, SWNTs had a large photoacoustic effect in suspension and would explode like a fire-cracker at nanoscale. Fluorescence imaging indicated that FA-SWNTs could selectively enter into cancer cells with a high target capability. Under the treatment of 1064-nm millisecond pulsed laser, SWNTs would explode in the cells and kill them, while 90% of the normal cells were kept alive because of the lack of SWNTs inside cells. Zhou

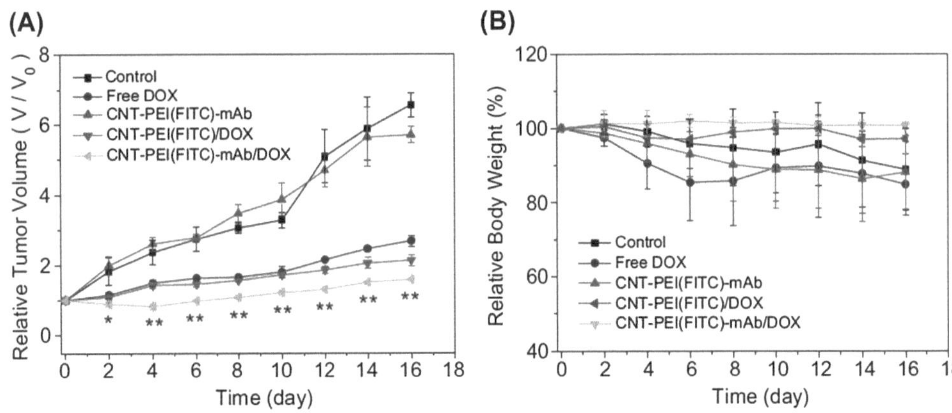

Figure 11.6 In vivo therapy effects of carbon nanotube-polyethylenimine-(fluorescein isothiocyanate)-monoclonal antibody/doxorubicin [CNT-PEI(FITC)-mAb/DOX] in nude mice-bearing PC-3 tumor. (A) Tumor growth curves for mice treated with free DOX, CNT-PEI(FITC)-mAb, CNT-PEI(FITC)/DOX and CNT-PEI(FITC)-mAb/DOX, in comparison with the blank control. The tumor volumes were normalized to their initial sizes. *$P < .05$, **$P < .01$ [CNT-PEI(FITC)-mAb/DOX group versus CNT-PEI(FITC)/DOX group]. (B) Body weight curves normalized to day 0 for different groups during the experiments *(22)*.

et al. *(28)* used a similar method for mitochondria-targeting photoacoustic therapy by using SWNTs. This method of using the photoacoustic properties of SWNTs provided a new way for therapeutic applications.

11.4.3 Photothermal Therapy

Photothermal therapy (PTT) is a therapeutic method whereby cancers are ablated by the heat generated from absorbed near-infrared (NIR) light energy. Liang et al. *(29)* used PEGylated SWNTs as the theranostic agent and demonstrated that imaging-guided PTT afforded effective destruction of primary tumors and cancer cells in sentinel lymph nodes and inhibited tumor metastasis. They found that mice with both primary tumors and popliteal sentinel lymph node ablated by SWNT-induced PTT showed greatly prolonged survival times compared with the three other groups (Fig. 11.7A and B). When

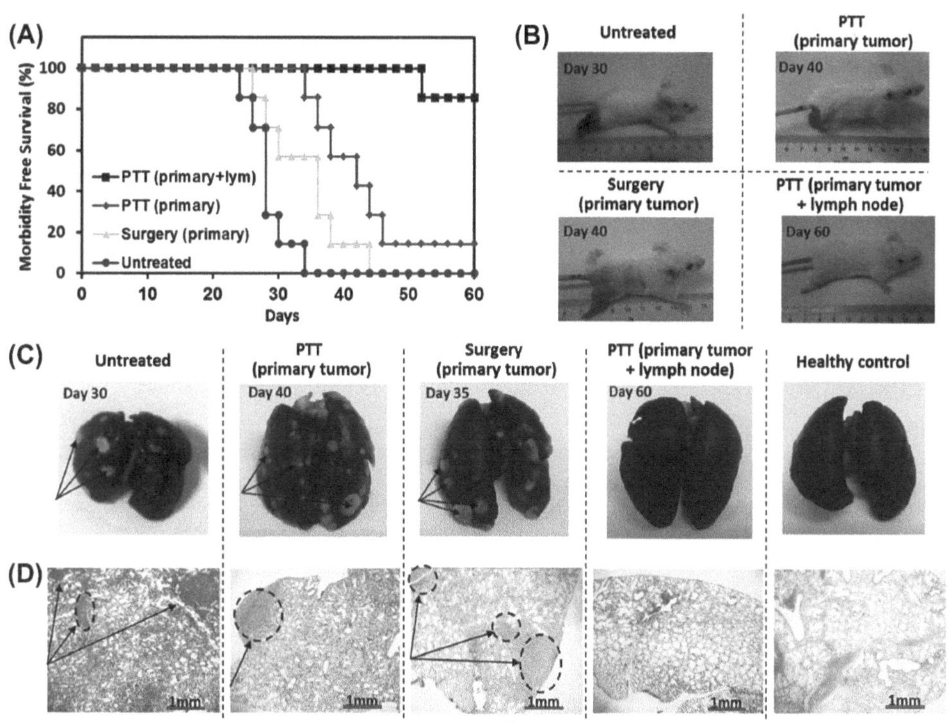

Figure 11.7 In vivo cancer therapy. (A) Morbidity-free survival of different groups of mice after various treatments indicated (seven mice per group). Mice with both primary tumors and popliteal sentinel lymph node ablated by single-walled carbon nanotube (SWNT)-induced photothermal therapy (PTT) showed greatly prolonged survival times compared to the three other groups. (B) Representative photos of mice in different groups. Photos were taken at different times as indicated. Photos of India ink-stained whole lungs (C) and microscopy images of H&E-stained lung slices (D) collected from different groups of mice after various treatments. Tumor metastasis sites are shown by *arrows (29)*.

the primary tumors and sentinel lymph nodes were photothermally ablated, all surviving mice had no pulmonary metastasis, while other methods were not sufficient to prevent cancer metastasis (Fig. 11.7C and D).

11.4.4 Photodynamic Therapy

SWNTs also showed great potential for photodynamic therapy (PDT). Wang et al. *(30)* synthesized two SWNTs covalently functionalized with PEI (SWNT-PEI) and noncovalently functionalized with PVPk30 (SWNT-PVPk30), and used them for PDT. The photodynamic effects of these two functionalized SWNTs were detected under visible light illumination both in vitro and in vivo. They found that the photodynamic effect was dependent not only on illumination time, but also on the modification method of the SWNTs. SWNT-PEI had a much stronger photodynamic effect than SWNT-PVPk30 both in vitro and in vivo.

Shi et al. *(31)* synthesized hyaluronic acid-derivatized carbon nanotubes (HA-CNTs). Hematoporphyrin monomethyl ether (HMME), a new PDT agent, was adsorbed onto HA-CNTs to develop HMME-HA-CNTs, which can combine local specific PDT with external NIR PTT. Fig. 11.8A and B shows tumor-bearing mice treated with 808 nm NIR laser and 532 nm laser, respectively. Experimental results indicated that the combined therapy method had a much higher therapeutic efficacy than the single methods (Fig. 11.8C).

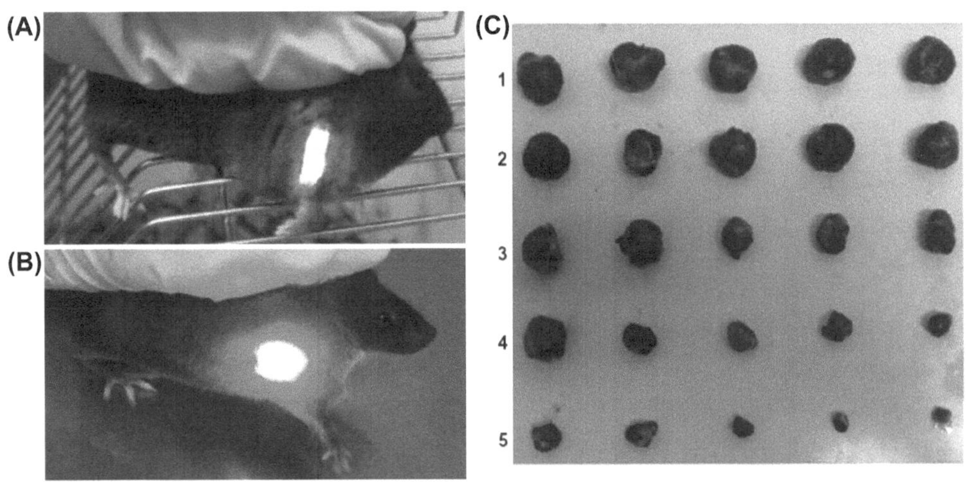

Figure 11.8 Tumor-bearing mice treated with (A) 808 nm near-infrared laser and (B) 532 nm laser. (C) Photo of tumors taken out of the saline group (1), the hyaluronic acid-derivatized carbon nanotubes (HA-CNTs; 808 nm) laser group (2), the hematoporphyrin monomethyl ether (HMME; 532 nm) laser group (3), the HMME-HA-CNT (532 nm) laser group (4), and the HMME-HA-CNT (532/808 nm) laser group after treatment for 8 days (5) *(31)*.

11.5 CARBON NANOTUBES IN TISSUE ENGINEERING

The development of tissue engineering requires combining biological science and material science. The rapid development of nanotechnology and nanomaterials gives a new chance for tissue engineering. CNTs not only have many excellent properties, but also can be assembled into various morphologies, such as fibers, films, and three-dimensional (3D) structures; all these make CNTs widely used in tissue engineering *(32)*.

Correa-Duarte et al. *(33)* synthesized CNT-based 3D networks as scaffolds for cell growth. The 3D interlocking resistive network of CNTs was prepared by treating a vertically aligned CNT array with exerting chemically induced capillary forces. This 3D network looked like a scaffold, which was suitable for cell growth. Fig. 11.9 shows scanning electron microscopy (SEM) images of L929 cells grown on the CNT network. Most of the L929 cells had adhered to the CNT surface as well as the walls of the cavities (Fig. 11.9A and B). Seven days later, the fibroblasts had completely covered almost the whole surface of the CNT-based 3D scaffolds (Fig. 11.9C and D).

Abarrategi et al. *(34)* developed MWNT/chitosan (CHI) composite scaffolds for tissue engineering. The scaffolds were composted of a major fraction of MWNTs (up to 89%)

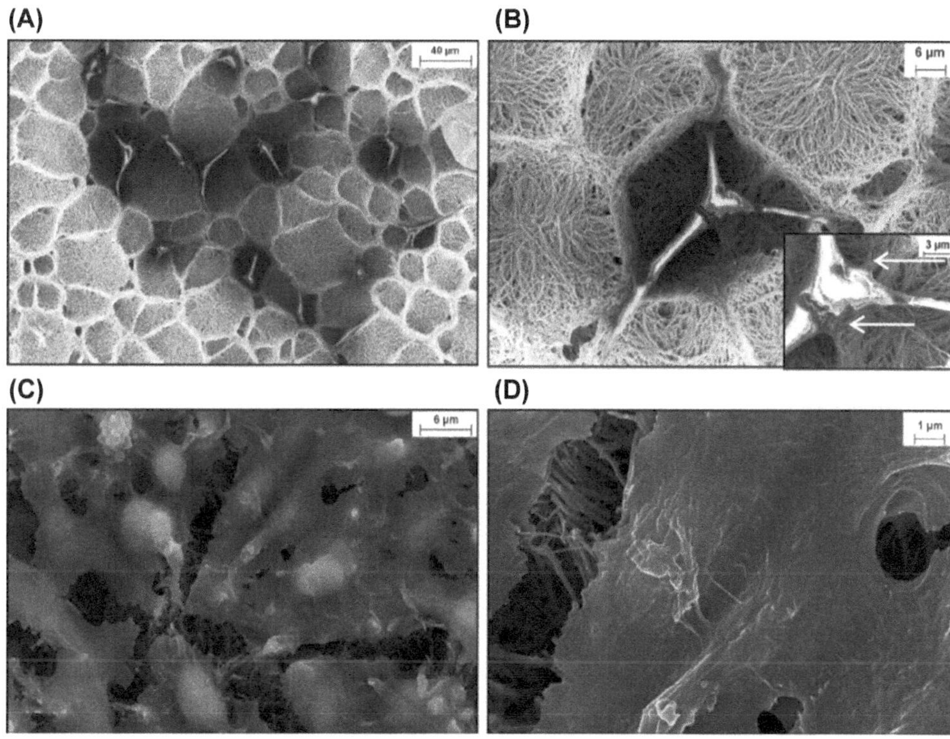

Figure 11.9 Scanning electron micrographs of L929 mouse fibroblasts growing on a multiwalled carbon nanotube (MWNT)-based network. (A and B) After 1 day. (C and D) After 7 days *(33)*.

and a minor one of CHI (11%). As shown in Fig. 11.10, the MWNT/CHI scaffolds had a well-defined microchannel porous structure, which was further used as a support for culture growth. The MWNT/CHI composite scaffolds showed remarkable biocompatible character in both in vitro and in vivo experiments. They also studied the evolution of the C_2C_{12} cell line toward an osteoblastic lineage in the presence of the recombinant human bone morphogenetic protein-2 (rhBMP-2) in vivo. As is shown in Fig. 11.10B, rhBMP-2-adhered MWNT/CHI scaffolds were implanted into subcutaneous pockets in the mouse back muscle tissue for 3 weeks. After 3 weeks, most of the MWNT/CHI scaffold structure was degraded and replaced by cells (Fig. 11.10C), and bone tissue regeneration was clearly observed. The remaining scaffold was further studied by optical micrograph after Masson's trichrome staining, as shown in Fig. 11.10D; the green-blue color was the characteristic of bone tissue, indicating high collagen expression. Plenty of fibroblasts (purple color in Fig. 11.10E) invaded the scaffold structure while collagen-expressing cells (blue-green color in Fig. 11.10E) filled degraded areas.

CNTs have also been used in nerve tissue engineering to probe cell behaviors, and to research the growth and organization of neural networks, finally to treat neural diseases. Keefer et al. *(35)* coated the conventional tungsten and stainless steel wire electrodes with

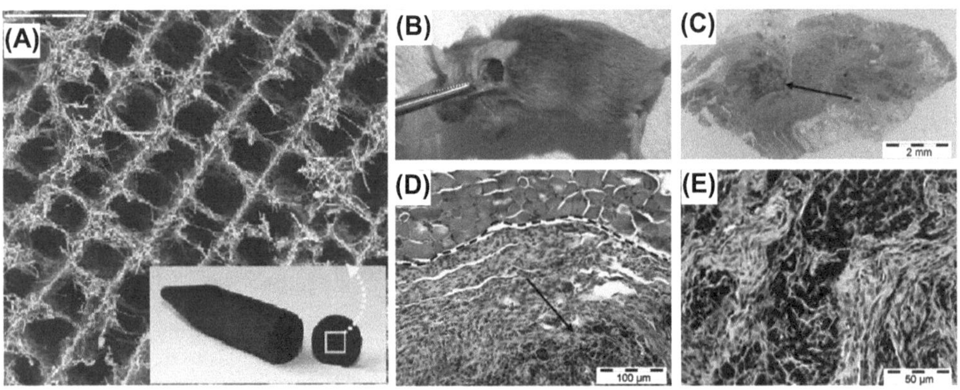

Figure 11.10 (A) Scanning electron microscopy (SEM) image of cross-sectioned multiwalled carbon nanotube/chitosan (MWNT/CHI) scaffold exhibiting a chamber-like structure. Bar is 20 mm. Inset shows a picture of the monolithic scaffolds resulting from the ice segregation-induced self-assembly (ISISA) process. (B) Photograph shows the surgery implantation of recombinant human bone morphogenetic protein-2 (rhBMP-2)-adsorbed MWNT/CHI scaffolds into mouse subcutaneous muscular pocket. (C) Optical microscope micrograph shows regenerated bone tissue and a minor fraction of remaining MWNT/CHI scaffold. (D) Optical micrograph shows detail of regenerated bone tissue (collagen expressing cells, blue-green (light gray in print versions)) after major disassembly of the MWNT/CHI scaffold, surrounded by muscle tissue (pink (gray in print versions)). Notable is the well-limited interface between adjacent tissues (*black dashed line*). The remaining MWNT/CHI scaffold (black) is shown by a *black arrow*. Optical micrograph (E) shows detail of remaining scaffold with plenty of fibroblasts (purple (black in print versions)), prior to its disassembly and colonization by collagen-expressing cells (blue-green (light gray in print versions)) *(34)*.

CNTs by electrochemical techniques under ambient conditions. The CNT coating improved both recording and electrical stimulation of neurons in culture, rats, and monkeys. Baranauskas et al. *(36)* found the CNT coating of neural microelectrodes enhanced the multiunit signal-to-noise ratio.

Cellot et al. *(37)* provided the mechanisms by which CNTs might affect the activity of cultured neuronal networks. They found that CNTs might improve neuronal performance by favoring electrical shortcuts between the proximal and distal compartments of the neuron. Electron microscopy analysis, single-cell electrophysiology techniques, and theoretical modeling were used to confirm their findings.

Lee et al. *(38)* pretreated rats with amine-modified SWNTs and found that the pretreatment can protect neurons and enhance the recovery of behavioral functions in rats with induced stroke. Less tissue damage was observed in SWNT-treated rats than in controls. The treated rats also took longer to fall from a rotating rod, which indicated that they had better motor functions after injury. Low levels of apoptotic, angiogenic, and inflammation markers showed that amine-modified SWNTs could protect the brains of treated rats from ischemic injury.

11.6 CARBON NANOTUBES FOR BIOSENSING

Biosensing is one of the important biomedical applications of CNTs, because CNTs can enhance electron transfer. There were already many reports of electrochemical biosensors made of CNTs *(39)*. These CNT-based biosensors have been used for detecting various biological structures such as glucose, antigens, viruses, DNA, and even whole cells. The sensing can be measured both in vitro and in vivo. We selectively introduced some works on biosensing in this chapter.

He et al. *(40)* prepared a CNT-DNA sensor by modified aligned CNT electrodes with single-strand DNA chains. This CNT-DNA sensor had high selectivity and sensitivity for probing complementary DNA and target DNA chains of specific sequences. As for the preparation of this CNT-DNA sensor, they first treated the gold-supported aligned CNT array with acetic acid. Then, single-strand DNA chains with an amino group were chemically bonded with the plasma-induced −COOH group on the CNTs via amide formation (Fig. 11.11). Ferrocenecarboxaldehyde (FCA) was labeled to complementary DNA chains and other noncomplementary DNA chains. The CNT-DNA can selectively detect the FCA-labeled complementary DNA chains but not the FCA-labeled noncomplementary DNA chains. This method can be used not only for sensing DNA chains but also for advanced device applications of CNTs *(40)*.

CNTs can also be used for living cell sensing. Sudibya et al. *(41)* reported functionalized CNTs with bioactive monosaccharides. They demonstrated that the functionalized CNT devices could biocompatibly interface with living cells and dynamically detect biomolecules from these living cells. Hu et al. *(42)* reported a hybrid film by

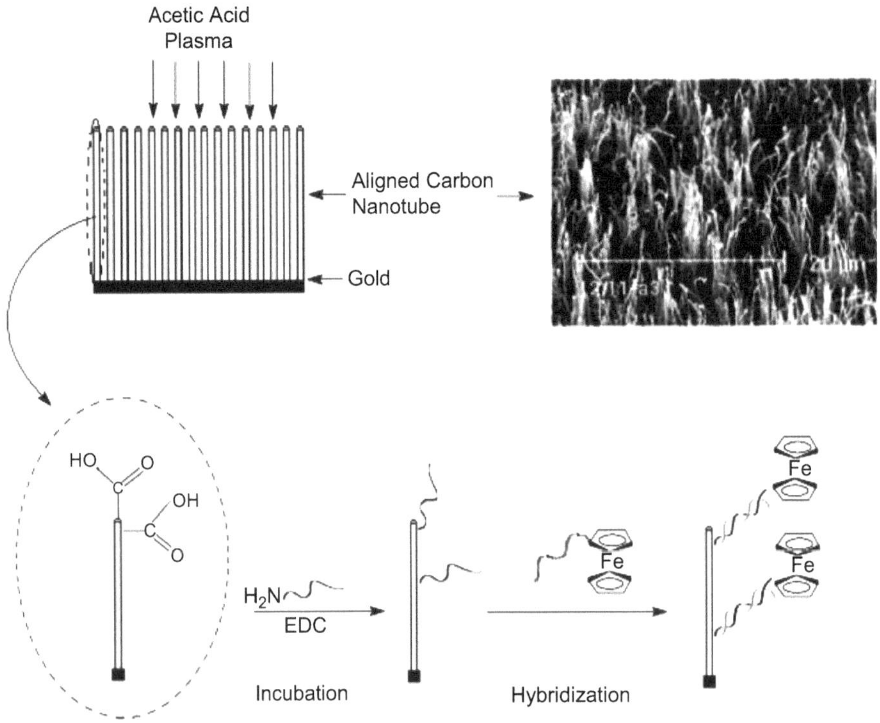

Figure 11.11 A schematic illustration of the aligned nanotube-DNA electrochemical sensor. The upper right scanning electron microscopy (SEM) image shows the aligned carbon nanotubes (CNTs) after having been transferred onto a gold foil. For reasons of clarity, only one of the many carboxyl groups is shown at the nanotube tip and wall, respectively *(40)*.

functionalizing the vertically aligned carbon nanotube/polyethylene (VACNT/PE) with DNA and $Mn_3(PO_4)_2$, which can provide high electroactivity. Living human breast carcinoma cells (MDA-MB-231) were directly grown on the DNA/ $Mn_3(PO_4)_2$ modified VACNT/PE hybrid film to form MDA-MB-231/Mn3(PO4) 2/DNA/VACNT/PE hybrid film. There was a very short diffusion distance between cells and reaction sites, resulting in a highly sensitive method for detecting reactive oxygen species (ROS) released from cancer cells under drug stimulations. Fig. 11.12 is the schematic illustration of the preparation procedure of the hybrid film and its living cell assay. The detection sensitivity of this in situ cell-attached hybrid film was more than six times those measured by cells in a culture medium, because the ROS molecules released from living cells on the hybrid film can be instantly captured by the smart structure and translated to electrical signal. This smart sensing platform delivered superior performance over the traditional method of detecting ROS molecules released by cells that are cultured in a Petri dish *(42)*.

Recently, CNTs for in vivo biosensing have also attracted attention. For example, Xiang et al. *(43)* used the VACNT-sheathed carbon fibers as pristine microelectrodes for selective

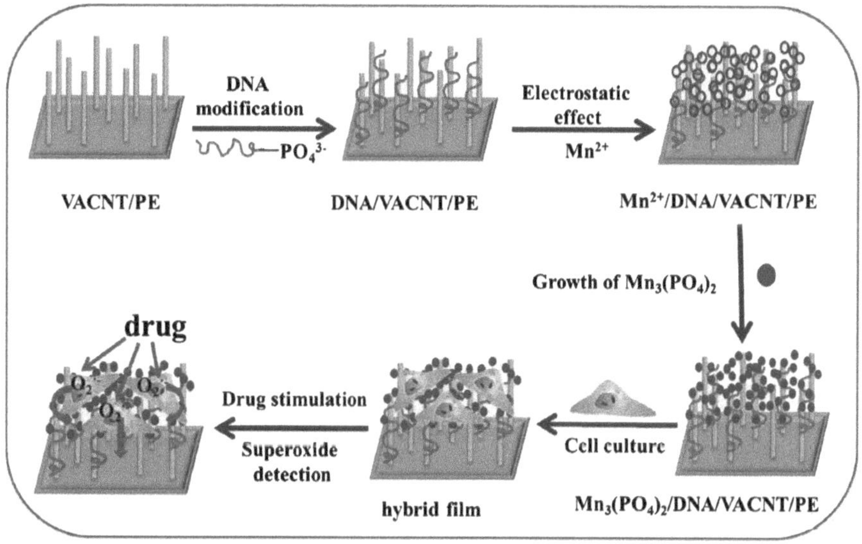

Figure 11.12 Schematic illustration of the preparation procedure for the MDA-MB-231/$Mn_3(PO_4)_2$/DNA/VACNT/PE hybrid film and its living cell assay *(42)*.

monitoring of ascorbate in vivo. Iverson et al. *(44)* have modified NIR fluorescent SWNTs with a PEG-ligated copolymer. The functionalized SWNTs were intravenously injected into mice for in vivo biosensing. The biosensor can selectively detect local nitric oxide concentration with a detection limit of 1 μm. They demonstrated that alginate-encapsulated SWNTs could function as implantable inflammation sensors for nitric oxide, without intrinsic immune reactivity or other adverse response for more than 400 days.

Another kind of sensor is the optical sensor. CNTs can also been use as optical sensors. Barone et al. *(45)* reported SWNT-based NIR optical sensors. They could modulate the emission of SWNTs in response to the adsorption of specific biomolecules. Jin et al. *(46)* used fluorescent SWNTs to detect single-molecule H_2O_2 signaling from epidermal growth factor receptor using fluorescent SWNTs. This platform provided a new method of understanding the signaling of ROS at the cellular level.

11.7 CARBON NANOTUBES FOR BIOIMAGING

SWNTs usually have some optical properties. For example, semiconducting SWNTs have NIR photoluminescence properties with low autofluorescence. The intrinsic photoluminescence can penetrate the deep tissue beyond 1 mm in the NIR region. These properties made SWNTs an ideal material for NIR photoluminescence imaging. CNTs can also be used for Raman or photoacoustic imaging.

Welshe et al. *(47)* synthesized phospholipid-polyethylene glycol (PL-PEG)-coated SWNTs by sonicating SWNTs with sodium cholate followed by surfactant exchange

(Fig. 11.13A). They thought this method was much better than directly sonicating SWNTs with PL-PEG, because the exchange process resulted in less damage.

Fig. 11.13B presents the NIR photoluminescence images of cholate-SWNT, exchange-SWNT, and direct-SWNT solutions excited at 808 nm at equal concentrations. The PL-PEG-coated SWNTs prepared by the exchange method were much brighter than the direct-SWNTs. Fig. 11.13C showed the photoluminescence versus excitation spectra for cholate-SWNT, exchange-SWNT, and direct-SWNT solutions.

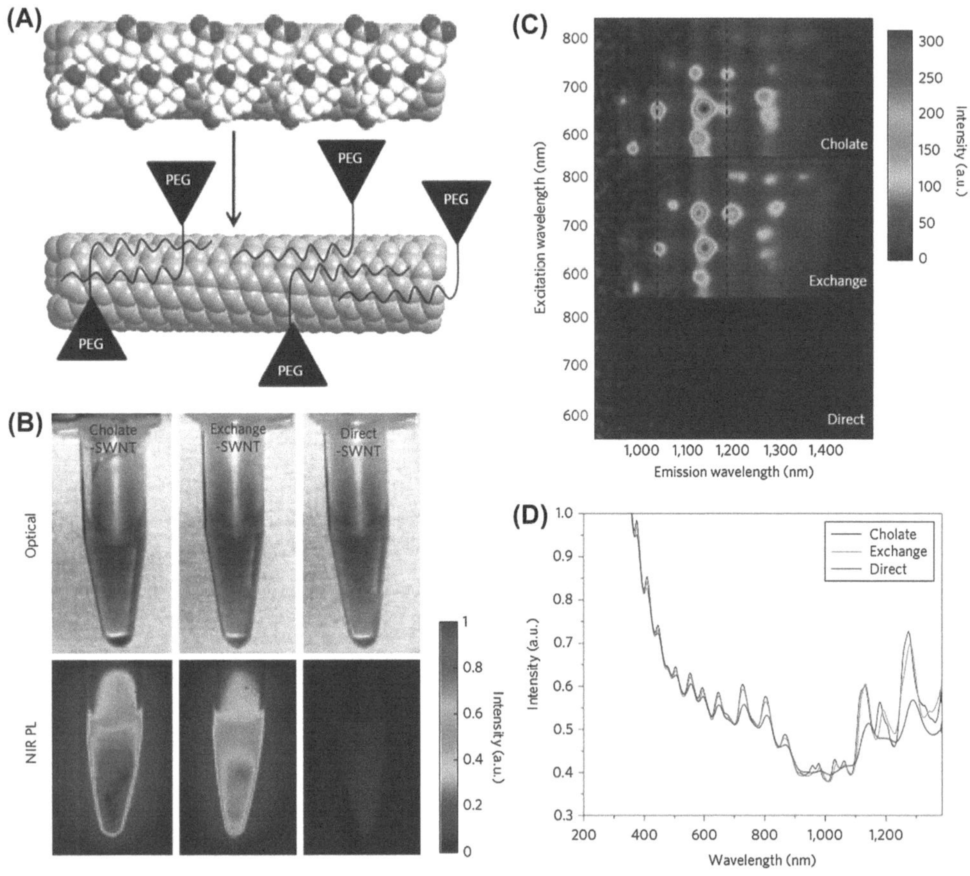

Figure 11.13 Comparison of exchange- and direct-single-walled carbon nanotubes (SWNTs). (A) Schematic of the exchange process. Cholate (*red* (gray in print versions) *and white balls*) on SWNTs (gray) is dialyzed and eventually replaced by phospholipid-polyethylene glycol (PL-PEG) to form biocompatible nanotubes without damaging the integrity of the nanotube sidewall. (B) Near-infrared (NIR) photoluminescence images of the three solutions excited at 808 nm at equal concentrations. Exchange-SWNTs show greater fluorescence yield than direct-SWNTs. (C) Photoluminescence versus excitation spectra show improved quantum yield in cholate and exchange samples. The *dotted lines* show how peaks are redshifted after exchange. (D) UV—Vis—NIR curves. Exchange- and cholate-SWNTs show sharp transition peaks; direct-SWNTs show very low and broad absorption features *(47)*.

The photoluminescence yield of exchange-SWNT was much higher than that of direct-SWNT. The sharp E11 absorption peaks (1000–1300 nm) in UV–Vis–NIR spectra also indicated exchange-SWNT high quantum yield. They further used the exchange-SWNT as in vivo NIR photoluminescence imaging agents for live mice. The high-resolution intravital image of tumor vessels beneath thick skin was achieved, because of the deep tissue penetration and low autofluorescence background.

Zerda et al. *(48)* synthesized cyclic Arg-Gly-Asp (RGD) peptide-functionalized SWNTs. As shown in Fig. 11.14A, SWNTs were first functionalized with PEG_{5000}, which were further incorporated with RGD at the top of PEF_{5000}. The SWNT-RGD can be used as a contrast agent for photoacoustic imaging of tumors. Two groups of mice bearing U87MG tumor xenografts were injected with plain SWNTs or SWNT-RGD at a concentration of 1.2 mM. Compared with control mice injected with plain SWNTs, those mice injected with SWNT-RGD showed a significant increase of photoacoustic signal in the tumor (Fig. 11.14B). These results were confirmed ex vivo using Raman microscopy. Photoacoustic imaging of targeted SWNTs may contribute to cancer imaging and therapy in living subjects.

Hong et al. *(49)* reported radionuclide-filled and glycosylated SWNTs and their usage as radioprobes. They first sealed $Na^{125}I$ inside SWNTs to form high-density

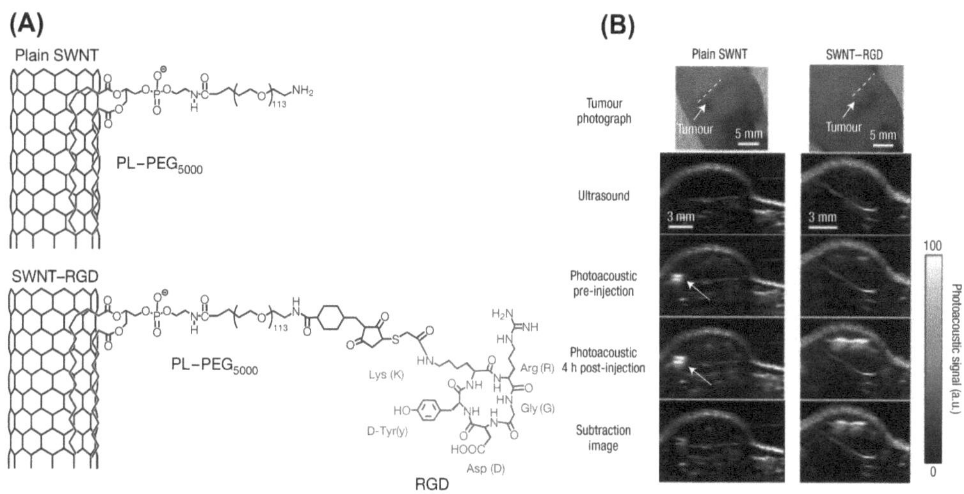

Figure 11.14 (A) Illustration of plain single-walled carbon nanotubes (plain SWNTs) and SWNT-Arg-Gly-Asp (RGD). The phospholipid binds to the sidewall of the SWNTs connecting the PEG_{5000} to the nanotubes. The RGD allows the SWNTs to bind to tumor integrins. (B) Ultrasound (gray) and photoacoustic (green (light gray in print versions)) images of one vertical slice (*white dotted line*) through the tumor. The ultrasound images show the skin and tumor boundaries. Subtraction images were calculated as the 4-h postinjection image minus the preinjection image. The high photoacoustic signal in the mouse injected with plain SWNTs (*white arrow*) is not seen in the subtraction image, suggesting that it is caused by a large blood vessel and not SWNTs *(48)*.

radioemitting crystals. Then, the outside surfaces of the metal halide-filled, sealed SWNTs were covalently modified with biantennary carbohydrates to improve their dispersibility and biocompatibility. Single-photon emission computed tomography/computed tomography (SPECT/CT) imaging was carried out immediately after tail-vein injection of the GlcNAc^{D}-Na^{125}I at SWNTs (Fig. 11.15A, left panel) or Na^{125}I (Fig. 11.15A, right panel) and after 4 h, 24 h and 7 days. GlcNAc^{D}-Na^{125}I at SWNTs predominantly accumulated in the lung, while the free Na^{125}I rapidly accumulated in

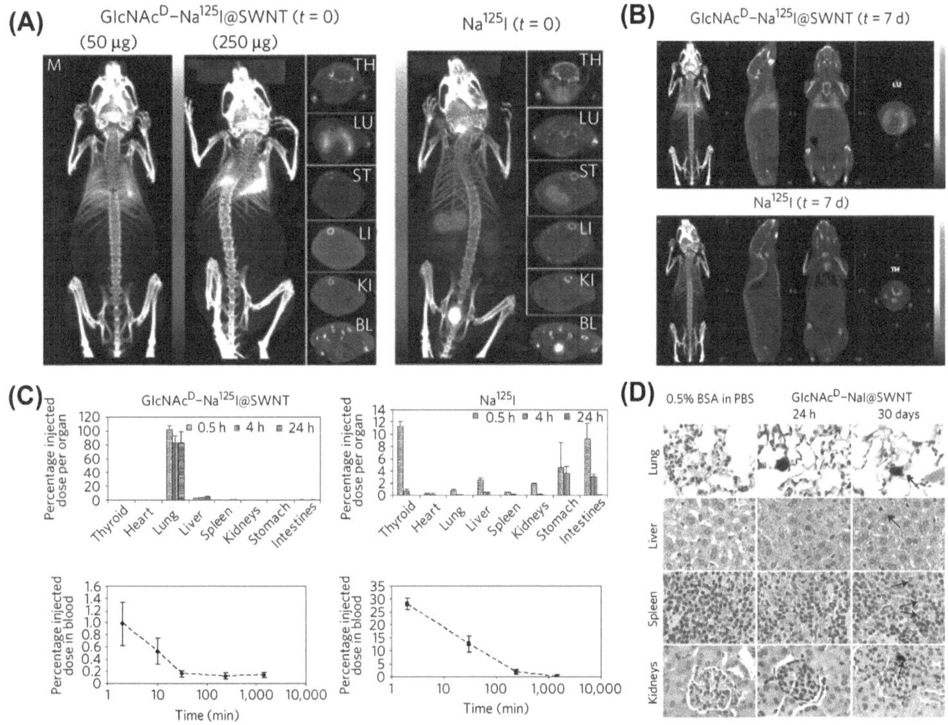

Figure 11.15 Whole-body single-photon emission-computed tomography (SPECT/CT) imaging was carried out immediately after tail-vein injection (A) and 7 days postinjection (B) of the GlcNAc^{D}-Na^{125}I at SWNTs (50 and 250 μg; 0.2 and 1.0 MBq, left-hand side in A and upper panels in B) or Na^{125}I (1.8 MBq, right hand-side in A and lower panels in B) with a scanning time of 40–60 min. Cross-sections of the thyroid (TH), lung (LU), stomach (ST), liver (LI), kidney (KI), and bladder at equivalent time points are shown. (C) Tissue biodistribution and blood-clearance profile of GlcNAc^{D}-Na^{125}I at SWNTs (50 μg, 0.2 MBq, left-hand panels) or Na^{125}I (0.36 MBq, right-hand panels) in mice after tail-vein injection. The percentage injected dose per organ was calculated at 30 min, 4 h, and 24 h after injection. Data are expressed as means ± s.d. ($n = 4$–6). (D) Hematoxylin- and eosin-stained sections of lung, liver, spleen, and kidney at 24 h and 30 days postadministration with "cold" GlcNAc^{D}-NaI at SWNTs (50 μg) or vehicle alone (0.5% BSA in PBS). Clearance to liver and kidney is consistent with single-walled carbon nanotubes (SWNTs) being slowly excreted through biliary and urinary routes (*black arrows* indicate nanotubes). All photomicrographs were captured at ×40 magnification *(49)*.

the thyroid and stomach. After 7 days postadministration, GlcNAc^{D}-Na^{125}I at SWNTs showed persistent lung accumulation, while only trace unencapsulated Na^{125}I remained in the thyroid (Fig. 11.15B). Gamma counting results indicated that 27% of free Na^{125}I entered in the blood 3 min after injection and cleared in the body within 24 h. However, very few GlcNAc^{D}-Na^{125}I at SWNTs were detected in the blood 3 min after administration and most of them remained in the lungs for up to 24 h (Fig. 11.15C). The results of histological examination of lungs, liver, spleen, and kidneys agreed well with SPECT/CT and gamma counting results (Fig. 11.15D).

Tong et al. *(50)* developed label-free imaging of both semiconducting and metallic CNTs in vitro and in vivo by using transient absorption microscopy. Their method could separate background signals generated by red blood cells and follow the movement of both types of nanotubes in cells, blood circulation, and organs of mice without any significant damaging cffccts.

11.8 SUMMARY AND OUTLOOK

In the past decades, CNTs have been widely investigated for biomedical applications because of their unique structure and excellent properties. Owing to their large surface area, CNTs can immobilize lots of biomolecules or load large amounts of drugs. The high conductivity of CNTs made them to transfer electron very quickly. The nanoscale diameters of CNTs enable them to penetrate the cell membrane and enter into cells. All these properties make CNTs promising materials for biosensing, drug delivery, therapy, and imaging.

One of the critical things related to their biomedical applications is the toxicity and safe of CNTs. It is necessary to understand the extensive toxicity of CNTs before developing their biomedical applications. The toxicity of CNTs depends on their structures, morphology, surface functional groups, and dose.

One of most used methods that make CNTs suitable for biomedical applications is surface modification. Surface modification can improve their hydrophilicity and solubility, as well as reducing toxicity. Surface modification of CNTs is performed first in almost all biomedical applications.

Our chapter has simply introduced recent biomedical applications of CNTs by choosing some excellent examples from the literature. Although many papers have used CNTs for drug delivery, therapy, biosensing, and bioimaging, more effort needs to be made to explore CNTs for biomedical applications in the future.

REFERENCES

1. Tasis, D.; Tagmatarchis, N.; Bianco, A.; Prato, M. Chemistry of Carbon Nanotubes. *Chemical Reviews* **2006,** *106* (3), 1105–1136.
2. Karousis, N.; Tagmatarchis, N.; Tasis, D. Current Progress on the Chemical Modification of Carbon Nanotubes. *Chemical Reviews* **2010,** *110* (9), 5366–5397.

3. Zhang, R. F.; Zhang, Y. Y.; Zhang, Q.; Xie, H. H.; Qian, W. Z.; Wei, F. Growth of Half-Meter Long Carbon Nanotubes Based on Schulz-Flory Distribution. *ACS Nano* **2013,** 7 (7), 6156–6161.
4. Iijima, S. Helical Microtubules of Graphitic Carbon. *Nature* **1991,** *354* (6348), 56–58.
5. Peer, D.; Karp, J. M.; Hong, S.; Farokhzad, O. C.; Margalit, R.; Langer, R. Nanocarriers as an Emerging Platform for Cancer Therapy. *Nature Nanotechnology* **2007,** *2* (12), 751–760.
6. Wu, H. C.; Chang, X.; Liu, L.; Zhao, F.; Zhao, Y. Chemistry of Carbon Nanotubes in Biomedical Applications. *Journal of Materials Chemistry* **2010,** *20,* 1036–1052.
7. Yang, W.; Thordarson, P.; Gooding, J. J.; Ringer, S. P.; Braet, F. Carbon Nanotubes for Biological and Biomedical Applications. *Nanotechnology* **2007,** *18.*
8. Lanone, S.; Andujar, P.; Kermanizadeh, A.; Boczkowski, J. Determinants of Carbon Nanotube Toxicity. *Advanced Drug Delivery Reviews* **2013,** *65* (15), 2063–2069.
9. Ryman-Rasmussen, J. P.; Cesta, M. F.; Brody, A. R.; Shipley-Phillips, J. K.; Everitt, J. I.; Tewksbury, E. W.; Moss, O. R.; Wong, B. A.; Dodd, D. E.; et al. Inhaled Carbon Nanotubes Reach the Subpleural Tissue in Mice. *Nature Nanotechnology* **2009,** *4* (11), 747–751.
10. Lam, C. W.; James, J. T.; McCluskey, R.; Hunter, R. L. Pulmonary Toxicity of Single-Wall Carbon Nanotubes in Mice 7 and 90 Days after Intratracheal Instillation. *Toxicological Sciences* **2004,** *77,* 126–134.
11. Mitchell, L. A.; Lauer, F. T.; Burchiel, S. W.; McDonald, J. D. Mechanisms for How Inhaled Multiwalled Carbon Nanotubes Suppress Systemic Immune Function in Mice. *Nature Nanotechnology* **2009,** *4* (7), 451–456.
12. Shvedova, A. A.; Castranova, V.; Kisin, E. R.; Schwegler-Berry, D.; Murray, A. R.; Gandelsman, V. Z.; Maynard, A.; Baron, P. Exposure to Carbon Nanotube Material: Assessment of Nanotube Cytotoxicity Using Human Keratinocyte Cells. *Journal of Toxicology and Environmental Health-Part A* **2003,** *66* (20), 1909–1926.
13. Bai, Y. H.; Zhang, Y.; Zhang, J. P.; Mu, Q. X.; Zhang, W. D.; Butch, E. R.; Snyder, S. E.; Yan, B. Repeated Administrations of Carbon Nanotubes in Male Mice Cause Reversible Testis Damage without Affecting Fertility. *Nature Nanotechnology* **2010,** *5* (9), 683–689.
14. Zhu, L.; Chang, D. W.; Dai, L.; Hong, Y. DNA Damage Induced by Multiwalled Carbon Nanotubes in Mouse Embryonic Stem Cells. *Nano Letters* **2007,** 7 (12), 3592–3597.
15. Schipper, M. L.; Nakayama-Ratchford, N.; Davis, C. R.; Kam, N. W.; Chu, P.; Liu, Z.; Sun, X.; Dai, H.; Gambhir, S. S. A Pilot Toxicology Study of Single-Walled Carbon Nanotubes in a Small Sample of Mice. *Nature Nanotechnology* **2008,** *3* (4), 216–221.
16. Jain, S.; Thakare, V. S.; Das, M.; Godugu, C.; Jain, A. K.; Mathur, R.; Chuttani, K.; Mishra, A. K. Toxicity of Multiwalled Carbon Nanotubes with End Defects Critically Depends on Their Functionalization Density. *Chemical Research in Toxicology* **2011,** *24,* 2028–2039.
17. Sayes, C. M.; Liang, F.; Hudson, J. L.; Mendez, J.; Guo, W.; Beach, J. M.; et al. Functionalization Density Dependence of Single-Walled Carbon Nanotubes Cytotoxicity in Vitro. *Toxicology Letters* **2006,** *161,* 135–142.
18. Xue, Y. H.; Chen, H. Synthesis and Cytotoxicity of POSS Modified Single Walled Carbon Nanotubes. *Journal of Nanomaterials* **2015,** 407437.
19. Chen, R. J.; Zhang, Y.; Wang, D.; Dai, H. Noncovalent Sidewall Functionalization of Single-Walled Carbon Nanotubes for Protein Immobilization. *Journal of the American Chemical Society* **2001,** *123* (16), 3838–3839.
20. Liu, Z.; Sun, X.; Nakayama-Ratchford, N.; Dai, H. Supramolecular Chemistry on Water-Soluble Carbon Nanotubes for Drug Loading and Delivery. *ACS Nano* **2007,** *1* (1), 50–56.
21. Ren, J.; Shen, S.; Wang, D.; Xi, Z.; Guo, L.; Pang, Z.; Qian, Y.; Sun, X.; Jiang, X. The Targeted Delivery of Anticancer Drugs to Brain Glioma by PEGylated Oxidized Multi-Walled Carbon Nanotubes Modified with Angiopep-2. *Biomaterials* **2012,** *33* (11), 3324–3333.
22. Wu, H.; Shi, H.; Zhang, H.; Wang, X.; Yang, Y.; Yu, C.; Hao, C.; Du, J.; Hu, H.; Yang, S. Prostate Stem Cell Antigen Antibody-Conjugated Multiwalled Carbon Nanotubes for Targeted Ultrasound Imaging and Drug Delivery. *Biomaterials* **2014,** *35* (20), 5369–5380.
23. Wong, B. S.; Yoong, S. L.; Jagusiak, A.; Panczyk, T.; Ho, H. K.; Ang, W. H.; Pastorin, G. Carbon Nanotubes for Delivery of Small Molecule Drugs. *Advanced Drug Delivery Reviews* **2013,** *65* (15), 1964–2015.

24. Pantarotto, D.; Singh, R.; McCarthy, D.; Erhardt, M.; Briand, J. P.; Prato, M.; Kostarelos, K.; Bianco, A. Functionalized Carbon Nanotubes for Plasmid DNA Gene Delivery. *Angewandte Chemie International Edition* **2004,** *43* (39), 5242–5246.
25. Tonelli, F. M. P.; Lacerda, S. M. S. N.; Silva, M. A.; Ávila, E. S.; Ladeira, L. O.; França, L. R.; Resende, R. R. Gene Delivery to Nile Tilapia Spermatogonial Stem Cells Using Carboxi-Functionalized Multiwall Carbon Nanotubes. *RSC Advances* **2014,** *4,* 37985–37987.
26. Liu, Z.; Fan, A. C.; Rakhra, K.; Sherlock, S.; Goodwin, A.; Chen, X.; Yang, Q.; Felsher, D. W.; Dai, H. Supramolecular Stacking of Doxorubicin on Carbon Nanotubes for in Vivo Cancer Therapy. *Angewandte Chemie International Edition* **2009,** *48* (11), 7668–7672.
27. Kang, B.; Yu, D.; Dai, Y.; Chang, S.; Chen, D.; Ding, Y. Cancer-Cell Targeting and Photoacoustic Therapy Using Carbon Nanotubes as "Bomb" Agents. *Small* **2009,** *5* (11), 1292–1301.
28. Zhou, F.; Wu, S.; Yuan, Y.; Chen, W. R.; Xing, D. Mitochondria-Targeting Photoacoustic Therapy Using Single-Walled Carbon Nanotubes. *Small* **2012,** *8* (10), 1543–1550.
29. Liang, C.; Diao, S.; Wang, C.; Gong, H.; Liu, T.; Hong, G.; Shi, X.; Dai, H.; Liu, Z. Tumor Metastasis Inhibition by Imaging-Guided Photothermal Therapy with Single-Walled Carbon Nanotubes. *Advanced Materials* **2014,** *26* (32), 5646–5652.
30. Wang, L.; Shi, J.; Liu, R.; Liu, Y.; Zhang, J.; Yu, X.; Gao, J.; Zhang, C.; Zhang, Z. Photodynamic Effect of Functionalized Single-Walled Carbon Nanotubes: A Potential Sensitizer for Photodynamic Therapy. *Nanoscale* **2014,** *6* (9), 4642–4651.
31. Shi, J.; Ma, R.; Wang, L.; Zhang, J.; Liu, R.; Li, L.; Liu, Y.; Hou, L.; Yu, X.; Gao, J.; et al. The Application of Hyaluronic Acid-Derivatized Carbon Nanotubes in Hematoporphyrin Monomethyl Ether-Based Photodynamic Therapy for In Vivo and In Vitro Cancer Treatment. *International Journal of Nanomedicine* **2013,** *8,* 2361–2373.
32. Harrison, B. S.; Atala, A. Carbon Nanotube Applications for Tissue Engineering. *Biomaterials* **2007,** *28* (2), 344–353.
33. Correa-Duarte, M. A.; Wagner, N.; Rojas-Chapana, J.; Morsczeck, C.; Thie, M.; Giersig, M. Fabrication and Biocompatibility of Carbon Nanotube-Based 3D Networks as Scaffolds for Cell Seeding and Growth. *Nano Letters* **2004,** *4* (11), 2233–2236.
34. Abarrategi, A.; Gutierrez, M. C.; Moreno-Vicente, C.; Hortiguela, M. J.; Ramos, V.; Lopez-Lacomba, J. L.; Ferrer, M. L.; del Monte, F. Multiwall Carbon Nanotube Scaffolds for Tissue Engineering Purposes. *Biomaterials* **2008,** *29* (1), 94–102.
35. Keefer, E. W.; Botterman, B. R.; Romero, M. I.; Rossi, A. F.; Gross, G. W. Carbon Nanotube Coating Improves Neuronal Recordings. *Nature Nanotechnology* **2008,** *3* (7), 434–439.
36. Baranauskas, G.; Maggiolini, E.; Castagnola, E.; Ansaldo, A.; Mazzoni, A.; Angotzi, G. N.; Vato, A.; Ricci, D.; Panzeri, S.; Fadiga, L. Carbon Nanotube Composite Coating of Neural Microelectrodes Preferentially Improves the Multiunit Signal-to-Noise Ratio. *Journal of Neural Engineering* **2011,** *8* (6), 066013.
37. Cellot, C.; Cilia, E.; Cipollone, S.; Rancic, V.; Sucapane, A.; Giordani, S.; et al. Carbon Nanotubes Might Improve Neuronal Performance by Favouring Electrical Shortcuts. *Nature Nanotechnology* **2009,** *4,* 126–133.
38. Lee, H. J.; Park, J.; Yoon, O. J.; Kim, H. W.; Lee do, Y.; Kim do, H.; Lee, W. B.; Lee, N. E.; Bonventre, J. V.; Kim, S. S. Amine-Modified Single-Walled Carbon Nanotubes Protect Neurons from Injury in a Rat Stroke Model. *Nature Nanotechnology* **2011,** *6* (2), 121–125.
39. Yang, W. R.; Ratinac, K. R.; Ringer, S. P.; Thordarson, P.; Gooding, J. J.; Braet, F. Carbon Nanomaterials in Biosensors: Should You Use Nanotubes or Graphene? *Angewandte Chemie-International Edition* **2010,** *49* (12), 2114–2138.
40. He, P.; Dai, L. Aligned Carbon Nanotube-DNA Electrochemical Sensors. *Chemical Communications* **2004,** *3,* 348–349.
41. Sudibya, H. G.; Ma, J.; Dong, X.; Ng, S.; Li, L. J.; Liu, X. W.; Chen, P. Interfacing Glycosylated Carbon-Nanotube-Network Devices with Living Cells to Detect Dynamic Secretion of Biomolecules. *Angewandte Chemie-International Edition* **2009,** *48* (15), 2723–2726.
42. Hu, F. X.; Kang, Y. J.; Du, F.; Zhu, L.; Xue, Y. H.; Chen, T.; Dai, L. M.; Li, C. M. Living Cells Directly Growing on a DNA/$Mn_3(PO_4)_2$-immobilized and Vertically Aligned CNT Array as a

Free-Standing Hybrid Film for Highly Sensitive in Situ Detection of Released Superoxide Anions. *Advanced Functional Materials* **2015,** *25,* 5924–5932.

43. Xiang, L.; Yu, P.; Hao, J.; Zhang, M. N.; Zhu, L.; Dai, L. M.; Mao, L. Q. Vertically Aligned Carbon Nanotube-Sheathed Carbon Fibers as Pristine Microelectrodes for Selective Monitoring of Ascorbate In Vivo. *Analytical Chemistry* **2014,** *86* (8), 3909–3914.
44. Iverson, N. M.; Barone, P. W.; Shandell, M.; Trudel, L. J.; Sen, S.; Sen, F.; Ivanov, V.; Atolia, E.; Farias, E.; McNicholas, T. P.; et al. In Vivo Biosensing via Tissue-Localizable Near-Infrared-Fluorescent Single-Walled Carbon Nanotubes. *Nature Nanotechnology* **2013,** *8* (11), 873–880.
45. Barone, P. W.; Baik, S.; Heller, D. A.; Strano, M. S. Near-Infrared Optical Sensors Based on Single-Walled Carbon Nanotubes. *Nature Materials* **2005,** *4* (1), 86–92.
46. Jin, H.; Heller, D. A.; Kalbacova, M.; Kim, J. H.; Zhang, J.; Boghossian, A. A.; Maheshri, N.; Strano, M. S. Detection of Single-Molecule H_2O_2 Signalling from Epidermal Growth Factor Receptor Using Fluorescent Single-Walled Carbon Nanotubes. *Nature Nanotechnology* **2010,** *5* (4), 302–309.
47. Welsher, K.; Liu, Z.; Sherlock, S. P.; Robinson, J. T.; Chen, Z.; Daranciang, D.; Dai, H. A Route to Brightly Fluorescent Carbon Nanotubes for Near-Infrared Imaging in Mice. *Nature Nanotechnology* **2009,** *4,* 773–780.
48. Zerda, A. D.; Zavaleta, C.; Keren, S.; Vaithilingam, S.; Bodapati, S.; Liu, Z.; et al. Carbon Nanotubes as Photoacoustic Molecular Imaging Agents in Living Mice. *Nature Nanotechnology* **2008,** *3,* 557–562.
49. Hong, S. Y.; Tobias, G.; Al-Jamal, K. T.; Ballesteros, B.; Ali-Boucetta, H.; Lozano-Perez, S.; et al. Filled and Glycosylated Carbon Nanotubes for In Vivo Radioemitter Localization and Imaging. *Nature Materials* **2010,** *9,* 485–490.
50. Tong, L.; Liu, Y.; Dolash, B. D.; Jung, Y.; Slipchenko, M. N.; Bergstrom, D. E.; Cheng, J. X. Label-Free Imaging of Semiconducting and Metallic Carbon Nanotubes in Cells and Mice Using Transient Absorption Microscopy. *Nature Nanotechnology* **2012,** *7,* 56–61.

CHAPTER 12

Carbon Nanotube Fibers for Wearable Devices

Y. Yao, N. Li, T. Lv, T. Chen
Tongji University, Shanghai, China

Contents

12.1 INTRODUCTION

To meet the rapid development of portable electronics, it is necessary to develop flexible energy conversion and/or storage devices with high performance *(1)*. Because of its unique structure, fiber-shaped (also called wearable) electronics (such as solar cells *(2)*, supercapacitors *(3)*, lithium-ion batteries *(4)*, and so on) not only show high flexibility, but also can be easily integrated into clothing, bags, or other textiles by using conventional woven technology. Generally, metal fibers or conducting material-coated plastic fibers were used as the fiber electrode in fiber-shaped energy devices *(2–4)*. However, the aforementioned fiber electrodes were often limited by their intrinsic defects, such as metal fibers being unstable or mismatched with active materials; also the conducting materials on the plastic substrate easily cracked as the devices were bent. Carbon

ISBN 978-0-323-41481-4

nanotubes (CNTs) exhibit excellent electrical, mechanical, optical, and electrochemical properties *(5)*, which can be easily assembled into macroscopic fibers with controlled structures and properties *(6,7)*. The developed CNT fibers showed highly electrical conductivity, mechanical strength, electrochemical property, and good flexibility, which represent an ideal fiber electrode for wearable fiber-shaped electronics. Up to now, great progress has been achieved on high-performance wearable electronics by using CNT fiber electrodes *(2)*. In this chapter, we first introduce the synthesis and properties of CNT fiber, then we focus on reviewing recent progress on its applications in various wearable electronics containing energy conversion, storage, and integrated systems.

12.2 PREPARATION AND PROPERTIES OF CARBON NANOTUBE FIBERS

The crucial condition to obtain high-performance CNT fiber is to make CNTs aligned along the axial direction of a fiber. Until now there have been four main approaches to prepare CNT fibers: (1) wet spinning from CNT solution *(8,9)*, (2) in situ spinning from a CNT aerogel directly synthesized in a chemical vapor deposition (CVD) reactor *(10,11)*, (3) dry spinning from a vertically aligned CNT array *(12,13)*, and (4) spinning from a CNT film *(14)*. The electrical and mechanical properties and morphology of the obtained CNT fibers strongly depend on the spinning processes used, which will be discussed next.

12.2.1 Wet Spinning CNT Fibers

The wet-spinning process has been successfully used to produce conventional polymer fibers, such as Kevlar fiber, poly(acrylonitrile) fiber, and so on. Typically, a polymer solution was extruded into a liquid bath in which the solvent in the polymer solution was soluble but the polymer was not *(8,15)*. By using a wet-spinning process, Vigolo et al. first assembled single-walled CNTs into macroscopic ribbon and fiber *(8)*. In this process, nanotubes were first dispersed with a 0.35 wt% concentration in an aqueous solutions of sodium dodecyl sulfate (SDS, 1.0 wt%), forming a homogeneous dispersion. Then, the nanotube dispersion was carefully injected into a coflowing stream of polymer solution containing 5 wt% polyvinyl alcohol (PVA) using a syringe needle or a thin glass capillary. In this system, amphiphilic PVA can presumably adsorb onto nanotubes displacing part of the SDS molecules, and flow-induced alignment of PVA solution provided the conditions for the formation of CNT ribbons or fibers. To remove the residue PVA and SDS, the as-prepared CNT ribbons or fibers can be washed and rinsed with pure water several times. By this simple method, fibers tens of centimeters long can be obtained by slowly pulling ribbons out of the water. CNT fiber with diameters ranging from a few micrometers to 100 μm can be fabricated by tuning the processing parameters, such as injection rate, flow conditions, and dimensions of the capillary tube. This CNT fiber showed a tensile strength of 300 MPa and a Young's modulus of 40 GPa, respectively, which were several times more than high-quality bucky paper of CNTs. In addition, the CNT fiber exhibited higher flexibility than traditional carbon fibers, which can be severely bent without breaking. This CNT fiber had a nonmetallic behavior as the temperature decreased, and showed an electrical conductivity of about 10 S/cm at room temperature.

The problems of the aforementioned method are: (1) the obtained CNT fiber is too week and (2) it is difficult to scale up because the extremely low preparing speed (about 1 cm/min) and the rotating bath are also not suitable to continuously spin long fibers. To resolve these problems, Razal et al. used lithium dodecyl sulfate as the surfactant to disperse nanotubes and continuously injected the CNT solution into the center of a cylindrical pipe with PVA coagulation solution flowing (Fig. 12.1A–D) *(16)*. The obtained CNT composite fiber not only showed high flexibility (Fig. 12.1D), but also exhibited highly mechanical properties, such as a tensile strength of 1.8 GPa and a Young's modulus of 80 GPa, and much higher energy-to-break than Kevar fiber and graphite fiber. The high mechanical properties can be attributed to the high loading of PVA (40%) in the resulting CNT composite fiber, but much electrical conductivity is lost at the same

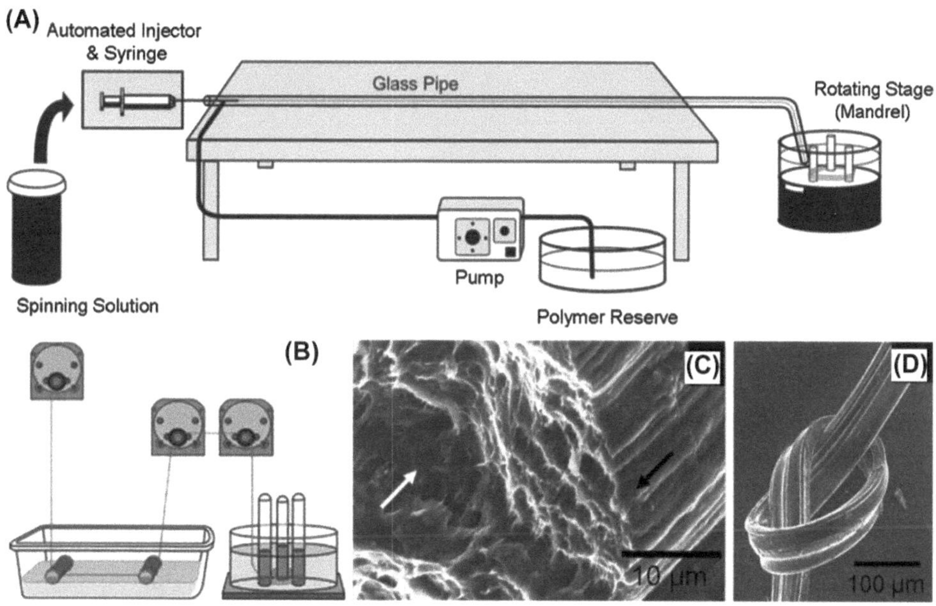

Figure 12.1 (A) Schematic illustration of the spinning apparatus and (B) setup for postspinning solvent treatment and draw. (C) Cross-sectional image of a carbon nanotube (CNT) composite fiber. (D) Scanning electron microscopy (SEM) image of fiber tied in a knot. *(Reproduced from Razal, J. M.; Coleman, J. N.; Munoz, E.; Lund, B.; Gogotsi, Y.; Ye, H.; Collins, S.; Dalton, A. B.; Baughman, R. H. Advanced Functional Materials* ***2007****, 17, 2918–2924 with permission. Copyright 2007 John Wiley and Sons.)*

time. To avoid using surfactant and PVA, Ericson et al. have produced a continuous length of macroscopic neat CNT fibers by extruding a homogeneous solution of 8 wt% purified single-walled CNTs in 102% sulfuric acid *(9)*. In superacid, single-walled CNTs will be charged and form ordered domains having the similar behavior to nematic liquid crystalline rod-like polymers at high concentration (>4 wt%), which make it possible to assemble nanotubes into CNT fiber. By using this modified method, 100 m-long continuous CNT composite fiber can be obtained at a rate of more than 70 cm/min. The tensile strength and Young's modulus of the resulting neat CNT fibers were 116 ± 10 MPa and 120 ± 10 GPa, respectively. After removing residues and annealing in a vacuum at 1100°C, the electrical conductivity of the neat CNT fiber reached 500 S/cm, which was two orders of magnitude higher than that of CNT fibers with PVA.

It is a great challenge to fabricate CNT fiber by using superacid, especially when scaled up in industry, because special protection equipment is required *(17)*. Therefore, a gentle environment in which to produce CNT fibers is more attractive for both academic and industrial fields. For example, Zhang et al. developed a simple approach to spin CNT fibers directly from the lyotropic liquid crystalline phase of CNTs formed in ethylene glycol solvent *(15)*. The residual solvent in as-prepared CNT fiber can be easily removed by annealing at much low temperature (280°C). The electrical conductivity along the pure CNT fiber axis was about 80 S/cm at room temperature, which was much higher than those of CNT/PVA composite fiber. However, because no polymer was introduced, the pure CNT fibers had lower mechanical properties, a Young's modulus of 69 ± 41 GPa, and tensile strength of 150 ± 60 MPa, respectively.

12.2.2 Direct Spinning of CNT Fiber From CVD Synthesis

Different to the aforementioned wet-spinning process, CNT fiber can also be achieved by directly spinning from a CVD reactor. Zhu et al. first reported the method of directly synthesizing long CNT strands by an optimized CVD technique in a vertical furnace *(18)*, where an *n*-hexane solution containing a given composition of ferrocene and thiophene was injected into the reactor with hydrogen as the gas carrier to grow CNT at the pyrolyzed temperature of *n*-hexane. CNT strands ranging from 10 to 20 cm in length can be obtained, the diameters of which can be tuned from 50 to 300 μm. The resulting CNT strands exhibited a metallic behavior with electrical conductivity from 2.0×10^3 to 1.4×10^3 S/cm as the temperature increased from 90K to 300K. The strength and stiffness of the CNT strands were about 1.0 and 100 GPa, respectively.

Li et al. developed a continuous spinning process to fabricate CNT fiber by winding aerogel formed in the CVD reactor onto a rotating rod *(10)*. They selected ethanol as the carbon source with a given concentration of ferrocene and thiophene for CNT growth. CNT aerogel was formed in the hot zone (1050–1200°C) as the solution was injected from the top of the furnace with a carrier gas of hydrogen, which was captured and spun continuously out of the hot zone by a rotating rod. The best electrical conductivity along the fiber was 8.3×10^3 S/cm, which is much higher than that of CNT fiber prepared by the wet-spinning process. The strength and stiffness of the obtained CNT fiber were 8.8 and 357 GPa, respectively. Zhong et al. further improved this method, which can continuously spin multilayered CNT fiber over several kilometers *(11)*. In brief, the multilayered CNT aerogel (also called sock) was densified with water as it came out of the reactor, followed by washing and further densification through a second spool of acetone; the resultant CNT fiber was wound onto the final spool after it was dried by an infrared heater (Fig. 12.2A–E). The electrical conductivity of the as-spun CNT fiber was about 5.0×10^3 S/cm as measured by a four-probe technique, which was higher than that of wetting-spun CNT fiber.

The properties of the CNT fibers greatly depend on parameters during synthesizing and spinning processes. During the CVD growth process, the content of iron played an important role on the structure of nanotubes and properties of the final fibers. The results showed that iron with a lower concentration led to a great

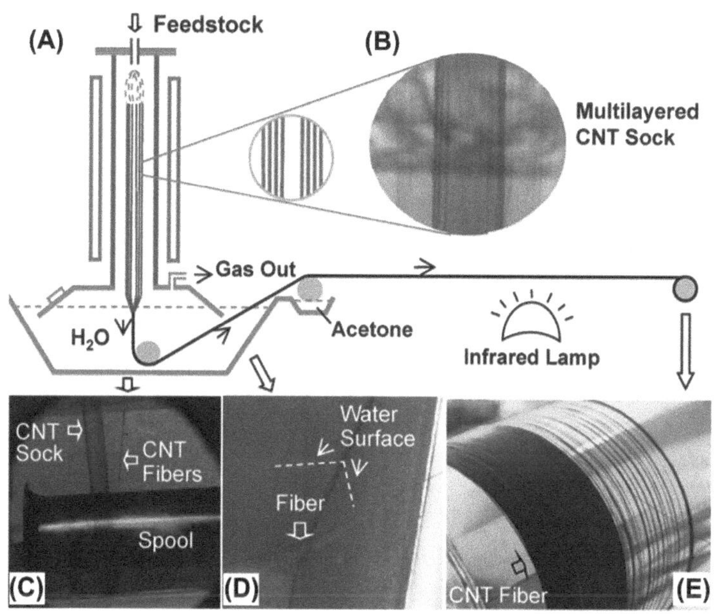

Figure 12.2 Setup of the direct-spinning process of CNT fiber from a chemical vapor deposition (CVD) reactor. (A) Schematic diagram of the synthesis and spinning setup for continuously spinning CNT fiber. (B) A photograph showing carbon nanotube (CNT) aerogel formed in the gas flow. Photographs of transformation of the CNT aerogel into a fiber (C), followed by water densification (D), and spinning of the resultant CNT fiber on a spool (E). *(Reproduced from Zhong, X. H.; Li, Y. L.; Liu, Y. K.; Qiao, X. H.; Feng, Y.; Liang, J.; Jin, J.; Zhu, L.; Hou, F.; Li, J. Y. Advanced Materials* ***2010****, 22, 692–696 with permission. Copyright 2010 John Wiley and Sons.)*

proportion of single-walled nanotubes; as a result the obtained CNT fiber exhibited higher strength and stiffness *(19)*. Meanwhile, the addition of sulfur can promote exceptionally long nanotube growth, resulting in mechanical integrity of the carbon plume, which enables the continuous spinning fiber *(20)*. The winding rate during the spinning process also had an obvious effect on the properties of CNT fiber, such as nanotube orientation, density, and mechanical property *(21)*. The experimental results showed that CNT fiber spun under higher winding rates (e.g., up to 20 m/min) possessed higher strength and stiffness because of higher density and higher alignment of nanotubes in the fiber *(21)*.

12.2.3 Dry Spinning of CNT Fiber From Highly Aligned CNT Array

Similar to drawing thread from a silk cocoon, in 2002 CNT fiber spun from a dry vertically aligned array was developed *(12,13)*. To start with, a superaligned CNT array should be first synthesized by the CVD method, in which CNTs are aligned parallel to each other and bundled together by van der Waals interactions *(12,13)*. An aligned CNT thin film with a controlled width can be directly drawn out from a CNT array (Fig. 12.3A) *(22)*, which can be spun into CNT fiber (Fig. 12.3B) by using a rotating probe. It has been estimated that about 10 m of CNT fiber can be produced by using a CNT array with an area of 1 cm^2 *(12)*. Furthermore, it is easily realized that the continuous spinning process is a simple setup. For instance, Liu et al. reported that a

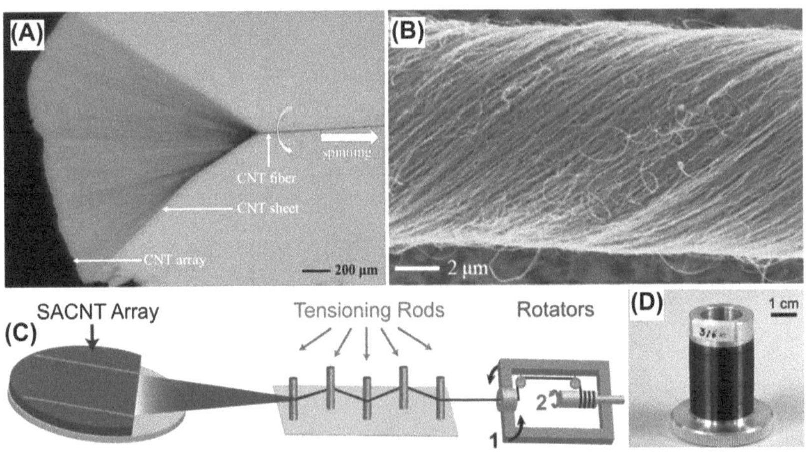

Figure 12.3 (A) A digital photograph of the process to spin a carbon nanotube (CNT) fiber from a CNT array. (B) A typical scanning electron microscopy (SEM) image of CNT fiber spun by the dry process. *(Reproduced from Chen, T.; Cai, Z.; Yang, Z.; Li, L.; Sun, X.; Huang, T.; Yu, A.; Kia, H. G.; Peng, H. Advanced Materials* ***2011****, 23, 4620–4625 with permission. Copyright 2011 John Wiley and Sons.)* (C) Schematic of the continuously spinning setup. (D) As-produced CNT fiber with a length of 316 m. *(Reproduced from Liu, K.; Zhu, F.; Liu, L.; Sun, Y.; Fan, S.; Jiang, K. Nanoscale* ***2012****, 4, 3389–3393. Copyright 2012 The Royal Society of Chemistry.) SACNT*, Superaligned carbon nanotube.

continuous 316 m-long CNT fiber can be obtained and collected on a roller (Fig. 12.3C and D) *(23)*. The structures and properties of dry-spun CNT fibers strongly depend on the length of the CNT array, parameters during the spinning process, and post-treatment.

Electrical resistance mainly comes from contact resistance between nanotubes, which means that less contact between nanotubes will reduce the electrical resistance of the resulting CNT fiber. Therefore the electrical conductivity of CNT fiber may be optimized by improving the alignment of the CNTs and increasing the lengths of individual nanotubes to minimize the contact resistance between nanotubes. Li et al. fabricated CNT fiber using CNT arrays with different heights and studied the dependence of electrical conductivity upon the length of nanotubes *(24)*. Their results showed that the electrical conductivity of CNT fibers increased from 465.3 to 596.5 S/cm as the heights of the CNT array increased from 0.3 to 1.0 mm. Further studies indicated that the CNT fibers exhibited a semiconducting behavior according to the dependence of electrical conductivity upon temperature. A three-dimensional hopping mechanism in the CNT fiber was verified according to Mott's variable range hopping model, which means that electrons can hop from one localized site to another or from one nanotube to another *(24)*. The length of nanotubes not only affects the electrical conductivity, but also has an obvious effect on the mechanical property of the CNT fibers. The results showed that the tensile strength increased from 320 to 850 MPa as the length of CNT array increased from 300 to 650 μm because of higher friction force at the interface of longer nanotubes *(25)*. A tensile strength as high as 1.91 GPa can be obtained by the post-spinning treatment of CNT fiber spun from a 650 μm-high CNT array, which can be attributed to the decrease in diameter and increase in the load-carrying capacity of the CNT fiber *(25)*.

12.3 CARBON NANOTUBE FIBER FOR WEARABLE ENERGY CONVERSION DEVICES

Because of its unique structure, excellent electrical conductivity, and mechanical property, CNT fiber has been widely used as electrodes in fiber-shaped energy conversion devices, such as solar cells and light emitting diodes (LEDs). The fiber-shaped energy conversion devices have similar structure to the devices with planar format, which can be easily fabricated based on one single-fiber electrode or two-fiber electrodes using a layer-by-layer method. Compared with planar devices, fiber-shaped energy conversion devices showed high flexibility, which can be easily integrated into other textiles for use as a power source for portable electronics by conventional woven techniques. Therefore fiber-shaped energy conversion devices have attracted intense attention over the last several years, and many interesting results have been achieved.

12.3.1 Fiber-Shaped Dye-Sensitized Solar Cells Based on CNT Fiber

It has been confirmed that CNTs can be used as efficient electrode materials for high-performance dye-sensitized solar cells (DSSCs) in conventional planar format *(26)*. CNT film with high transmittance and electrical conductivity can be used as a transparent electrode, replacing the expensive fluorine-doped tin oxide (FTO) electrode, which can also be used as a highly efficient counter electrode to replace the conventional platinum catalyst for redox reactions between I^- and I_3^-. Therefore as a one-dimensional macroscopic assemble, CNT fiber with CNTs as building blocks is a promising candidate electrode for use as efficient fiber electrodes in fiber-shaped DSSCs, which has been widely investigated.

CNT fiber prepared by the dry-spinning process exhibited semiconductive behavior, and may be used as semiconductive material to absorb dye molecules because of its high surface area and strong $\pi-\pi$ interaction between bipyridyl-based dye molecules and CNTs. Chen et al. first attempted to apply the semiconductive CNT fiber in DSSCs, in which an aligned CNT fiber transferred onto the FTO conductive glass was used as the anode after absorbed photosensitive dyes and a Pt-coated FTO conductive glass was used as the cathode *(27)*. In this system, the electrons generated by dyes can rapidly transport along aligned CNTs to the anodic FTO electrode, which can efficiently avoid charge recombination. As a result, a power conversion efficiency (η) of 2.2% and an incident photo-to-electron conversion efficiency of 90% were obtained when one single CNT fiber was used. This work opened up a new field of application of CNT fibers in wearable solar cells. A real fiber-shaped DSSC has been developed by using a dye-sensitized CNT fiber as the photoanode electrode and a poly(vinylidene fluoride) (PVDF)-coated CNT fiber as the photocathode *(28)*. A porous PVDF layer outside of the CNT fiber can efficiently avoid a short circuit between two CNT fiber electrodes and enable ions or electrons to transport through. The results showed that both CNT fiber and PVDF-coated CNT fiber had a highly catalytic property toward the redox action between I^- and I_3^-, but low energy conversion efficiency was obtained for the fiber-shaped DSSC. The poor performance of this type of fiber-shaped DSSC may be ascribed to the limited ability of CNT fiber to absorb enough photosensitive dyes and mismatch of bandgap between CNTs and dyes.

To resolve these problems, it is required to introduce highly efficient photoanode materials, such as TiO_2. In 2012, Chen et al. coated a 5−10 μm-thick TiO_2 layer outside of the CNT fiber by using a simple dip-coating method, resulting in a CNT/TiO_2 composite fiber, which was used as the photoanode after absorbing photosensitive dyes (Fig. 12.4A and B) *(29)*. As a result, all CNT fiber electrode-based fiber-shaped DSSCs have been developed by twisting this photoanode and a bare CNT fiber as the photocathode (Fig. 12.4C and D). An energy conversion efficiency of about 3% (Fig. 12.4F) can be obtained for one single-fiber-shaped DSSC, which is much high than that of

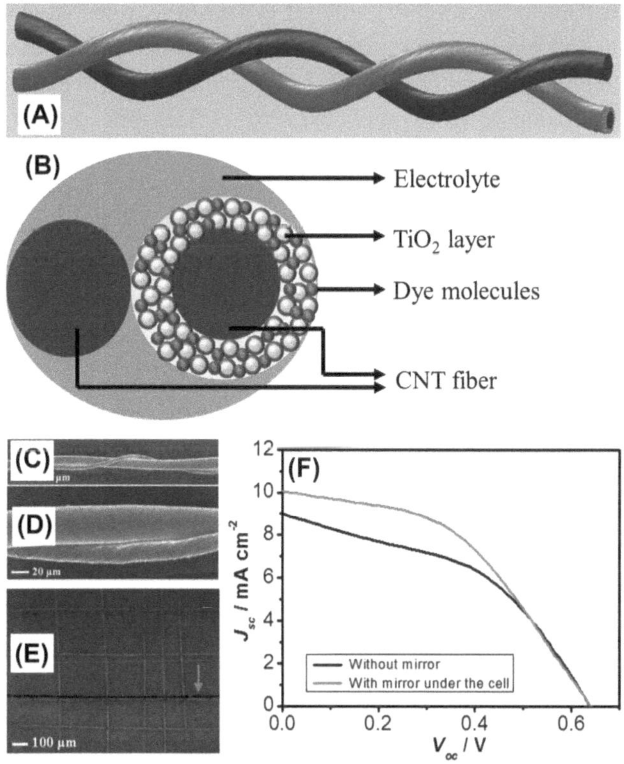

Figure 12.4 (A and B) Schematic structures of a fiber-shaped dye-sensitized solar cell (DSSC) based on two carbon nanotube (CNT) fiber electrodes. (C and D) Scanning electron microscopy (SEM) images of a fiber-shaped DSSC with different magnifications. (E) A digital photograph of a fiber-shaped DSSC (*red arrow* (gray in print versions)) being woven into a textile. (F) The current density–voltage curves of a fiber-shaped DSSC with and without a mirror under the AM1.5 solar light. *(Reprinted with permission from Chen, T.; Qiu, L.; Cai, Z.; Gong, F.; Yang, Z.; Wang, Z.; Peng, H. Nano Letters* ***2012****, 12, 2568–2572. Copyright 2012 American Chemical Society.)*

devices without using the TiO_2 layer *(28)*. Interestingly, almost all the parameters (e.g., open-circuit voltage, short-circuit current density, and fill factor) of the devices varied very slightly as the length of the light incident angles or devices changed *(29)*. Importantly, the obtained fiber-shaped DSSCs were flexible and wearable, and could be easily woven into textiles for further applications (Fig. 12.4E).

An efficient route to improve the performance of DSSCs is decreasing the transmission distance of photogenerated electrons, which can be realized by using electrodes with aligned structure. Therefore a Ti wire radially grown with an aligned TiO_2 nanotube array has been prepared and used to chemisorb photosensitive dyes *(30)*. The fiber-shaped DSSCs have been developed by twisting the Ti wire supported with aligned TiO_2 nanotube with a bare aligned CNT fiber as the photocathode. Because of the highly aligned structure of both CNTs and TiO_2 nanotube, photogenerated charges can efficiently separate and transport to corresponding electrodes in the device. Therefore an energy conversion efficiency as high as 4.6% can be achieved, which is higher than that of fiber-shaped DSSCs based on CNT networks *(30,31)*. The efficiency of the fiber-shaped DSSCs can be improved to 4.85% when a CNT fiber is modified with platinum nanoparticles *(32)*. For all the aforementioned fiber-shaped DSSCs, a conventional I^-/I_3^- redox couple was used as electrolyte, and it has been confirmed that it was efficient for the Pt-based counter electrode, but may not be suitable for the carbon materials-based counter electrode because there may be a mismatch of bandgap between the electrolyte and counter electrode. Therefore Pan et al. designed a CNT fiber-based fiber-shaped DSSC using an organic thiolate/disulfide redox couple as electrolyte *(33)*. An energy conversion efficiency of 7.33% has been achieved, which was much higher (5.97%) than that of devices based on the I^-/I_3^- redox couple *(33)*. Up to now, the highest efficiency of fiber-shaped DSSCs was 8.54%, because a porous graphene fiber modified with Pt nanoparticles was used as the counter electrode *(34)*, very close to that of DSSCs in planar format.

12.3.2 Fiber-Shaped Polymer Solar Cells Based on CNT Fiber

Great achievements on liquid electrolyte-based fiber-shaped DSSCs have been achieved over the last several years. However, the problem of fiber-shaped DSSCs using liquid electrolyte is also remarkable, such as toxicity and corrosivity of organic solvents and leakage and vaporization of liquid electrolytes, which are the main obstacles for industrial applications. To resolve these issues, efforts have been made to replace liquid electrolytes with gel electrolyte *(35)* or quasi-solid-state electrolyte *(36)*. The problems with gel electrolytes are their instability and poor mechanical property; meanwhile, the quasi-solid-state electrolytes have problems of low electrical conductivity, high viscosity, and poor contact with the TiO_2 layer. Fiber-shaped solar cells based on polymer photoactive materials that do not need liquid electrolyte and could be easily fabricated by the simple dip-coating method can overcome the problems of liquid electrolyte-based DSSCs.

Lee et al. first developed a fiber-shaped polymer solar cell (PSC), in which a steel wire was used as the conductive substrate and a composite of poly(3-hexyl-2,5-thiophene) (P3HT) and [(6,6)-phenyl-C_{61} butyric acid methyl ester] ($PC_{61}BM$) was used as the photoactive layer *(37)*. An energy conversion efficiency of 3.87% has been achieved for the fiber-shaped PSC with a two-electrode twisted structure. Compared with steel wire, CNT fiber is more lightweight, flexible, and stable, which represents an ideal candidate for use as a conductive electrode in fiber-shaped PSCs. Chen et al. tried to fabricate a fiber-shaped PSC in which a CNT fiber electrode was twisted with a Ti wire supported with anodized TiO_2 nanotubes where a composite of P3HT and $PC_{61}BM$ was coated as photoactive materials *(38)*. A relative low efficiency of 0.15% has been obtained, which

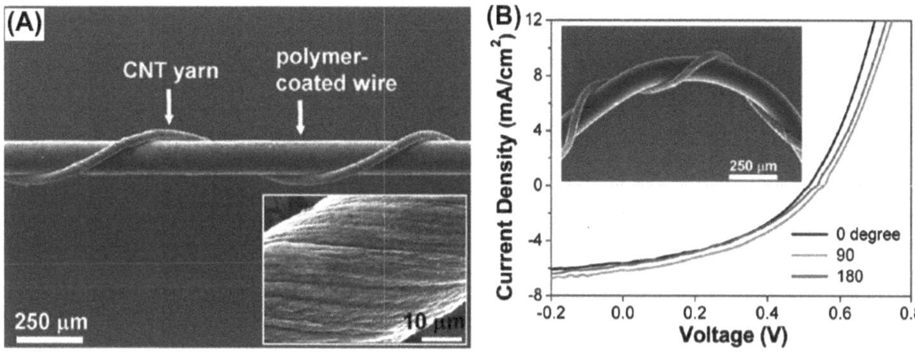

Figure 12.5 (A) Scanning electron microscopy (SEM) image of a fiber-shaped polymer solar cell (PSC) twisted by a carbon nanotube (CNT) fiber (the inset image) electrode. (B) The current density–voltage curves of a fiber-shaped PSC under different bending states. The inset SEM image showed a device being bent to 90 degrees. *(Reprinted with permission from Liu, D.; Zhao, M.; Li, Y.; Bian, Z.; Zhang, L.; Shang, Y.; Xia, X.; Zhang, S.; Yun, D.; Liu, Z.; Cao, A. ACS Nano* ***2012****, 6, 11027–11034. Copyright 2012 American Chemical Society.)*

can be ascribed to the ineffective infiltration of photoactive materials into TiO_2 nanotubes. To overcome this defect, Zhang et al. modified the interface between TiO_2 layer and photoactive layer, and an energy conversion efficiency of 1.78% can be achieved *(39)*. Liu et al. reported a fiber-shaped PSC using steel wire coated with a photoactive layer of P3HT and $PC_{61}BM$ as the primary electrode and CNT fiber as the counter electrode (Fig. 12.5A), in which a layer of ZnO was introduced to function as hole-blocking layer *(40)*. The resulting fiber-shaped PSC based on one single CNT fiber showed an energy conversion efficiency of 2.11%, comparable with that of the device using CNT film as the counter electrode. The performance of this fiber-shaped PSC without using liquid electrolyte changed very slightly under different bending states (Fig. 12.5B), and exhibited high flexibility. The challenge for fiber-shaped PSCs is their lower efficiency compared to planar counterparts, which should be optimized further in the future.

12.3.3 Fiber-Shaped Perovskite Solar Cells Based on CNT Fiber

A new class of solar cell based on mixed organic–inorganic halide perovskites has attracted great attention *(41)* because of its unique characteristics of ultrahigh energy conversion efficiency (close to 20%), low cost, ease of fabrication, and it is liquid electrolyte free. In perovskite solar cells, the most involved perovskite material was methylammonium lead triiodide ($CH_3NH_3PbI_3$), and mixed halides ($CH_3NH_3PbI_{3-x}Cl_x$ and $CH_3NH_3PbI_{3-x}Br_x$) have also been investigated widely. At the same time, highly flexible and stable perovskite solar cells have also been developed, whose energy conversion efficiency can approach as high as 12%, much higher than flexible PSCs *(42)*. Therefore it is reasonable to predict that high-performance fiber-shaped perovskite solar cells can be

rather lower than their counterparts with planar structure. The poor performance of fiber-shaped perovskite solar cells can be ascribed to the following factors: (1) the mismatch of bandgap between the fiber electrode and photoactive perovskite materials and (2) the ineffective contact at interfaces, such as electron transport layer/perovskite material, perovskite materials/hole transport layer. Therefore it is promising to obtained high-performance fiber-shaped perovskite solar cells if these problems are resolved.

12.3.4 Stretchable Fiber-Shaped Solar Cells Based on CNT Composite Fiber

Stretchable electronics can maintain their properties even though some unpredictable tensile or compressive force has been loaded. *(46,47)*. To this end, various stretchable electronics (such as transistors *(48)*, solar cells *(49)*, LEDs *(50)*, supercapacitors *(51)*, and so on) have been developed by preparing these devices on a stretchable substrate (e.g., polydimethylesiloxane, PDMS). Previous studies showed that the stretchability of electronics greatly depended on the combined properties of the substrate used. However, it was a great challenge to develop fiber-shaped devices because conventional fiber substrates either have high electrical conductivity with poor stretchability (e.g., metal fiber) or have high stretchability with low conductivity (e.g., polymeric elastic fiber). Therefore a fiber substrate with highly electrical conductivity and stretchable property is crucial for the development of high-performance fiber-shaped electronics.

Yang et al. developed a fiber-shaped DSSC (Fig. 12.7A–C) in which a CNT sheet-wrapped elastic composite fiber was used as the cathode and a spring-like Ti wire with perpendicularly grown aligned TiO_2 nanotubes was used as the anode electrode *(52)*. The elastic CNT composite fiber was inserted into the working electrode, transferred into a stretchable and transparent polyethylene tube, and injected with redox electrolyte to produce the stretchable fiber-shaped DSSC. The high stretchability of the elastic CNT composite fiber and spring-like Ti wire enabled the obtained fiber-shaped DSSC with high stretchability (Fig. 12.7D–G). The performance of the fiber-shaped DSSC did not changed obviously when the device was stretched by 30% strain, and maintained 90% after stretching for 20 cycles (Fig. 12.7H). Based on a similar idea, a stretchable fiber-shaped PSC has also been developed, which also showed high stretchability during stretching from 0% to 30% strain *(53)*.

12.3.5 Wearable Fiber-Shaped Polymer Light-Emitting Diode Based on CNTs

Like the fiber-shaped solar cells mentioned earlier, fiber-shaped polymer light-emitting diodes (PLEDs) with a coaxial structure have also been realized by using all-solution-based processes *(54,55)*. In a typical fiber-shaped PLED, a modified steel wire was used as the core cathode, a layer of CNT sheet was used as the outer anode, and a layer of electroluminescent polymer was sandwiched in between (Fig. 12.8A) *(54)*. Unlike conventional planar PLEDs, the fiber-shaped structure enabled fiber device luminance

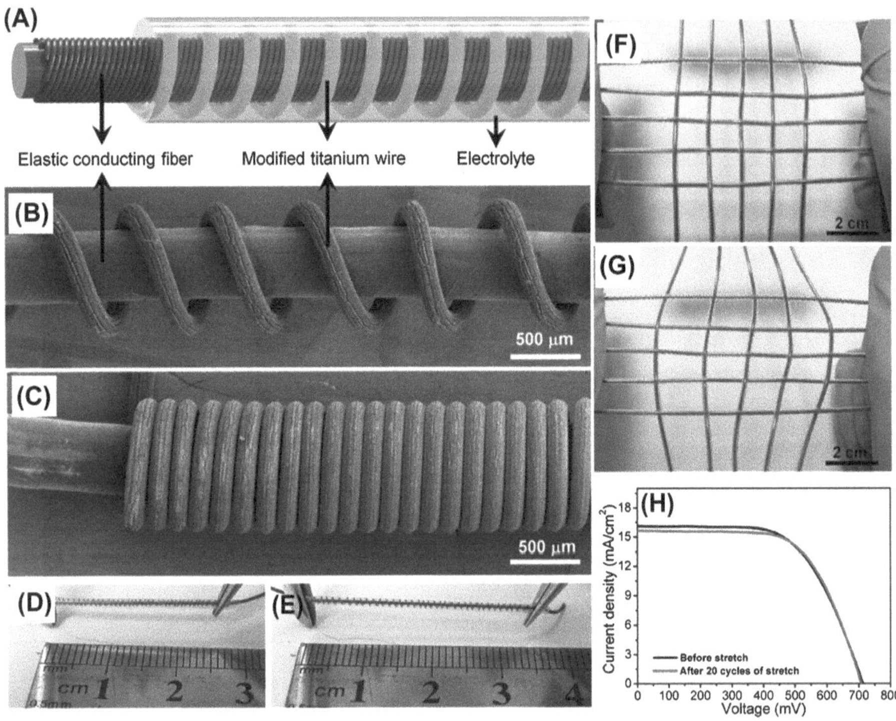

Figure 12.7 (A) Schematic illustration of a stretchable dye-sensitized solar cells (DSSC). (B and C) Scanning electron microscopy (SEM) images of stretchable DSSCs with pitch distances of 560 and 164 μm, respectively. (D and E) Digital photographs of a fiber-shaped DSSC without and with stretching by 30%. (F and G) Digital photographs of a photovoltaic textile without and with stretching. (H) Current density–voltage curves of a fiber-shaped DSSC before and after stretching for 20 cycles. *(Reproduced from Yang, Z.; Deng, J.; Sun, X.; Li, H.; Peng, H. Advanced Materials* ***2014****, 26, 2643–2647 with permission. Copyright 2014 John Wiley and Sons.)*

from all of the outer surface, and the brightness did not depend on the observation angles (Fig. 12.8B). Because of its unique structure and ease of fabrication, it was convenient to integrate a series of devices with different emitting colors into one single-fiber substrate by changing the electroluminescent polymer materials during preparation processes. As a simple model, two PLEDs with blue and yellow luminance were fabricated on one fiber electrode; selected lighting can be easily realized according to different application requirements (Fig. 12.8C–E). The fiber-shaped PLEDs maintained over 90% of their maximum brightness after 100 cycles of bending with a radius of curvature of 6 mm, indicating high flexibility. Like other fiber-shaped electronics, the fiber-shaped PLEDs can be easily woven into other textiles (Fig. 12.8F), and PLEDs with different colors can be selectively lit to produce the required configurations (Fig. 12.8G–I).

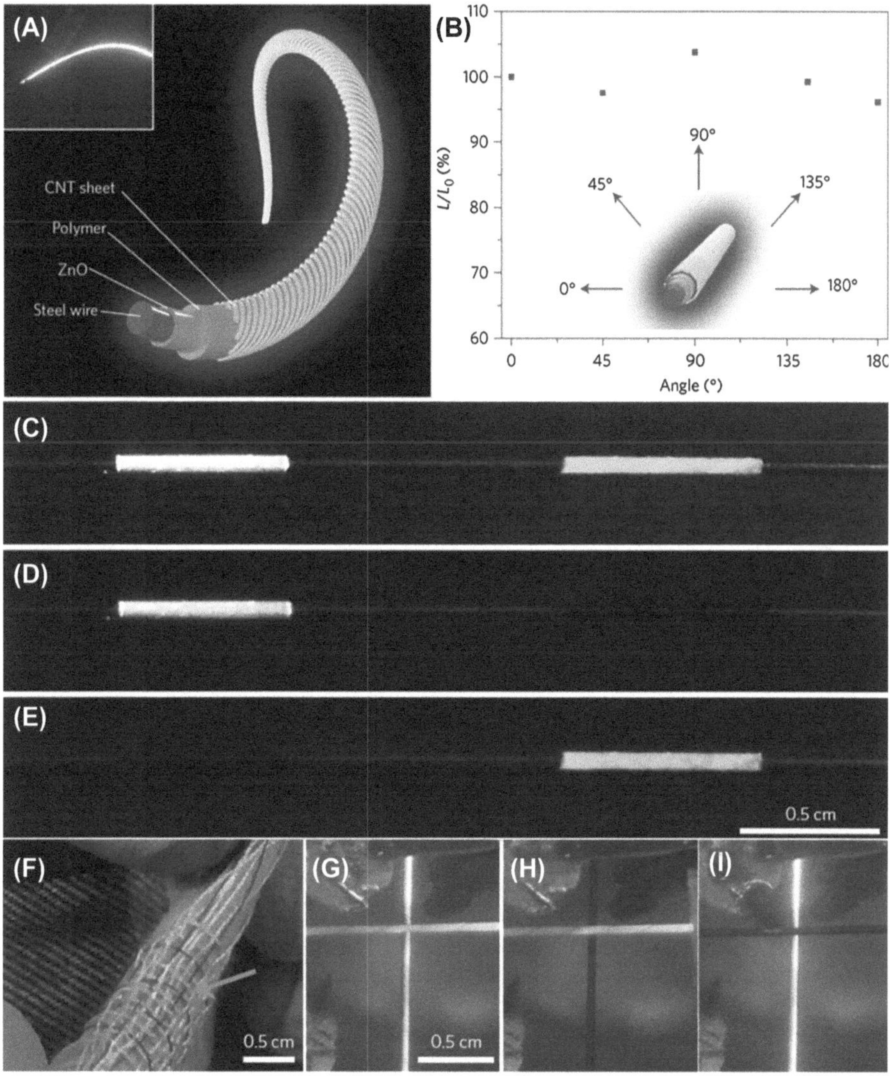

Figure 12.8 (A) Schematic illustration of a coaxial fiber-shaped polymer light-emitting diode (PLED). Inset image is a digital photograph of a device biased at 10 V. (B) Dependence of luminance on the angle of a fiber-shaped PLED. L_0 and L correspond to luminance measured at 0 degrees and the other angle, respectively. (C—E) Fiber-shaped PLED displaying blue (white in print versions) and/or yellow (light gray in print versions) light when biased at 9 V. (F) A digital photograph of a fiber-shaped PLED woven into a textile. (G and H) Two fiber-shaped PLEDs with different colors being selectively illuminated biased at 10 V. *(Reprinted from Zhang, Z.; Guo, K.; Li, Y.; Li, X.; Guan, G.; Li, H.; Luo, Y.; Zhao, F.; Zhang, Q.; Wei, B. Nature Photonics **2015**, 9, 233—238 with permission. Copyright 2014 Nature Publishing Group.)*

12.4 CARBON NANOTUBE FIBER FOR WEARABLE ENERGY STORAGE DEVICES

Energy storage devices, such as supercapacitors and batteries, are also required in various areas ranging from hybrid vehicles, even all-electric vehicles, to consumer electronics. The energy storage devices can overcome intermittent and geographical limitations of energy conversion systems *(56)*. Great progress on energy storage systems has been achieved in the last 20 years, especially based on the electrodes of new emerging carbon nanomaterials (e.g., graphene *(57)* and CNT *(58)*). Meanwhile, based on CNT fiber with high electrical conductivity and flexibility, wearable fiber-shaped energy storage devices containing supercapacitors and batteries have also been widely investigated *(59)*.

12.4.1 Wearable Supercapacitors Based on CNT Fiber

Generally, wearable supercapacitors are fabricated based on a one-fiber-electrode core–shell structure or two-fiber-electrode twisted structure using a polymer gel as both electrolyte and separator, typically aqueous solution of phosphoric acid and PVA. In 2012, Ren et al. reported a wearable two-electrode-based fiber-shaped supercapacitor by arranging two CNT fibers in parallel without a separator or twisting two CNT fibers coated with a separator of PVDF *(60)*. The resulting solid-state supercapacitor exhibited a specific capacitance of 3.53 mF/cm^2 based on bare CNT fiber electrodes, which was rather lower than that of supercapacitors with planar format. To enhance the performance of the fiber-shaped supercapacitors, ordered mesoporous carbon (OMC) with high specific surface area was introduced into the CNT fiber, resulting in a CNT/OMC composite fiber (Fig. 12.9A and B) *(61)*. Using CNT fibers modified with OMC as electrodes, the fiber-shaped supercapacitors (Fig. 12.9C) showed a specific capacitance of 39.7 mF/cm^2, responding to a length specific capacitance of 1.9 mF/cm (Fig. 12.9D). The obtained fiber-shaped supercapacitors exhibited long cycling stability and good flexibility, and can be easily woven into other textiles.

Apart from two-electrode twisted devices, fiber-shaped supercapacitors with a core–shell structure (Fig. 12.9E) have also been developed based on one single-fiber substrate by Chen et al. *(62)* An as-spun CNT fiber was directly used as the core electrode, after gel electrolyte was infiltrated into the fiber and coated on its surface; a CNT sheet was then wrapped outside of the fiber, resulting in the core–shell structural supercapacitor. The fiber device was dipped into the electrolyte solution again to improve the contact between the electrolyte and the top CNT sheet electrode. From the scanning electron microscopy (SEM) images (Fig. 12.9F–H), the core and shell CNT electrodes can be observed clearly. The core–shell fiber-shaped supercapacitors exhibited a specific capacitance of 59 F/g, and had a good cycling retention (Fig. 12.9I). In this core–shell fiber-shaped supercapacitor, two electrodes were soundly connected by the gel electrolyte, so that there was no risk that two fiber electrodes were separated

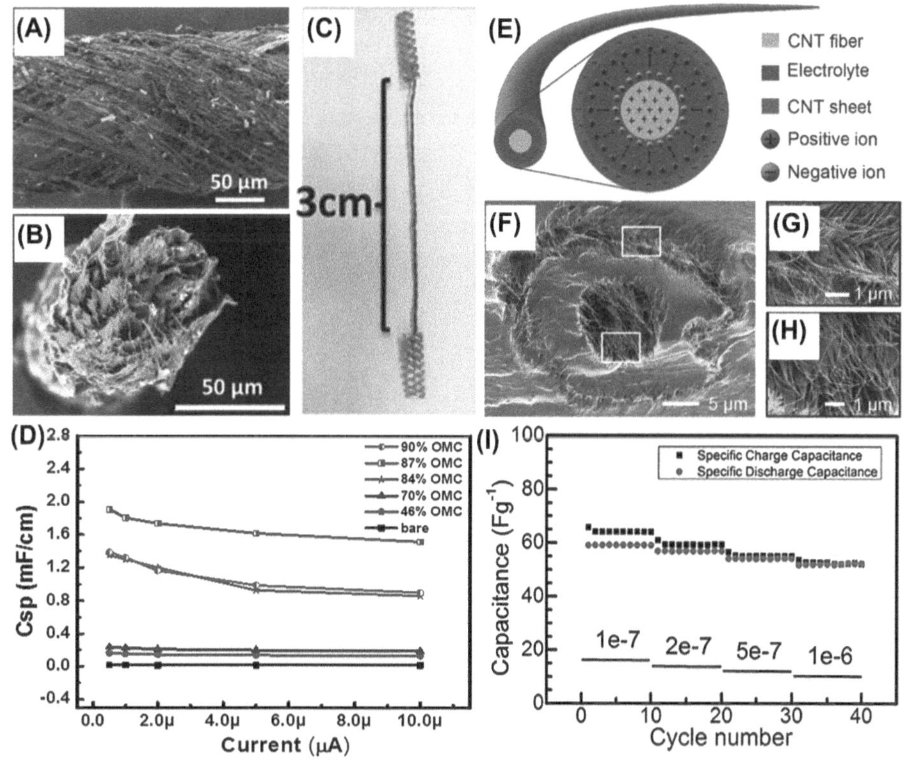

Figure 12.9 (A and B) Scanning electron microscopy (SEM) image of carbon nanotube/mesoporous carbon (CNT/OMC) composite fiber with an OMC weight percentage of 87%. (C) A digital photograph of a typical fiber-shaped supercapacitor based on CNT/OMC composite fiber. (D) Dependence on current of specific capacitance (Csp) of the fiber-shaped devices. *(Reproduced from Ren, J.; Bai, W.; Guan, G.; Zhang, Y.; Peng, H. Advanced Materials* ***2013****, 25, 5965–5970 with permission. Copyright 2013 John Wiley and Sons.)* (E) Schematic diagram of the cross-sectional structure and mechanism of the coaxial fiber-shaped supercapacitor. (F–H) Cross-sectional SEM image of a coaxial device. (I) Cycling performance of a device as the electric current increased. *(Reproduced from Chen, X.; Qiu, L.; Ren, J.; Guan, G.; Lin, H.; Zhang, Z.; Chen, P.; Wang, Y.; Peng, H. Advanced Materials* ***2013****, 25, 6436–6441 with permission. Copyright 2014 John Wiley and Sons.)*

when the device was bent. As predicted, the core–shell fiber-shaped supercapacitor showed good flexibility and stability, which almost had no obvious damage during bending with large curvature.

Conductive polymers and metal oxides that have excellent pseudocapacitance have been widely used to improve specific capacitance and energy density. However, both conductive polymers and metal oxides often suffer from their limited stability during cycling when they are used as bulk materials in supercapacitors. Therefore pseudocapacitive materials were often composited with carbon nanomaterials used to achieve supercapacitors with high specific capacitance, energy density, and long cycling stability.

To this end, Wang et al. deposited a polyaniline (PANI) nanowire array outside of a CNT fiber by an in situ polymerization process, resulting in a CNT/PANI composite fiber (Fig. 12.10A–C) *(63)*. Based on the CNT/PANI composite fiber, a wearable fiber-shaped supercapacitor with a twisted structure has been developed, which shows an area specific capacitance of 38 mF/cm^2 (Fig. 12.10D), comparable to that of devices based on CNT/OMC composite fibers *(61)*. This fiber-shaped supercapacitor had a good retention after 150 bending cycles, indicating high flexibility, so that it can be easily woven into a textile (Fig. 12.10E). Choi et al. reported a flexible fiber-shaped

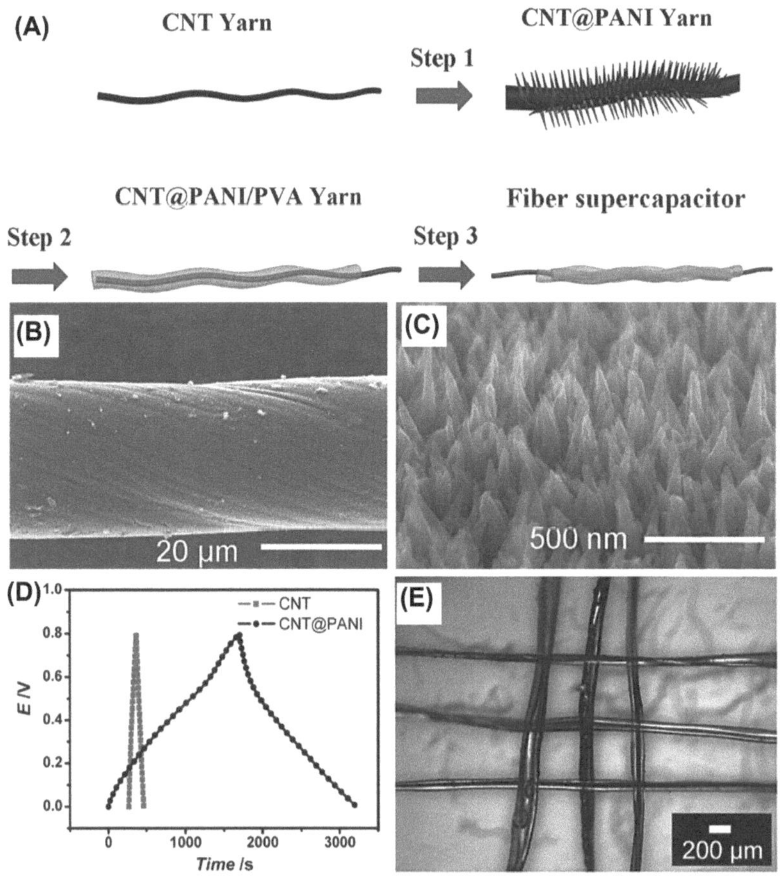

Figure 12.10 (A) Schematic illustration of preparation processes for the fiber-shaped supercapacitors based on carbon nanotube/polyaniline (CNT/PANI) composite fibers. (B and C) Scanning electron microscopy (SEM) images of CNT/PANI nanowire composite fiber. (D) Galvanostatic charge/discharge curves of fiber-shaped supercapacitors using bare CNT and CNT/PANI composite fibers. (E) A digital photograph of a model textile by weaving six fiber-shaped supercapacitors. *(Reproduced from Wang, K.; Meng, Q.; Zhang, Y.; Wei, Z.; Miao, M. Advanced Materials* ***2013****, 25, 1494–1498 with permission. Copyright 2014 John Wiley and Sons.)*

supercapacitor based on CNT/MnO_2 composite fiber, which was prepared by electrochemically depositing MnO_2 on a bare CNT fiber *(64)*. The obtained fiber-shaped supercapacitor showed a specific capacitance of 3.7 mF/cm^2.

An efficient approach to preparing CNT composite fiber with high void volume guest materials loading can be realized by a biscrolled spinning technique *(65)*. Lee et al. first deposited conductive poly(3,4-ethylenedioxythiophene) (PEDOT) on a CNT sheet by a vapor phase polymerization method to prepare a polymer-coated nanomembrane, followed by twisting the nanomembrane into a CNT composite fiber (Fig. 12.11A and B) *(65)*. The loading of PEDOT in the composite fiber can reach 75%. Based on the CNT/PEDOT composite fiber, supercapacitors based on both liquid and solid electrolytes have been developed. The results showed that the specific capacitance of the twisted fiber-shaped devices using gel electrolyte (Fig. 12.11C) was about 145 mF/cm *(2)* at the scanning rate of 1.0 V/s, which was comparable to that

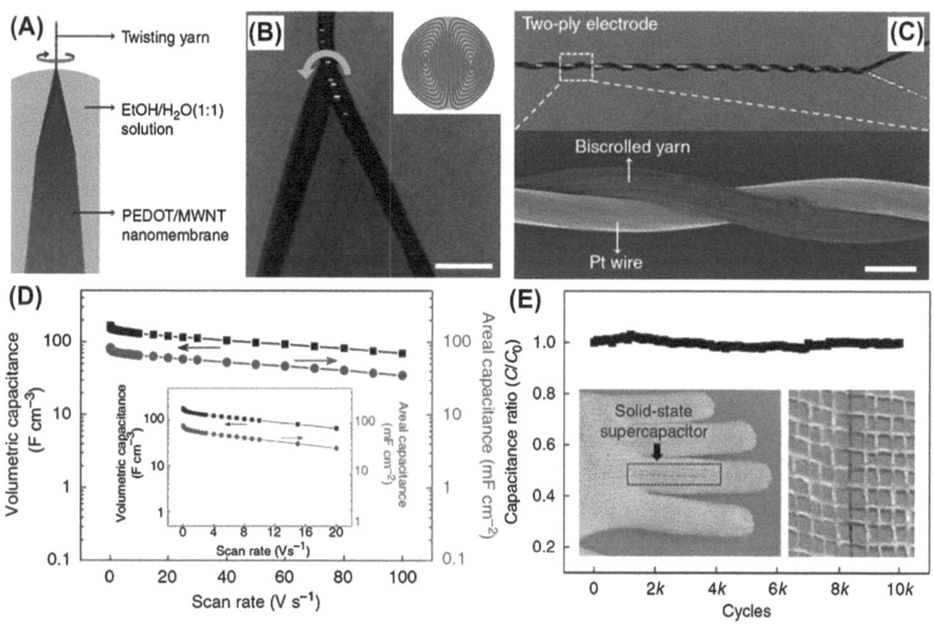

Figure 12.11 (A) Schematic illustration to fabricate a biscrolled carbon nanotube/poly(3,4-ethylenedioxythiophene (CNT/PEDOT) composite fiber. (B) Optical microscope image of the spinning wedge. The inset schematically showed the wedge edges being twisted to form a dual-Archimedean scroll fiber. Scale bar, 50 μm. (C) Two scanning electron microscopy (SEM) images of a CNT/PEDOT biscrolled fiber twisted with a platinum wire with 25 μm diameter. Scale bar, 40 μm. (D) Dependence of volumetric and areal capacitances on the scan rate for the twisted fiber-shaped supercapacitor based on liquid electrolyte and solid electrolyte (inset). (E) Cycling stability of a fiber-shaped supercapacitor woven into a glove. *(Reprinted from Lee, J. A.; Shin, M. K.; Kim, S. H.; Cho, H. U.; Spinks, G. M.; Wallace, G. G.; Lima, M. D.; Lepró, X.; Kozlov, M. E.; Baughman, R. H. Nature Communications* ***2013****, 4, 1970–1977 with permission. Copyright 2013 Nature Publishing Group.)*

(147 mF/cm^2) of devices based on liquid electrolyte (Fig. 12.11D), and much higher than that of previously mentioned counterparts *(61–63)*. The developed fiber-shaped supercapacitors were stable in ambient atmosphere and showed good retention after 10,000 cycles even when they were bent or woven into a glove. The power density and energy density of this fiber-shaped supercapacitor can reach 40 W/cm^3 and 1.4 mWh/cm^3, respectively. The extraordinary combination of high power density and energy density for these devices can be explained by the continuous pathway provided by the scroll structure, which can enable both ions and electrons to transport from surface to center of the fiber.

Because of its excellent electrical and mechanical properties and large surface area, graphene has showed great potential for use as an efficient electrode for high-performance energy storage devices *(57)*. Graphene fiber can be easily fabricated by the wet-spinning process, which has been widely used as an electrode to achieve high-performance fiber-shaped supercapacitors *(66,67)*. Therefore it may be an efficient way to enhance the property of fiber-shaped supercapacitors based on CNT fiber by the introduction of graphene materials. Yu et al. developed a continuous approach to prepare a large-scale CNT/graphene composite fiber (Fig. 12.12A and B) *(68)*. The obtained CNT/graphene composite fiber exhibited excellent electrical conductivity, high packing density, and a large ion-accessible surface area (Fig. 12.12C–E). Based on the resultant CNT/graphene composite fibers, fiber-shaped supercapacitors using solid electrolyte showed an area specific capacitance of 116.3 mF/cm^2 and a volumetric energy density of 6.3 mWh/cm^3 (Fig. 12.12F), respectively. The capacitance of the fiber-shaped supercapacitor retained 93% of its initial state after 10,000 charge/discharge cycles, and showed high stability with long cycling life. The fiber-shaped supercapacitor maintained more than 97% of its initial capacitance after bending 1000 cycles at 90 degrees (Fig. 12.12G), indicating high flexibility.

12.4.2 Multifunctional Wearable Supercapacitors Based on CNT Fibers

Multifunctional wearable supercapacitors that can realize other functions are not only being used as energy sources but have also attracted considerable attention. These devices can be achieved easily by introducing other functional substrates or materials. For instance, stretchable, self-healing, and shape-memory wearable supercapacitors have been developed by using stretchable, self-healing, and shape-memory fiber substrates, respectively *(69–71)*. In addition, electrochromic wearable supercapacitors have also been obtained by introducing conductive polymers *(72)*.

Xu et al. fabricated a two-electrode twisted fiber-shaped supercapacitor, then wrapped the fiber devices on a prestrained stretchable fiber substrate (Spandex), followed by releasing Spandex fiber, resulting in buckled fiber-shaped devices *(73)*. The obtained fiber-shaped supercapacitor showed an area specific capacitance of 4.28 mF/cm^2.

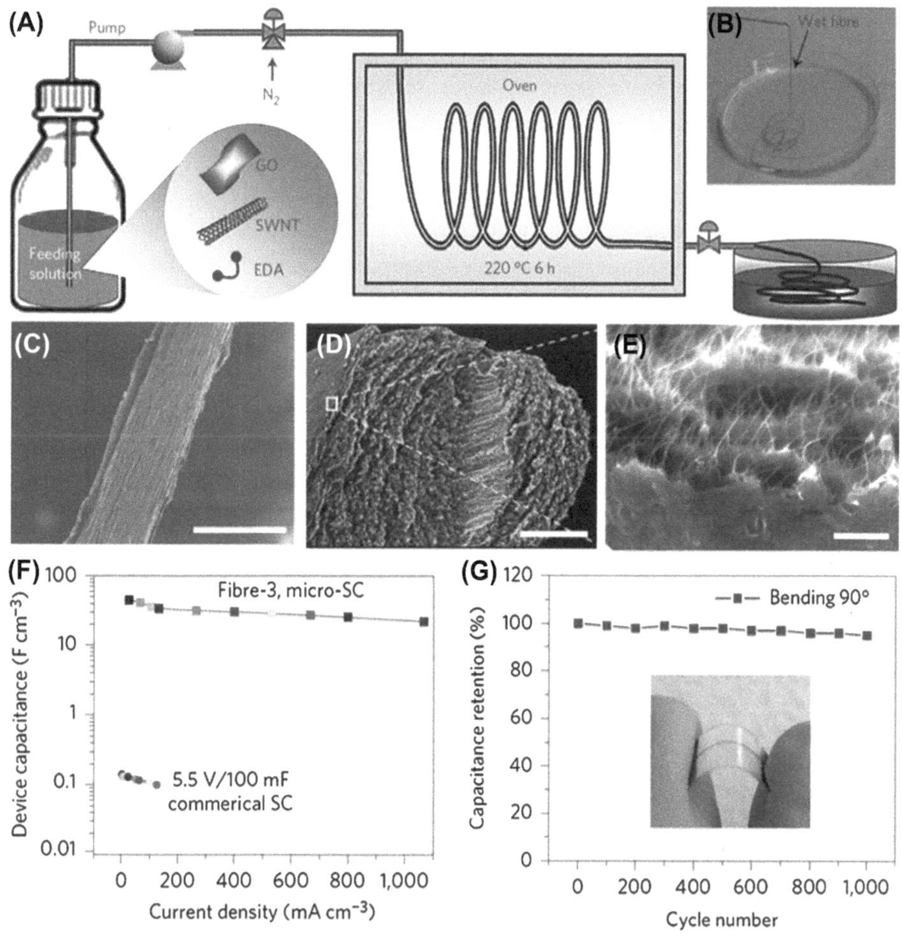

Figure 12.12 (A) Schematic illustration of continuously preparing a carbon nanotube (CNT)/graphene composite fiber. (B) A digital photograph of the as-prepared fiber collected in water. (C) Scanning electron microscopy (SEM) image of an obtained CNT/graphene composite fiber from top view. (D and E) The cross-sectional SEM images of a fiber, the square area in D is highlighted as E. (F) Volumetric capacitance as a function of current densities. (G) Capacitance retention after bending at 90 degrees for 1000 cycles. Inset: Photograph of a bent supercapacitor. *EDA*, ethylene diamine; *GO*, graphene oxide; *SWNT*, single-walled carbon nanotube. *(Reprinted from Yu, D.; Goh, K.; Wang, H.; Wei, L.; Jiang, W.; Zhang, Q.; Dai, L.; Chen, Y. Nature Nanotechnology* ***2014****, 9, 555–562 with permission. Copyright 2014 Nature Publishing Group.)*

The electrochemical performance of this fiber-shaped supercapacitor did not decrease but improved slightly when a tensile strain up to 100% or after 20 stretching–releasing cycles was loaded, which can be attributed to the increased pressure between CNT fiber electrodes at kink sites of buckled fiber-shaped supercapacitor, which induced better wettability of the CNT fiber electrode *(73)*. A singe fiber substrate-based core–shell fiber-shaped supercapacitor has been developed by Yang et al., in which a layer of CNT sheet, a layer of gel electrolyte, and a top CNT sheet electrode were fabricated outside of an elastic rubber fiber by a layer-by-layer technique *(69)*. From the SEM image (Fig. 12.13A–D), two layers of CNT sheet with gel electrolyte in between can be clearly observed. The specific capacitance of 19.2 F/g was obtained based on bare CNT sheet 330 nm thick, which increased to 41.4 F/g when composite layers of CNT sheet and OMC were used as electrodes. It can be seen that capacitance/voltage (Fig. 12.13E) curves of the fiber-shaped supercapacitors changed slightly as the tensile strain increased from 0% to 75%, indicated high stretchability. The stretchability of this core–shell fiber-shaped supercapacitor can be improved to 400% using a superelastic rubber fiber as substrate *(74)*.

To make a wrinkled (or wavy, or buckled) structure has been proved as an efficient approach to obtain highly stretchable electrodes *(47)*, which can be easily realized by transferring or directly coating conducting materials onto a prestrained substrate,

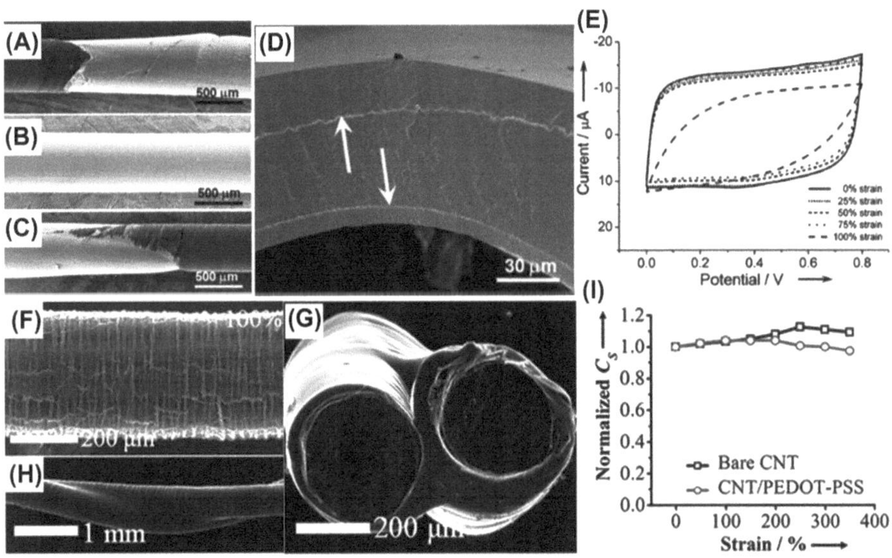

Figure 12.13 (A) Scanning electron microscopy (SEM) image of the inner carbon nanotube (CNT) sheet electrode. (B and C) SEM images of the middle and right parts of a coaxial fiber-shaped supercapacitor. (D) Cross-sectional SEM image of a fiber device. (E) Capacitance/voltage curves of a fiber device recorded at different strains. *(Reproduced from Yang, Z.; Deng, J.; Chen, X.; Ren, J.; Peng, H. Angewandte Chemie International Edition* ***2013****, 52,* 13453–13457 *with permission. Copyright 2013 John Wiley and Sons.)* (F) SEM image of wrinkled CNTs on an elastic fiber with 100% prestrain during preparation. (G and H) SEM images of a twisted structural fiber-shaped supercapacitor from cross-section and top view. (I) Dependence of normalized capacitance on different tensile strain. *(Reproduced from Chen, T.; Hao, R.; Peng, H.; Dai, L. Angewandte Chemie International Edition* ***2015****, 54, 618–622 with permission. Copyright 2015 John Wiley and Sons.) MWCNT,* multiwalled carbon nanotube.

followed by releasing the substrate. To this end, Chen et al. fabricated a CNT composite fiber with a wrinkled surface (Fig. 12.13F) by wrapping CNT sheets onto a prestrained elastic fiber, then releasing the tensile strain *(75)*. The electrical resistance of the CNT composite fiber with the wrinkled surface only increased by 30% when a tensile strain of 300% was applied, indicating high stretchability. Based on this stretchable fiber electrode, two-electrode twisted fiber-shaped supercapacitors (Fig. 12.13G and H) have been developed, which can be stretched to 370% strain without damage (Fig. 12.13I).

A self-healing fiber-shaped supercapacitor has been developed by using self-healing conductive CNT composite fibers fabricated by wrapping CNTs and silver nanowires around self-healing polymer fibers *(70)*. The obtained fiber-shaped supercapacitor exhibited a high specific capacitance of 140 F/g (or 1.34 mF/cm), which can recover to 92% after self-healing a broken fiber-shaped device, indicating high self-healing property. Fig. 12.14A–C showed three fiber-shaped supercapacitors connected in series to power an LED before breaking and after healing. Similarly, shape-memory fiber-shaped supercapacitors can be achieved by using shape-memory CNTs composite fiber substrates *(71)*. As Fig. 12.14D shows, the fiber-shaped supercapacitor deformed and shaped when the temperature was higher than the thermal transition temperature of shape-memory polymer (T_{trans}), followed by cooling to room temperature to induce a physical cross-linking in the polymer. The deformed and shaped supercapacitors recovered as the temperature increased beyond T_{trans}. Fig. 12.14E shows that capacitance/voltage curves of a shape-memory fiber-shaped supercapacitor were almost unchanged before and after deformation, indicating good shape-memory property. Because a conductive polymer of PANI was introduced into the CNT composite electrode, an electrochromic fiber-shaped supercapacitor can be achieved, which can be attributed to the redox reaction of PANI occurring during the charge/discharge process *(72)*. Color changes (Fig. 12.14F) can be obviously observed as the fiber-shaped supercapacitor was charged with different states. Interestingly, different electrochromic patterns can be realized during the charge/discharge process by weaving the fiber-shaped supercapacitors into textiles (Fig. 12.14G and H).

12.4.3 Wearable Fiber-Shaped Batteries Based on CNT Fiber

As well as fiber-shaped supercapacitors, wearable electrochemical ion batteries (such as Li ion, Li sulfur) have also been attracting intense interest because of their higher energy density compared to supercapacitors *(4)*. Ren et al. demonstrated a fiber-shaped Li ion battery by twisting a positive electrode of aligned CNT/MnO_2 composite and a negative electrode of Li wire with PVDF as the separator *(60)*. A solution of $LiPF_6$ in a mixture solvent of ethylene carbonate, diethyl carbonate, and dimethyl carbonate (weight ratios of 1/1/1) with a concentration of 1 M was used as electrolyte. For the fiber-shaped Li ion battery based on CNT/MnO_2 composite fiber, a specific capacity

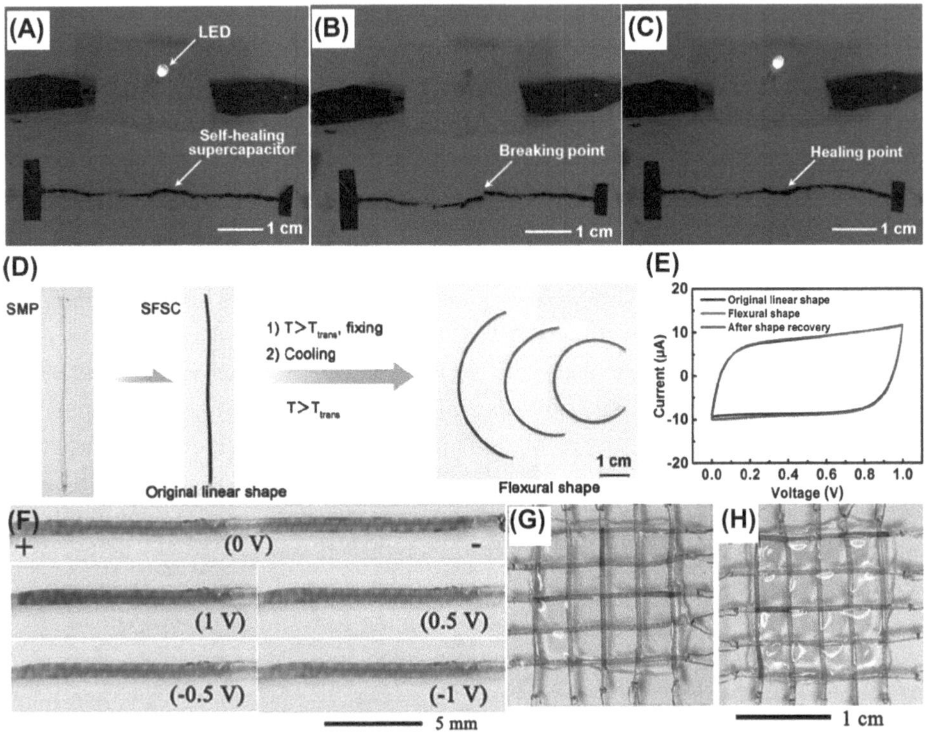

Figure 12.14 (A–C) Digital photographs of three fiber-shaped self-healing supercapacitors connected in series to power a light-emitting diode (LED) before breaking (A), after breaking (B), and after healing (C). *(Reproduced from Sun, H.; You, X.; Jiang, Y.; Guan, G.; Fang, X.; Deng, J.; Chen, P.; Luo, Y.; Peng, H. Angewandte Chemie International Edition* ***2014****, 53, 9526–9531 with permission. Copyright 2014 John Wiley and Sons.)* (D) Photographs of a fiber-shaped shape-memory supercapacitor at different states. *SFSC*, fiber-shaped supercapacitor; *SMP*, shape-memory polyurethane. (E) Capacitance/voltage curves of a fiber device before and after deformation and shape recovery. *(Reproduced from Deng, J.; Zhang, Y.; Zhao, Y.; Chen, P.; Cheng, X.; Peng, H. Angewandte Chemie International Edition* ***2015****.* http://dx.doi.org/10.1002/anie.201508293 *with permission. Copyright 2015 John Wiley and Sons.)* (F) Digital photographs of a fiber-shaped electrochromic supercapacitor with different colors at different potentials. (G and H) Photographs of textiles woven by using fiber-shaped electrochromic supercapacitors showing different patterns. *(Reproduced from Chen, X.; Lin, H.; Deng, J.; Zhang, Y.; Sun, X.; Chen, P.; Fang, X.; Zhang, Z.; Guan, G.; Peng, H. Advanced Materials* ***2014****, 26, 8126–8132 with permission. Copyright 2014 John Wiley and Sons.)*

of 218.32 mAh/g (or 109.62 mAh/cm^3) can be achieved at a current of 5×10^{-4} mA; the energy density and power density were 35.74 mWh/cm^3 and 2.43 W/cm^3, respectively. To improve performance of the fiber-shaped Li ion battery, Lin et al. first coated Si onto a CNT sheet, followed by twisting the Si-coated CNT sheet into a CNT/Si composite fiber for use as anode electrodes in the fiber-shaped batteries (Fig. 12.15A) *(76)*. A high specific capacity of 1670 mAh/g can be obtained because the weight percentage of Si was 38.1% in the CNT/Si composite fiber (Fig. 12.15B). The specific capacity of the half-cell only changed 6% before and after

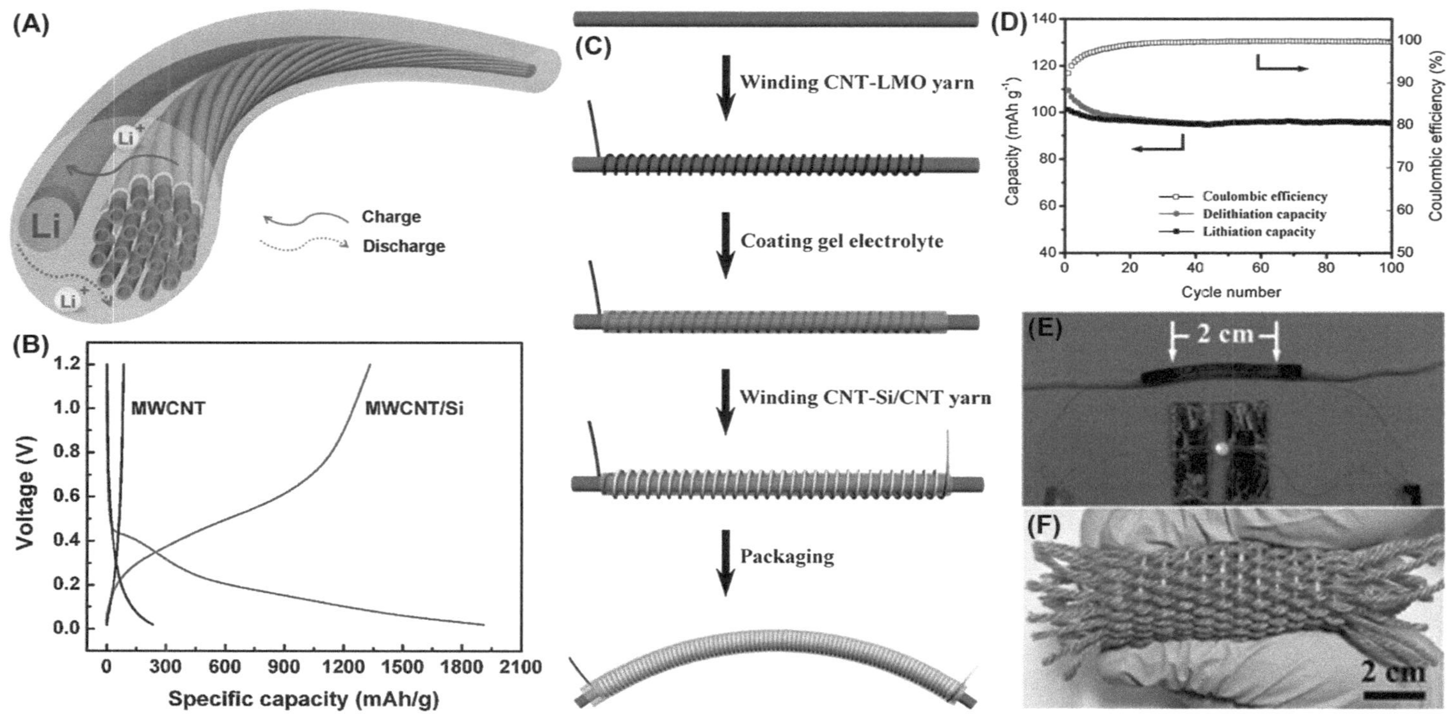

Figure 12.15 (A) Schematic illustration of a half Li ion battery using a carbon nanotube (CNT)/Si composite fiber as the working electrode and a Li wire as both counter and reference electrodes. (B) Charge/discharge curves of a half Li ion battery based on bare CNT fiber and CNT/Si composite fiber with a weight percentage of 38.1% at 1 A/g. *(Reproduced from Lin, H.; Weng, W.; Ren, J.; Qiu, L.; Zhang, Z.; Chen, P.; Chen, X.; Deng, J.; Wang,Y.; Peng, H. Advanced Materials* ***2014****, 26, 1217–1222 with permission. Copyright 2014 John Wiley and Sons.)* (C) Schematic illustration to prepare the coaxial fiber-shaped full Li ion battery. (D) Cycling performance of a fiber-shaped Li ion battery between 2.0 and 4.3 V at 1°C. (E) Digital photograph of a fiber-shaped battery to light an LED. (F) Fiber-shaped full Li ion battery being woven into a textile. *(Reprinted with permission from Weng, W.; Sun, Q.; Zhang, Y.; Lin, H.; Ren, J.; Lu, X.; Wang, M.; Peng, H. Nano Letters* ***2014****, 14, 3432–3438. Copyright 2014 American Chemical Society.)*

bending for 100 cycles, indicating good flexibility, which can be attributed to the unique twisted and aligned structure of the CNT composite fiber.

The aforementioned fiber-shaped Li ion batteries were based on liquid electrolyte, which had problems of leakage of electrolyte and complex packaging process. To overcome these problems, Weng et al. designed a core—shell full fiber-shaped Li ion battery by winding two CNT composite fibers onto a cotton fiber with a gel electrolyte as separator at the same time (Fig. 12.15C) *(77)*. In this full fiber-shaped Li ion battery, a CNT/Si composite was used as anode electrode and a CNT/$LiMn_2O_4$ (CNT/LMO) was used as the cathode electrode. For the full fiber-shaped Li ion battery, a specific discharge capacity of 106.5 mAh/g can be achieved at the first cycle based on the cathode (Fig. 12.15D), the linear capacity density and gravimetric energy density were 0.22 mAh/cm and 27.7 mWh/g, respectively. One single-fiber device can successfully light an LED (Fig. 12.15E). Like other fiber-shaped devices, this fiber-shaped Li ion battery can also be easily woven into a flexible textile (Fig. 12.15F) for more applications.

Highly stretchable fiber-shaped Li ion batteries have also been realized by fabricating the Li ion battery on stretchable substrate and gel electrolyte *(78,79)*. Ren and Zhang et al. developed a stretchable fiber-shaped Li ion battery by simultaneously twisting two composite fibers onto a heat-shrinkable tube, where a CNT/$Li_4Ti_5O_{12}$ (LTO) composite fiber and a CNT/LMO composite fiber were used as anode and cathode (Fig. 12.16A), respectively *(79)*. The resulting full fiber-shaped Li ion battery showed high stretchability, the specific capacity of which remained at 90% as it was stretched by 100% and maintained over 80% after 200 cycles of stretching by 100% (Fig. 12.16B—D). When the Li ion battery was fabricated on a PDMS fiber substrate, a superstretchy fiber-shaped Li ion battery with 600% strain can be obtained *(78)*.

12.5 CARBON NANOTUBE FIBER FOR WEARABLE INTEGRATED DEVICE OF ENERGY CONVERSION AND STORAGE

Integrated energy devices that can simultaneously store the electric energy generated by an energy conversion device (e.g., solar cell) into an energy storage device (e.g., supercapacitor and battery) have attracted great interest *(80,81)*, and are showing promise for use as self-powering systems for portable electronic equipment. Compared with planar structure, fiber-shaped integrated energy devices are lighter and smaller, and can be easily woven into other portable textiles by conventional techniques *(82,83)*. As demonstrated earlier, CNT fiber can be used as an efficient electrode for high-performance solar cells, supercapacitors, and batteries, which shows great potential for use as a fiber electrode for the integrated devices of energy conversion and storage. To this end, Chen et al. have developed a wearable fiber-shaped integrated energy device by fabricating a DSSC and a supercapacitor on a Ti fiber grown with a TiO_2 nanotube array *(84)*. In this

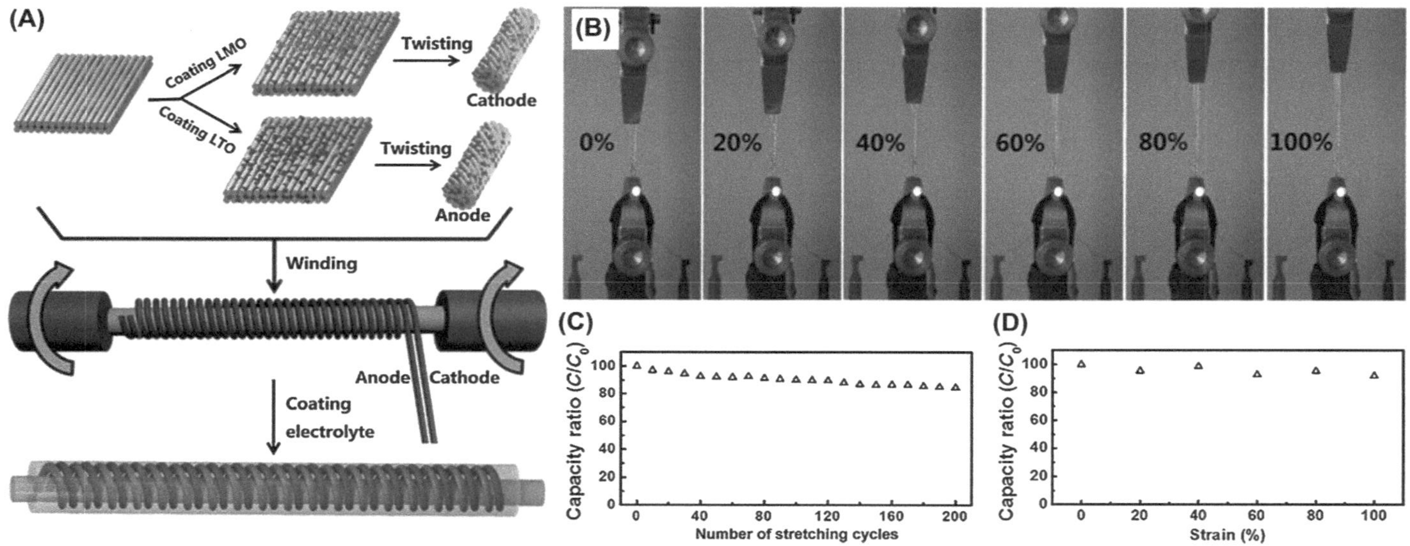

Figure 12.16 (A) Schematic illustration to prepare stretchable Li ion battery by using carbon nanotube (CNT)/LMO and CNT/LTO composite electrodes, where LMO and LTO represent $LiMn_2O_4$ and $Li_4Ti_5O_{12}$, respectively. *(Reproduced from Ren, J.; Zhang, Y.; Bai, W.; Chen, X.; Zhang, Z.; Fang, X.; Weng, W.; Wang, Y.; Peng, H. Angewandte Chemie* ***2014****, 126, 7998–8003. Copyright 2014 The Royal Society of Chemistry.)* (B) Digital photographs of a fiber-shaped Li ion battery being stretched from 0% to 100%. (C) Normalized capacity as a function of stretching cycles with 100% strain. (D) Dependence of normalized capacity on the tensile strains. *(Reproduced from Zhang, Y.; Bai, W.; Ren, J.; Weng, W.; Lin, H.; Zhang, Z.; Peng, H. Journal of Materials Chemistry A* ***2014****, 2, 11054–11059 with permission. Copyright 2014 John Wiley and Sons.)*

fiber-shaped integrated device, both DSSC and supercapacitor can work individually, which shows an energy conversion efficiency of 2.2% and a specific capacitance of 0.024 mF/cm, respectively. Significantly, photogenerated electric energy can be simultaneously stored in the supercapacitor part when the DSSC part is under the light and two parts are connected (Fig. 12.17A and B). The voltage of the supercapacitor rapidly charged to 0.6 V (Fig. 12.17C), close to the open-circuit voltage of DSSC, indicating that the energy of the supercapacitor part came from the DSSC part. The discharging process can be obtained as a discharging current is loaded when the energy conversion and storage parts are disconnected. In this photocharge/discharge process, the entire

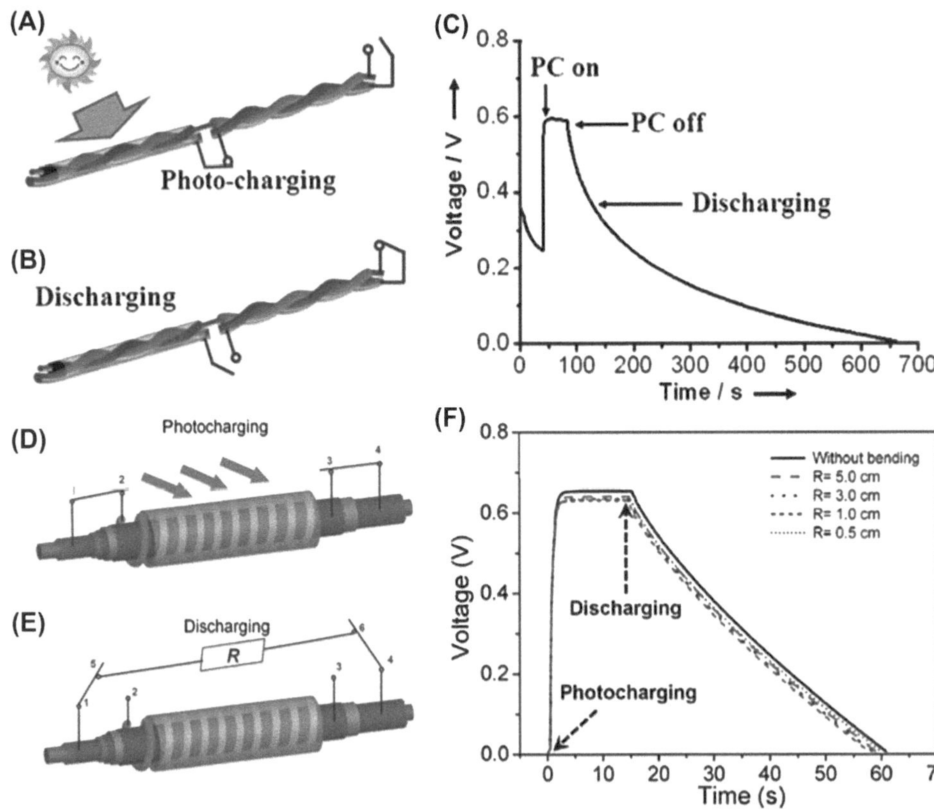

Figure 12.17 (A and B) Schematic illustration of photocharging/discharging of a fiber-shaped integrated energy conversion and storage device. (C) Photocharging/discharging curve of an integrated device. *(Reproduced from Chen, T.; Qiu, L.; Yang, Z.; Cai, Z.; Ren, J.; Li, H.; Lin, H.; Sun, X.; Peng, H. Angewandte Chemie International Edition* ***2012****, 51, 11977–11980 with permission. Copyright 2012 John Wiley and Sons.)* (D and E) Schematic illustration of a core–shell fiber-shaped integrated device. (F) Photocharging/discharging curves of an integrated device being bent with different states. *(Reproduced from Yang, Z.; Deng, J.; Sun, H.; Ren, J.; Pan, S.; Peng, H. Advanced Materials* ***2014****, 26, 7038–7042 with permission. Copyright 2014 John Wiley and Sons.)*

photoelectric conversion and storage efficiency was about 1.5%. An all-solid-state fiber-shaped integrated energy conversion and storage device has also been realized based on a PSC and a supercapacitor *(85)*.

By using CNT composite fiber and CNT sheet electrodes, a core—shell fiber-shaped integrated device (Fig. 12.17D and E) has been developed, in which an all-solid-state supercapacitor in the core was used as energy storage device while a DSSC in the shell was used as energy conversion device *(86)*. In brief, the route to obtain the integrated device was as follows: (1) a supercapacitor with core—shell structure was fabricated through a layer-by-layer method on an elastic rubber fiber, which was then inserted into an plastic tube to separate it to the DSSC part; (2) a layer of CNT sheet was wrapped outside of the plastic tube for use as the cathode electrode for the shell DSSC; (3) the aforementioned core device was inserted into a helical Ti fiber grown with TiO_2 nanotube absorbed with photosensitive dyes that functioned as the anode in the DSSC part; and (4) the resulting device was encapsulated in a transparent plastic tube after injecting with redox electrolyte. The DSSC part showed a power conversion efficiency of 6.47%, while the inner supercapacitor had a specific capacitance of about 20 F/g. This core—shell fiber-shaped integrated device exhibited a high energy conversion and storage efficiency (1.83%), which also showed highly flexible (Fig. 12.17F) and stretchable properties.

Furthermore, integrated energy conversion and storage devices (Fig. 12.18A) prepared by stacking textile electrodes have also been developed *(87)*. In the energy conversion part (a DSSC, Fig. 12.18B), a textile woven from CNT fiber was used as the counter electrode, and a textile of anodized Ti wire absorbed with dye molecules was used as the working electrode *(87,88)*. A power conversion efficiency of 2.2% can be achieved for the textile DSSC based on gel electrolyte. The energy storage part (a supercapacitor, Fig. 12.18C) was fabricated by stacking two CNT (or CNT/PANI composite) fiber textiles with gel electrolyte sandwiched in between, which showed a specific capacitance of 272.7 F/g by using CNT fiber textiles with a weight percentage of 50% PANI *(87)*. The integrated device was prepared by simply stacking the energy conversion part and energy storage part. Fig. 12.18D showed a photocharging/discharging curve for an integrated device based on all electrolytes, according to which a total energy conversion and storage efficiency of 1.5% was obtained.

12.6 CONCLUSION

Because of their complex structure and properties, CNTs must be assembled into macroscopic formats with controllable structure and properties for further applications. As one of the most important macroscopic assembles, CNT fiber can be fabricated through various methods and can be easily scaled up to produce continuously, which is promising for use in the fields of engineering fiber, electrical conductor, wearable electronics, and so on.

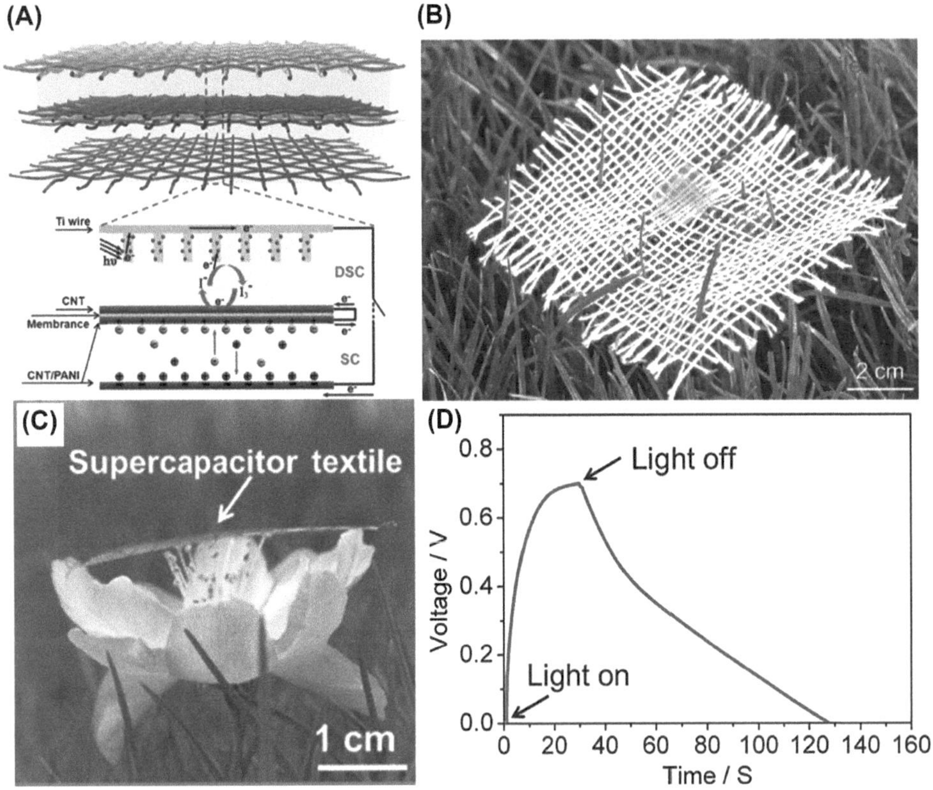

Figure 12.18 (A) Schematic illustration of the structure and working mechanism of the integrated device based on textile electrodes. *DSC*, dye-sensitized solar cell; *SC*, supercapacitor. *(Reproduced from Pan, S.; Lin, H.; Deng, J.; Chen, P.; Chen, X.; Yang, Z.; Peng, H. Advanced Energy Materials* ***2015****, 5, 1401438 with permission. Copyright 2015 John Wiley and Sons.)* (B) A digital photograph of a dye-sensitized solar cell (DSSC) textile. *(Reproduced from Pan, S.; Yang, Z.; Chen, P.; Deng, J.; Li, H.; Peng, H. Angewandte Chemie* ***2014****, 126, 6224–6228. with permission. Copyright 2014 John Wiley and Sons.)* (C) A digital photograph of a supercapacitor textile. (D) Photocharging and galvanostatic discharging curves of an all-solid-state integrated energy textile based on carbon nanotube/polyaniline (CNT/PANI) composite fiber textile electrodes. *(Reproduced from Pan, S.; Lin, H.; Deng, J.; Chen, P.; Chen, X.; Yang, Z.; Peng, H. Advanced Energy Materials* ***2015****, 5, 1401438. with permission. Copyright 2015 John Wiley and Sons.)*

Because of its highly electrical conductivity, mechanical property, and excellent flexibility, CNT fiber has showed its unique advantages for use as fiber electrodes in wearable electronics. However, there still remain some challenges for wearable electronics based on CNT fiber, although great achievements have been obtained so far. For instance, the performance, stability, and packaging process of wearable electronics should be improved further. On the other hand, how the molecules absorbed on the CNTs affect the transport properties of CNT fiber and the derived wearable electronics should be further investigated, which is crucial to developing devices with high performance. Furthermore, to achieve commercial production, continuously fabricating processes maintaining high performance at a large scale should be resolved. There is still much room for scientific research and development of wearable electronics based on CNT fiber for various applications. There exist many reasons why wearable electronics are very promising for use in our daily lives.

ACKNOWLEDGMENTS

The authors would like to acknowledge the support from NSFC (51503152), Science & Technology Commission of Shanghai Municipality (14DZ2261100) and High-level Linghang Program of the foundational subjects in Tongji University.

REFERENCES

1. Li, L.; Wu, Z.; Yuan, S.; Zhang, X.-B. *Energy & Environmental Science* **2014,** *7,* 2101–2122.
2. Chen, T.; Qiu, L.; Yang, Z.; Peng, H. *Chemical Society Reviews* **2013,** *42,* 5031–5041.
3. Cai, X.; Peng, M.; Yu, X.; Fu, Y.; Zou, D. *Journal of Materials Chemistry C* **2014,** *2,* 1184–1200.
4. Lee, S.-Y.; Choi, K.-H.; Choi, W.-S.; Kwon, Y. H.; Jung, H.-R.; Shin, H.-C.; Kim, J. Y. *Energy & Environmental Science* **2013,** *6,* 2414–2423.
5. De Volder, M. F.; Tawfick, S. H.; Baughman, R. H.; Hart, A. J. *Science* **2013,** *339,* 535–539.
6. Lu, W.; Zu, M.; Byun, J. H.; Kim, B. S.; Chou, T. W. *Advanced Materials* **2012,** *24,* 1805–1833.
7. Behabtu, N.; Green, M. J.; Pasquali, M. *Nano Today* **2008,** *3,* 24–34.
8. Vigolo, B.; Penicaud, A.; Coulon, C.; Sauder, C.; Pailler, R.; Journet, C.; Bernier, P.; Poulin, P. *Science* **2000,** *290,* 1331–1334.
9. Ericson, L. M.; Fan, H.; Peng, H.; Davis, V. A.; Zhou, W.; Sulpizio, J.; Wang, Y.; Booker, R.; Vavro, J.; Guthy, C. *Science* **2004,** *305,* 1447–1450.
10. Li, Y.-L.; Kinloch, I. A.; Windle, A. H. *Science* **2004,** *304,* 276–278.
11. Zhong, X. H.; Li, Y. L.; Liu, Y. K.; Qiao, X. H.; Feng, Y.; Liang, J.; Jin, J.; Zhu, L.; Hou, F.; Li, J. Y. *Advanced Materials* **2010,** *22,* 692–696.
12. Jiang, K.; Li, Q.; Fan, S. *Nature* **2002,** *419,* 801.
13. Zhang, M.; Atkinson, K. R.; Baughman, R. H. *Science* **2004,** *306,* 1358–1361.
14. Ma, W.; Liu, L.; Yang, R.; Zhang, T.; Zhang, Z.; Song, L.; Ren, Y.; Shen, J.; Niu, Z.; Zhou, W.; Xie, S. *Advanced Materials* **2009,** *21,* 603–608.
15. Zhang, S.; Koziol, K. K.; Kinloch, I. A.; Windle, A. H. *Small* **2008,** *4,* 1217–1222.
16. Razal, J. M.; Coleman, J. N.; Munoz, E.; Lund, B.; Gogotsi, Y.; Ye, H.; Collins, S.; Dalton, A. B.; Baughman, R. H. *Advanced Functional Materials* **2007,** *17,* 2918–2924.
17. Kozlov, M. E.; Capps, R. C.; Sampson, W. M.; Ebron, V. H.; Ferraris, J. P.; Baughman, R. H. *Advanced Materials* **2005,** *17,* 614–617.
18. Zhu, H.; Xu, C.; Wu, D.; Wei, B.; Vajtai, R.; Ajayan, P. *Science* **2002,** *296,* 884–886.
19. Motta, M.; Li, Y.-L.; Kinloch, I.; Windle, A. *Nano Letters* **2005,** *5,* 1529–1533.
20. Sundaram, R. M.; Koziol, K. K.; Windle, A. H. *Advanced Materials* **2011,** *23,* 5064–5068.
21. Koziol, K.; Vilatela, J.; Moisala, A.; Motta, M.; Cunniff, P.; Sennett, M.; Windle, A. *Science* **2007,** *318,* 1892–1895.
22. Chen, T.; Cai, Z.; Yang, Z.; Li, L.; Sun, X.; Huang, T.; Yu, A.; Kia, H. G.; Peng, H. *Advanced Materials* **2011,** *23,* 4620–4625.
23. Liu, K.; Zhu, F.; Liu, L.; Sun, Y.; Fan, S.; Jiang, K. *Nanoscale* **2012,** *4,* 3389–3393.
24. Li, Q.; Li, Y.; Zhang, X.; Chikkannanavar, S. B.; Zhao, Y.; Dangelewicz, A. M.; Zheng, L.; Doorn, S. K.; Jia, Q.; Peterson, D. E. *Advanced Materials* **2007,** *19,* 3358–3363.
25. Zhang, X.; Li, Q.; Tu, Y.; Li, Y.; Coulter, J. Y.; Zheng, L.; Zhao, Y.; Jia, Q.; Peterson, D. E.; Zhu, Y. *Small* **2007,** *3,* 244–248.
26. Batmunkh, M.; Biggs, M. J.; Shapter, J. G. *Small* **2015,** *11,* 2963–2989.

27. Chen, T.; Wang, S.; Yang, Z.; Feng, Q.; Sun, X.; Li, L.; Wang, Z. S.; Peng, H. *Angewandte Chemie International Edition* **2011,** *50,* 1815–1819.
28. Cai, F.; Chen, T.; Peng, H. *Journal of Materials Chemistry* **2012,** *22,* 14856–14860.
29. Chen, T.; Qiu, L.; Cai, Z.; Gong, F.; Yang, Z.; Wang, Z.; Peng, H. *Nano Letters* **2012,** *12,* 2568–2572.
30. Chen, T.; Qiu, L.; Kia, H. G.; Yang, Z.; Peng, H. *Advanced Materials* **2012,** *24,* 4623–4628.
31. Zhang, S.; Ji, C.; Bian, Z.; Liu, R.; Xia, X.; Yun, D.; Zhang, L.; Huang, C.; Cao, A. *Nano Letters* **2011,** *11,* 3383–3387.
32. Zhang, S.; Ji, C.; Bian, Z.; Yu, P.; Zhang, L.; Liu, D.; Shi, E.; Shang, Y.; Peng, H.; Cheng, Q. *ACS Nano* **2012,** *6,* 7191–7198.
33. Pan, S.; Yang, Z.; Li, H.; Qiu, L.; Sun, H.; Peng, H. *Journal of the American Chemical Society* **2013,** *135,* 10622–10625.
34. Yang, Z.; Sun, H.; Chen, T.; Qiu, L.; Luo, Y.; Peng, H. *Angewandte Chemie* **2013,** *125,* 7693–7696.
35. Li, H.; Guo, J.; Sun, H.; Fang, X.; Wang, D.; Peng, H. *ChemNanoMat* **2015,** *1,* 399–402.
36. Sun, H.; Li, H.; You, X.; Yang, Z.; Deng, J.; Qiu, L.; Peng, H. *Journal of Materials Chemistry A* **2014,** *2,* 345–349.
37. Lee, M. R.; Eckert, R. D.; Forberich, K.; Dennler, G.; Brabec, C. J.; Gaudiana, R. A. *Science* **2009,** *324,* 232–235.
38. Chen, T.; Qiu, L.; Li, H.; Peng, H. *Journal of Materials Chemistry* **2012,** *22,* 23655–23658.
39. Zhang, Z.; Yang, Z.; Wu, Z.; Guan, G.; Pan, S.; Zhang, Y.; Li, H.; Deng, J.; Sun, B.; Peng, H. *Advanced Energy Materials* **2014,** *4,* 1301750.
40. Liu, D.; Zhao, M.; Li, Y.; Bian, Z.; Zhang, L.; Shang, Y.; Xia, X.; Zhang, S.; Yun, D.; Liu, Z.; Cao, A. *ACS Nano* **2012,** *6,* 11027–11034.
41. Green, M. A.; Ho-Baillie, A.; Snaith, H. J. *Nature Photonics* **2014,** *8,* 506–514.
42. Kaltenbrunner, M.; Adam, G.; Głowacki, E. D.; Drack, M.; Schwödiauer, R.; Leonat, L.; Apaydin, D. H.; Groiss, H.; Scharber, M. C.; White, M. S. *Nature Materials* **2015,** *14,* 1032–1039.
43. Qiu, L.; Deng, J.; Lu, X.; Yang, Z.; Peng, H. *Angewandte Chemie International Edition* **2014,** *53,* 10425–10428.
44. Li, R.; Xiang, X.; Tong, X.; Zou, J.; Li, Q. *Advanced Materials* **2015,** *27,* 3831–3835.
45. He, S.; Qiu, L.; Fang, X.; Guan, G.; Chen, P.; Zhang, Z.; Peng, H. *Journal of Materials Chemistry A* **2015,** *3,* 9406–9410.
46. Rogers, J. A.; Someya, T.; Huang, Y. *Science* **2010,** *327,* 1603–1607.
47. Cheng, T.; Zhang, Y.; Lai, W. Y.; Huang, W. *Advanced Materials* **2015,** *27,* 3349–3376.
48. Sekitani, T.; Noguchi, Y.; Hata, K.; Fukushima, T.; Aida, T.; Someya, T. *Science* **2008,** *321,* 1468–1472.
49. Lipomi, D. J.; Tee, B. C. K.; Vosgueritchian, M.; Bao, Z. *Advanced Materials* **2011,** *23,* 1771–1775.
50. Sekitani, T.; Nakajima, H.; Maeda, H.; Fukushima, T.; Aida, T.; Hata, K.; Someya, T. *Nature Materials* **2009,** *8,* 494–499.
51. Yu, C.; Masarapu, C.; Rong, J.; Wei, B.; Jiang, H. *Advanced Materials* **2009,** *11,* 4793–4797.
52. Yang, Z.; Deng, J.; Sun, X.; Li, H.; Peng, H. *Advanced Materials* **2014,** *26,* 2643–2647.
53. Zhang, Z.; Yang, Z.; Deng, J.; Zhang, Y.; Guan, G.; Peng, H. *Small* **2015,** *11,* 675–680.
54. Zhang, Z.; Guo, K.; Li, Y.; Li, X.; Guan, G.; Li, H.; Luo, Y.; Zhao, F.; Zhang, Q.; Wei, B. *Nature Photonics* **2015,** *9,* 233–238.
55. Zhang, Z.; Zhang, Q.; Guo, K.; Li, Y.; Li, X.; Wang, L.; Luo, Y.; Li, H.; Zhang, Y.; Guan, G. *Journal of Materials Chemistry C* **2015,** *3,* 5621–5624.
56. Simon, P.; Gogotsi, Y. *Nature Materials* **2008,** *7,* 845–854.
57. Raccichini, R.; Varzi, A.; Passerini, S.; Scrosati, B. *Nature Materials* **2015,** *14,* 271–279.
58. Chen, T.; Dai, L. *Journal of Materials Chemistry A* **2014,** *2,* 10756–10775.
59. Pan, S.; Zhang, Z.; Weng, W.; Lin, H.; Yang, Z.; Peng, H. *Materials Today* **2014,** *17,* 276–284.
60. Ren, J.; Li, L.; Chen, C.; Chen, X.; Cai, Z.; Qiu, L.; Wang, Y.; Zhu, X.; Peng, H. *Advanced Materials* **2013,** *25,* 1155–1159.
61. Ren, J.; Bai, W.; Guan, G.; Zhang, Y.; Peng, H. *Advanced Materials* **2013,** *25,* 5965–5970.
62. Chen, X.; Qiu, L.; Ren, J.; Guan, G.; Lin, H.; Zhang, Z.; Chen, P.; Wang, Y.; Peng, H. *Advanced Materials* **2013,** *25,* 6436–6441.

63. Wang, K.; Meng, Q.; Zhang, Y.; Wei, Z.; Miao, M. *Advanced Materials* **2013,** *25,* 1494–1498.
64. Choi, C.; Lee, J. A.; Choi, A. Y.; Kim, Y. T.; Lepró, X.; Lima, M. D.; Baughman, R. H.; Kim, S. J. *Advanced Materials* **2014,** *26,* 2059–2065.
65. Lee, J. A.; Shin, M. K.; Kim, S. H.; Cho, H. U.; Spinks, G. M.; Wallace, G. G.; Lima, M. D.; Lepró, X.; Kozlov, M. E.; Baughman, R. H. *Nature Communications* **2013,** *4,* 1970–1977.
66. Kou, L.; Huang, T.; Zheng, B.; Han, Y.; Zhao, X.; Gopalsamy, K.; Sun, H.; Gao, C. *Nature Communications* **2014,** *5,* 3754–3763.
67. Meng, Y.; Zhao, Y.; Hu, C.; Cheng, H.; Hu, Y.; Zhang, Z.; Shi, G.; Qu, L. *Advanced Materials* **2013,** *25,* 2326–2331.
68. Yu, D.; Goh, K.; Wang, H.; Wei, L.; Jiang, W.; Zhang, Q.; Dai, L.; Chen, Y. *Nature Nanotechnology* **2014,** *9,* 555–562.
69. Yang, Z.; Deng, J.; Chen, X.; Ren, J.; Peng, H. *Angewandte Chemie International Edition* **2013,** *52,* 13453–13457.
70. Sun, H.; You, X.; Jiang, Y.; Guan, G.; Fang, X.; Deng, J.; Chen, P.; Luo, Y.; Peng, H. *Angewandte Chemie International Edition* **2014,** *53,* 9526–9531.
71. Deng, J.; Zhang, Y.; Zhao, Y.; Chen, P.; Cheng, X.; Peng, H. *Angewandte Chemie International Edition* **2015;** http://dx.doi.org/10.1002/anie.201508293.
72. Chen, X.; Lin, H.; Deng, J.; Zhang, Y.; Sun, X.; Chen, P.; Fang, X.; Zhang, Z.; Guan, G.; Peng, H. *Advanced Materials* **2014,** *26,* 8126–8132.
73. Xu, P.; Gu, T.; Cao, Z.; Wei, B.; Yu, J.; Li, F.; Byun, J. H.; Lu, W.; Li, Q.; Chou, T. W. *Advanced Energy Materials* **2014,** *4,* 1300759.
74. Zhang, Z.; Deng, J.; Li, X.; Yang, Z.; He, S.; Chen, X.; Guan, G.; Ren, J.; Peng, H. *Advanced Materials* **2015,** *27,* 356–362.
75. Chen, T.; Hao, R.; Peng, H.; Dai, L. *Angewandte Chemie International Edition* **2015,** *54,* 618–622.
76. Lin, H.; Weng, W.; Ren, J.; Qiu, L.; Zhang, Z.; Chen, P.; Chen, X.; Deng, J.; Wang, Y.; Peng, H. *Advanced Materials* **2014,** *26,* 1217–1222.
77. Weng, W.; Sun, Q.; Zhang, Y.; Lin, H.; Ren, J.; Lu, X.; Wang, M.; Peng, H. *Nano Letters* **2014,** *14,* 3432–3438.
78. Zhang, Y.; Bai, W.; Ren, J.; Weng, W.; Lin, H.; Zhang, Z.; Peng, H. *Journal of Materials Chemistry A* **2014,** *2,* 11054–11059.
79. Ren, J.; Zhang, Y.; Bai, W.; Chen, X.; Zhang, Z.; Fang, X.; Weng, W.; Wang, Y.; Peng, H. *Angewandte Chemie* **2014,** *126,* 7998–8003.
80. Wee, G.; Salim, T.; Lam, Y. M.; Mhaisalkar, S. G.; Srinivasan, M. *Energy & Environmental Science* **2011,** *4,* 413–416.
81. Xu, J.; Chen, Y.; Dai, L. *Nature Commun* **2015,** *6,* 8103–8109.
82. Chen, T.; Yang, Z.; Peng, H. *ChemPhysChem* **2013,** *14,* 1777–1782.
83. Bae, J.; Park, Y. J.; Lee, M.; Cha, S. N.; Choi, Y. J.; Lee, C. S.; Kim, J. M.; Wang, Z. L. *Advanced Materials* **2011,** *23,* 3446–3449.
84. Chen, T.; Qiu, L.; Yang, Z.; Cai, Z.; Ren, J.; Li, H.; Lin, H.; Sun, X.; Peng, H. *Angewandte Chemie International Edition* **2012,** *51,* 11977–11980.
85. Zhang, Z.; Chen, X.; Chen, P.; Guan, G.; Qiu, L.; Lin, H.; Yang, Z.; Bai, W.; Luo, Y.; Peng, H. *Advanced Materials* **2014,** *26,* 466–470.
86. Yang, Z.; Deng, J.; Sun, H.; Ren, J.; Pan, S.; Peng, H. *Advanced Materials* **2014,** *26,* 7038–7042.
87. Pan, S.; Lin, H.; Deng, J.; Chen, P.; Chen, X.; Yang, Z.; Peng, H. *Advanced Energy Materials* **2015,** *5,* 1401438.
88. Pan, S.; Yang, Z.; Chen, P.; Deng, J.; Li, H.; Peng, H. *Angewandte Chemie* **2014,** *126,* 6224–6228.

CHAPTER 13

Growth of Aligned Carbon Nanotubes and Their Applications

L.X. Kang, D. Li, Z.Z. Yong, X.H. Zhang, Q. Li
Chinese Academy of Sciences, Suzhou, China

Contents

13.1 INTRODUCTION

In assembling carbon nanotubes (CNTs), alignment plays an important role in optimizing the performance of single-walled CNT (SWCNT)-based devices by avoiding the tube—tube overlap and entanglement, and in facilitating the mechanical and electrical transport between the CNTs in a CNT-based structural material *(1—3)*. In different applications, the alignment can be realized in a horizontal way where the parallelized CNTs lie on a flat substrate or in a vertical way where the CNTs "stand up" on a substrate to form a CNT forest.

There have been a handful of experimental technologies to align CNTs, which can be categorized into two groups. The first one is a posttreatment process, where the randomly distributed CNTs are aligned by fluidics *(4,5)*, electrophoresis *(6,7)*, surface forces *(8)*, surface acoustic waves *(9)*, or self-assembling *(10)*. However, as the alignment can just be improved by a certain level, the improvement on device performance is still quite limited. In contrast, the direct achievement of high alignment during CNT growth

Industrial Applications of Carbon Nanotubes
ISBN 978-0-323-41481-4

has become a more promising strategy because high-quality and well-aligned CNT horizontal or vertical arrays can be obtained. It is also an in situ alignment control during chemical vapor deposition (CVD).

In this chapter, by focusing on such in situ strategy, we first review the fundamental structures and unique properties of aligned CNTs, including the horizontal array, vertical forest, and a special forest that is spinnable for fiber production. Then, different applications based on aligned CNTs are discussed, that is, nanoelectronics and functional thin-film and fiber-shaped devices. Further, the development of high-performance composite materials because of the high alignment is also provided. At the end, we briefly discuss the challenge and future development of alignment control in CVD growth.

13.2 IN SITU ALIGNMENT CONTROL

Because of the one-dimensional feature or the extremely high aspect ratio, a high level of alignment (orientation) allows the full utilization of a CNT's electrical and mechanical properties along its longitudinal direction. The avoidance of tube—tube overlap is also very promising for high-performance applications. Furthermore, especially for electronics and optics application, it is very necessary to control the morphology of each CNT in an array, the spatial arrangement among them, and the uniformity across the entire array.

In situ approaches to produce aligned CNTs, that is, during the CNT growth process, are straightforward and simple and are also the most common way today. Among the complicated growth parameters, the type, structure, activity, lift time, and particle distribution of catalysts are key factors that influence aligned growth. To discuss the in situ alignment control, horizontal array and vertical forest are deliberated separately according to their different growth procedures and different interrelations between the CNTs.

13.2.1 Horizontal CNT Array

Highly aligned SWCNT arrays are favorable for large-scale device integration *(11,12)*, and controlled growth is obviously the first step toward practical applications. Since their discovery in 1993 *(13,14)*, various methods including arc-discharge *(15)*, laser ablating method *(16)*, and the CVD approach *(17)* have been developed to produce SWCNTs. CVD is the most promising one because it can produce SWCNTs at large scale and low cost *(18)*. As horizontal SWCNT arrays are grown on substrates, the fabrication of CNT-based nanoelectronics can thus be greatly facilitated. As schematically illustrated in Fig. 13.1A, the CVD process for array growth is a catalytic conversion of a gaseous precursor at high temperatures into a solid structure on the surface of catalyst particles *(11,19)*. The influencing factors to the tube structure include gas, catalyst, and substrate, as shown in Fig. 13.1B. These factors are interrelated and affect the structures of SWCNTs together.

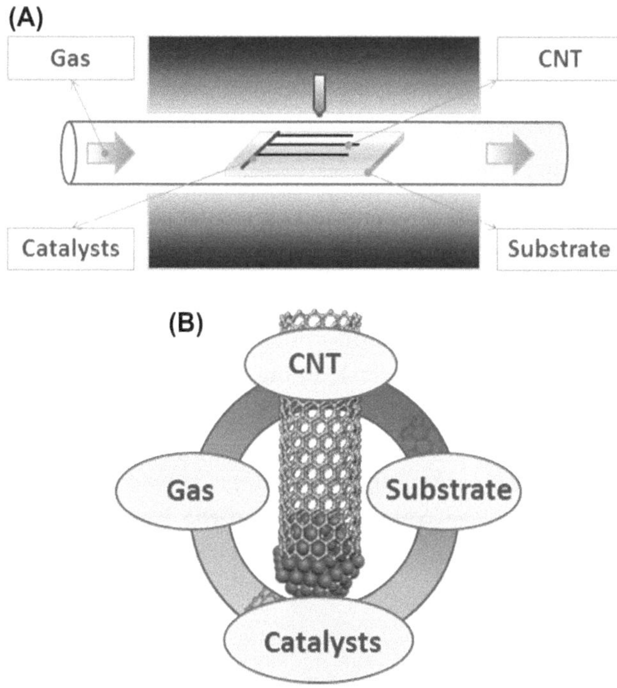

Figure 13.1 Illustration of the chemical vapor deposition (CVD) process of single-walled CNT (SWCNT) array growth on a surface. (A) General growth process of SWNTs in CVD. (B) The interaction of various factors in the growth of SWNTs.

There are many ways to control CNT growth, providing us with many opportunities to realize a high level of alignment. For the precursors, gaseous or liquid small-weight molecules *(20,21)*, that is, carbon monoxide, methane, acetylene, ethylene, alcohol, benzene, and cyclohexane, can be used for the growth of SWCNTs. The catalysts can be either metallic or nonmetallic *(22–24)*, like Fe, Co, Ni, TiO_2, Al_2O_3, and so on. Further, the widely used substrates for array growth even include single crystals *(25)* (sapphire, ST-cut quartz) and amorphous types *(26)* (silicon). Besides these choices of raw materials, control of temperature, atmosphere, and pressure can be realized in a wide range.

According to the control techniques, the aligned growth can be classified into three categories: gas flow directed *(27)*, lattice directed *(28,29)*, and external field assisted *(30)*. In each method, catalysts play an extremely important role in affecting the tube structure. Iron-family elements such as Fe, Co, Ni, and their alloys are the most commonly used catalysts for the growth of SWCNTs because of their catalytic function of graphite formation *(22)*. Although the binding energy between carbon atoms and metal nanoparticles is different, noble metals are also active catalysts inside which carbon atoms can be dissolved at high temperatures. A number of studies on nonmetallic nanoparticles have indicated a new efficient way to design the catalysts. The first success was achieved with 6H-SiC surface in 2002 *(31)*. The resulting SWCNTs formed a highly ordered network with a narrow diameter distribution, both on and below the substrate surface. In the following, semiconducting nanoparticles (SiC, Ge, and Si) *(32)* and many metal oxides (SiO_2 *(28,33)*, Al_2O_3 *(23)*, and TiO_2 *(24)*) are reported as catalysts to grow SWCNTs.

However, to satisfy the requirements of nanodevice fabrication, there are still many important issues to be solved. First, for a (n,m) SWCNT ($n \geq m$), there is a one-third possibility that the tube is metallic, which is damaging for device applications. Considering the diversity in tube diameter, chiral angles, and even handedness, there could be significant a difference in bandgap for the semiconducting tubes. Second, the growth mechanism is still controversial with several ambiguous hypotheses *(33–35)*. Although vapor–liquid–solid and vapor–solid mechanisms have been proposed to illustrate the cap formation–elongation–termination process, the detailed elemental reactions are still not clear to scientists. These include the interactions between the carbon fragment and catalyst, the actual roles of the nanoparticles, and the key factor to chiral structures *(36)*. A well-established relationship among CVD conditions, the used catalysts, and the chiral structures of the grown SWCNTs is still lacking *(37)*.

13.2.2 Vertical CNT Forest

Since the first success in 1998 *(38)*, the vertically aligned CNT forests have gained great attention because of their superb assembly structure and broad applications. The merits of this ordered structure are high purity, high quality, and quasi-unidirectional alignment. Therefore CNT forests can be used in thin-film harvesting devices *(39)*, incandescent displays *(40)*, transparent loudspeakers *(41)*, and super-strong CNT fibers and composites *(41)*. Importantly, there is a special type of CNT forest, the spinnable forest, which is distinguished from ordinary forests because it is capable of being converted into continuous CNT sheets by a simple drawing process in a dry state *(42)*. Nevertheless, there are quite a lot of common issues in growing both the ordinary and the spinnable CNT forests.

In a typical process to grow a CNT forest, catalysts are first prepared by physically depositing a layer of catalyst film on a substrate. Then, annealing treatment at high temperatures with diluted hydrogen is performed to form catalyst nanoparticles on the substrate. Hydrocarbon gases such as acetylene, ethylene, or methane are decomposed at high temperatures into active carbon atoms, which can diffuse into the catalyst nanoparticles, reach saturation, and precipitate into CNTs *(43,44)*. The first CNT forest was grown on Ni-coated glass using a plasma-enhanced CVD method below 666°C (Fig. 13.2A and B) *(38)*. Acetylene gas was used as the carbon source and ammonia gas was used as a catalyst and dilution gas. Nanotubes with controllable diameters from 20 to 400 nm and lengths from 0.1 to 50 μm were obtained. After this, numerous efforts

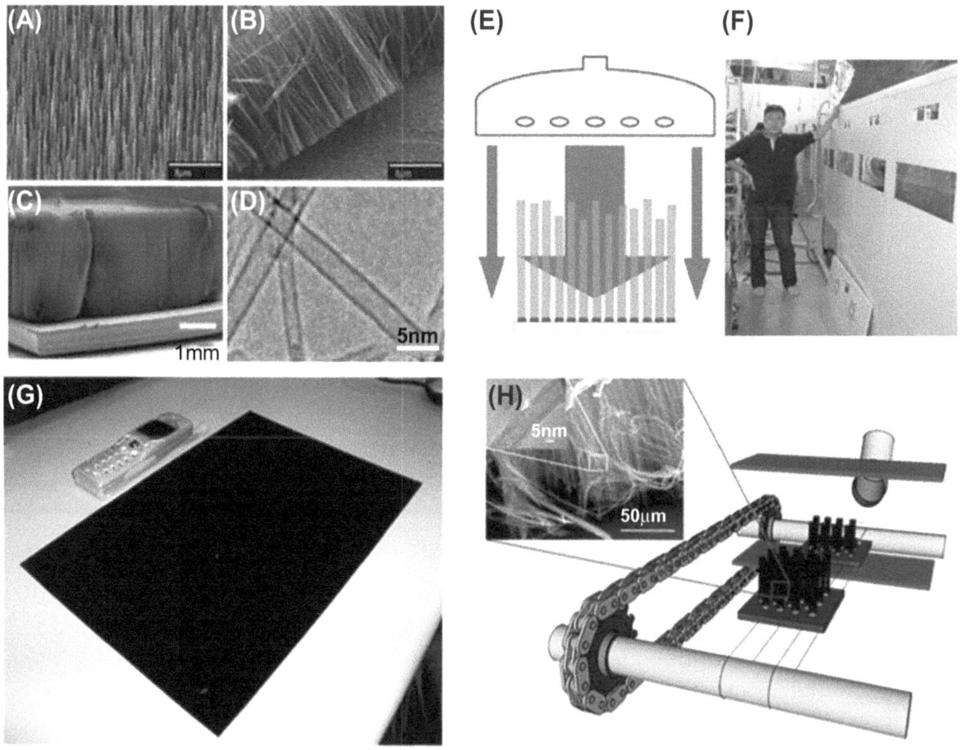

Figure 13.2 (A and B) Scanning electron microscopy (SEM) image of carbon nanotubes (CNTs) aligned perpendicular to the substrate. *(Reproduced with permission from Ren, Z. F.; Huang, Z. P.; Xu, J. W.; Wang, J. H.; Bush, P.; Siegal, M. P.; Provencio, P. N. Synthesis of Large Arrays of Well-Aligned Carbon Nanotubes on Glass. Science* ***1998,*** *282 (5391), 1105–1107.)* (C) SEM image of a 2.5 mm-tall single-walled CNT (SWNT) array and (D) transmission electron microscopy (TEM) image of the SWNTs. *(Reproduced with permission from Hata, K.; Futaba, D. N.; Mizuno, K.; Namai, T.; Yumura, M.; Iijima, S. Water-Assisted Highly Efficient Synthesis of Impurity-Free Single-Walled Carbon Nanotubes. Science* ***2004,*** *306 (5700), 1362–1364.)* (E) Schematic of a shower growth system showing gases delivered from the top of the CNT array. (F) Photographs of constructed 6 m × 1.5 m lateral batch chemical vapor deposition (CVD) furnace. (G) Photograph of a CNT array grown on an A4-size substrate. *(Reproduced with permission from Yasuda, S.; Futaba, D. N.; Yamada, T.; Satou, J.; Shibuya, A.; Takai, H.; Arakawa, K.; Yumura, M.; Hata, K. Improved and Large Area Single-Walled Carbon Nanotube Forest Growth by Controlling the Gas Flow Direction. ACS Nano* ***2009,*** *3 (12), 4164–4170.)* (H) Cold-wall reactor system developed for continuous growth of vertically aligned nanotube arrays. *(Reproduced with permission from Guzmán De Villoria, R.; Hart, A. J.; Wardle, B. L. Continuous High-Yield Production of Vertically Aligned Carbon Nanotubes on 2D and 3D Substrates. ACS Nano* ***2011,*** *5 (6), 4850–4857.)*

have been devoted to fabricate CNT forests with tunable tube diameter, tube length, covering areas, improved alignment, and even well-designed patterns *(45,46)*.

For example, with an efficient "super-growth" method, a 2.5 mm-high SWCNT forest was grown in 10 min by introducing a trace amount of water vapor to the growth ambient (Fig. 13.2C and D) *(47)*. The water vapor acted as a weak oxidizer to etch the amorphous carbon coated on the catalyst particles, and thus prolonged their catalytic activity. It could also inhibit the aggregation of catalyst particles on the substrates *(48)*. Further studies showed that a series of oxygen-containing molecules are capable of enhancing the CVD growth when used together with a carbon source without containing oxygen *(49)*. Besides these enhancers, the introduction of an alumina buffer layer between the silicon substrate and catalyst film *(47)* can sustain the activity of the catalyst by inhibiting the diffusion into the substrate *(50,51)*. By further optimizing the growth condition, centimeter-long CNT forests were produced in a large area *(52–55)*. For example, by using a shower system to deliver the gases that contained water vapor (Fig. 13.2E), in a constructed 6 m × 1.5 m lateral batch CVD furnace (Fig. 13.2F), CNT forests were grown on A4 size substrates (Fig. 13.2G) *(53)*. To further increase the growth area, a continuous production of CNT forests was demonstrated with a movable quartz stage on which substrates containing an Fe/Al_2O_3 film were placed (Fig. 13.2H) *(56,57)*. At a moving speed of 2.4 mm/s millimeter-high CNT forests were produced.

There is no doubt that the performance of CNT forest-based functional devices and composites highly depend on the structure of nanotubes. Therefore controlled growth of vertical CNT forest is of great importance. However, obtaining a cost-effective and scalable process for the synthesis of high-quality CNT forest is still in its infancy. Thus massive growth of CNT forests and precise control of their structure are regarded as important points to be further studied in the future.

13.3 SPINNABLE CNT FORESTS

For the purpose of engineering applications where the remarkable mechanical, electrical, and thermal properties of individual CNTs can be translated to a macroscopic scale, the self-assembly of aligned CNTs is the most efficient way. One important breakthrough was reported in 2002, where macroscopically continuous CNT fibers were produced by drawing the aligned CNTs out of a spinnable CNT forest *(42)*. From then on, forest-based dry spinning has become a hot topic in CNT research. Besides fiber spinning *(58)*, the drawn-out aligned CNT sheet can be used as transparent conducting films *(59)*, loudspeakers *(60)*, and artificial muscles *(61)*.

13.3.1 Synthesis Methods

The growth of spinnable CNT forests is very similar to those of ordinary forests, but the growth condition is much narrower. Fixing catalyst nanoparticles on a substrate is one key point to synthesize spinnable forests. There are three stages that affect the catalyst structures, that is, the deposition of catalyst film, annealing treatment, and CNT growth. First, Fe catalyst film is deposited on a silicon wafer with an SiO_2 layer by electron beam evaporation or magnetron sputtering. To improve the lifetime of the catalyst and to fix the catalyst nanoparticles, an Al_2O_3 buffer layer can also be introduced before the catalyst deposition. Then, the wafer is treated at high temperatures in a diluted hydrogen atmosphere. Without taking the wafer out, acetylene or ethylene is introduced into the furnace, and the forest growth will start in a base-growth mode at appropriate temperatures.

The first reported spinnable CNT forest was produced by using acetylene as carbon source *(42)*. The CNTs had a high level of alignment, clean surfaces, a high density, and perfect spinnability, as shown in Fig. 13.3A and B. Notice that the acetylene-based growth usually generates multiwalled CNTs (MWCNTs) with a diameter of 10–15 nm *(62)*. To reduce the number of walls and modify the spinnability, ethylene was found to be more efficient than acetylene *(63–68)*.

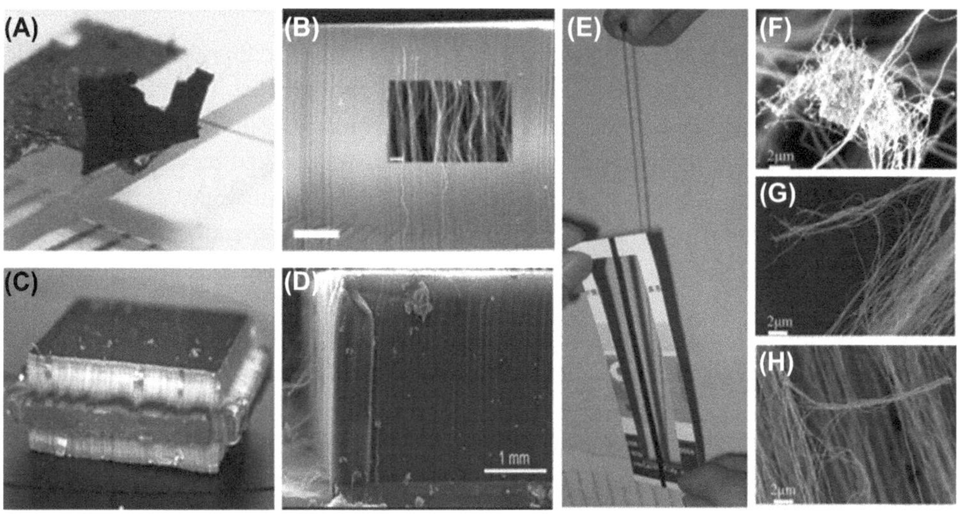

Figure 13.3 (A) A carbon nanotube (CNT) yarn being continuously pulled out from a free-standing CNT array. (B) Scanning electron microscopy (SEM) images of a CNT array grown on a silicon substrate, showing the superalignment of CNTs (scale bars: 100 μm; inset, 200 nm). *(Reproduced with permission from Jiang, K.; Wang, J.; Li, Q.; Liu, L.; Liu, C.; Fan, S. Superaligned Carbon Nanotube Arrays, Films, and Yarns: A Road to Applications. Advanced Materials* ***2011,*** *23 (9), 1154–1161.)* (C) Photograph of vertically aligned CNT (VACNT) sample synthesized chloride-mediated chemical vapor deposition (CM-CVD). VACNTs were grown not only on the top surface but also on the side and back surfaces. (D) SEM image of 2.1 mm-long VACNTs. *(Reproduced with permission from Inoue, Y.; Kakihata, K.; Hirono, Y.; Horie, T.; Ishida, A.; Mimura, H. One-Step Grown Aligned Bulk Carbon Nanotubes by Chloride Mediated Chemical Vapor Deposition. Applied Physics Letters* ***2008,*** *92 (21), 213113–213115.)* (E) Spun CNT yarn from the CNT array produced by the floating catalyst process. (F–H) Typical connection types of CNTs in the CNT yarn. *(Reproduced with permission from Zhang, Q.; Wang, D.-G.; Huang, J.-Q.; Zhou, W.-P.; Luo, G.-H.; Qian, W.-Z.; Wei, F. Dry Spinning Yarns from Vertically Aligned Carbon Nanotube Arrays Produced by an Improved Floating Catalyst Chemical Vapor Deposition Method. Carbon* ***2010,*** *48 (10), 2855–2861.)*

Besides the wafer-supported CVD growth, chloride-mediated CVD on a bare quartz surface *(69)* and the floating catalyst CVD *(70)* were also developed to synthesize spinnable CNT forests. In these two methods, one can avoid the use of high-cost silicon wafers and it is not necessary to predeposit a catalyst film. In the former method, only iron chloride ($FeCl_2$) powder and acetylene were required to grow spinnable CNT forests, and the substrate can be of various types. As the growth was based on the reaction of $FeCl_2$ in vapor phase with acetylene, the forests were finally grown not only on the top surface but also on the side and back surfaces (Fig. 13.3C and D). For the latter, the growth was performed as a multistep interim reactant supply reaction to produce a double-layered CNT forest, in which the top CNT layer was unspinnable and the second CNT layer was spinnable (Fig. 13.3E–H). Because of the avoidance of predeposition treatment, both the chloride-mediated CVD and the floating catalyst CVD are more convenient for mass production of CNT forests.

13.3.2 Tuning the Wafer-Based Growth

For wafer-based growth, the key issue to obtain highly spinnable forests is the proper distribution density and uniform particle size of catalyst nanoparticles. Therefore the thickness of the predeposited Fe film and the annealing treatment play important roles, and the tube diameter, number of walls, and tube length can be controlled by carefully controlling these two pregrowth stages *(64–66,68)*. Further, the spinnability control is quite sensitive to the growth parameters, including the growth temperature, gas flow rate, and reaction time *(66,67)*. For example, by introducing a small amount of hydrogen or oxygen during the growth, the overall morphology of the grown forests can be tuned in a wide range: highly aligned forests were obtained with hydrogen while wavy CNTs were grown by oxygen assistance (Fig. 13.4A–C) *(71)*. As a result, different levels of spinnability were obtained. Thermal treatment on the predeposited Fe film can also influence spinnability *(72)*. The shorter the treatment time, the higher the spinnability, because the prolonged treatment resulted in coarsening of catalyst particles (Fig. 13.4D).

The best way to improve spinnability is to introduce a buffer layer between the Fe catalyst film and the substrate (usually the SiO_2-coated silicon wafer) *(52)*. Because the buffer layer can reduce the surface roughness and avoid the interdiffusion of catalyst nanoparticles, the particle size was no larger than ~25 nm and the areal density was significantly as high as ~700 μm^2, according to a study by the research team at Suzhou Institute of Nano-Tech and Nano-Bionics (SINANO) *(73)*. Although the grown CNTs were multiwalled, the average diameter was just 10 nm and the wall number was less than 8 *(52)*. Most importantly, as the life time of catalyst was improved, the forest was as high as more than 3 mm (Fig. 13.4E).

Recent development in spinnable CNT forests is mainly concentrated on the further decrease in number of walls and tube diameter, rather than on the increase in forest height. Such achievement was realized recently by the SINANO team by carefully

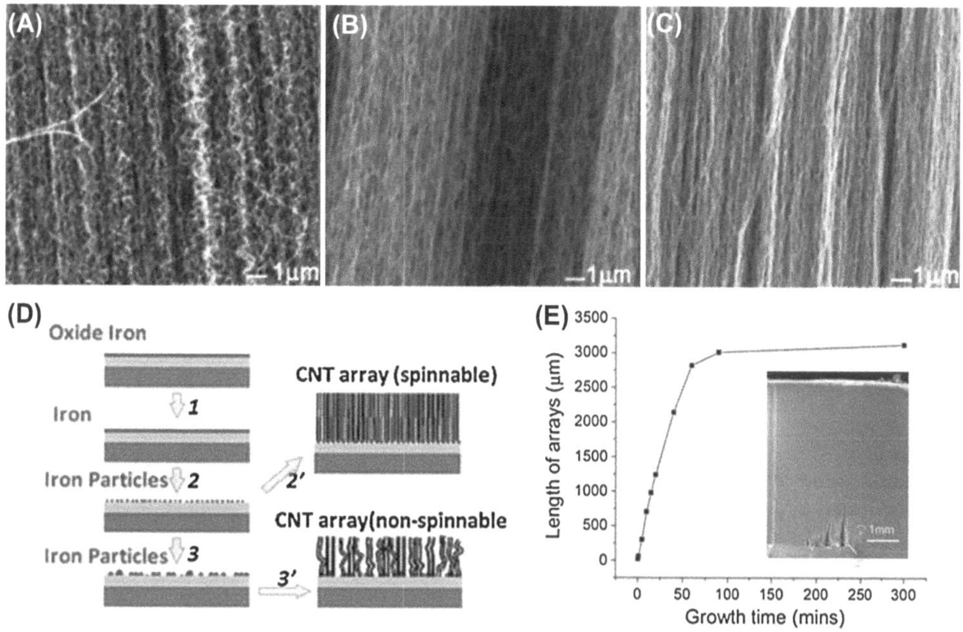

Figure 13.4 (A) Typical scanning electron microscopy (SEM) images of the oxygen assistance-grown carbon nanotube (CNT) arrays (oxygen concentration 150 ppm, no hydrogen introduced). (B) The normally grown arrays (without introducing hydrogen or oxygen). (C) The hydrogen assistance-grown CNT arrays (2% hydrogen, no oxygen intentionally introduced). *(Reproduced with permission from Zheng, L.; Sun, G.; Zhan, Z. Tuning Array Morphology for High-Strength Carbon-Nanotube Fibers. Small* ***2010,*** *6 (1), 132–137.)* (D) Schematic illustration shows long catalyst pretreatment time leads to coarsening of catalyst particles, and the resulting CNT array is not uniform and not spinnable. *(Reproduced with permission from Zhang, Y.; Zou, G.; Doorn, S. K.; Htoon, H.; Stan, L.; Hawley, M. E.; Sheehan, C. J.; Zhu, Y.; Jia, Q. Tailoring the Morphology of Carbon Nanotube Arrays: From Spinnable Forests to Undulating Foams. ACS Nano* ***2009,*** *3 (8), 2157–2162.)* (E) Evolution of CNT array length with time at 750°C with no water vapor and with ethylene and carrier-gas flow rates at 100 sccm. The inset was an SEM image of 4.7 mm-long CNT arrays grown on the catalyst SiO_2/Al_2O_3 (10 nm)/Fe (1 nm) with the assistance of water vapor. The dense Al_2O_3 buffer layer (10 nm) was deposited on an Si substrate using ion beam-assisted deposition. *(Reproduced with permission from Li, Q. W.; Zhang, X. F.; DePaula, R. F.; Zheng, L. X.; Zhao, Y. H.; Stan, L.; Holesinger, T. G.; Arendt, P. N.; Peterson, D. E.; Zhu, Y. T. Sustained Growth of Ultralong Carbon Nanotube Arrays for Fiber Spinning. Advanced Materials* ***2006,*** *18 (23), 3160–3163.)*

optimizing the growth parameters: (1) ethylene was used as carbon source; (2) the Al_2O_3 buffer layer was about 20 nm and the Fe film was 2 nm thick; (3) the furnace temperature was heated up to 660–750°C in a flow of Ar gas within 10 min to pretreat the catalyst film; and (4) hydrogen gas and ethylene were added to the Ar gas to start the forest growth *(74)*. Under such narrowed growth conditions, the CNTs were examined to be 2–3 walled and ≤6 nm in diameter. Furthermore, wafer size can be as large as up to 4 or 6 inches in diameter.

13.3.3 Mass and Low-Cost Production

Although the chloride-mediated CVD and the floating catalyst CVD have advantages in potential mass and low-cost production, the many-walled CNTs were very thick in the tube wall and thus had a diameter larger than 30 nm. Therefore it is still necessary to improve the production ability for wafer-based growth. To meet this aim, a batch growth on 8-inch wafers (Fig. 13.5A) was reported in 2010 *(75)*. In each growth run, at least eight pieces of wafers were used to grow spinnable forests. However, the detailed growth techniques were still not discussed. The SINANO team also succeeded in batch growing spinnable forests on eight pieces of 4-inch wafers where the CNTs were well controlled to be mainly double walled (Fig. 13.5B and C).

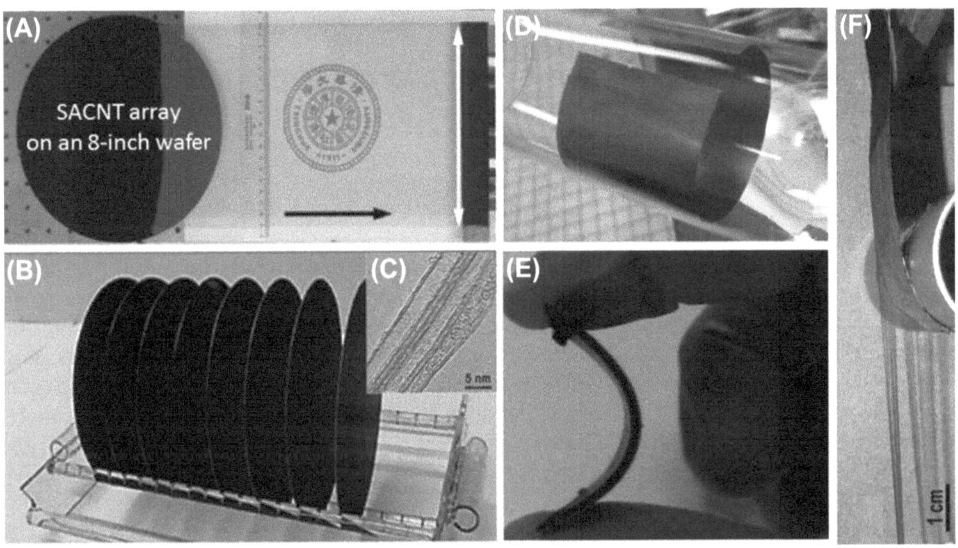

Figure 13.5 (A) Free-standing superaligned carbon nanotube (SACNT) film drawn out from a 230 μm-high SACNT array on an 8-inch silicon wafer. *(Reproduced with permission from Feng, C.; Liu, K.; Wu, J.-S.; Liu, L.; Cheng, J.-S.; Zhang, Y.; Sun, Y.; Li, Q.; Fan, S.; Jiang, K. Flexible, Stretchable, Transparent Conducting Films Made from Superaligned Carbon Nanotubes. Advanced Functional Materials* ***2010,*** *20 (6), 885–891.)* (B) Batch growth of 4-inch spinnable carbon nanotube (CNT) wafers. (C) Transmission electron microscopy (TEM) image of double-walled CNTs in a spinnable array. *(Reproduced with permission from Ren, Z. F.; Huang, Z. P.; Xu, J. W.; Wang, J. H.; Bush, P.; Siegal, M. P.; Provencio, P. N. Synthesis of Large Arrays of Well-Aligned Carbon Nanotubes on Glass. Science* ***1998,*** *282 (5391), 1105–1107.)* (D) A stainless steel foil bent on the inner surface of the quartz tube (inner diameter, c. 65 mm) used as a reactor chamber before being submitted to the chemical vapor deposition (CVD) thermal process. (E) Piece of forest growth on a stainless steel foil being elastically bent. (F) Photograph of a 5 cm-wide CNT sheet being pulled from a forest growth on a bent stainless steel substrate. *(Reproduced with permission from Lepró, X.; Lima, M. D.; Baughman, R. H. Spinnable Carbon Nanotube Forests Grown on Thin, Flexible Metallic Substrates. Carbon* ***2010,*** *48 (12), 3621–3627.)*

The high cost of silicon wafer is another hurdle to industrial scale-up. To solve this problem, recycling of silicon wafers is the simplest way, and experiments have shown its availability: after the forest removal, a new forest was regrown on the same substrate even after five cycles *(76)*. This is because the catalyst particles had strong interaction with the substrate and can be reactivated. The avoidance of catalyst predeposition was another way, as discussed earlier, at the expense of increased wall number and tube diameter. A breakthrough in reducing the price was reported in 2010, where spinnable forests were grown on thin and flexible metallic substrates (Fig. 13.5D–F) *(77)*. The retained mechanical flexibility of the stainless steel sheets after the CNT drawable forest synthesis and its low cost (at least 60 times cheaper per area than Si wafers) were important considerations for their use as belts for continuous or semicontinuous CNT yarn and sheet production from forests. This technique was significant for the upscaling of forest-based yarn and sheet production. Following this idea, an industrial company adopted a similar technique, where a roll-to-roll CVD system was developed for the mass production of spinnable forests on metal foils *(78)*.

13.3.4 Mechanism of Spinnability

The narrow growth condition for spinnable CNT forests indicates that there should be special features in the forest's assembly structure. It was suggested that disordered regions on the top and bottom of the forests induced the spinnability *(59)*. However, later characterization on the forest bottom did not show strong CNT entanglements *(62)*. Thus the top/bottom entanglement should just be a possible but not prerequisite reason for spinnability. For example, by removing the disordered top regions with etching, the remaining forests were still highly spinnable *(68)*.

In situ scanning electron microscopy (SEM) observation of the drawing process can serve as a strong tool to understand the spinnability *(79)*. During the pulling process, an entangled structure was formed at the end of CNTs when they were drawn out, as schematically shown in Fig. 13.6A. This means that self-entangling between the CNTs played a key role in making the CNTs connect in an end-to-end way. A theoretical study based on similar experimental observations showed a suggestive spinning model *(80)*. In this model the interconnections between neighboring CNT bundles are the source of the pulling-out forces. When a bundle is being pulled out, the interconnections are concentrated at the forest top or bottom, and thus make it possible to pull the next bundle out (Fig. 13.6B). Because the interconnection determines the pulling-out forces, there should be a suitable forest height that is dependent on the bundle size and the interconnection density. Obviously, this model can qualitatively explain why taller spinnable CNT forests produce denser CNT sheets.

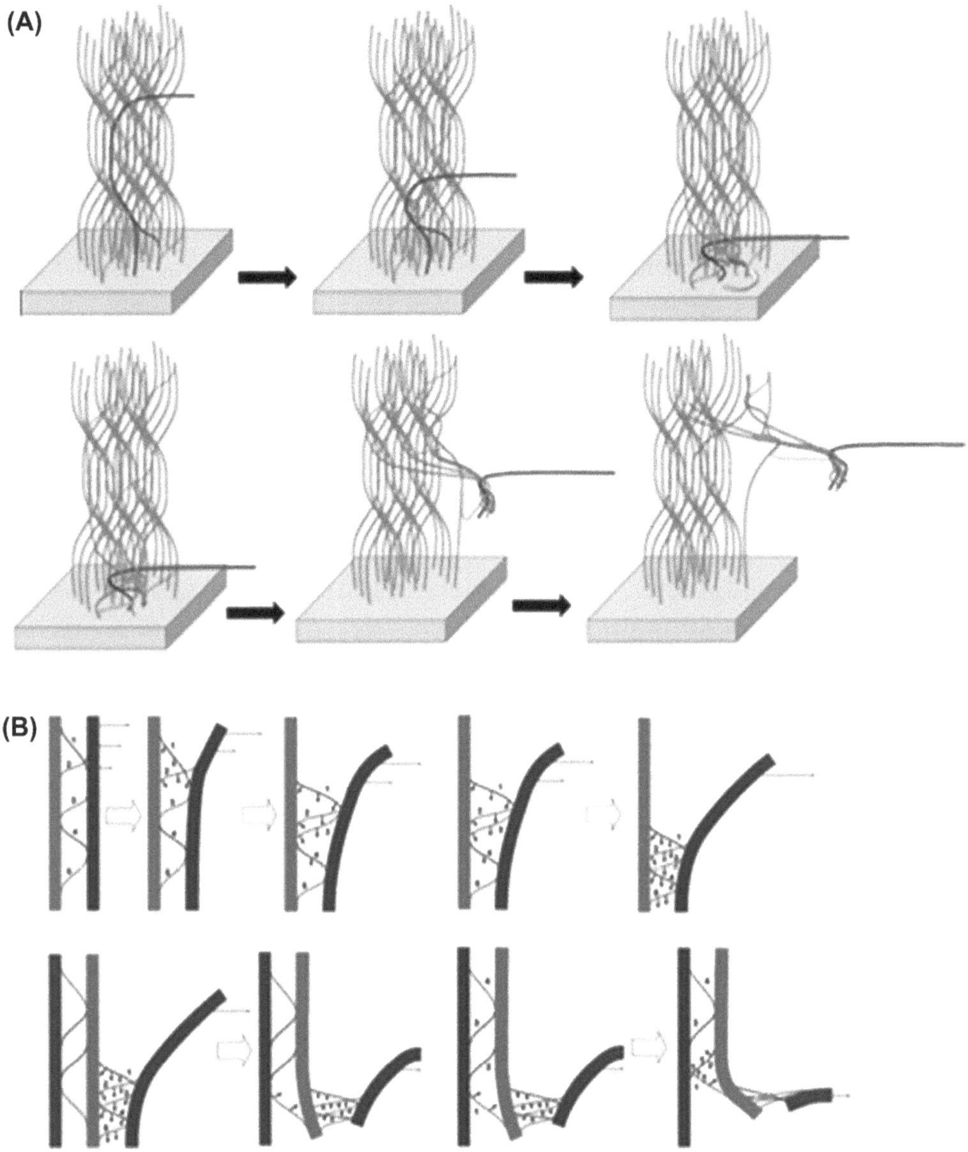

Figure 13.6 (A) Self-entanglement mechanism for continuous pulling of carbon nanotube (CNT) yarns. *(Reproduced with permission from Zhu, C.; Cheng, C.; He, Y. H.; Wang, L.; Wong, T. L.; Fung, K. K.; Wang, N. A Self-Entanglement Mechanism for Continuous Pulling of Carbon Nanotube Yarns. Carbon* ***2011,*** *49 (15), 4996–5001.)* (B) Scheme of the process of pulling out a forest tree from the forest. *Blue* (dark gray in print versions) *and red* (gray in print versions) *bars* represent forest trees or bundles of CNTs. *Red* (gray in print versions) *arrows* show the external force applied to the first bundle. *Small black arrows* show the direction of the net movement of the interconnections. *(Reproduced with permission from Kuznetsov, A. A.; Fonseca, A. F.; Baughman, R. H.; Zakhidov, A. A. Structural Model for Dry-Drawing of Sheets and Yarns from Carbon Nanotube Forests. ACS Nano* ***2011,*** *5 (2), 985–993.)*

13.4 APPLICATIONS OF ALIGNED CNTs

The aligned feature provides a new way to develop transistors, harvesting and energy storage devices, and super-strong composite materials. By considering the ways to align CNTs, the horizontal array and vertical forest have also great differences in their applications.

13.4.1 Electronic Transistors

Parallel arrays of nanotubes on surfaces can find various applications in advanced electronics. They are better candidates for digital electronics than traditional silicon because of their high carrier mobility and current-carrying capacity. The superior electrostatics were experimentally verified via the fabrication of sub-10 nm gate length transistors (Fig. 13.7A) that show superior transport properties to that of silicon devices, while operating at substantially lower voltages *(81)*. In terms of manipulation and integration of nanotubes for device applications, aligned nanotube arrays possess an apparent advantage. Rogers et al. *(82)* reported the use of dense, perfectly aligned arrays of long, perfectly linear SWNTs as an effective thin-film semiconductor suitable for integration into transistors with a top gate (Fig. 13.7B). The large number of SWNTs enables excellent device-level performance characteristics and good device-to-device uniformity, even with SWCNTs that are electronically heterogeneous. Though these SWCNT electron devices still leave much to be desired, complex SWCNT field-effect transistor (FET)-based digital systems can now be realized. For example, a general-purpose fully programmable processor has been experimentally demonstrated *(83)*. An image of such a full processor is shown in Fig.13.7C. It consists of 178 nanotube FETs, each comprising 10–200 SWCNTs (depending on the FET sizes), and is created using the highly scalable fabrication techniques and design flows. This SWCNT computer can be used to execute any arbitrary program and it was a significant development for the advancement of CNT circuits.

Scientists at the IBM research center, being a longtime leader of carbon nanoelectronics, have clearly claimed that the desired SWNT arrays should have a density of 125 nanotubes/μm and metallic impurity of less than 0.0001% *(84)*. Without doubt, this ultimate target for SWCNT sample preparation presents tremendous challenges, and the essential prerequisite is how to realize the orientation-controlled synthesis of SWNTs.

13.4.2 Thin-Film Devices

Besides the horizontal CNT arrays that can be grown directly on substrates, 50 nm-thick CNT sheets can be drawn out from spinnable forest and then laid on various substrates, rigid or flexible, opaque or transparent, and flat or rough. The drawn-out CNT sheet is also electronically conductive because CNT entanglement still exists (see the discussion of forest spinnability). Therefore the forest-based CNT sheets can serve as transparent

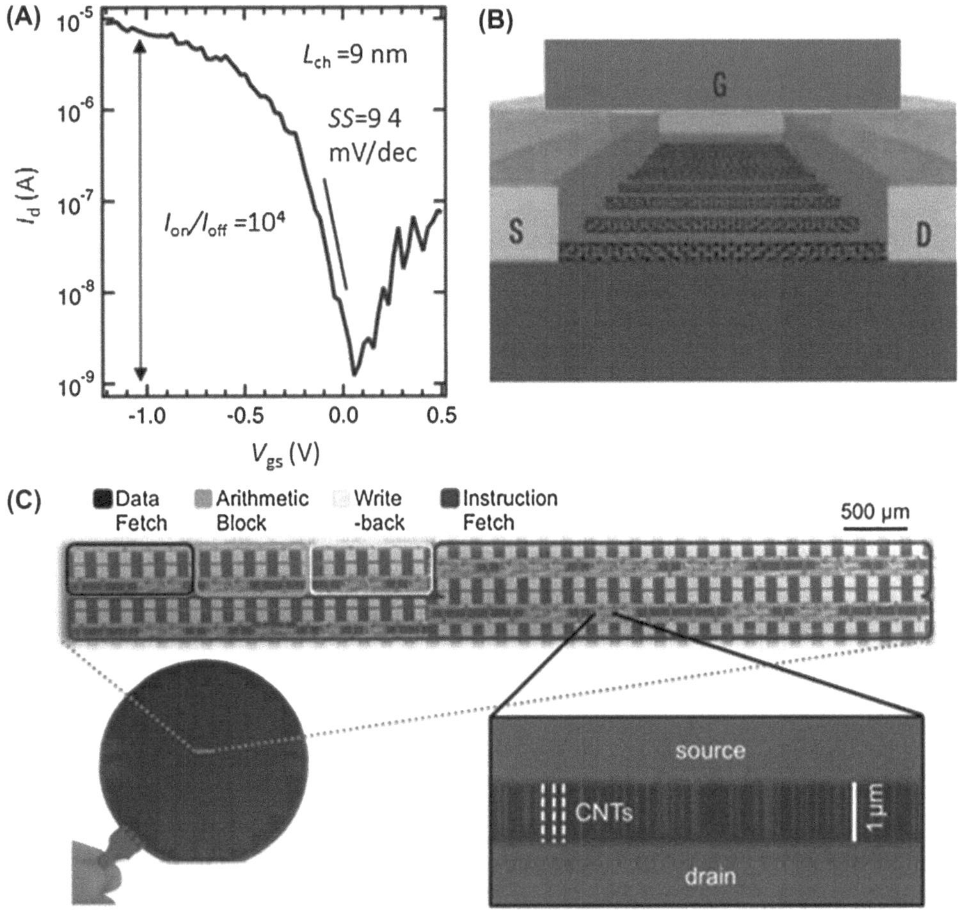

Figure 13.7 (A) Representative *I*–*V* curve of a 9-nm device showing excellent switching behavior. *(Reproduced with permission from Franklin, A. D.; Luisier, M.; Han, S.-J.; Tulevski, G.; Breslin, C. M.; Gignac, L.; Lundstrom, M. S.; Haensch, W. Sub-10 nm Carbon Nanotube Transistor. Nano Letters* ***2012,*** *12 (2), 758–762.)* (B) Schematic illustration of a dual-gated transistor that incorporates horizontally aligned arrays of single-walled CNTs (SWNTs) as the channel. The device uses source (S), drain (D) and gate (G) electrodes. *(Reproduced with permission from Kang, S. J.; Kocabas, C.; Ozel, T.; Shim, M.; Pimparkar, N.; Alam, M. A.; Rotkin, S. V.; Rogers, J. A. High-Performance Electronics Using Dense, Perfectly Aligned Arrays of Single-Walled Carbon Nanotubes. Nature Nanotechnology* ***2007,*** *2 (4), 230–236.)* (C) Optical image of the 4-inch wafer where multiple instances of the full carbon nanotube (CNT) field-effect transistor-based processor are fabricated. *(Reproduced with permission from Shulaker, M. M.; Hills, G.; Patil, N.; Wei, H.; Chen, H.-Y.; Wong, H. S. P.; Mitra, S. Carbon Nanotube Computer. Nature* ***2013,*** *501 (7468), 526–530.)*

electrodes to replace the resource-limited and brittle indium tin oxide (ITO) electrodes for high-performance optoelectronic devices *(41)*. They can also act as conducting scaffolds with large surface area for semiconductors to prepare functional devices.

Fig. 13.8A shows a way to use forest-based CNT sheets to develop solar cells *(39)*, where a CNT sheet was directly laid on silicon wafer with the interfacial contact strengthened with ethanol treatment. The aligned CNTs formed a high CNT-Si junction density and provided efficient charge-transport paths. As a result, the power conversion efficiency reached 10.5%, even higher than that of solar cells fabricated using pristine

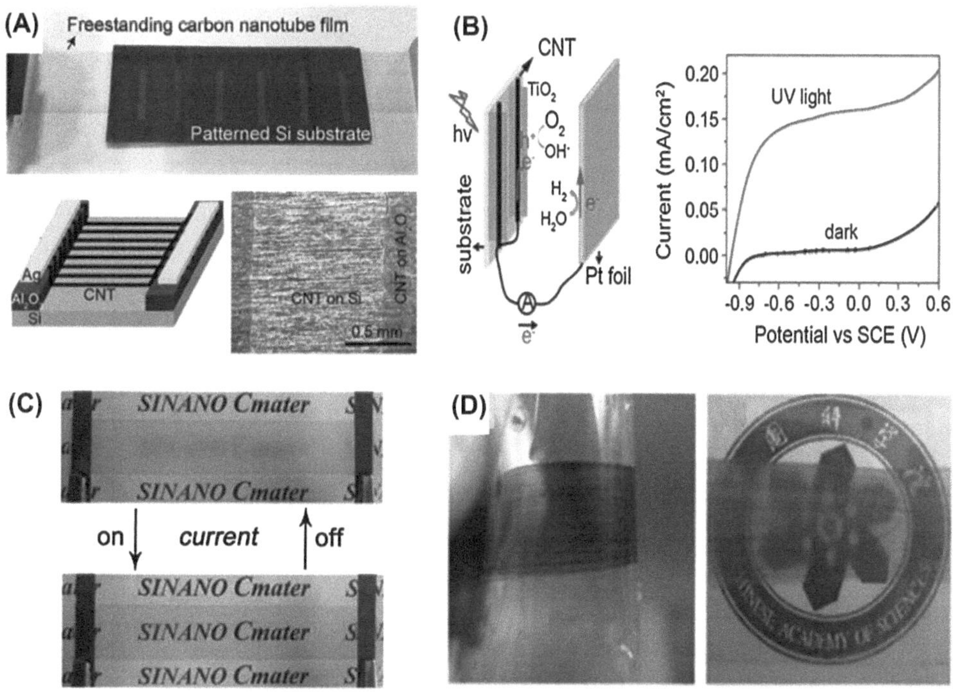

Figure 13.8 (A) Preparation of the carbon nanotube (CNT)-Si solar cells using aligned CNT drawn from spinnable nanotube arrays. Schematic showing the structure of the device and optical image showing aligned CNTs on Si. *(Reproduced with permission from Ren, Z. F.; Huang, Z. P.; Xu, J. W.; Wang, J. H.; Bush, P.; Siegal, M. P.; Provencio, P. N. Synthesis of Large Arrays of Well-Aligned Carbon Nanotubes on Glass. Science* ***1998,*** *282 (5391), 1105–1107.)* (B) Schematic illustrating the structure of a water-splitting solar cell containing a nanotube/TiO_2 film and corresponding current–potential curves of a nanotube/TiO_2 film under dark and UV illumination. *(Reproduced with permission from Di, J.; Yong, Z.; Yao, Z.; Liu, X.; Shen, X.; Sun, B.; Zhao, Z.; He, H.; Li, Q. Robust and Aligned Carbon Nanotube/Titania Core/Shell Films for Flexible TCO-Free Photoelectrodes. Small* ***2013,*** *9 (1), 148–155.)* (C) An opaque CNT/polyurethane (PU) composite film is connected to two copper electrodes (top), and becomes transparent when an electric current is passing through the film (bottom). *(Reproduced with permission from Meng, F.; Zhang, X.; Xu, G.; Yong, Z.; Chen, H.; Chen, M.; Li, Q.; Zhu, Y. Carbon Nanotube Composite Films with Switchable Transparency. ACS Applied Materials & Interfaces* ***2011,*** *3 (3), 658–661.)* (D) A colored WO_3/CNT film on a PET film demonstrating the sheet's flexibility and on the logo of the Chinese Academy of Sciences demonstrating the sheet's transparency. *(Reproduced with permission from Yao, Z.; Di, J.; Yong, Z.; Zhao, Z.; Li, Q. Aligned Coaxial Tungsten Oxide-Carbon Nanotube Sheet: A Flexible and Gradient Electrochromic Film. Chemical Communications* ***2012,*** *48 (66), 8252–8254.)*

and random CNT networks. Besides the photovoltaic property, these aligned CNTs were also applied in photoelectrochemical cells *(85)*. By coating TiO_2 on a free-standing CNT sheet, a core−shell hybrid structure was formed. As shown in Fig. 13.8B, the CNTs provided fast transport paths for photogenerated charges in the oxide, and thus induced a high photon-to-electron conversion ratio of up to 32%. Because of the lattice mismatch, when ZnO nanoparticles were deposited on the aligned CNT sheets, the defect types of ZnO were significantly limited to the planar types rather than the point defects *(86)*. Such planar defect-rich ZnO nanoparticles exhibited a nearly monochromatic ultraviolet luminescence within a wavelength range of 373−376 nm, and acted also as highly efficient photocatalysts.

The ability to conduct electric current opened a way to design a smart window with switchable transparency *(87)*. When the aligned CNTs were fabricated inside a polyurethane (PU) film, the electroheating effect caused a rapid crystal melting of the soft segments in the PU. As a result, the film can be switched from opaque to transparent in just several seconds, and also can reverse to opaque after tuning off the voltage (Fig. 13.8C). In another study, a tungsten oxide (WO_3)/CNT hybrid film showed electrochromic behavior (Fig. 13.8D) when a current passed through the CNT sheet *(88)*. Compared with the WO_3/ITO film, the WO_3/CNT film showed a larger charge density for proton intercalation.

13.4.3 High-Strength Composite Materials

The development of the spinnable CNT forest opened a new way to design high-strength nanocomposites, because the super alignment provides a way to fully utilize the extraordinary mechanical property of CNT. The CNT sheets can be first assembled into a preform and then infiltrated with epoxy resin *(89,90)*. The CNTs with a loading up to 16.5 wt% were homogeneously dispersed and highly aligned in the matrix. Because of the reinforcement, the Young's modulus and tensile strength of the composites reached 20.4 GPa and 231.5 MPa, respectively. Although it was already 160% stronger than the pure epoxy, the low CNT content still hindered the future mechanical strengthening. To overcome this problem, the aligned CNT sheets were layer-by-layer stacked with the aid of solution spraying *(91)*. The as-drawn CNT sheets were continuously wound onto a rotating mandrel under the spray of a polymer solution. Various polymers can be used to act as interfacial strengthening agents, such as poly(vinyl alcohol) (PVA), nylon, epoxy, and polyimide. By tuning the polymer content in the range of 0−40 wt%, aligned CNT/polymer composites with different tensile properties were obtained. For example, the CNT/PVA composite film exhibited a strength of 1.8 GPa at a CNT content of 65 wt%. Furthermore, because of the alignment, the composite film had also a high modulus (40−96 GPa) and toughness (38−100 J/g), and was electrically conductive (conductivity of 780 S/cm). By using a high-performance polyimide, bismaleimides, the composite film can be even as strong as 3.8 GPa with a modulus of 293 GPa *(92)*, showing the superb advantages of the aligned CNTs.

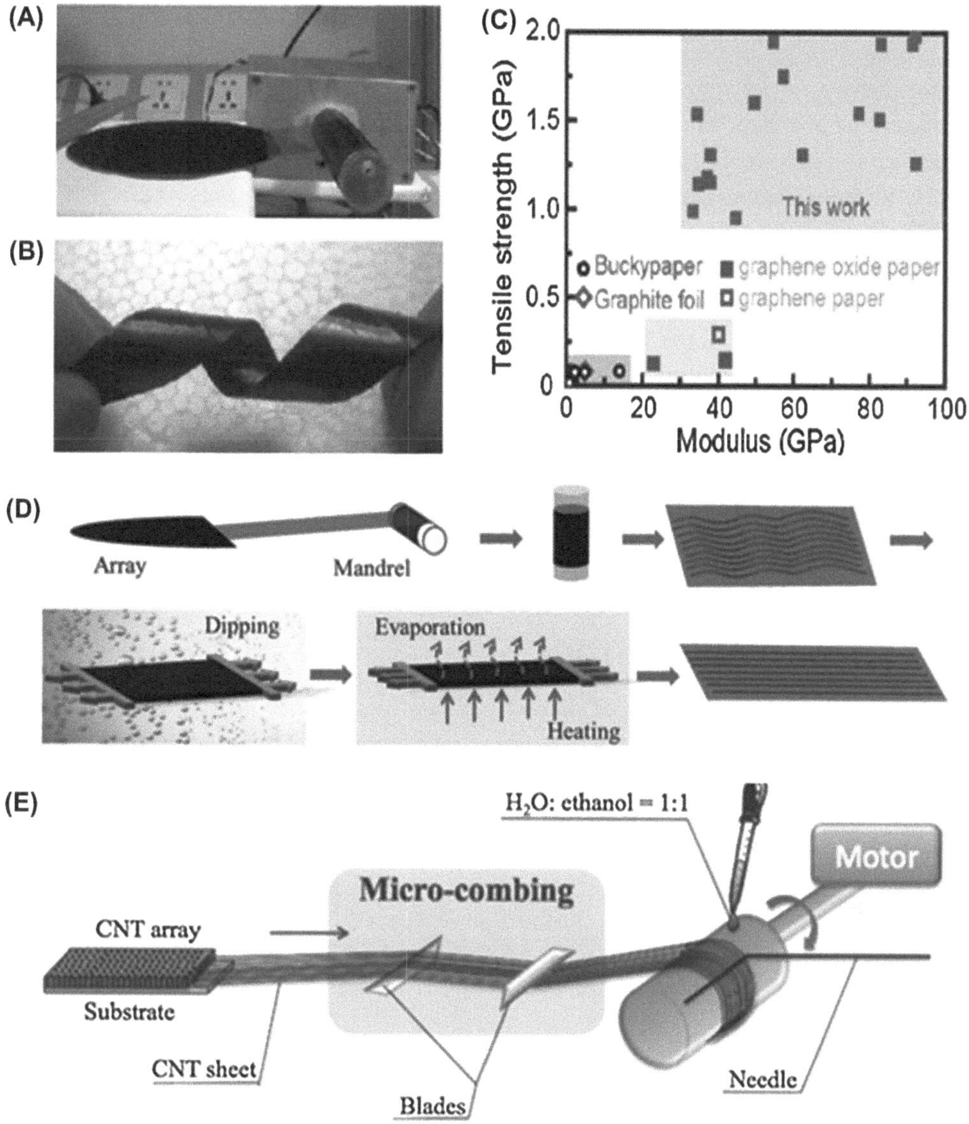

Figure 13.9 (A) Setup for winding aligned carbon nanotube (CNT) films and (B) an as-prepared CNT film. (C) Comparison of aligned CNT films in terms of tensile strength and modulus with the reported buckypapers, graphite foil, and graphene papers by wet chemistry methods. *(Reproduced with permission from Di, J.; Hu, D.; Chen, H.; Yong, Z.; Chen, M.; Feng, Z.; Zhu, Y.; Li, Q. Ultrastrong, Foldable, and Highly Conductive Carbon Nanotube Film. ACS Nano* ***2012,*** *6 (6), 5457–5464.)* (D) Preparation procedure of the stretch-dip-drying method. *(Reproduced with permission from Wang, Y.; Li, M.; Gu, Y.; Zhang, X.; Wang, S.; Li, Q.; Zhang, Z. Tuning Carbon Nanotube Assembly for Flexible, Strong and Conductive Films. Nanoscale* ***2015,*** *7 (7), 3060–3066.)* (E) Schematics of the microcombing experimental setup. *(Reproduced with permission from Zhang, L.; Wang, X.; Xu, W.; Zhang, Y.; Li, Q.; Bradford, P. D.; Zhu, Y. Strong and Conductive Dry Carbon Nanotube Films by Microcombing. Small* ***2015,*** *11 (31), 3830–3836.)*

The layer-by-layer stacking allowed a more uniform penetration of polymer molecules compared to the prepreg method and thus could improve the interfacial properties under the lowest polymer loadings. Furthermore, this approach can produce composite structures in one step, can be controlled by the rotating motion, and thus is easy to be scaled up for industrial production.

By further reducing the tube diameter and number of walls, the self-packing between CNTs allows a fabrication of strong CNT films without introducing polymer matrix. For example, when 2–3-walled, 4–6 nm-diameter CNTs were drawn out from their spinnable forests, they can be densely packed just with the aid of ethanol wetting *(93)*. As shown in Fig. 13.9A–C, the CNT films had a smooth and shining surface, were flexible, and exhibited tensile strengths of 1.1–1.9 GPa and Young's modulus of 40–90 GPa. By further introducing a stretching treatment in a liquid environment on the dry-stacked CNT films *(94)* or by microcombing *(95)*, as shown in Fig. 13.9D–E, the last wavy CNTs can be aligned to further improve the film's mechanical performance. These two methods improved both the tensile strength up to ~3.2 GPa and the modulus up to 124–172 GPa, respectively.

13.5 SUMMARY AND OUTLOOK

The CVD-based in situ growth of aligned CNTs provides an efficient way to utilize the unique properties of individual CNTs. The catalyst type, size, and distribution play very important roles in obtaining the aligned CNTs, and the growth condition becomes quite narrow to precisely control the CNT structure. Nevertheless, exciting developments have been developed in recent years, including chirality control and mass production. The aligned CNTs have also been successfully applied in electronic devices and engineering materials, based on the high utilization of the CNT's electrical and mechanical properties.

ACKNOWLEDGMENTS

The authors thank financial supports from the National Natural Science Foundation of China (11302241, 11404371, 21273269, 21473238, 21503267, 51561145008), Natural Science Foundation of Jiangsu Province (BK20140390), Suzhou Industrial Science and Technology Program (ZXG201416), and the Youth Innovation Promotion Association of the Chinese Academy of Sciences (2015256, Grant to X.Z.).

REFERENCES

1. Huang, L.; Jia, Z.; O'Brien, S. Orientated Assembly of Single-Walled Carbon Nanotubes and Applications. *Journal of Materials Chemistry* **2007,** *17* (37), 3863–3874.
2. Xiao, J.; Dunham, S.; Liu, P.; Zhang, Y.; Kocabas, C.; Moh, L.; Huang, Y.; Hwang, K.-C.; Lu, C.; Huang, W.; et al. Alignment Controlled Growth of Single-Walled Carbon Nanotubes on Quartz Substrates. *Nano Letters* **2009,** *9* (12), 4311–4319.

3. Li, Z.; Liu, Z.; Sun, H.; Gao, C. Superstructured Assembly of Nanocarbons: Fullerenes, Nanotubes, and Graphene. *Chemical Reviews* **2015,** *115* (15), 7046–7117.
4. Deheer, W. A.; Bacsa, W. S.; Châtelain, A.; Gerfin, T.; Humphrey-Baker, R.; Forro, L.; Ugarte, D. Aligned Carbon Nanotube Films: Production and Optical and Electronic Properties. *Science* **1995,** *268* (5212), 845–847.
5. Yan, Y. H.; Li, S.; Chen, L. Q.; Chan-Park, M. B.; Zhang, Q. Large-Scale Submicron Horizontally Aligned Single-Walled Carbon Nanotube Surface Arrays on Various Substrates Produced by a Fluidic Assembly Method. *Nanotechnology* **2006,** *17* (22), 5696–5701.
6. Huang, Y.; Duan, X.; Wei, Q.; Lieber, C. M. Directed Assembly of One-Dimensional Nanostructures into Functional Networks. *Science* **2001,** *291* (5504), 630–633.
7. Oh, S. J.; Cheng, Y.; Zhang, J.; Shimoda, H.; Zhou, O. Room-Temperature Fabrication of High-Resolution Carbon Nanotube Field-Emission Cathodes by Self-Assembly. *Applied Physics Letters* **2003,** *82* (15), 2521–2523.
8. Rao, S. G.; Huang, L.; Setyawan, W.; Hong, S. Nanotube Electronics: Large-Scale Assembly of Carbon Nanotubes. *Nature* **2003,** *425* (6953), 36–37.
9. Seemann, K. M.; Ebbecke, J.; Wixforth, A. Alignment of Carbon Nanotubes on Pre-structured Silicon by Surface Acoustic Waves. *Nanotechnology* **2006,** *17* (17), 4529–4532.
10. Diao, P.; Liu, Z. Vertically Aligned Single-Walled Carbon Nanotubes by Chemical Assembly–Methodology, Properties, and Applications. *Advanced Materials* **2010,** *22* (13), 1430–1449.
11. Chen, Y. B.; Zhang, Y. Y.; Hu, Y.; Kang, L. X.; Zhang, S. C.; Xie, H. H.; Liu, D.; Zhao, Q. C.; Li, Q. W.; Zhang, J. State of the Art of Single-Walled Carbon Nanotube Synthesis on Surfaces. *Advanced Materials* **2014,** *26* (34), 5898–5922.
12. Chen, Y. B.; Zhang, J. Chemical Vapor Deposition Growth of Single-Walled Carbon Nanotubes with Controlled Structures for Nanodevice Applications. *Accounts of Chemical Research* **2014,** *47* (8), 2273–2281.
13. Bethune, D. S.; Kiang, C. H.; Devries, M. S.; Gorman, G.; Savoy, R.; Vazquez, J.; Beyers, R. Cobalt-Catalyzed Growth of Carbon Nanotubes with Single-Atomic-Layerwalls. *Nature* **1993,** *363* (6430), 605–607.
14. Iijima, S.; Ichihashi, T. Single-Shell Carbon Nanotubes of 1-nm Diameter. *Nature* **1993,** *363* (6430), 603–605.
15. Ebbesen, T. W.; Ajayan, P. M. Large-Scale Synthesis of Carbon Nanotubes. *Nature* **1992,** *358* (6383), 220–222.
16. Lebedkin, S.; Hennrich, F.; Skipa, T.; Kappes, M. M. Near-Infrared Photoluminescence of Single-Walled Carbon Nanotubes Prepared by the Laser Vaporization Method. *Journal of Physical Chemistry B* **2003,** *107* (9), 1949–1956.
17. Endo, M.; Takeuchi, K.; Igarashi, S.; Kobori, K.; Shiraishi, M.; Kroto, H. W. The Production and Structure of Pyrolytic Carbon Nanotubes (PCNTs). *Journal of Physics and Chemistry of Solids* **1993,** *54* (12), 1841–1848.
18. Zhang, Q.; Huang, J.-Q.; Qian, W.-Z.; Zhang, Y.-Y.; Wei, F. The Road for Nanomaterials Industry: A Review of Carbon Nanotube Production, Post-treatment, and Bulk Applications for Composites and Energy Storage. *Small* **2013,** *9* (8), 1237–1265.
19. Li, Y.; Cui, R. L.; Ding, L.; Liu, Y.; Zhou, W. W.; Zhang, Y.; Jin, Z.; Peng, F.; Liu, J. How Catalysts Affect the Growth of Single-Walled Carbon Nanotubes on Substrates. *Advanced Materials* **2010,** *22* (13), 1508–1515.
20. Kong, J.; Cassell, A. M.; Dai, H. J. Chemical Vapor Deposition of Methane for Single-Walled Carbon Nanotubes. *Chemical Physics Letters* **1998,** *292* (4–6), 567–574.
21. Mayne, M.; Grobert, N.; Terrones, M.; Kamalakaran, R.; Ruhle, M.; Kroto, H. W.; Walton, D. R. M. Pyrolytic Production of Aligned Carbon Nanotubes from Homogeneously Dispersed Benzene-Based Aerosols. *Chemical Physics Letters* **2001,** *338* (2–3), 101–107.
22. Hamilton, J. C.; Blakely, J. M. Carbon Segregation to Single-Crystal Surfaces of Pt, Pd and Co. *Surface Science* **1980,** *91* (1), 199–217.
23. Huang, S.; Cai, Q.; Chen, J.; Qian, Y.; Zhang, L. Metal-Catalyst-Free Growth of Single-Walled Carbon Nanotubes on Substrates. *Journal of the American Chemical Society* **2009,** *131* (6), 2094–2095.

24. Kang, L.; Hu, Y.; Liu, L.; Wu, J.; Zhang, S.; Zhao, Q.; Ding, F.; Li, Q.; Zhang, J. Growth of Close-Packed Semiconducting Single-Walled Carbon Nanotube Arrays Using Oxygen-Deficient TiO_2 Nanoparticles as Catalysts. *Nano Letters* **2015,** *15* (1), 403–409.
25. Chen, Y.; Hu, Y.; Fang, Y.; Li, P.; Feng, C.; Zhang, J. Lattice-Directed Growth of Single-Walled Carbon Nanotubes with Controlled Geometries on Surface. *Carbon* **2012,** *50* (9), 3295–3297.
26. Yang, F.; Wang, X.; Zhang, D. Q.; Yang, J.; Luo, D.; Xu, Z. W.; Wei, J. K.; Wang, J.-Q.; Xu, Z.; Peng, F.; et al. Chirality-Specific Growth of Single-Walled Carbon Nanotubes on Solid Alloy Catalysts. *Nature* **2014,** *510* (7506), 522–524.
27. Huang, S.; Maynor, B.; Cai, X.; Liu, J. Ultralong, Well-Aligned Single-Walled Carbon Nanotube Architectures on Surfaces. *Advanced Materials* **2003,** *15* (19), 1651–1655.
28. Su, M.; Li, Y.; Maynor, B.; Buldum, A.; Lu, J. P.; Liu, J. Lattice-Oriented Growth of Single-Walled Carbon Nanotubes. *Journal of Physical Chemistry B* **2000,** *104* (28), 6505–6508.
29. Ismach, A.; Segev, L.; Wachtel, E.; Joselevich, E. Atomic-Step-Templated Formation of Single Wall Carbon Nanotube Patterns. *Angewandte Chemie-International Edition* **2004,** *43* (45), 6140–6143.
30. Zhang, Y. G.; Chang, A. L.; Cao, J.; Wang, Q.; Kim, W.; Li, Y. M.; Morris, N.; Yenilmez, E.; Kong, J.; Dai, H. J. Electric-Field-Directed Growth of Aligned Single-Walled Carbon Nanotubes. *Applied Physics Letters* **2001,** *79* (19), 3155–3157.
31. Derycke, V.; Martel, R.; Radosvljevic, M.; Ross, F. M. R.; Avouris, P. Catalyst-Free Growth of Ordered Single-Walled Carbon Nanotube Networks. *Nano Letters* **2002,** *2* (10), 1043–1046.
32. Takagi, D.; Hibino, H.; Suzuki, S.; Kobayashi, Y.; Homma, Y. Carbon Nanotube Growth from Semiconductor Nanoparticles. *Nano Letters* **2007,** *7* (8), 2272–2275.
33. Tessonnier, J. P.; Su, D. S. Recent Progress on the Growth Mechanism of Carbon Nanotubes: A Review. *ChemSusChem* **2011,** *4* (7), 824–847.
34. Page, A. J.; Ohta, Y.; Irle, S.; Morokuma, K. Mechanisms of Single-Walled Carbon Nanotube Nucleation, Growth, and Healing Determined Using QM/MD Methods. *Accounts of Chemical Research* **2010,** *43* (10), 1375–1385.
35. Jourdain, V.; Bichara, C. Current Understanding of the Growth of Carbon Nanotubes in Catalytic Chemical Vapour Deposition. *Carbon* **2013,** *58*, 2–39.
36. Lamouroux, E.; Serp, P.; Kalck, P. Catalytic Routes Towards Single Wall Carbon Nanotubes. *Catalysis Reviews – Science and Engineering* **2007,** *49* (3), 341–405.
37. Chiang, W.-H.; Mohan Sankaran, R. Linking Catalyst Composition to Chirality Distributions of As-Grown Single-Walled Carbon Nanotubes by Tuning Ni_xFe_{1-x} Nanoparticles. *Nature Materials* **2009,** *8* (11), 882–886.
38. Ren, Z. F.; Huang, Z. P.; Xu, J. W.; Wang, J. H.; Bush, P.; Siegal, M. P.; Provencio, P. N. Synthesis of Large Arrays of Well-Aligned Carbon Nanotubes on Glass. *Science* **1998,** *282* (5391), 1105–1107.
39. Di, J.; Yong, Z.; Zheng, X.; Sun, B.; Li, Q. Aligned Carbon Nanotubes for High-Efficiency Schottky Solar Cells. *Small* **2013,** *9* (8), 1367–1372.
40. Zhang, M.; Fang, S.; Zakhidov, A. A.; Lee, S. B.; Aliev, A. E.; Williams, C. D.; Atkinson, K. R.; Baughman, R. H. Materials Science: Strong, Transparent, Multifunctional, Carbon Nanotube Sheets. *Science* **2005,** *309* (5738), 1215–1219.
41. Jiang, K.; Wang, J.; Li, Q.; Liu, L.; Liu, C.; Fan, S. Superaligned Carbon Nanotube Arrays, Films, and Yarns: A Road to Applications. *Advanced Materials* **2011,** *23* (9), 1154–1161.
42. Jiang, K.; Li, Q.; Fan, S. Nanotechnology: Spinning Continuous Carbon Nanotube Yarns. *Nature* **2002,** *419* (6909), 801.
43. Helveg, S.; López-Cartes, C.; Sehested, J.; Hansen, P. L.; Clausen, B. S.; Rostrup-Nielsen, J. R.; Abild-Pedersen, F.; Nørskov, J. K. Atomic-Scale Imaging of Carbon Nanofibre Growth. *Nature* **2004,** *427* (6973), 426–429.
44. Stadermann, M.; Sherlock, S. P.; In, J. B.; Fornasiero, F.; Park, H. G.; Artyukhin, A. B.; Wang, Y.; De Yoreo, J. J.; Grigoropoulos, C. P.; Bakajin, O.; et al. Mechanism and Kinetics of Growth Termination in Controlled Chemical Vapor Deposition Growth of Multiwall Carbon Nanotube Arrays. *Nano Letters* **2009,** *9* (2), 738–744.
45. Lan, Y.; Wang, Y.; Ren, Z. F. Physics and Applications of Aligned Carbon Nanotubes. *Advances in Physics* **2011,** *60* (4), 553–678.

46. Wei, B. Q.; Vajtai, R.; Jung, Y.; Ward, J.; Zhang, R.; Ramanath, G.; Ajayan, P. M. Microfabrication Technology: Organized Assembly of Carbon Nanotubes. *Nature* **2002,** *416* (6880), 495–496.
47. Hata, K.; Futaba, D. N.; Mizuno, K.; Namai, T.; Yumura, M.; Iijima, S. Water-Assisted Highly Efficient Synthesis of Impurity-Free Single-Walled Carbon Nanotubes. *Science* **2004,** *306* (5700), 1362–1364.
48. Amama, P. B.; Pint, C. L.; McJilton, L.; Kim, S. M.; Stach, E. A.; Murray, P. T.; Hauge, R. H.; Maruyama, B. Role of Water in Super Growth of Single-Walled Carbon Nanotube Carpets. *Nano Letters* **2008,** *9* (1), 44–49.
49. Futaba, D. N.; Goto, J.; Yasuda, S.; Yamada, T.; Yumura, M.; Hata, K. General Rules Governing the Highly Efficient Growth of Carbon Nanotubes. *Advanced Materials* **2009,** *21* (47), 4811–4815.
50. De Los Arcos, T.; Wu, Z. M.; Oelhafen, P. Is Aluminum a Suitable Buffer Layer for Carbon Nanotube Growth? *Chemical Physics Letters* **2003,** *380* (3–4), 419–423.
51. Amama, P. B.; Pint, C. L.; Kim, S. M.; McJilton, L.; Eyink, K. G.; Stach, E. A.; Hauge, R. H.; Maruyama, B. Influence of Alumina Type on the Evolution and Activity of Alumina-Supported Fe Catalysts in Single-Walled Carbon Nanotube Carpet Growth. *ACS Nano* **2010,** *4* (2), 895–904.
52. Li, Q.; Zhang, X.; DePaula, R. F.; Zheng, L.; Zhao, Y.; Stan, L.; Holesinger, T. G.; Arendt, P. N.; Peterson, D. E.; Zhu, Y. T. Sustained Growth of Ultralong Carbon Nanotube Arrays for Fiber Spinning. *Advanced Materials* **2006,** *18* (23), 3160–3163.
53. Yasuda, S.; Futaba, D. N.; Yamada, T.; Satou, J.; Shibuya, A.; Takai, H.; Arakawa, K.; Yumura, M.; Hata, K. Improved and Large Area Single-Walled Carbon Nanotube Forest Growth by Controlling the Gas Flow Direction. *ACS Nano* **2009,** *3* (12), 4164–4170.
54. Puretzky, A. A.; Eres, G.; Rouleau, C. M.; Ivanov, I. N.; Geohegan, D. B. Real-Time Imaging of Vertically Aligned Carbon Nanotube Array Growth Kinetics. *Nanotechnology* **2008,** *19* (5), 055605–055609.
55. Hasegawa, K.; Noda, S. Millimeter-Tall Single-Walled Carbon Nanotubes Rapidly Grown with and without Water. *ACS Nano* **2011,** *5* (2), 975–984.
56. Guzmán De Villoria, R.; Hart, A. J.; Wardle, B. L. Continuous High-Yield Production of Vertically Aligned Carbon Nanotubes on 2D and 3D Substrates. *ACS Nano* **2011,** *5* (6), 4850–4857.
57. Guzmán De Villoria, R.; Figueredo, S. L.; Hart, A. J.; Steiner, S. A., III; Slocum, A. H.; Wardle, B. L. High-Yield Growth of Vertically Aligned Carbon Nanotubes on a Continuously Moving Substrate. *Nanotechnology* **2009,** *20* (40), 405611–405618.
58. Zhang, X.; Li, Q.; Tu, Y.; Li, Y.; Coulter, J. Y.; Zheng, L.; Zhao, Y.; Jia, Q.; Peterson, D. E.; Zhu, Y. Strong Carbon-Nanotube Fibers Spun from Long Carbon-Nanotube Arrays. *Small* **2007,** *3* (2), 244–248.
59. Zhang, M.; Fang, S.; Zakhidov, A. A.; Lee, S. B.; Aliev, A. E.; Williams, C. D.; Atkinson, K. R.; Baughman, R. H. Strong, Transparent, Multifunctional, Carbon Nanotube Sheets. *Science* **2005,** *309* (5738), 1215–1219.
60. Xiao, L.; Chen, Z.; Feng, C.; Liu, L.; Bai, Z.-Q.; Wang, Y.; Qian, L.; Zhang, Y.; Li, Q.; Jiang, K.; et al. Flexible, Stretchable, Transparent Carbon Nanotube Thin Film Loudspeakers. *Nano Letters* **2008,** *8* (12), 4539–4545.
61. Aliev, A. E.; Oh, J.; Kozlov, M. E.; Kuznetsov, A. A.; Fang, S.; Fonseca, A. F.; Ovalle, R.; Lima, M. D.; Haque, M. H.; Gartstein, Y. N.; et al. Giant-Stroke, Superelastic Carbon Nanotube Aerogel Muscles. *Science* **2009,** *323* (5921), 1575–1578.
62. Zhang, X.; Jiang, K.; Feng, C.; Liu, P.; Zhang, L.; Kong, J.; Zhang, T.; Li, Q.; Fan, S. Spinning and Processing Continuous Yarns from 4-Inch Wafer Scale Super-Aligned Carbon Nanotube Arrays. *Advanced Materials* **2006,** *18* (12), 1505–1510.
63. Zhang, M.; Atkinson, K. R.; Baughman, R. H. Multifunctional Carbon Nanotube Yarns by Downsizing an Ancient Technology. *Science* **2004,** *306* (5700), 1358–1361.
64. Kim, J.-H.; Jang, H.-S.; Lee, K. H.; Overzet, L. J.; Lee, G. S. Tuning of Fe Catalysts for Growth of Spin-Capable Carbon Nanotubes. *Carbon* **2010,** *48* (2), 538–547.
65. Kim, J.-H.; Lee, K. H.; Burk, D.; Overzet, L. J.; Lee, G. S. The Effects of Pre-annealing in Either H_2 or He on the Formation of Fe Nanoparticles for Growing Spin-Capable Carbon Nanotube Forests. *Carbon* **2010,** *48* (15), 4301–4308.

66. Huynh, C. P.; Hawkins, S. C. Understanding the Synthesis of Directly Spinnable Carbon Nanotube Forests. *Carbon* **2010,** *48* (4), 1105–1115.
67. Cui, Y.; Wang, B.; Zhang, M. Optimizing Reaction Condition for Synthesizing Spinnable Carbon Nanotube Arrays by Chemical Vapor Deposition. *Journal of Materials Science* **2013,** *48* (21), 7749–7756.
68. Liu, K.; Sun, Y.; Chen, L.; Feng, C.; Feng, X.; Jiang, K.; Zhao, Y.; Fan, S. Controlled Growth of Super-Aligned Carbon Nanotube Arrays for Spinning Continuous Unidirectional Sheets with Tunable Physical Properties. *Nano Letters* **2008,** *8* (2), 700–705.
69. Inoue, Y.; Kakihata, K.; Hirono, Y.; Horie, T.; Ishida, A.; Mimura, H. One-Step Grown Aligned Bulk Carbon Nanotubes by Chloride Mediated Chemical Vapor Deposition. *Applied Physics Letters* **2008,** *92* (21), 213113–213115.
70. Zhang, Q.; Wang, D.-G.; Huang, J.-Q.; Zhou, W.-P.; Luo, G.-H.; Qian, W.-Z.; Wei, F. Dry Spinning Yarns from Vertically Aligned Carbon Nanotube Arrays Produced by an Improved Floating Catalyst Chemical Vapor Deposition Method. *Carbon* **2010,** *48* (10), 2855–2861.
71. Zheng, L.; Sun, G.; Zhan, Z. Tuning Array Morphology for High-Strength Carbon-Nanotube Fibers. *Small* **2010,** *6* (1), 132–137.
72. Zhang, Y.; Zou, G.; Doorn, S. K.; Htoon, H.; Stan, L.; Hawley, M. E.; Sheehan, C. J.; Zhu, Y.; Jia, Q. Tailoring the Morphology of Carbon Nanotube Arrays: From Spinnable Forests to Undulating Foams. *ACS Nano* **2009,** *3* (8), 2157–2162.
73. Di, J. T.; Yong, Z. Z.; Yang, X. J.; Li, Q. W. Structural and Morphological Dependence of Carbon Nanotube Arrays on Catalyst Aggregation. *Applied Surface Science* **2011,** *258* (1), 13–18.
74. Li, S.; Zhang, X.; Zhao, J.; Meng, F.; Xu, G.; Yong, Z.; Jia, J.; Zhang, Z.; Li, Q. Enhancement of Carbon Nanotube Fibres Using Different Solvents and Polymers. *Composites Science and Technology* **2012,** *72* (12), 1402–1407.
75. Feng, C.; Liu, K.; Wu, J.-S.; Liu, L.; Cheng, J.-S.; Zhang, Y.; Sun, Y.; Li, Q.; Fan, S.; Jiang, K. Flexible, Stretchable, Transparent Conducting Films Made from Superaligned Carbon Nanotubes. *Advanced Functional Materials* **2010,** *20* (6), 885–891.
76. Huynh, C. P.; Hawkins, S. C.; Redrado, M.; Barnes, S.; Lau, D.; Humphries, W.; Simon, G. P. Evolution of Directly-Spinnable Carbon Nanotube Growth by Recycling Analysis. *Carbon* **2011,** *49* (6), 1989–1997.
77. Lepró, X.; Lima, M. D.; Baughman, R. H. Spinnable Carbon Nanotube Forests Grown on Thin, Flexible Metallic Substrates. *Carbon* **2010,** *48* (12), 3621–3627.
78. Hitachi Zosen Corporation. *What's New,* 2013. http://www.hitachizosen.co.jp/english/news/2013/01/000850.html.
79. Zhu, C.; Cheng, C.; He, Y. H.; Wang, L.; Wong, T. L.; Fung, K. K.; Wang, N. A Self-Entanglement Mechanism for Continuous Pulling of Carbon Nanotube Yarns. *Carbon* **2011,** *49* (15), 4996–5001.
80. Kuznetsov, A. A.; Fonseca, A. F.; Baughman, R. H.; Zakhidov, A. A. Structural Model for Dry-Drawing of Sheets and Yarns from Carbon Nanotube Forests. *ACS Nano* **2011,** *5* (2), 985–993.
81. Franklin, A. D.; Luisier, M.; Han, S.-J.; Tulevski, G.; Breslin, C. M.; Gignac, L.; Lundstrom, M. S.; Haensch, W. Sub-10 nm Carbon Nanotube Transistor. *Nano Letters* **2012,** *12* (2), 758–762.
82. Kang, S. J.; Kocabas, C.; Ozel, T.; Shim, M.; Pimparkar, N.; Alam, M. A.; Rotkin, S. V.; Rogers, J. A. High-Performance Electronics Using Dense, Perfectly Aligned Arrays of Single-Walled Carbon Nanotubes. *Nature Nanotechnology* **2007,** *2* (4), 230–236.
83. Shulaker, M. M.; Hills, G.; Patil, N.; Wei, H.; Chen, H.-Y.; Wong, H. S. P.; Mitra, S. Carbon Nanotube Computer. *Nature* **2013,** *501* (7468), 526–530.
84. Franklin, A. D. Electronics: The Road to Carbon Nanotube Transistors. *Nature* **2013,** *498* (7455), 443–444.
85. Di, J.; Yong, Z.; Yao, Z.; Liu, X.; Shen, X.; Sun, B.; Zhao, Z.; He, H.; Li, Q. Robust and Aligned Carbon Nanotube/Titania Core/Shell Films for Flexible TCO-Free Photoelectrodes. *Small* **2013,** *9* (1), 148–155.
86. Zhu, Y.; Zhang, X.; Li, R.; Li, Q. Planar-Defect-Rich Zinc Oxide Nanoparticles Assembled on Carbon Nanotube Films as Ultraviolet Emitters and Photocatalysts. *Scientific Reports* **2014,** *4,* 4728–4734.
87. Meng, F.; Zhang, X.; Xu, G.; Yong, Z.; Chen, H.; Chen, M.; Li, Q.; Zhu, Y. Carbon Nanotube Composite Films with Switchable Transparency. *ACS Applied Materials & Interfaces* **2011,** *3* (3), 658–661.

88. Yao, Z.; Di, J.; Yong, Z.; Zhao, Z.; Li, Q. Aligned Coaxial Tungsten Oxide-Carbon Nanotube Sheet: A Flexible and Gradient Electrochromic Film. *Chemical Communications* **2012,** *48* (66), 8252–8254.
89. Cheng, Q. F.; Wang, J. P.; Wen, J. J.; Liu, C. H.; Jiang, K. L.; Li, Q. Q.; Fan, S. S. Carbon Nanotube/Epoxy Composites Fabricated by Resin Transfer Molding. *Carbon* **2010,** *48* (1), 260–266.
90. Cheng, Q.; Wang, J.; Jiang, K.; Li, Q.; Fan, S. Fabrication and Properties of Aligned Multiwalled Carbon Nanotube-Reinforced Epoxy Composites. *Journal of Materials Research* **2008,** *23* (11), 2975–2983.
91. Liu, W.; Zhang, X.; Xu, G.; Bradford, P. D.; Wang, X.; Zhao, H.; Zhang, Y.; Jia, Q.; Yuan, F. G.; Li, Q.; et al. Producing Superior Composites by Winding Carbon Nanotubes onto a Mandrel under a Poly(Vinyl Alcohol) Spray. *Carbon* **2011,** *49* (14), 4786–4791.
92. Wang, X.; Yong, Z. Z.; Li, Q. W.; Bradford, P. D.; Liu, W.; Tucker, D. S.; Cai, W.; Wang, H.; Yuan, F. G.; Zhu, Y. T. Ultrastrong, Stiff and Multifunctional Carbon Nanotube Composites. *Materials Research Letters* **2012,** *1* (1), 19–25.
93. Di, J.; Hu, D.; Chen, H.; Yong, Z.; Chen, M.; Feng, Z.; Zhu, Y.; Li, Q. Ultrastrong, Foldable, and Highly Conductive Carbon Nanotube Film. *ACS Nano* **2012,** *6* (6), 5457–5464.
94. Wang, Y.; Li, M.; Gu, Y.; Zhang, X.; Wang, S.; Li, Q.; Zhang, Z. Tuning Carbon Nanotube Assembly for Flexible, Strong and Conductive Films. *Nanoscale* **2015,** *7* (7), 3060–3066.
95. Zhang, L.; Wang, X.; Xu, W.; Zhang, Y.; Li, Q.; Bradford, P. D.; Zhu, Y. Strong and Conductive Dry Carbon Nanotube Films by Microcombing. *Small* **2015,** *11* (31), 3830–3836.

CHAPTER 14

Safety of Carbon Nanotubes

B. Wan, J. Hou, L.-H. Guo
Chinese Academy of Science, Beijing, China

Contents

14.1 INTRODUCTION

In recent years, engineered nanomaterials (ENMs) and nanotechnologies have improved the quality of life, contributed to the economic growth, and increased the competitiveness of industry globally. Carbon nanotubes (CNTs), because of their unique physical, chemical, and electronic properties including high specific surface area, strength, conductivity, magnetic susceptibility, and catalytic activity *(1,2)*, have become one of the shining stars among nanomaterials. They are perhaps among the most dynamic nanomaterials, and are undergoing the most rapid pace of development.

ENMs are also a double-edged sword in that they do not only possess beneficial properties but in some cases can also cause harmful effects to humans or the environment. In fact, it is their unique physicochemical properties that likely make them more potent in causing adverse effects. During the life cycle of ENMs, they can be released into environment compartments including surface water, the atmosphere, and soil during the

Industrial Applications of Carbon Nanotubes
ISBN 978-0-323-41481-4

production, transportation, application, and disposal of these materials. There is an increasing concern about the potential risks of CNTs to the environment and human health. A quantitative risk assessment is based on the equation: hazard × exposure = risk. However, lack of systematic data on CNT exposure in different environments, or on their effects on human health, makes the assessment of the potential risks and safety of CNTs a huge challenge. The amount of data on exposure in the workplace and the general population is even more limited, which makes the risk assessment of CNT exposure practically infeasible. In the following sections, the biological effects of CNTs will be dealt with, beginning with CNT exposure, and then the interactions between CNTs and living systems as well as the toxicity to a few potential target systems will be discussed.

14.2 POTENTIAL EXPOSURE OF CARBON NANOTUBES

The increasing scope of CNTs in industrial applications includes electronics, lithium-ion batteries, solar cells, supercapacitors, thermoplastics, polymer composites, coatings, adhesives, biosensors, enhanced microscopy imaging, and inks. Additionally, CNTs are also used in pharmaceutical/biomedical devices for bone grafting, tissue repair, drug delivery, and disease diagnostics *(3,4)*. CNT producers are now capable of offering CNTs in commercial-scale quantities—in kilograms, tons, and even hundreds of tons, depending on the specific grade—at more affordable and competitive unit prices. The global commercial market in 2013 was approximately 2000 tons for multiwalled carbon nanotubes (MWCNTs) and 6 tons for single-walled carbon nanotubes (SWCNTs), and is expected to exceed 20,000 tons (SWCNTs and MWCNTs) with a total value estimated to be as high as USD 3.42 billion by 2022 *(5)*. The governments of China, Japan, the United States, and India have increased research funds to promote domestic nanotechnology, which is expected to open new avenues for the CNT market over the next several years. An ever-increasing range of applications and higher production rates of CNTs will undoubtedly result in increased environmental and human exposure.

Human exposure to CNTs can potentially occur in facilities involved with CNTs throughout the CNT's life cycle, ranging from laboratory research, manufacturing, incorporation of CNTs into products, manipulation of CNT-containing products, and finally through disposal and recycling of CNTs *(6–8)*. Maynard et al. *(9)* investigated the effect of handling unrefined CNT in a laboratory production facility and found about 0.2–6 mg per hand of CNT deposits on gloves, indicating that dermal exposure occurred on unprotected body regions. Therefore the use of protective clothing was recommended to minimize dermal exposure to CNTs.

Respiratory tract inhalation is the most likely route of CNT exposure during production and use. Compared with larger respirable particles at the same mass dose, nanosized particles (<0.1 μm in diameter) have a higher deposition efficiency in the deep pulmonary regions (Fig. 14.1) *(10)*. There is sufficient evidence to support the notion that

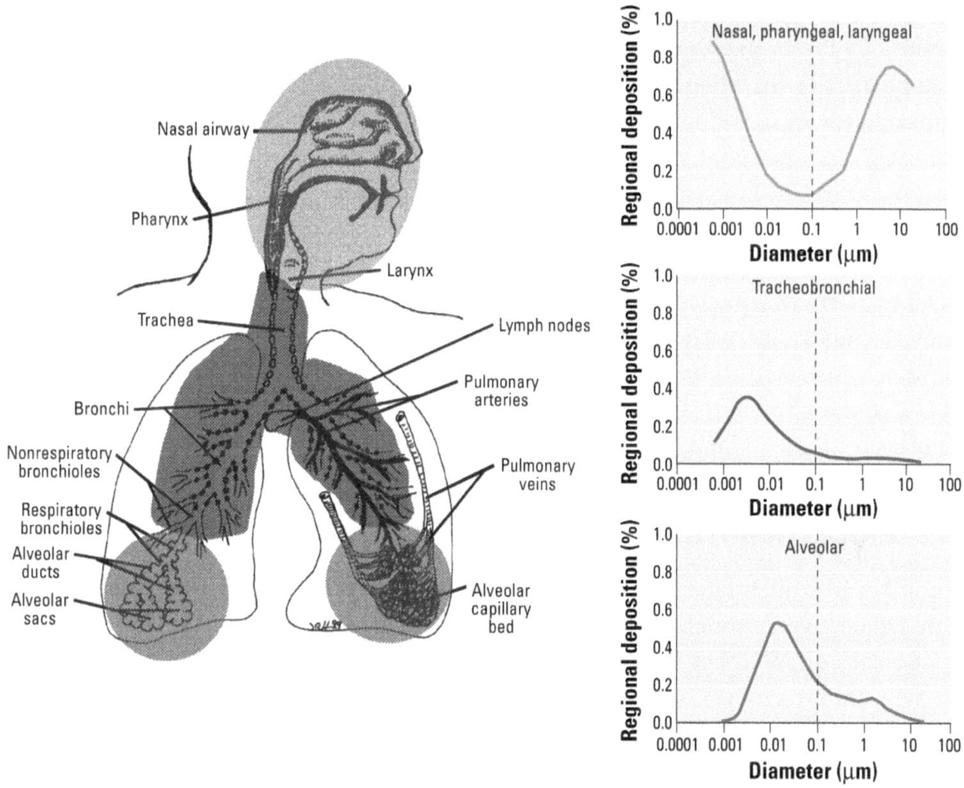

Figure 14.1 Predicted fractional deposition of inhaled particles in the nasopharyngeal, tracheobronchial, and alveolar region of the human respiratory tract during nose breathing *(12)*. *(Reproduced with permission from Environmental Health Perspectives.)*

ultrafine ambient air particles pose considerable health risks if they gain access to the airways. Agitation such as weighing, transferring, mixing, blending, or sonication can lead to aerosolized CNTs in the workplace. Although CNTs may form loose agglomerates with a much greater aerodynamic diameter than the singlet particles because of their low density, they linger in workplace air for a considerable period of time, and a large fraction of these structures are respirable *(11)*. Inhaled CNTs can pass through epithelia of the respiratory tract into the interstitium and access the blood circulation directly or via lymphatic pathways and be transported to other organs and tissues including the central nervous system, liver, spleen and kidneys, where they cause adverse effects *(13)*. Therefore airborne CNT particles should be avoided in workplaces or protective measurements should be taken to limit inhalation exposure to CNTs in occupational settings.

Proposed occupational air exposure limits for specific types of CNTs range from approximately 2 to 50 mg/m^3 based on the estimated risk of noncancerous adverse lung effects (inflammation and fibrosis) *(7)*. The main work practice recommendations described are: (1) wearing masks during the processes that could result in the release of airborne CNTs in the workplace; (2) avoiding handling CNTs in a "free particle" state (e.g., powder form) whenever possible; (3) performing dispensing, weighing, or sonication processes on well-designed ventilated benches; and (4) performing initial and periodic area and personal exposure monitoring *(7)*. Medical surveillance of workers involved with CNTs must include initial and periodic evaluations of the respiratory system, and exposure registries should be developed for all workers with potential exposure to CNTs *(7)*. Although the injection route is not the main route of exposure for CNTs at present, this route is suggested to be relevant for human exposure when CNTs are used in medical applications, particularly drug delivery for cancer therapy *(14)*. Oral exposure of CNTs is considered to be a minor route of consumer or environmental exposure, but CNTs were indicated to be absorbed from the gastrointestinal tract (GIT) to the blood circulation in mice *(15)* and rats *(16)*.

14.3 BIODISTRIBUTION OF CARBON NANOTUBES

A number of studies have investigated the biodistribution of CNTs after labeling them with radioactive isotopes. For example, diethylenetriaminepentaacetic acid (DTPA)-functionalized SWCNTs were labeled with indium to form ^{111}In-DTPA-SWCNT-3 (no free amino groups) and ^{111}In-DTPA-SWCNT-5 (40% free amino groups), and exposed to mice through intravenous injection *(17)*. Soon after injection (0.5 h), both types of SWCNT were detected in most of the organs. The overall radioactive levels reached the highest within 0.5 h and then decreased rapidly, suggesting that the SWCNTs were eliminated through the renal excretion route. A slightly different phenomenon was seen in mice exposed to ^{13}C SWCNTs by a single tail vein injection *(18)*. SWCNTs in the bloodstream were quickly distributed to many organs including the brain, and the clearance was relatively slow in the lungs and livers or even gradually increased in the spleen. The accumulation of CNTs in livers and spleens was also seen for ^{64}Cu-labeled SWCNT-PEG2000 (short-chain) and SWCNT-PEG5400 (long-chain) in U87KG tumor-bearing nude mice *(19)*, ^{86}Y-DOTA-SWCNTs in male athymic nude mice *(20)*, and ^{111}I-DOTA-SWCNTs in BALB/c mice *(21)*.

Wang et al. *(15)* compared the biodistribution patterns of ^{125}I-SWCNTols (OH-SWCNTs) in mice after exposure of CNTs by intravenous, intraperitoneal, or subcutaneous injection, or gavage. No significant difference was observed among these routes of exposure, with most SWCNTs being excreted via urine and feces, and the excretion percentage was 94% via urine and 6% via feces. It is worth noting that SWCNTs can be readily accumulated in bones and retained for a long time, indicating a potential risk of CNTs to the immune system, where immune cells originated in mammals. By contrast, the accumulation of CNTs in organs was much less when high-pressure carbon monoxide (HiPco)-SWCNTs were singly injected at 75 mg into rabbits through implanted jugular vein catheters *(22)*. No SWCNTs were detected in the liver, spleen, or kidney after a single intratracheal instillation of HiPco-SWCNTs *(23)*. Biocompatible SWCNTs could be eliminated through the kidney and bile ducts *(24,25)* of animals receiving intraperitoneal or intravenous injection of CNTs.

Generally speaking, blood clearance of CNTs was rapid and accumulation in the spleen and liver, and in some cases kidneys, was higher after intravenous than after intraperitoneal injection. In addition to feces and biliary pathways, renal excretion of CNTs is the main pathway for the clearance of CNTs from animals. Brain and bones are potential target organs of CNT exposure, which deserve serious attention because the adverse effects in these sites would cause significant consequences to the health of exposed individuals. There is still some controversy regarding the influence of exposure routes on the biodistribution of CNTs. The insignificant difference in CNT biodistribution patterns after various exposure routes *(15)* might be because CNTs could be actively transported throughout the bodies via the bloodstream or by phagocytes *(13)*.

14.4 TRANSMEMBRANE TRANSPORTATION OF CARBON NANOTUBES

Nanoparticle uptake usually begins with an initial adhesion of the nanoparticle to the cell and interactions with the lipids, proteins, and other components of the cell membrane. This is followed by the activation of an energy-dependent uptake mechanism *(26–28)*, which allows the nanoparticles to be internalized into the cell and further trafficked to different subcellular locations, typically ending in lysosomal accumulation *(26,29,30)*. The interactions between nanoparticles and living systems are commonly mediated by what adsorbs to the nanoparticles in the biological environment rather than the pristine surface. As a result, adhesion of nanoparticles on cells is strongly reduced in comparison to that of bare materials *(31)*. The lowered adhesion in the presence of proteins thereby causes a concomitant decrease in nanoparticle uptake efficiency.

There are four types of endocytosis: clathrin mediated, caveolae mediated, macropinocytosis, and phagocytosis *(32)*. Phagocytosis is the main mechanism by which particulate matters larger than 1 μm (with a maximum phagocytosis rate of 2–3 μm) in diameter are internalized by macrophages *(33)*. Macropinocytosis is an actin-driven process that occurs in the highly ruffled regions of the plasma membrane to form a pocket, which then pinches off into the cell to form a vesicle (0.5–5 μm in diameter) filled with a large volume of extracellular fluid and molecules. The vesicle then moves into the cytosol and fuses with other vesicles, such as endosomes and lysosomes *(34)*. Macropinocytosis can be transiently induced in most cells and might have a role in modulating cell migration and stimulating immune surveillance in antigen-presenting cells *(35,36)*. Clathrin-dependent endocytosis is a receptor-mediated endocytosis that begins with ligand binding to plasma membrane-spanning receptors. A signal is then sent, leading to membrane coating, and the receptor and its ligand are opsonized in the clathrin-coated vesicles

(37,38). Nanoparticles less than 200 nm (optimal size is ~25 nm) are usually deposited in the vesicles, which fuse with the early endosomes and are released into the cytoplasm *(39,40)*.

The detailed mechanism of CNT endocytosis has not been reported yet, but evidences of particle internalization showed that nonspherical particles with lower length-diameter ratios could be readily internalized via several different mechanisms of endocytosis, mostly phagocytosis and clathrin-mediated endocytosis. In the case of particles or fibers with high length-diameter ratios, macropinocytosis plays a major role in their internalization *(41)*. Therefore it was speculated that CNTs, with high length–diameter ratio, would most likely be endocytosized through macropinocytosis. A study by Jin et al. *(42)* showed that DNA-wrapped SWCNTs with an average length of 320 nm had the highest uptake in NIH-3T3 cells when compared with the other DNA-wrapped SWCNTs (660, 430, and 130 nm). Although the influence of this length-selective uptake of CNTs on their cytotoxicity is unknown, CNTs with length long enough (>20 μm) cannot be internalized efficiently by cells, leading to frustrated phagocytosis and serious mechanical damage on cell membrane and the induction of inflammation *(43,44)*. These findings suggest that the relationship between transmembrane transportation of CNTs and their length is not monotonically linear. More evidence showed that all CNTs with various surface modifications could penetrate cell membranes in multiple cell models, indicating that the transmembrane transportation of CNTs is independent of surface functional groups *(44,45)*.

14.5 BIODEGRADATION OF CARBON NANOTUBES

Once inside organisms, the physical and chemical properties of nanomaterials usually undergo a series of alterations that either increase or reduce their toxicity. CNTs were considered to be resistant to chemical damage because of their rigid and perfect chemical structure, which rendered them immune to biodegradations. The bioaccumulation of CNTs in organs and tissues was usually high, and as a consequence, the nondegradable nanotubes are more prone to cause toxic effects. However, recent studies reported that after functionalization that introduces defect sites on the sidewall and tips of the CNTs, they were degraded by peroxidases or under oxidative conditions. Allen et al. *(46,47)* for the first time investigated the biodegradation of CNTs by using a variety of techniques including transmission electron microscopy, thermogravimetric analysis, gel electrophoresis, and UV–Vis–NIR spectroscopy, and demonstrated that acid-treated SWCNTs were degraded by plant-derived horse radish peroxidase (HRP) in the presence of low concentrations of hydrogen peroxide. Thorough disintegration of the CNTs was achieved in 10 days, and oxidized aromatic fragments were detected in the course of the nanotube degradation, which eventually evolved to carbon dioxide *(46,47)*. A study investigated the degradation of SWCNTs and MWCNTs in artificial fluids mimicking phagolysosome content, and found that carboxylated SWCNTs were degraded over time along with the fracture of CNTs and disaggregation of bundles, further proving the feasibility of biodegradation of CNTs *(48,49)*.

In mammalian immune systems, neutrophils and monocytes/macrophages can be activated as they come into contact with different bacteria or fungi, and the oxygen consumption increases significantly along with the rapid release of reactive oxygen species (ROS; superoxide radicals and hydrogen peroxide), which is defined as respiratory burst. Respiratory burst is a crucial reaction that occurs in phagocytes to degrade internalized particles and bacteria, and thus plays an important role in the immune system. The whole reaction is initiated with the activation of nicotinamide adenine dinucleotide phosphate-oxidase that reduces O_2 to superoxide anion and then H_2O_2. Neutrophils and monocytes utilize myeloperoxidase (MPO) to further combine H_2O_2 with Cl^- to produce hypochlorite, which plays a role in destroying bacteria as well as internalized particles such as SWCNTs *(50,51)*. Kagan and colleagues *(51)* compared the degradation of SWCNTs with different functional groups in neutrophil and macrophage models. Their results found that carboxylated SWCNTs were the most readily degraded and the biodegraded nanotubes did not generate inflammatory response when aspirated into the lungs of mice, while pristine SWCNTs were resistant to degradation. Molecular modeling suggests that there are two binding sites of MPO, and interactions of basic amino acids of the enzyme with the carboxyls on the CNTs position the nanotubes near the catalytic site, facilitating the attack of the free radicals to the defect sites of the nanotubes (Fig. 14.2). However, it is uncertain as to whether or to what extent MPO would degrade SWCNTs in vivo (e.g., in the lung) because polymorphonuclear neutrophil (PMN) recruitment after SWCNT exposure is a transient rather than persistent response, and there is no strong evidence for SWCNT phagocytosis by PMNs. Instead, alveolar macrophages are the most likely cells engulfing CNTs, but their degradation efficiency is much lower than PMNs because of a relatively lower level of MPOs present in the former cells *(52)*.

These studies illustrated the possible pathways and mechanisms for the biodegradation of SWCNTs, which showed close relationship between the surface structure and the biodegradation. Table 14.1 lists the degradation of CNTs with different surface modifications and oxidation conditions. For both SWCNTs and MWCNTs, it seems that carboxylation on the CNT surface is a prerequisite for CNT degradation. The possible reason is because the carboxylation on CNTs might break the π ring of CNTs and create two reactive sites that are potential sites for oxidative attack *(53)*.

14.6 TOXIC EFFECTS OF CARBON NANOTUBES

14.6.1 Respiratory System

Inhaled particles with aerodynamic diameters less than 0.1 μm have much higher deposition efficiency in the pulmonary region compared with larger particles. Nanomaterials

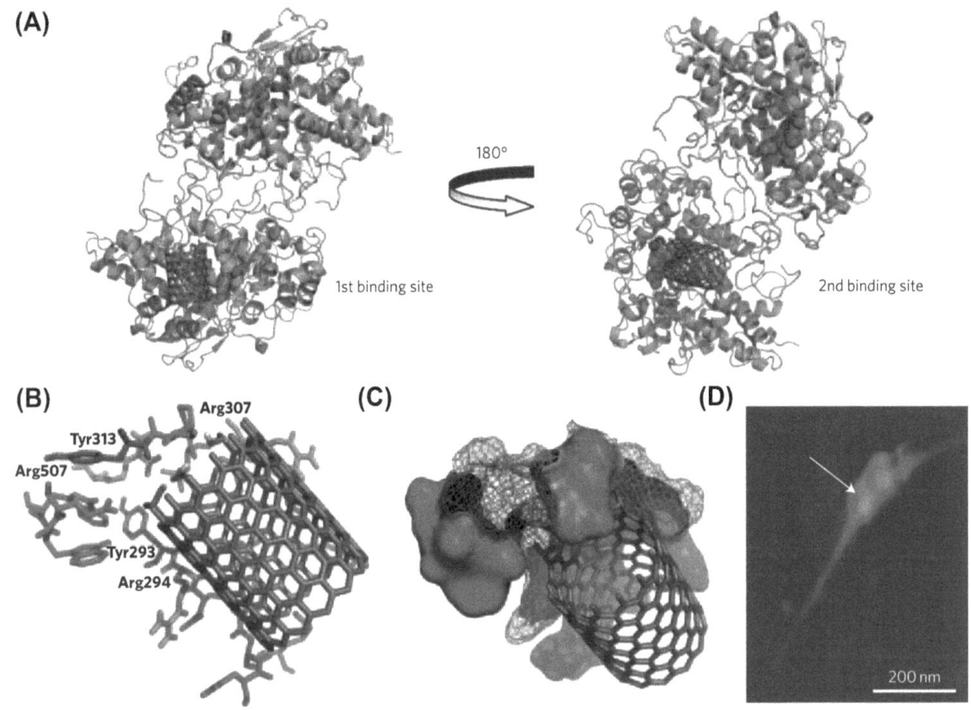

Figure 14.2 Molecular modeling demonstrating possible nanotube interaction sites on human myeloperoxidase (hMPO). (A) Two putative binding sites for nanotubes on the hMPO monomer surface are located on each side of the protein. The best model site with lowest binding energy is shown for carboxylated short-cut nanotubes (*left*) and pristine nanotubes (*right*). (B) Stabilization of the carboxylated ends of the nanotubes by arginine (*yellow*) and tyrosine (*red*) residues in hMPO in the first binding pocket (*left*). All residues shown are within 5 Å of nanotubes and are predicted to participate in the degradative catalysis of the nanotubes. They include Tyr293, Arg294, Arg307, Tyr313, and Arg507. (C) As (B), using space-fill representation. (D) Atomic force microscopy (AFM) image confirming the binding of an hMPO dimer with a single nanotube *(51)*.

Table 14.1 Influence of surface functionalization on the oxidative degradation of CNTs

Materials	Functional groups	Oxidation conditions	Degradation
SWCNT/MWCNT	Pristine	HRP/PSF	No
SWCNT	COOH	HRP/MPO	Yes
SWCNT	COOH/taurine	MPO	No
SWCNT	COOH/phosphatidyl serine	MPO	Yes
SWCNT	COOH/phosphatidyl choline	MPO	No
SWCNT	COOH	PSF	Yes
SWCNT/MWCNT	Aryl-sulfonation	PSF	No
MWCNT	COOH	HRP/PSF	Yes
MWCNT	COOH	HRP	Yes
MWCNT	N-doped	HRP	Yes
MWCNT	$CONH\text{-}(CH_2CH_2O)_2CH_2CH_2\text{-}NH^{3+}$	HRP/PSF/MPO	Yes
MWCNT	COOH+1,3-dipolar cycloaddition	HRP/PSF	Yes

HRP, horseradish peroxidase; *MPO*, myeloperoxidase; *MWCNT*, multiwalled carbon nanotubes; *PSF*, Phagolysosomal Simulant Fluid; *SWCNT*, single-walled carbon nanotubes.

by definition are within this range, therefore airborne nanoparticles are readily inhaled into airways and deposit in lungs *(54)*. The only published report on the direct real-world human exposure to CNTs was from Maynard and coworkers *(9)*, who investigated the release of particles from unrefined SWNTs into the air and its effect on workers at a small-scale production facility. The results showed very low airborne particle concentrations were generated as a consequence of handling unrefined materials with no effects. Although health effects of CNTs on humans are not well characterized, there is some preliminary evidence that exposure to CNTs can induce severe adverse effects in rodents. Individual CNTs have structural characteristics resembling fibers and their extreme aspect ratio (length:diameter ratio) and low solubility in aqueous media suggest toxic properties similar to those observed with other fibrous particles such as asbestos, raising concerns that their widespread use may lead to pleural fibrosis and/or mesothelioma. Short or tangled CNTs could move rapidly through to the pleural space and drain easily into the lymphatic system, whereas the movement of long, fibrous CNT aggregates is impeded at the drainage points of pleural tissue, where they may then exert pathogenic effects (e.g., inflammation) over time (Fig. 14.3).

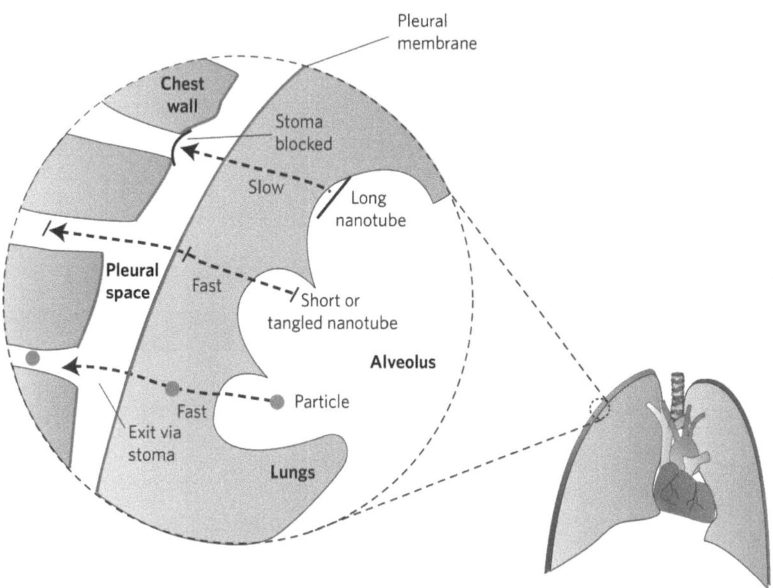

Figure 14.3 The lung and chest cavity are surrounded by thin membranes called pleura (*red lines*). Between the two membranes is the pleural space, which contains fluid that lubricates the movement of the lung. Small particles (*blue circles*) and short or tangled nanotubes (*black lines*) can move rapidly through to the pleural space and drain easily into the lymphatic system through pores in the chest wall (3–8 μm diameter) whereas long, fibrous nanotubes cannot. Long nanotubes that lodge at the drainage points and block them can cause inflammation, leading to pathological changes over time *(55)*.

A number of studies explored lung toxicity of different types of CNTs (SWCNT or MWCNT, intact or ground MWCNT, purified or not) in varying animals species (rat, mouse, or guinea pig), and demonstrated that CNTs in lungs were retained mainly within the alveolar septa and subpleural tissues inducing injury, inflammation, and fibrosis. Warheit *(56)* and Lam *(57)* investigated the pulmonary lesions caused by intratracheal instillation of unpurified SWCNTs in rodents. Histopathological examinations on lungs revealed that all CNT particles induced dose-dependent epithelioid granulomas and/or interstitial inflammation in the animals. These lesions persisted and were more pronounced after 90-day exposure. The lungs of some animals also revealed peribronchial inflammation and necrosis that had extended into the alveolar septa (Fig. 14.4). The CNTs used in these studies contained a high percentage of Fe and Ni (20−30%). Shvedova et al. *(52)* demonstrated that pharyngeal aspiration of acid-purified SWCNTs elicited unusual pulmonary effects in C57BL/6 mice, including a robust but acute inflammation, progressive fibrosis, and granulomas. SWCNT-induced granuloma was mainly associated with hypertrophied epithelial cells surrounding SWCNT aggregates but diffuse interstitial fibrosis and alveolar wall thickening were found associated with dispersed SWCNTs. In addition, SWCNTs also induced a dose-dependent increase in bronchoalveolar lavage fluid (BALF) protein content, lactate dehydrogenase and γ-glutamyl transferase activities, accumulation of 4-hydroxynonenal (an oxidative biomarker), and depletion of glutathione. The elevation of proinflammatory cytokines (TNF-α, IL-1β) was observed, which was followed by neutrophil accumulation, lymphocyte, and macrophage influx over time *(52)*. Another study performed by the same group investigated the pulmonary response of C57BL/6 mice to repeated exposures to SWCNTs and showed in great details that granuloma formed near bronchioles and

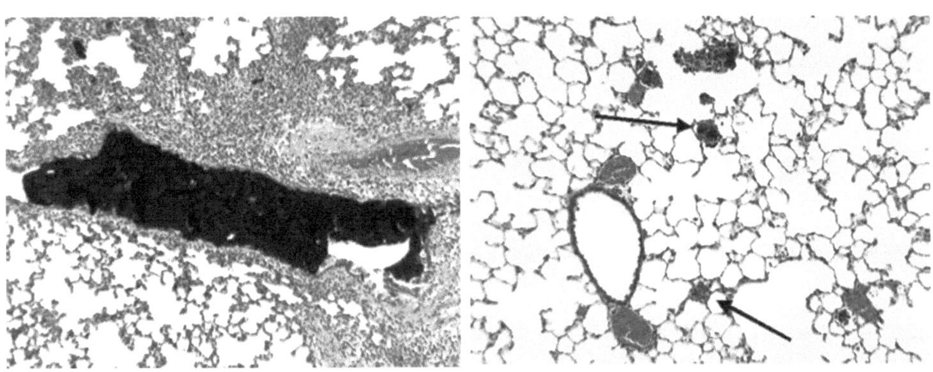

Figure 14.4 (*Left*) Light micrograph of lung tissues from a rat exposed to 5 mg/kg SWCNT (a few hours after exposure). The major airways are mechanically blocked by the SWCNT instillation. (*Right*) Light micrograph of lung tissues from a rat exposed to single-walled carbon nanotubes (SWCNTs) (1 mg/kg) for 1 week. Note the early development of lesions surrounding the instilled SWCNT (*arrows*) and the nonuniformity of CNTs *(56)*.

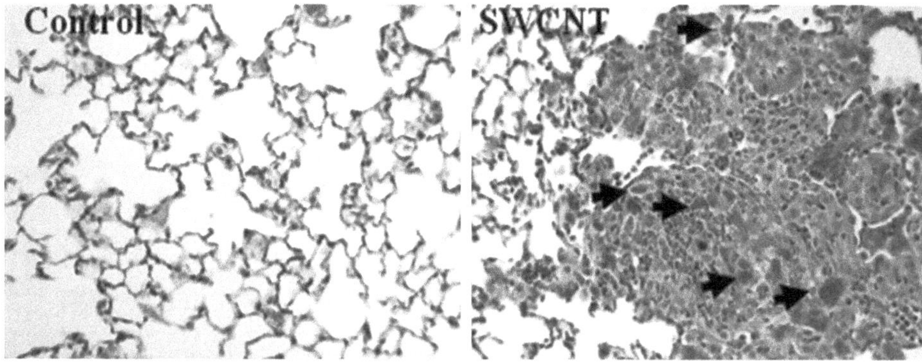

Figure 14.5 Histopathology of lung from the single-walled carbon nanotube (SWCNT) repeated aspiration study (40 mg per mouse, twice a week, for 3 weeks): SWCNTs caused granuloma formation near bronchioles and adjacent alveoli. Fibrosis found within these granulomas extended for a short distance into surrounding septal walls. *Arrows* indicate agglomerates of SWCNTs within granulomas *(58)*.

adjacent areas contained clusters of large macrophages and multinucleated giant cells with SWCNT agglomerates observed within (Fig. 14.5). Many of these granulomas contained numerous polymorphonuclear cells (pyogranulomas), and some extended from the alveoli into the wall and lumen of the bronchiole (*58*).

It is worth noting that studies showed the occurrence of lung toxicity including persistent fibrotic granulomas, cell injury, and interstitial inflammation after CNT exposure, but the toxic effects are most often not dose dependent, which could be caused by the aggregation/agglomeration of CNTs being clumped in larger airways that did not allow for adequate dispersion of the delivered CNTs in the lungs. Muller et al. *(54)* ground the CNTs into shorter fibers and compared the distribution and effects with the intact ones. The results showed that the ground CNTs were better dispersed in lungs (Fig. 14.6) and caused dose-dependent toxic effects and the lesions were more evenly distributed, suggesting that CNTs are intrinsically toxic to lungs. Exposure to well-dispersed MWCNTs in mice via pharyngeal aspiration resulted in dose- and time-dependent pulmonary inflammation *(59–64)*.

Similarly, N-doped (N-)MWCNTs were less toxic than nondoped pristine nanotubes after intrathecal injection, which might be caused by their improved dispersibility *(65)*. Elgrabli et al. *(66)* used bovine serum albumin to improve the dispersibility of MWCNTs and observed no inflammatory, physiological, or histological pathologies. It appears that the agglomeration of CNTs in the airways is the primary cause of morbidity and granuloma formation.

Most of the studies were performed on animal models using the intratracheal instillation technique, which is not a realistic way of exposure. The nonphysiological rapid delivery of the CNT particles, the possible delivery of aggregates, and the bypass of the nose-filtering mechanism can induce artifacts of so-called "foreign-body

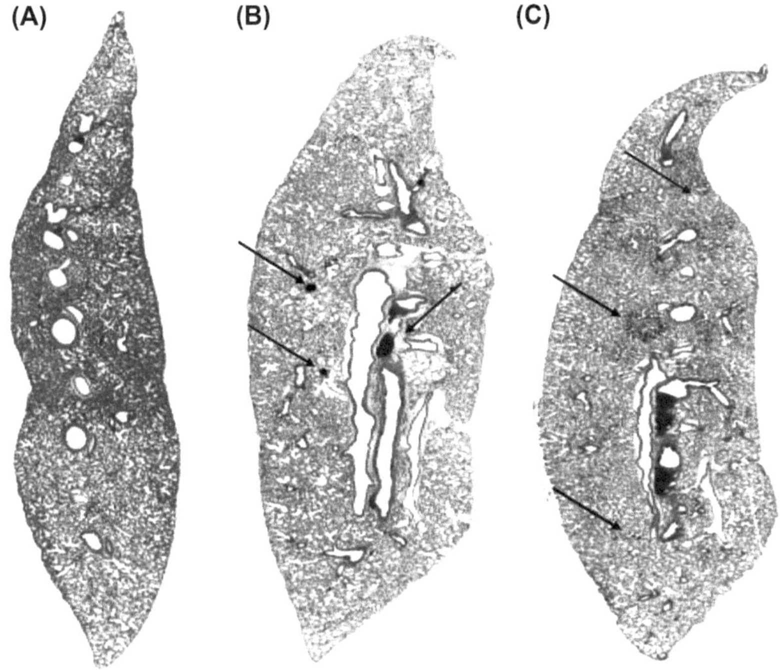

Figure 14.6 Sections of control rat lung (A) and immediately after intratracheal instillation of 2 mg/animal ground MWCNTs (C, *arrows*) which are much better dispersed in the lung than intact MWCNTs, which mainly remained clumped in large airways (B, *arrows*) *(54)*.

granulomas" *(54,67)*. Most of the literature agrees that chronic inflammation and oxidative stress observed during and after exposure to CNT can induce adverse health effects such as genotoxicity and cancer.

14.6.2 Cardiovascular System

Nanoparticles inhaled in lungs may penetrate the alveolar membrane and reach blood circulation. The unique physical characteristics of CNTs raise concerns that they may have not only pulmonary toxicity but also be associated with extrapulmonary toxicity including heart and blood vessels. In vitro, CNTs induced dose-dependent mitochondria DNA damage and oxidative modification of low density lipoproteins in human aortic endothelial cells *(68)*. Walker et al. *(69)* examined the effects of both purified SWCNTs and MWCNTs and found marked actin filament and vascular endothelial-cadherin disruption, cytotoxicity, and reduced tubule formation in these cells. In vivo, a single intrapharyngeal instillation of SWCNTs to HO-1 reporter transgenic mice induced activation of heme oxygenase-1, a marker of oxidative insults, in aorta and heart tissues. Furthermore, C57BL/6 mice exposed to SWCNTs (10 and 40 μg/mouse) developed aortic mitochondria DNA damage and accumulation of protein carbonyl levels at 7,

28, and 60 days after exposure. Repeated exposure to SWCNTs (20 μg/mouse once every other week for 8 weeks) stimulated the progression of atherosclerosis, a major pathogenic factor for the development of cardiovascular diseases and resulted in accelerated plaque formation in aortas and the brachiocephalic arteries of $ApoE^{(-/-)}$ mice *(68,70)*. Castranova et al. *(11)* showed that pulmonary exposure to SWCNTs induced oxidative stress in aortic tissues and increased plaque formation in an atherosclerotic mouse model. In addition, MWCNTs depressed the ability of coronary arterioles to respond to dilators. These cardiovascular effects may result from neurogenic signals from sensory irritant receptors in the lung.

Semberova et al. *(71)* reported that CNTs activated complement cascade by directly binding with plasma proteins such as fibrinogens and apolipoproteins. Activated complement serum proteins are important for phagocyte stimulation and lead to lytic attack on cell membranes. CNTs in blood have also been reported to activate transforming growth factor-β1 (TGF-β1) along with the induction of platelet aggregation and prothrombotic changes in hepatic microvasculature of healthy rodents. In addition, they accelerated the rate of vascular thrombosis in rat carotid artery *(72)*. Fe-rich and Fe-depleted SWCNTs increased levels of creatine kinase in serum, endothelin-1, a potent vasoconstrictive substance, in BALF and plasma, and angiotensin-converting enzyme in plasma *(73)*. Histopathological changes were induced by both types of SWCNTs in the heart. Decreased heart rate and number of baroreflex sequences were observed in rats intratracheally instilled with SWCNTs, and SWCNTs were suggested to affect autonomic cardiovascular control regulation and alter baroreflex function *(74)*. But when pristine or hyaluronate-functionalized SWCNTs were singly injected into tooth sockets or the intraperitoneal cavity of rats, no effects were observed on cardiac function including systolic or diastolic tension, coronary flow, or heart rate *(75)*. These findings suggest that SWCNTs have direct or indirect oxidative effects on vascular system through mitochondria oxidative modifications and accelerating atheroma formation. SWCNTs also affected the autonomic cardiovascular control regulation and altered the baroreflex function.

14.6.3 Skin

For humans, skin contact is a very likely route that may cause an inflammatory response via dermal toxicity. There is a lack of information on whether CNTs can penetrate the corneous layer or not. Huczko et al. *(76)* found no association between working with soot containing CNTs and the risk from possible skin allergies as assessed by various dermatological tests on human volunteers and rabbits *(77)*. However, exposure of mice skin to unpurified CNTs caused oxidative stress, depletion of glutathione, an increase of dermal cell number, localized alopecia, and skin thickening *(78)*. Rouse et al. *(79)* investigated the dermal absorption of amino acid-derivatized fullerenes using dermatomed porcine skin, and the results showed that fullerenes in the unflexed skin were mostly localized in the epidermal layer, while those in the flexed areas penetrated deeper in the epidermal and even the dermal layer, where fullerenes could potentially be absorbed through capillaries.

After exposure of human epidermal keratinocytes (HaCaT) to SWCNTs, oxidative stress and cellular toxicity were indicated by formation of free radicals, accumulation of peroxidative products, antioxidant depletion, and loss of cell viability. Exposure to SWCNTs also resulted in ultrastructural and morphological changes in cultured skin cells. These data indicate that dermal exposure to unrefined SWCNTs may lead to dermal toxicity caused by accelerated oxidative stress in the skin of exposed workers *(80)*. In addition, 6-aminohexanoic acid-derivatized SWCNTs decreased the viability of HEKs cells and induced the expression of interleukin (IL)-6 and IL-8 *(81)*. CNTs were most often localized in intracytoplasmic vacuoles in HEKs, causing lower cell viability and increased expression of inflammation cytokines *(82)*. It is speculated that CNTs' distribution profile and thus dermal toxicity will be highly dependent on the physicochemical characteristics of the nanotubes, the dispersing agents, coating materials, or functional group and their length. More research is needed to investigate the toxic effects of CNTs on skin both in vitro and in vivo.

14.6.4 Gastrointestinal Tract

The GIT is a selective mucosal barrier that represents a considerable surface area for potential interaction with nanoparticles. In addition to direct ingestion, a proportion of inhaled particulate materials are eventually removed via the GIT, after being mobilized up the trachea via the mucociliary escalator *(6)*. Once inside the GIT, CNTs are readily absorbed onto mucosa because of their smaller size but cause no significant adverse effects on the GIT histopathologically. Perhaps the most observed effect is the disruption of COOH-SWCNTs on tight junctions in the GIT through perturbation of zonulin-1 distribution *(83)*, which could enhance paracellular permeability. In human epithelial colorectal adenocarcinoma cells (Caco-2), SWCNT-COOH also inhibited efflux pump activity through interaction with the P-glycoprotein efflux system, with increased cellular accumulation of the pump substrate, rhodamine-123 *(84)*. When male mice were exposed to CNTs through drinking water, there was a significant change in the fine structure of the villi of the small intestine and proliferation of epithelial cells. The effect was most pronounced in animals exposed for up to 2 months *(85)*. Carboxylated MWCNTs induced hepatotoxicity in Swiss—Webster mice through oxidative stress activation *(86)*. Moreover, Ji et al. found that acid-oxidized MWCNTs (O-MWCNTs) exposure through intravenous injection caused spotty necrosis, inflammatory cell infiltration and mitochondria swelling and lysis in hepatocytes of Kunming mice. Liver damage of the T-MWCNT (Tween-80-dispersed) group was more severe than that of the O-MWCNT, but was recovered after 60 days *(87)*.

14.6.5 Immune System

Inhaled CNTs through the airway tend to deposit in the alveolar, where they either are engulfed by macrophages or, at significant amounts, induce oxidative stress and cause direct damage to cell membrane, inducing cytotoxicity and immune responses *(88)*. When CNTs are not efficiently recognized and internalized by alveolar macrophages, they can also diffuse into the circulatory system and accumulate in a variety of organs, where they interact with the immune cells and molecules. SWCNTs with a low content of metal impurities were shown to suppress chemotaxis of primary human monocytes. Moreover, macrophage engulfment of apoptotic target cells was significantly impaired following preincubation with SWCNTs at noncytotoxic concentrations *(89)*. The same effects were observed in mice primary peritoneal macrophages *(90)*. In addition, acid-functionalized SWCNTs were demonstrated to inhibit the antigen-presenting function and altered cytokine production profile, resulting in reduced T-cell proliferation and Th1 differentiation. These data suggest that SWCNT exposure might affect the cell-mediated immunity and promote Th1-associated autoimmune diseases. The ingestion of microorganisms and tissue homeostasis may be compromised by CNTs because of suppressive effects on macrophages that would ultimately increase the susceptibility of host-to-pathogen infections.

To examine the effect of CNTs on allergic immune responses, Balb/cA mice were exposed to SWCNTs and MWCNTs through subcutaneous administration in the footpad during sensitization with the allergen ovalbumin (OVA). Increased serum levels of OVA-specific IgE, the number of eosinophils in BALF, and the secretion of Th2-associated cytokines in the mediastinal lymph node cells were observed. MWCNTs also increased IgG2a level, neutrophil cell number, and tumor necrosis factor-alpha and monocyte chemoattractant protein-1 levels in BALF. These findings indicate that CNTs promote allergic immune responses in mice *(91)*.

14.6.6 Genotoxicity and Carcinogenicity

Genotoxicity refers to the property of chemical agents that damages the genetic material within a cell causing mutations, which may lead to cancers. DNA damage includes single- and double-strand breaks, loss of excision repair, cross-linking, alkali-labile sites, point mutations, and structural and numerical chromosomal aberrations (CAs). A number of studies have demonstrated that SWCNTs and MWCNTs possess the potential to be genotoxic, although the findings are somewhat contradictory probably because of the use of different CNTs and biological systems. In vitro studies with mammalian cells showed that many types of CNTs induced DNA strand breakage, DNA base modification, and chromosomal damage. By using Comet assay, the genotoxicity of a mixture of CNTs (>50% SWCNTs and 40% other CNTs) was evaluated in human bronchial epithelial BEAS 2B cells in vitro, and the results found that CNTs induced a dose-dependent increase in DNA damage at all treatment times (24, 48, or 72 h), even at the lowest dose (~3.8 μg/mL) tested. However, there is no dose-dependent effect in the micronucleus assay, although at some time points a significant increase in micronucleated cells was seen. This activity may be caused by the fibrous nature of these carbon nanomaterials with a possible contribution by catalyst metals present in the materials, for example, Co and Mo in CNTs (<5 wt%) *(92)*. In addition, SWCNTs were shown to induce a dose-dependent increase in chromosome-type CAs and chromatid-type CAs *(93)*. The occurrence of CAs was the most significant after 48-h exposure because CNTs need to access to the nucleus during the first mitosis and about two cell cycles are needed for CA induction.

In vivo, SWCNTs and MWCNTs have been reported to cause aortic mitochondria DNA damage in mice and rats *(68,94)*. Rats orally exposed to SWCNTs (0.064 or 0.64 mg/kg) suspended in both saline and corn oil were found to have higher levels of 8-oxo-7,8-dihydro-20-deoxyguanosine in the lungs and liver *(16)*. SWCNTs increased the frequency of K-ras gene mutations in the lung of female C57BL/6 mice after inhalation *(95)* and pharyngeal aspiration *(96)*. Micronucleated cells were also increased in lungs after inhalation *(96)*. An increased level of DNA damage in BALF cells was detected by Comet assay after intratracheal instillation *(97)*.

However, in a study by Naya et al. *(98)*, a micronucleus test of supergrowth (SG)-SWCNTs was performed by oral gavage of SG-SWCNTs to mice *(98)* at relatively high dosages (60 or 200 mg/kg/day for 2 days). No increases in micronucleated immature erythrocytes were observed. Similarly, Comet assay was performed by using lung tissues of rats exposed to N-SWCNTs singly at 0.2 or 1.0 mg/kg or repeatedly (once a week for 5 weeks) at 0.04 or 0.2 mg/kg. No effect was obtained, although pulmonary inflammation was observed at both doses *(99)*. Neither did this type of SWCNT induce increases in micronucleated immature erythrocytes of mice *(100)*. No change in the levels of oxidative damage of DNA by Comet assay was observed in the lungs of female $ApoE^{-/-}$ mice subjected to intratracheal instillation *(101)*. These findings suggest that CNTs induced DNA damage in the lung and aorta by airway exposure and in the lung and liver by oral exposure. By contrast, CNTs did not induce an increased frequency of micronucleated immature erythrocytes or DNA damage in the lung cells of rats instilled even at doses that elicited inflammation.

The influence of the physical–chemical properties including length, surface area, density, shape, agglomeration, aggregation, and impurity remains unclear. But rigid CNTs with a high aspect ratio have been of particular concern since they may be able to induce lung cancer and mesothelioma in a manner similar to asbestos fibers *(102)*. In a dose–response study using p53 heterozygous mice exposed (intraperitoneal to MWCNTs, a cumulative mesothelioma incidence was obtained *(103)*. All mice in the low-dose (3 μg/mice) group that survived the experiment displayed atypical mesothelial hyperplasia, along with an accumulation of lymphocytes and macrophages. Some of the findings were considered to be mediated by frustrated phagocytosis. Long CNTs deposited in the distal alveoli could migrate to the intrapleural space, causing mesothelial proliferation and inflammatory response in the pleural cavity. The mechanism for the formation of mesothelioma was considered to be caused by the incomplete phagocytosis of CNTs by macrophages in pleural space. This would elicit a modest proinflammatory cytokine response, which would amplify further proinflammatory responses in mesothelial cells *(104)*. Based on these findings, two main modes of action can be proposed for the genotoxicity of CNTs: (1) generation of ROS by the particles themselves or particle-elicited inflammation and (2) mechanical interference with cellular components, either direct interaction with the DNA or indirectly as a consequence of interaction with a non-DNA target such as the spindle apparatus or DNA repair enzymes *(105)*. Oxidative stress and inflammation can ultimately lead to carcinogenesis.

14.7 FACTORS DETERMINING CNT TOXICITY

The bioactivity/toxicity of CNTs varies with their physicochemical properties including impurities, structure, length, aggregate/agglomerate status, surface chemistry, and possibly exposure route. There is great interest in elucidating the relationships between these properties of CNTs and their bioactivity. Understanding such relationships would greatly benefit the design and safe use of CNTs with lower toxicity, better biocompatibility, and higher efficiency.

14.7.1 Metal Impurities

Increasing evidence suggests that contaminating catalytic metals in CNTs appear to play an important role in dictating CNT toxicity. The metal contaminants are introduced into CNTs during the synthesis process and are most often wrapped with a layer of graphene shell so that they are very difficult to remove completely. But they can be released into cell culture or human bodies in varying degrees, and frequently introduce variations in experiments. It is well known that transition metals are a definite source of free radicals, which are important in ultrafine particle toxicity *(106)*. Kagan et al. *(51)* demonstrated that Fe-rich SWCNTs induced more generation of ROS and nitric oxide in RAW264.7 cells than did Fe-depleted ones. Metal impurities contribute to the cytotoxicity partially by inhibiting glutathione and inducing oxidative stress in cells *(107)*.

However, metal contaminants do not appear to cause different pulmonary responses in vivo *(57)*. Both metal-containing and metal-depleted SWCNTs are able to penetrate the cell membrane and induce similar degrees of granulomatous reactions. Similarly, the inflammatory response to aspiration of raw (17.7% iron) versus purified (0.23% iron) SWCNTs was not significantly different *(52,95)*.

14.7.2 Aggregate/Agglomerate

The degree of aggregation/agglomeration dictates the shape and surface area of CNT bundles, which is considered an important factor influencing the toxicity of CNTs. Wick et al. *(108)* have investigated the cytotoxicity of CNTs at varying degrees of agglomeration in human MSTO-211H cells and found that rope-like agglomerates induced more pronounced cytotoxic effects than suspended CNT bundles at the same concentrations. Agglomerated MWCNTs were reported to induce a greater degree of inflammation and stronger granulomatous response *(94,109)*. In contrast, well-dispersed SWCNTs and MWCNTs were more fibrogenic in vitro than agglomerated CNTs, causing greater proliferation and collagen formation with fibroblasts and greater TGF-β1 production by lung epithelial cells *(110–112)*. Aspiration of a well-dispersed SWCNT preparation also caused more transient inflammation and persistent interstitial fibrosis than poorly dispersed SWCNTs *(113)*. Well-dispersed CNT suspension might facilitate interactions between CNTs and cells, resulting in higher toxicity *(114)*.

14.7.3 Surface Chemistry

Successful application of CNTs most often begins with the control of dispersion. SWCNTs have a strong tendency to aggregate into microscopic bundles. Therefore to extend their applications, considerable efforts have been made to improve the dispersion of SWCNTs in water and organic solvents *(115,116)*. In most cases, improved hydrophilicity of CNTs is accomplished with the addition of functional groups such as carboxyl and hydroxyl on their surface.

The surface groups on one side influence the biodistribution dynamics of CNTs, and on the other side alter the toxicity of CNTs. ^{125}I-labeled hydroxylated SWCNTs were quickly cleared from the bloodstream and distributed in many organs including bones, and ^{14}C-taurine-MWCNTs were mainly accumulated in livers and retained for over 3 months *(17,117)*. However, ^{111}In-DTPA-SWCNTs were rapidly excreted through renal pathway without any residence of CNTs in the liver and spleen *(118)*. Carboxylation of MWCNTs (MWCNT-COOH) reduced their capability of inducing TGF-β1 production in bronchial epithelial cells and stimulating IL-1β in THP-1 cells *(119)*. Similarly, MWCNT-COOH mitigated pulmonary inflammation, lung damage, and IL-1β production after aspiration *(120)*. On the other hand, surface function groups of SWCNTs may also contribute to enhanced toxicity. The results of cell culture studies have demonstrated that the toxicity of CNTs increased significantly when carbonyl, carboxyl, hydroxyl, and/or epoxy groups were present on their surface *(121,122)*.

14.7.4 Fiber Length

The similar fibrous shape of CNTs to that of asbestos raises concern that they may pose an asbestos-like inhalation hazard. Direct instillation of long and short CNTs into the pleural cavity of mice generated very different side effects. It is the long CNTs but not the short ones that induced acute inflammation, leading to progressive fibrosis in the parietal pleura of mice, because short CNTs were readily cleared from the pleural space *(123)*. Similarly, Poland et al. *(124)* demonstrated that intraperitoneal injection of long MWCNTs were inflammatory and induced granulomatous lesions on the diaphragm. Among various types of CNTs, long MWCNTs were shown to induce the strongest DNA damage, inflammatory responses, and highest total cell number in abdominal lavage fluid (BALF) in mice *(125)*. These effects may result from the higher retention of long MWCNTs in organs than the shorter ones *(126)* and frustrated phagocytosis of long fibers by macrophages *(43)*.

14.8 SUMMARY AND PERSPECTIVES

Nanomaterials and nanotechnology extend rapidly into our daily life, offering obvious benefits to socioeconomic development. With the increased production and use of nanomaterials, the safety issues of these nanomaterials have raised great concerns. The risk of their potential negative impacts on human health and the environment has to be investigated and communicated for a sustainable development of nanotechnologies. As we have discussed, a risk may occur only if exposure and hazard coexist. Information on the real-world human exposure to CNTs is very rare, probably because there are a limited number of techniques available for specific measurement of CNT levels in the environment. The release of CNTs into the environment could occur at any stage of their life cycle, including synthesis, processing, handling, downstream use, and disposal. In addition, by considering the swift growth of CNT markets, it can be reasonably expected that CNT exposure would occur more frequently and seriously. Therefore effective measurements need to be taken to protect and/or prevent CNT exposure to workers and consumers. Understanding how the exposure occurs and what CNTs will do to human health would definitely benefit these efforts. In this chapter, we comprehensively reviewed the interactions between CNTs and living systems, from CNT exposure to the potential influences CNTs may have on human health. It seems that CNT nanoparticles present deleterious effects on multiple organs and systems, including pulmonary, skin, GIT, cardiovascular system, and immune system, through inhalation, digesting, and contact exposure to CNTs. Because of their unique physicochemical properties, CNTs also have the potential to travel through these compartments, inducing genotoxicity. A few studies showed the carcinogenicity of CNTs. It has to be emphasized that a variety of factors including metal contaminants, aggregate/agglomerate status, surface chemistry as well as fiber length modulate the toxicity profile of CNTs.

With future applications of CNTs in new forms, the increased toxicity of CNTs will be revealed and needs to be investigated closely. Although the interactions between CNTs and living systems are discussed in this chapter, the dynamic processes including endocytosis and exocytosis of CNTs as well as the corresponding mechanisms are largely unknown, partially caused by limited methods for quantitative measurement of CNTs within population cells/organs. In addition, CNTs released into the environment tend to coexist with various contaminants including organic and inorganic chemicals as well as natural compounds such as humic acids, which largely modified the behavior of CNTs within living systems. More effort is definitely required in these fields for a better understanding of the toxicity of CNTs and underlying mechanisms, and, most importantly, it may provide more efficient intervention measurements for controlling CNT exposure and toxicity.

REFERENCES

1. Walters, A.; Ericson, L. M.; Casavant, M. J.; Liu, J.; Colbert, D. T.; Smith, K. A.; Smalley, R. E. Elastic Strain of Freely Suspended Single-Walled Carbon Nanotube Ropes. *Applied Physics Letters* **1999,** *74* (25), 3803–3805.
2. Yu, M. F.; Louire, O.; Dyer, M. J.; Moloni, K.; Kelly, T. F.; Ruoff, R. S. The Strength and Breaking Mechanism of Multiwalled Carbon Nanotubes Under Tensile Load. *Science* **2000,** *287* (5453), 637–640.
3. Pantarotto, D.; Singh, R.; McCarthy, D.; Erhardt, M.; Briand, J. P.; Prato, M.; Kostarelos, K.; Bianco, A. Functionalized Carbon Nanotubes for Plasmid DNA Gene Delivery. *Angewandte Chemie-International Edition* **2004,** *43,* 5242–5246.
4. Milne, W. I.; Mann, M.; Dijon, J.; Bachmann, P.; McLaughlin, J.; Robertson, J.; Teo, K. B. K.; Lewalter, A.; de, Souza, M.; Boggild, P.; Briggs, A.; Mogensen, K. B.; Gabriel, J.-C. P.; Roche, S.; Baptist, R. Carbon Nanotubes. *E-Nano Newsletter* **2008,** *13,* 3–5.
5. Hexa Reports: Carbon Nanotubes (CNT) Market Analysis By Product (Single Walled Carbon Nanotubes (SWCNT), Multi Walled Carbon Nanotubes (MWCNT)), By Application (Polymers, Energy, Electrical & Electronics) And Segment Forecasts to 2022. http://hexareports.com/report/carbon-nanotubes-cnt-market/details.
6. Lamberti, M.; Zappavigna, S.; Sannolo, N.; et al. Advantages and Risks of Nanotechnologies in Cancer Patients and Occupationally Exposed Workers. *Expert Opinion on Drug Delivery* **2014,** *11,* 1087–1101.
7. Schulte, P. A.; Kuempel, E. D.; Zumwalde, R. D.; et al. Focused Actions to Protect Carbon Nanotubes Workers. *American Journal of Industrial Medicine* **2012,** *55,* 395–411.
8. Maynard, A. D.; Kuempel, E. Airborne Nanostructured Particles and Occupational Health. *Journal of Nanoparticle Research* **2005,** 7 (6), 587–614.
9. Maynard, A. D.; Baron, P. A.; Foley, M.; Shvedova, A. A.; Kisin, E. R.; Castranova, V. Exposure to Carbon Nanotube Material: Aerosol Release During the Handling of Unrefined Single-Walled Carbon Nanotube Material. *Journal of Toxicology and Environmental Health, Part A: Current Issues* **2004,** *67,* 87–107.
10. Anderson, P. J.; Wilson, J. D.; Hiller, F. C. Respiratory Tract Deposition of Ultrafine Particles in Subjects With Obstructive or Restrictive Lung Disease. *Chest* **1990,** *97,* 1115–1120.
11. Castranova, V.; Schulte, P. A.; Zumwalde, R. D. Occupational Nanosafety Considerations for Carbon Nanotubes and Carbon Nanofibers. *Accounts of Chemical Research* **2013,** *46,* 642–649.
12. Oberdörster, G.; Oberdörster, E.; Oberdörster, J. Nanotoxicology: An Emerging Discipline Evolving From Studies of Ultrafine Particles. *Environmental Health Perspectives* **2005,** *113* (7), 823–839.
13. Lamberti, M.; Pedata, P.; Sannolo, N.; Porto, S.; De, Rosa, A.; Caraglia, M. Carbon Nanotubes: Properties, Biomedical Applications, Advantages and Risks in Patients and Occupationally-Exposed Workers. *International Journal of Immunopathology and Pharmacology* **2015,** *28* (1), 4–13.
14. Singh, C.; Song, W. Carbon Nanotube Structure, Synthesis, and Application. In *The Toxicology of Carbon Nanotubes;* Donaldson, K., Poland, C., Duffin, R., Bonner, J., Eds.; Cambridge University Press: Cambridge, UK, 2012; pp 1–37.

15. Wang, H.; Wang, J.; Deng, X.; Sun, H.; Shi, Z.; Gu, Z.; Liu, Y.; Zhao, Y. Biodistribution of Carbon Single-Wall Carbon Nanotubes in Mice. *Journal of Nanoscience and Nanotechnology* **2004,** *4,* 1019–1024.
16. Folkmann, J. K.; Risom, L.; Jacobsen, N. R.; Wallin, H.; Loft, S.; Møller, P. Oxidative Damaged DNA in Rats Exposed by Oral Gavage to C60 Fullerenes and Single-Walled Carbon Nanotubes. *Environmental Health Perspectives* **2009,** *117,* 703–708.
17. Singh, R.; Pantarotto, D.; Lacerda, L.; Pastorin, G.; Klumpp, C.; Prato, M.; Bianco, A.; Kostarelos, K. Tissue Biodistribution and Blood Clearance Rates of Intravenously Administered Carbon Nanotube Radiotracers. *Proceedings of the National Academy of Sciences of the United States of America* **2006,** *103,* 3357–3362.
18. Yang, S.; Guo, W.; Lin, Y.; Deng, X.; Wabg, H.; Sun, H.; Liu, Y.; Wang, X.; Wang, W.; Chen, M.; et al. Biodistribution of Pristine Single-Walled Carbon Nanotubes in Vivo. *Journal of Physical Chemistry C* **2007,** *111,* 17761–17764.
19. Liu, Z.; Cai, W.; He, L.; Nakayama, N.; Chen, K.; Sun, X.; Chen, X.; Dai, H. In Vivo Biodistribution and Highly Efficient Tumour Targeting of Carbon Nanotubes in Mice. *Nature Nanotechnology* **2007,** *2,* 47–52.
20. McDevitt, M. R.; Chattopadhay, D.; Jaggi, J. S.; Finn, R. D.; Zanzonico, P. B.; Villa, C.; Rey, D.; Mendenhall, J.; Batt, C. A.; Njardarson, J. T.; et al. PET Imaging of Soluble Yttrium-86-Labeled Carbon Nanotubes in Mice. *PLoS One* **2007a,** *2* (9), e907.
21. McDevitt, M. R.; Chattopadhay, D.; Kappel, B. J.; Jaggi, J. S.; Schiffman, S. R.; Antczak, C.; Njardarson, J. T.; Brentjens, R.; Scheunberg, D. A. Tumor Targeting With Antibody-Functionalized, Radiolabeled Carbon Nanotubes. *Journal of Nuclear Medicine* **2007b,** *48,* 1180–1189.
22. Cherukuri, P.; Gannon, C. J.; Leeuw, T.; Schmidt, H. K.; Smalley, R. E.; Curley, S. A.; Weisman, R. B. Mammalian Pharmacokinetics of Carbon Nanotubes Using Intrinsic Near-Infrared Fluorescence. *Proceedings of the National Academy of Sciences of the United States of America* **2006,** *103,* 18882–18886.
23. Faraj, A. A.; Besaad, A.; Cieslar, K.; Lacroix, G.; Canet-Soulas, E.; Cremillieux, Y. Long-Term Follow-up of Lung Biodistribution and Effect of Instilled SWCNTs Using Multiscale Imaging Techniques. *Nanotechnology* **2010,** *21* (17), 175103.
24. Kolosnjaj-Tabi, J.; Hartman, K. B.; Boudjemaa, S.; Ananta, J. S.; Morgant, G.; Szwarc, H.; Wilson, L. J.; Moussa, F. In Vivo Behavior of Large Doses of Ultrashort and Full-Length Single-Walled Carbon Nanotubes After Oral and Intraperitoneal Administration to Swiss Mice. *ACS Nano* **2010,** *4,* 1481–1492.
25. Liu, Z.; Davis, C.; Cai, W.; He, L.; Chen, X.; Dai, H. Circulation and Long-Term Fate of Functionalized, Biocompatible Single-Walled Carbon Nanotubes in Mice Probed by Raman Spectroscopy. *Proceedings of the National Academy of Sciences of the United States of America* **2008,** *105,* 1410–1415.
26. Rejman, J.; Oberle, V.; Zuhorn, I. S.; Hoekstra, D. Size-Dependent Internalization of Particles via the Pathways of Clathrin- and Caveolae-Mediated Endocytosis. *Biochemical Journal* **2004,** *377,* 159–169.
27. Chithrani, B. D.; Ghazani, A. A.; Chan, W. C. Determining the Size and Shape Dependence of Gold Nanoparticle Uptake Into Mammalian Cells. *Nano Letters* **2006,** *6,* 662–668.
28. Dausend, J.; Musyanovych, A.; Dass, M.; Walther, P.; Schrezenmeier, H.; Landfester, K.; Mailänder, V. Uptake Mechanism of Oppositely Charged Fluorescent Nanoparticles in HeLa Cells. *Macromolecular Bioscience* **2008,** *8,* 1135–1143.
29. Salvati, A.; Åberg, C.; dos Santos, T.; Varela, J.; Pinto, P.; Lynch, I.; Dawson, K. A. Experimental and Theoretical Comparison of Intracellular Import of Polymeric Nanoparticles and Small Molecules: toward Models of Uptake Kinetics. *Nanomedicine: Nanotechno Biology, and Medicine* **2011,** *7,* 818–826.
30. Shapero, K.; Fenaroli, F.; Lynch, I.; Cottell, D. C.; Salvati, A.; Dawson, K. A. Time and Space Resolved Uptake Study of Silica Nanoparticles by Human Cells. *Molecular BioSystems* **2011,** *7,* 371–378.
31. Lesniak, A.; Salvati, A.; Santos-Martinez, M. J.; Radomski, M. W.; Dawson, K. A.; Åberg, C. Nanoparticle Adhesion to the Cell Membrane and Its Effect on Nanoparticle Uptake Efficiency. *Journal of the American Chemical Society* **2013,** *135* (4), 1438–1444.
32. Rupper, A.; Cardelli, J. Regulation of Phagocytosis and Endo-Phagosomal Trafficking Pathways in Dictyosteliumdiscoideum. *Biochimica et Biophysica Acta* **2001,** *1525* (3), 205–216.

33. Xiao, Y.; Forry, S. P.; Gao, X.; Holbrook, R. D.; Telford, W. G.; Tona, A. Dynamics and Mechanisms of Quantum Dot Nanoparticle Cellular Uptake. *Journal of Nanobiotechnology* **2010,** *8,* 13.
34. Falcone, S.; Cocucci, E.; Podini, P.; Kirchhausen, T.; Clementi, E.; Meldolesi, J. Macropinocytosis: Regulated Coordination of Endocytic and Exocytic Membrane Traffic Events. *Journal of Cell Science* **2006,** *119,* 4758−4769.
35. Ridley, A. J. Rho Proteins: Linking Signaling With Membrane Trafficking. *Traffic* **2001,** *2* (5), 303−310.
36. Mellman, I.; Steinman, R. M. Dendritic Cells: Specialized and Regulated Antigen Processing Machines. *Cell* **2001,** *106* (3), 255−258.
37. Howe, C. L. Modeling the Signaling Endosome Hypothesis: Why a Drive to the Nucleus Is Better Than a (Random) Walk. *Theoretical Biology and Medical Modelline* **2005,** *2,* 43.
38. Kholodenko, B. N. Four-Dimensional Organization of Protein Kinase Signaling Cascades: The Roles of Diffusion, Endocytosis and Molecular Motors. *Journal of Experimental Biology* **2006,** *206,* 2073−2082.
39. Prokop, A.; Davidson, J. M. Nanovehicular Intracellular Delivery Systems. *Journal of Pharmaceutical Sciences* **2008,** *97* (9), 3518−3590.
40. Liu, B. R.; Lo, S. Y.; Liu, C. C.; Chyan, C. L.; Huang, Y. W.; Aronstam, R. S.; et al. Endocytic Trafficking of Nanoparticles Delivered by Cell-Penetrating Peptides Comprised of Nona-Arginine and Penetration Accelerating Sequence. *PLoS One* **2013,** *8* (6), e67100.
41. Gratton, S. E.; Ropp, P. A.; Pohlhaus, P. D.; Luft, J. C.; Madden, V. J.; Napier, M. E.; DeSimone, J. M. The Effect of Particle Design on Cellular Internalization Pathways. *Proceedings of the National Academy of Sciences of the United States of America* **2008,** *105* (33), 11613−11618.
42. Jin, H.; Heller, D. A.; Sharma, R.; Strano, M. S. Size-Dependent Cellular Uptake and Expulsion of Single-Walled Carbon Nanotubes: Single Particle Tracking and a Generic Uptake Model for Nanoparticles. *ACS Nano* **2009,** *3* (1), 149−158.
43. Donaldson, K.; Murphy, F. A.; Duffin, R.; Poland, C. A. Asbestos, Carbon Nanotubes and the Pleural Mesotheliomia Review of the Hypothesis Regarding the Role of Long Fibre Retention in the Parietal Pleura, Inflammation and Mesothelioma. *Partical and Fibre Toxicology* **2010,** *7,* 5.
44. Chou, C. C.; Hsiao, H. Y.; Hong, Q. S.; Chen, C. H.; Peng, Y. W.; Chen, H. W.; Yang, P. C. Single-Walled Carbon Nanotubes Can Induce Pulmonary Injury in Mouse Model. *Nano Letters* **2008,** *8* (2), 437−445.
45. Kostarelos, K.; Lacerda, L.; Pastorin, G.; Wu, W.; Wieckowski, S.; Luangsivilay, J.; Godefroy, S.; Pantarotto, D.; Briand, J. P.; Muller, S.; et al. Cellular Uptake of Functionalized Carbon Nanotubes Is Independent of Functional Group and Cell Type. *Nature Nanotechnology* **2007,** *2* (2), 108−113.
46. Allen, B. L.; Kichambare, P. D.; Gou, P.; Vlasova, I. I.; Kapralov, A. A.; Konduru, N.; Kagan, V. E.; Star, A. Biodegradation of Single-Walled Carbon Nanotubes Through Enzymatic Catalysis. *Nano Letters* **2008,** *8* (11), 3899−3903.
47. Allen, B. L.; Kotchey, G. P.; Chen, Y.; Yanamala, N. V.; Klein-Seetharaman, J.; Kagan, V. E.; Star, A. Mechanistic Investigations of Horseradish Peroxidase-Catalyzed Degradation of Single-Walled Carbon Nanotubes. *Journal of the American Chemical Society* **2009,** *131* (47), 17194−17205.
48. Russier, J.; Ménard-Moyon, C.; Venturelli, E.; Gravel, E.; Marcolongo, G.; Meneghetti, M.; Doris, E.; Bianco, A. Oxidative Biodegradation of Single- and Multi-walled Carbon Nanotubes. *Nanoscale* **2011,** *3* (3), 93−96.
49. Liu, X.; Hurt, R. H.; Kane, A. B. Biodurability of Single-Walled Carbon Nanotubes Depends on Surface Functionalization. *Carbon* **2010,** *48* (7), 1961−1969.
50. Inoguchi, T.; Sonta, T.; Tsubouchi, H.; Etoh, T.; Kakimoto, M.; Sonoda, N.; Sato, N.; Sekiguchi, N.; Kobayashi, K.; Sumimoto, H.; et al. Protein Kinase C-dependent Increase in Reactive Oxygen Species (ROS) Production in Vascular Tissues of Diabetes: Role of Vascular NAD(P)H Oxidase. *Journal of the American Society of Nephrology* **2003,** *14* (8 Suppl. 3), S227−S232.
51. Kagan, V. E.; Konduru, N. V.; Feng, W.; Allen, B. L.; Conroy, J.; Volkov, Y.; Vlasova, I. I.; Belikova, N. A.; Yanamala, N.; Kapralov, A.; et al. Carbon Nanotubes Degraded by Neutrophil Myeloperoxidase Induce Less Pulmonary Inflammation. *Nature Nanotechnology* **2010,** *5* (5), 354−359.

52. Shvedova, A. A.; Kisin, E. R.; Mercer, R.; Murray, A. R.; Johnson, V. J.; Potapovich, A. I.; Tyurina, Y. Y.; Gorelik, O.; Arepulli, S.; Schwegler-Berry, D.; et al. Unusual Inflammatory and Fibrogenic Pulmonary Responses to Single Walled Carbon Nanotubes in Mice. *American Journal of Physiology: Lung Cellular and Molecular Physiology* **2005,** *289,* L698–L708.
53. Zhao, Y.; Allen, B. L.; Star, A. Enzymatic Degradation of Multiwalled Carbon Nanotubes. *Journal of Physical Chemistry A* **2011,** *115* (34), 9536–9544.
54. Muller, J.; Huaux, F.; Lison, D. Respiratory Toxicity of Carbon Nanotubes: How Worried Should We Be? *Carbon* **2006,** *44,* 1048–1056.
55. Donaldson, K.; Poland, C. A. Nanotoxicology: New Insights Into Nanotubes. *Nature Nanotechnology* **2009,** *4* (11), 708–710.
56. Warheit, D. B.; Laurence, B. R.; Reed, K. L.; Roach, D. H.; Reynolds, G. A.; Webb, T. R. Comparative Pulmonary Toxicity Assessment of Single-Wall Carbon Nanotubes in Rats. *Toxicological Sciences* **2004,** 77 (1), 117–125.
57. Lam, C. W.; James, J. T.; McCluskey, R.; Hunter, R. L. Pulmonarytoxicity of Single-Walled Carbon Nanotubes in Mice 7 and 90 Days After Intratracheal Instillation. *Toxicological Sciences* **2004,** *77,* 126–134.
58. Teeguarden, J. G.; Webb-Robertson, B. J.; Waters, K. M.; Murray, A. R.; Kisin, E. R.; Varnum, S. M.; Jacobs, J. M.; Pounds, J. G.; Zanger, R. C.; Shvedova, A. A. Comparative Proteomics and Pulmonary Toxicity of Instilled Single-Walled Carbon Nanotubes, Crocidolite Asbestos, and Ultrafine Carbon Black in Mice. *Toxicological Sciences* **2011,** *120* (1), 123–135.
59. Han, S. G.; Andrews, R.; Gairola, G.; Bhalia, D. Acute Pulmonary Effects of Combined Exposure to Carbon Nanotubes and Ozone in Mice. *Inhalation Toxicology* **2008,** *20* (4), 391–398.
60. Wolfarth, M. G.; Porter, D. W.; Hubbs, A. F.; Leonard, S.; Battelli, L.; Andrews, M.; Castranova, V. Pulmonary Toxicity of Multi-walled Carbon Nanotubes. *The Toxicologist* **2009,** *108,* A2196.
61. Hubbs, A. F.; Mercer, R. R.; Coad, J. E.; Battelli, L. A.; Willard, P.; Sriram, K.; Wolfarth, M.; Castranova, V.; Porter, D. Persistent Pulmonary Inflammation, Airway Mucous Metaplasia and Migration of Multi-walled Carbon Nanotubes From the Lung after Subchronic Exposure. *The Toxicologist* **2009,** *108,* A2193.
62. Han, S. G.; Andrews, R.; Gairola, C. G. Acute Pulmonary Response of Mice to Multi-wall Carbon Nanotubes. *Inhalation Toxicology* **2010,** *22* (4), 340–347.
63. Porter, D. W.; Hubbs, A. F.; Mercer, R. R.; Wu, N.; Wolfarth, M. G.; Sriram, K.; Leonard, S. S.; Battelli, L.; Schwegler- Berry, D.; Friend, S.; et al. Mouse Pulmonary Dose- and Time Course-Responses Induced by Exposure to Multi-walled Carbon Nanotubes. *Toxicology* **2010,** *269* (2–3), 136–147.
64. Mercer, R. R.; Hubbs, A. F.; Scabilloni, J. F.; Wang, L.; Battelli, L. A.; Friend, S.; Castranova, V.; Porter, D. W. Pulmonary Fibrotic Response to Aspiration of Multi-walled Carbon Nanotubes. *Particle and Fibre Toxicology* **2011,** *8* (1), 21.
65. Carrero-Sanchez, J. C.; Elías, A. L.; Mancilla, R.; Arrellín, G.; Terrones, H.; Laclette, J. P.; Terrones, M. Biocompatibility and Toxicological Studies of Carbon Nanotubes Doped With Nitrogen. *Nano Letters* **2006,** *6* (8), 1609–1616.
66. Elgrabli, D.; Abella-Gallart, S.; Robidel, F.; Rogerieux, F.; Boczkowski, J.; Lacroix, G. Induction of Apoptosis and Absence of Inflammation in Rat Lung After Intratracheal Instillation of Multiwalled Carbon Nanotubes. *Toxicology* **2008,** *253* (1–3), 131–136.
67. Driscoll, K. E.; Costa, D. L.; Hatch, G.; Henderson, R.; Oberdorster, G.; Salem, H.; et al. Intratracheal Instillation as an Exposure Technique for the Evaluation of Respiratory Tract Toxicity: Uses and Limitations. *Toxicological Sciences* **2000,** *55,* 24–35.
68. Li, Z.; Salmen, R.; Hulderman, T.; et al. Pulmonary Carbon Nanotube Exposure and Oxidative Status in Vascular System. *Free Radical Biology and Medicine* **2004,** *37,* S142–S143.
69. Walker, V. G.; Li, Z.; Hulderman, T.; et al. Potential in Vitro Effects of Carbon Nanotubes on Human Aortic Endothelial Cells. *Toxicology and Applied Pharmacology* **2009,** *236* (3), 319–328.
70. Li, Z.; Hulderman, T.; Salmen, R.; et al. Cardiovascular Effects of Pulmonary Exposure to Single-Wall Carbon Nanotubes. *Environmental Health Perspectives* **2007b,** *115* (3), 377–382.

71. Semberova, J.; De Paoli Lacerda, S. H.; Simakova, O.; et al. Carbon Nanotubes Activate Blood Platelets by Inducing Extracellular Ca^{2+} Influx Sensitive to Calcium Entry Inhibitors. *Nano Letters* **2009,** *9,* 3312–3317.
72. Guidetti, G. F.; Consonni, A.; Cipolla, L.; et al. Nanoparticles Induce Platelet Activation in Vitro Through Stimulation of Canonical Signalling Pathways. *Nanomedicine* **2012,** *8,* 1329–1336.
73. Ge, C.; Meng, L.; Xu, L.; Bai, R.; Du, J.; Zhang, L.; Li, Y.; Chang, Y.; Zkao, Y.; Chen, C. Acute Pulmonary and Moderate Cardiovascular Responses of Spontaneous Hypertensive Rats After Exposure to Single-Walled Carbon Nanotubes. *Nanotoxicology* **2012,** *6* (5), 526–542.
74. Legramante, J. M.; Valentini, F.; Magrini, A.; Palleschi, G.; Sacco, S.; Iavicoli, I.; Pallante, M.; Moscone, D.; Galante, A.; Bergamaschi, E.; et al. Cardiac Autonomic Regulation after Lung Exposure to Carbon Nanotubes. *Human & Experimental Toxicology* **2009,** *28,* 369–375.
75. Joviano-Santos, J. V.; Sa, M. A.; De, Maria, M. L.; Almeida, T. C. S.; Geraldo, V.; Oliveira, S.; Ladeira, L. O.; Ferreira, A. J. Evaluation of Cardiovascular Toxicity of Carbon Nanotubes Functionalized With Sodium Hyaluronate in Oral Regenerative Medicine. *Brazilian Journal of Medical and Biological Research* **2014,** *47,* 560–566.
76. Huczko, A.; Lange, H. Carbon Nanotubes: Experimental Evidence for a Null Risk of Skin Irritation and Allergy. *Fullerene Science and Technology* **2001,** *9,* 247–250.
77. Ali-Boucetta, H.; Kostarelos, K. Pharmacology of Carbon Nanotubes: Toxicokinetics, Excretion and Tissue Accumulation. *Advanced Drug Delivery Reviews* **2013,** *65* (15), 2111–2119.
78. Murray, A. R.; Kisin, E.; Leonard, S. S.; Young, S. H.; Kommineni, C.; Kagan, V. E.; Castranova, V.; Shvedova, A. A. Oxidative Stress and Inflammatory Response in Dermal Toxicity of Single Walled Carbon Nanotubes. *Toxicology* **2009,** *257,* 161–171.
79. Rouse, J. G.; Yang, J.; Ryman-Rasmussen, J. P.; Barron, A. R.; Monteiro-Riviere, N. A. Effects of Mechanical Flexion on the Penetration of Fullerene Amino Acid-Derivatized Peptide Nanoparticles Through Skin. *Nano Letters* **2007,** 7 (1), 155–160.
80. Shvedova, A. A.; Castranova, V.; Kisin, E. R.; Schwegler-Berry, D.; Murray, A. R.; Gandelsman, V. Z.; Maynard, A.; Baron, P. Exposure to Carbon Nanotube Material: Assessment of Nanotube Cytotoxicity Using Human Keratinocyte Cells. *Journal of Toxicology and Environmental Health, Part A: Current Issues* **2003,** *66,* 1909–1926.
81. Zhang, L. W.; Zeng, L.; Barron, A. R.; Monteiro-Riviere, N. A. Biological Interactions of Functionalized Single-Wall Carbon Nanotubes in Human Epidermal Keratinocytes. *Internet Journal of Toxicology* **2007,** *26* (2), 103–113.
82. Monteiro-Riviere, N. A.; Nemanich, R. J.; Inman, A. O.; Wang, Y. Y.; Riviere, J. E. Multi-walled Carbon Nanotube Interactions With Human Epidermal Keratinocytes. *Toxicology Letters* **2005,** *155,* 377–384.
83. Lai, X.; Blazer-Yost, B. L.; Clack, J. W.; et al. Protein Expression Profiles of Intestinal Epithelial Co-cultures: Effect of Functionalised Carbon Nanotube Exposure. *International Journal of Biomedical Nanoscience and Nanotechnology* **2013,** *3,* 1–2.
84. Jos, A.; Pichardo, S.; Puerto, M.; et al. Cytotoxicity of Carboxylic Acid Functionalized Single Wall Carbon Nanotubes on the Human Intestinal Cell Line Caco-2. *Toxicology In Vitro* **2009,** *23,* 1491–1496.
85. Beliaeva, N. N.; Mikhaĭlova, R. I.; Sycheva, L. P. Effect of Multi-walled Carbon Nanotubes on the Morphofunctional State of the Small Intestine Cells of Mice. *Gigiena I Sanitariya* **2012,** *6,* 58–61.
86. Patlolla, A. K.; Berry, A.; Tchounwou, P. B. Study of Hepatotoxicity and Oxidative Stress in Male Swiss-Webster Mice Exposed to Functionalized Multiwalled Carbon Nanotubes. *Molecular and Cellular Biochemistry* **2011,** *358,* 189–199.
87. Ji, Z.; Zhang, D.; Li, L.; et al. The Hepatotoxicity of Multi-walled Carbon Nanotubes in Mice. *Nanotechnology* **2009,** *20,* 445101.
88. Owen, R.; Depledge, M. Nanotechnology and the Environment: Risks and Rewards. *Marine Pollution Bulletin* **2005,** *50* (6), 609–612.
89. Witasp, E.; Shvedova, A. A.; Kagan, V. E.; Fadeel, B. Single-Walled Carbon Nanotubes Impair Human Macrophage Engulfment of Apoptotic Cell Corpses. *Inhalation Toxicology* **2009,** *21,* 131–136.

90. Dong, P. X.; Wan, B.; Wang, Z. X.; Guo, L. H.; Yang, Y.; Zhao, L. Exposure of Single-Walled Carbon Nanotubes Impairs the Functions of Primarily Cultured Murine Peritoneal Macrophages. *Nanotoxicology* **2013,** 7 (5), 1028–1042.
91. Nygaard, U. C.; Hansen, J. S.; Samueisen, M.; Alberg, T.; Marioara, C. D.; Løvik, M. Single-Walled and Multi-walled Carbon Nanotubes Promote Allergic Immune Responses in Mice. *Toxicological Sciences* **2009,** *109,* 113–123.
92. Lindberg, H. K.; Falck, G. C.; Suhonen, S.; Vippola, M.; Vanhala, E.; Catalán, J.; Savolainen, K.; Norppa, H. Genotoxicity of Nanomaterials: DNA Damage and Micronuclei Induced by Carbon Nanotubes and Graphite Nanofibres in Human Bronchial Epithelial Cells in Vitro. *Toxicology Letters* **2009,** *186* (3), 166–173.
93. Catalán, J.; Järventaus, H.; Vippola, M.; Savolainen, K.; Norppa, H. Induction of Chromosomal Aberrations by Carbon Nanotubes and Titanium Dioxide Nanoparticles in Human Lymphocytes in Vitro. *Nanotoxicology* **2012,** *6,* 825–836.
94. Li, J. G.; Li, W. X.; Xu, J. Y.; Cai, X. Q.; Liu, R. L.; Li, Y. J.; Zhao, Q. F.; Li, Q. N. Comparative Study of Pathological Lesions Induced by Multiwalled Carbon Nanotubes in Lungs of Mice by Intratracheal Instillation and Inhalation. *Environmental Toxicology* **2007,** *22,* 415–421.
95. Shvedova, A. A.; Kisin, E.; Murray, A. R.; Johnson, V. J.; Gorelik, O.; Arepali, S.; Hubbs, A. F.; Mercer, R. R.; Keohavong, P.; Sussman, N.; et al. Inhalation Versus Aspiration of Single Walled Carbon Nanotubes in C57BL/6 Mice: Inflammation, Fibrosis, Oxidative Stress and Mutagenesis. *American Journal of Physiology: Lung Cellular and Molecular Physiology* **2008,** *295,* L552–L565.
96. Shvedova, A. A.; Yanamala, N.; Kisin, E. R.; Murray, A. R.; Hubbs, A.; Chirila, M. M.; Keohavong, P.; Sycheva, L. P.; Kagan, V. E.; Castranova, V. Long-term Effects of Carbon Containing Engineered Nanomaterials and Asbestos in the Lung: One Year Postexposure Comparisons. *American Journal of Physiology: Lung Cellular and Molecular Physiology* **2014,** *306* (2), L170–L182.
97. Jacobsen, N. R.; Møller, P.; Jensen, K. A.; Vogel, U.; Ladefoged, O.; Loft, S.; Wallin, H. Lung Inflammation and Genotoxicity Following Pulmonary Exposure to Nanoparticles in $ApoE^{-/-}$ Mice. *Particle and Fibre Toxicology* **2009,** *6,* 2.
98. Naya, M.; Kobayashi, N.; Mizuno, K.; Matsumoto, K.; Ema, M.; Nakanishi, J. Evaluation of the Genotoxic Potential of Single-Walled Carbon Nanotubes by Using a Battery of in Vitro and in Vivo Genotoxicity Assays. *Regulatory Toxicology and Pharmacology* **2011,** *61,* 192–198.
99. Naya, M.; Kobayashi, N.; Endoh, S.; Maru, J.; Honda, K.; Ema, M.; Tanaka, J.; Fukumuro, M.; Hasegawa, K.; Nakajima, M.; et al. In Vivo Genotoxicity Study of Single-Wall Carbon Nanotubes Using Comet Assay Following Intratracheal Instillation in Rats. *Regulatory Toxicology and Pharmacology* **2012,** *64,* 124–129.
100. Ema, M.; Imamura, T.; Suzuki, H.; Kobayashi, N.; Naya, M.; Nakanishi, J. Genotoxicity Evaluation for Single-Walled Carbon Nanotubes in Battery on in Vitro and in Vivo Assays. *Journal of Applied Toxicology* **2013,** *33,* 933–939.
101. Vesterdal, L. K.; Jantzen, K.; Sheykhzade, M.; Roursgaard, M.; Folkman, J. K.; Loft, S.; Møller, P. Pulmonary Exposure to Particles From Diesel Exhaust, Urban Dust or Single-Walled Carbon Nanotubes and Oxidatively Damaged DNA and Vascular Function in $ApoE^{-/-}$ Mice. *Nanotoxicology* **2014,** *8,* 61–71.
102. Takagi, A.; Hirose, A.; Nishimura, T.; Fukumori, N.; Ogata, A.; Ohashi, N.; Kitajima, S.; Kanno, J. Induction of Mesothelioma in $P53^{+/-}$ Mouse by Intraperitoneal Application of Multi-wall Carbon Nanotube. *Toxicological Sciences* **2008,** *33* (1), 105–116.
103. Takagi, A.; Hirose, A.; Futakuchi, M.; Tsuda, H.; Kanno, J. Dose-Dependent Mesothelioma Induction by Intraperitoneal Administration of Multi-wall Carbon Nanotubes in P53 Heterozygous Mice. *Cancer Science* **2012,** *103* (8), 1440–1444.
104. Murphy, F. A.; Schinwald, A.; Poland, C. A.; Donaldson, K. The Mechanism of Pleural Inflammation by Long Carbon Nanotubes: Interaction of Long Fibres With Macrophages Stimulates Them to Amplify Pro-inflammatory Responses in Mesothelial Cells. *Particle and Fibre Toxicology* **2012,** *9,* 8.
105. Donner, M.; Tran, L.; Muller, J.; Vrijhof, H. Genotoxicity of Engineered Nanomaterials. *Nanotoxicology* **2010,** *4* (4), 345–346.

106. MacNee, W.; Donaldson, K. Particulate Air Pollution: Injurious and Protective Mechanisms in the Lungs. In *Air Pollution and Health;* Holgate, S. T., Samet, J. M., Koren, H. S., Maynard, R. L., Eds.; Academic Press: London, UK, 1999; pp 653–672.
107. Chen, Z.; Meng, H.; Xing, G. M.; Chen, C. Y.; Zhao, Y. L. Toxicological and Biological Effects of Nanomaterials. *International Journal of Nanotechnology* **2007,** *4,* 179–196.
108. Wick, P.; Manser, P.; Limbach, L. K.; Dettlaff-Weglikowska, U.; Krumeich, F.; Roth, S.; Stark, W. J.; Bruinink, A. The Degree and Kind of Agglomeration Affect Carbon Nanotube Cytotoxicity. *Toxicology Letters* **2007,** *168* (2), 121–131.
109. Pauluhn, J. Subchronic 13-week Inhalation Exposure of Rats to Multiwalled Carbon Nanotubes: Toxic Effects Are Determined by Density of Agglomerate Structures, Not Fibrillar Structure. *Toxicological Sciences* **2010,** *113,* 226–242.
110. Wang, L.; Castranova, V.; Mishra, A.; Chen, B.; Mercer, R. R.; Schwegler-Berry, D.; Rojanasakul, Y. Dispersion of Single-Walled Carbon Nanotubes by a Natural Lung Surfactant for Pulmonary in Vitro and in Vivo Toxicity Studies. *Particle and Fibre Toxicology* **2010,** *7,* 31.
111. Wang, L.; Mercer, R. R.; Rojanasakul, Y.; Qiu, A.; Lu, Y.; Scabilloni, J. F.; Wu, N.; Castranova, V. Direct Fibrogenic Effects of Dispersed Single-Walled Carbon Nanotubes on Human Lung Fibroblasts. *Journal of Toxicology and Environmental Health, Part A: Current Issues* **2010,** *73,* 410–422.
112. Wang, X.; Xia, T.; Ntim, S. A.; Ji, Z.; George, S.; Meng, H.; Zhang, H.; Castranova, V.; Mitra, S.; Nel, A. E. Quantitative Techniques for Assessing and Controlling the Dispersion State and Biological Effects of Multi-walled Carbon Nanotubes in Mammalian Tissue Culture Cells. *ACS Nano* **2010,** *4,* 7241–7252.
113. Mercer, R. R.; Scabilloni, J.; Wang, L.; Kisin, E.; Murray, A. R.; Schwegler-Berry, D.; Shvedova, A. A.; Castranova, V. Alteration of Deposition Pattern and Pulmonary Response as a Result of Improved Dispersion of Aspirated Single Walled Carbon Nanotubes in a Mouse Model. *American Journal of Physiology: Lung Cellular and Molecular Physiology* **2008,** *294,* L87–L97.
114. Oberdörster, G. Significance of Particle Parameters in the Evaluation of Exposure-Dose-Response Relationships of Inhaled Particles. *Inhalation Toxicology* **1996,** *8,* 73–89.
115. Saxena, R. K.; Williams, W.; Mcgee, J. K.; Daniels, M. J.; Boykin, E.; Gilmour, M. I. Enhanced in Vitro and in Vivo Toxicity of Poly-Dispersed Acid-Functionalized Single-Wall Carbon Nanotubes. *Nanotoxicology* **2007,** *1* (4), 291–300.
116. Davis, V. A.; Parra-Vasquez, A. N. G.; Green, M. J.; Rai, P. K.; Behabtu, N.; Prieto, V.; Booker, R. D.; Schmidt, J.; Kesselman, E.; Zhou, W.; et al. True Solutions of Single-Walled Carbon Nanotubes for Assembly Into Macroscopic Materials. *Nature Nanotechnology* **2009,** *4,* 830–834.
117. Deng, X. Y.; Jia, G.; Wang, H. F.; Liu, Y. F.; et al. Translocation and Fate of Multi-walled Carbon Nanotubes. *Carbon* **2007,** *45,* 1419–1424.
118. Lopez, C. F.; Nielsen, S. O.; Moore, P. B.; Klein, M. L. Understanding Nature's Design for a Nanosyringe. *Proceedings of the National Academy of Sciences of the United States of America* **2004,** *101* (13), 4431–4434.
119. Wang, X.; Xia, T.; Ntim, S. A.; Ji, Z.; Lin, S.; Meng, H.; Chung, C.-H.; George, S.; Zhang, H.; Wang, M.; et al. The Dispersal State of Multi-walled Carbon Nanotubes Elicits Pro-fibrogenic Cellular Responses That Correlate With Fibrogenesis Biomarkers and Fibrosis in the Murine Lung. *ACS Nano* **2012,** *5,* 9772–9787.
120. Sager, T. M.; Wolfarth, M.; Porter, D.; Castranova, V.; Wu, N.; Holian, A. Effect of Surface Modification on the Bioavailability and Inflammatory Potential of Multi-walled Carbon Nanotubes. *The Toxicologist* **2011,** *120,* A1178.
121. Magrez, A.; Kasas, S.; Salicio, V.; Pasquier, N.; Seo, J. W.; Celio, M.; Catsicas, S.; Schwaller, B.; Forró, L. Cellular Toxicity of Carbon-Based Nanomaterials. *Nano Letters* **2006,** *6* (6), 1121–1125.
122. Sayes, C. M.; Liang, F.; Hudson, J. L.; Mendez, J.; Guo, W.; Beach, J. M.; Moore, V. C.; Doyle, C. D.; West, J. L.; Billups, W. E.; et al. Functionalization Density Dependence of Single-Walled Carbon Nanotubes Cytotoxicity in Vitro. *Toxicology Letters* **2006,** *161* (2), 135–142.
123. Murphy, F. A.; Poland, C. A.; Duffin, R.; Al-Jamal, K. T.; Ali-Boucetta, H.; Nunes, A.; Byrne, F.; Prina-Mello, A.; Volkov, Y.; Li, S.; et al. Length-Dependent Retention of Carbon Nanotubes in the

Pleural Space of Mice Initiates Sustained Inflammation and Progressive Fibrosis on the Parietal Pleura. *American Journal of Pathology* **2011,** *178* (6), 2587–2600.

124. Poland, C. A.; Duffin, R.; Kinloch, I.; Maynard, A.; Wallace, W. A.; Seaton, A.; Stone, Y.; Brown, S.; MacNee, W.; Donaldson, K. Carbon Nanotubes Introduced into the Abdominal Cavity of Mice Show Asbestos-Like Pathogenicity in a Pilot Study. *Nature Nanotechnology* **2008,** *3,* 423–428.
125. Yamashita, K.; Yoshioka, Y.; Higashisaka, K.; Morishita, Y.; Yoshida, T.; Fujimura, M.; Kayamuro, H.; Nabeshi, H.; Yamashita, T.; et al. Carbon Nanotubes Elicit DNA Damage and Inflammatory Response Relative to Their Size and Shape. *Inflammation* **2010,** *33* (4), 276–280.
126. Muller, J.; Delos, M.; Panin, N.; Rabolli, V.; Huaux, F.; Lison, D. Absence of Carcinogenic Response to Multi Wall Carbon Nanotubes in a 2-year Bioassay in the Peritoneal Cavity of the Rat. *Toxicological Sciences* **2009,** *110,* 442–447.

CHAPTER 15

Challenge and Opportunities of Carbon Nanotubes

M.Q. Jian, H.H. Xie, K.L. Xia, Y.Y. Zhang
Tsinghua University, Beijing, People's Republic of China

Contents

As stated in previous chapters, carbon nanotubes (CNTs) have been intensively explored over the past 25 years because of their unique structure, superior properties, and wide applications, and remarkable progress and enormous achievements have been made. Carbon nanotube production capacity exceeds several 1000 tons per year *(1)*. Advances in CNT controlled growth, purification, modification, and processing are showing great advantages in many aspects of everyday life. Carbon nanotubes have been integrated into various commercial products *(2)*, ranging from rechargeable batteries, mobile screens, conductive additives, sporting goods, and water filters.

However, challenges and opportunities remain. For example, the properties of the mass-produced CNT powders or CNT macrostructures including fibers, yarns, and

Industrial Applications of Carbon Nanotubes
ISBN 978-0-323-41481-4

sheets are far from the ideal properties of CNTs, which hinders the large-scale application of CNTs.

In addition, it is well known that the discovery and development of graphene has an obvious impact on the applications of CNTs. The progress and development of CNTs provide good reference for the research of graphene, ranging from synthesis and characterization to applications. In particular, promising materials combining CNTs and graphene that integrate the outstanding properties of these two-carbon allotropes are on the rise. The increase of CNT/graphene hybrids has expanded the research and application of carbon nanomaterials, providing new opportunities for the application of CNTs.

In this chapter, we first summarize the challenges and opportunities that exist in the synthesis and application of CNTs and then describe the synthesis and application of CNT/graphene hybrid materials.

15.1 CHALLENGES AND OPPORTUNITIES IN THE SYNTHESIS AND PROCESSING OF CNTs

In general, CNTs are synthesized by laser ablation, arc discharge, and chemical vapor deposition (CVD). Among these techniques, CVD has the advantages of high yield, desirable controllability, and low cost, and has been the dominant approach to obtaining high-volume CNTs production and synthesizing specific CNTs. Currently, mass production of CNTs based on the CVD method usually contains contaminants, which negatively influence the properties of CNTs and have to be removed by thermal or chemical treatment. The purification process inevitably introduces defects to the CNTs, resulting in adverse effects on the properties and potential applications.

It has been shown that the CVD approach enables preferential growth of semiconducting or metallic single-wall CNTs (SWNTs) *(3,4)*, even specific-chirality SWNTs *(5–8)* and half-meter long CNTs *(9)*. However, the concentration of semiconducting SWNTs (s-SWNTs) in the as-obtained SWNTs cannot meet the high requirement for fabrication of high-performance electronic devices, which should be more than 99.9999% *(10)*. Furthermore, the baseline target of s-SWNT density on surfaces is 125 CNTs per μm *(10)*. To realize these goals, it is crucial to understand the growth mechanisms and to design novel catalysts so as to shed light on how catalyst composition and growth conditions influence the diameter, length, chirality, and purity of CNTs *(1)*. Moreover, efforts are still needed to realize the chirality-controlled synthesis of CNTs, which is of great significance for applications in electronic devices.

Alternatively, methods for selective enrichment of s-SWNTs or metallic SWNTs (*m*-SWNTs) produced from CNT mixtures have been developed, such as the smart scotch tape separation technique *(11)*, sodium dodecyl sulfonate washing method *(12)*, UV irradiation *(13)*, thermocapillary-resist approach *(14–16)*, etc. The high purity of specific-chirality SWNTs can be separated from powders by solution methods, including density-gradient ultracentrifugation *(17)*, gel chromatography *(18)*, selective surfactant, or polymer wrapping *(19,20)*. With these approaches, although high-purity SWNTs can be obtained, the price is too high for commercial applications *(2)*. In addition, stable and uniform CNTs suspensions are usually achieved with the assistance of chemical modification of CNTs or surfactants, which influences the properties and applications of CNTs.

In addition, long and aligned CNT materials including fibers, ribbons, or sheets, which have excellent properties, can be synthesized without liquid dispersion. Horizontally *(21–23)* or vertically *(24)* aligned CNT arrays on substrates can be obtained by CVD. Furthermore, CNT sheets and fibers, which possess excellent mechanical and electrical properties, can be directly prepared from vertically aligned CNT arrays *(25)* or floating-catalyst CVD systems *(26–28)*. However, the properties of these CNT assemblies remain far lower than the intrinsic properties of individual CNTs. Continuing efforts to understand the growth mechanisms and to optimize the growth conditions should be made to obtain high-purity CNTs with desired structures and properties for the practical applications of CNTs.

15.2 CHALLENGES AND OPPORTUNITIES IN THE APPLICATION OF CNTs

Bulk CNT powders are used as electrically conductive fillers in polymers with low concentration. Generally, the mechanical properties of the matrix, such as the strength and modulus, can be increased. The diameter, length, and dispersion of CNTs and the interfacial interaction will affect the performance of the composites. In the long run, CNTs may compete with carbon fibers in application in high-end composite materials *(1)*. Currently, the strength of CNT fibers (or yarns) reaches 8.8 GPa when prepared from CNT solution *(29)*, while the strength of CNT fibers from a CVD system achieves 9.6 GPa *(28)*. Unlike carbon fibers, CNTs possess high surface area and flexibility, which shows their potential for use in energy conversion and storage devices, conductive textiles, etc. However, scaling up these techniques while maintaining low cost and high quality is a big challenge.

In addition, CNT films prepared from solutions or CVD systems are used as multifunctional coating materials, such as anticorrosion coating *(30)* and electromagnetic shielding films *(31)*. The most promising application of CNT film is as transparent conducting film *(32)*, an indispensable part of display and photovoltaic devices. Carbon nanotube films with flexibility and foldability show promise in replacing indium tin oxide (ITO) films. However, the performance of CNT films including its transparency and sheet resistivity is still slightly lower than ITO. Alternatively, CNT films can be integrated with inorganic metal nanowires. Single-wall CNT film can also be used as flexible thin-film transistors (TFTs) with a low-cost printing method *(33)*. However, the enrichment of s-SWNTs is essential for practical application of CNT films in TFTs, which requires further understanding of the surface chemistry of SWNTs *(1)*.

High-quality SWNTs have been applied to transistors due to their supreme electrical property and unique bandgap *(34)*. In 2013, the first CNT computer came out, which was a milestone in the development of CNT-based electronic devices. Carbon nanotubes were being considered as highly energy-efficient electronic systems *(35)*. However, despite the high-performance of individual devices, integration of massive high-performance SWNT devices remains a challenge due to the difficulties in controlling the diameter, chirality, density, and location of CNTs.

Furthermore, CNTs show broad prospects in environmental applications, sensors, and biotechnology. However, potential toxicity remains a concern, which may hinder their application in these fields. More research is needed on the safety issues in these applications.

15.3 EXTENDED APPLICATIONS OF CNTs WITH INCORPORATION OF GRAPHENE

Carbon nanotubes are one-dimensional (1D) materials composed of sp^2 carbon and, graphene is a two-dimensional (2D) material also composed of sp^2 carbon. Both of them have aroused tremendous interest since their discovery due to outstanding properties such as large specific surface area, electronic conductivity, thermal stability, and excellent mechanical properties. In particular, since Geim et al. *(36)* reported single atomic layer graphene in 2004, graphene has drawn great interest in both the scientific and industrial worlds. Graphene is a magic material that has a high intrinsic mobility of 200,000 $cm^{-2}/(V \cdot s)$, high optical transmittance (97.7% for monolayer graphene), high thermal conductivity of 5300 $W/(m \cdot K)$, high Young's modulus of 1.0 TPa, and a high theoretical specific surface area of 2630 m^2/g. Researchers have attempted to combine CNTs with graphene to obtain better performance than when using CNTs alone, or to fabricate new devices. Carbon nanotubes/graphene hybrid materials are expected to integrate the merits of both, opening new opportunities for exploring new properties and devices. Research has shown that hybridization of CNTs and graphene have high performance in hydrogen storage *(37)*, electric transport *(37)*, and thermal transport *(38)*.

In the past decade, great efforts have been applied to the controlled synthesis of CNT/graphene hybrids as well as to the exploration of their applications. It has been shown that the construction of CNT/graphene is an efficient and versatile strategy to explore the synergistic effects while making the most of the advantages of each. In view of hybrids, CNTs have the advantage of preventing graphene from stacking and aggregating, resulting in higher electrical conductivity and larger specific surface area, while graphene sheets facilitate charge transfer and offer larger contact area with other materials. Therefore CNTs and graphene have significant synergistic effects in electrical properties. Furthermore, when combining CNTs and graphene in fiber, the gravimetric toughness has been shown to be far more than spider silk and Kevlar *(39)*. Similarly, 3D CNT/graphene hybrids have shown synergistic effects in superelasticity *(40)* and mechanical energy storage *(41)*. Fabrication of hybrids enables the integration of the advantages of CNTs and graphene, resulting in tremendous applications in many aspects, such as electronic devices, transparent conductive films (TCFs), sensors and actuators, energy conversion and storage, etc. The application of CNTs, with the incorporation of graphene, has been further widened due to the enhanced performance of hybrids. In this section, the fundamentals of graphene will be briefly introduced and then the state of the art for the fabrication of CNT/graphene will be discussed, followed by a discussion on the applications of CNT/graphene hybrids.

15.3.1 Graphene: Structure, Synthesis, and Properties

15.3.1.1 Structure of Graphene

Graphene has a honeycomb 2D crystal lattice, which is composed of sp^2 hybridized carbon atoms that are close-packed and bonded together with σ bonds. The strong σ bonds in the graphene lattice give the high mechanical properties of graphene. Each carbon atom has a π orbital that constitutes a delocalized electrons network. The π electrons can move freely within the graphene crystal plane, thus the graphene has excellent electrical conductivity *(42)*. Graphene have a thickness of 0.335 nm. It is a 2D material and can be seen as the basic unit of all other dimensional carbon materials. For example, it can be wrapped up into zero-dimensional fullerene and rolled into 1D carbon nanotubes or stacked into 3D graphite (Fig. 15.1A) *(43)*.

Graphene's structure is very stable and the connection between the carbon atoms is extremely flexible. Observed by transmission electron microscope (TEM), graphene is not perfectly flat, but is intrinsically micro-uneven *(44)*. Under TEM, wrinkles and curls can be clearly seen within the microlevel monolayer graphene sheets. The surface and edges of monolayer graphene also have nanoscale wrinkles and these wrinkles exhibit $\pm$ 0.5 nm changes in the vertical direction and show 10 nm changes in the lateral dimensions *(45)*. These fluctuations in 3D can cause static electricity in the graphene plane, leading to aggregation of graphene at the macroscale.

15.3.1.2 Synthesis of Graphene

Since Geim et al. *(36)* first obtained monolayer graphene by mechanical exfoliation of highly oriented-pyrolytic-graphite (HOPG) in 2004, the study of graphene has become a significant trend in the scientific world. Currently, the preparation of graphene is no longer confined to mechanical exfoliation. Various preparation methods have been explored and developed. Each method has its own advantages and limitations. For practical applications, choosing the right approach is very important.

In the following, several major methods to prepare graphene, including mechanical exfoliation, chemical vapor deposition (CVD), thermal decomposition of SiC and other

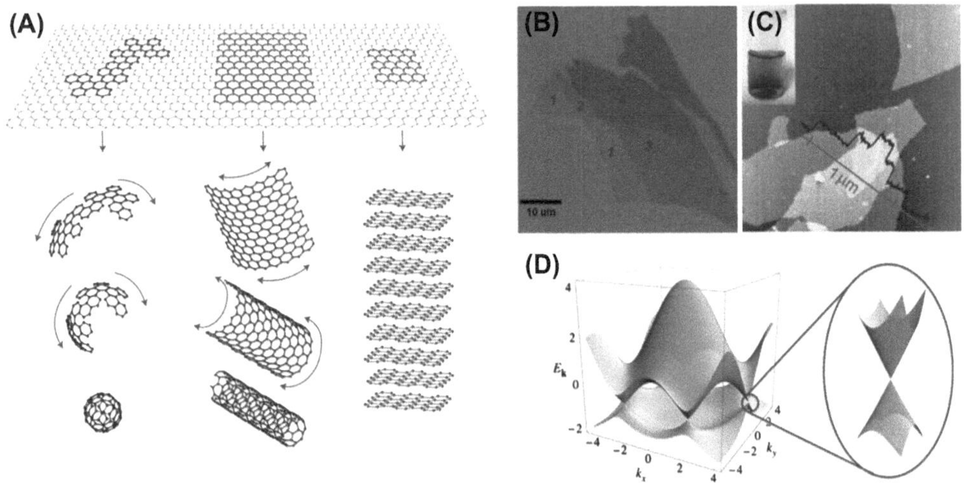

Figure 15.1 (A) The basic unit of all graphitic form. Graphene is a 2D building material for all other dimensional carbon materials. *(Adapted with permission from Geim, A. K.; Novoselov, K. S. The Rise of Graphene. Nature Materials* ***2007,*** *6, 183–191. Copyright 2007 Nature Publishing Group.)* (B) Optical image of mechanical exfoliation graphene with different layers. *(Adapted with permission from Ni, Z.; Wang, H.; Kasim, J.; Fan, H.; Yu, T.; Wu, Y.; Feng, Y.; Shen, Z. Graphene Thickness Determination Using Reflection and Contrast Spectroscopy. Nano Letters* ***2007,*** *7, 2758–2763. Copyright 2007 American Chemical Society.)* (C) AFM image of GO sheets deposited onto a mica substrate from an aqueous dispersion (inset). *(Adapted with permission from Stankovich, S.; Dikin, D. A.; Dommett, G. H.; Kohlhaas, K. M.; Zimney, E. J.; Stach, E. A.; Piner, R. D.; Nguyen, S. T.; Ruoff, R. S. Graphene-Based Composite Materials. Nature* ***2006,*** *442, 282–286. Copyright 2006 Nature Publishing Group.)* (D) Electronic dispersion in graphene lattice. Left: energy spectrum for finite values; right: zoom in of the energy bands closed to one of the Dirac points. *(Adapted with permission from Neto, A. C.; Guinea, F.; Peres, N.; Novoselov, K. S.; Geim, A. K. The Electronic Properties of Graphene. Reviews of Modern Physics* ***2009,*** *81, 109. Copyright 2009 American Physical Society.)*

substrates, oxidation-reduction methods, electrochemical methods, and the important advances of these approaches, will be introduced. These fundamental aspects of graphene are closely related to the fabrication of CNT/graphene hybrids.

15.3.1.2.1 Mechanical Exfoliation

The basic strategy of mechanical exfoliation is using adhesive tapes to stick on the two sides of a graphite sheet and repeatedly stripe to get thin graphene layers *(36,46,47)*. Detailed operations are as follows.

First, oxygen plasma is used to dry etch the surface of HOPG and to prepare graphite squares of various size ranging from 20 μm to 2 mm and with a thickness about 5 μm. Then, the structure is pressed against a glass substrate with a thick, fresh photoresist layer, followed by baking. Thirdly, scotch tape is used to repeatedly peel off graphite flakes from the mesas and then the photoresist is dissolved with acetone along with sonication to release the remaining thin layer of graphite. Finally, a silicon wafer is dipped into acetone to pick up the thin layer of graphite. Graphite with layers less than 10 (or called multilayer graphene) will be adsorbed on the silicon surface due to the van der Waals and capillary forces (Fig. 15.1B and C).

In addition, there are other mechanical exfoliation methods such as rubbing the graphite on a solid surface to leave thin graphene layers on the surface (48), applying direct current voltage to the graphite surface and discharging on a mica substrate to obtain the electrostatic deposition of graphene *(49)*, etc.

Graphene prepared from mechanical exfoliation of HOPG has the best electrical properties. By far, the best-published performance of graphene device was obtained from mechanical exfoliation graphene. However, the size of the graphene obtained from this approach is generally between several to tens of micrometer, and the probability of getting monolayer is quite low. Therefore graphene prepared by mechanical exfoliation is suitable for laboratory use but do not for large-scale application.

15.3.1.2.2 Chemical Vapor Deposition

As mentioned in previous chapters, CVD is a widely used method of preparing CNTs. Similar technology is also successful in the synthesis of graphene, which enables the large-area preparation of graphene. Nickel (Ni) was the first catalytic material used for the production of graphene using CVD *(50)*. Generally, Ni foil or SiO_2/Si substrates covered with a thin film of Ni is used as the substrate for CVD *(51–53)*. When Ni is exposed in a mixed gas of hydrogen and methane at high temperature, Ni and carbon atoms will form a solid solution. Since the solubility of C atoms in Ni is related to temperature, carbon atoms will precipitate from Ni substrates during the cooling process, leading to the formation of graphene. Since the solubility of carbon atoms in Ni substrates is relatively large (>0.1 at %), it is easy to separate out excess carbon, resulting in the formation of multilayer graphene.

Compared to Ni, the carbon solubility in Cu is relatively low (<0.001 at %). Li et al. first grew large-area, high-quality single-layer graphene on copper (Cu) foils *(54)*. Since the dissolution of carbon is very limited in Cu, the formation mechanism of graphene on Cu substrates is based on the surface reaction, which is a real vapor deposition process rather than the largely dissolved carbon atoms precipitation process from metal substrate like Ni. Li et al. verified the different growth mechanism of graphene on Cu and Ni using isotope labeling *(55)*. The entire region of graphene on the Ni surface was uniformly mixed with ^{12}C and ^{13}C. In contrast, graphene on Cu surface consisted of orderly arranged ^{12}C and ^{13}C region. The Cu-catalyzed growth process was based on surface reaction and thus Cu is considered to be the most promising substrate for preparation of large-area and high-quality graphene by CVD *(17)*. Theoretically, graphene has high thermal conductivity, electrical conductivity, and other properties, but the actual obtained graphene cannot provide the superior properties shown by theoretical simulations due to the grain boundaries and defects that exist in the graphene prepared by CVD.

Grain boundaries can reduce the migration velocity of electrons, so the thermal and electrical conductivity and mechanical properties will be affected. Therefore people are trying to grow larger single crystals graphene film on Cu substrate. The methods used include heating the Cu substrate only in inert atmosphere, extending the annealing time in hydrogen at high temperature, and increasing the growth temperature. Further, the pretreated Cu surface by chemical or mechanical etching can also help to reduce the growth of nucleation sites. For example, Lee et al. recently reported synthesizing large-area seamless stitching of hexagonal on single-crystal Cu substrate through annealing for a long time with large hydrogen partial pressure *(56)*.

In addition, other metal substrates such as Ru *(57)*, Ir *(58)*, Co *(59)*, and Pt *(60)* have also been investigated for the CVD growth of graphene.

CVD is considered to be the most promising method to controllably produce large-scale high-quality graphene, which includes large single crystals and single, double, and multilayer graphene films. These graphene samples can meet the quality requirement of a variety of electronic devices and have broad application potential. Of course, compared to mechanical exfoliation graphene, the structure of graphene grown by CVD is not perfect and needs further improvement.

15.3.1.2.3 Thermal Decomposition of SiC and Other Substrates

Graphene can be obtained on silicon carbide (SiC) substrate through annealing SiC under ultrahigh vacuum condition (UHV). Berger et al. obtained mono- and few-layer graphene through heating SiC and studied its properties *(61,62)*. When heating the SiC substrate under UHV, silicon atoms sublimate and the carbon atoms rearrange on the surface to form graphene. The number of graphene layers depends on the annealing time and temperature. Although the graphene obtained by this method is uniform, it is still a great challenge to control the number of graphene layers. Large-scale production is also difficult. In addition to SiC substrates, other carbon-containing metals such as Ru *(63)* have also been used for the preparation of graphene by annealing.

15.3.1.2.4 Oxidation Reduction Graphene From Graphite

Oxidation reduction is a widely used method for large-scale synthesis of graphene. The preparation process can be summarized in two steps: (1) using a mixture of strong oxidant and acid-treated graphite to obtain graphene oxide (GO); (2) utilizing chemical reduction, thermal reduction, or catalytic reduction treatment methods to transform GO into reduction graphene oxide (rGO). Due to the simple preparation conditions, which are feasible for the laboratory, raw materials readily available, and high yield, oxidation reduction is one of the most popular methods for large-scale preparation of graphene.

GO was first prepared by Brodie who treated graphite repetitively with nitric acid and potassium chlorate *(64)*. Later, this method was modified by several groups, including Staudenmaier et al. *(65)* who used a mixture of nitric acid and sulfuric with potassium chlorate, Hummers et al. *(66)* who utilized a mixture of sodium acid and sodium nitrate with potassium permanganate, and other investigators and other modified methods *(67–71)*. Hummers' method is less dangerous and more efficient and commonly used for the oxidation of graphene. In addition, in order to make graphite oxidize sufficiently, a pretreating process can be used to cause graphite to swell *(72,73)*. Later, an improved Hummers' method could be used to oxidize the expanded graphite, with the following process: (1) mixing the expansion graphite with potassium permanganate, sodium nitrate, and sulfuric acid and then holding 24 h at 0–4°C to let the oxidizing agent fully penetrate into the graphite interlayer; (2) stirring the mixture at 35°C for 30 min, and diluting it, then stirring for 15 min at 95°C; and (3) adding hydrogen peroxide to remove the residual oxidant, then treating with deionized water until the solution is rendered neutral. Finally, GO solution is obtained.

During the oxidation process, graphite is strongly destroyed. Oxygen atoms enter into the graphite interlayers and form a large amount of oxygen-containing functional groups such as hydroxyl, carboxyl, carbonyl, etc. These groups damage the structure and the π bond in graphene. The electrical conductivity of graphene is drastically reduced, as well as the mechanical properties. Thus in order to restore the conductivity and mechanical properties of graphene, these functional groups need to be removed by reduction. There are three main methods to reduce GO. The first approach is using reductant directly to reduce GO under high-temperature and high-pressure environment. Commonly used reductants contain hydrazine hydrate *(74,75)*, sodium borohydride *(76,77)*, hydroquinone *(78,79)*, etc. The second approach is heating GO at 200°C to let the oxygen-containing functional groups leave the GO in the form of water and carbon dioxide under an inert atmosphere *(80,81)*. The third approach is mixing catalyst, such as titanium dioxide, into GO to induce the reduction of GO with optical exposure or high-temperature conditions *(82)*.

The oxidation-reduction method is simple and easy for scale up. However, the oxidation process is harsh, which strongly damages the graphene's structure. Even after the reduction process, it is still difficult to obtain graphene with good conductivity.

15.3.1.2.5 Electrochemical Method

Electrochemical approaches have also been developed for the large-scale synthesis of graphene. The electrochemical device generally comprises a graphite-working electrode, a graphene auxiliary electrode, electrolyte solution (including aqueous solution, organic solvent, and ionic liquid), and a direct-current (DC) stabilized power. The preparation process can be summarized in three steps: (1) apply a DC voltage across the electrodes to form an electrical field. Under the electric forces, the electrolyte will be dissociated into cations and anions and move to the cathode and the anode, respectively; (2) these ions will be embedded into the graphite layers, leading to expansion and deformation of graphite; (3) eventually, graphene layers will peel from the graphite electrode to form graphene. The electrochemical methods for the preparation of graphene include anodic oxidation, cathodic reduction, and electrochemical reduction approaches.

Anodic oxidation is a common electrochemical method that promotes the dissociated anion embedded into the graphite anode layers through controlling the applied potential, leading to the expansion and desquamation of graphite. Wang et al. used poly(sodium-4-styrenesulfonate) as the electrolyte and electrolyzed the graphite anode under 5 V constant voltage to obtain graphene sheets *(83)*. The other electrolytes include H_2SO_4, a mixed solution of H_2SO_4 and KOH *(84)*, and aqueous ammonium hydroxide solution *(85)*. The anodic oxidation method, which uses the electric force as the power to produce graphene, has the advantages of low cost, simple process, and large-scale production. However, this method is constrained by experimental conditions due to the fact that it is necessary to strictly control the voltage, type of electrolytes, and other parameters.

Cathodic reduction, similar to anodic oxidation, also uses electric forces to exfoliate the graphite. For anodic oxidation, oxygen promotes the expansion of anode and anions intercalate into graphite layers while for cathodic reduction, hydrogen facilitates the expansion of cathode and cations are intercalated graphite layers. Cathodic reduction reaction mainly happens in the cathode. The major factors for cathodic reduction reaction are the electrolyte and the adding of surfactants. Wang et al. used a Li^+ and propylene carbonate mixed solution as the electrolyte and obtained well-dispersed graphene in the cathode without going through other processes such as oxidation and strong acid treatment *(86)*. The Li^+ plays an important role in promoting the polymerization of propylene carbonate and destroying the graphite intercalation force. Compared to the anodic oxidation method, cathodic reduction not only achieves the high yield and high purity of graphene but also avoids the production of GO.

In addition, electrochemical reduction is another green and large-scale method for the preparation of graphene. Compared to the cathodic reduction process, this approach does not involve the process of cation embedding into graphite, but directly transforms the graphene oxide made from Hummers' method into graphene *(87,88)*.

Currently, people are continuously looking for alternative oxidation and reduction agents for electrochemical approaches to achieve improved products or to make the process safer and greener.

15.3.1.3 Properties of Graphene

Because of its unique structure, graphene has shown superior properties in many aspects important for its application in CNT/graphene hybrid materials. In the following, we will briefly introduce the electrical, optical, thermal, and mechanical properties of graphene.

15.3.1.3.1 Electrical Properties

Graphene is a zero-gap semiconductor. The Brillouin zone of graphene has two points, K and K′, which are called Dirac points, at the corners of the graphene (Fig. 15.1D) *(89,90)*.

The valance band and conduction band in graphene all have the conical shape and intersect at a point. The electrons at the top of the valance band can easily access the bottom of the low-energy conduction band. Electrons in the graphene can be seen as relativistic particles *(91)*, showing a half-integer quantum Hall effect *(92,93)*. Graphene has a unique characteristic carrier and special linear spectral characteristics and has great potential in transistor, microelectronics, batteries, etc.

As noted before, graphene lattice is composed of sp^2 hybridized carbon atoms. Each carbon atom in the graphene lattice is bonded together with σ bonds and has a π orbital, which constitutes a delocalized electron network. Due to the strong interatomic force and the freely moved π electrons, graphene has excellent electrical conducting performance. The electric conductivity of graphene can reach 10^6 S/m *(94)*, higher than the conductivity of silver. The electron mobility of graphene can reach 2×10^5 $cm^2/(V \cdot s)$ *(95)*, which is 140 times that of silicon.

15.3.1.3.2 Optical Properties

Graphene has excellent optical properties. The high-frequency conductivity for Dirac fermions in graphene is equal to $\pi e^2/2h$ *(96,97)*. Under visible light incidence, the optical transmittance (T) can be expressed as $T = (1 + 1/2\pi\alpha)^{-2}$, and the reflectance R is given by $R = 1/4\pi^2\alpha^2 T$, where $\alpha = 2\pi e^2/hc \approx 1/137$ (e is the electron charge, c is the light speed, and h is the Planck constant). Thus the opacity of monolayer graphene is $(1 - T) \approx \pi\alpha \approx 2.3\%$. Geim et al. measured the monolayer graphene opacity, which is 2.3%, and negligible reflectance (<0.1%) using suspended mechanical exfoliation graphene *(98)*. The opacity spectroscopy of graphene is independent of wavelength, λ. Furthermore, the opacity is increased with the membranes' thickness and each graphene layer increases another 2.3%. Thus it is a facile method to estimate graphene layers according to the visible light opacity of graphene film.

Theoretical and experimental results all show that the optical characteristic of graphene is linear with the thickness of graphene. Due to the outstanding electrical and optical properties, graphene film can be used as transparent conductor which possesses excellent conductivity and good transparency comparable with ITO *(94,99)*. Furthermore, in infrared spectroscopy, graphene has interband and gate-dependent optical transitions. The low density of states near the Dirac points in graphene contribute to a significant change in transmission, with a shift of the Fermi lever when the gate causes a significant variation in charge density *(100)*. Therefore graphene can be used in ultrafast photonics *(101)* and optoelectronics *(102,103)*, such as fiber laser *(104)*.

15.3.1.3.3 Thermal Properties

The thermal conductance (κ) of graphene is dominated by phonon transport, which has diffusive conduction at high temperature and ballistic conduction at sufficiently low temperature *(105)*. By molecular dynamic (MD) simulation *(106)*, the κ of defect-free graphene has a relationship with T shown as $\kappa \propto 1/T$. The κ of a suspended monolayer graphene is predicted to be about 6000 W/(m·K) at room temperature. Balandin et al. measured the thermal conductivity of monolayer graphene near 5300 W/(m·K) through a no-contact optical method at room temperature *(107)*. This value is higher than that of CNTs and diamond, and 10 times of that of copper [401 W/(m·K)] at room temperature. For graphene on a substrate, Seol et al. *(108)* measured the thermal conductivity of mechanical exfoliation graphene on a SiO_2 substrate, which was 600 W/(m·K). This value is higher than that of Cu but lower than that of suspended graphene because of the phonons leaking across the graphene–substrate interface and the strong interface scattering *(107,109)*. Graphene's excellent thermal conductivity could be retained at high temperature and thus could be used as an efficient radiator material.

15.3.1.3.4 Mechanical Properties

The mechanical properties of monolayer graphene have been studied theoretically and experimentally. The carbon atoms in graphene lattice are arranged in sp^2 hybridization. Reddy et al. investigated the Young's modulus of suspended few-layer graphene strips over trenches using atomic force microscopy (AFM) through force-displacement measurement *(110)*. Kern et al. studied the Young' modulus and fracture strength of defect-free monolayer graphene by a similar AFM indentation method, which were 1.0 TPa and 130 GPa, respectively, *(111,112)*. The ultimate strength of plain steel is in the range of 250–1200 MPa, and the ideal graphene is approximately 100 times stronger than plain steel.

In addition, graphene is very flexible. When an external force is applied, the bending deformation of the graphene plane will happen so that the carbon atoms do not have to rearrange to accommodate the foreign force, thus maintaining a stable structure.

The excellent mechanical properties make graphene an ideal 2D reinforcement material, which has a wide range of applications in the field of composite materials. It can also be used as paper-like thin and ultralight aircraft fuselage material, supertough material for bulletproof vests, and even as space elevator cables.

15.3.2 Synthesis of CNT/Graphene Hybrids

Over the past several years, several methods have been explored to fabricate CNT/graphene hybrid. These methods are categorized as postgrowth assembly and in situ CVD growth.

15.3.2.1 Postgrowth Assembly of CNTs and Graphene

Postgrowth assembly is the most convenient approach to realize the hybridization of CNTs and graphene. As the fabrication techniques depend on the form of the two components, the postgrowth assembly approaches can be considered as wet routes and dry routes.

15.3.2.1.1 Wet Routes

Pioneering works on the fabrication of CNT/graphene hybrids through wet processes are based on the formation of stable precursor solution with the two components. As CNTs and graphene have similar atomic structures, a stable dispersion solution could be prepared with similar processes *(113)*. There are three major strategies for the dispersion of CNTs or graphene, including (1) directly dispersing raw CNTs or graphene in neat solvents, such as *N,N*-dimethylformamid (DMF), *N*-methyl-2-pyrrolidone (NMP), and anhydrous hydrazine; (2) noncovalently functionalizing them by adding certain agents, such as surfactants and polymers, then dispersing them; and (3) covalently functionalizing CNTs or graphene sheets, then dispersing them.

In addition, graphene sheets can also come from rGO, which is obtained by exfoliation of graphite oxide and the following reduction. Since the hydrophilic oxygen groups attach to the graphene sheets in GO, GO can form stable solutions in several solvents. Therefore GO is preferably selected for hybridization with CNTs in the solution process, while extra-reduction treatment is needed to restore the conductivity of graphene to achieve CNT/graphene hybrids with good electrical conductivity.

Finally, macroscopic CNT/graphene architectures with uniform dispersion can be produced from the uniform dispersion of CNT/graphene with appropriate assembling routes. According to the type of the interactions that drive the assembly of CNTs and graphene, wet processes can be classified as $\pi-\pi$ interactions, noncovalently electrostatic interactions by layer-by-layer (LBL) deposition, electrophoretic deposition (EPD), etc.

Hybridization by $\pi-\pi$ Interactions $\pi-\pi$ interaction is a particular type of dispersion force from van der Waals forces, which is established between unsaturated (poly)cyclic molecules *(114)*. Carbon nanotubes and graphene have the same hexatomic ring of carbon atoms and therefore can be spontaneously stacked with each other by the $\pi-\pi$ interactions *(115)*. In this case, stable CNT/graphene solutions would be formed by adding one to the other, from which CNT/graphene hybrid materials could be prepared.

The early fabrication of CNT/graphene hybrids was reported by Cai et al. *(116)* CNT/graphene hybrid films were prepared through casting the GO/DMF and CNTs/DMF mixture on a cover slip. The CNTs were stacked parallel to the surface of GO nanoplates (as shown in Fig. 15.2A). Carbon nanotubes functionalized with aqueous cationic polyelectrolytes were also used for the preparation of hybrids, which consisted of intercalated layer of CNTs and GO *(117)*. In addition, GO could also be used as the dispersant of CNTs based on the strong $\pi-\pi$ stacking interactions between them *(118)*. Tung et al. reported the preparation of CNT/graphene hybrids with 86% optical transmittance by dispersing GO powder and slightly oxidized CNTs in anhydrous hydrazine *(119)*. Moreover, the prepared CNT/graphene hybrid films or papers from this approach were thick and strong enough to be freestanding *(120,121)*.

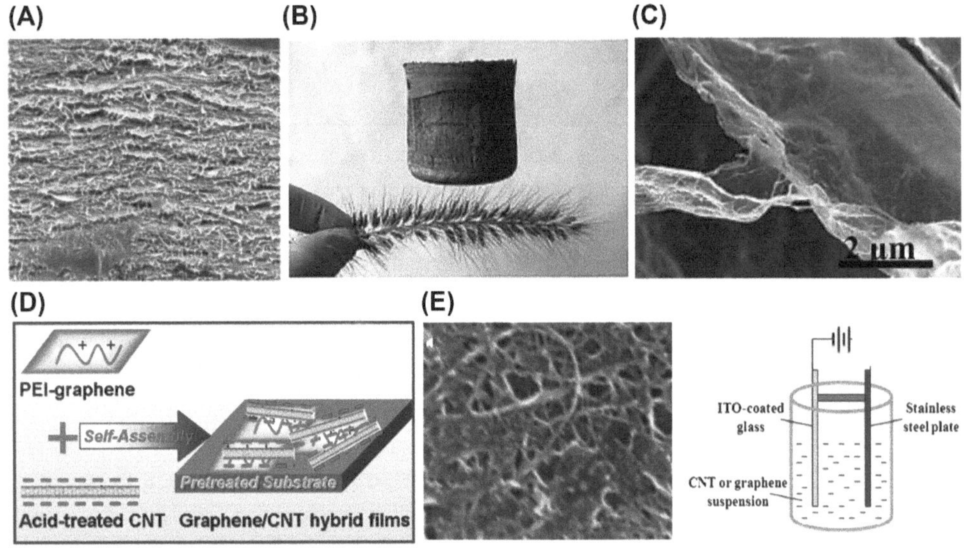

Figure 15.2 (A—C) SEM image and macroscopic and microscopic structures of CNT/graphene aerogels fabricated by EM interactions. *(The image in panel (A) was adapted with permission from Cai, D.; Song, M.; Xu, C. Highly Conductive Carbon-Nanotube/Graphite-Oxide Hybrid Films. Advanced Materials* ***2008,*** *20, 1706—1709. Copyright 2008 John Wiley and Sons. The images in panels (B and C) were adapted with permission from Sun, H.; Xu, Z.; Gao, C. Multifunctional, Ultra-Flyweight, Synergistically Assembled Carbon Aerogels. Advanced Materials* ***2013,*** *25, 2554—2560. Copyright 2013 John Wiley and Sons.)* (D—E) CNT/graphene hybrid materials fabricated by noncovalent electrostatic force. (D) Illustration of hybridization of PEI-modified graphene and acid-treated CNT with LBL method. (E) SEM image. *(Reprinted with permission from Yu, D.; Dai, L. Self-Assembled Graphene/Carbon Nanotube Hybrid Films for Supercapacitors. The Journal of Physical Chemistry Letters* ***2009,*** *1, 467—470. Copyright 2010 American Chemical Society.)* (F) Diagram of EPD process. *(Reprinted with permission from Hong, X.; Chen, Y.; Wu, P. Z.; Zheng, H. Simple, Effective Fabrication of Layered Carbon Nanotube/Graphene Hybrid Field Emitters by Electrophoretic Deposition. Journal of Vacuum Science & Technology B* ***2015,*** *33, 011802. Copyright 2015 AIP Publishing LLC.)*

The three-dimensional architecture of graphene and CNTs, which is ultralight porous nanomaterials with large surface area, could also be prepared by this approach *(122)*. Fig. 15.2B shows a CNT/graphene hybrid aerogel prepared by a template-free method via direct cryodesiccation from the aqueous solution of CNTs and giant GO flakes *(123)*. The nanostructure of CNT/graphene hybrid aerogels are constructed with graphene cell walls covered by spaghetti-like CNT networks (as shown in Fig. 15.2C), leading the synergistic effect between CNTs and graphene walls.

The hybridization of CNTs and graphene via $\pi-\pi$ interactions is efficient and inexpensive. The structures of the hybrids depend on the substrates, solution concentration, and fabrication techniques. However, the stacking structure of graphene and CNTs cannot been well controlled, and aggregated graphene or CNTs exists in the CNT/graphene hybrid nanostructure.

Layer-by-Layer Deposition LBL deposition is one of the fabrication techniques for hybrid films via sequential, repeated deposition of alternate layers of complementarily charged or functionalized materials *(124)*. The process of deposition can be realized with various film-coating methods, such as dip coating, spin coating, spray coating, electromagnetism, etc. LBL deposition is an efficient method to fabricate CNT/graphene hybrid films with alternate CNT and graphene layers *(125–128)*. GO sheets could be functionalized with groups, such as carboxylic acids, on the surface and edge, inducing negative charge ($GO\text{-}COO^-$), while CNTs could be positively charged with introduction of amine groups ($CNT\text{-}NH_3^+$). Kim et al. prepared ultrathin transparent CNT/graphene hybrid films with double-layer structure via LBL deposition (125). In addition, graphene sheets could be positively charged with poly (ethyleneimine) (PEI), and CNTs could be negatively charged with carboxylic acids by acid treatment. Hybrid films were fabricated by sequential self-assembly of PEI-modified graphene with acid-oxidized CNTs on various substrates (as shown in Fig. 15.2D), such as silicon and ITO glass. The hybrid carbon films had interconnected networks with well-defined nanopores (as shown in Fig. 15.2E), which showed promise as supercapacitor electrodes.

LBL deposition is a convenient and efficient method to fabricate CNT/graphene hybrids with sandwich structures by integrating CNT layers and graphene layer. However, the functionalization groups on the graphene and CNTs would destroy their intrinsic structures and influence the performance and application of the obtained hybrids.

Electrophoretic Deposition EPD is a versatile processing technique for coating without the requirement of binders. In EPD, charged colloidal particles dispersed or suspended in a liquid medium are attracted and deposited onto an electrically conductive substrate with opposite charges under an electric field *(129)*. In order to hybridize CNTs and graphene via EPD, the preparation of stable mixture suspensions, where both CNTs and graphene have the same charge, is key.

For example, Bon et al. fabricated transparent and conducting CNT/graphene thin films on conducting ITO-coated polyethylene terephthalate (PET) substrates via EPD using carboxyl-functionalized CNTs and graphene sheets dispersed in 1-ethyl-2-pyrrolidinone anhydrous *(130)*. In addition, Su et al. used methyl violet (MV) as both an efficient dispersant and a good charging agent *(131)*. Koh et al. *(132)* prepared graphene and CNTs solutions in ethanol with $Al(NO_3)_3$ as the charger agent. Similarly, with the introduction of $Mg(NO_3)_2$, Hong et al. fabricated CNT/graphene sandwich structures with alternate CNT layers and graphene layers via EPD *(133)*. Fig. 15.2F shows the experimental setup.

EPD is a low-cost, high-efficient method to fabricate CNT/graphene films on various substrates, the thickness of which depends on the parameters used in the EPD and the morphology of which depends on the deposited substrates. However, in order to realize EPD, the substrates need to be conductive, which restricts the types of applied substrates. In addition, the colloids or particles in the suspensions need to be charged via specific functionalization, which leads to destruction of the intrinsic structure of CNTs and graphene or introduction of contaminations in the hybrid films.

15.3.2.1.2 Dry Routes

The obvious difference between wet routes and dry routes is whether the graphene and CNTs need to be dispersed into solvents before the assembly of CNT/graphene hybrid. In a dry process, the assemblies of graphene and CNTs could be prepared through direct growth followed by dry-route assembly, in which case the materials could preserve their intrinsic properties.

For example, Li et al. reported a CNT/graphene hybrid with a CNT network patched with graphene by a solid-phase layer-stacking approach (as shown in Fig. 15.3A) *(134)*. Carbon nanotube films were prepared by floating CVD, while graphene films were

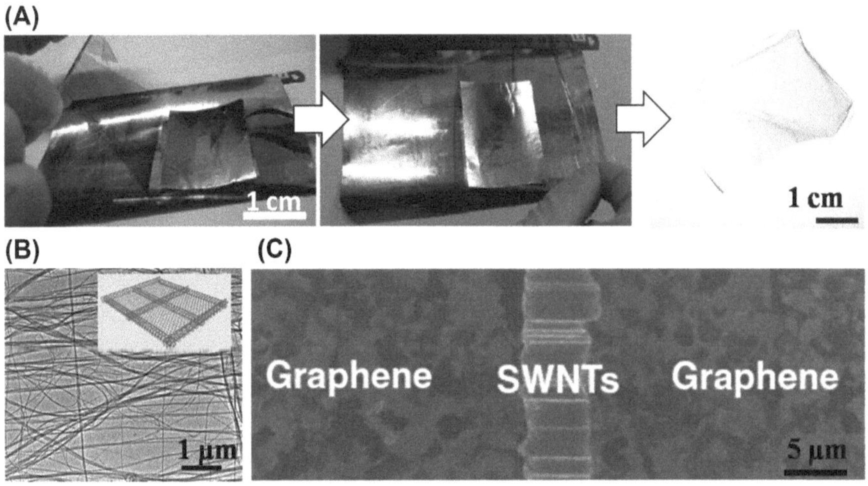

Figure 15.3 CNT/graphene hybrid materials assembled by solid-phase layer-by-layer stacking. (A) The process to coat CNT film on graphene and a freestanding CNT/graphene hybrid film floating on the water. *(Reprinted with permission from Li, C.; Li, Z.; Zhu, H.; Wang, K.; Wei, J.; Li, X.; Sun, P.; Zhang, H.; Wu, D. Graphene Nano-"Patches" on a Carbon Nanotube Network for Highly Transparent/Conductive Thin Film Applications. The Journal of Physical Chemistry C* ***2010,*** *114, 14008–14012. Copyright 2010 American Chemical Society.)* (B) TEM image and schematic illustration (inset) of a CNT/graphene membrane with graphene film supported by CNT vein. *(Reprinted with permission from Lin, X.; Liu, P.; Wei, Y.; Li, Q.; Wang, J.; Wu, Y.; Feng, C.; Zhang, L.; Fan, S.; Jiang, K. Development of an Ultra-Thin Film Comprised of a Graphene Membrane and Carbon Nanotube Vein Support. Nature Communications* ***2013,*** *4, 2920. Copyright 2013 Nature Publishing Group.)* (C) An SEM image of the monolithic carbon hybrid with aligned CNTs between two graphene electrodes. *(Adapted with permission from Hong, S. W.; Du, F.; Lan, W.; Kim, S.; Kim, H. S.; Rogers, J. A. Monolithic Integration of Arrays of Single-walled Carbon Nanotubes and Sheets of Graphene. Advanced Materials* ***2011,*** *23, 3821–3826. Copyright 2011 John Wiley and Sons.)*

grown on the copper foils. The CNT film was transferred onto the graphene followed by ethanol treatment. In addition, inspired by the vein-membrane structure in nature, Lin et al. prepared CNT/graphene hybrid via overlaying graphene film onto a porous vein-like CNT network (Fig. 15.3B) *(135)*. Kholmanov et al. fabricated CNT/graphene hybrid films with CNT/graphene/CNT or graphene/CNT/graphene sandwich structures *(136)*. Furthermore, with transfer printing strategies, Hong et al. fabricated all-carbon transistors via incorporating CNT arrays as the semiconductor and graphene sheets as the source, drain, and gate electrodes (Fig. 15.3C) *(137)*. Liu et al. fabricated a planer atomically thin CNT/graphene hybrid via transferring graphene film onto SWNT films on the silicon substrate *(138)*.

The above shows that, in a dry process, as the assemblies of separated CNTs and graphene are done by direct growth without any additive treatment, the CNTs and graphene could preserve their inherent properties, enabling enhanced properties. However, dry routes also have limitations, such as difficulty for large scale, weak interactions between the CNTs and the graphene, etc.

15.3.2.2 In Situ CVD Growth of CNT/Graphene Hybrid

Although hybridization of separated CNTs and graphene is accessible and feasible, the structures of the hybrid films are simple, with physically stacked CNTs and graphene films without the formation of covalent C—C bonding between them. Accordingly, in situ growth routes provide possibilities for fabrication of novel structures with the formation of C—C bonds between CNTs and graphene.

Chemical vapor deposition, which is the typical growth method used to grow both CNTs and graphene, could be directly used for the fabrication of CNT/graphene hybrid. In situ growth of CNT/graphene hybrids by CVD can be divided into three categories: growth of CNTs or graphene on the other one (which is not prepared by CVD), growth of CNT/graphene through two separated CVD processes, and growth of CNT/graphene by a one-step CVD process.

15.3.2.2.1 Growth of CNTs or Graphene by CVD on the Other Component

The catalysts for the growth of CNTs are in the form of nanoparticles, while graphene could be catalytically grown on a flat substrate. A convenient strategy to realize the growth of CNT/graphene hybrids is to synthesize one of the building blocks on the other through CVD. For example, CNTs can be grown and intercalated between graphene layers or GO layers by CVD (as shown in Fig. 15.4A). For the first time, Fan et al. *(139)* reported the preparation of CNT/graphene 3D hybrids by CVD, with CNT pillars grown in between graphene layers. $Co(NO_3)_2 \cdot 6H_2O$ was added into the GO solution to work as the catalyst precursor for the growth of CNTs. Following that, Zhang et al. *(140)* prepared hybrid materials with most of the CNTs grown perpendicularly to the graphene platelets (as shown in Fig. 15.4B). Three-dimensional

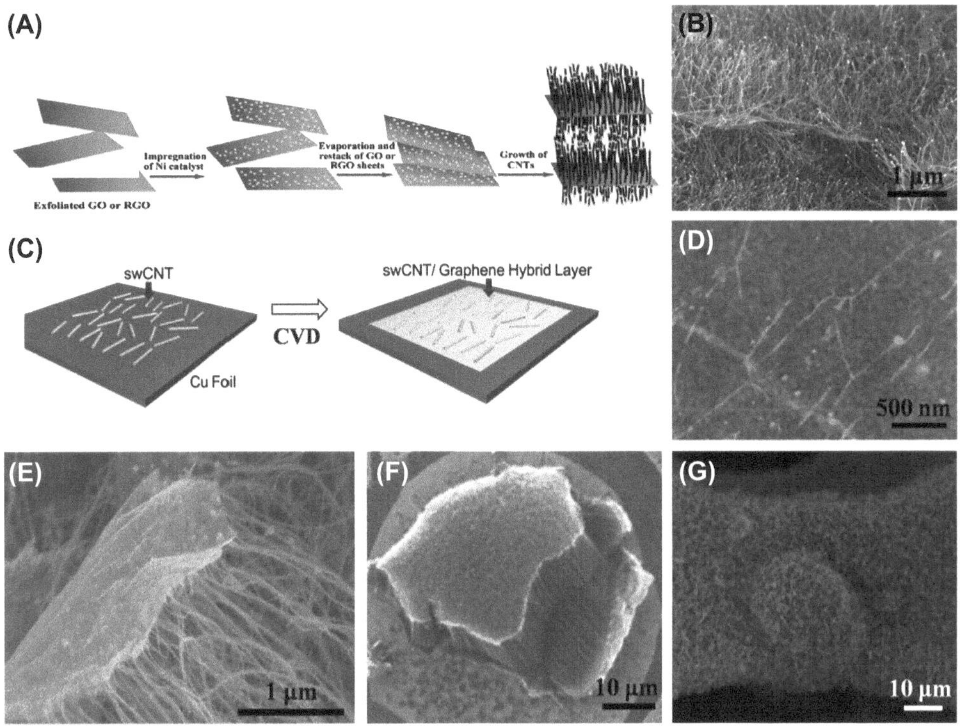

Figure 15.4 (A) Illustration of the formation of hybrid materials with CNTs grown in between graphene nanosheets. (B) SEM image of the CNT/graphene hybrid with CNTs grown on the reduced GO. *(The images (A) and (B) were reprinted with permission from Zhang, L. L.; Xiong, Z.; Zhao, X. S. Pillaring Chemically Exfoliated Graphene Oxide with Carbon Nanotubes for Photocatalytic Degradation of Dyes under Visible Light Irradiation. ACS Nano* ***2010,*** *4, 7030–7036. Copyright 2010 American Chemical Society.)* (C) Schematic diagram for synthesis of planer CNT/graphene hybrids. (D) AFM image of the planer hybrid film. *(The images (C) and (D) were reprinted with permission from Kim, S. H.; Song, W.; Jung, M. W.; Kang, M. A.; Kim, K.; Chang, S. J.; Lee, S. S.; Lim, J.; Hwang, J.; Myung, S.; et al. Carbon Nanotube and Graphene Hybrid Thin Film for Transparent Electrodes and Field Effect Transistors. Advanced Materials* ***2014,*** *26, 4247–4252. Copyright 2014 John Wiley and Sons.)* (E–G) SEM images of the CNT/graphene hybrids grown with two separated CVD processes. *(The image (E) was adapted with permission from Zhang, W.; Xie, H.; Zhang, R.; Jian, M.; Wang, C.; Zheng, Q.; Wei, F.; Zhang, Y. Synthesis of Three-Dimensional Carbon Nanotube/Graphene Hybrid Materials by a Two-Step Chemical Vapor Deposition Process. Carbon* ***2015,*** *86, 358–362. Copyright 2015 Elsevier. The image (F) was adapted with permission from Zhu, Y.; Li, L.; Zhang, C.; Casillas, G.; Sun, Z.; Yan, Z.; Ruan, G.; Peng, Z.; Raji, A. R.; Kittrell, C.; et al. A Seamless Three-Dimensional Carbon Nanotube Graphene Hybrid Material. Nature Communications* ***2012,*** *3, 1225. Copyright 2012 Nature Publishing Group. The image (G) was adapted with permission from Dong, X.; Ma, Y.; Zhu, G.; Huang, Y.; Wang, J.; Chan-Park, M. B.; Wang, L.; Huang, W.; Chen, P. Synthesis of Graphene–Carbon Nanotube Hybrid Foam and Its Use as a Novel Three-Dimensional Electrode for Electrochemical Sensing. Journal of Materials Chemistry* ***2012,*** *22, 17044. Copyright 2012 Royal Society of Chemistry.)*

CNT/graphene hybrids can also be prepared by directly growing vertically aligned CNTs on graphene substrates. Lee et al. reported a versatile hybrid composed of vertical CNT arrays grown on mechanically compliant graphene films *(141)*. Fe catalyst nanoparticles were deposited on graphene for the growth of CNTs. It was found that the junctions between the CNT and graphene were ohmic electrical contacts. Du et al. created 3D pillared CNT/graphene architectures by intercalated growth of vertically aligned CNTs catalyzed by pyrolysis of FePc into thermally expanded HOPG *(142)*. In addition, Li et al. prepared vertical CNT arrays on free-standing graphene papers by deposition of Al_2O_3 and Fe for the CVD growth of CNTs *(143)*.

In addition, planar CNT/graphene hybrids were synthesized via the CVD growth of graphene on the Cu or Ni substrates coated with CNT networks (as shown in Fig. 15.4C). For example, Kim et al. synthesized CNT/graphene hybrids with the copper foil spin coated with CNTs by CVD with methane source (as shown in Fig. 15.4D) *(144)*. Nguyen et al. grew multilayer graphene films on a Ni substrate coated with CNT networks using cellulose acetate (CA) member as the carbon precursor in the CVD process *(145)*. In addition, Yan et al. proposed the synthesis of CNT/graphene hybrids with CNTs networks without introducing extraneous carbon sources for the growth of graphene *(146)*. It was reported that CNT acted as reinforcing bar (rebar) in the hybrid films, through both $\pi-\pi$ stacking and covalent bonding with graphene films. Furthermore, novel hybrid structure with graphene film embroidered by CNT networks was produced by Shi et al. *(147)*. High-density and interconnected CNT spider webs acted as porous template for the growth of graphene on the copper substrate.

Direct CVD growth of CNTs or graphene on the other component enabling the fabrication of hybrid with novel structures, such as 3D structures, layered structures, and structures with aligned CNTs, has led to novel or enhanced performance.

15.3.2.2.2 Growth of CNTs and Graphene by CVD

The graphene and CNTs grown by CVD usually have high quality and fewer defects. To synthesize hybrids with good and neat structure, CNTs and graphene can be grown by CVD, using either two separated CVD processes or a one-step CVD process.

For example, Kim et al. synthesized CNT/graphene hybrids with short CNT arrays on graphene *(148)*. Graphene films were grown on a copper foil and transferred onto a quartz substrate for the growth of CNTs. In addition, Zhang et al. reported a CNT/graphene hybrid with one end of vertically aligned CNTs connected tightly with graphene film (Fig. 15.4E) *(149)*. The graphene was grown on a copper substrate, then transferred onto a SiO_2/Si substrate deposited with Al_2O_3 and Fe. The vertically aligned CNTs were to be grown below the graphene films and the graphene was lifted up on the end of CNTs. Carbon nanotube/graphene hybrids could also be prepared without the transferring process. Zhu et al. synthesized a seamless 3D CNT/graphene hybrid material on copper foils with one end of the CNTs covalently bonded with graphene films (as shown in Fig. 15.4F) *(150)*. Similar work was performed by Wang et al., where graphene films were rooted between CNT and CNTs *(151)*. Graphene film between the copper and CNTs facilitated the charge transfer, which inspired the binder-free electrodes in lithium-ion batteries.

In addition, 3D hybrids can also be grown with the appropriate templates. For example, Dong et al. synthesized 3D CNT/graphene hybrid foam with a two-step CVD process using nickel foam as the template (Fig. 15.4G) *(152)*. The 3D monolithic carbon hybrids have great potential in fabrication of versatile electrodes for sensors and energy devices. Moreover, Yan et al. reported multifunctional 3D hybrids with graphene and vertically aligned CNTs grown separately on a porous Ni substrate without etching Ni *(153)*. The graphene in the hybrid acted as a linking agent between the CNTs and metal interfaces, facilitating the fabrication of high-performance field-emitter devices and double-layer capacitors.

The growth of CNT/graphene hybrid could also be done with a one-step CVD process. The process needs to realize simultaneous catalytic growth of graphene and CNTs. For example, Paul et al. synthesized hybrids with vertically aligned CNTs and graphene films in a one-batch CVD process *(154)*. The substrates were SiO_2/Si wafers with sequentially deposited 500-nm-thick copper film for growth of graphene and 5-nm-thick iron catalysts for growth of CNTs. The growth of graphene and CNTs simultaneously occurred at 750°C. A planer hybrid film could also be synthesized using a one-step CVD process. Copper foils spin coated with silicon nanoparticles were used to grow graphene and CNTs in a one-step CVD process *(155)*. In addition, Zhao et al. synthesized CNT/graphene hybrids on 2D lamellar FeMgAl layered double hydroxide (LDH) by one-step CVD with methane as the carbon resource *(156)*. Similar CNT/graphene hybrids can be prepared on the exfoliated vermiculites impregnated with Fe/Mo *(157)*.

Growth of CNT/graphene hybrid material through CVD processes is an effective approach to combine the two high-quality building blocks with covalent bonds, enabling good coupling between CNTs and graphene. However, the cost of this approach is relatively high and the scale-up needs more effort than other common postgrowth processes.

15.3.2.3 Summary

Great efforts have been continuously applied to controllably assemble the hybrid during the past few years. Postgrowth assembly of CNT/graphene hybrid is a simple and easy approach that develops quickly with the mass production of CNTs and graphene. However, postgrowth organization methods can only construct simple nanostructures in which CNTs are stacked together with graphene films with van der Waals interaction or electrostatic force. Based on the postgrowth process, it is hard to control the architectures at nanometer scale and to obtain covalent C—C bonding connections between the two components. In comparison, in situ CVD assembly is an effective method to prepare high-quality hybrids with novel structures. Among these, pillared graphene, which is a CNT/graphene hybrid with graphene surfaces covalently connected by the ends of CNTs connected with pure C—C bonds, have gained enormous momentum. However, its process is complicated and not easy for low-cost and large-scale production of the hybrids. It is still a great challenge to explore facile methods for the preparation of high-quality CNT/graphene hybrids.

15.3.3 Applications of CNT/Graphene Hybrids

Due to the superior properties of CNT/graphene hybrids resulting from the synergetic effect of CNTs and graphene, their applications have been widely explored in many fields, such as electronic devices, TCFs, sensors and actuators, energy conversion and storage, composites, catalysts, adsorption and desalination, etc.

15.3.3.1 Electronic Devices

In the past few decades, the silicon-based semiconductor industry has created devices that are more compact, efficient, stable, denser, and faster, which has revolutionized almost all aspects of our life. However, due to continuous miniaturization of devices based on silicon technology, limits such as short-channel effects and tunneling current have created bottlenecks for further development *(158)*. Therefore pursuit of other materials to replace silicon is needed to address these problems and to improve the performance of devices. At the same time, development of all-carbon electronic devices is becoming one of the most attractive research fields to realize completely flexible, transparent, and stretchable electronic devices. Carbon nanotubes and graphene are the most promising materials due to their extremely high mobility, electrical current capacity, and other fascinating properties *(159)*. Most importantly, they are flexible and stretchable and show great potential as future flexible electronic devices. Therefore all-carbon electronic devices consisting of graphene acting as source, drain electrodes, and CNTs used as channel material can be fabricated to achieve the extraordinary flexibility and stretchability and even transparent devices with high performance.

For example, SWNTs/graphene could be used as channel-active materials to fabricate transparent field effect transistors (FETs) *(144)*. These devices have high transmittance at 550 nm (96.4%) and high mobility ($394.46 \pm 176.27\ cm^2/V \cdot s$). Importantly, the on-state current increases due to the existence of graphene (Fig. 15.5A—C). Moreover, SWNTs/graphite (thicker graphene) was synthesized by a two-step CVD and the graphite acted as source and drain electrodes *(160)*. This FET demonstrated that graphite and aligned CNTs have good electrical contact; the effective mobility was about $20\ cm^2/V \cdot s$. Similarly, rGO could also be used as electrodes *(161)*.

Moreover, all-carbon electronic devices that are flexible and stretchable can be fabricated. W. J. Yu et al. used monolayer graphene to replace metal electrodes and found that the FET device hysteresis was narrowed because graphene dramatically reduced

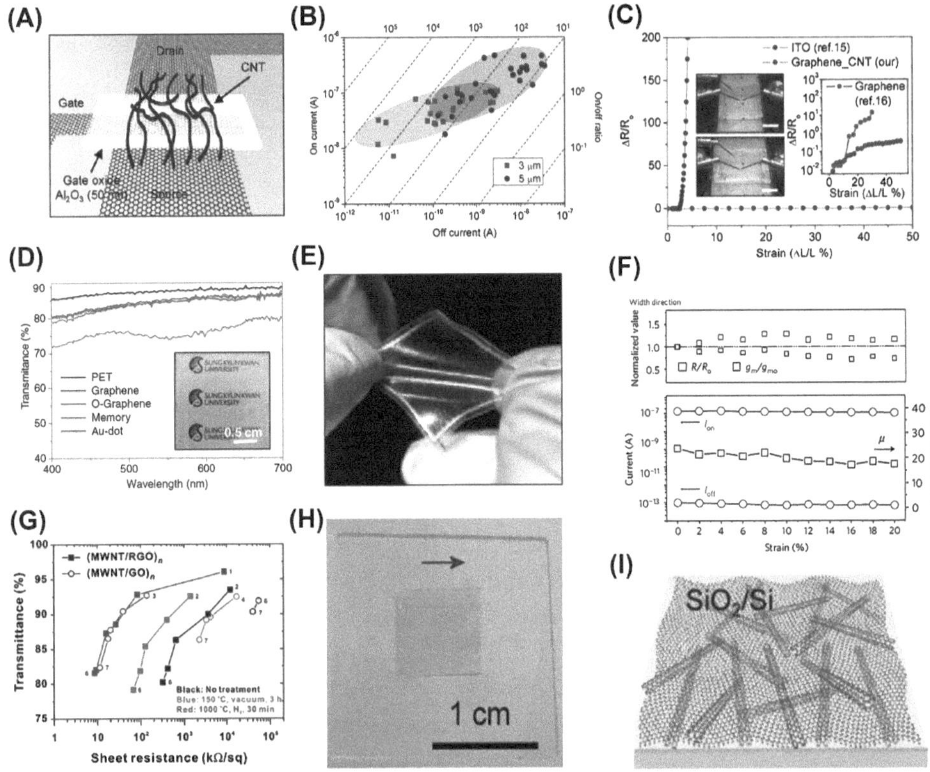

Figure 15.5 Electronic devices and transparent conductive films based on CNT/graphene hybrids. (A) The schematic diagram of CNT/graphene field effect transistors (FETs) composed of single-layer graphene and CNTs thin-film channels. (B) Electrical properties of CNT/graphene FETs with different channel widths. (C) Resistance variation of devices against a strain compared with ITO electrodes. *(The images (A–C) were adapted with permission from Yu, W.J.; Lee, S. Y.; Chae, S. H.; Perello, D.; Han, G. H.; Yun, M.; Lee, Y. H. Small Hysteresis Nanocarbon-Based Integrated Circuits on Flexible and Transparent Plastic Substrate. Nano Letters* ***2011,*** *11, 1344–1350. Copyright 2011 American Chemical Society.)* (D) Transmittance of memory device array on a polyethylene terephthalate (PET) film. *(Reproduced with permission from Yu, W. J.; Chae, S. H.; Lee, S. Y.; Duong, D. L.; Lee, Y. H. Ultra-Transparent, Flexible Single-Walled Carbon Nanotube Non-Volatile Memory Device with an Oxygen-Decorated Graphene Electrode. Advanced Materials* ***2011,*** *23, 1889–1893. Copyright 2011 John Wiley and Sons.)* (E) Optical image of stretching the devices. (F) Normalized properties with respect to values at zero strain along the width direction. *(Reproduced with permission from Chae, S. H.; Yu, W. J.; Bae, J. J.; Duong, D. L.; Perello, D.; Jeong, H. Y.; Ta, Q. H.; Ly, T. H.; Vu, Q. A.; Yun, M. Transferred Wrinkled Al_2O_3 for Highly Stretchable and Transparent Graphene–Carbon Nanotube Transistors. Nature Materials* ***2013,*** *12, 403–409. Copyright 2013 Nature Publishing Group.)* (G) Sheet resistance and transmittance changes of CNT/graphene hybrid films after thermal treatment. *(Adapted with permission from Hong, T.K.; Lee, D.W.; Choi, H. J.; Shin, H. S.; Kim, B.-S. Transparent, Flexible Conducting Hybrid Multilayer Thin Films of Multiwalled Carbon Nanotubes with Graphene Nanosheets. ACS Nano* ***2010,*** *4, 3861–3868. Copyright 2010 American Chemical Society.)* (H) Schematic illustration of CNTs embroidered graphene film. *(Reproduced with permission from Kholmanov, I. N.; Magnuson, C. W.; Piner, R.; Kim, J. Y.; Aliev, A. E.; Tan, C.; Kim, T. Y.; Zakhidov, A. A.; Sberveglieri, G.; Baughman, R. H.; et al. Optical, Electrical, and Electromechanical Properties of Hybrid*

the interface charge trap *(162)*. In addition, the FET was transparent (Fig. 15.5D) and flexible, and there were no obvious changes of performance during the bending test. These SWNTs/graphene FETs could be used to fabricate logic circuit array composed of inverters and different gates devices. Furthermore, the SWNTs/graphene FET device can be used to fabricate programmable memory devices *(163)*. The device showed high mobility (44 $cm^2/V \cdot s$) and a fast response speed of 100 ns. In the case of wrinkled Al_2O_3 used as dielectric layer, the transparent SWNTs/graphene FET device was highly stretchable (Fig. 15.5E) *(164)*. The performances of the on/off ratio, on/off current, and mobility showed almost no change during specific strain (Fig. 15.5F). After stretching and releasing for 1000 times, there was no significant change in device performance. This device could be integrated with other flexible devices to obtain completely flexible and stretchable electronic devices.

15.3.3.2 Transparent Conductive Film

TCFs are widely used in industry, including in touch panels, displays, and optoelectronic devices *(32,165)*. Currently, ITO is the most commonly used TCF because it has superior optical and electrical properties (sheet resistance, 10 Ω/sq and transmittance, 90%) *(113)*. However, indium resources are nonrenewable and it has been difficult to satisfy the increasing demand, and ITO lacks flexibility and cracks easily so that it does not meet the needs of flexible, wearable, stretchable, low-cost TCFs. Therefore new TCFs with flexibility, stretchability, high optically transparent, and electrically conductivity are needed.

In practice, CNTs and graphene are generally considered to be ideal materials to replace ITO due to their excellent conductivity, mechanical property, stability, and flexibility. Carbon nanotubes/graphene hybrid film could also be high-performance TCFs. Carbon nanotube/graphene TCFs can be fabricated by electrostatic self-assembly *(126)*, solution mixing *(119)*, CVD method *(166,167)*, etc. Graphene sheets can cover a majority of surfaces and CNTs can link the graphene sheets together so that the sheet resistance can be reduced while the transmittance is maintained. For example, CNTs and rGO fabricated via LBL assembly followed by thermal treatment showed a sheet resistance of ~8 kΩ/sq with a transmittance of 81% (Fig. 15.5G) *(126)*. Carbon nanotube/rGO films prepared through solution routes with anhydrous hydrazine solutions showed sheet resistance of 240 Ω/sq at 86% transmittance *(119)*. Moreover, monolayer CVD-grown

Graphene/Carbon Nanotube Films. Advanced Materials ***2015,*** *27, 3053–3059. Copyright 2015 John Wiley and Sons.)* (I) Optical image of aligned CNT/graphene film on a glass substrate. *(Reprinted with permission from Shi, E.; Li, H.; Yang, L.; Hou, J.; Li, Y.; Li, L.; Cao, A.; Fang, Y. Carbon Nanotube Network Embroidered Graphene Films for Monolithic All-Carbon Electronics. Advanced Materials* ***2015,*** *27, 682–688. Copyright 2015 John Wiley and Sons.)*

graphene can be combined with aligned CNTs sheets to achieve highly transparent and flexible sheets *(136)*. In addition, the hybrid film could be obtained by CVD method (Fig. 15.5H and I) *(147)*. Moreover, incorporating CNT/graphene with other materials is a way to prepare high-performance TCFs *(168)*.

In view of practical applications, the sheet resistance of hybrid film needs to be further reduced while maintaining high optical transmittance. In addition, flexibility and stretchability are becoming indispensable. Therefore more research should be carried out, such as on using dopant and improving the quality of graphene and CNTs, so as to obtain high-performance CNT/graphene TCFs to meet today's requirements.

15.3.3.3 Energy Conversion and Storage

15.3.3.3.1 Solar Cells

The energy crisis is becoming one of the world's most severe problems. Thus it is of great importance to develop clean and renewable energy, such as wind, solar energy, tidal, etc. Traditional silicon-based solar cells cannot satisfy the increasing demand for solar energy devices. New types of solar cells, such as dye-sensitized solar cells (DSSCs), polymer solar cells (PSCs), and perovskite solar cells (PVCs), can be fabricated by simple processes at low cost *(169,170)*.

Carbon nanotubes/graphene can be utilized as both conductive electrode and electron/hole transport layers due to their high mobility in new types of solar cells. Kim et al. added a certain amount of SWNTs into GO layer to fabricate PSCs *(171)*. The hybrid film dramatically improved the fill factor (FF) and short-circuit current (J_{SC}) because it reduced the resistance and increased the hole transport. Moreover, functioned graphene/CNTs hybrid electrode could tune the work functions due to doping alkali carbonate salt *(172)*. This electrode in PSCs showed higher capability of collecting electron so that the power conversion efficiency (PCE) could be improved. Spinnable aligned CNTs can be combined with graphene for use as counter electrode in DSSCs *(173)*. This hybrid structure significantly improved the catalytic behavior and the PCE, which rose to 7.55% beyond the 4.65% with the use of graphene alone. The hybrid film can be prepared by CVD, and the DSSCs fabricated by this film had high performance, with a PCE of 7.9% *(174)*. Other studies in this field have also obtained excellent results *(175,176)*.

In addition, fiber-shaped solar cells have drawn wide attention due to their 1D configuration, good flexibility, and knittability *(177)*. Moreover, they are able to integrate with other wearable electronic devices, such as sensors and energy storages *(178,179)*. Generally, conductive materials, including CNTs, graphene, and metallic nanowires, are used as the fibers and act as the electrodes. Carbon nanotube films, which were drawn out of CNT arrays, have been combined with GO to fabricate CNT/rGO hybrid fibers (Fig. 15.6A and B) *(180,181)*, where CNTs could offer high electric conductivity while graphene would provide high catalytic performance. The DSSCs fabricated by CNTs/rGO fibers showed a high PCE (8.50%).

Carbon nanotube/graphene hybrids are promising materials to obtain high-performance solar cells. However, the fabrication process needs to be simplified and the quality of materials must be improved. Therefore effort is still needed in this field to push the applications of CNT/graphene in solar cells forward.

15.3.3.3.2 Supercapacitors

Compared with energy conversion, energy storage is equally or more important. With the advancement in portable electronic equipment, new energy-storage devices with higher power density and energy density are urgently needed. Supercapacitors and lithium batteries are widely used for energy-storage systems, which typically consist of anode, cathode, electrolyte, and separator. Improvement in their performance in devices mainly relies on the optimization of electrode materials. Carbon materials, such as activated carbon, have had an important influence on the development of supercapacitors and lithium batteries. Currently, CNTs, graphene, and carbon allotropes have been applied in energy-storage devices due to their large specific surface area, high conductivity, and excellent mechanical, catalytic properties.

As for supercapacitors, CNTs and graphene, which possess large specific surface area, can offer high specific capacitance due to their effective charge separation. Moreover, the hybrids of CNTs and graphene can prevent agglomeration as CNTs act as the spacer for graphene sheets. For example, a CNTs/rGO film was prepared by the solution method, which is normally usually used to fabricate supercapacitors *(115,182–184)*. Three-dimensional porous graphene/CNTs hybrid papers, which were porous, conductive, and flexible, had a high specific capacitance of 294 F/g *(185)*. Carbon nanotubes could apparently improve the electrical property and mesh on the rGO offer the fast ion diffusion, so that the supercapacitors delivered high specific capacitance (Fig. 15.6C), and an excellent rate capability as well as outstanding cycling performance *(186)*. With the exception of the solution process, CNT/graphene hybrids prepared by CVD were also applied to fabricate energy-storage devices *(139,150,187,188)*. Carbon nanotube/graphene hybrid film prepared on Ni foam was used to fabricate supercapacitors. The device showed a high specific capacitance of 286 F/g and outstanding cycling performance (capacitance retention of 99.34% after 85,000 cycles) *(188)*. Moreover, CNT/graphene/metal oxide and CNT/graphene-conducting polymer have been used as supercapacitors to obtain further improved electrochemical performance *(189–192)*.

Similar to solar cells, fiber-shaped supercapacitors have become one of the most attractive research topics in the past few years. Yu and coworkers synthesized a hierarchical structure composed of SWNTs and N-doped rGO, which showed excellent conductivity and high specific volumetric capacity (305 F/cm^3), as well as outstanding flexibility (Fig. 15.6D and E) *(193)*. Foroughi et al. combined CNTs sheets with

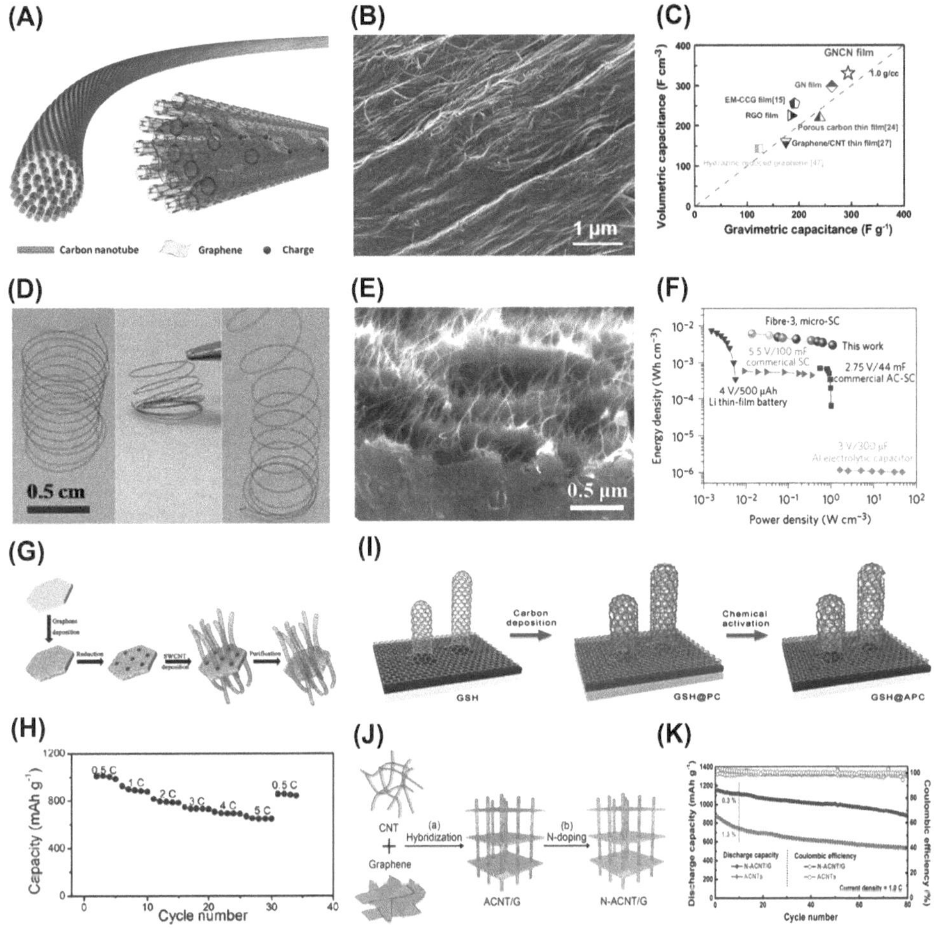

Figure 15.6 Energy conversion and storage devices based on CNT/graphene hybrids. (A) Schematic illustration of the aligned CNT/graphene composite fiber. (B) SEM image of CNT/graphene composite fiber. *(The images (A) and (B) were reprinted with permission from Sun, H.; You, X.; Deng, J.; Chen, X.; Yang, Z.; Ren, J.; Peng, H. Novel Graphene/Carbon Nanotube Composite Fibers for Efficient Wire-Shaped Miniature Energy Devices. Advanced Materials* ***2014,*** *26, 2868–2873. Copyright 2014 John Wiley and Sons.)* (C) Comparison of specific capacitances of CNT/graphene hybrid electrode with other carbon material electrodes. *(Reprinted with permission from Jiang, L.; Sheng, L.; Long, C.; Fan, Z. Densely Packed Graphene Nanomesh-Carbon Nanotube Hybrid Film for Ultra-High Volumetric Performance Supercapacitors. Nano Energy* ***2015,*** *11, 471–480. Copyright 2015 Elsevier.)* (D) Compressed and stretched CNT/graphene fibers. (E) Microstructures of CNT/graphene hybrid fibers. (F) Comparison of energy and power densities of supercapacitors with commercially available energy-storage systems. *(The images (D–F) were adapted with permission from Yu, D.; Goh, K.; Wang, H.; Wei, L.; Jiang, W.; Zhang, Q.; Dai, L.; Chen, Y. Scalable Synthesis of Hierarchically Structured Carbon Nanotube-Graphene Fibres for Capacitive Energy Storage. Nature Nanotechnology* ***2014,*** *9, 555–562. Copyright 2014 Nature Publishing Group.)* (G) Schematic illustration of SWNTs/graphene hybrids on layered double hydroxides (LDH) flakes. (H) Rate performance of the SWNTs/graphene hybrids for LSBs. *(The images (G and H) were adapted with*

graphene flakes, and the hybrid was twisted into yarn, which had electrical conductivity as high as 900 S/cm *(194)*. This yarn had high specific capacitance, more than 111 F/g, which was higher than the pristine CNTs sheets by a factor of 4.

15.3.3.3.3 Lithium Batteries

Lithium batteries, primarily lithium-ion batteries (LIBs), are widely used in our daily life, such as in mobile phones, tablets, and cameras. In the LIB system, carbon materials usually replace the metal lithium to store and release lithium ions in the electrochemical process. In theory, carbon can deliver a specific capacity of 372 mAh/g, but the actual performance closely relies on the structure and crystallinity *(195)*. Carbon nanotubes and graphene are considered as promising materials for fabrication of high performance LIBs, due to their large specific surface area and high electrical conductivity. Similar to the advantages mentioned earlier, CNT/graphene hybrids can provide a larger specific surface area as CNTs can act as the spacer to prevent graphene flakes from agglomerating.

Li and coworkers grew vertically aligned CNTs on graphene papers for use as anode in LIBs, and this device had excellent rate capability and good cycling performance *(143)*. Carbon nanotube/graphene hybrids prepared by the CVD technique, which have seamless connections between CNTs and graphene, facilitate the charge transfer and show high reversible capacity (900 mAh/g) *(151)*. In CNT/graphene hybrids, the length of the CNTs had an important influence on electrochemical performance *(196)*. In addition, 3D graphene structure from nickel foam has been mixed with SWNTs, and the hybrids provided an excellent reversible capacity of 2640 mAh/g due to the decoupling of lithium-storage reactions *(197)*. Carbon nanotube/graphene hybrids prepared through solution mixing of CNTs and graphene were also used in LIBs *(119,166)*. If GO was chosen to blend with CNTs, oxygen functional groups on the GO played a significant role in the performance of devices *(198)*. Furthermore, CNTs and graphene have often been integrated with other materials, such as metal oxide, conducting polymer, and

permission from Zhao, M.-Q.; Liu, X. F.; Zhang, Q.; Tian, G. L.; Huang, J. Q.; Zhu, W.; Wei, F. Graphene/Single-Walled Carbon Nanotube Hybrids: One-Step Catalytic Growth and Applications for High-rate Li–S Batteries. ACS Nano ***2012,*** *6, 10759–10769. Copyright 2012 American Chemical Society.)* (I) Schematic illustration of the growth of CNT/graphene/carbon hybrids. *(Reprinted with permission from Peng, H. J.; Huang, J. Q.; Zhao, M. Q.; Zhang, Q.; Cheng, X. B.; Liu, X. Y.; Qian, W. Z.; Wei, F. Nanoarchitectured Graphene/CNT@Porous Carbon with Extraordinary Electrical Conductivity and Interconnected Micro/Mesopores for Lithium-Sulfur Batteries. Advanced Functional Materials* ***2014,*** *24, 2772–2781. Copyright 2014 John Wiley and Sons.)* (J) Sketch map of the design of N-CNT/graphene hybrids. (K) The cycling performance of LSBs based on *N*-CNT/graphene hybrids at a current density of 1.0°C. *(The images (J) and (K) were reprinted with permission from Tang, C.; Zhang, Q.; Zhao, M. Q.; Huang, J. Q.; Cheng, X. B.; Tian, G. L.; Peng, H. J.; Wei, F. Nitrogen-Doped Aligned Carbon Nanotube/Graphene Sandwiches: Facile Catalytic Growth on Bifunctional Natural Catalysts and Their Applications as Scaffolds for High-Rate Lithium-Sulfur Batteries. Advanced Materials* ***2014,*** *26, 6100–6105. Copyright 2014 John Wiley and Sons.)*

transition-metal dichalcogenide, to improve the practical performance of devices *(199,200)*.

With the exception of LIBs, lithium—sulfur batteries (LSBs) have been getting increased attention because of their high capacity (1675 mAh/g) and energy density *(201,202)*. In this device system, metallic lithium is employed as anode, while sulfur is used as cathode. Due to the isolation nature of sulfur, a second component is essential to support this cathode material. The second component must be conductive and have a general high affinity with sulfur, and have some microstructures. Similarly, CNT/graphene hybrids are good choices to incorporate with sulfur because they have high intrinsic conductivity and abundant nanostructure. For example, SWNTs/graphene hybrids were prepared by a facile CVD method on the LDH, as noted in Section 15.3.2.2 (Fig. 15.6G) *(156)*. The hybrids provided high conductivity of 3130 S/cm, which allowed the battery to work well at a high current rate (Fig. 15.6H). In order to accommodate sulfur, nanostructured porous carbon was introduced to the CNT/graphene interlinked structure (Fig. 15.6I) *(203)*, leading to LSBs with a high specific capacity of 1121 mAh/g (at 0.5°C). Even at 10°C, the LSBs also delivered specific capacity of 809 mAh/g, showing a good high-rate performance. Moreover, *N*-CNT/graphene could be prepared, as illustrated in Fig. 15.6J *(157)*. The sandwich structure could prevent graphene from stacking and provided a larger surface area to store sulfur. The LSBs had an excellent specific capacity (Fig. 15.6K) and good rate performance.

15.3.3.3.4 Summary

In summary, efficient energy conversion and storage have strongly affected many aspects of our daily life. Novel carbon materials, including CNTs and graphene, especially hybrids of CNTs and graphene, have been widely studied in solar cells, supercapacitors, and lithium batteries and are considered the materials that will push high-performance devices forward due to their large specific surface area and high electrical conductivity. Moreover, CNTs and graphene can be combined with metal oxides or conducting polymers to offer better overall system performance. Furthermore, fiber-shaped energy conversion and storage devices have drawn widespread attention due to being light weight, flexible, and wearable. Continuous CNTs and graphene fibers have received more attention in this field because of their high mechanical strength, electrical conductivity, and excellent flexibility.

In addition, integrating energy-conversion devices and storage devices is becoming more and more important. For example, solar cells can be connected with supercapacitors or LIBs. However, this integration usually has relatively poor performance because of its difficulty in control and fabrication with better optimization. Therefore more research is needed on the integration of CNT/graphene-based energy conversion and storage equipment.

15.3.3.4 Sensors and Actuators

15.3.3.4.1 Sensors

Carbon nanotubes and graphene have been widely used as chemical sensors to detect gases. The detection mechanism is that the electrical conductivity of CNTs and graphene will change while gas molecules are absorbed on them. Hence, the absorbed amount of gas can be obtained by detecting the change in electrical conductivity. Carbon nanotubes and graphene have the advantages of large specific surface area, enabling improved detection accuracy and sensitivity. Carbon nanotube/graphene hybrids have played a significant role in this field. For example, combined with Pt nanoparticles, CNTs and graphene could be employed as hydrogen sensor for excellent sensitivity at room temperature *(204)*. A flexible NO_2 sensor has been demonstrated by using CNT/rGO films on polyimide substrates *(205)*. The sensor can be bent due to the rGO film and presented high sensitivity that benefited from the vertically aligned CNTs array. Moreover, the hybrids can act as a hydrogen peroxide (H_2O_2) sensor because of their excellent electrocatalytic activity. In this case, the composite can catalyze the reduction of H_2O_2, which will induce the change of anodic and cathodic peak currents from cyclic voltammogram (CV) *(206)*. In addition, 3D graphene and CNTs hybrids have also been used as H_2O_2 sensors. Generally, these sensors possess a high sensitivity and low detection limit (1 μM) *(152)*.

Carbon nanotube/graphene hybrids also have potential applications in biosensors. Various biomolecules, such as dopamine *(152)*, glucose *(207)*, and some drugs *(208)*, were detected in low amounts with the hybrids. Among them, some sensors were fabricated on flexible substrates so that the sensors could be interfaced with human body, live plants, or insects.

Furthermore, with the development of wearable electronic devices, body-motion sensors have aroused considerable interest. Due to high conductivity and flexibility, CNT/graphene hybrids have shown outstanding potential applications. Carbon nanotube/graphene hybrids on flexible substrates have been used to fabricate pressure sensors *(209)*. In addition, transparent pressure sensors have been demonstrated based on monolayer graphene and SWNTs covered on thermal plastic polyurethane pyramids *(210)*. The sensors had a transparency of 91% and could detect load distribution.

15.3.3.4.2 Actuators

Actuators, which are important building blocks in bionic robots, microsensors, and artificial muscle, should be sensitive and able to achieve complex actions. Since conventional actuators (shape memory alloy, piezoelectric ceramic, etc.) cannot satisfy the demands of technology, electroactive polymer (EAP) actuators, including electronic actuators and ionic actuators, have been developed. EAP actuators are usually driven by electrical field with high-strain actuation and easy controllability. Electronic actuators have the advantages of large strain with quick response, but they are worked at high electric field intensity (150 MV/m) with a voltage usually higher than 1 kV. In comparison, ionic actuators can achieve large strain with low voltage (<10 V). In order to improve stability and actuation performance, CNTs and graphene have been applied because of their high electrical conductivity, outstanding stability, and flexibility *(211)*. Carbon nanotubes/rGO could be employed as hybrid electrodes. As previously noted, CNTs could effectively stop graphene from stacking or accumulating and increase electrical conductivity. The hybrids could significantly improve the electrochemical process, allowing the actuator to possess a quick response, large displacement, and high stability (a million times) *(212)*.

In the past few years, electrothermal and photothermal actuators have triggered increasing research interest in this field. Electrothermal actuators only need low driven voltage and do not need electrolytes. Photothermal actuators are driven by light, so they can achieve large-strain actuations by remote stimulus. Among them, one type of actuator uses conductive polymer to realize extension or shrinkage. The other type is composed of conductive materials and polymer film. When the current or light is introduced, the composites will bend due to the huge difference between the conductive materials and the polymer in coefficient of thermal expansion. Carbon nanotube/graphene hybrids have been used widely as conductive materials in electrothermal and photothermal actuators. When the carbon nanomaterials are mixed with polymer, the actuators generally have small actuations *(213,214)*. However, with a double-layer structures, they usually have large-strain bending with low voltage or low energy density of light *(215–218)*. Moreover, if CNTs are mixed with rGO, the hybrid film is endowed with high electrical conductivity. When photothermal actuators are fabricated by CNTs/rGO and poly(dimethylsiloxane) (PDMS), the rGO could offer strong interface with PDMS so that the thermal energy absorbed by CNTs/rGO layer can effectively transport to PDMS, inducing large expansion *(219)*. The actuators could bend from 84 to 563 degree in only 3.6 s when driven by visible light. Based on this, intelligent devices, such as roller-shutter, light-driven robots can be designed. In summary, CNT/graphene hybrids have played a significant role in the design of intelligent actuators and will have increasing potential applications.

15.3.3.5 Composites

Due to their superior electrical, optical, thermal, and mechanical properties, both CNTs and graphene are suited as additives used to improve the performance of polymer. The stacking of graphene can be effectively avoided by introducing CNTs. Carbon nanotubes can bridge adjacent graphene sheets and prevent graphene from aggregation, resulting in higher electrical conductivity and higher contact area than only using one component. After a CNT/graphene hybrid was introduced into epoxy, the tensile strength of composites was 35.4% higher than the original epoxy *(220)*. Meanwhile, the thermal conductivity of composites was increased by 146.9%, compared with the pristine epoxy *(220)*.

However, if the amount of CNTs was too high, the phonon scattering increased, resulting in reduction of thermal conductivity *(221)*. Moreover, CNT/graphene/epoxy composites can be used as thermal interface materials, essential constituents for thermal management *(222)*.

15.3.3.6 Catalysts

CNT/graphene hybrids also have great potential for applications in both of oxygen reduction reaction (ORR) and oxygen evolution reaction (OER), which are generally involved in water splitting, fuel cells, and metal-air batteries. Carbon nanotube/graphene hybrids have been extensively researched in these aspects *(223–225)*. In addition, other catalytic reactions have been explored with CNT/graphene, including hydrogen evolution reaction (HER) *(226)*, photocatalytic reactions *(158)*, etc., demonstrating that CNT/graphene delivers high catalytic results.

15.3.3.7 Adsorption and Desalination

Our earth is facing major environmental issue such as contamination of waters by aromatic compounds or heavy metal ions. One of the most popular methods to remove these pollutants is adsorption, which can remove trace amounts of undesired chemicals. The most common adsorbent is activated carbon. Carbon nanotubes and graphene can be used to adsorb metal ions or dyes because of their large specific surface area and catalytic performance. It was shown that SWNTs/graphene has significant synergistic effects on the absorption of aromatic organic compounds *(227)*. In addition, 3D hybrid of graphene and CNTs could be used to remove oils and organic compounds from water *(228)*. Carbon nanotube/rGO composites, which have large specific surfaces (352 m^2/g), exhibited superior performance in photocatalytic degradation dye due to their porous structure and electrons transport *(140)*. Similarly, the heavy-metal ions could be absorbed and removed *(229)*. Furthermore, CNT/graphene hybrids could be applied in water desalination *(122)*.

15.3.3.8 Other Applications

In addition to those mentioned above, CNT/graphene also showed promise in other aspects. For example, CNT/graphene hybrids could be used as cell culture *(230)* and gene transfection *(231)*. Moreover, the hybrid could be utilized to fabricate optical devices *(148)* and mechanical resonators *(232)*, etc.

In summary, CNT/graphene hybrids have large specific surface area and superior electrical, thermal, and mechanical properties, providing hybrids promising materials with tremendous potential in many fields, such as electronic devices, TCFs, energy-conversion and storage, sensors and actuators, catalysts. It is believed that CNTs, when they are combined with graphene, will play more important and more significant roles in the development of new materials and new technology, which will revolutionize many aspects of our daily life.

15.4 PERSPECTIVE

In the past 25 years, tremendous advancements and outstanding achievements have been achieved from both fundamental research and practical applications of CNTs. However, some urgent issues still need to be addressed, representing the opportunities to promote the process of research and applications.

Controlled synthesis of CNTs with specific structures (diameter, chiral angle, etc.) and morphologies (length, density, purity, etc.) is of extreme significance for both research and industrial applications, especially in the field of electronic devices due to the fact that CNT can extend Moore's law. However, it is still very challenging to grow CNTs with specific structures or high density or superpurity. Thus continuing efforts to understand the growth mechanisms and to design novel catalysts should be made to realize the controlled growth of CNTs in order to obtain high-purity CNTs with specific structures for the fabrication of high-performance CNT devices. In addition, the reasons the properties of CNT macrostructures, including fibers, yarns, and sheets, are far from the intrinsic properties of individual CNTs should be understood to facilitate the large-scale and high-end applications of CNTs. Success in controlled synthesis of CNTs will greatly promote fundamental research of CNTs and provide valuable references for the research and application of other nanomaterials, and even for all nanoscience and nanotechnology *(233)*.

In addition, the practical application of CNTs is another focus of attention. Due to its superior properties, CNTs are expected to revolutionize many aspects of daily life. However, the development of large-scale and high-performance electronic devices based on CNTs has been delayed, which has had an effect on the practical application of CNTs in nanoelectronics. Currently, CNTs are used as electrically conductive fillers in composites and electrodes due to their superior electrical conductivity, replacing traditional additives. Carbon nanotube-reinforced composites with excellent properties may take the spotlight in some fields. In addition, in the development of controlled growth and postprocessing CNTs show great potential in electronic devices, thin-film transistors, TCFs, energy conversion and storage, chemical and biological sensors, catalysts, etc. Compared with carbon fibers, CNTs have high surface area, high mechanical strength, and flexibility, and show great potential in application in newly developed flexible devices. It is worth noting that the above potential applications are all closely related to the specific structures of CNTs *(233)*, further emphasizing the importance of the controlled growth of CNTs with desired structures, morphologies, and properties.

In addition, the development of CNTs provides valuable information for the research of graphene. In particular, the rise of CNT/graphene hybrids, which integrate the outstanding properties of these two-carbon allotropes, will greatly expand the research and applications of carbon nanomaterials. Due to the large specific surface area, high electrical conductivity, high mechanical strength, and good flexibility of CNT/graphene hybrids, they hold great potential in many fields, such as flexible electronic devices, TCFs, sensors, energy conversion and storage devices, adsorption, etc. Currently, CNT/graphene hybrids can be synthesized by a postgrowth process and in situ CVD growth. However, these approaches, especially in situ CVD growth, are complicated and not easy for low-cost and large-scale production of the hybrids. It is still a great challenge to explore facile methods for the preparation of high-quality and large-scale CNT/graphene hybrids. Once success is achieved in the controlled growth of CNT/graphene hybrids with high-quality and large-scale preparation, the full potential practical applications of CNT/graphene hybrids in many fields will be realized.

REFERENCES

1. De Volder, M. F.; Tawfick, S. H.; Baughman, R. H.; Hart, A. J. Carbon Nanotubes: Present and Future Commercial Applications. *Science* **2013,** *339,* 535–539.
2. Zhang, Q.; Huang, J. Q.; Qian, W. Z.; Zhang, Y. Y.; Wei, F. The Road for Nanomaterials Industry: A Review of Carbon Nanotube Production, Post-Treatment, and Bulk Applications for Composites and Energy Storage. *Small* **2013,** *9,* 1237–1265.
3. Kang, L.; Hu, Y.; Liu, L.; Wu, J.; Zhang, S.; Zhao, Q.; Ding, F.; Li, Q.; Zhang, J. Growth of Close-Packed Semiconducting Single-Walled Carbon Nanotube Arrays Using Oxygen-Deficient TiO_2 Nanoparticles as Catalysts. *Nano Letters* **2014,** *15,* 403–409.
4. Harutyunyan, A. R.; Chen, G.; Paronyan, T. M.; Pigos, E. M.; Kuznetsov, O. A.; Hewaparakrama, K.; Kim, S. M.; Zakharov, D.; Stach, E. A.; Sumanasekera, G. U. Preferential Growth of Single-walled Carbon Nanotubes with Metallic Conductivity. *Science* **2009,** *326,* 116–120.
5. Yu, X.; Zhang, J.; Choi, W.; Choi, J. Y.; Kim, J. M.; Gan, L.; Liu, Z. Cap Formation Engineering: from Opened C60 to Single-walled Carbon Nanotubes. *Nano Letters* **2010,** *10,* 3343–3349.
6. Yang, F.; Wang, X.; Zhang, D.; Yang, J.; Luo, D.; Xu, Z.; Wei, J.; Wang, J. Q.; Xu, Z.; Peng, F. Chirality-Specific Growth of Single-Walled Carbon Nanotubes on Solid Alloy Catalysts. *Nature* **2014,** *510,* 522–524.
7. Yao, Y.; Feng, C.; Zhang, J.; Liu, Z. "Cloning" of Single-walled Carbon Nanotubes via Open-End Growth Mechanism. *Nano Letters* **2009,** *9,* 1673–1677.
8. Yang, F.; Wang, X.; Zhang, D.; Qi, K.; Yang, J.; Xu, Z.; Li, M.; Zhao, X.; Bai, X.; Li, Y. Growing Zigzag (16, 0) Carbon Nanotubes with Structure-Defined Catalysts. *Journal of the American Chemical Society* **2015,** *137,* 8688–8691.
9. Zhang, R.; Zhang, Y.; Zhang, Q.; Xie, H.; Qian, W.; Wei, F. Growth of Half-Meter Long Carbon Nanotubes Based on Schulz–Flory Distribution. *ACS Nano* **2013,** *7,* 6156–6161.
10. Franklin, A. D. Electronics: The Road to Carbon Nanotube Transistors. *Nature* **2013,** *498,* 443–444.
11. Hong, G.; Zhou, M.; Zhang, R.; Hou, S.; Choi, W.; Woo, Y. S.; Choi, J. Y.; Liu, Z.; Zhang, J. Separation of Metallic and Semiconducting Single-Walled Carbon Nanotube Arrays by "Scotch Tape". *Angewandte Chemie International Edition* **2011,** *50,* 6819–6823.
12. Hu, Y.; Chen, Y.; Li, P.; Zhang, J. Sorting Out Semiconducting Single-Walled Carbon Nanotube Arrays by Washing off Metallic Tubes Using SDS Aqueous Solution. *Small* **2013,** *9,* 1306–1311.
13. Zhang, Y.; Zhang, Y.; Xian, X.; Zhang, J.; Liu, Z. Sorting Out Semiconducting Single-Walled Carbon Nanotube Arrays by Preferential Destruction of Metallic Tubes Using Xenon-Lamp Irradiation. *The Journal of Physical Chemistry C* **2008,** *112,* 3849–3856.
14. Du, F.; Felts, J. R.; Xie, X.; Song, J.; Li, Y.; Rosenberger, M. R.; Islam, A. E.; Jin, S. H.; Dunham, S. N.; Zhang, C. Laser-Induced Nanoscale Thermocapillary Flow for Purification of Aligned Arrays of Single-Walled Carbon Nanotubes. *ACS Nano* **2014,** *8,* 12641–12649.

15. Xie, X.; Jin, S. H.; Wahab, M. A.; Islam, A. E.; Zhang, C.; Du, F.; Seabron, E.; Lu, T.; Dunham, S. N.; Cheong, H. I. Microwave Purification of Large-area Horizontally Aligned Arrays of Single-Walled Carbon Nanotubes. *Nature Communications* **2014,** *5.*
16. Jin, S. H.; Dunham, S. N.; Song, J.; Xie, X.; Kim, J-h.; Lu, C.; Islam, A.; Du, F.; Kim, J.; Felts, J. Using Nanoscale Thermocapillary Flows to Create Arrays of Purely Semiconducting Single-Walled Carbon Nanotubes. *Nature Nanotechnology* **2013,** *8,* 347–355.
17. Arnold, M. S.; Green, A. A.; Hulvat, J. F.; Stupp, S. I.; Hersam, M. C. Sorting Carbon Nanotubes by Electronic Structure Using Density Differentiation. *Nature Nanotechnology* **2006,** *1,* 60–65.
18. Liu, H.; Nishide, D.; Tanaka, T.; Kataura, H. Large-Scale Single-Chirality Separation of Single-Wall Carbon Nanotubes by Simple Gel Chromatography. *Nature Communications* **2011,** *2,* 309.
19. Tu, X.; Manohar, S.; Jagota, A.; Zheng, M. DNA Sequence Motifs for Structure-Specific Recognition and Separation of Carbon Nanotubes. *Nature* **2009,** *460,* 250–253.
20. Pochorovski, I.; Wang, H.; Feldblyum, J. I.; Zhang, X.; Antaris, A. L.; Bao, Z. H-Bonded Supramolecular Polymer for the Selective Dispersion and Subsequent Release of Large-Diameter Semiconducting Single-Walled Carbon Nanotubes. *Journal of the American Chemical Society* **2015,** *137,* 4328–4331.
21. Xie, H.; Zhang, R.; Zhang, Y.; Zhang, W.; Jian, M.; Wang, C.; Wang, Q.; Wei, F. Graphene/Graphite Sheet Assisted Growth of High-Areal-Density Horizontally Aligned Carbon Nanotubes. *Chemical Communications* **2014,** *50,* 11158–11161.
22. Xie, H.; Zhang, R.; Zhang, Y.; Li, P.; Jin, Y.; Wei, F. Growth of High-Density Parallel Arrays of Ultralong Carbon Nanotubes with Catalysts Pinned by Silica Nanospheres. *Carbon* **2013,** *52,* 535–540.
23. Xie, H.; Zhang, R.; Zhang, Y.; Yin, Z.; Jian, M.; Wei, F. Preloading Catalysts in the Reactor for Repeated Growth of Horizontally Aligned Carbon Nanotube Arrays. *Carbon* **2016,** *98,* 157–161.
24. Hata, K.; Futaba, D. N.; Mizuno, K.; Namai, T.; Yumura, M.; Iijima, S. Water-Assisted Highly Efficient Synthesis of Impurity-Free Single-Walled Carbon Nanotubes. *Science* **2004,** *306,* 1362–1364.
25. Zhang, M.; Atkinson, K. R.; Baughman, R. H. Multifunctional Carbon Nanotube Yarns by Downsizing an Ancient Technology. *Science* **2004,** *306,* 1358–1361.
26. Li, Y. L.; Kinloch, I. A.; Windle, A. H. Direct Spinning of Carbon Nanotube Fibers from Chemical Vapor Deposition Synthesis. *Science* **2004,** *304,* 276–278.
27. Wang, J.; Luo, X.; Wu, T.; Chen, Y. High-Strength Carbon Nanotube Fibre-like Ribbon with High Ductility and High Electrical Conductivity. *Nature Communications* **2014,** *5.*
28. Xu, W.; Chen, Y.; Zhan, H.; Wang, J. N. High-Strength Carbon Nanotube Film from Improving Alignment and Densification. *Nano Letters* **2016,** *16,* 946–952.
29. Koziol, K.; Vilatela, J.; Moisala, A.; Motta, M.; Cunniff, P.; Sennett, M.; Windle, A. High-Performance Carbon Nanotube Fiber. *Science* **2007,** *318,* 1892–1895.
30. Wei, H.; Ding, D.; Wei, S.; Guo, Z. Anticorrosive Conductive Polyurethane Multiwalled Carbon Nanotube Nanocomposites. *Journal of Materials Chemistry A* **2013,** *1,* 10805–10813.
31. Yang, Y.; Gupta, M. C.; Dudley, K. L.; Lawrence, R. W. Novel Carbon Nanotube-Polystyrene Foam Composites for Electromagnetic Interference Shielding. *Nano Letters* **2005,** *5,* 2131–2134.
32. Hecht, D. S.; Hu, L.; Irvin, G. Emerging Transparent Electrodes Based on Thin Films of Carbon Nanotubes, Graphene, and Metallic Nanostructures. *Advanced Materials* **2011,** *23,* 1482–1513.
33. Chen, P.; Fu, Y.; Aminirad, R.; Wang, C.; Zhang, J.; Wang, K.; Galatsis, K.; Zhou, C. Fully Printed Separated Carbon Nanotube Thin Film Transistor Circuits and Its Application in Organic Light Emitting Diode Control. *Nano Letters* **2011,** *11,* 5301–5308.
34. Tans, S. J.; Verschueren, A. R.; Dekker, C. Room-temperature Transistor Based on a Single Carbon Nanotube. *Nature* **1998,** *393,* 49–52.
35. Shulaker, M. M.; Hills, G.; Patil, N.; Wei, H.; Chen, H. Y.; Wong, H. S. P.; Mitra, S. Carbon Nanotube Computer. *Nature* **2013,** *501,* 526–530.
36. Novoselov, K. S.; Geim, A. K.; Morozov, S.; Jiang, D.; Zhang, Y.; Dubonos, S. V.; Grigorieva, I.; Firsov, A. Electric Field Effect in Atomically Thin Carbon Films. *Science* **2004,** *306,* 666–669.
37. Dimitrakakis, G. K.; Tylianakis, E.; Froudakis, G. E. Pillared Graphene: a New 3-D Network Nanostructure for Enhanced Hydrogen Storage. *Nano Letters* **2008,** *8,* 3166–3170.

38. Varshney, V.; Patnaik, S. S.; Roy, A. K.; Froudakis, G.; Farmer, B. L. Modeling of Thermal Transport in Pillared-graphene Architectures. *ACS Nano* **2010,** *4,* 1153–1161.
39. Shin, M. K.; Lee, B.; Kim, S. H.; Lee, J. A.; Spinks, G. M.; Gambhir, S.; Wallace, G. G.; Kozlov, M. E.; Baughman, R. H.; Kim, S. J. Synergistic Toughening of Composite Fibres by Self-alignment of Reduced Graphene Oxide and Carbon Nanotubes. *Nature Communications* **2012,** *3,* 650.
40. Kim, K. H.; Oh, Y.; Islam, M. Graphene Coating Makes Carbon Nanotube Aerogels Superelastic and Resistant to Fatigue. *Nature Nanotechnology* **2012,** *7,* 562–566.
41. Tang, C.; Zhang, Q.; Zhao, M. Q.; Tian, G. L.; Wei, F. Resilient Aligned Carbon Nanotube/Graphene Sandwiches for Robust Mechanical Energy Storage. *Nano Energy* **2014,** *7,* 161–169.
42. Mermin, N. D. Crystalline Order in Two Dimensions. *Physical Review* **1968,** *176,* 250.
43. Geim, A. K.; Novoselov, K. S. The Rise of Graphene. *Nature Materials* **2007,** *6,* 183–191.
44. Meyer, J. C.; Geim, A. K.; Katsnelson, M.; Novoselov, K.; Booth, T.; Roth, S. The Structure of Suspended Graphene Sheets. *Nature* **2007,** *446,* 60–63.
45. Brar, V. W.; Zhang, Y.; Yayon, Y.; Ohta, T.; McChesney, J. L.; Bostwick, A.; Rotenberg, E.; Horn, K.; Crommie, M. F. Scanning Tunneling Spectroscopy of Inhomogeneous Electronic Structure in Monolayer and Bilayer Graphene on SiC. *Applied Physics Letters* **2007,** *91,* 122102.
46. Blake, P.; Hill, E.; Neto, A. C.; Novoselov, K.; Jiang, D.; Yang, R.; Booth, T.; Geim, A. Making Graphene Visible. *Applied Physics Letters* **2007,** *91,* 063124.
47. Ni, Z.; Wang, H.; Kasim, J.; Fan, H.; Yu, T.; Wu, Y.; Feng, Y.; Shen, Z. Graphene Thickness Determination Using Reflection and Contrast Spectroscopy. *Nano Letters* **2007,** *7,* 2758–2763.
48. Novoselov, K.; Jiang, D.; Schedin, F.; Booth, T.; Khotkevich, V.; Morozov, S.; Geim, A. Two-Dimensional Atomic Crystals. *Proceedings of the National Academy of Sciences of the United States of America* **2005,** *102,* 10451–10453.
49. Sidorov, A. N.; Yazdanpanah, M. M.; Jalilian, R.; Ouseph, P.; Cohn, R. W.; Sumanasekera, G. Electrostatic Deposition of Graphene. *Nanotechnology* **2007,** *18,* 135301.
50. Somani, P. R.; Somani, S. P.; Umeno, M. Planer Nano-Graphenes from Camphor by CVD. *Chemical Physics Letters* **2006,** *430,* 56–59.
51. Chae, S. J.; Gunes, F.; Kim, K. K.; Kim, E. S.; Han, G. H.; Kim, S. M.; Shin, H. J.; Yoon, S. M.; Choi, J. Y.; Park, M. H. Synthesis of Large-Area Graphene Layers on Poly-Nickel Substrate by Chemical Vapor Deposition: Wrinkle Formation. *Advanced Materials* **2009,** *21,* 2328–2333.
52. Cao, H.; Yu, Q.; Colby, R.; Pandey, D.; Park, C.; Lian, J.; Zemlyanov, D.; Childres, I.; Drachev, V.; Stach, E. A. Large-Scale Graphitic Thin Films Synthesized on Ni and Transferred to Insulators: Structural and Electronic Properties. *Journal of Applied Physics* **2010,** *107,* 044310.
53. Reina, A.; Jia, X.; Ho, J.; Nezich, D.; Son, H.; Bulovic, V.; Dresselhaus, M. S.; Kong, J. Large Area, Few-Layer Graphene Films on Arbitrary Substrates by Chemical Vapor Deposition. *Nano Letters* **2008,** *9,* 30–35.
54. Li, X.; Cai, W.; An, J.; Kim, S.; Nah, J.; Yang, D.; Piner, R.; Velamakanni, A.; Jung, I.; Tutuc, E. Large-Area Synthesis of High-quality and Uniform Graphene Films on Copper Foils. *Science* **2009,** *324,* 1312–1314.
55. Li, X.; Cai, W.; Colombo, L.; Ruoff, R. S. Evolution of Graphene Growth on Ni and Cu by Carbon Isotope Labeling. *Nano Letters* **2009,** *9,* 4268–4272.
56. Nguyen, V. L.; Shin, B. G.; Duong, D. L.; Kim, S. T.; Perello, D.; Lim, Y. J.; Yuan, Q. H.; Ding, F.; Jeong, H. Y.; Shin, H. S. Seamless Stitching of Graphene Domains on Polished Copper (111) Foil. *Advanced Materials* **2015,** *27,* 1376–1382.
57. Sutter, P. W.; Flege, J.-I.; Sutter, E. A. Epitaxial Graphene on Ruthenium. *Nature Materials* **2008,** *7,* 406–411.
58. Coraux, J.; N'Diaye, A. T.; Busse, C.; Michely, T. Structural Coherency of Graphene on Ir (111). *Nano Letters* **2008,** *8,* 565–570.
59. Liu, N.; Fu, L.; Dai, B.; Yan, K.; Liu, X.; Zhao, R.; Zhang, Y.; Liu, Z. Universal Segregation Growth Approach to Wafer-Size Graphene from Non-Noble Metals. *Nano Letters* **2010,** *11,* 297–303.
60. Sutter, P.; Sadowski, J. T.; Sutter, E. Graphene on Pt (111): Growth and Substrate Interaction. *Physical Review B* **2009,** *80,* 245411.

61. Berger, C.; Song, Z.; Li, T.; Li, X.; Ogbazghi, A. Y.; Feng, R.; Dai, Z.; Marchenkov, A. N.; Conrad, E. H.; First, P. N. Ultrathin Epitaxial Graphite: 2D Electron Gas Properties and a Route toward Graphene-Based Nanoelectronics. *The Journal of Physical Chemistry B* **2004,** *108,* 19912–19916.
62. Berger, C.; Song, Z.; Li, X.; Wu, X.; Brown, N.; Naud, C.; Mayou, D.; Li, T.; Hass, J.; Marchenkov, A. N. Electronic Confinement and Coherence in Patterned Epitaxial Graphene. *Science* **2006,** *312,* 1191–1196.
63. Pan, Y.; Zhang, H.; Shi, D.; Sun, J.; Du, S.; Liu, F.; Gao, H. Highly Ordered, Millimeter-Scale, Continuous, Single-crystalline Graphene Monolayer Formed on Ru (0001). *Advanced Materials* **2009,** *21,* 2777–2780.
64. Brodie, B. C. On the Atomic Weight of Graphite. *Philosophical Transactions of the Royal Society of London* **1859,** 249–259.
65. Staudenmaier, L. Verfahren zur darstellung der graphitsäure. *Berichte der deutschen chemischen Gesellschaft* **1898,** *31,* 1481–1487.
66. Hummers, W. S., Jr.; Offeman, R. E. Preparation of Graphitic Oxide. *Journal of the American Chemical Society* **1958,** *80,* 1339.
67. Hamdi, H. Zur Kenntnis der kolloidchemischen Eigenschaften des Humus. *Fortschrittsberichte über Kolloide und Polymere* **1943,** *54,* 554–634.
68. Nakajima, T.; Mabuchi, A.; Hagiwara, R. A New Structure Model of Graphite Oxide. *Carbon* **1988,** *26,* 357–361.
69. Kovtyukhova, N. I.; Ollivier, P. J.; Martin, B. R.; Mallouk, T. E.; Chizhik, S. A.; Buzaneva, E. V.; Gorchinskiy, A. D. Layer-by-Layer Assembly of Ultrathin Composite Films from Micron-Sized Graphite Oxide Sheets and Polycations. *Chemistry of Materials* **1999,** *11,* 771–778.
70. Niyogi, S.; Bekyarova, E.; Itkis, M. E.; McWilliams, J. L.; Hamon, M. A.; Haddon, R. C. Solution Properties of Graphite and Graphene. *Journal of the American Chemical Society* **2006,** *128,* 7720–7721.
71. Stankovich, S.; Dikin, D. A.; Dommett, G. H.; Kohlhaas, K. M.; Zimney, E. J.; Stach, E. A.; Piner, R. D.; Nguyen, S. T.; Ruoff, R. S. Graphene-Based Composite Materials. *Nature* **2006,** *442,* 282–286.
72. Schniepp, H. C.; Li, J. L.; McAllister, M. J.; Sai, H.; Herrera-Alonso, M.; Adamson, D. H.; Prud'homme, R. K.; Car, R.; Saville, D. A.; Aksay, I. A. Functionalized Single Graphene Sheets Derived from Splitting Graphite Oxide. *The Journal of Physical Chemistry B* **2006,** *110,* 8535–8539.
73. McAllister, M. J.; Li, J. L.; Adamson, D. H.; Schniepp, H. C.; Abdala, A. A.; Liu, J.; Herrera-Alonso, M.; Milius, D. L.; Car, R.; Prud'homme, R. K. Single Sheet Functionalized Graphene by Oxidation and Thermal Expansion of Graphite. *Chemistry of Materials* **2007,** *19,* 4396–4404
74. Dikin, D. A.; Stankovich, S.; Zimney, E. J.; Piner, R. D.; Dommett, G. H.; Evmenenko, G.; Nguyen, S. T.; Ruoff, R. S. Preparation and Characterization of Graphene Oxide Paper. *Nature* **2007,** *448,* 457–460.
75. Stankovich, S.; Dikin, D. A.; Piner, R. D.; Kohlhaas, K. A.; Kleinhammes, A.; Jia, Y.; Wu, Y.; Nguyen, S. T.; Ruoff, R. S. Synthesis of Graphene-Based Nanosheets via Chemical Reduction of Exfoliated Graphite Oxide. *Carbon* **2007,** *45,* 1558–1565.
76. Si, Y.; Samulski, E. T. Synthesis of Water Soluble Graphene. *Nano Letters* **2008,** *8,* 1679–1682.
77. Shen, J.; Hu, Y.; Shi, M.; Lu, X.; Qin, C.; Li, C.; Ye, M. Fast and Facile Preparation of Graphene Oxide and Reduced Graphene Oxide Nanoplatelets. *Chemistry of Materials* **2009,** *21,* 3514–3520.
78. Bourlinos, A. B.; Gournis, D.; Petridis, D.; Szabó, T.; Szeri, A.; Dékány, I. Graphite Oxide: Chemical Reduction to Graphite and Surface Modification with Primary Aliphatic Amines and Amino Acids. *Langmuir* **2003,** *19,* 6050–6055.
79. Wang, G.; Yang, J.; Park, J.; Gou, X.; Wang, B.; Liu, H.; Yao, J. Facile Synthesis and Characterization of Graphene Nanosheets. *The Journal of Physical Chemistry C* **2008,** *112,* 8192–8195.
80. Wang, X.; Zhi, L.; Müllen, K. Transparent, Conductive Graphene Electrodes for Dye-Sensitized Solar Cells. *Nano Letters* **2008,** *8,* 323–327.
81. Becerril, H. A.; Mao, J.; Liu, Z.; Stoltenberg, R. M.; Bao, Z.; Chen, Y. Evaluation of Solution-Processed Reduced Graphene Oxide Films as Transparent Conductors. *ACS Nano* **2008,** *2,* 463–470.

82. Williams, G.; Seger, B.; Kamat, P. V. TiO_2-Graphene Nanocomposites. UV-Assisted Photocatalytic Reduction of Graphene Oxide. *ACS Nano* **2008,** *2,* 1487–1491.
83. Wang, G.; Wang, B.; Park, J.; Wang, Y.; Sun, B.; Yao, J. Highly Efficient and Large-scale Synthesis of Graphene by Electrolytic Exfoliation. *Carbon* **2009,** *47,* 3242–3246.
84. Su, C. Y.; Lu, A. Y.; Xu, Y.; Chen, F. R.; Khlobystov, A. N.; Li, L. J. High-quality Thin Graphene Films from Fast Electrochemical Exfoliation. *ACS Nano* **2011,** *5,* 2332–2339.
85. Chang, L. C.; Hsieh, Y. C.; Chen, Y. M.; Wu, P. W.; Lee, J. F. Fabrication of Graphene by Electrochemical Exfoliation in Alkaline Electrolytes. *ECS Transactions* **2014,** *58,* 33–38.
86. Wang, J.; Manga, K. K.; Bao, Q.; Loh, K. P. High-Yield Synthesis of Few-layer Graphene Flakes through Electrochemical Expansion of Graphite in Propylene Carbonate Electrolyte. *Journal of the American Chemical Society* **2011,** *133,* 8888–8891.
87. Guo, H. L.; Wang, X. F.; Qian, Q. Y.; Wang, F. B.; Xia, X. H. A Green Approach to the Synthesis of Graphene Nanosheets. *ACS Nano* **2009,** *3,* 2653–2659.
88. Shao, Y.; Wang, J.; Engelhard, M.; Wang, C.; Lin, Y. Facile and Controllable Electrochemical Reduction of Graphene Oxide and its Applications. *Journal of Materials Chemistry* **2010,** *20,* 743–748.
89. Neto, A. C.; Guinea, F.; Peres, N.; Novoselov, K. S.; Geim, A. K. The Electronic Properties of Graphene. *Reviews of Modern Physics* **2009,** *81,* 109.
90. Wallace, P. R. The Band Theory of Graphite. *Physical Review* **1947,** *71,* 622.
91. Avouris, P.; Chen, Z.; Perebeinos, V. Carbon-Based Electronics. *Nature Nanotechnology* **2007,** *2,* 605–615.
92. Zhang, Y.; Tan, Y. W.; Stormer, H. L.; Kim, P. Experimental Observation of the Quantum Hall Effect and Berry's Phase in Graphene. *Nature* **2005,** *438,* 201–204.
93. Novoselov, K.; Geim, A. K.; Morozov, S.; Jiang, D.; Katsnelson, M.; Grigorieva, I.; Dubonos, S.; Firsov, A. Two-Dimensional Gas of Massless Dirac Fermions in Graphene. *Nature* **2005,** *438,* 197–200.
94. Kim, K. S.; Zhao, Y.; Jang, H.; Lee, S. Y.; Kim, J. M.; Kim, K. S.; Ahn, J. H.; Kim, P.; Choi, J. Y.; Hong, B. H. Large-Scale Pattern Growth of Graphene Films for Stretchable Transparent Electrodes. *Nature* **2009,** *457,* 706–710.
95. Bolotin, K. I.; Sikes, K.; Jiang, Z.; Klima, M.; Fudenberg, G.; Hone, J.; Kim, P.; Stormer, H. Ultrahigh Electron Mobility in Suspended Graphene. *Solid State Communications* **2008,** *146,* 351–355.
96. Peres, N.; Guinea, F.; Neto, A. C. Electronic Properties of Disordered Two-dimensional Carbon. *Physical Review B* **2006,** *73,* 125411.
97. Gusynin, V.; Sharapov, S.; Carbotte, J. Unusual Microwave Response of Dirac Quasiparticles in Graphene. *Physical Review Letters* **2006,** *96,* 256802.
98. Nair, R.; Blake, P.; Grigorenko, A.; Novoselov, K.; Booth, T.; Stauber, T.; Peres, N.; Geim, A. Fine structure Constant Defines Visual Transparency of Graphene. *Science* **2008,** *320,* 1308.
99. Bae, S.; Kim, H.; Lee, Y.; Xu, X.; Park, J.-S.; Zheng, Y.; Balakrishnan, J.; Lei, T.; Kim, H. R.; Song, Y. I. Roll-to-Roll Production of 30-inch Graphene Films for Transparent Electrodes. *Nature Nanotechnology* **2010,** *5,* 574–578.
100. Wang, F.; Zhang, Y.; Tian, C.; Girit, C.; Zettl, A.; Crommie, M.; Shen, Y. R. Gate-Variable Optical Transitions in Graphene. *Science* **2008,** *320,* 206–209.
101. Bao, Q.; Zhang, H.; Yang, J.x.; Wang, S.; Tang, D. Y.; Jose, R.; Ramakrishna, S.; Lim, C. T.; Loh, K. P. Graphene–Polymer Nanofiber Membrane for Ultrafast Photonics. *Advanced Functional Materials* **2010,** *20,* 782–791.
102. Gu, T.; Petrone, N.; McMillan, J. F.; van der Zande, A.; Yu, M.; Lo, G.-Q.; Kwong, D.-L.; Hone, J.; Wong, C. W. Regenerative Oscillation and Four-Wave Mixing in Graphene Optoelectronics. *Nature Photonics* **2012,** *6,* 554–559.
103. Bonaccorso, F.; Sun, Z.; Hasan, T.; Ferrari, A. Graphene Photonics and Optoelectronics. *Nature Photonics* **2010,** *4,* 611–622.
104. Zhang, H.; Bao, Q.; Tang, D.; Zhao, L.; Loh, K. Large Energy Soliton Erbium-Doped Fiber Laser with a Graphene-Polymer Composite Mode Locker. *Applied Physics Letters* **2009,** *95,* 141103.
105. Yu, C.; Shi, L.; Yao, Z.; Li, D.; Majumdar, A. Thermal conductance and Thermopower of an Individual Single-Wall Carbon Nanotube. *Nano Letters* **2005,** *5,* 1842–1846.

106. Berber, S.; Kwon, Y.-K.; Tománek, D. Unusually High Thermal Conductivity of Carbon Nanotubes. *Physical Review Letters* **2000,** *84,* 4613.
107. Balandin, A. A.; Ghosh, S.; Bao, W.; Calizo, I.; Teweldebrhan, D.; Miao, F.; Lau, C. N. Superior thermal Conductivity of Single-Layer Graphene. *Nano Letters* **2008,** *8,* 902–907.
108. Seol, J. H.; Jo, I.; Moore, A. L.; Lindsay, L.; Aitken, Z. H.; Pettes, M. T.; Li, X.; Yao, Z.; Huang, R.; Broido, D. Two-dimensional Phonon Transport in Supported Graphene. *Science* **2010,** *328,* 213–216.
109. Cai, W.; Moore, A. L.; Zhu, Y.; Li, X.; Chen, S.; Shi, L.; Ruoff, R. S. Thermal Transport in Suspended and Supported Monolayer Graphene Grown by Chemical Vapor Deposition. *Nano Letters* **2010,** *10,* 1645–1651.
110. Frank, I.; Tanenbaum, D. M.; Van der Zande, A.; McEuen, P. L. Mechanical Properties of Suspended Graphene Sheets. *Journal of Vacuum Science & Technology B* **2007,** *25,* 2558–2561.
111. Wu, J.; Becerril, H. A.; Bao, Z.; Liu, Z.; Chen, Y.; Peumans, P. Organic Solar Cells with Solution-processed Graphene Transparent Electrodes. *Applied Physics Letters* **2008,** *92,* 263302.
112. Gómez-Navarro, C.; Burghard, M.; Kern, K. Elastic Properties of Chemically Derived Single Graphene Sheets. *Nano Letters* **2008,** *8,* 2045–2049.
113. Du, J.; Pei, S.; Ma, L.; Cheng, H. M. 25th Anniversary Article: Carbon Nanotube- and Graphene-based Transparent Conductive Films for Optoelectronic Devices. *Advanced Materials* **2014,** *26,* 1958–1991.
114. Perez, E. M.; Martin, N. Pi-pi Interactions in Carbon Nanostructures. *Chemical Society Reviews* **2015,** *44,* 6425–6433.
115. Zhang, D.; Yan, T.; Shi, L.; Peng, Z.; Wen, X.; Zhang, J. Enhanced Capacitive Deionization Performance of Graphene/Carbon Nanotube Composites. *Journal of Materials Chemistry* **2012,** *22,* 14696.
116. Cai, D.; Song, M.; Xu, C. Highly Conductive Carbon-Nanotube/Graphite-Oxide Hybrid Films. *Advanced Materials* **2008,** *20,* 1706–1709.
117. Tristan-Lopez, F.; Morelos-Gomez, A.; Vega-Diaz, S. M.; García-Betancourt, M. L.; Perea-López, N.; Elias, A. L.; Muramatsu, H.; Cruz-Silva, R.; Tsuruoka, S.; Kim, Y. A. Large Area Films of Alternating Graphene–Carbon Nanotube Layers Processed in Water. *ACS Nano* **2013,** *7,* 10788–10798.
118. Dong, X.; Xing, G.; Chan-Park, M. B.; Shi, W.; Xiao, N.; Wang, J.; Yan, Q.; Sum, T. C.; Huang, W.; Chen, P. The Formation of a Carbon Nanotube–Graphene Oxide Core–Shell Structure and Its Possible Applications. *Carbon* **2011,** *49,* 5071–5078.
119. Tung, V. C.; Chen, L. M.; Allen, M. J.; Wassei, J. K.; Nelson, K.; Kaner, R. B.; Yang, Y. Low temperature Solution Processing of Graphene–Carbon Nanotube Hybrid Materials for High-performance Transparent Conductors. *Nano Letters* **2009,** *9,* 1949–1955.
120. Tang, Y.; Gou, J. Synergistic Effect on Electrical Conductivity of Few-layer Graphene/Multi-walled Carbon Nanotube Paper. *Materials Letters* **2010,** *64,* 2513–2516.
121. King, P. J.; Khan, U.; Lotya, M.; De, S.; Coleman, J. N. Improvement of Transparent Conducting Nanotube Films by Addition of Small Quantities of Graphene. *ACS Nano* **2010,** *4,* 4238–4246.
122. Sui, Z.; Meng, Q.; Zhang, X.; Ma, R.; Cao, B. Green synthesis of Carbon Nanotube–Graphene Hybrid Aerogels and Their Use as Versatile Agents for Water Purification. *Journal of Materials Chemistry* **2012,** *22,* 8767–8771.
123. Sun, H.; Xu, Z.; Gao, C. Multifunctional, Ultra-Flyweight, Synergistically Assembled Carbon Aerogels. *Advanced Materials* **2013,** *25,* 2554–2560.
124. Chen, W.; McCarthy, T. J. Layer-by-Layer Deposition: a Tool for Polymer Surface Modification. *Macromolecules* **1997,** *30,* 78–86.
125. Kim, Y. K.; Min, D. H. Durable Large-Area Thin Films of Graphene/Carbon Nanotube Double Layers as a Transparent Electrode. *Langmuir: the ACS Journal of Surfaces and Colloids* **2009,** *25,* 11302–11306.
126. Hong, T. K.; Lee, D. W.; Choi, H. J.; Shin, H. S.; Kim, B.-S. Transparent, Flexible Conducting Hybrid Multilayer Thin Films of Multiwalled Carbon Nanotubes with Graphene Nanosheets. *Acs Nano* **2010,** *4,* 3861–3868.

127. Byon, H. R.; Lee, S. W.; Chen, S.; Hammond, P. T.; Shao-Horn, Y. Thin Films of Carbon Nanotubes and Chemically Reduced Graphenes for Electrochemical Micro-capacitors. *Carbon* **2011,** *49,* 457−467.
128. Yu, D.; Dai, L. Self-Assembled Graphene/Carbon Nanotube Hybrid Films for Supercapacitors. *The Journal of Physical Chemistry Letters* **2009,** *1,* 467−470.
129. Besra, L.; Liu, M. A Review on Fundamentals and Applications of Electrophoretic Deposition (EPD). *Progress in Materials Science* **2007,** *52,* 1−61.
130. Bittolo Bon, S.; Valentini, L.; Kenny, J. M.; Peponi, L.; Verdejo, R.; Lopez-Manchado, M. A. Electrodeposition of Transparent and Conducting Graphene/Carbon Nanotube Thin Films. *Physica Status Solidi (A)* **2010,** *207,* 2461−2466.
131. Su, Y.; Zhitomirsky, I. Electrophoretic Deposition of Graphene, Carbon Nanotubes and Composite Films Using Methyl Violet Dye as a Dispersing Agent. *Colloids and Surfaces A: Physicochemical and Engineering Aspects* **2013,** *436,* 97−103.
132. Koh, A. T. T.; Chen, T.; Pan, L.; Sun, Z.; Chua, D. H. C. Effective Hybrid Graphene/Carbon Nanotubes Field Emitters by Electrophoretic Deposition. *Journal of Applied Physics* **2013,** *113,* 174909.
133. Hong, X.; Chen, Y.; Wu, P. Z.; Zheng, H. Simple, Effective Fabrication of Layered Carbon Nanotube/Graphene Hybrid Field Emitters by Electrophoretic Deposition. *Journal of Vacuum Science & Technology B* **2015,** *33,* 011802.
134. Li, C.; Li, Z.; Zhu, H.; Wang, K.; Wei, J.; Li, X.; Sun, P.; Zhang, H.; Wu, D. Graphene Nano-"Patches" on a Carbon Nanotube Network for Highly Transparent/Conductive Thin Film Applications. *The Journal of Physical Chemistry C* **2010,** *114,* 14008−14012.
135. Lin, X.; Liu, P.; Wei, Y.; Li, Q.; Wang, J.; Wu, Y.; Feng, C.; Zhang, L.; Fan, S.; Jiang, K. Development of an Ultra-Thin Film Comprised of a Graphene Membrane and Carbon Nanotube Vein Support. *Nature Communications* **2013,** *4,* 2920.
136. Kholmanov, I. N.; Magnuson, C. W.; Piner, R.; Kim, J. Y.; Aliev, A. E.; Tan, C.; Kim, T. Y.; Zakhidov, A. A.; Sberveglieri, G.; Baughman, R. H.; et al. Optical, Electrical, and Electromechanical Properties of Hybrid Graphene/Carbon Nanotube Films. *Advanced Materials* **2015,** *27,* 3053−3059.
137. Hong, S. W.; Du, F.; Lan, W.; Kim, S.; Kim, H. S.; Rogers, J. A. Monolithic Integration of Arrays of Single-walled Carbon Nanotubes and Sheets of Graphene. *Advanced Materials* **2011,** *23,* 3821−3826.
138. Liu, Y.; Wang, F.; Wang, X.; Wang, X.; Flahaut, E.; Liu, X.; Li, Y.; Wang, X.; Xu, Y.; Shi, Y.; et al. Planar Carbon Nanotube-graphene Hybrid Films for High-Performance Broadband Photodetectors. *Nature Communications* **2015,** *6,* 8589.
139. Fan, Z.; Yan, J.; Zhi, L.; Zhang, Q.; Wei, T.; Feng, J.; Zhang, M.; Qian, W.; Wei, F. A Three-Dimensional Carbon Nanotube/graphene Sandwich and its Application as Electrode in Supercapacitors. *Advanced Materials* **2010,** *22,* 3723−3728.
140. Zhang, L. L.; Xiong, Z.; Zhao, X. S. Pillaring Chemically Exfoliated Graphene Oxide with Carbon Nanotubes for Photocatalytic Degradation of Dyes under Visible Light Irradiation. *ACS Nano* **2010,** *4,* 7030−7036.
141. Lee, D. H.; Kim, J. E.; Han, T. H.; Hwang, J. W.; Jeon, S.; Choi, S. Y.; Hong, S. H.; Lee, W. J.; Ruoff, R. S.; Kim, S. O. Versatile Carbon Hybrid Films Composed of Vertical Carbon Nanotubes Grown on Mechanically Compliant Graphene Films. *Advanced Materials* **2010,** *22,* 1247−1252.
142. Du, F.; Yu, D.; Dai, L.; Ganguli, S.; Varshney, V.; Roy, A. K. Preparation of Tunable 3D Pillared Carbon Nanotube−Graphene Networks for High-Performance Capacitance. *Chemistry of Materials* **2011,** *23,* 4810−4816.
143. Li, S.; Luo, Y.; Lv, W.; Yu, W.; Wu, S.; Hou, P.; Yang, Q.; Meng, Q.; Liu, C.; Cheng, H.-M. Vertically Aligned Carbon Nanotubes Grown on Graphene Paper as Electrodes in Lithium-Ion Batteries and Dye-Sensitized Solar Cells. *Advanced Energy Materials* **2011,** *1,* 486−490.
144. Kim, S. H.; Song, W.; Jung, M. W.; Kang, M. A.; Kim, K.; Chang, S. J.; Lee, S. S.; Lim, J.; Hwang, J.; Myung, S.; et al. Carbon Nanotube and Graphene Hybrid Thin Film for Transparent Electrodes and Field Effect Transistors. *Advanced Materials* **2014,** *26,* 4247−4252.

145. Nguyen, D. D.; Tiwari, R. N.; Matsuoka, Y.; Hashimoto, G.; Rokuta, E.; Chen, Y. Z.; Chueh, Y. L.; Yoshimura, M. Low Vacuum Annealing of Cellulose Acetate on Nickel towards Transparent Conductive CNT-Graphene Hybrid Films. *ACS Applied Materials & Interfaces* **2014,** *6,* 9071–9077.
146. Yan, Z.; Peng, Z.; Casillas, G.; Lin, J.; Xiang, C.; Zhou, H.; Yang, Y.; Ruan, G.; Raji, A.-R. O.; Samuel, E. L. Rebar Graphene. *ACS Nano* **2014,** *8,* 5061–5068.
147. Shi, E.; Li, H.; Yang, L.; Hou, J.; Li, Y.; Li, L.; Cao, A.; Fang, Y. Carbon Nanotube Network Embroidered Graphene Films for Monolithic All-Carbon Electronics. *Advanced Materials* **2015,** *27,* 682–688.
148. Kim, U. J.; Lee, I. H.; Bae, J. J.; Lee, S.; Han, G. H.; Chae, S. J.; Gunes, F.; Choi, J. H.; Baik, C. W.; Kim, S. I.; et al. Graphene/Carbon Nanotube Hybrid-Based Transparent 2D Optical Array. *Advanced Materials* **2011,** *23,* 3809–3814.
149. Zhang, W.; Xie, H.; Zhang, R.; Jian, M.; Wang, C.; Zheng, Q.; Wei, F.; Zhang, Y. Synthesis of Three-Dimensional Carbon Nanotube/Graphene Hybrid Materials by a Two-Step Chemical Vapor Deposition Process. *Carbon* **2015,** *86,* 358–362.
150. Zhu, Y.; Li, L.; Zhang, C.; Casillas, G.; Sun, Z.; Yan, Z.; Ruan, G.; Peng, Z.; Raji, A. R.; Kittrell, C.; et al. A Seamless Three-Dimensional Carbon Nanotube Graphene Hybrid Material. *Nature Communications* **2012,** *3,* 1225.
151. Wang, W.; Ruiz, I.; Guo, S.; Favors, Z.; Bay, H. H.; Ozkan, M.; Ozkan, C. S. Hybrid Carbon Nanotube and Graphene Nanostructures for Lithium Ion Battery Anodes. *Nano Energy* **2014,** *3,* 113–118.
152. Dong, X.; Ma, Y.; Zhu, G.; Huang, Y.; Wang, J.; Chan-Park, M. B.; Wang, L.; Huang, W.; Chen, P. Synthesis of Graphene–Carbon Nanotube Hybrid Foam and Its Use as a Novel Three-Dimensional Electrode for Electrochemical Sensing. *Journal of Materials Chemistry* **2012,** *22,* 17044.
153. Yan, Z.; Ma, L.; Zhu, Y.; Lahiri, I.; Hahm, M. G.; Liu, Z.; Yang, S.; Xiang, C.; Lu, W.; Peng, Z. Three-Dimensional Metal–Graphene–Nanotube Multifunctional Hybrid Materials. *ACS Nano* **2012,** *7,* 58–64.
154. Paul, R. K.; Ghazinejad, M.; Penchev, M.; Lin, J.; Ozkan, M.; Ozkan, C. S. Synthesis of a Pillared Graphene Nanostructure: a Counterpart of Three-dimensional Carbon Architectures. *Small* **2010,** *6,* 2309–2313.
155. Dong, X.; Li, B.; Wei, A.; Cao, X.; Chan-Park, M. B.; Zhang, H.; Li, L. J.; Huang, W.; Chen, P. One-step Growth of Graphene–Carbon Nanotube Hybrid Materials by Chemical Vapor Deposition. *Carbon* **2011,** *49,* 2944–2949.
156. Zhao, M.-Q.; Liu, X. F.; Zhang, Q.; Tian, G. L.; Huang, J. Q.; Zhu, W.; Wei, F. Graphene/Single-Walled Carbon Nanotube Hybrids: One-Step Catalytic Growth and Applications for High-Rate Li–S Batteries. *ACS Nano* **2012,** *6,* 10759–10769.
157. Tang, C.; Zhang, Q.; Zhao, M. Q.; Huang, J. Q.; Cheng, X. B.; Tian, G. L.; Peng, H. J.; Wei, F. Nitrogen-Doped Aligned Carbon Nanotube/Graphene Sandwiches: Facile Catalytic Growth on Bifunctional Natural Catalysts and Their Applications as Scaffolds for High-Rate Lithium-Sulfur Batteries. *Advanced Materials* **2014,** *26,* 6100–6105.
158. Biswas, C.; Lee, Y. H. Graphene Versus Carbon Nanotubes in Electronic Devices. *Advanced Functional Materials* **2011,** *21,* 3806–3826.
159. Sun, D. M.; Liu, C.; Ren, W. C.; Cheng, H. M. A Review of Carbon Nanotube- and Graphene-Based Flexible Thin-Film Transistors. *Small* **2013,** *9,* 1188–1205.
160. Lee, K.; Park, J.; Lee, M. S.; Kim, J.; Hyun, B. G.; Kang, D. J.; Na, K.; Lee, C. Y.; Bien, F.; Park, J. U. In-situ Synthesis of Carbon Nanotube-Graphite Electronic Devices and Their Integrations onto Surfaces of Live Plants and Insects. *Nano Letters* **2014,** *14,* 2647–2654.
161. Li, B.; Cao, X.; Ong, H. G.; Cheah, J. W.; Zhou, X.; Yin, Z.; Li, H.; Wang, J.; Boey, F.; Huang, W.; et al. All-Carbon Electronic Devices Fabricated by Directly Grown Single-Walled Carbon Nanotubes on Reduced Graphene Oxide Electrodes. *Advanced Materials* **2010,** *22,* 3058–3061.
162. Yu, W. J.; Lee, S. Y.; Chae, S. H.; Perello, D.; Han, G. H.; Yun, M.; Lee, Y. H. Small Hysteresis Nanocarbon-Based Integrated Circuits on Flexible and Transparent Plastic Substrate. *Nano Letters* **2011,** *11,* 1344–1350.

163. Yu, W. J.; Chae, S. H.; Lee, S. Y.; Duong, D. L.; Lee, Y. H. Ultra-Transparent, Flexible Single-Walled Carbon Nanotube Non-Volatile Memory Device with an Oxygen-Decorated Graphene Electrode. *Advanced Materials* **2011,** *23,* 1889–1893.
164. Chae, S. H.; Yu, W. J.; Bae, J. J.; Duong, D. L.; Perello, D.; Jeong, H. Y.; Ta, Q. H.; Ly, T. H.; Vu, Q. A.; Yun, M. Transferred Wrinkled Al_2O_3 for Highly Stretchable and Transparent Graphene–Carbon Nanotube Transistors. *Nature Materials* **2013,** *12,* 403–409.
165. Zheng, Q.; Li, Z.; Yang, J.; Kim, J. K. Graphene Oxide-Based Transparent Conductive Films. *Progress in Materials Science* **2014,** *64,* 200–247.
166. Nguyen, D. D.; Tai, N. H.; Chen, S. Y.; Chueh, Y. L. Controlled Growth of Carbon Nanotube-graphene Hybrid Materials for Flexible and Transparent Conductors and Electron Field Emitters. *Nanoscale* **2012,** *4,* 632–638.
167. Wu, S.; Shi, E.; Yang, Y.; Xu, W.; Li, X.; Cao, A. Direct Fabrication of Carbon Nanotube-graphene Hybrid Films by a Blown Bubble Method. *Nano Research* **2015,** *8,* 1746–1754.
168. Liu, Y.-T.; Feng, Q.-P.; Xie, X.-M.; Ye, X.-Y. The Production of Flexible and Transparent Conductive Films of Carbon Nanotube/Graphene Networks Coordinated by Divalent Metal (Cu, Ca or Mg) Ions. *Carbon* **2011,** *49,* 3371–3375.
169. Sun, H.; Deng, J.; Qiu, L.; Fang, X.; Peng, H. Recent Progress in Solar Cells Based on One-Dimensional Nanomaterials. *Energy & Environmental Science* **2015,** *8,* 1139–1159.
170. Yang, Z.; Ren, J.; Zhang, Z.; Chen, X.; Guan, G.; Qiu, L.; Zhang, Y.; Peng, H. Recent Advancement of Nanostructured Carbon for Energy Applications. *Chemical Reviews* **2015,** *115,* 5159–5223.
171. Kim, J.; Tung, V. C.; Huang, J. Water Processable Graphene Oxide:Single Walled Carbon Nanotube Composite as Anode Modifier for Polymer Solar Cells. *Advanced Energy Materials* **2011,** *1,* 1052–1057.
172. Huang, J.-H.; Fang, J.-H.; Liu, C.-C.; Chu, C.-W. Effective Work Function Modulation of Graphene/Carbon Nanotube Composite Films as Transparent Cathodes for Organic Optoelectronics. *ACS Nano* **2011,** *5,* 6262–6271.
173. Velten, J.; Mozer, A. J.; Li, D.; Officer, D.; Wallace, G.; Baughman, R.; Zakhidov, A. Carbon Nanotube/Graphene Nanocomposite as Efficient Counter Electrodes in Dye-Sensitized Solar Cells. *Nanotechnology* **2012,** *23,* 085201.
174. Qiu, L.; Wu, Q.; Yang, Z.; Sun, X.; Zhang, Y.; Peng, H. Freestanding Aligned Carbon Nanotube Array Grown on a Large-Area Single-layered Graphene Sheet for Efficient Dye-Sensitized Solar Cell. *Small* **2015,** *11,* 1150–1155.
175. Yang, Z.; Liu, M.; Zhang, C.; Tjiu, W. W.; Liu, T.; Peng, H. Carbon Nanotubes Bridged with Graphene Nanoribbons and Their Use in High-efficiency Dye-Sensitized Solar Cells. *Angewandte Chemie International Edition* **2013,** *52,* 3996–3999.
176. Ma, J.; Zhou, L.; Li, C.; Yang, J.; Meng, T.; Zhou, H.; Yang, M.; Yu, F.; Chen, J. Surfactant-free Synthesis of Graphene-Functionalized Carbon Nanotube Film as a Catalytic Counter Electrode in Dye-Sensitized Solar Cells. *Journal of Power Sources* **2014,** *247,* 999–1004.
177. Chen, T.; Qiu, L.; Yang, Z.; Peng, H. Novel Solar Cells in a Wire Format. *Chemical Society Reviews* **2013,** *42,* 5031–5041.
178. Yang, Z.; Li, L.; Luo, Y.; He, R.; Qiu, L.; Lin, H.; Peng, H. An Integrated Device for Both Photoelectric Conversion and Energy Storage Based on Free-Standing and Aligned Carbon Nanotube Film. *Journal of Materials Chemistry A* **2013,** *1,* 954–958.
179. Zhang, Z.; Chen, X.; Chen, P.; Guan, G.; Qiu, L.; Lin, H.; Yang, Z.; Bai, W.; Luo, Y.; Peng, H. Integrated Polymer Solar Cell and Electrochemical Supercapacitor in a Flexible and Stable Fiber Format. *Advanced Materials* **2014,** *26,* 466–470.
180. Fang, X.; Yang, Z.; Qiu, L.; Sun, H.; Pan, S.; Deng, J.; Luo, Y.; Peng, H. Core-sheath Carbon Nanostructured Fibers for Efficient Wire-shaped Dye-Sensitized Solar Cells. *Advanced Materials* **2014,** *26,* 1694–1698.
181. Sun, H.; You, X.; Deng, J.; Chen, X.; Yang, Z.; Ren, J.; Peng, H. Novel Graphene/Carbon Nanotube Composite Fibers for Efficient Wire-Shaped Miniature Energy Devices. *Advanced Materials* **2014,** *26,* 2868–2873.

182. Beidaghi, M.; Wang, C. Micro-supercapacitors Based on Interdigital Electrodes of Reduced Graphene Oxide and Carbon Nanotube Composites with Ultrahigh Power Handling Performance. *Advanced Functional Materials* **2012,** *22,* 4501–4510.
183. Yang, C.; Shen, J.; Wang, C.; Fei, H.; Bao, H.; Wang, G. All-solid-state Asymmetric Supercapacitor Based on Reduced Graphene Oxide/Carbon Nanotube and Carbon Fiber Paper/polypyrrole Electrodes. *Journal of Materials Chemistry A* **2014,** *2,* 1458–1464.
184. Zeng, F.; Kuang, Y.; Zhang, N.; Huang, Z.; Pan, Y.; Hou, Z.; Zhou, H.; Yan, C.; Schmidt, O. G. Multilayer Super-Short Carbon Nanotube/Reduced Graphene Oxide Architecture for Enhanced Supercapacitor Properties. *Journal of Power Sources* **2014,** *247,* 396–401.
185. Fan, W.; Miao, Y. E.; Huang, Y.; Tjiu, W. W.; Liu, T. Flexible Free-standing 3D Porous N-Doped Graphene–Carbon Nanotube Hybrid Paper for High-performance Supercapacitors. *RSC Advances* **2015,** *5,* 9228–9236.
186. Jiang, L.; Sheng, L.; Long, C.; Fan, Z. Densely Packed Graphene Nanomesh-Carbon Nanotube Hybrid Film for Ultra-High Volumetric Performance Supercapacitors. *Nano Energy* **2015,** *11,* 471–480.
187. Lin, J.; Zhang, C.; Yan, Z.; Zhu, Y.; Peng, Z.; Hauge, R. H.; Natelson, D.; Tour, J. M. 3-Dimensional Graphene Carbon Nanotube Carpet-Based Microsupercapacitors with High Electrochemical Performance. *Nano Letters* **2013,** *13,* 72–78.
188. Liu, K.; Hong, X.; Zhou, Q.; Jin, C.; Li, J.; Zhou, W.; Liu, J.; Wang, E.; Zettl, A.; Wang, F. High-throughput Optical Imaging and Spectroscopy of Individual Carbon Nanotubes in Devices. *Nature Nanotechnology* **2013,** *8,* 917–922.
189. Gupta, T. K.; Singh, B. P.; Mathur, R. B.; Dhakate, S. R. Multiwalled Carbon Nanotube-Graphene-Polyaniline Multiphase Nanocomposite with Superior Electromagnetic Shielding Effectiveness. *Nanoscale* **2014,** *6,* 842–851.
190. Liu, J.; Zhang, L.; Wu, H. B.; Lin, J.; Shen, Z.; Lou, X. W. D. High-Performance Flexible Asymmetric Supercapacitors Based on a New Graphene Foam/Carbon Nanotube Hybrid Film. *Energy & Environmental Science* **2014,** *7,* 3709–3719.
191. Zhu, G.; He, Z.; Chen, J.; Zhao, J.; Feng, X.; Ma, Y.; Fan, Q.; Wang, L.; Huang, W. Highly Conductive Three-dimensional MnO_2-Carbon Nanotube-Graphene-Ni Hybrid Foam as a Binder-Free Supercapacitor Electrode. *Nanoscale* **2014,** *6,* 1079–1085.
192. Sun, Y.; Cheng, Y.; He, K.; Zhou, A.; Duan, H. One-step Synthesis of Three-Dimensional Porous Ionic Liquid–Carbon Nanotube–Graphene Gel and MnO_2–Graphene Gel as Freestanding Electrodes for Asymmetric Supercapacitors. *RSC Advances* **2015,** *5,* 10178–10186.
193. Yu, D.; Goh, K.; Wang, H.; Wei, L.; Jiang, W.; Zhang, Q.; Dai, L.; Chen, Y. Scalable Synthesis of Hierarchically Structured Carbon Nanotube-Graphene Fibres for Capacitive Energy Storage. *Nature Nanotechnology* **2014,** *9,* 555–562.
194. Foroughi, J.; Spinks, G. M.; Antiohos, D.; Mirabedini, A.; Gambhir, S.; Wallace, G. G.; Ghorbani, S. R.; Peleckis, G.; Kozlov, M. E.; Lima, M. D.; et al. Highly Conductive Carbon Nanotube-Graphene Hybrid Yarn. *Advanced Functional Materials* **2014,** *24,* 5859–5865.
195. Bavykin, D. V.; Friedrich, J. M.; Walsh, F. C. Protonated Titanates and TiO_2 Nanostructured Materials: Synthesis, Properties, and Applications. *Advanced Materials* **2006,** *18,* 2807–2824.
196. Chen, S.; Chen, P.; Wang, Y. Carbon Nanotubes Grown in Situ on Graphene Nanosheets as Superior Anodes for Li-Ion Batteries. *Nanoscale* **2011,** *3,* 4323–4329.
197. Cohn, A. P.; Oakes, L.; Carter, R.; Chatterjee, S.; Westover, A. S.; Share, K.; Pint, C. L. Assessing the Improved Performance of Freestanding, Flexible Graphene and Carbon Nanotube Hybrid Foams for Lithium Ion Battery Anodes. *Nanoscale* **2014,** *6,* 4669–4675.
198. Byon, H. R.; Gallant, B. M.; Lee, S. W.; Shao-Horn, Y. Role of Oxygen Functional Groups in Carbon Nanotube/Graphene Freestanding Electrodes for High Performance Lithium Batteries. *Advanced Functional Materials* **2013,** *23,* 1037–1045.
199. Zhang, B.; Zheng, Q. B.; Huang, Z. D.; Oh, S. W.; Kim, J. K. SnO_2–Graphene–Carbon Nanotube Mixture for Anode Material with Improved Rate Capacities. *Carbon* **2011,** *49,* 4524–4534.
200. Zhu, C.; Mu, X.; van Aken, P. A.; Maier, J.; Yu, Y. Fast Li Storage in MoS_2-Graphene-Carbon Nanotube Nanocomposites: Advantageous Functional Integration of 0D, 1D, and 2D Nanostructures. *Advanced Energy Materials* **2015,** *5,* 1401170.

201. Fang, X.; Peng, H. A Revolution in Electrodes: Recent Progress in Rechargeable Lithium-Sulfur Batteries. *Small* **2015,** *11,* 1488−1511.
202. Manthiram, A.; Chung, S. H.; Zu, C. Lithium-Sulfur Batteries: Progress and Prospects. *Advanced Materials* **2015,** *27,* 1980−2006.
203. Peng, H. J.; Huang, J. Q.; Zhao, M. Q.; Zhang, Q.; Cheng, X. B.; Liu, X. Y.; Qian, W. Z.; Wei, F. Nanoarchitectured Graphene/CNT@Porous Carbon with Extraordinary Electrical Conductivity and Interconnected Micro/Mesopores for Lithium-Sulfur Batteries. *Advanced Functional Materials* **2014,** *24,* 2772−2781.
204. Kaniyoor, A.; Imran Jafri, R.; Arockiadoss, T.; Ramaprabhu, S. Nanostructured Pt Decorated Graphene and Multi Walled Carbon Nanotube Based Room Temperature Hydrogen Gas Sensor. *Nanoscale* **2009,** *1,* 382−386.
205. Jeong, H. Y.; Lee, D. S.; Choi, H. K.; Lee, D. H.; Kim, J. E.; Lee, J. Y.; Lee, W. J.; Kim, S. O.; Choi, S. Y. Flexible Room-Temperature NO_2 Gas Sensors Based on Carbon Nanotubes/reduced Graphene Hybrid Films. *Applied Physics Letters* **2010,** *96,* 213105.
206. Woo, S.; Kim, Y. R.; Chung, T. D.; Piao, Y.; Kim, H. Synthesis of a Graphene−Carbon Nanotube Composite and its Electrochemical Sensing of Hydrogen Peroxide. *Electrochimica Acta* **2012,** *59,* 509−514.
207. Shahriary, L.; Ghourchian, H.; Athawale, A. A. Graphene-Multiwalled Carbon Nanotube Hybrids Synthesized by Gamma Radiations: Application as a Glucose Sensor. *Journal of Nanotechnology* **2014,** *2014,* 1−10.
208. Chen, X.; Zhu, J.; Xi, Q.; Yang, W. A High Performance Electrochemical Sensor for Acetaminophen Based on Single-Walled Carbon Nanotube−Graphene Nanosheet Hybrid Films. *Sensors and Actuators B: Chemical* **2012,** *161,* 648−654.
209. Janczak, D.; Sloma, M.; Wroblewski, G.; Mlozniak, A.; Jakubowska, M. Screen-printed Resistive Pressure Sensors Containing Graphene Nanoplatelets and Carbon Nanotubes. *Sensors (Basel)* **2014,** *14,* 17304−17312.
210. Zhang, X.; Hu, S.; Wang, M.; Yu, J.; Khan, Q.; Shang, J.; Ba, L. Continuous Graphene and Carbon Nanotube Based High Flexible and Transparent Pressure Sensor Arrays. *Nanotechnology* **2015,** *26,* 115501.
211. Zhao, Y.; Song, L.; Zhang, Z.; Qu, L. Stimulus-Responsive Graphene Systems towards Actuator Applications. *Energy & Environmental Science* **2013,** *6,* 3520.
212. Lu, L.; Liu, J.; Hu, Y.; Zhang, Y.; Randriamahazaka, H.; Chen, W. Highly Stable Air Working Bimorph Actuator Based on a Graphene Nanosheet/Carbon Nanotube Hybrid Electrode. *Advanced Materials* **2012,** *24,* 4317−4321.
213. Wang, W.; Sun, X.; Wu, W.; Peng, H.; Yu, Y. Photoinduced Deformation of Crosslinked Liquid-Crystalline Polymer Film Oriented by a Highly Aligned Carbon Nanotube Sheet. *Angewandte Chemie International Edition* **2012,** *51,* 4644−4647.
214. Jiang, W.; Niu, D.; Liu, H.; Wang, C.; Zhao, T.; Yin, L.; Shi, Y.; Chen, B.; Ding, Y.; Lu, B. Photoresponsive Soft-Robotic Platform: Biomimetic Fabrication and Remote Actuation. *Advanced Functional Materials* **2014,** *24,* 7598−7604.
215. Ji, M.; Jiang, N.; Chang, J.; Sun, J. Near-Infrared Light-Driven, Highly Efficient Bilayer Actuators Based on Polydopamine-Modified Reduced Graphene Oxide. *Advanced Functional Materials* **2014,** *24,* 5412−5419.
216. Zhang, X.; Yu, Z.; Wang, C.; Zarrouk, D.; Seo, J. W.; Cheng, J. C.; Buchan, A. D.; Takei, K.; Zhao, Y.; Ager, J. W.; et al. Photoactuators and Motors Based on Carbon Nanotubes with Selective Chirality Distributions. *Nature Communications* **2014,** *5,* 2983.
217. Chen, L.; Weng, M.; Zhou, Z.; Zhou, Y.; Zhang, L.; Li, J.; Huang, Z.; Zhang, W.; Liu, C.; Fan, S. Large-deformation Curling Actuators Based on Carbon Nanotube Composite: Advanced-structure Design and Biomimetic Application. *ACS Nano* **2015,** *9,* 12189−12196.
218. Li, Q.; Liu, C.; Lin, Y.-H.; Liu, L.; Jiang, K.; Fan, S. Large-strain, Multiform Movements from Designable Electrothermal Actuators Based on Large Highly Anisotropic Carbon Nanotube Sheets. *ACS Nano* **2015,** *9,* 409−418.
219. Hu, Y.; Wu, G.; Lan, T.; Zhao, J.; Liu, Y.; Chen, W. A Graphene-Based Bimorph Structure for Design of High Performance Photoactuators. *Advanced Materials* **2015,** *27,* 7867−7873.

220. Yang, S. Y.; Lin, W. N.; Huang, Y. L.; Tien, H. W.; Wang, J. Y.; Ma, C. C. M.; Li, S. M.; Wang, Y.-S. Synergetic Effects of Graphene Platelets and Carbon Nanotubes on the Mechanical and Thermal Properties of Epoxy Composites. *Carbon* **2011,** *49,* 793–803.
221. Im, H.; Kim, J. Thermal conductivity of a Graphene Oxide–Carbon Nanotube Hybrid/epoxy Composite. *Carbon* **2012,** *50,* 5429–5440.
222. Shahil, K. M.; Balandin, A. A. Graphene-multilayer Graphene Nanocomposites as Highly Efficient Thermal Interface Materials. *Nano Letters* **2012,** *12,* 861–867.
223. Chung, H. T.; Won, J. H.; Zelenay, P. Active and Stable Carbon Nanotube/Nanoparticle Composite Electrocatalyst for Oxygen Reduction. *Nature Communications* **2013,** *4,* 1922.
224. Jafri, R. I.; Arockiados, T.; Rajalakshmi, N.; Ramaprabhu, S. Nanostructured Pt Dispersed on Graphene-Multiwalled Carbon Nanotube Hybrid Nanomaterials as Electrocatalyst for PEMFC. *Journal of the Electrochemical Society* **2010,** *157,* B874.
225. Li, Y.; Zhou, W.; Wang, H.; Xie, L.; Liang, Y.; Wei, F.; Idrobo, J. C.; Pennycook, S. J.; Dai, H. An Oxygen Reduction Electrocatalyst Based on Carbon Nanotube-Graphene Complexes. *Nature Nanotechnology* **2012,** *7,* 394–400.
226. Youn, D. H.; Han, S.; Kim, J. Y.; Kim, J. Y.; Park, H.; Choi, S. H.; Lee, J. S. Highly Active and Stable Hydrogen Evolution Electrocatalysts Based on Molybdenum Compounds on Carbon Nanotube–Graphene Hybrid Support. *ACS Nano* **2014,** *8,* 5164–5173.
227. Dichiara, A. B.; Sherwood, T. J.; Benton-Smith, J.; Wilson, J. C.; Weinstein, S. J.; Rogers, R. E. Free-standing Carbon Nanotube/graphene Hybrid Papers as Next Generation Adsorbents. *Nanoscale* **2014,** *6,* 6322–6327.
228. Dong, X.; Chen, J.; Ma, Y.; Wang, J.; Chan-Park, M. B.; Liu, X.; Wang, L.; Huang, W.; Chen, P. Superhydrophobic and Superoleophilic Hybrid Foam of Graphene and Carbon Nanotube for Selective Removal of Oils or Organic Solvents from the Surface of Water. *Chemical Communications (Cambridge)* **2012,** *48,* 10660–10662.
229. Wang, C.; Cao, M.; Wang, P.; Ao, Y.; Hou, J.; Qian, J. Preparation of Graphene–Carbon Nanotube–TiO_2 Composites with Enhanced Photocatalytic Activity for the Removal of Dye and Cr (VI). *Applied Catalysis A: General* **2014,** *473,* 83–89.
230. Sebaa, M.; Nguyen, T. Y.; Paul, R. K.; Mulchandani, A.; Liu, H. Graphene and Carbon Nanotube–Graphene Hybrid Nanomaterials for Human Embryonic Stem Cell Culture. *Materials Letters* **2013,** *92,* 122–125.
231. Hollanda, L. M.; Lobo, A. O.; Lancellotti, M.; Berni, E.; Corat, E. J.; Zanin, H. Graphene and Carbon Nanotube Nanocomposite for Gene Transfection. *Materials Science & Engineering. C, Materials for Biological Applications* **2014,** *39,* 288–298.
232. Eichler, A.; Moser, J.; Chaste, J.; Zdrojek, M.; Wilson-Rae, I.; Bachtold, A. Nonlinear Damping in Mechanical Resonators Made from Carbon Nanotubes and Graphene. *Nature Nanotechnology* **2011,** *6,* 339–342.
233. Liu, C.; Cheng, H. M. Carbon Nanotubes: Controlled Growth and Application. *Materials Today* **2013,** *16,* 19–28.

INDEX

'*Note*: Page numbers followed by "f" indicate figures and "t" indicate tables.'

D

E

G

M

Q

R

S

T

U

V

MIX
Papier | Fördert
gute Waldnutzung
FSC® C083411

Kolibri 360 GmbH
Ferdinand-Jühlke-Straße 7
99095 Erfurt, Deutschland
produktsicherheit@kolibri360.de